Renewable Energy in the Process Industry

Renewable Energy in the Process Industry

Eric van Steen, Nicholas S. Featherstone, and Linda H. Callanan

WILEY-VCH

Authors

Prof. Eric van Steen
University of Cape Town
Private Bag X3
Rondebosch 7701
Cape Town
South Africa

Prof. Nicholas S. Featherstone
University of Cape Town
Private Bag X3
Rondebosch 7701
Cape Town
South Africa

Dr. Linda H. Callanan
University of Cape Town
Private Bag X3
Rondebosch 7701
Cape Town
South Africa

Cover Design: Wiley
Cover Image: © bombermoon/Shutterstock

Typesetting: Straive, Chennai, India
Druck und Bindung: CPI Group (UK) Ltd, Croydon, CR0 4YY

Library of Congress Card No.: applied for

British Library Cataloguing-in-Publication Data A catalogue record for this book is available from the British Library.

Bibliographic information published by the Deutsche Nationalbibliothek
The Deutsche Nationalbibliothek lists this publication in the Deutsche Nationalbibliografie; detailed bibliographic data are available on the Internet at <http://dnb.d-nb.de>.

Print ISBN: 978-3-527-35446-7
ePDF ISBN: 978-3-527-84956-7
ePub ISBN: 978-3-527-84955-0
oBook ISBN: 978-3-527-84957-4

C9783527354467_120126

Contents

About the Authors

Eric van Steen is a professor in the Department of Chemical Engineering at the University of Cape Town (UCT), South Africa, where he holds the South African Research Chair in Reaction Engineering (SARChI). He holds an MSc from Eindhoven University of Technology in the Netherlands, and a Dr.-Ing. from Karlsruhe University of Technology in Germany. He is a Fellow of the South African Institute of Chemical Engineering and of the South African Academy of Engineering. His commitment to student development was recognised when he received the NRF Champion of Research Capacity Development and Transformation Award in 2024.

Prof. van Steen is part of the Catalysis Institute, in the Department of Chemical Engineering at the University of Cape Town, where the focus is on the development of novel materials and processes towards a sustainable society. His specific research focuses on heterogeneous catalysis, particularly applying principles in heterogeneous catalysis to the areas of renewable energy like CO/CO_2 hydrogenation, and photo-catalytic water splitting, but also in alternative ways to obtain building blocks for the chemical industry using the selective oxidation of methane.

Nicholas S. Featherstone holds a BSc Eng. (Chemical) from the University of Cape Town and is currently a lecturer in the Department of Chemical Engineering at the University of Cape Town, South Africa. He is part of the Catalysis Institute, in the Department of Chemical Engineering at the University of Cape Town, where the focus of his research is on the direct hydrogenation of CO_2 to liquid fuels.

Linda H. Callanan holds a B.Sc. (Eng.) (Chemical) and a Ph.D. from the University of Cape Town. After studying, she held a postdoctoral position in the Department of Chemical Technology II (Lehrstuhl für Technische Chemie II) at the Technische Universität München, Germany. She was then a senior lecturer at the Department of Chemical Engineering, Stellenbosch University, South Africa. Her research focus is on reaction kinetics, particularly in the field of heterogeneous catalysis.

Preface

'Renewable energy' is a buzzword with the potential to change the world. This may be achieved through reducing greenhouse gas emissions, thereby mitigating climate change in a drive towards a more sustainable society. Society is energy-driven, and the increasing population, together with the drive for a large part of the world population to improve their economic situation, will increase the global demand for energy. The process industry, which is expected to expand to cater for a growing world economy, is a major energy consumer and emitter of greenhouse gases. Hence, the process industry must look at minimising its energy usage and greenhouse gas emissions. Further, it must also actively engage in the discussion on how to change and adapt its processes towards more sustainable production. Electrification of the process industry using energy supplied from renewable resources will reduce emissions, but this will not be entirely sufficient, and other aspects will need to be considered.

The idea for this book originated from a conversation with one of my postgraduate students, Masana Mhinga, in early 2019, about topics she felt were lacking in the chemical engineering curriculum offered at the University of Cape Town. Although renewable energy was discussed in the research arena, the 'why, what and how' of renewable energy was missing in our undergraduate curriculum. Hence, the thought arose to introduce a course for senior undergraduate chemical engineering students, which would introduce them to this field. A course suitable for our final year students was developed following several interesting discussions with my colleagues from within the department (interrupted by the COVID-19 pandemic). In 2023, I was joined by the enthusiastic Nicholas Featherstone, and we decided to take the deep dive and run a course on 'renewable energy in the process industry', and a larger-than-expected group of eager students enrolled for this course. At that stage, we were still producing the course material and delivering the course simultaneously (with associated course notes being released after completing a lecture). Running the course for the 2nd time in 2024, we decided to produce a full set of course notes, and we asked Linda Callanan to take the crude, original notes produced in 2023 and transform them into a consistent set of notes. This set of notes has now been completely re-worked and extended to give a more in-depth view of the energy scene worldwide, electricity generation (focusing on electricity generation from renewable resources) and energy usage in industry. We have tried to cover a diverse set of industries, ranging from basic industries, including cement and steel manufacture, all the way through to the producers of consumer goods. In the course, we asked the enrolled students to research on how renewable energy can be incorporated in a specific industry, which they then presented. Their contribution is gratefully acknowledged.

This book aims to give the reader an overview of the challenges associated with the generation and storage of renewable energy, the usage of energy in the process industry at large and the examples of industries with their specific requirements. We wrote this book to educate and open the eyes of process engineers working in industry about the opportunities and challenges ahead associated with the penetration of renewable energy into society. It should be realised that the process industry does not stand on its own as it is embedded within the community in which it operates. Renewable energy in the process industry requires stronger grid connections, extension of the grid together with its stability and good communication between grid operators and the process industry with the ability to trade or 'wheel' renewable energy over the grid. Therefore, grid operators may need to learn more about the process industry's requirements. Furthermore, the opportunity to distribute heat generated by the process industry 'over the fence' could result in the development of new businesses. Engineers in those facilities should be familiar with the industries delivering the heat so that they understand possible intermittency/interruptions in the heat supply that can arise when renewable energy is used.

We hope that other engineers will find this book useful as well. Though it emerged from a course, this book is not a textbook. However, the material in the book can be used to teach a course on renewable energy in the process industry (and the authors can supply ideas for tutorials associated with such a course).

We are grateful to Adj. Professor Dave Wright (University of Cape Town), who went over the material critically (always with the caveat that the mistakes are ours).

Books written by multiple authors will always have some variation in style, which hopefully will only have a minimal effect on the readability. We, the authors, have learned a lot in putting the material for this book together, and we hope that, in turn, the readers will learn a lot as well.

June 2025

On behalf of the authors
Eric van Steen
Department of Chemical Engineering
University of Cape Town
Cape Town, South Africa

Nomenclature

Symbols

Symbol	Name	Units
a	acceleration	m/s^2
a	activity coefficient	
A	area	m^2
a	heat transfer area	m^2
$A_{orifice}$	area of nozzle	m^2
B	exergy	kJ
C	capacitance	F
C	cost (of electrolyser plant)	
C_D	drag coefficient	
CF	capacity factor	
CF	concentrating factor	
Cp	coefficient of performance (efficiency)	
C_p	heat capacity (constant pressure)	kJ/kg/°C
CR	compression ratio	
C_v	heat capacity (constant volume)	J/kg/°C
d_w	wire diameter	mm
E	energy	J (alt: kWh)
	energy density	MJ/l
E	specific energy	MJ/kg
E_0	maximum specific energy	MJ/kg
E_{stored}	stored energy	MJ
E	potential difference	
E_{CO_2}	associated CO_2 emissions	tonne CO_2/tonne material
ΔE^0	electric potential (reference state)	
E_{eq}	equilibrium potential	
E_{rms}	electric field strength	V/m

Symbol	Name	Units
E	(cell) potential	V
e	elementary charge	$1.602176634 \times 10^{-19}$ C
F	force	N
f	frequency of the incoming radiation	
f_i	fugacity of species i	
$\mathcal{F}$	Faraday constant	96 485 A s/mol
FF	fill factor	
ff	time factor	
$F_{\text{tubes-flame}}$	view factor of the tubes area of heat transfer	m^2
g	gravitational constant (acceleration)	9.81 m/s^2
ΔG	Gibbs free energy of reaction	kJ/mol
H	enthalpy	kJ
ΔH_{rxn}	heat (enthalpy) of reaction	kJ/mol
ΔH^0_{rxn}	heat of reaction at reference conditions	kJ/mol
h	heat transfer coefficient	$W/m^2/K$
h	height	m
H_{rms}	strength of magnetic field	T (N/Am)
I	current per unit surface area	A
I	moment of inertia	kg m^2
I	rate of discharge	
i_0	exchange current density	
I_{rms}	root mean square of the applied AC current	
I_t	investment cost	
I_W	Wobbe index	MJ/Nm^3
J	current density	A/m^2
K	shape factor (rotor)	
k	thermal conductivity	W/m/°C
ΔL	thermal expansion	
L_w	wire length	m
$\dot{m}$	mass flow rate	kg/s
n	mole fraction	
n	moles of electrons transferred in a reaction	
n	number of moles transferred in a reaction	
N_A	Avogadro's number	$6.022 \cdot 10^{-19}$ 1/mol
N_C	number of carbon atoms in a chain	
Nu	Nusselt number	
P	power	kWh
P	pressure	bar
P^{vap}	vaporisation pressure	bar
pc	Peukert's constant	A
Pr	Prandtl number	

Symbol	Name	Units
Q	heat (energy required for heating)	kWh
q	stored charge	
$Q_{H2,t}$	hydrogen production rate	
Q_r	equilibrium quotient	
r	discount rate	
R	gas constant	8.314 J/mol K
r	rate of reaction	
R	ratio of the pressure after the nozzle	
ΔS	change in entropy upon reaction	J/mol K
S	area swept out by the blades	m^2
T	temperature	°C (alt: K)
$T_{ambient}$	ambient (outside) temperature	°C
t_{final}	lifetime of project	years
V	number of installations	
V_0	number of installations in a reference year	
$\dot{V}$	volumetric flow rate	m^3/s
v	velocity	m/s
v_{air}	wind speed	m/s
V	voltage	V
V	volume	m^3
V_{OC}	open-circuit voltage	V
V_{rms}	root mean square of the applied AC voltage	
W	work	kWh
Y	expansion coefficient	
$\bar{y}$	mean wetness	
Z	compressibility factor (of hydrogen)	
Z	height change	
α	temperature dependency of the specific resistance	
α	absorptivity	
α	chain growth probability	
α	scaling factor	
α	thermal diffusivity	m^2/s
α	transfer coefficient	
β	learning factor	
β	ratio of the diameter of the orifice relative to d_0	
γ	heat capacity ratio (C_p/C_v)	
δ	penetration depth	
ε	electric polarisability	
ε	emissivity	
ε_0	electric permittivity in vacuum	

Symbol	Name	Units
ε_{∞}	permittivity at infinite frequency	
$\varepsilon^{\varepsilon}_{r,eff}$	relative effective dielectric loss	
η_{Carnot}	Carnot efficiency	
η_{Dry}	efficiency operating under dry conditions	
η_{SO}	Shockley-Queisser efficiency	
η_{Wet}	thermal efficiency under slightly wet conditions	
η^{energy}	energy (energetic) efficiency	
η^{exergy}	exegetic efficiency	
η	overpotential	
θ_i	ratio of the components in the effluent relative to the feed of the combustible component	
λ	actual air to methane ratio relative to the stoichiometric air to methane ratio	
λ	air-to-fuel ratio	
λ	wavelength	m
λ_{max}	wavelength with the highest intensity	m
μ	fluid viscosity	Pa·s
μ_0	magnetic permeability in vacuum	
μ_{JT}	Joule-Thomson coefficient	
μ^{ε}_{r}	relative effective magnetic permeability	
ρ	density	kg/m^3
ρ	reflectance	
ρ_0	specific resistance	
σ	Stefan-Boltzmann constant	$5.67 \cdot 10^{-8}$ W $//m^2/K^4$
σ	tensile strength	
τ	relaxation time	s
ω	rotational speed	

Acronyms

AD	anaerobic digester
AEL	alkaline electrolysis
AEM	alkaline exchange membrane electrolysis
AFE	annualised furnace efficiency
APIs	active pharmaceutical ingredients
ASF	Anderson-Schulz-Flory
AST	activated sludge tank
ATR	autothermal reforming
BCC	body-centred-cubic
BCT	body-centred tetragonal

BDT	dibenzyl toluene
BFP	bulk Fermentation Process
C.O.P.	coefficient of performance
CAES	compressed air energy storage
CAGR	compound annual growth rate
CaL	calcium looping
CBP	Chorleywood Bread Process
CCS	carbon capture and sequestration
CCUS	carbon capture, utilisation and storage
CHP	combined heat and power
COGC	Combined Open Gas Cycle
CSP	concentrated solar power
CTMP	chemi-thermo-mechanical pulping
DME	dimethyl ether
DMFC	direct methanol fuel cell
DOD	depth of discharge
EMA	European Medicines Agency
ETC	evacuated tube collectors
$F\&E_t$	costs of feedstock and energy input
FCC	face-centred-cubic
FDA	Food and Drug Administration (U.S.A.)
FOG	fats, oil and grease
FPF	forming, pressing and finishing
(HT)FT	(high temperature) Fischer-Tropsch
GSR	Generator-to-System area
GWP	Global Warming Potential
HAWT	horizontal axis wind turbine
HHV	higher heating value
IEA	International Energy Agency
IR	infrared radiation
IS	individual section
LAES	liquid air energy storage
LCA	life cycle analysis
LCOE	levelised cost of electricity
LCOH	levelised cost of hydrogen
LFL	lower flammability limit
LHV	lower heating value
LOHC	liquid organic hydrogen carriers
LPG	liquefied petroleum gas
LRs	lignocellulosic residues
MCH	methyl cyclohexane
MG-Si	metallurgical-grade silicon
MTBE	methyl-t-butyl ether

MTG	methanol to gasoline
MTJ	methanol to jet
MTO	methanol to olefins
NPV	net present value
$O\&M_t$	operations and maintenance costs
ODC	oxygen-depolarised cathode cell
ODP	ozone depletion potential
PATs	Process Analytical Technologies
PEM	proton exchange membrane electrolysis (acid electrolysis)
PFD	process flow diagram
PGM	platinum group metals
PH	process heat
PHES	pumped hydro-energy storage
PS	primary sedimentation
PtL	power-to-liquid
PV	photovoltaic
REEs	rare earth elements
RH	relative humidity
RMP	refiner mechanical pulping
SCMs	supplementary cementitious materials
SDG	sustainable development goal
SG-Si	solar-grade silicon
SI	solar irradiation
SMR	steam methane reforming
SNG	synthetic natural gas
SOC	state of charge
SOE	solid oxide electrolysis
SOFC	solid oxide fuel cells
SS	secondary sedimentation
STES	solar thermal energy storage
TAME	tertiary amyl methyl ether
TMP	thermomechanical pulping
TOF	turnover frequency
UFL	upper flammability limit
UPS	uninterrupted power supplies
VAWT	vertical axis wind turbine
(R)WGS	(reverse) water gas shift
WHO	World Health Organisation
WWTP	wastewater treatment plant

1

Setting the Energy Scene

1.1 Introduction

Humankind has embarked on the third energy transition (see Figure 1.1). In the pre-industrial time, biomass, for example in the form of wood and peat, was used as the main source of energy with limited use of fossil fuels. The introduction of steam engines, which heralded the start of the first industrial revolution, increased the demand for easily transportable energy carriers, such as coal. For a long time, coal remained the main energy carrier (and in various countries still is by far the main energy carrier) but was gradually replaced by oil as an energy carrier when the need for transportation fuels increased. Going into the energy-demanding fourth industrial revolution, fossil fuels are being replaced with energy carriers derived from renewable resources. It is often argued that the driving force for the replacement of fossil fuels as the main energy carrier is driven by the depletion of these resources. However, this is incorrect. It has been stated in various forms: '*The Stone Age came to an end, not because we had a* ***lack of stones****, and the oil age will come to an end not because we have a* ***lack of oil***' [1] implying that the change is brought about by alternative technologies becoming more attractive than the continued use of existing technologies. This may be brought about by lowering the cost of alternative technologies but also by changing the view of costs associated with existing technologies (through implementation of environmental laws or through carbon tax).

The demand for primary energy has increased dramatically in the period 1820–2018 (see Figure 1.2), which can be ascribed in part to the growing world population, but also to the increased energy usage per person. It can be estimated that annual primary energy demand per person increased almost sixfold in that period from 16.7 GJ per person per annum (0.5 kW per person) in 1820 to 95.3 GJ per person per annum (3.0 kW) in 2018. The increase in the primary energy demand between 1850 and 2020 is stronger than would be expected based on a simple exponential growth pattern, possibly due to the presence of two independent factors driving the energy demand, viz. the increased population as well as the higher energy demand per capita. The power demand varies widely across the world, with developed countries having higher-than-average power consumption (though it should be noted that comparing the energy demand at a national level is difficult

Renewable Energy in the Process Industry, First Edition.
Eric van Steen, Nicholas S. Featherstone, and Linda H. Callanan.

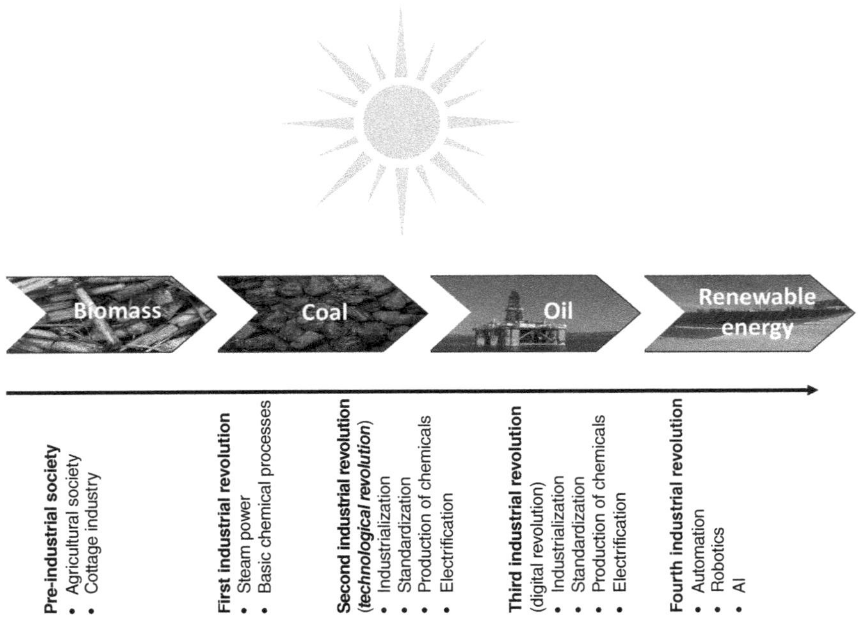

Figure 1.1 Energy carriers used in human history with the main technological areas associated with them (image by Pixabay).

due to the local differences in energy consumption and energy accounting, which often ignores the import/export of goods with their respective energy contents). The global average demand for primary energy per capita is still comparatively low at 2.4–3.0 kW per person. If it is assumed that the worldwide primary power demand per capita will come in line with that of the EU (4.6 ± 0.3 kW per person), the global primary energy demand will double by 2086 (1515 EJ per annum vs. ca. 719 EJ [2] or 604 EJ [3] per annum in 2018; note that different resources may yield different values for the total energy demand due to the different conversion factors used [10]). The change in the primary energy mix, which is currently taking place, would reduce the primary energy demand via improved efficiency for power generation from renewable resources. However, novel technologies such as AI may push the energy demand higher [11].

The industrialisation and population growth over the past decades have been accompanied by a steady increase in the emission of carbon dioxide (CO_2) into the atmosphere as a logical consequence of the dominance of fossil fuels as the energy carrier in this period, whose energy is being released by its combustion with CO_2 being the final product. The COVID-19 pandemic resulted in a (small) dip in the rate of CO_2 emissions, most likely because of a decrease in energy use for transportation, but the rate of emission has been climbing ever since then. The emission of other greenhouse gases such as methane has also steadily increased, although the emission of methane does not show the same rapid increase as seen for the CO_2 emission from 1950 onwards. Methane emissions into the atmosphere are

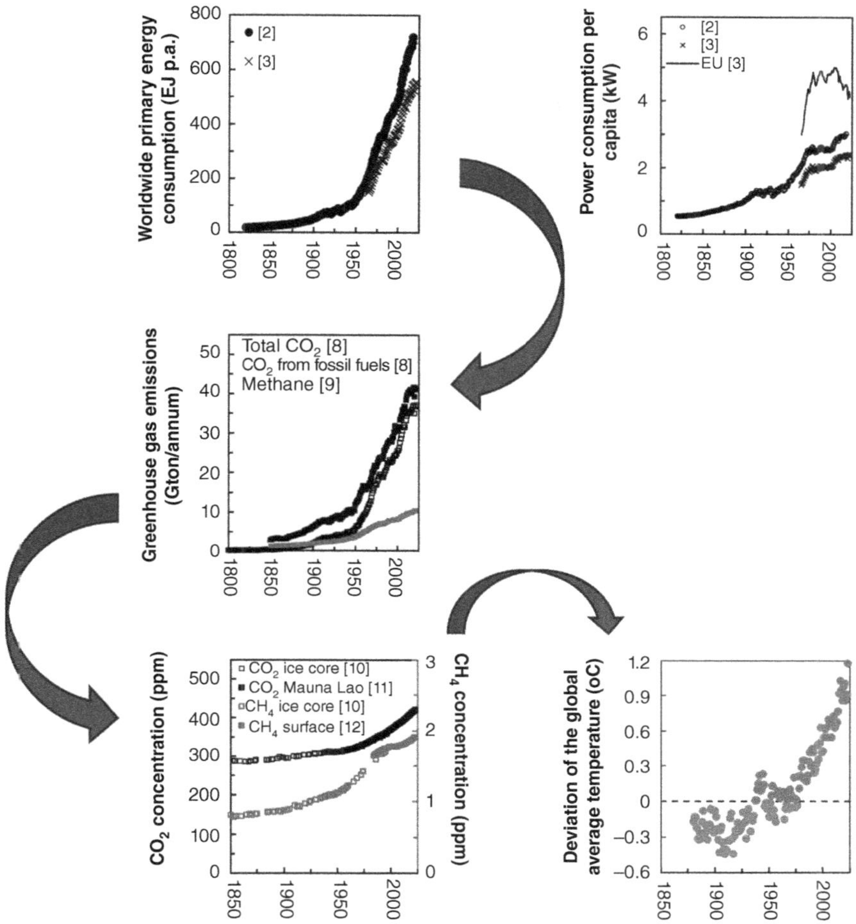

Figure 1.2 Historical trend of the primary energy consumption [2, 3], the power consumption per capita [2, 3], the emission of the greenhouse gases CO_2 (data from Global Carbon Project [4]) and CH_4 [5], the average concentration of CO_2 and CH_4 in the atmosphere on a dry air basis (from ice core data [6] and CO_2-concentrations measured at Mauna Lao [7] and the globally averaged methane concentration at marine level [8]) and the average temperature deviation from the average value in the period 1880–2018 (data from Lindsey and Dahlman [9]).

partly associated with anthropogenic activities, such as the recovery of coal and oil, and from agricultural sources and landfill sites [12] with the latter being associated with the size and living standards of the world population. Whilst greenhouse gases have natural sinks (such as growth of biomass consuming CO_2), larger emissions of these greenhouse gases will lead to an increase in their concentration in the atmosphere. As a result, the rate of uptake will need to be adjusted to account for the increased rate of release and, owing to the increased concentration, will also need to be done at a higher atmospheric CO_2 concentration than before.

The increase in the concentration of a greenhouse gas in the atmosphere will result in greater light absorption, which leads to the heating up of the atmosphere (it was already noted in 1938 that the average temperature would increase by 0.003 °C per annum due to CO_2 emissions at that time [13]). Solar light enters the earth atmosphere and is partially reflected by the clouds (ca. 30%) with the remainder travelling freely through the atmosphere (except the ultraviolet [UV]-part of the spectrum which is absorbed in the ozone layer protecting us from this type of radiation) and heats up the earth. Warm objects, and thus also the earth surface, give off radiation in the form of infrared (IR) radiation, which can be absorbed by greenhouse gases in the atmosphere (as well as water vapour). An increase in the concentration of these greenhouse gases will thus increase the absorption of light in the infrared range. The absorbed energy is released by emitting IR radiation in all directions (also back to earth) or is transformed directly into kinetic energy (via internal relaxation processes and quenching of the excited states in a collision with other molecules); the increased average kinetic energy of the molecules in the gas phase corresponds to a higher average temperature of the atmosphere.

The impact of the various greenhouse gases on the global average temperature is compound-specific and is often summarised in the Global Warming Potential (GWP), which represents the radiative forcing (difference between energy entering the atmosphere and energy leaving the atmosphere) integrated over a time frame relative to the radiative forcing induced by CO_2 over the same time frame [14]. The GWP is thus time-dependent (often an arbitrary timescale of 100 years is invoked) due to the decomposition of the compound in the atmosphere. For instance, the average lifetime of methane in the atmosphere is ca. 12 years [14]. Hence, the GWP of methane over a period of 20 years is 84 times higher than the potential of CO_2 and is over a period of 100 years 28 times higher. The total radiative forcing of methane is ca. 25% that of CO_2 as the high GWP is off set by the low concentration of methane (ca. 1.9 ppm). Ammonia has a relatively short average lifetime in the atmosphere (due to reaction and uptake by particles) [15], and hence its direct impact is small (although ammonia oxidation contributes significantly to the presence of nitrous oxide in the atmosphere [16]). Water vapour is, in principle, also a greenhouse gas, but its concentration in the atmosphere is controlled by the temperature due to its relatively low vapour pressure. However, the increasing water concentration in the atmosphere due to a higher average global temperature will amplify global warming (Table 1.1).

Table 1.1 Global Warming Potential (GWP) over 20 and 100 years for different gases (data from Myhre et al. [14]).

Compound	Chemical formula	$GWP_{20\ years}$	$GWP_{100\ years}$
Carbon dioxide	CO_2	1	1
Methane	CH_4	84	28
Nitrous oxide	N_2O	264	265

The increase in the average temperature of the atmosphere poses a problem as it may affect long-term weather patterns and thus climate both locally and globally (linking short-term changes in the weather to changes in the atmosphere is much more difficult thus giving rise to different scenarios [17–19]). The changing weather patterns may affect food supply and physical safety. This will affect various regions on the planet differently and thus the overall societal stability. The linkage between greenhouse gas emissions and the global average temperature requires a drastic reduction in the rate of emission of these gases, posing thus the significant challenge of reducing the emission of greenhouse gases in a society where the use of primary energy may increase. This requires a critical look at not only new processes but also at the replacement/modification of current processes.

1.2 Energy Carriers

The term 'energy' is often used loosely, but it represents the ability to perform work in a conversion process or for heating purposes (such as chemical energy powering a car or electricity powering a heater). The term 'energy' is thus often used as an indication of the ability to degrade an energy carrier, so that the energy given off in the degradation process can be used for driving another process via heating or delivering work.

There are many different energy carriers. In cells of the human body, the energy carrier is adenosine triphosphate (ATP), which is converted into adenosine diphosphate (ADP) upon the release of energy (which was formally stored in the phosphate bond). ADP can be transformed back into ATP using another energy carrier (e.g. carbohydrates, which are added to the body in the form of food). Other energy carriers are, for example, petrol or diesel, which releases energy upon combustion; batteries, which release electrochemical energy; or radioactive elements, which release energy through nuclear processes; or even simple water, which can release gravitational energy by bringing it down from a storage at a height.

1.2.1 Solar Energy

The sun produces an approximately constant amount of power from the hydrogen fusion taking place in its core (estimated power generated: 385 yottawatts = 385×10^{24} W [20]). The power generated in the sun radiates outwards into space, reducing the amount of power per unit area further away from the sun. At the earth, we receive ca. 173 000 TW of the power generated by the sun (i.e. only a tiny fraction of the total amount of energy released by the sun). This corresponds to an annual received amount of energy from the sun on the earth of 5.47 YJ or 5.47 million EJ. This would be sufficient to power the needs of humanity now and in the future, as the current consumption of primary energy carriers worldwide is ca. 555 EJ per annum [3] or 719 EJ per annum [2] (this corresponds to 19.1 or 22.8 TW, respectively, which represents only 0.011–0.013% of all incoming solar

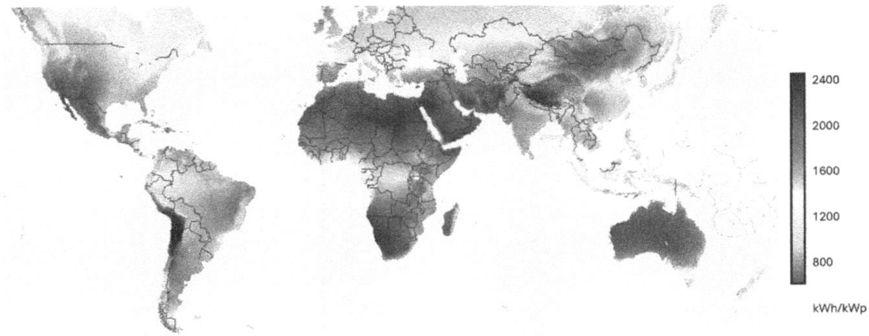

Figure 1.3 Solar irradiance over the world in kWh/day/m^2 (Global Solar Atlas [21]; map from Global Solar Atlas 2.0, Global Solar Atlas).

energy). Even the anticipated doubling of the energy demand will not alter this picture significantly.

The amount of solar irradiation reaching the top of the atmosphere can be estimated from the total energy released by the sun relative to the area of the earth shading the sunlight and amounts to 1360 W/m^2 (this corresponds to an average irradiance on the surface of the earth of ca. 340 W/m^2 – the decrease by a factor of 4 originating from the sphericity of the earth). The solar irradiance is often expressed as a daily average of 8.2 kWh/m^2 or an annual average of 2981 kWh/m^2. Some of the light scatters back into space before reaching sea level, reducing the amount of light at sea level depending on cloud cover and the local climate.

Solar energy received at sea level is, by its nature, intermittent due to the rotation of the earth with the intensity changing with time as a consequence of the changing elevation of the sun (i.e. the angle between the sun and horizon). Furthermore, the intensity of solar radiation reaching the surface is impacted by the local weather conditions, the latitude and seasonal changes due to the tilt of the earth (thus changing the area being covered by sunlight). Hence, the amount of the incoming irradiance strongly depends on the location (see Figure 1.3), with Chile, Northern Africa, South Africa/Namibia, the Middle East, the Himalayas and Australia having regions where more than two-thirds of the incoming irradiance at the top of the atmosphere reaches sea level.

1.2.2 Wind Energy

Wind energy is the energy associated with the movement of air originating from pressure differences in the atmosphere, which are induced by variations in temperature. Hence, wind energy can be regarded as the first derivative of solar energy, which is transformed into kinetic energy in the atmosphere.

The amount and speed (and direction) of the wind vary over the earth with the average wind speed being the highest over the seas and in coastal regions. High aver-

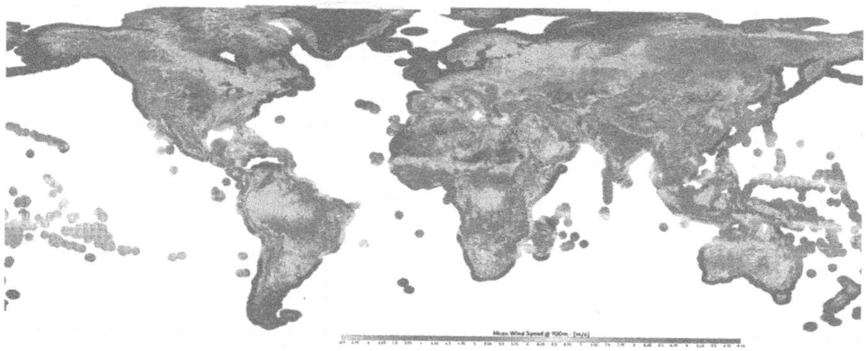

Figure 1.4 Average wind speed at a height of 100 m worldwide (Davis et al. [22]; Map obtained from the Global Wind Atlas version 3.3, a free, web-based application developed, owned and operated by the Technical University of Denmark (DTU). The Global Wind Atlas version 3.3 is released in partnership with the World Bank Group, utilising data provided by Vortex, using funding provided by the Energy Sector Management Assistance Program (ESMAP). For additional information: https://globalwindatlas.info).

age wind speeds in the (near) land region have been recorded in the coastal regions of Patagonia, Alaska, Eastern Canada, Greenland, North-West Europe, Mauritania, Western Sahara, Horn of Africa, Japan, Australia and New Zealand (see Figure 1.4).

The nature of wind results in a strong variability of wind both in direction and in wind speed. Even in windy areas, the wind speed may drop to close to zero during certain times. Wind energy has an effect on large bodies of water, such as the ocean, where wind is one of the primary causes for the formation of waves, and thus wave energy.

1.2.3 Hydropower

The evaporation and condensation of water are natural processes initiated by the presence of solar irradiation. The collection of condensed water at elevated positions allows the recovery of a (very small) part of the energy, as the gravitational energy of liquid water at a height of 1000 m above sea level is less than 0.5% of the energy required to evaporate the water at sea level.

The recovery of energy in the form of hydropower requires the release of stored water over a height difference. For instance, the stored energy in the Hoover Dam in the Colorado River (USA) with a capacity of 19.55 km^3 and a hydraulic height of 180 m corresponds to 3.5 PJ (or 0.0035 EJ). Hydropower is a useful and reliable energy carrier, which should be used, if possible, but it requires a large area in order to store water at a sufficient elevation, attached to a suitable catchment area. However, it can be affected by drought or lack of rain if the dam volume or stored water volume is not sufficiently replenished in order to maintain the flow required to generate the power.

Hydropower is thus a form of solar energy, where the energy is stored as gravitational energy. Another form of gravitational energy that can be captured is tidal energy. The tides are caused by the gravitational pull of the moon and the sun, and the energy released by the movement of the tides can be captured directly in places with the appropriate geography.

1.2.4 Chemical Energy Carriers

Chemicals have been used by humankind for a long time as energy carriers. This started with the use of wood, followed by coal, oil and natural gas as energy carriers; even biofuels are chemical energy carriers (and even the food we eat is a chemical energy carrier). The energy in these substances is transformed into heat/work by converting them mostly in the presence of oxygen or air into CO_2 and water. Energy carriers, such as coal, oil and natural gas (in large reservoirs) are non-renewable as they can only be replenished on a timescale covering many millions of years. Biofuels, on the other hand, are renewable chemical energy carriers as the carbon cycle can be closed; the combustion of biofuels will release CO_2 into the atmosphere, but the same amount of CO_2 will be taken up when re-growing plants, which will be converted into biofuels.

Figure 1.5 shows the distribution of fossil fuel reserves over the world for national reserves larger than 100 EJ (this corresponds 95.2% of all proven coal reserves, 93.2% of all proven oil reserves and 82.9% of all natural gas reserves). Large coal reserves are found in the United States, China, Russia, India and Australia with significant reserves in Ukraine, Poland, Germany and South Africa. Oil reserves are

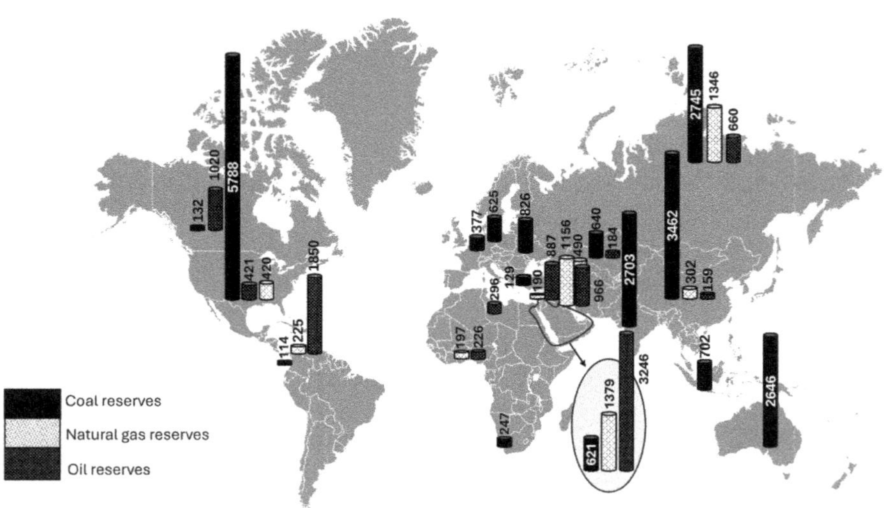

Figure 1.5 Proven reserves of natural gas, oil and coal with a reserve of over 100 EJ (value indicates proven reserves [3] in EJ calculated using a heating value for natural gas of 36 EJ/10^9 m^3; oil of 6.119 EJ/1000 Mio barrel; coal of 0.025 EJ/Mio tonne for hard coal and 0.0105 EJ/Mio tonne for lignite and sub-bituminous coal).

mainly located in the Middle East, which including Iran and Iraq accounts for 48% of all proven oil reserves worldwide. Other regions with large oil reserves include Venezuela, Canada, the United States and Russia. Natural gas reserves can be found in the oil-rich Middle East, Russia, Turkmenistan and the United States with many smaller reserves found all over the world.

These fossil fuels have a large effect on the changing composition of the atmosphere owing to their large international usage and high CO_2 output from combustion; this is the driving force to change towards the implementation of renewable or green energy carriers worldwide. Chemicals will be used as an energy carrier going forward, as they are a convenient way to store (and thus transport) energy and the energy content can be released on demand in a controlled manner. The amount of energy carried by a chemical is its heat of combustion, if the energy is released in a combustion reaction, or by the electrochemical potential, if the energy is released in an electrochemical cell. For instance, the lower heating value of hydrogen is 241.8 kJ/mol, implying that 241.8 kJ/mol or 120.9 MJ/kg will be released upon reacting hydrogen with oxygen forming water in the gas phase. However, the electrochemical potential for the electrochemical conversion of hydrogen to water is 1.23 V, which corresponds to an energy content of 237.1 kJ/mol or 118.6 MJ/kg.

The energy content of chemicals can be defined in terms of the specific energy (the energy released per unit mass of the chemical – for combustion reactions, the lower heating value is used) or in terms of the energy density (the energy per unit volume of the chemical). The energy is normalised to obtain a direct relationship between the amount of energy required in a process and the required amount of the energy carrier for this process. A high specific energy is desirable as it means that more energy can be obtained from a smaller quantity (in terms of mass or the number of molecules) of the substance. For instance, the required mass of wood (with a specific energy of ca. 15 MJ/kg) is eight times larger than the required mass of hydrogen (with a specific energy of 120.9 MJ/kg) for the same process (see Table 1.2).

However, chemicals are energy carriers that need to be transported to the required process, and the volume of the chemical to be transported becomes important (this is of particular importance in the transportation industry, where the energy carrier must be carried whilst being consumed). Diesel and petrol are well-known energy carriers, which have a reasonable specific energy of 45–47 MJ/kg and, due to their relatively high density, have a high energy density of 35.8 and 31.7 MJ/l (see Figure 1.6), respectively. Hydrogen is much less dense, and its energy density is much lower than that of diesel and petrol, despite its high specific energy of 120.9 MJ/kg. Compressing hydrogen to 700 bar or even liquefying hydrogen brings up the energy density to 4.9 and 8.5 MJ/l, respectively. Methane, which has a similar specific energy to diesel and petrol, also needs to be compressed or even liquefied to obtain energy densities approaching those of diesel and petrol. The specific energy and the energy density of other alternative energy carriers such as ammonia, ethanol and methanol are also lower than those of diesel and petrol as shown in Figure 1.6.

Chemical energy is not only released upon combustion but can also be stored in electrochemical devices. The maximum specific energy which can be stored in

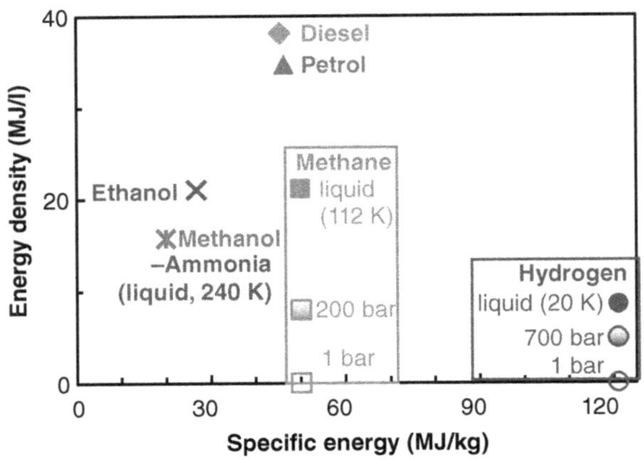

Figure 1.6 Specific energy and energy density for different gaseous and liquid energy carriers.

a certain type of battery can be estimated from the equilibrium cell potential, E_0 (which relates to the Gibbs free energy of the reaction – $\Delta_{rxn}\underline{G} = -n \cdot \mathcal{F} \cdot E_0$), and the mass of the electrolytes involved in the electrochemical conversion. Batteries have a relatively low specific energy in comparison to other chemical energy carriers (see Table 1.2). It should be noted that the actual specific energy of batteries is lower than the quoted value for the specific energy as the weight of the electrodes and other internals will also contribute to the weight of the battery. Batteries do have the advantage that the reaction can be more readily reverted by changing the potential.

Table 1.2 Specific energy of some energy carriers.

Compound	Specific energy (MJ/kg)	
Hydrogen	120.9	$\Delta\underline{H}^{comb} = -241.8$ kJ/mol
	117.6	In an electrochemical cell at 298.15 K
Methane	50.0	$\Delta\underline{H}^{comb} = -802.6$ kJ/mol
Petrol	46.4	Approximate value [23]
Diesel	45.6	Approximate value [23]
Ethanol	26.8	$\Delta\underline{H}^{comb} = -1235$ kJ/mol
Wood	15.0	Dry oak wood [24]
Anthracite	32.6	Approximate value [23]
Lignite	14.0	Approximate value [23]
Li-Co battery	2.1	$E_0 = 3.6$ V; $LiC_6 + CoO_2 \rightleftharpoons LiCoO_2 + 6\,C$
Lead-acid battery	0.8	$E_0 = 2.0$ V[a)]; $Pb + PbO_2 + 4H^+ \rightleftharpoons 2PbSO_4 + 2\,H_2O$[b)]
V-flow battery	0.3	$E_0 = 1.26$ V[a)]; $VO_2^+ + V^{2+} + 2H^+ \rightleftharpoons VO^{2+} + V^{3+} + H_2O$[b)]

a) At standard conditions (298.15 K, 1 M ideal solution).
b) Cations balanced by sulphate ions.

1.2.5 Nuclear Energy

The nuclei of atoms contain protons and neutrons which are kept together by the strong nuclear force, which is opposed by the Coulombic repulsion between the protons. The nucleus of elements with the atomic numbers of around 26–28 is most stable, and energy can thus be obtained by either the fusion of atoms with a low atomic number or the fission of atoms with a high atomic number. The obvious advantage of nuclear energy is that the conversion and utilisation of this energy carrier is without the emission of greenhouse gases.

Radioactive decay of heavy elements has been known since the beginning of the 1900s. In the fission of ^{235}U, ca. 94% of all energy (202.79 MeV) is in the form of kinetic energy of the fragments, which can be recovered as heat. Hence, the specific energy of ^{235}U is 78.2×10^6 MJ/kg, i.e. more than a million times larger than the chemical energy carriers. Natural uranium consists of ca. 0.72% fissile uranium (^{235}U) and hence the specific energy of natural uranium is 56.3×10^4 MJ/kg.

The total uranium reserves around the world are estimated at ca. 6 million tonnes of uranium with more than 50% of the reserves located in Australia, Kazakhstan and Canada (see Table 1.3). Hence, the total known reserves of uranium represent a total energy of ca. 3400 EJ or ca. 6–7 years of the total annual primary energy demand in the world.

The use of nuclear fission as the primary energy carrier generates radioactive waste, which must be placed in a controlled environment to avoid radiation damage to the environment. This is often denoted as storage of nuclear waste, but the natural decay of the different radioactive nuclides will continue to take place gradually reducing the irradiation intensity. The high thermal output of spent nuclear fuel requires the initial storage of this material in pools of water to facilitate cooling for a period of up to 5–10 years [26], following which the material is typically stored in dry caskets. Subsequently, the material is supposed to enter long-term (over 1000 years) storage in a suitable geological location, but currently not sufficient facilities have been designated for this. This is aggravated by the great uncertainty regarding the reactions taking place during the storage at the geological location and (possible) changes of the geological site (and in particular water/brine penetration) [26].

Even when not used directly as an energy source by humans, all energy sources are ultimately derivatives of nuclear energy. One derivative found occurring naturally on earth is geothermal energy, and subterranean heat is partially generated by radioactive decay, which heats up water. The generated, pressurised hot water or steam can be used as an energy carrier if the resource can be accessed in a reasonable manner.

On the other end of the spectrum of nuclear energy is nuclear fusion using small atoms. The sun irradiates energy because of the H + H fusion reaction yielding ^4He (and 26.73 MeV). Hence, the specific energy of hydrogen in the fusion reaction corresponds to 630×10^6 MJ/kg. The controlled fusion reaction would be a game-changer in the energy field, and some significant advances have been made in this field recently [27], but there is still a long road to make this technology reality.

Table 1.3 Uranium reserves across the world in tonnes (in brackets the relative amount to the total world reserves) (data from World Nuclear Association [25]).

Country	Reserve	
Australia	1 684 100	(28%)
Kazakhstan	815 200	(13%)
Canada	588 500	(10%)
Russia	480 900	(8%)
Namibia	470 100	(8%)
South Africa	320 900	(5%)
Niger	311 100	(5%)
Brazil	276 800	(5%)
China	223 900	(4%)
Mongolia	144 600	(2%)
Uzbekistan	131 300	(2%)
Ukraine	107 200	(2%)
Botswana	87 200	(1%)
USA	59 400	(1%)
Tanzania	58 200	(1%)
Jordan	52 500	(1%)
Rest of the world	266 600	(5%)
Total	6 078 500	

1.3 Current Usage of Energy Carriers

Figure 1.7 shows the global annual energy consumption as well as the contributions of the various energy carriers to the total consumption in the period 1965–2022.

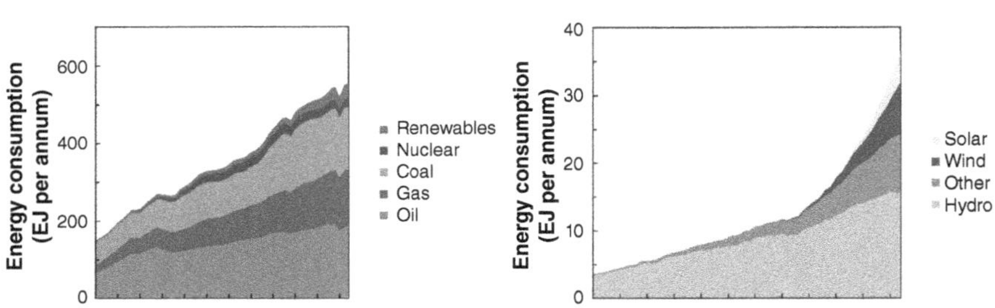

Figure 1.7 Use of the different energy carriers worldwide in the period 1965–2022 (other refers to biofuels/biomass and geothermal energy) (data from The Energy Institute [3]).

Oil dominated the energy scene in the 1960s and 1970s, and in 1973 it contributed 51.2% to the total primary energy use. Its relative contribution to the overall energy mix and even the total oil consumption dropped rapidly between 1979 and 1985, as the global economic recession in the early 1980s resulted in a drop in the overall energy demand, which together with the high price of oil enforced the use of other energy carriers. From 1985 onwards, the oil consumption increased, although its relative contribution to the total energy mix kept falling, and it stood at 34.4% in 2022.

Natural gas is seen as a better alternative to coal when it comes to CO_2 emissions for power generation (power generation with an efficiency of 40% from methane with a specific energy of 50 MJ/kg will generate 0.138 kg CO_2/MJ; a similar calculation using anthracite with a specific energy of 32.6 MJ/kg will generate 0.281 kg CO_2/MJ – these estimates are in line with the CO_2 emissions for coal and natural gas as indicated by [28]). The availability of methane through pipelines and the replacement of coal power stations with natural gas power stations may have led to an increase in the use of natural gas as the primary energy carrier [29]. In 2022, natural gas made up 25.6% of the total primary energy palette (down from a high of 26.6% in 2020, which may have been caused by the reduced delivery of Russian gas to the European market in 2021 [30] aggravated by the Russia–Ukraine war).

The use of coal as the primary energy carrier shows an interesting trend, with an almost linear increase from 58.1 EJ per annum in 1965 to 93.1 EJ per annum in 1988, after which the coal consumption remained almost constant for a decade (with coal consumption equivalent to 95.3 EJ per annum in 1999). A decade later, the coal consumption had increased by more than 50% to 158.1 EJ per annum in 2011, which is attributed to the rapid economic expansion in Asia during that time [31]. The coal consumption virtually did not change the following decade, and it remained at 158.3 ± 2.5 EJ per annum (excluding the COVID-19 year in which the coal consumption was 4% lower). This was because of the move away from coal by some countries (especially the United States) and the balance taken up by new consumers (e.g. India). The contribution of coal to the overall primary energy mix has been remarkably stable from 1973 to 2022 ($28.8\% \pm 1.7\%$).

The contribution of nuclear energy increased significantly up to 2001, where it reached 27.2 EJ per annum, representing 7.0% of the total primary energy demand in that year. The development of new nuclear energy power plants stagnated at that point as the public perception of this technology changed due to accidents such as the partial nuclear meltdown in the plant on Three Mile Island (1979) and the explosion at the Chernobyl nuclear power plant (1986). Following the nuclear accident at the Fukushima power plant, when a tsunami hit the nuclear power plant in 2011, Germany shut down all its nuclear plants by 2020. Worldwide, there has been a net decrease in nuclear energy as a primary energy carrier, and in 2022, nuclear energy contributed only 24.1 EJ to the overall primary energy mix (4.1% of the total demand mainly due to the increased total primary energy demand).

Renewable energy (hydropower, geothermal energy, biomass, wind energy and solar energy) has slowly become an important contributor to the overall primary energy mix with a total contribution in 2022 of 36.6 EJ or 6.6% of the total primary energy consumption. The contribution of renewable energy, and in particular hydro-electric energy, wind energy and solar energy, is even more remarkable as the energy contribution of these renewable energy carriers contributes directly to the available power. The efficiency of power generation using fossil fuels is typically at ca. 40% implying a much larger impact of the introduction of these renewable energy carriers to the primary energy mix.

Hydropower has a long history and is still the dominant renewable energy carrier. Hydropower is expected to be overtaken by the other renewable energy carriers (specifically wind and solar energy), despite its steady increase over the period from 1965 to 2022.

Geothermal energy is rather small, and the total installed capacity for geothermal energy is ca. 16 GW (which corresponds to a maximum annual energy consumption of 0.5 EJ) [32]. The plants are typically located in volcanically active regions of the earth with easy access to the thermal source (viz. relatively close to Earth's surface). The plants built so far are typically small with sizes that are typically less than 1 MW.

Biomass has been used for a long time as an energy carrier, and the use of biomass as an energy carrier in the formal primary energy market has increased to ca. 8 EJ per annum in 2022. Biomass is a term encompassing all kinds of biological materials (including animal waste), which can be converted into a more refined energy carrier, biofuel, for example ethanol or biodiesel. Growing biomass for the purpose of creating a renewable energy carrier competes with food supply for natural resources, such as land and water. Even in regions where water supply is not a primary issue, clearing land to grow crops to generate energy carriers not only has a detrimental effect on the environment, but will also have a sociological impact, as was seen with the cultivation of palm oil for biodiesel production in Indonesia [33]. This type of analysis is convoluted as the product can have a dual purpose, i.e. food or biofuel [34]. The use of biowaste is less controversial, but multiple processing steps are required to convert this into biofuels [35].

Wind energy contributed 7.6 EJ to the overall primary energy consumption in 2022 and its contribution has been growing substantially. At the end of the 1990s, the amount of energy recovered from wind energy grew by more than 30% annually. The growth rate has since then slowed down, and in the period 2014–2022, the growth rate of the energy captured from wind energy has grown by 13.7% ± 2.9% per annum.

Solar energy has seen a similar growth to wind energy, even though it only began contributing more than 0.1 EJ to the primary energy demand in 2010, more than a decade later than wind energy. The growth rate of solar energy in the period 2014–2022 is 28.0% ± 6.6% per annum. This period includes 2020, the year of COVID-19, which caused a significant slowdown of all economic activity and the year-on-year growth of solar energy in 2020 was 'only' 20.7%.

1.3.1 Interconversion of Energy Carriers

Although energy carriers are treated as interchangeable, the different energy carriers are used for different purposes. Figure 1.8 shows a Sankey diagram of the energy carriers worldwide to visualise the flow of various primary energy carriers to the

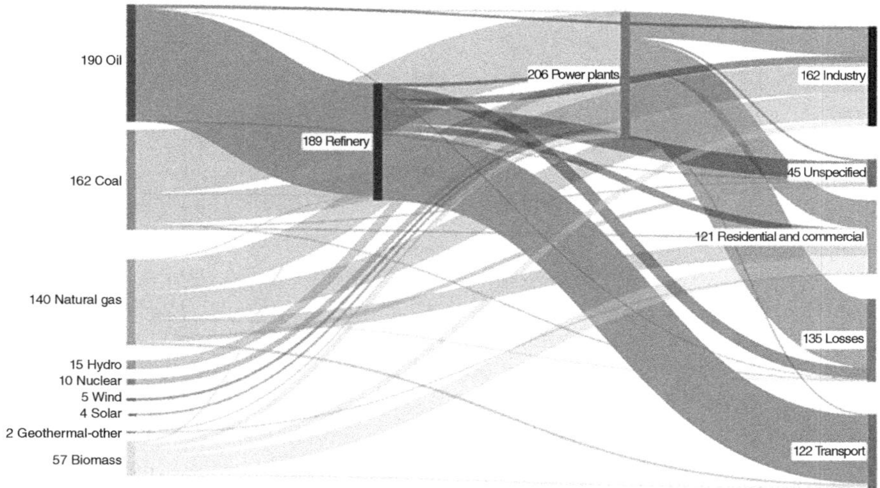

Figure 1.8 Sankey diagram of the primary energy consumption in 2019 (redrawn from Pathak et al. [36], using 'Sankeymatic', 25 February 2024).

different end users in 2019 [36]. Some of the energy carriers are transformed (e.g. in power plants) and the difference in the primary energy supply and the energy consumption indicates an overall efficiency of energy conversion of 76.6%, but not all processes are equally efficient.

Oil produced is mainly being refined (95%) with the balance being used for power generation, in industry, or in other non-specified sectors. Most of the coal (63%) is used for power generation, with industry taking ca. 30%. The use of natural gas is split between power generation (38%), industry (30%) and the residential and commercial sector (21%), where it is used for heating. A small part of natural gas is used in the transportation sector and for non-energy use (for instance, as a chemical feedstock). Hydropower, nuclear energy, wind energy and solar energy feed only into the power systems. Geothermal energy is again used in the residential and commercial sector (heating) and to some extent for power generation. Biomass shows a more varied palette of applications, viz. the residential and commercial sector (51%), industry (19%), power generation (16%) and even transport (7%).

In oil refineries, primary energy carriers (primarily oil, 96% with small contributions of biomass, natural gas and coal) are converted into refined products. The overall efficiency is high (89.9%), mostly due to the fact that the energetic efficiency of the refining of oil is high compared to the conversion of biomass, natural gas [37] or coal [38] into refined products. The refined products are subsequently mainly used for transportation (59%), with 15% going into the non-energy usage (as a feedstock for the chemical industry). The industry and the residential and commercial sectors each use ca. 6% of all refined products, with only 3% being used for power generation.

It should be noted that overall, 23.5% of the primary energy supply is lost by the time it gets to its final usage, which is caused by inefficiencies in power plants in the conversion of heat into power as well as losses in the refineries. Though it should be noted here that the energetic losses in these two sectors, which are currently the biggest, will become smaller when renewable energy sources are used to a greater extent.

Electric power is currently being generated from a variety of different primary energy carriers with coal making up ca. 50% of the total energy mix in 2019 and natural gas 26%. The remainder comes from hydropower (7%), nuclear energy (5%), biomass (4%), oil and refined products (3%), wind energy (2%) and solar energy (2%). The power generation has an overall efficiency of 46.1%, affected primarily by the relatively low efficiency of turning heat into power as expected for the conversion of coal, gas, nuclear energy, biomass and oil and its refined products into power. An energy transition with more power being generated from renewable sources such as hydropower, wind energy and solar energy should improve this efficiency. Power is being used mainly in the commercial and residential sector (48%) and industry (45%). In industry, the power is used mainly to drive machinery and appliances (ca. 87%) with only a small amount used for heating purposes (13%). In 2019, only a small fraction of electric power was used in the transportation sector.

Energy is being used throughout society with the industry using 36% of the final, total consumed energy. The energy consumed in the industry comes in the form of coal (30%), electric power (27%) and natural gas 26% (with the balance coming from oil and oil products and biomass).

Transportation and the residential and commercial sectors are the next largest users of the energy with each using ca. 27% of all consumed energy. Energy usage in the transportation sector was in 2019 still dominated by oil products (92%) with a much smaller contribution from natural gas, biomass and electric power. The residential and commercial sector (in 2019) used a variety of different energy carriers, i.e. electric power (38%), natural gas (24%) and biomass (24%).

Figure 1.9 shows the historical trend of the primary energy used for power generation as extracted from the Energy Institute data [3] (note that the relative contribution of power generation appears to be low from this set of data, possibly as a consequence of differences in scaling [10]). Electricity generation (and thus the associated consumption of primary energy) has been increasing over the period 1985–2022. The increase in electric power generation is stronger than the increase

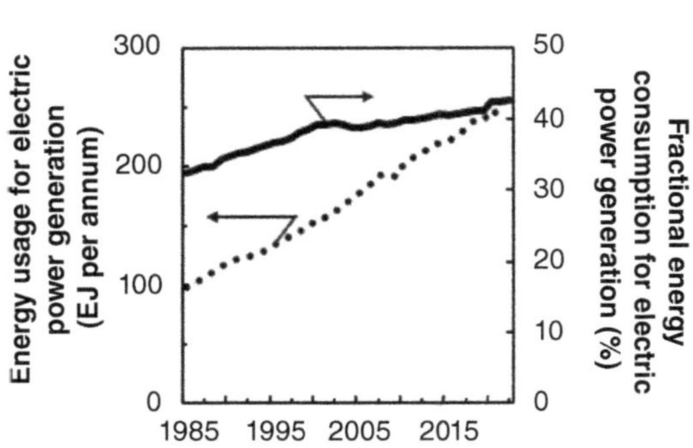

Figure 1.9 Primary energy consumption in EJ per annum for the generation of electricity and the amount used for electricity generation relative to the total amount of energy consumed in the period 1985–2022 (data from The Energy Institute [3]).

in primary energy demand, and thus the fraction of primary energy used to generate electric power has been increasing steadily with time as well. The ongoing fourth industrial revolution will push the demand for electric power up further, which may push towards a further increase in the energy demand going forward, as well as increase the utilisation of electric power.

1.4 Outlook

The world population has increased fivefold over the last century, and there is a general demand for a more equitable world in terms of economic opportunities. This would require a strong economic growth in the world overall and specifically in the developing world. Whilst there is a link between economic growth and increased energy consumption, the relationship is highly complex [39]. In open economies, energy consumption is also related to factors such as industrialisation, the presence of natural resources, energy intensity and urbanisation [40], as well as the export of energy-intensive materials [41]. Achieving this with the current energy mix will increase the concentration of greenhouse gases in the atmosphere, thus increasing the global average temperature.

Hence, the world has set itself on a path to reduce the anthropogenic emission of greenhouse gases to attempt to limit global warming. In the Paris Agreement (COP21) in 2015, the participants set as a target 'to limit global temperature increase to well below 2 °C, while pursuing efforts to limit the increase to 1.5 °C by reducing the anthropogenic emission of greenhouse gases, but also preserving natural sinks for these compounds' [42]. This legally binding agreement is supported by nationally determined contributions (NDCs), which are being monitored to track the progress biannually from 2024 onwards. The ultimate goal is to get to a net-zero society, where anthropogenic emissions are balanced by anthropogenic sinks (e.g. carbon capture and sequestration); the rate at which this is achieved will determine the expected increase in the global average temperature and the impact on global climate. Different scenarios have been developed to determine the course of the global average temperature upon reducing the net anthropogenic emission of greenhouse gases [43]. By limiting the increase of the global average temperature relative to pre-industrial levels to 1.5 °C with limited overshoot, the net anthropogenic greenhouse gas emissions should decrease by ca. 45% from the 2010 level by 2030 [36]. Higher levels of emission of greenhouse gases in 2030 will push back the time at which the net-zero state will be achieved [36, 43] and increase the likelihood of a temperature overshoot.

The net-zero scenario requires a strong and rapid reduction of the dependency of the world on fossil fuels. Electric power, which is mainly produced from fossil fuels (78% in 2019), could be replaced cost effectively [36] with renewable energy carriers such as wind and solar energy, provided sufficient storage and excess capacity is built-in [44], reducing CO_2 emissions by 40% relative to 2019 emissions and 36% relative to 2010 emissions (based on data from the Energy Institute [3]). Transportation consumes ca. 27% of all energy, and decarbonisation of this sector is necessary to achieve a net-zero situation, although some of its sub-sectors, such as aviation, may be difficult to decarbonise [45].

Industry is the largest sectorial converter of energy and will have to transform if a net-zero society is to be achieved. It should be noted that most of the greenhouse gas emissions per unit income are associated with the basic industries, whereas the derivative industries have much lower greenhouse emission per unit income [46]. For instance, the equivalent CO_2 emission per unit income in the chemical industry is 4 times higher than the emissions from the derivative group of fast-moving consumer groups; similar comparisons can be made for steel production versus automobile manufacturing (4.5 times higher), cement industry versus construction industry (46 times higher) and agriculture versus food industry (3.9 times higher). This is a consequence of the energy intensity associated with the base industries, indicating a dilution of the monetary carbon footprint down the product value chain. However, it also implies that decarbonisation efforts in industry should have a strong focus on the base industries.

Industry can classify its emissions into emissions directly emanating from its own operation (so-called Scope 1 emissions), emissions associated with its electricity usage (Scope 2 emissions), as well as the more difficult to quantify Scope 3 emissions, which are associated with all activities surrounding the production process (from the transport and production of the raw materials to getting the product to the client). Table 1.4 shows the contribution of the various sources to the Scope 1 and Scope 2 emissions of greenhouse gases by the industry in 2019 [47]. The direct emission of greenhouse gases by the industry in 2019 amounted to ca. 14.1 Gton, of which 49.4% arose from heating by fuel combustion, 22.2% was emitted as a product of the chemical processes and 16.5% was emitted as a waste greenhouse gas. The indirect emission of greenhouse gases from using electricity and imported heating sources (Scope 2 emissions) amounted to ca. 5.9 Gton in 2019 (with electricity contributing 71.9% to the indirect emissions).

A part of the transformation of the industry will be relatively invisible as greening of electric power upon the introduction of renewable energy to the power grid will reduce the Scope 2 emissions associated with the production process and thus indirectly reduce the carbon footprint of the industry. However, electric power makes up only 21% of the total energy utilisation in industry, and further measures will need to be implemented to move towards a net-zero society, such as further improvement in process efficiency, electrification of processes or even changing the materials used in the various processes.

The implementation of novel technologies to use renewable energy can be driven by an economic value proposition, as well as social factors. The economics of a process is determined by its operating and capital costs, which are a determining factor when considering the replacement of processes in industry. Policy-driving carbon pricing forms a part of these considerations as they contribute to the operating costs.

Carbon pricing can take the form of a carbon tax or emission trading systems (carbon credits) to be paid at the point of emission [48] and has been introduced in several countries as a measure to incentivise industry to reduce their emissions. Carbon pricing can be considered as a sin tax (much like taxation on tobacco and

Table 1.4 Industrial greenhouse gas emissions in 2019 (data from Bashmakov et al. [47]).

Source		**2019 emissions MiO tonne CO_2-eq.**	
Direct CO_2 emissions (fuel combustion)	Iron and steel	2481	(12.4%)
	Non-metallic minerals	1148	(5.7%)
	Chemical and petrochemical	977	(4.9%)
	Non-ferrous metals	163	(0.8%)
	Paper, pulp and printing	150	(0.7%)
	Food and tobacco	265	(1.3%)
	Other	1797	(9.0%)
CO_2 emission (chemical process)	Non-metallic minerals	2008	(10.0%)
	Chemical and petrochemical	720	(3.6%)
	Metallurgy	391	(2.0%)
	Other	25	(0.1%)
Industrial product use GHG		204	(1.0%)
Other non-CO_2 GHG		1470	(7.3%)
Waste GHG		2327	(11.6%)
Indirect emissions	Electricity	4236	(21.2%)
	Heating	1663	(8.3%)
Total		**20 025**	

gambling), which adds to the cost of manufacture and will thus increase the cost of living. Although it can be argued that the income from carbon pricing should contribute to the mitigation of the impact of greenhouse gas emissions or shelter the most affected by the increase in the cost of living, the effects of the greenhouse gas emissions and the associated effect of the global temperature are not bound by the borders of nations.

Carbon tax is often based on the amount of carbon in the fuel purchased for a process, whereas companies have to purchase carbon credits for their emissions, when going beyond their allocated carbon credits. Governments will have a better handle on emissions nationwide through the restrictive release of these credits. The carbon credits can be traded with the marketplace setting the price. This may result in fluctuations in the price of carbon credits (e.g. EU carbon credits were traded in 2022–2023 between €54.2 and €104.8 per tonne $CO_{2,eq}$ [49]). Carbon tax, on the other hand, is more predictive in nature and gives the industry more control over its production costs. It has been argued [48] that carbon tax and emission credit systems should come to the same carbon pricing, although carbon tax is determined in the political arena and carbon credits in the open market.

Carbon pricing schemes have been implemented in the EU, in various nations and even in some sub-nation regions such as California. However, in some countries, less than 30% of all greenhouse gas emissions are subject to carbon pricing, whilst the remainder is exempt [48]. Furthermore, the actual carbon price varies across the nations which have implemented carbon tax and/or an emission trading system between US$3 and 100/tonne $CO_{2,eq}$.

The variability in the carbon pricing worldwide skews the economic costs of operation and may result in the movement of industries with a large carbon footprint to regions with lower or no carbon pricing (termed carbon leakage). To counter this behaviour, carbon tariffs will be put in place by countries with a high carbon pricing [50], in principle safeguarding their own industry from competition operating in a different cost environment. In the world-trade scenario, it implies that products produced by industries operating outside the boundaries of the carbon tariff zone will have an added cost factor. A clear distinction between carbon tax and carbon tariffs is the receiver of these funds, with carbon tax being paid at the point of emission and carbon tariffs being paid at the border. The introduction of carbon tariffs by large markets should also provide an incentive for industry targeting these markets to reduce their emissions whilst not being part of that system and thus minimising the impact of the carbon tariffs on the cost of the product at the marketplace. Direct emissions, so-called Scope 1 emissions, can be addressed readily by the affected industry, but it will have less influence on so-called Scope 2 emissions, which depend on the local power grid. The effect of carbon tariffs on the world market and especially on the division between the developed and the developing world will have to be considered [51].

Industries are not operating in isolation, but form part of a society. Hence, the acceptance of their operation by society has become an important aspect in decision-making in industry, which has been termed a 'social license to operate' (SLO) [52]. The concept of SLO finds its origin in the resource extraction industry, where the acceptance of their operations by the local communities has slowly become an integral part of the decision-making process. The concept can be extended to all industries, as their products are offered to a global marketplace with competitors and willing and informed consumers who may prefer products with a lower environmental impact and greenhouse gas emissions.

There are sufficient driving forces to implement renewable energy in the process industry and reduce their associated greenhouse gas emissions, but there are also challenges in the implementation. These challenges will need to be addressed by the industry, but the involvement of other role players, such as power suppliers for grid connectivity, local government for local infrastructure, national government for supporting policies and financing sector, is also required. Here, we will look mainly at the technical and techno-economic aspects of changing the energy carrier in the process industry.

References

1 Quote Investigator (2018). The stone age did not end because the world ran out of stones, and the oil age will not end because we run out of oil, *Quote Investigator*, 7 January 2018 [Online]. Available: https://quoteinvestigator.com/2018/01/07/stone-age/#note-17685-1 (accessed 14 April 2021).

2 Malanima, P. (2020). World energy consumption – a database 1820–2018 (revision 2020) [Online]. Available: https://histecon.fas.harvard.edu/energyhistory/DATABASE%20World%20Energy%20Consumption.pdf (accessed 7 February 2024).

3 The Energy Institute (2023). Statistical review of world energy [Online]. Available: https://www.energyinst.org/statistical-review/about (accessed 7 February 2024).

4 Friedlingstein, P., O'Sullivan, M., Jones, M. et al. (2023). Global carbon budget 2023. *Earth System Science Data* 15 (2023): 5301–5369.

5 Jones, M.W., Peters, G.P., Gasser, T. et al. (2023). National contributions to climate change due to historical emissions of carbon dioxide, methane, and nitrous oxide since 1850. *Science Data* 10: 155.

6 MacFarling Meure, C., Etheridge, D., Trudinger, C. et al. (2006). Law dome CO_2, CH_4 and N_2O ice core records extended to 2000 years BP. *Geophysical Research Letters* 33 (14): L14810.

7 Lan, X. and Keeling, R. (2024). Trends in CO_2, CH_4, N_2O, SF_6 [Online]. Available: https://gml.noaa.gov/ccgg/trends/data.html (accessed February 2024).

8 Lan, X., Thoning, K.W., and Dlugokencky, E.J. (2024). Trends in globally-averaged CH_4, N_2O, and SF_6 determined from NOAA Global Monitoring Laboratory measurements. Version 2024–02 [Online]. Available: https://doi.org/10.15138/P8XG-AA10 (accessed 7 February 2024).

9 Lindsey, R. and Dahlman, L. (2024). Climate change: Global temperature [Online]. Available: https://www.climate.gov/news-features/understanding-climate/climate-change-global-temperature (accessed 7 February 2024).

10 Lightfoot, H. (2007). Understand the three different scales for measuring primary energy and avoid errors. *Energy* 32: 1478–1483.

11 de Vries, A. (2023). The growing energy footprint of artificial intelligence. *Joule* 7 (10): 2191–2194.

12 Lau, B. (2022). What are the major sources of methane in the atmosphere? [Online]. Available: https://earth.org/sources-of-methane-in-the-atmosphere (accessed 8 February 2024).

13 Callendar, G.S. (1938). The artificial production of carbon dioxide and its influence on temperature. *Quarterly Journal of the Meteorological Society* 64 (275): 223–240.

14 Myhre, G., Shindell, D., Bréon, F.-M. et al. (2013). Anthropogenic and natural radiative forcing. In: *Climate Change 2013: The Physical Science Basis.*

Contribution of Working Group I to the Fifth Assessment Report of the Intergovernmental Panel on Climate Change (ed. T.F. Stocker, D. Qin, G. Plattner, et al.), 659–740. Cambridge, UK: Cambridge University Press.

15 Nair, A.A. and Yu, F. (2020). Quantification of atmospheric ammonia concentrations: a review of its measurement and modeling. *Atmosphere* 11: 1092.

16 Pai, S.J., Heald, C.L., and Murphy, J.G. (2021). Exploring the global importance of atmospheric ammonia oxidation. *ACS Earth and Space Chemistry* 5 (7): 1674–1685.

17 Carvalho, D., Rafael, S., Monteiro, A. et al. (2022). How well have CMIP3, CMIP5 and CMIP6 future climate projections portrayed the recently observed warming. *Scientific Reports* 12: 11983.

18 Hausfather, Z., Marvel, K., Schmidt, G.A. et al. (2022). Climate simulations: recognize the 'hot model' problem. *Nature* 605: 26–29.

19 Randall, D.A., Wood, R.A., Bony, S. et al. (2007). Climate models and their evaluation. In: *Climate Change 2007: The Physical Science Basis. The Contribution of Working Group I to the Fourth Assessment Report of the Intergovernmental Panel on Climate Change* (ed. S. Solomon, D. Qin, M. Manning, et al.), 589–662. Cambridge, UK: Cambridge University Press.

20 Larson, R.B. and Bromm, V. (2004). The first stars in the universe. *Scientific America* 14 (4): 4–11.

21 Global Solar Atlas (2025). Global Solar Atlas 2.0 [Online]. Available: https://globalsolaratlas.info (accessed 28 August 2025).

22 Davis, N.N., Badger, J., Hahmann, A.N. et al. (2023). The global wind atlas: a high-resolution dataset of climatologies and associated web-based application. *Bulletin of the American Meteorological Society* 104 (8): E1507–E1525.

23 Engineering Toolbox (2003). Higher and lower calorific calues [online]. Available: https://www.engineeringtoolbox.com/fuels-higher-calorific-values-d_169.html (accessed 16 February 2024).

24 Engineering Toolbox (2003). Wood - combustion heat values [Online]. Available: https://www.engineeringtoolbox.com/wood-combustion-heat-d_372.html (accessed 16 February 2024).

25 World Nuclear Association (2023). What is uranium? How does it work? [Online]. Available: https://www.world-nuclear.org/information-library/nuclear-fuel-cycle/introduction/what-is-uranium-how-does-it-work.aspx#Who%20Uses%20Nuclear%20Power? (accessed 20 February 2024).

26 Ewing, R. (2015). Long-term storage of spent nuclear fuel. *Nature Materials* 14: 252–257.

27 Tollefson, J. and Gibney, E. (2022). Nuclear-fusion lab achieves 'ignition': what does it mean? *Nature* 612: 597–598.

28 Schlömer, S., Bruckner, T., Fulton, L. et al. (2014). Annex III: Technology-specific cost and performance parameters. In: *Climate Change 2014: Mitigation of Climate Change. Contribution of Working Group III to the Fifth Assessment Report of the Intergovernmental Panel on Climate Change* (ed. O. Edenhofer, R. Pichs-Madruga, Y. Sokona, et al.), 1329–1356. Cambridge, UK: Cambridge University Press.

29 Ritchie, H. (2021). How have the world's energy sources changed over the last two centuries? http://OurWorldInData.org, 1 December 2021 [Online]. Available: https://ourworldindata.org/global-energy-200-years (accessed 19 February 2024).

30 IEA (2022). Russian supplies to global energy markets, IEA, Paris, Licence: CC BY 4.0 [Online]. Available: https://www.iea.org/reports/russian-supplies-to-global-energy-markets (accessed 19 February 2024).

31 Alvarez, C. and Arnold, F. (2020). What the past decade can tell us about the future of coal, IEA, 02 December 2020 [Online]. Available: https://www.iea.org/commentaries/what-the-past-decade-can-tell-us-about-the-future-of-coal (accessed 19 February 2024).

32 Lu, M. (2023). Energy transition: Explainer: this is how geothermal energy works, World Economic Forum, 04 January 2023 [Online]. Available: https://www.weforum.org/agenda/2023/01/geothermal-energy-infographics-data-renewable (accessed 20 February 2024).

33 Obidzinski, K., Andriani, R., Komarudin, H., and Andrianto, A. (2021). Environmental and social impacts of oil palm plantations and their implications for biofuel production in Indonesia. *Ecology and Society* 17: 25.

34 Meijaard, E., Brooks, T., Carlson, K. et al. (2020). The environmental impacts of palm oil in context. *Nature Plants* 6: 1418–1426.

35 Osman, A., Mehta, N., Elgarahy, A. et al. (2021). Conversion of biomass to biofuels and life cycle assessment: a review. *Environmental Chemistry Letters* 19: 4075–4118.

36 Pathak, M., Slade, R., Shukla, P. et al. (2022). Technical summary. In: *Climate Change 2022: Mitigation of Climate Change. Contribution of Working Group III to the Sixth Assessment Report of the Intergovernmental Panel on Climate Change* (ed. P. Shukla, J. Skea, R. Slade, et al.), 51–147. Cambridge, UK: Cambridge University Press.

37 Ermolaev, I., Ermolaev, V., and Mordkovich, V. (2014). Efficiency of gas-to-liquids technology with different synthesis gas production methods. *Industrial & Engineering Chemistry Research* 53 (7): 2758–2763.

38 Li, Y.-M., Khurshid, A., and Khan, K. (2023). Optimization of coal-to-liquid processes; a way forward towards carbon neutrality, high economic returns and effective resource utilization. Evidences from China. *Fuel* 344: 128082.

39 Ozturk, I., Aslan, A., and Kalyoncu, H. (2010). Energy consumption and economic growth relationship: evidence from panel data for low and middle income countries. *Energy Policy* 38: 4422–4428.

40 Azam, M., Gohar, A., and Bekun, F. (2022). Estimating the energy consumption function: evidence from across the globe. *Environmental Science and Pollution Research* 29: 59060–59075.

41 Wu, X., Guo, J., Meng, J., and Chen, G. (2019). Energy use by globalized economy: total-consumption-based perspective via multi-region input-output accounting. *Science of the Total Environment* 662: 65–76.

42 UN Climate Change. Key aspects of the Paris agreement [Online]. Available: https://unfccc.int/process-and-meetings/the-paris-agreement (accessed 22 February 2024).

43 Rogelj, J., Popp, A., Calvin, K. et al. (2018). Scenarios towards limiting global mean temperature increase below 1.5 °C. *Nature Climate Change* 8: 325–332.
44 Tong, D., Farnham, D., Duan, L. et al. (2021). Geophysical constraints on the reliability of solar and wind power worldwide. *Nature Communications* 12: 6146.
45 Jaramillo, P., Ribeiro, S.K., Newman, P. et al. (2022). Transportation. In: *IPCC, 2022: Climate Change 2022: Mitigation of Climate Change. Contribution of Working Group III to the Sixth Assessment Report of the Intergovernmental Panel on Climate Change* (ed. P. Shukla, J. Skea, R. Slade, et al.), 1049–1160. Cambridge, UK: Cambridge University Press.
46 Gomez, F., Kolomanska-van Iperen, J., Sandström, J. et al. (2021). Implementing low-carbon emitting technologies in the chemical industry: a way forward. World Economic Forum, 2021 [Online]. Available: https://www3.weforum.org/docs/WEF_Implementing_Low_Carbon_Emitting_Technologies_in_the_Chemical_Industry_A_Way_Forward_2021.pdf (accessed 21 May 2024).
47 Bashmakov, I., Nilsson, L., Acquaye, A. et al. (2022). Industry. In: *IPCC, 2022: Climate Change 2022: Mitigation of Climate Change. Contribution of Working Group III to the Sixth Assessment Report of the Intergovernmental Panel on Climate Change* (ed. P. Shukla, J. Skea, R. Slade, et al.), 1161–1243. Cambridge, UK: Cambridge University Press.
48 Parry, I., Black, S., and Zhunussova, K. (2022). Carbon taxes or emissions trading systems? Instrument choice and design. International Monetary Fund, July 2022 [Online]. Available: https://www.imf.org/en/Publications/staff-climate-notes/Issues/2022/07/14/Carbon-Taxes-or-Emissions-Trading-Systems-Instrument-Choice-and-Design-519101 (accessed 21 May 2024).
49 Trading Economics (2024). Carbon credits [Online]. Available: https://tradingeconomics.com/commodity/carbon#:~:text=EU%20carbon%20permit%20prices%20have%20stabilized%20at%20around,demand%20offset%20the%20recovery%20in%20natural%20gas%20prices (accessed 21 May 2024).
50 Keen, M., Parry, I., and Roaf, J. (2021). Border carbon adjustments: rationale, design and impact. International Monetary Fund, September 2021 [Online]. Available: https://www.imf.org/en/Publications/WP/Issues/2021/09/24/Border-Carbon-Adjustments-Rationale-Design-and-Impact-466176 (accessed 21 May 2024).
51 Böhringer, C., Fischer, C., Rosendahl, K., and Rutherford, T. (2022). Potential impacts and challenges of border carbon adjustments. *Nature Climate Change* 12: 22–29.
52 Moffat, K., Lacey, J., Zhang, A., and Leipold, S. (2016). The social licence to operate: a critical review. *Forestry* 89: 477–488.

2

Power Generation – Harnessing Renewable Energy

2.1 Introduction

The contribution of electric power to the overall energy palette has been increasing steadily over the last 50 years (see Figure 1.9). Electric power has been generated from different primary energy carriers (see Table 2.1 and Figure 2.1) with coal being the major contributor (35% in 2022). Renewable energy carriers made up 31% of the total primary energy used to generate electric power in 2022, of which the majority was generated using hydropower (56%), but wind and solar energy contribute significantly at 27% and 17%, respectively, of the renewable energy. Natural gas was the third largest pillar for power generation in 2022, contributing 23% to the overall power generation, followed by nuclear (9%). A small fraction (4%) is generated using hydro-storage, non-renewable waste, from production in the chemical industry, and other sources.

Power generation is not evenly distributed over the world. Furthermore, the world uses a variety of primary energy carriers to generate power (see Table 2.1). Coal or natural gas is, in most countries, the dominant primary energy carrier used to generate power. There are some exceptions, e.g. power generation in France is dominated by nuclear energy, whilst hydropower is dominant in Canada and Brazil.

North America, with 6% of the world's population, generates more than 19% of all the electric power in the world. The United States (having 67% of the North American population) is the dominant player in this region, generating almost 82% of the total power. Hence, the sources of primary energy used to generate power in North America strongly mimic the power generation in the United States (slightly shifted towards the generation using hydropower due to dominance of hydropower in Canada). Power generation in North America is now dominated by natural gas (38%) which has replaced coal as the major energy carrier for power generation [2]. Nuclear contributes 16% to the power generation in the region, with hydropower contributing 13%. Power generation from renewables, in particular wind, has been growing strongly in this region and accounted for 9% of the total power generation in 2022.

Renewable Energy in the Process Industry, First Edition.
Eric van Steen, Nicholas S. Featherstone, and Linda H. Callanan.

Table 2.1 Power generation in 2022 in selected countries and the source of the primary energy (data sourced from The Energy Institute [1]).

	Population Million	Annual generation TWh	Source in % of total annual generation						
			Coal	Gas	Nuclear	Hydro	Wind	Solar	Other
North America	*504*	*5548*	*17.3*	***37.7***	*16.4*	*12.5*	*9.0*	*4.2*	*3.0*
USA	338	4548	19.9	**39.9**	17.9	5.7	9.7	4.5	2.4
Canada	38	660	5.2	12.3	13.1	**60.4**	5.7	0.9	2.4
South & Central America	*533*	*1410*	*4.0*	*16.8*	*1.6*	***52.9***	*8.4*	*3.9*	*12.5*
Argentina	46	151	1.4	**53.3**	5.0	15.8	9.4	1.9	13.1
Brazil	215	677	2.4	6.2	2.1	**63.1**	12.1	4.4	9.6
Europe	*821*	*5068*	*16.6*	***25.7***	*19.0*	*15.1*	*11.0*	*4.9*	*7.6*
France	65	468	0.7	10.0	**63.0**	9.5	8.1	4.3	4.3
Germany	83	577	**31.3**	13.8	6.0	3.0	21.7	10.5	13.6
Italy	59	287	6.1	**54.4**	0.0	9.8	7.2	9.6	12.9
Poland	40	179	**71.1**	6.5	0.0	1.1	10.9	4.5	5.9
Russia	145	1167	16.5	**45.8**	19.2	16.9	0.4	0.2	1.1
Spain	48	294	3.2	**30.4**	20.0	6.2	21.4	11.5	7.4
UK	68	326	1.7	**38.4**	14.6	1.6	24.6	4.3	14.7
Middle East	*279*	*1365*	*1.4*	***72.0***	*2.0*	*0.9*	*0.2*	*1.7*	*21.8*
Iran	89	348	0.2	**86.2**	1.9	2.1	0.4	0.2	9.0
Saudi Arabia	36	407	0.0	**67.1**	0.0	0.0	0.0	0.2	32.7
Africa	*1427*	*893*	*26.5*	***40.7***	*1.1*	*17.6*	*2.7*	*2.0*	*9.4*
Egypt	111	201	0.0	**79.3**	0.0	6.9	2.6	2.5	8.8
South Africa	60	235	**84.0**	0.0	4.3	1.3	4.1	2.6	3.6
Asia/Australia	*4411*	*14 881*	***55.1***	*11.1*	*5.0*	*13.2*	*5.0*	*6.1*	*4.5*
Australia	26	274	**47.9**	16.9	0.0	6.3	11.6	14.2	3.2
China	1426	8849	**61.0**	3.3	4.7	14.7	8.6	4.8	2.8
India	1417	1858	**74.3**	2.5	2.5	9.4	3.8	5.1	2.4
Indonesia	276	333	**61.6**	16.8	0.0	8.2	0.1	0.1	13.2
Japan	26	274	**47.9**	16.9	0.0	6.3	11.6	14.2	3.2
South Korea	52	620	**33.6**	27.9	28.4	0.6	0.5	4.3	4.6
Vietnam	98	260	**38.8**	10.7	0.0	36.9	3.1	10.1	0.4

Values in bold: dominant primary energy carrier for power generation; values in italics: dominant primary energy carrier on a (sub-)continent.

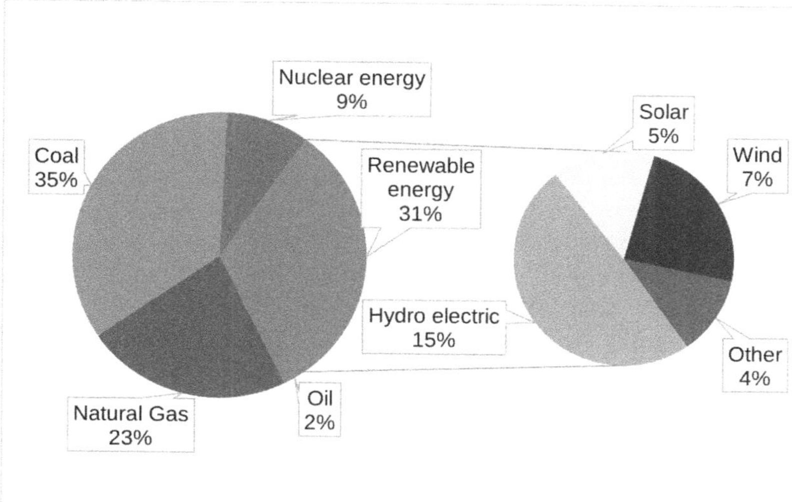

Figure 2.1 Worldwide usage of energy carriers to generate electricity in 2022 (data from The Energy Institute [1]).

South America, that in terms of population is somewhat larger than North America, generates only 5% of the electric power in the world. Hydropower is the main energy source for power generation with 57% of all hydropower in this region being generated in Brazil.

Power generation in Europe (including the Russian Federation) amounted to 17% of all the electric power generated in the world in 2022 (whilst accounting for 10% of the world population). A more diverse palette of energy carriers is being used to generate power, although most power is still being generated from natural gas (26%). Nuclear energy, due to the dominance of power generation in France, contributes 19% to the power generation. Renewable energy (hydropower, wind energy and solar energy) contributes 31%. It is interesting to note that wind energy contributes more than 20% to the power generation in the United Kingdom, Germany and Spain, but also solar energy contributes more than 10% in Germany and Spain.

The Middle East is a relatively small region (with 3% of the world population), which generates 5% of the power in the world. Power generation in this region is dominated by natural gas, although a relatively high proportion is being generated using oil as well (denoted as others in Table 2.1).

Africa, with a rapidly growing population (currently 18% of the world population), can be classified as a power-poor region. It generates only ca. 3% of all the electric power in the word (i.e. less than the Middle East). It can be noted that 49% of the power on this continent was generated in two countries, viz. Egypt and South Africa, each of them using their local resources (natural gas and coal, respectively) to generate power. Power generation in Africa can be expected to grow significantly going forward.

Most people live in Asia and Oceania (55% of the world's population). Power generation in this part of the world has shown a strong growth accompanying the economic growth in the region [3] and stands, in 2022, at 51% of the total power generation in the world. Further economic growth is expected to fuel growth in the electric power demand as well. Coal is the main source used to generate power in this region across all countries, but most notably China and India, the two largest power producers (and largest populations), though it is worth noting that despite a similar population China produces 4.8 times the amount of power that is produced in India. The continued growth of the number of coal-fired power stations in this region is poorly understood [4], but might be explained by a mixture of factors ranging from the need of cheap and stable supply of electricity to the use of local resources and the ability to store the energy carrier.

2.2 Thermal Power Generation Processes

Most of the electricity is generated using a heat source, by burning a fuel or from a nuclear reaction. The fuel can be a carbon-containing fuel, such as coal, natural gas or municipal waste, or (much less commonly) ammonia. The conversion of a heat source into electric power is thermodynamically limited by the Carnot efficiency, which is the maximum efficiency by which work can be extracted from this system. As an example, receiving the heat at 600 °C and giving it off at 100 °C limits the efficiency for thermal power generation to 57.3%, whilst decreasing the temperature at which heat is given off to 35 °C increases the Carnot efficiency to 65.9% (see Figure 2.2). Hence, the drive is to maximise the thermal efficiency by increasing the difference between the temperature at which heat is received and the temperature at which heat is given off.

2.2.1 Rankine Cycle

Electric power from heat is generated using the so-called Rankine cycle (see Figure 2.3), although other more direct conversion technologies could be developed, such as thermoelectric devices [5], although the maximum efficiency of these

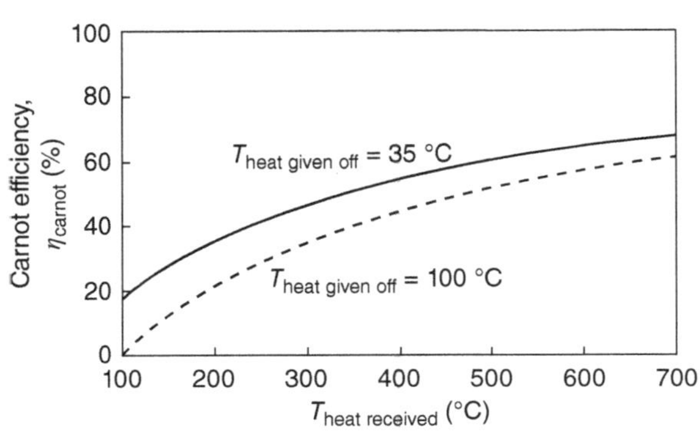

Figure 2.2 Carnot efficiency as a function of the temperature at which heat is received with heat being given off at 35 °C or 100 °C ($\eta_{Carnot} = 1 - \frac{T_C}{T_H}$).

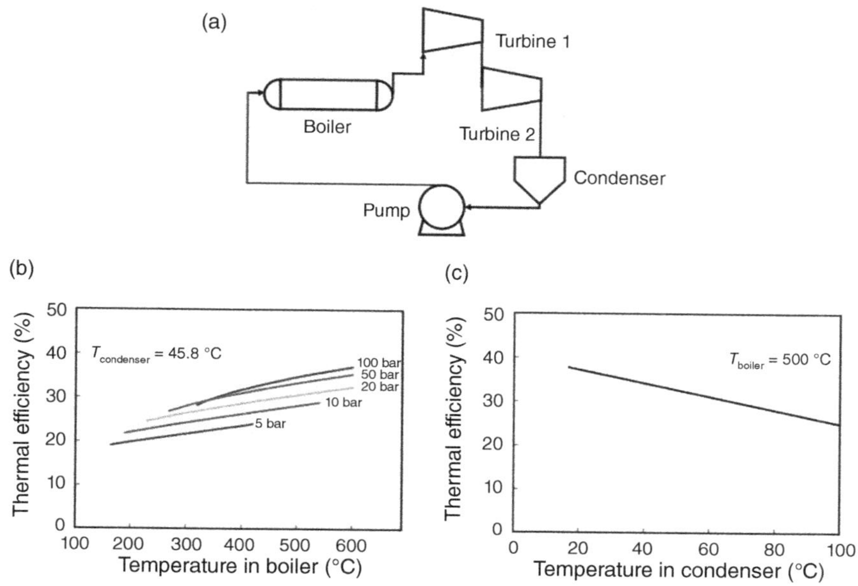

Figure 2.3 Power generation using a Rankine cycle using steam as the working fluid operating with two turbines in series to achieve maximum efficiency. (a) process flow diagram (PFD), (b) efficiency as a function of the operating conditions, temperature and pressure, in the boiler and (c) efficiency as a function of the temperature in the condenser. In all cases, an efficiency of the pump and the turbines under dry conditions of 0.9 is assumed.

devices is low with currently used materials [6]. In the Rankine cycle, a compressed working fluid (typically water/steam) is heated in a boiler to the required temperature. The fluid is subsequently expanded over a turbine or a set of turbines causing a decrease in the temperature. This may be accompanied by partial condensation of the working fluid. Heat is subsequently removed from the working fluid by passing it through a condenser yielding a saturated liquid. The pressure in the condenser (and thus the outlet pressure of the compressor) is determined by the temperature in the condenser. Subsequently, the saturated liquid is pumped up to the required pressure.

The low pressure in the condenser will result in the formation of wet steam at the exit of the turbine(s), if water is used as the working fluid. The appearance of droplets in the expanded steam may cause blade erosion and corrosion damage [7], but also in a loss in the thermal efficiency due to the so-called wetness losses [8]. It is, therefore, more efficient from a maintenance viewpoint to have multiple turbines in series with only the last turbine operating in the wet steam regime, i.e. entering with a saturated steam and expanding it to the desired pressure.

The typical operating variables for a Rankine cycle are the temperature at which heat is received in the boiler, the pressure of the working fluid in the boiler and the temperature at which heat is being given off to the surroundings. The latter will control the thermal efficiency of the last turbine. The decrease in the thermal efficiency of the turbine operating under slightly wet conditions (η_{wet})

can be empirically correlated to the efficiency of the turbine operating under dry conditions (η_{dry}) with so-called Baumann relationship [8]:

$$\eta_{wet} = \eta_{dry} \cdot (1 - \alpha \cdot \bar{y})$$

with $\bar{y}$ the mean wetness of the stream and α an empirical factor between 0.4 and 2.5.

Figure 2.3 shows the efficiency of a Rankine cycle operating with 2 turbines in series (each operating with a 90% efficiency under dry conditions) as a function of the various operating variables, viz. temperature and pressure in the boiler, and temperature in the condenser. The effluent leaving the 1st turbine is a slightly superheated steam, and the effluent leaving the condenser a saturated liquid in order to maximise the overall efficiency. A higher efficiency can be obtained by reducing the quality of the steam leaving the turbine, which can be achieved by increasing the number of turbines.

The range of operating temperatures, when operating at a modest pressure of 5 bar, is restricted (see Figure 2.3b) with the low temperature limit given by the saturation point of steam (151.83 °C), and the high temperature limit (410 °C) given by the point at which there is no resulting condensation in the turbine and thus, multiple turbines will not increase the overall thermal efficiency. Increasing the pressure in the boiler will shift the operating range for two turbines in series towards higher temperatures, whilst at the same time increasing the thermal efficiency of the Rankine cycle. Furthermore, the thermal efficiency increases with increasing pressure in the boiler. The quality of steam generated in the second turbine from steam at a given pressure decreases with increasing temperature in the boiler – countering the effect of improved efficiency with this parameter. Hence, the thermal efficiency of a 2-turbine system with the boiler operating at 100 bar and below 330 °C is less than with the boiler operating at 50 bar. It should however be noted that the Rankine cycle will not operate at these conditions as the quality of the steam is too low, resulting in damage to the turbine blades, and multiple low-pressure turbines will be used to minimise the condensate formation in the turbines, thereby improving the efficiency (at the cost of a higher capital expenditure).

The efficiency for electric power generation from heat thus increases with increasing temperature and pressure in the boiler, thus driving the development of boilers which can operate at these conditions (see Table 2.2). In coal-based power plants, the combustion of pulverised coal is used to generate steam at ca. 500 °C and 100 bar [9, 10]. Similar operating conditions can be obtained when using natural gas, LPG, light fuel oil or heavy fuel oil as the fuel [11]. Increasing the pressure to beyond the critical pressure of water (220.64 bar) will improve the efficiency of the steam generation in a coal power plant by ca. 1.6% [9]. Going to ultra-supercritical conditions can increase the efficiency to over 50% [12] (under these conditions, the Carnot efficiency increases to 70%, assuming heat is received at 750 °C and given off at 35 °C). These conditions cannot be achieved, when converting heat generated by combustion of solid municipal waste, due to corrosion concerns in the boiler [13].

An important parameter in the conversion of heat into power is the temperature at which heat is removed (see Figure 2.3c). Being able to lower the temperature (and thus the pressure in the condenser) improves the thermal efficiency, which starts

Table 2.2 Typical temperatures and pressures in steam boilers for different power plants.

Heat source	Operation	*T* (°C)	*P* (bar)	Reference
Coal power plant		510	100	[9, 10]
	Supercritical	540–600	250–300	[9, 12]
	Ultra-supercritical	760	350	[14]
Waste incineration		400	400	[13]
Nuclear power plant	Boiling water plant	288	70	[15]
	High pressure reactor	326	144	[15]

getting closer to the Carnot efficiency at these conditions. This is typically achieved by using cooling water, and hence, locality and the availability of water do affect the efficiency of power generation.

2.2.2 Brayton Cycle

The combustion of fuels at a constant pressure can be used to generate the heat used for the extraction of work from compressed air, in the so-called Brayton cycle. In this process, compressed air is mixed with the fuel gas at elevated pressure and combusted adiabatically. The pressurised fuel gas is then expanded over a turbine. Heat is then rejected from the process in the form of the effluent gas. This process is often used in describing open gas turbines (hence it is also referred to as the 'open cycle') and jet engines.

The efficiency of the Brayton cycle, defined as the ratio of the net amount of work extracted from the process relative to the heat of combustion of the fuel, depends on variables, such as pressure at which the combustion takes place, the air to fuel ratio and the type of fuel. Figure 2.4 shows the efficiency of the Brayton process utilising methane as a fuel; it passes a maximum as a function the parameter λ (defined as the actual air to methane ratio relative to the stoichiometric air to methane ratio for complete combustion). The efficiency of the process is determined by the change in the amount of work required by the compressors and the work delivered by the turbine. The work required by the compressors increases linearly with the amount of air being compressed. The work delivered by the turbine also increases linearly with the amount of air fed to the system but shows a discontinuity at $\lambda = 1$. The work delivered by the turbine is a function of the amount of gas going through the turbine and the temperature of the gas mixture leaving the combustion zone. The gas flow increases linearly with the air to methane ratio but the temperature of the gas leaving the combustion zone passes a maximum at $\lambda = 1$.

A higher pressure in the combustion zone results in a higher maximum efficiency for the power generation, but the efficiency drops off more rapidly upon increasing the amount of excess air ($\lambda > 1$). The work required for the compressors increases roughly linearly with increasing pressure; the work generated by the turbine is related to the pressure drop over the turbine, as the pressure differential determines the temperature gradient over the turbine. The change in the amount of work delivered by the turbine with changing excess ratio, and thus with the total molar flow rate, increases with increasing pressure. This can be readily seen by considering the maximum amount of work that can be obtained by expansion of an ideal gas, with a constant heat capacity, from a pressure p_{high} to atmospheric pressure, p_{at}:

$$W_{turbine}^{max} = \dot{n} \cdot C_P \cdot (T_{adiabatic} - T_{flue\ gas}) = \dot{n} \cdot C_P \cdot T_{adiabatic} \cdot \left(1 - \left(\frac{p_{at}}{p_{high}}\right)^{\frac{R}{C_P}}\right)$$

It can be deduced from Figure 2.4 that the variation in the amount of work delivered by the turbine as a function of pressure, whilst keeping the amount of excess air constant, will result in a maximum at a particular pressure.

A large fraction of the original thermal energy gained in the system can remain in the effluent stream leaving the Brayton cycle. The temperature of the effluent gas follows the adiabatic temperature and increases initially with the amount of air fed to the system. The slope of the temperature as a function of the parameter λ shows a discontinuity at $\lambda = 0.75$; after this air to methane ratio CO_2 will be formed stoichiometrically. The highest effluent temperature at a given pressure is obtained for a stoichiometric mixture of air and methane for the formation of CO_2. Mixtures with excess air yield lower effluent temperatures.

Additional work can be obtained from the effluent by using the effluent stream to generate steam for power generation in a Rankine cycle. This then becomes the so-called Combined Cycle Gas Turbine (CCGT) and results in enhanced thermal efficiency. It should be noted that the power plant now needs to be optimised with respect to both the pressure at which the combustion takes place, as well as the pressure and temperature in the boiler of the Rankine cycle. Figure 2.5 illustrates the improved efficiency obtained in the CCGT with, as constraints, a minimum temperature of the flue gas leaving the boiler of 400 °C and the Rankine cycle operating with steam leaving the boiler at 400 °C and 50 bar (the temperature at which the effluent leaves the boiler for the Rankine cycle is a design variable as well, with a higher temperature resulting in a higher amount of work produced in the Rankine cycle, but

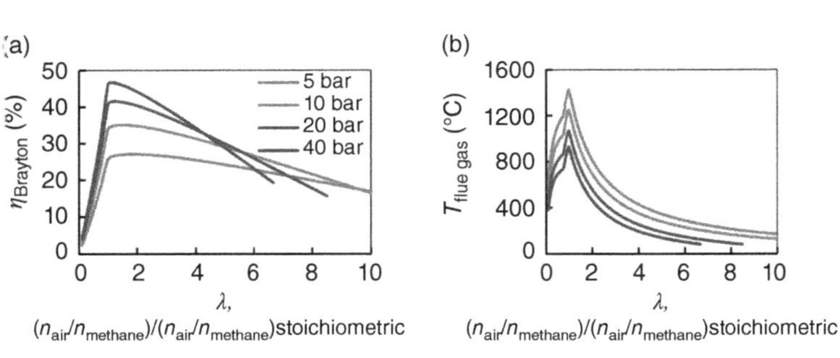

Figure 2.4 Efficiency of power generation in a Brayton cycle (a) and flue gas temperature (b) as a function of the air to methane ratio, λ and the pressure in the combustion chamber assuming an efficiency of the compressors and turbine of 0.9 (pressure ratio in compressors: ca. 3; gas phase modelled using SRK-EOS; Gibbs reactor with methane combusted to CO or CO_2; the outlet temperature of the turbine is set to at least 80 °C and 1 bar).

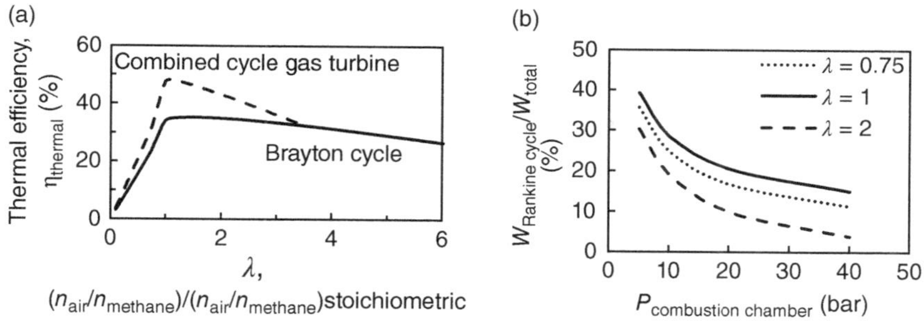

Figure 2.5 Efficiency of Combined Cycle Gas Turbine (CCGT) assuming an efficiency of the compressors and turbine of 0.9 (pressure ratio in compressors: ca. 3; gas phase modelled using SRK-EOS; Gibbs reactor with methane combustion to CO or CO_2; the outlet temperature of the turbine is set to at least 400 °C; Rankine cycle operating with boiler at 400 °C and 50 bar – $\eta_{Rankine} = 31.9\%$). (a) Comparing the efficiency of the CCGT with a Brayton cycle as a function of the air to methane ratio, λ. Pressure in the combustion chamber: 10 bar. (b) Contribution of Rankine cycle to work generated in the CCGT as a function of the pressure in the combustion chamber.

less work from the Brayton cycle). The CCGT must be run at much closer to a stoichiometric feed to achieve a high thermal efficiency. At these conditions, the flue gas temperature is high (see Figure 2.4b), and more heat can be converted into work using the Rankine cycle. The contribution of the additional Rankine cycle to the overall power generation passes a maximum as a function of the excess air fed to the Brayton cycle because of the variation of the flue gas temperature as a function of the excess air fed to the combustion chamber. Furthermore, the additional Rankine cycle contributes less to the overall work generated in the CCGT. This is because a higher pressure of the combustion gas results in a larger temperature reduction over the turbine and thus less heat being available for the Rankine cycle.

The pressure in the combustion chamber, as well as the pressure and the temperature of the steam leaving the boiler in the Rankine cycle are the prime variables that can be used to increase the efficiency of the CCGT further [16], with the efficiency of the CCGT being potentially increased to well above 50%.

2.2.3 Environmental Aspects

The environmental impact of power generation should factor in waste generation during fuel provision, arising from the conversion of fuel into power, as well as arising from the construction and decommissioning of the plant [17]. Hence, the environmental impact thus becomes not only strongly dependent on the applied technology, but also on the time and location of construction, and duration of operation.

The generation of heat using nuclear fuel generates ca. 0.22–1.1 m^3 spent nuclear fuel per TWh of thermal energy depending on the design of the nuclear plant in addition to contaminated concrete, steel and cooling material [18]; this corresponds to ca. 0.7–3.4 m^3 spent nuclear fuel per TWh of electrical power taking into account the typical thermal efficiencies in nuclear power plants [19, 20]. The waste needs to be stored for extended times with the highly radioactive material being initially stored in a water pool for a period up to 5–15 years, and all materials required to be stored for millennia, a time far exceeding the lifetime of a nuclear plant.

Other power plants burn organic matter to generate heat generating waste streams (typically either solid or gaseous waste), which must be dealt with. The source of the waste is primarily associated with the elements present in the fuel and the oxidant. An indication of the typical waste streams generated in the power generation from the combustion of coal, natural gas, municipal waste and biomass, is given in Table 2.3. The flue gas from these processes is enriched in nitrogen, with the oxygen content determined by the excess air used in the combustion process. Carbon dioxide makes up ca. 4–15% of the flue gas on a dry basis depending on the excess air used in the combustion process (a dry basis is chosen here as water generated in the combustion process will condense and precipitate out, and the process stream will typically be saturated with water).

NO_x in the flue gas is due to the presence of nitrogen in the air as well as nitrogen in the fuel. Sulphur present in the feed will result in the formation of SO_2/SO_3, which may react further with water to sulphurous and sulphuric acid. Halides can be present specifically in (unsorted) municipal waste and its combustion may result in the formation of the corresponding acidic hydrogen halides.

Solid particles are formed by the presence of transition metals and metalloids in the feed (fly ash), as well as incomplete combustion of the feed.

2.2.3.1 Carbon Capture

The relatively high concentration of CO_2 in the effluent from power plants (point source) renders it an ideal opportunity to capture carbon from entering the atmosphere directly and store it (carbon capture and sequestration [CCS]) or utilise it further (carbon capture and utilisation [CCU]/carbon capture utilisation and storage [CCUS]). The minimum amount of work required for the separation of CO_2 from the effluent can be determined by considering a simple system for the splitting of the

Table 2.3 Indication of the flue gas composition from power plants fired with different feedstock.

Power-plant	O_2 (vol.%)	CO_2 (vol.%)	CO (ppm)	NO_x (ppm)	SO_x (ppm)	Particle (ppm)	Reference
Brown coal	4–6.5	12–15	~100	237–1750	215–457		[21, 22]
Hard coal	9.4	14.2		108–216	216–432	30	[23]
Natural gas	12–18	1.0–5.3	5–315	20–250	<110	<25	[24]
Municipal waste	8.8–10.3	9–14	10	120–210	24		[25, 26]
Industrial incinerator	7.0–17	7–13		<150	<75		[26, 27]
Biomass	6–15	4–7	180–560			25	[28, 29]

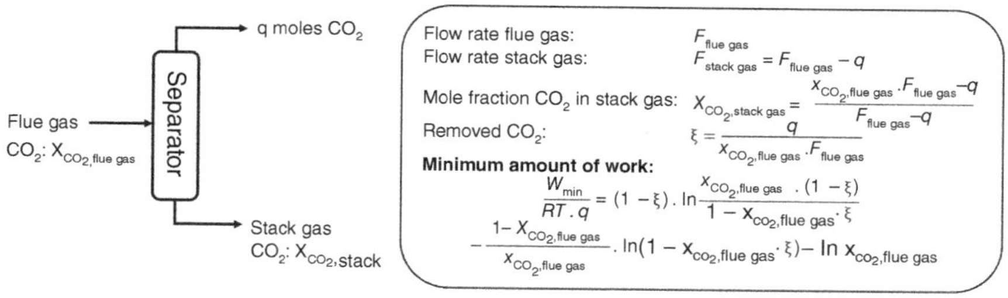

Figure 2.6 Minimum amount of work required for the separation of CO_2 from flue gas.

stream into a CO_2-rich stream and a stack gas relatively poorer in CO_2:

$$\frac{W_{\min}}{RT} = \sum_i n_{i,\mathrm{CO_2-rich}} \cdot \ln x_{i,\mathrm{CO_2-rich}} + \sum_i n_{i,\mathrm{stack}} \cdot \ln x_{i,\mathrm{stack}} - \sum_i n_{i,\mathrm{feed}} \cdot \ln x_{i,\mathrm{feed}}$$

The estimation of the minimum amount of work normalised with respect to the amount of CO_2 removed is only dependent on the fraction of CO_2 in the feed gas and the fraction of CO_2 removed from this stream, if the CO_2-rich stream only contains CO_2 (see Figure 2.6).

The minimum work normalised by the amount of CO_2 removed increases with the fraction of CO_2 removed and decreases with the concentration of CO_2 in the feed. It can thus be estimated that the amount of work for the direct air capture (removing 90% from a stream containing 400 ppm CO_2) will require ca. 0.13 MWh/tonne of CO_2 (21.2 kJ/mol), whereas the removal of 90% of CO_2 from the effluent of a power station (15% CO_2) would require ca. 0.04 MWh/tonne of CO_2 (6.37 kJ/mol) (see also [30]) and removal from even more concentrated streams would require even less energy. This is thus an incentive to create a flue gas with a high CO_2-content from which the separation of CO_2 would be less energetically costly.

The actual energy cost for the removal of CO_2 is typically much higher, as the processes typically use separation through either absorption or adsorption [31] with desorption induced by raising the temperature or decreasing the pressure (membrane technology could be used as an energetically more favourable separation, but this process is still in development phase and not yet ready for industrial application [32]). Table 2.4 gives an overview of the required energy costs to separate CO_2 from an effluent. High-pressure storage of CO_2 at ca. 90 bar would require an additional 0.1 MWh/tonne of CO_2.

CO_2 can, in principle, be removed by scrubbing the effluent with water. However, the amount of CO_2 dissolved in water is relatively low, and a large amount of liquid must be pumped around to achieve the desired level of CO_2-removal from the effluent, increasing the loss of water in the process through evaporation.

Acidic gases can be removed by using a basic absorbent, for instance, an aqueous ammonia solution. Ammonia is volatile and may result in slip of ammonia into the flue gas, which may be minimised by reducing the operating temperature. However, ammonium carbonate may precipitate out at low temperature and high concentration of ammonia.

Table 2.4 Reported energetic costs to extract CO_2 from flue gas.

Technology	CO_2 content in feed (vol.%)	CO_2 recovery (%)	Energy (MWh/tonne CO_2)	Reference
Water wash	40	82[a)]	0.3	[37]
Chilled ammonia process	[b)]	75	0.7	[38]
	32	—[c)]	1.0	[39]
Amine scrubbing				
Mono ethanol amine	—[c)]	—[c)]	0.6	[34]
	2.5–20.5	ca. 90	0.9–1.2	[30]
	32	—[c)]	1.4	[39]
Dimethyl carbonate	32	—[c)]	0.6	[39]
Rectisol (methanol solvent)	32	—[c)]	0.5[d)]	[39]
Temperature swing adsorption (zeolite NaUSY)	15	84	1.3	[40]
Pressure swing adsorption (zeolite 13X)	33	[a)]	0.5	[41]
Calcium looping	12	90	0.6	[42]

a) Process optimised for purification of methane.
b) Flue gas from power station.
c) Data not available.
d) Low energy cost due to heat integration with cryogenic air separation unit.

Amine scrubbing (see Figure 2.7), e.g. using mono ethanol amine (MEA), is currently the leading technology for CO_2 capture from concentrated streams [33]. The flue gas is passed through an amine-containing solution and reacts with the amine in solution. The chemical reaction is limited by chemical equilibrium and thus the amount of amine in the solution. The saturated solution is then sent to a stripper to release high-purity CO_2, where the CO_2-containing solution is heated up, boiling off some of the water. A benefit of the amine scrubbing process is its ability to run at high CO_2 capture (>90%), but there are a few drawbacks [34] and the presence of metals and NO_x in the flue gas would affect the process. The process is associated with a high thermal energy demand of at least 0.6 MWh/tonne of CO_2, which corresponds to 22.4% of the heat of combustion of coal. Hence, this carbon capture technology adapted to a coal-based power station using the Rankine cycle with an efficiency of 40% would require 55% of the generated power to remove CO_2 from the flue gas. The picture is slightly better when using natural gas as the energy carrier and the required energy for the CO_2 separation amounts to ca. 11% of the heat of combustion of methane and thus ca. 20% of the power generated in a CCGT (with an assumed efficiency of 55%) would be required to separate out CO_2. However, most of this energy is used in the stripper for boiling off water, which can be integrated with the medium temperature streams in various locations on an existing plant, as the required temperatures are still relatively low [35].

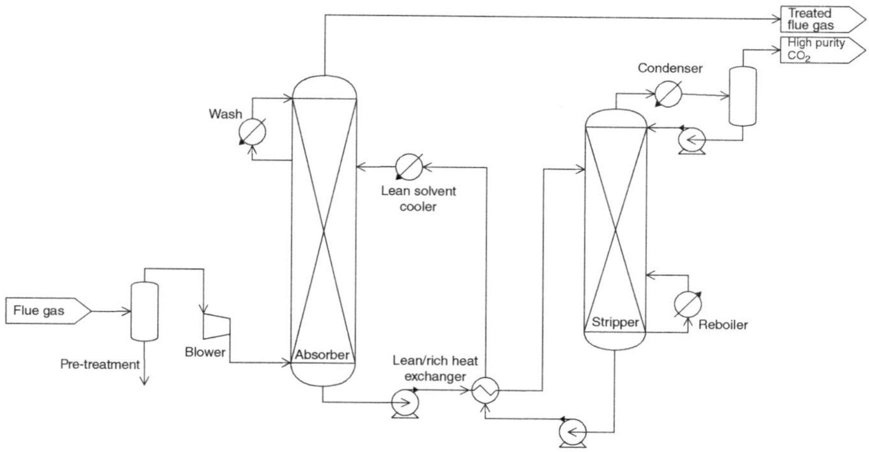

Figure 2.7 CO_2 removal using amine scrubbing process (redrawn from Mangiaracina et al. [36]).

Physical absorption of CO_2 can also be used for CO_2 removal. This requires solvents with a high affinity for CO_2, such as dimethyl carbonate or methanol. CO_2 is absorbed at relatively low temperature (ca. −50 °C for the so-called Rectisol process using methanol as the solvent, which requires thus multiple refrigeration loops). In this absorption process, the CO_2 loading of the solvent increases roughly linearly with increasing pressure (in contrast to a chemical adsorption, where the CO_2 uptake as a function of pressure is limited) and thus a relatively high pressure is required to obtain a high CO_2 loading of the solvent. CO_2 is released at low pressure whilst minimising the loss of the solvent (hence, solvents with a high boiling point are preferred).

Solids, such as zeolites, can likewise be used to selectively adsorb CO_2 from a flue gas. CO_2 can desorb from these materials by either increasing the temperature or decreasing the pressure. In contrast to the processes using a liquid absorbent, where the absorbent is pumped around, these processes operate with switching columns. The flue gas passes over the column containing the adsorbent for a certain period, after which the flue gas is diverted to a fresh column and the loaded column is either depressurised or heated up to remove the adsorbed CO_2.

CO_2 can also be removed by reacting CO_2 in the flue gas with calcium hydroxide, $Ca(OH)_2$ or CaO in a process called calcium looping. In this process, CO_2 is mineralised in the form of calcium carbonate ($CaCO_3$). Alternatively, a relatively pure stream of CO_2, albeit at low pressure, can be obtained by heating $CaCO_3$ to high temperatures (>950 °C to form a stream with a partial pressure of CO_2 of larger than 1 bar) regenerating CaO.

2.2.3.2 Desulphurisation

The presence of sulphur in the fuel will result in the formation of sulphur dioxide, which can be removed in an alkali wash. For instance, contacting the sulphur-containing flue gas with a slurry containing limestone, $CaCO_3$, or hydrated lime, $Ca(OH)_2$, will result in the formation of calcium sulphite, $CaSO_3$. This product can be monetised by oxidising it to form gypsum, $CaSO_4 \cdot 2H_2O$.

The alkali wash is a low temperature process, which will need to precede the DeNO$_x$ operation, as the addition of ammonia to a sulphur-containing flue gas may result in the formation of the corrosive ammonium bisulphate, which may be deposited downstream of the addition point [43]. Before the alkali wash, particulates in the flue gas will need to be removed using electro-static precipitators.

2.2.3.3 DeNO$_x$

Combustion processes may result in the thermal formation of NO_x, which may be formed from nitrogen present in the fuel (fuel NO_x), in the high temperature combination of oxygen and nitrogen in the Zeldovich reaction (thermal NO_x), and in the initial stage of combustion (prompt NO_x) initiated by the interaction of radicals present in the flame and nitrogen in the air. Figure 2.8 shows the temperature regions in which the various formation pathways are dominant. Fuel-NO_x is dominant at relatively low temperatures in the combustion chamber and is, of course, determined by the amount of nitrogen in the feed, with natural gas as a feed having a low nitrogen content, whereas coal and heavy oil are richer in nitrogen. At high temperature under fuel lean conditions, and especially for a fuel with a low nitrogen content, thermal NO_x becomes the dominant reaction pathway. The contribution of prompt-NO_x is typically minor to the overall NO_x formation. The formation of prompt-NO_x can be further minimised by intensive mixing of fuel with the combustion air. The formation of NO_x is further controlled by the air to fuel ratio, with the highest amount typically formed under close to stoichiometric conditions.

NO_x may result in the formation of ground-level ozone upon interaction with volatile organic compounds (VOCs) and exposure to UV-A light [45], which will then affect the respiratory system. Furthermore, NO_x may result in the formation of acid rain upon reaction with water (and oxygen) in the atmosphere. N_2O, a strong greenhouse gas, is only expected at rather low temperatures (see Figure 2.8). The

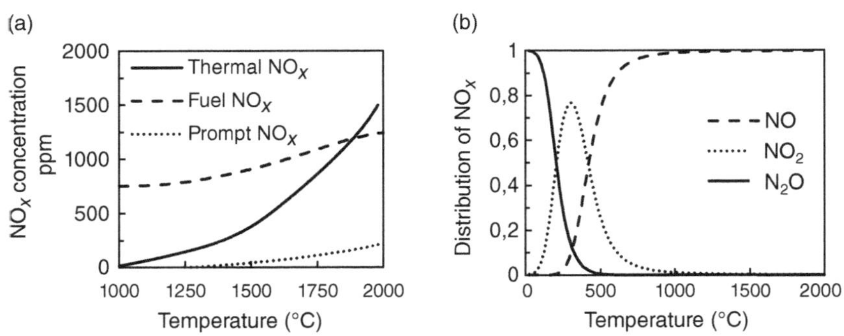

Figure 2.8 (a) NO_x formation attributed to the different reaction pathways in the effluent of pulverised coal combustion as a function of the maximum temperature in the combustion chamber (redrawn according to Schrod et al. [44]); (b) Equilibrium distribution of NO_x as a function of temperature ($p_{O2} = 0.15$ bar; $p_{NOx,total} = 0.1$ mbar).

environmental and health effects association with NO_x has resulted in the regulation of its emission. The NO_x-concentration in the flue gas can be reduced by the reduction of NO with ammonia:

$$NO + NH_3 + \frac{1}{4}O_2 \rightarrow N_2 + \frac{3}{2}H_2$$

or urea. In the Selective Non-Catalytic Reduction (SNCR) process, anhydrous gaseous ammonia or urea is injected into the hot flue gas stream (900–1100 °C) with the ammonia slip being controlled by allowing the conversion of NO_x to be limited to 50–75% [46]. A higher degree of NO_x-removal with a reduced consumption of the reductant, ammonia or urea, can be achieved using Selective Catalytic Reduction (SCR), which typically takes place over a vanadia supported on titania catalyst or a copper-containing zeolite at temperatures in the range of 300–400 °C [46].

2.3 Power Generation from Geothermal Energy

Electric power can also be generated from heat available directly from the Earth, especially where the Earth's crust is relatively thin. Heat from the Earth's core is conducted through the crust, augmented by convective heat transfer through magma, and hot springs. This type of heat is most readily available in volcanic regions, such as the region surrounding the Pacific plate (also known as the 'Pacific ring of fire', surrounded by the West coast of the America's and the East coast of Asia/Oceania), Iceland (on the border between the North American plate and the Eurasian plate) and the area between the Eurasian plate, the African plate and the Arabian plate (i.e. the region around the Mediterranean Sea).

2.3.1 Principle

Thermal energy in geothermal environments can be classified according to the temperature and the amount of water/steam present to carry the energy to the Earth's surface [47]. Magma has the highest temperature (650–1300 °C) but contains very little water. When magma cools, it will result in dry, warm rock. Cooled magma will result in dry, warm rock and the presence of water may generate saturated rock with a variable porosity. Heat may be recovered by the transport of hot water/steam depending on the depth below the surface, facilitated by the injection of cold liquid water (see Figure 2.9), but this is dependent on the porosity and permeability of the rock.

The simplest process is when the pumped water is heated by the geothermally heated rock underground to generate steam, which is then recovered at the wellhead. Dry steam can be passed through a turbine to generate power. A condenser may then be used to recapture liquid water which can be pumped back down to the subsurface rock. This is a typical example of a Rankine cycle where the boiler is the heated sub-surface rock fired by the Earth. There are various requirements for the rock to make this process work, viz. the rock needs to be permeable enough for the water

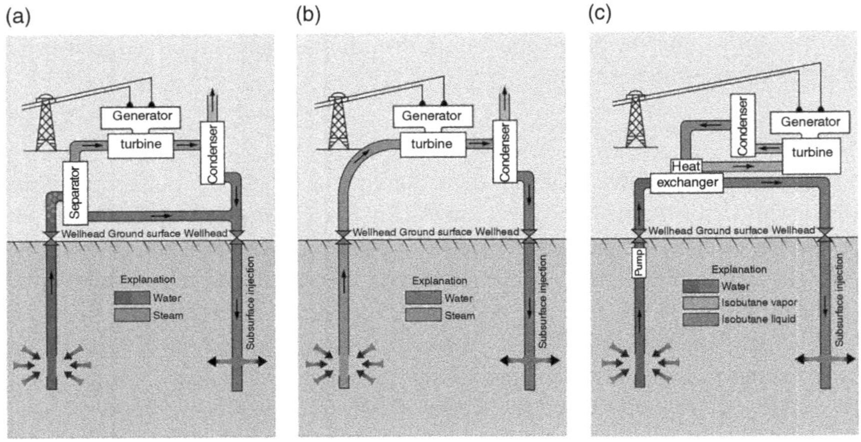

Figure 2.9 Power generation using geothermal energy (from Duffield and Sass [47], report in public domain). (a) Flash steam process. (b) Dry-steam process. (c) Binary cycle process.

to get from the injection point to the hot well and the thermal energy needs to be sufficient to generate dry steam – see Figure 2.9 middle.

Often wet steam is generated (i.e. a water/steam mixture) and appears at the wellhead, thus requiring upfront separation of water from steam in a flash tank (see Figure 2.9a). Hence, this process is named the flash steam process. For the rest, this process operates identically to the dry steam process. The flash process is typically used with a water/steam feed with a specific enthalpy between 800 and 2550 kJ/kg [48]. This process can be modified for feed with an enthalpy in the range 750–1900 kJ/kg by applying a double flash with two generators [48].

An alternative process utilises two cycles (hence its name 'binary cycle' process) shown in Figure 2.9b. The heat from geothermal sources is often of a low quality, which is insufficient to evaporate water but warm enough to exchange heat with an alternative working fluid for the power generation cycle. Water is pumped into the subsurface rock and recovered at the wellhead. At this stage, it is still liquid water and not steam. Hot water can then be used to evaporate a low-boiling working fluid in a counter-current heat exchange. Cooled water is recycled back to recover heat from the subsurface area. The evaporated, low-boiling working fluid passes through a turbine to extract work and is recovered as a liquid in the condenser. Alternatively, power may be generated using the Kalina cycle, in which heat is exchanged with an ammonia–water mixture [49].

2.3.2 Efficiency

Geothermal energy is typically operated at much lower temperatures (<250 °C) than fired power stations, due to the temperature of the energy reservoir. The maximum efficiency of power generation from geothermal energy, the Carnot efficiency, for a process receiving heat at 160 °C (corresponding to saturated steam at ca. 6 bar) and

Table 2.5 The top 10 countries with the largest installed capacity of geothermal power generation, their annualised production in TWh (1 TWh = 0.0036 EJ) and the annualised capacity factor of geothermal power generation (CF_{geo}) in 2020/2021 (data from Gutiérrez-Negrín [50]).

Country	Capacity (GW)	Production (TWh)	CF_{geo} [a)] (%)
USA	3.92	18.70	54.5
Indonesia	2.38	16.59	79.4
Philippines	1.95	11.67	68.3
Turkey	1.72	10.84	72.1
New Zealand	1.05	7.82	84.6
Mexico	1.00	4.51	51.4
Kenya	0.97	5.59	65.6
Italy	0.92	5.92	73.8
Iceland	0.76	5.79	87.5
Japan	0.55	2.66	55.7
Total world	*16.32*	*96.55*	*67.5*

a) Capacity factor $CF_{geo} = \frac{\text{annual production}}{\text{installed capacity} \cdot 24 \cdot 365} \cdot 100\%$

rejecting it at 30 °C is only ca. 30%. The efficiency reported for the dry steam and single flash process is in the range between 6% and 21% [48] depending on the enthalpy of the emerging stream. The efficiency for the binary cycle process lies between 1% and 9% depending on the temperature of water received from the wellhead.

2.3.3 Installations

The total capacity for geothermal power generation in 2020/2021 stood at 16.32 GW [50] (see Table 2.5) producing globally 96.55 TWh per annum. The United States accounts for 24% of all installed capacity, followed by Indonesia, the Philippines, Turkey, New Zealand and Mexico. All these countries have at least 1 GW of installed geothermal power. Almost 53% of the total geothermal power is generated using the flash steam process, whereas 25% of the installed capacity utilises the binary cycle process.

The annualised capacity factor for power generation from geothermal energy is high with a global average of 67.5%, and in New Zealand and Iceland the capacity factor is even 84–88%, thus highlighting the greatest benefit of this technology, viz. its high capacity factor.

2.3.4 Environmental Aspects of Geothermal Power

The environmental impact of geothermal power generation is typically considered to be low with a CO_2 emission over the lifetime of the plant 20–200 times lower than that of a coal-fired power plant (4–40 kg CO_2/MWh versus ~900 kg CO_2/MWh [51]) with the main emissions related to construction and maintenance. Only 1% of the CO_2 emissions are associated with the operation and thus the carbon footprint will be further reduced when society is moving towards a low carbon economy.

The construction of a geothermal plant consumes significant amounts of water in the drilling phase [52]. Water is also used in the operational phase to extract geothermal energy, and in some cases to improve the efficiency of the power generation cycle by cooling the working fluid. In the latter case, water consumption can be significant [52]. The water usage during extraction can be negative, as ground water ingress into the reservoir can also be recovered for the extraction of geothermal energy [51].

Extraction of geothermal energy may result in land subsidence depending on the geological environment [51]. The injection and extraction of water/steam from the wells of geothermal plants may cause seismicity in the area because of increased stress underground, with increasing fluid flow rate enhancing the risk of induced seismic activity.

The stream coming out of the wellhead may contain (dissolved) gases, mainly CO_2, but also H_2S as well as other minor components [52, 53]. In closed-loop plants, waste gas is re-injected into the ground after heat recovery. In open-loop systems, the gas will be emitted into the atmosphere when cooling the water/steam mixture. The gas composition and the amounts of gas released differ between different fields, and even between different wells, but the amount of gas is relatively small in comparison to fossil-fuel-fired power stations.

2.4 Hydropower

Approximately 15% of the electric power worldwide is generated using hydropower through fast flowing water passing over a turbine. The power generation from flowing water is a logical extension of the old water mills used to drive different machineries over the centuries.

2.4.1 Principle

Hydropower converts gravitational potential energy associated within stored water into electrical power by flowing through a turbine. Water in a reservoir is let out through the penstock, passes through a turbine and is discharged into a river (Figure 2.10).

The amount of work ($\dot{W}$) obtained from falling water with a mass flow rate of $\dot{m}$ can be obtained by combining a mass and energy balance for this non-steady-state system (neglecting the kinetic energy associated with the stream(s) refilling the reservoir, but including the kinetic energy of the tail water):

$$\frac{\dot{W}}{\dot{m}} = \eta \cdot \left(g \cdot \Delta h - \frac{1}{2} \cdot \left(\frac{\dot{m}}{\rho_{\text{water}} \cdot A_{\text{outflow}}} \right)^2 \right) \approx \eta \cdot g \cdot \Delta h$$

Figure 2.10 Schematic sketch of hydropower plant.

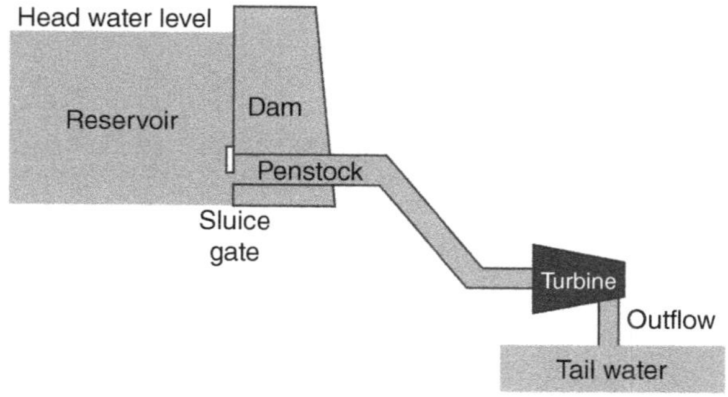

with gravitational constant $g = 9.81 \cdot \frac{m}{s^2}$; density of water $\rho_{water} = 10^3 \cdot \frac{kg}{m^3}$; $A_{outflow}$: cross-sectional area of the outflow pipe.

The change in the height of the reservoir as well as the kinetic energy of the discharged water can also typically be neglected. The head, the height difference between the level of the reservoir and the turbine, is the main determining factor in the work obtained from a hydropower plant. The maximum amount of work that can be extracted from a flow of 1 tonnne/s of water is thus 9.81 kW/m of height difference. The amount of electric power generated from the plant can be regulated by controlling the mass flow rate of water coming down the penstock through partial closure of the sluice gate.

In the turbine, the water will turn the blades, which can then be transformed into electrical power. The turbines used can be classified into reaction turbines and impulse turbines. In reaction turbines, such as the Kaplan or Francis turbines, the pressure of the water changes when it passes through the turbine. In impulse turbines, the water passes through a nozzle before contacting the turbine blades and the passage over the turbine blades takes place with no noticeable pressure drop. Examples of impulse turbines are the overflow turbine, the cross-flow turbine, the Pelton turbine and the Turgo turbine.

Figure 2.11 shows the different turbines, and their operating windows. The Pelton and Turgo turbines would be most suited for operations demanding a high-power delivery, but a relatively low flow rate (and thus a high head), whereas a Francis turbine or Kaplan turbine would be more appropriate when dealing with large flow rates (and relatively low head for a given power output).

2.4.2 Efficiency

The efficiency of the hydropower generation is typically determined by the efficiency of the turbine, although the energy loss through the kinetic energy of the water leaving the turbine can be significant, in particular when the height difference is too small, the mass velocity of water is too high and/or the cross-sectional area of the outflow is too small:

$$\eta_{flow} = 1 - \frac{1}{2 \cdot g \cdot \Delta h} \cdot \left(\frac{\dot{m}}{\rho_{water} \cdot A_{outflow}} \right)^2$$

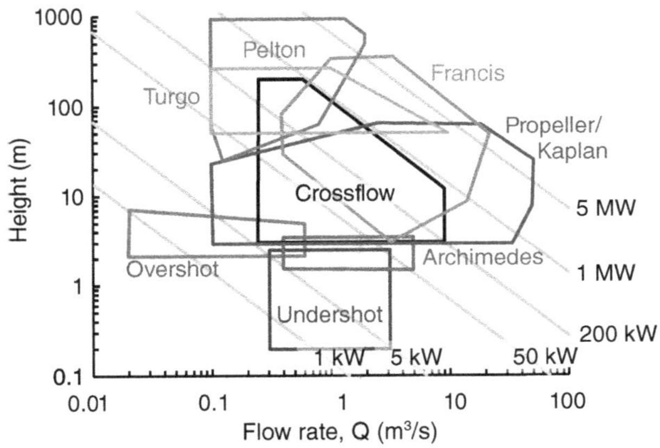

Figure 2.11 Operating regimes for different turbines (redrawn from Williamson et al. [54]).

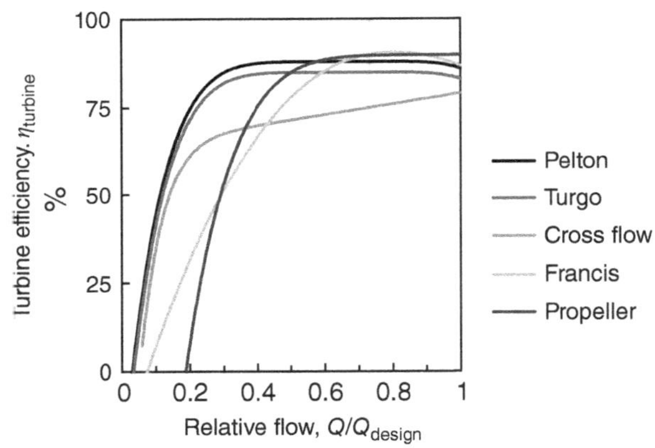

Figure 2.12 Turbine efficiency as a function of the flow rate relative to the design flow rate (curves based on correlations given by Minister of Natural Resources [55]).

The efficiency of the turbine can be estimated using the correlations given by literature [55] (the calculated efficiencies are minimally dependent on the chosen design flow rate, and only slightly dependent on the height difference over which hydropower is being generated). The attained efficiency as a function of the flow rate relative to the design flow rate is shown in Figure 2.12. A high efficiency of larger than 80% can be obtained with the various turbines; the efficiency of the propeller turbine and the Francis turbine can even reach 90%. The highest efficiency is not always attained at the design flow rate but typically at ca. 80–90% of the design flow rate. It can further be noted that the efficiency of these turbines drops off quite rapidly upon reducing the flow rate relative to the design flow rate. A wider working range is obtained with the Pelton and Turgo turbines, but the ability to deliver power over a wider range comes at a cost of a slightly lower maximum efficiency of 88% and 85%, respectively.

2.4.3 Installations

The total installed capacity of hydropower worldwide in 2020 amounted to 1330 GW, of which 159 GW was installed as hydro-storage, which generated 15% of the world's electric power (see Table 2.6). The capacity factor, taken as the ratio of the annual energy generation from hydropower over the installed hydropower capacity assuming continuous operation over the year, is 37.5% worldwide.

China has the largest installed single hydropower facility, the Three Gorges Dam on the Yangtze River. The reservoir has a capacity of 39.3 km^3 and a surface area of 1084 km^2. The installed generators have a design capacity of 22.5 GW. The power produced varies seasonally and over the years with the water inflow into the dam [56]. The actual capacity factor varies between 40% and 57% on an annual basis and can vary strongly from month-to-month and between seasons. The next largest hydropower facility is the Itaipu power plant, located on the border between Brazil and Paraguay. The reservoir has a capacity of 29 km^3 and a surface area of 1350 km^2. The facility has an installed capacity of 14 GW with a capacity factor between 62% and 84% (calculated based on the annual generation data given by [57]). The largest hydroelectric facility in North America is the Grand Coulee Dam (Washington State, USA) attached to a reservoir containing 12 km^3 of water and a surface area of 324 km^2. The installed capacity of 6.08 GW has an annualised capacity factor of $33.7 \pm 4.6\%$. The monthly capacity factor varied in 2021–2022 between 19% and 58% (calculated based on data from [58]). The generation of hydroelectricity has a

Table 2.6 The top 10 countries with the largest installed capacity of hydropower, hydropower capacity reserved for hydro-storage, their annualised production in TWh (1 TWh = 0.0036 EJ) and the annualised capacity factor of hydropower plants (CF_{hydro}) in 2020 (data from HA - International Hydropower Association [59]).

Country	Capacity (GW)	Hydro-storage (GW)	Production (TWh)	CF_{hydro} [a)] (%)
China	370.2	31.5	1355	41.8
Brazil	109.3	0.0	410	42.8
USA	102.0	22.9	291	32.6
Canada	82.0	0.2	383	53.3
India	50.5	4.8	155	35.0
Japan	50.0	27.6	89	20.4
Russia	49.9	1.4	196	44.8
Norway	33.0	1.4	142	49.0
Turkey	31.0	—	77	28.5
France	25.5	5.8	65	29.1
Rest of the world	*426.7*	*63.9*	*1207*	*32.3*

a) Capacity factor $CF_{hydro} = \frac{\text{annual production}}{\text{installed capacity} \cdot 24 \cdot 365} \cdot 100\%$

seasonal and annual variation depending on the local geographical features, as well as rainfall within its catchment area.

Dams have been constructed for a variety of different purposes, viz. irrigation, drinking water supply, hydroelectricity, flood control, inland navigation and even recreation. The construction of a dam with a hydroelectric power station often can thus assist attaining other objectives at the same time.

2.4.4 Environmental Aspects of Hydropower

Hydroelectricity is generally seen as a green energy source. However, the environmental impact caused by greenhouse gas emissions over its lifetime, loss in carbon sequestration by the now submerged land, acidification, etc. will need to be considered [60]. Construction of the dam has a greenhouse gas contribution of ca. 19 kg $CO_{2,eq}$/MWh [61]. However, greenhouse gases are produced from the flooded land, as the landscape changes from a carbon sink to a carbon source. The submerged biomass is converted into CO_2 and in an abiotic process into CH_4 with the rate of production depending on the type of original vegetation, the temperature, precipitation, variation in the water level in the reservoir, mixing in the reservoir and the age of the reservoir [62]. The emission of greenhouse gases from this source can be substantial; Scherer and Pfister [63] reported a surprisingly high, average carbon footprint of ca. 273 kg $CO_{2,eq}$/MWh, albeit with a large variation in the estimated carbon footprint of different hydroelectric plants, due to uncertainty in the underlying processes and the variation in the data underpinning the model. Ubierna et al. [64] estimated the global median life cycle carbon footprint associated with hydropower of 23 kg $CO_{2,eq}$/MWh using the GRes tool [65].

2.5 Wind Energy

Wind, the movement of air, can be used as a source of energy. It is readily available in a wide range of areas, and specifically in the coastal regions of Western Europe, eastern parts of the United States and Canada, north-western Africa, South Africa, Australia and New Zealand. Wind energy has been utilised by humankind for a long time.

The disadvantage of wind as an energy source is its variability. Figure 2.13 shows the variability in the wind speed in the year 2021 at the coast of North Holland (the Netherlands). The wind speed varies strongly on a daily basis, but it also shows seasonal variations with less wind in the period May–September than in the autumn and winter months. This implies that power generation in the summer months from wind energy will be less than during the autumn and winter months and the seasonal fluctuations will need to be buffered. The Germans invented a new word to describe the period with a reduced power generation through meteorological phenomena as a 'Dunkelflaute' combining the words 'Dunkelheit' (darkness) and 'Windflaute' (low wind) [66].

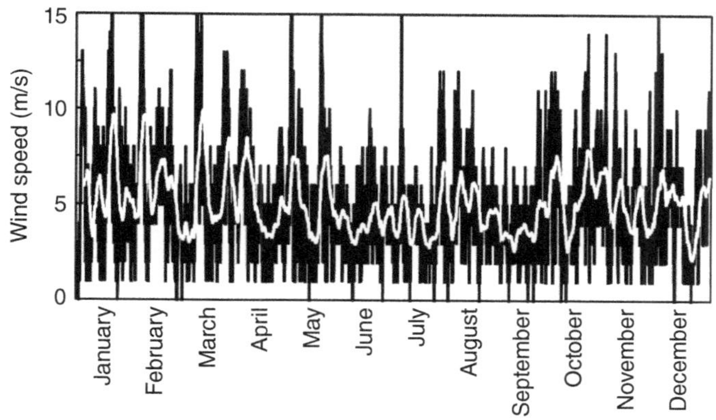

Figure 2.13 Hourly (black line) and averaged wind speed (white line) in 2021 at the weather station De Kooy in the Netherlands (data from KNMI [67]).

2.5.1 Principle

Energy can be extracted from the movement of air (wind) by passing the wind over wind turbine blades. This results in a drag force on the blades parallel to the flowing air and a lift force on the blades perpendicular to the flowing air, induced by the pressure variations around the blades, which causes the blade to turn. Rotation of the blades due to the lift force is more efficient in power generation than rotation due to the drag force [68]. The blades then turn a rotor, which is connected by a gearbox to the high-speed rotor in the generator.

In general, the power obtained from a wind turbine can be expressed as

$$P_{\text{wind turbine}} = \text{Cp} \cdot \frac{1}{2} \cdot \rho_{\text{air}} \cdot S \cdot v_{\text{air}}^3$$

with Cp the coefficient of performance (or efficiency) of the wind turbine, S the area swept out by the blades of the turbine and v_{air} the wind speed. The power obtained is thus proportional to the area swept by the blades, and turbines with larger blades will deliver a higher nominal power and hence wind turbines with a nominal power of 2 MW have a diameter of the swept area of over 110 m [69, 70].

There are in principle two variations, with a rotor perpendicular to the ground (vertical axis wind turbine – VAWT), and with a rotor parallel to the ground (horizontal axis wind turbine – HAWT) with the HAWT being the most commonly used, as it has a better rotor control through pitch and yaw control [68].

2.5.2 Efficiency

Air flows over the area, A_S, swept by the blades (see Figure 2.14). The transfer of kinetic energy from the air to the blades will drop the velocity of the air from v_1 (before the blades) to v_2 (after the blades) with the velocity of air at the blades being equal to v_S. The mass balance implies that

$$\dot{m}_1 = \dot{m}_S = \dot{m}_2$$

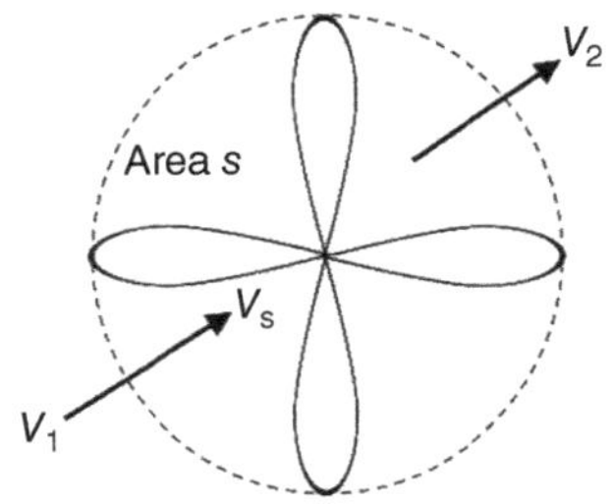

Figure 2.14 Change in air speed upon passing the blades of a wind turbine.

with the mass flow rate related to the linear velocity of the air via the density and the area through which the air is flowing:

$$\rho_{air} \cdot A_1 \cdot v_1 = \rho_{air} \cdot A_S \cdot v_S = \rho_{air} \cdot A_2 \cdot v_2$$

The power associated with wind, due to the kinetic energy of the air is:

$$\dot{E}_{kin} = \frac{1}{2} \cdot \dot{m} \cdot v^2$$

and over the wind turbine, the velocity drops from v_1 to v_2, and the power given off to the wind turbine can be estimated from the energy balance and equals the change in the kinetic energy of the air:

$$P_{\text{wind turbine}} = \frac{1}{2} \cdot \dot{m} \cdot \left(v_1^2 - v_2^2\right)$$

The work done on the blades can also be estimated from the force exerted by the air on the blades:

$$F = m \cdot a = m \cdot \frac{dv}{dt} = m \cdot \frac{\Delta v}{\Delta t} = \dot{m} \cdot (v_1 - v_2)$$

and the power on the wind turbine

$$P = \frac{dE}{dt} = F \cdot \frac{dx}{dt} = F \cdot v_S = \dot{m} \cdot v_S \cdot (v_1 - v_2)$$

Both expressions must be equal and hence $v_S = \frac{1}{2} \cdot (v_1 + v_2)$, or alternatively the velocity at the rotating blade is the average between the velocity of the wind before the blades and the velocity after the blades.

Thus, the power given off to the wind turbine can be expressed in terms of the relative reduction in the wind speed:

$$P_{\text{wind turbine}} = \frac{1}{4} \cdot \rho_{air} \cdot S \cdot v_1^3 \cdot \left(1 + \frac{v_2}{v_1} - \left(\frac{v_2}{v_1}\right)^2 - \left(\frac{v_2}{v_1}\right)^3\right)$$

which is a maximum at a relative velocity of $\frac{v_2}{v_1} = \frac{1}{3}$. At this point, the power given off to the turbine equals $\frac{16}{27}$ or 59.3% of the original kinetic energy flow of the air ($\frac{1}{2} \cdot \rho_{air} \cdot S \cdot v_1^3$). This is known as the 'Betz Limit', a theoretical upper limit for wind turbine efficiency. The occurrence of a maximum can be rationalised by the two opposing factors affecting the power of the wind turbine, viz. the change in the wind speed, which increases with decreasing velocity, v_2, as well as the mass velocity passing through the area swept by the blades, which decreases with decreasing velocity, v_2. The model used here assumes a constant linear velocity and hence no entropic losses on the blades from drag and vortex shedding. However, the entropic losses, as well as efficiencies in converting rotational motion into electrical work will further reduce the efficiency [68]. Hence, the overall performance of a wind turbine can be summarised as:

$$P_{\text{wind turbine}} = \text{Cp} \cdot \frac{1}{2} \cdot \rho_{\text{air}} \cdot S \cdot v_1^3$$

with $\text{Cp} = \eta_{\text{entropic}} \cdot \eta_{\text{electrical}} \cdot \eta_{\text{Betz}}$ and $\eta_{\text{Betz}} = 1 + \frac{v_2}{v_1} - \left(\frac{v_2}{v_1}\right)^2 - \left(\frac{v_2}{v_1}\right)^3$

The entropic losses (and thus the efficiency of the rotor) depend on the ratio of the speed of the tip of the rotor to the free-flowing wind speed (tip speed ratio). Wind flows unperturbed through the area swept by the blades; if they rotate too slowly, then energy transfer is not efficient. However, the area starts to appear as a disk for the wind coming through the area, if the blades rotate too quickly. This will increase the drag on the blades, and thus the turbulence around the blades, which reduces the efficiency [71]. Furthermore, the wind may change direction due to the interaction with the blades; the presence of severe velocity gradients will reduce the efficiency even further. These losses can be minimised by improving the design of the blades as well as the number of blades on the wind turbine.

The typical efficiency of windmills and wind turbines is given in Figure 2.15. The classical American and Dutch windmills have efficiencies of less than 20% as the drag forces play an important role in these designs. The modern wind turbines (exploiting lift forces) have a much higher efficiency, with a 2-blade HAWT having a maximum efficiency of 46% at a tip speed ratio of 5.6–6 [71]. Increasing the number of blades increases the maximum efficiency achieved, at a lower tip speed ratio [72], which is an attractive option for areas with a lower average wind speed. The typical number of blades is 3, showing an increase in efficiency (ca. 50%) compared to the 2-blade, and a lower cost than a 4-blade system. The optimal number of blades is thus a trade-off between the increased cost and the (limited) increase in efficiency (and thus income).

The wind speed needs to be in the correct operating window for the turbines to operate most efficiently. Wind turbines can either operate in the variable speed mode, where the rotor speed is given by the wind speed, or in a constant speed mode (see Figure 2.16). The latter option is preferred as turbines operated in this manner are mechanically robust, stable and reliable, with lower cost for maintenance [73]. No power is generated below the so-called cut-in speed (Region I), which is the wind speed required to turn the wind turbines at a sufficient speed to produce electricity. In Region II, the power generation is proportional to the cube of the wind speed. At the rated wind speed (coming into Region III), the speed of the high-speed rotor is reduced by applying brakes, to protect the mechanical integrity of the wind turbine, and the efficiency of transfer of the wind energy is reduced. Above the cut-out speed (Region IV), the brakes are applied fully so that the wind turbine does not rotate. No power is being generated above the cut-out speed to protect the wind turbine from mechanical failure.

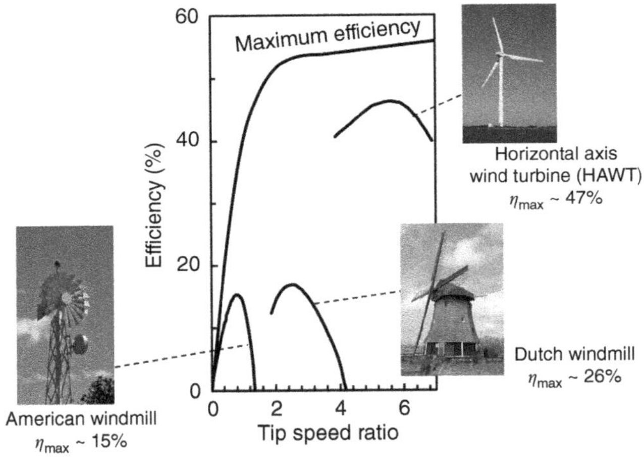

Figure 2.15 Efficiency of windmills and wind turbines as a function of the tip speed ratio (Photo 1: Linnaea Mallette/PublicDomainPictures.Net/Public Domain; Photo 2: Petr Kratochvil/PublicDomainPictures.Net/Public Domain; Photo 3: X posid/PublicDomain Pictures.Net/Public Domain).

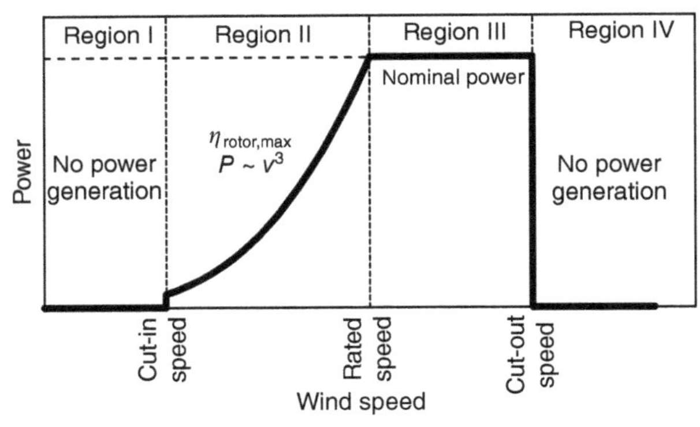

Figure 2.16 Generated power in a constant speed HAWT wind turbine (redrawn from Adeseye et al. [72]).

2.5.2.1 Efficiency of a Wind Park

Multiple wind turbines are typically put together in a wind park to generate sufficient electrical power efficiently. However, each wind turbine creates its own wake, thus lowering the wind velocity coming to the next turbine. Hence, a wind park efficiency can be defined as a function of the wind speed (see Figure 2.16), the type of turbine employed (see Figure 2.15) and the configuration of the wind park. Modelling the wake behind the turbine results in a relatively complex model, but short-cut methods do exist to estimate the overall efficiency of a wind park [74, 75]. The distance between the wind turbines on an evenly distributed wind park is at least 6 times more than the diameter of the area swept by the blades [74], resulting in efficiencies of the wind park of 80% and higher depending on the average wind speed and the number of wind turbines in the wind park (with a higher number of turbines needing to be spread out over a larger land area).

2.5.3 Installations

Wind energy has grown in popularity over the last 2 decades with a significant growth in the installed capacity of wind turbines; the installed capacity grew on average by 13.7% ± 2.9% per annum in the period 2014–2022 and reached an installed capacity of 899 GW worldwide in 2022 (see Table 2.7). Most installations are onshore, with only 6.5% of all installed capacity being off shore [76] (although this may become a growth market). The top 15 countries with the largest installed capacity, listed in Table 2.7 represent 89% of the installed capacity in the world, with China and the United States on their own representing already 41% and 16% of the global installed capacity, respectively.

The installed capacity differs (significantly) from the obtained power from these installations, due to variations in the wind speed (see Figure 2.14), wind direction and down time, but also more subtle variations in wind park design and the specifics of the wind turbine. The installed global capacity delivered 2104.8 TWh of electric power in 2022 (or 7.58 EJ). The total amount that could have been produced from the installed capacity assuming it was running 24/7 throughout the whole year was 28.3 EJ. The capacity factor (CF_{wind}), defined as the ratio of the energy obtained from wind power in a year relative to the maximum amount of energy that could have been obtained in a year based on the installed capacity, stood worldwide in 2022 at 26.7% (see Table 2.7). The capacity factor of wind turbines and wind parks appears to be a strong function of the location of the installations as it

Table 2.7 The top 15 countries with the largest installed capacity of wind power, their annualised production in TWh (1 TWh = 0.0036 EJ) and the capacity factor in 2022 (data from The Energy Institute [1]).

	Capacity (GW)	Production (TWh)	CF_{wind} [a] (%)		Capacity (GW)	Production (TWh)	CF_{wind} [a] (%)
China	366.0	762.7	23.8	Canada	15.3	37.5	28.0
USA	140.9	439.2	35.6	Sweden	14.6	32.6	25.6
Germany	66.3	125.3	21.6	Italy	11.8	20.7	20.0
India	41.9	70.0	19.1	Turkey	11.4	35.1	35.2
Spain	29.3	62.7	24.4	Australia	10.1	31.6	35.7
United Kingdom	28.5	80.2	32.1	Netherlands	9.3	21.2	25.9
Brazil	24.2	81.6	38.6	Poland	8.0	19.4	27.8
France	21.1	38.0	20.5	***World total***	*898.8*	*2104.8*	*26.7*

a) Capacity factor $CF_{wind} = \frac{\text{annual production}}{\text{capacity} \cdot 24 \cdot 365} \cdot 100\%$

varies between 19.1% (India in 2022) to 38.6% (Brazil in 2022) implying that Brazil obtains twice as much electric power per unit of installed capacity of wind power than India.

The capacity factor can be taken as an indication of the required installed power to obtain on average a target power. For instance, replacing a 400 MW thermal power plant would require 400 3 MW wind turbines if the capacity factor is 33.3%. However, the actual delivered power will vary with time due to the variation in the wind speed (and thus the variability in the efficiency, if the wind speed is below rated velocity or over the cut-off speed).

2.5.4 Environmental Aspects of Harnessing Wind Energy

Harnessing wind power to generate electric power is seen as a green technology as the process of transforming wind energy into electricity does not generate greenhouse gases (and CO_2 in particular) directly. Life cycle analyses indicate that capturing wind energy onshore is associated with less than 7 kg $CO_{2,eq}$/MWh, and for off-shore operation less than 11 kg $CO_{2,eq}$/MWh with the main contribution coming from the materials used in the construction of wind turbines and taking it over the 20 year lifespan of wind turbines [77, 78] (increasing the life of wind turbines to beyond the 20 year period would thus almost proportionally decrease the carbon footprint). Working Group III of the 5th IPCC considered a somewhat larger CO_2-footprint for wind energy of 24 kg $CO_{2,eq}$/MWh [61]. Comparing these values with the typical carbon footprint of power generated from natural gas in a combined cycle power plant (490 kg $CO_{2,eq}$/MWh [61]) indicates that the wind power has a relatively small carbon footprint, which can be further reduced by increasing the nominal power of the wind turbines [79].

There have been some environmental concerns raised over the use of wind turbines [78, 80]. An increased mortality of birds and bats upon colliding with wind turbines has been reported, which may be reduced by increasing the cut-in speed (as bats appear to be affected more at low wind speed [78]), cutting out rotor movement in the migratory season and increasing the visibility of wind turbines for birds and bats. The presence of wind turbines and the greater human activity around wind turbines (in particular during the construction and commissioning phase) may also displace some animals from their natural habitat [80].

The passing of the blades past the tower, the movement of the blades through the air, both the in-flow turbulence involving the leading edge of the blades and the trailing edge flow, and the tip vortex all produce noise (together with the gearbox) [81]. This low-frequency noise can lead to annoyance of people and animals living in the neighbourhood. Improved designs of the blades could reduce the noise levels during operation [82].

The end-of-life of wind turbines, and especially the blades of modern wind turbines, is an environmental concern [83]. Wind blades are made of non-biodegradable glass/carbon fibre reinforced polymer and have an estimated lifetime of 20–25 years. Alternative uses/recycle schemes will have to be devised going forward [84].

2.6 Solar Power

The Earth continuously receives energy from the sun. The total solar irradiation (TSI) received at the top of the atmosphere by the Earth visible to the sun is 1367 W/m^2; the average irradiation over the whole Earth visible to the sun amounts thus to 342 W/m^2 (a fourfold reduction as the Earth is spherical rather than circular). The elliptic nature of the orbit of the Earth around the sun results in a 3.3% variation of the mean value of the received solar irradiation [85].

The extraterrestrial solar radiation spectrum (referred to as AM0) is shown in Figure 2.17. The spectrum compares well with that of a black body radiating at 5800 K. Deviations of the extraterrestrial spectrum from the ideal black body radiation is ascribed to absorption in the outer, cooler regions of the sun [85]. The solar light at the top of the atmosphere comprises of UV-light ($\lambda < 400$ nm; 7.6% of the total power received), visible light ($\lambda = 400$–800 nm; 48.8% of the total power received) and infrared light ($\lambda > 800$ nm; 43.6% of the total power received).

Some of the light is absorbed and scattered upon travelling through the atmosphere. Light scattering (Rayleigh scattering) is dependent on the wavelength of light. Figure 2.17 shows the direct irradiation received directly from the sun arriving through the atmosphere (referred to as AM1.5d). The passage through the atmosphere results in absorption of solar light with specific wavelengths. For example, ozone reduces the UV radiation reaching Earth in the form of direct solar irradiation quite significantly and UV radiation at the Earth's surface makes up only 3.4% of the direct solar irradiation. Furthermore, absorption bands in the near infrared region due to the presence of water, oxygen and, to a lesser extent, CO_2 in the atmosphere are clearly visible.

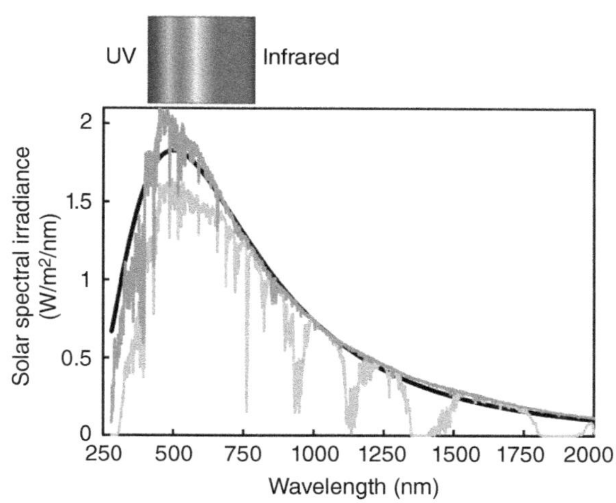

Figure 2.17 Solar irradiation spectrum at the top of the atmosphere (AM0, upper curve), the corresponding black body radiation of the sun at 5800 K (solid black curve) and the direct solar irradiation at a zenith angle of 48.2° (AM1.5d; lower curve) (data from PV Lighthouse [86]).

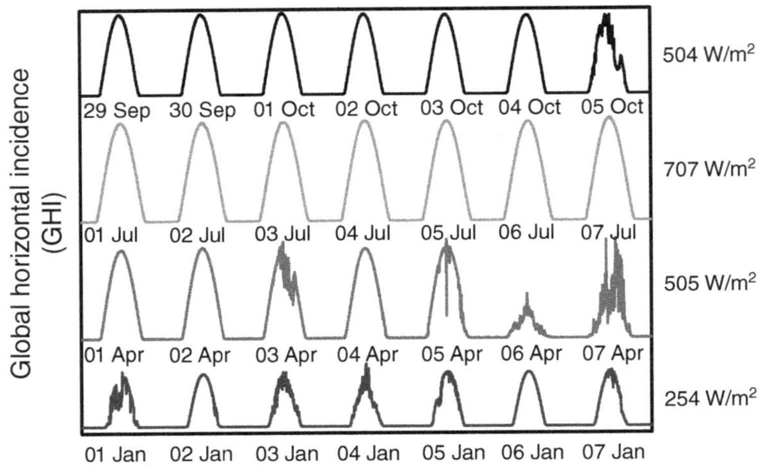

Figure 2.18 Intermittency in solar energy (GHI) at various times in 2020 in Desert Rock, Nevada USA (average GHI during daytime over the reported period of seven days indicated as well; calculated based on data reported by NOAA [87–89]).

The total irradiance from the sun received on a horizontal surface parallel to the Earth (global horizontal irradiance [GHI]) depends on the diffuse horizontal irradiance (DHI – light scattered by the atmosphere), the direct normal irradiance (DNI) and the solar zenith angle, and is thus dependent on the time of the day, the season and meteorological conditions.

Solar light is intermittent in nature due to the Earth's rotation and the tilt of the Earth. Figure 2.18 shows the global horizontal incidence in Desert Rock (Nevada, USA; latitude: 36.624° north) [87, 88] over 7-day periods in the four seasons over the year. The day–night pattern in the received solar radiation is readily identified, but also a reduced solar irradiation due to the weather pattern can be seen overlaying the daily variation. The seasonal variation in the average solar irradiation received during daytime is significant with the average solar irradiation (GHI) received during daytime in summer being more than 2.5 times the received solar irradiation in winter.

2.6.1 PhotoVoltaics

2.6.1.1 Principle

Photovoltaic devices utilise (doped) semi-conductors in contact with each other [90]. Semi-conductors are characterised by the band gap between the valence band and the conduction band, with its Fermi-energy located halfway between the valence and conduction bands. Doping the semi-conductor with a material with more valence electrons will result in the formation of an n-type semi-conductor and the Fermi-level of this material will be shifted towards the conduction band. On the other hand, doping the semi-conductor with materials with less valence electrons will result in the formation of a p-type semi-conductor, whose Fermi-level is shifted towards the energy level of the valence band. Upon contact of an n-type semi-conductor with a p-type of semi-conductor, a (small) amount of electrons flow from the n-type of semi-conductor to the p-type of semi-conductor, thus reducing the chemical potential of the system. The flow will be countered by the electric field generated by the positive charged ions left behind in the n-type semi-conductor and the additional electrons in the p-type of conductor resulting in a potential difference at the p–n junction.

The absorption of light by a semi-conductor may result in the formation of an excited electron in the conduction band and a hole in the valence band, if the energy of the light is larger than the band gap energy of the semi-conductor. The rate of charge carrier generation decreases exponentially with the thickness of material and thus thin films are used. In doped semi-conductors, the absorption of light will affect the concentration of the so-called minority charge carrier (electrons in a p-type semi-conductor and holes in an n-type semi-conductor), whereas the concentration of the majority charge carrier remains almost constant.

The electron taken from the n-type semi-conductor can be taken through an external circuit before recombining with a hole in the p-type of semi-conductor (see Figure 2.19). Not all generated electrons will flow through the external circuit, and the voltage and current emanating from the solar cell can be modelled as a simple electrical circuit with some losses through the diode current, I_D, shunt resistance and resistance in series. The resulting I–V (current–voltage) curve for a PV cell depends on variables such as the resistances in the solar cell, and the diode current, which depends on the obtained voltage and temperature. It shows that the highest voltage occurs when the current tends to zero (the so-called open-circuit voltage, V_{OC}), and the highest current is obtained when the voltage tends to zero (the so-called short-circuit current, I_{SC}).

The open-circuit voltage of a PV-cell containing a classical p–n junction depends on the difference between the quasi Fermi-level of the n-type and the p-type

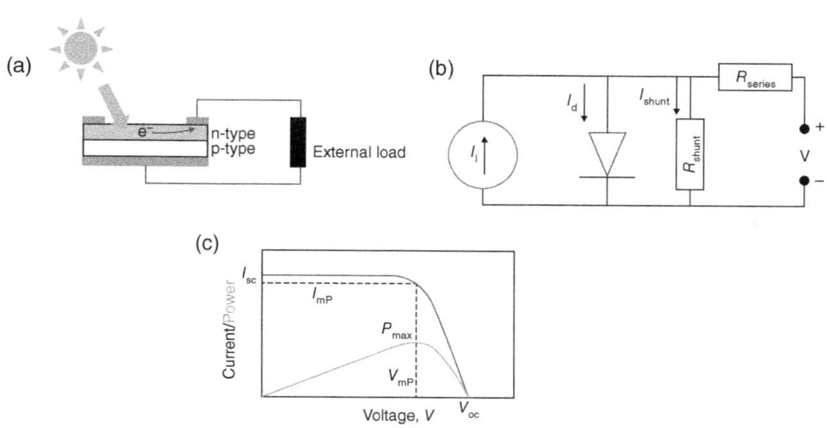

Figure 2.19 Principal functioning of a PV-cell (a), its electrical equivalent model (b) and the typical I–V curve for PV-cells (c).

semi-conductors, which, for a given material depends on the temperature [90]. In organic PV-cells, the open-circuit current voltage also depends on the light intensity [91]. The short-circuit current depends on the light intensity and the recombination rate of the charge carriers, which is strongly dependent on the presence of defects.

The maximum power obtained from a solar cell can be obtained by plotting the obtained power as a function of the voltage. The so-called fill factor (FF), the ratio of the maximum power to the product of the short-circuit current and the open-circuit voltage, is a measure of the efficiency with which the generated electrons are used for work in an external circuit: $\text{FF} = \frac{P_{\text{max}}}{V_{\text{OC}} \cdot I_{\text{SC}}}$ (in the region of 80–86% for Si-based PV-cells [92]).

A solar panel typically comprises of several solar cells joint together in series to increase the voltage delivered by a solar panel.

2.6.1.2 Efficiency of Converting Solar Light into Power

The theoretical, thermodynamic efficiency of capturing energy from solar irradiation, can be obtained from an energy and entropy balance around a device transforming solar irradiation into work [93, 94] (see Figure 2.20).

For the sun radiating at 5800 K, and the device operating at 300 K, the maximum thermodynamic efficiency for the conversion of solar radiation into work is thus 93.10% (the inclusion of radiative pressure effects increases the maximum efficiency only marginally, whereas the inclusion of the geometric factor and dilution of the incoming black body radiation will decrease the efficiency [93]). The maximum efficiency for a device operating at 300 K is less than the Carnot efficiency, which would amount to 94.83% if the sun is a heat source operating at 5800 K.

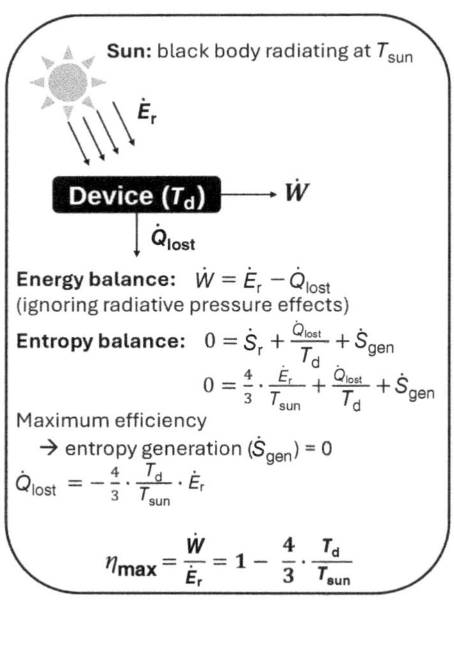

Figure 2.20 Maximum efficiency of capturing solar irradiation ignoring radiative pressure.

The maximum efficiency of photovoltaic devices is greatly reduced using semi-conductors to capture solar energy, as only light with an energy larger than the energy of the band gap will result in the formation of electrons and holes. Furthermore, the freshly generated electrons and holes may recombine, thus reducing the efficiency of the PV-cell. The Shockley-Queisser limit [95, 96] for the efficiency can be derived assuming that the rate of radiative recombination is proportional to the energy difference between the quasi Fermi-energy of the electron and the hole (set equal to the voltage obtained from the PV-cell). The voltage at maximum power (V_{mP}) increases with increasing band gap, whereas the current at maximum power (I_{mP}) decreases with increasing band gap (the latter due to the reduced number of photons which can create the electron–hole pair).

The Shockley-Queisser efficiency (η_{SQ}) passes a maximum as a function of the band gap energy of the semi-conductor (see Figure 2.21), because with increasing energy of the band gap, the number of photons from solar irradiation with an energy larger than the band gap energy decreases, but at the same time the rate of recombination decreases as well implying that the generated electrons are more likely to be used in the external circuit. Silicon, the most used semi-conductor for PV-cells, has a band gap energy of ca. 1.1 eV. The calculated open-circuit voltage (V_{OC}) of silicon upon exposure to solar irradiation (AM1.5 g) is determined as 0.89 V and

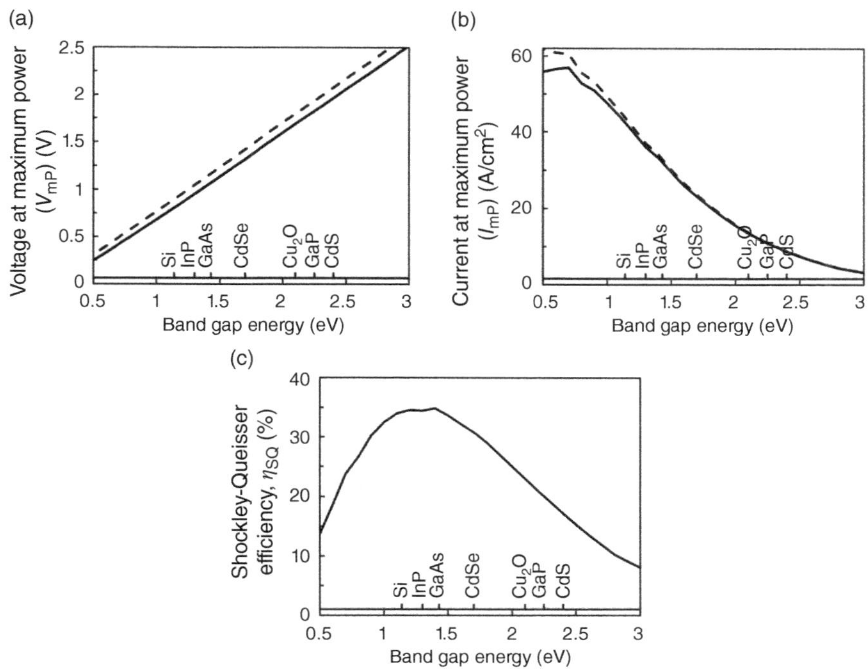

Figure 2.21 Efficiency of PV-cells as a function of the band gap energy of the semi-conductor upon exposure to solar irradiation (AM1.5 g). (a) Voltage at maximum power (dotted line: open-circuit voltage, V_{OC}). (b) Current at maximum power (dotted line: short-circuit current, I_{SC}). (c) Shockley-Queisser efficiency of PV-cells.

the short-circuit current (I_{SC}) is 45.5 mA/cm^2. Silicon will yield maximum power of 339 W/m^2 (at which point the voltage is 0.77 V, and the current is 44.0 mA/cm^2), and thus a fill factor (FF) of 0.868. The maximum efficiency that can be achieved with silicon upon exposure to solar irradiation (AM1.5 g) (which can be determined from the maximum power relative to the energy content in the solar irradiation) is 34%, substantially lower than the thermodynamic efficiency. More than 80% of the energy of the solar spectrum is light with energy greater than 1.1 eV indicating that recombination contributes significantly to the reduction in the efficiency.

The efficiency of PV-cells is further reduced via ohmic losses in the transport of electrons out of the cell. Furthermore, the performance of PV-cells decreases with time due to exposure to the environment. The loss in performance is affected by mechanical shocks (and thus impact of wind, rain, snow and hail), as well as dust accumulation, humidity, temperature and irradiance (in particular, UV radiation) [97], causing an intrinsic loss in efficiency through the formation of cracks or defects in the semi-conductor, a loss in the number of photons reaching the semi-conductor (through discolouration or delamination and bubble formation in the protective layer over the semi-conductor) or an increase in ohmic losses through corrosion. The loss in performance of PV-cells due to exposure to the environment has been reported to be between 0.6% and 1.1% per annum for monocrystalline and polycrystalline silicon-based PV-cells [98]. An annual loss in performance of 0.8% per annum on a PV-cell with an initial efficiency of 20% would result in an efficiency of 16% after 25 years and an average efficiency over a 25-year lifetime of the PV-cell of 18%.

The amount of solar irradiation is proportional to the area and large tracts of land will need to be covered to generate sufficient power. Hence, PV-cells are being put together in large-scale PV-plants. The land usage can be characterised by an efficiency, or land occupation factor [99, 100] as well, where the efficiency is defined by the ratio of the minimum area required for the required power generation relative to the actual area occupation by the PV-plant:

$$\eta_{\text{land}} = \frac{A_{\text{minimum}}}{A_{\text{actually occupied}}}$$

The minimum area can be defined as the area fully covered by PV-cells operating with an efficiency of $\eta_{\text{PV-panel}}$. The land efficiency depends on the packing efficiency (η_{packing}), i.e. the ratio of the area of the PV-panels to the ground area required to avoid excessive self-shading, and the so-called Generator-to-System area (GSR), representing the ratio of the area covered by the PV-panels including the area between them to avoid self-shading (the generation area) to the total ground area (system area).

$$\eta_{\text{land}} = \eta_{\text{packing}} \cdot \text{GSR}$$

The packing efficiency depends on the latitude of the PV-installation and a packing factor of 60–65% can be achieved at a latitude of 30° (in the northern hemisphere: Houston, USA; Cairo, Egypt; Basra, Iraq and Shanghai, China; in the southern hemisphere: Porto Alegre, Brazil; Durban, South Africa, and Perth, Australia), whereas at a latitude of 40° (Salt Lake City, USA; New York, USA; Madrid, Spain; Ankara, Turkey; Beijing, China), the packing efficiency is less than 45–50% [101]. The GSR is typically between 0.7 and 0.85, with larger plants closer to 0.7 as it becomes more difficult to utilise all land for larger operations [100].

The land efficiency has a strong effect on the amount of land required. The minimum area required related to the required power generation, the average solar irradiation (SI; taken here as the yearly average) and the efficiency of the solar panel to convert solar energy into electrical power ($\eta_{\text{PV-panel}}$):

$$A_{\text{minimum}} = \frac{E_{\text{required}}}{\eta_{\text{PV-panel}} \cdot \text{SI}} = \frac{P_{\text{required}} \cdot 12 \cdot 365}{\eta_{\text{PV-panel}} \cdot \text{SI}}$$

(taking into account that the average exposure time to solar irradiation is 12 hours per day, though this value varies greatly depending on the latitude and the time of year).

Hence, the minimum amount of land required for a PV-power plant with a nameplate capacity of 1 GW located in Salt Lake City (USA, 40° north) with a GHI of 1720 kWh/m^2/year can be estimated to be 17.0 km^2 (assuming an efficiency of the PV-cells of 15%). A PV-field with fixed tilt angles optimised for the maximum yearly production at the latitude of Salt Lake City allowing for no shading at noon will have a packing efficiency of 50%. Hence, the actual required area for a 1 GW plant would be 45.3 km^2. In comparison, Mauro and Lughi [102] estimated the required area of 28.7 and 35.2 km^2/GW for PV utilising a tracking system and a non-tracking system, respectively, located in Lecce (Italy, 40° north). A similar sized plant in the Kalahari Desert (border of Namibia and South Africa; latitude 29° south with a GHI of 2500 kWh/m^2/year) requires a minimum area of 11.7 km^2 and an actual area of 24.0 km^2 (with a packing efficiency of 65% and thus a land efficiency of 48.8%).

2.6.1.3 Installations

Generating power from solar light using PV-installations has seen a strong growth over the last two decades [1]. The installed capacity of PV-power stations has grown on average by 21.6% ± 1.5% per annum in the period 2018–2022 with an even stronger growth in the decade preceding this period. In 2022, the global installed capacity was 1053 GW (see Table 2.8). The top 15 countries, with the largest installed capacity, listed in Table 2.8 represent 87% of the installed capacity in the world, with China and the United States together representing 48% of the global installed capacity. This will change with time as the growth rate of PV-installations in areas more conducive for capturing solar energy, such as Brazil, the Middle East, India and Africa, is much higher.

The installed global capacity delivered 1332.6 TWh of electric power in 2022, and PV-generated power contributed 4.6% to the global power generation in that year [1]. It should be realised that in some countries, such as Australia, Spain, Vietnam and, surprisingly, Germany and the Netherlands, PV-generated power contributes more than 10% to the total power generation.

Table 2.8 The top 15 countries with the largest installed capacity of PV-power, their annualised production in TWh (1 TWh = 0.0036 EJ) and the capacity factor in 2022 (data from The Energy Institute [1]).

	Capacity (GW)	Production (TWh)	CF_{PV} [a)] (%)		Capacity (GW)	Production (TWh)	CF_{PV} [a)] (%)
China	393.1	427.7	12.4	Netherlands	22.6	17.7	8.9
USA	113.0	206.2	20.8	South Korea	21.0	27.0	14.7
Japan	78.8	102.4	14.8	Spain	20.5	33.8	18.8
Germany	66.6	60.8	10.4	Vietnam	18.5	26.4	16.3
India	63.1	95.2	17.2	France	17.4	20.1	13.2
Australia	26.8	38.8	16.6	United Kingdom	14.4	13.9	11.0
Italy	25.1	27.5	12.5	Poland	11.2	8.1	8.3
Brazil	24.1	30.1	14.3	***World total***	*1053.1*	*1322.6*	*14.3*

a) Capacity factor $CP_{PV} = \frac{\text{annual production}}{\text{installed capacity} \cdot 24 \cdot 365} \cdot 100\%$.

Theoretically, the total amount of electric energy that could have been produced from the global installed capacity was 9225 TWh (31.6 EJ), but realistically only 50% of this can be obtained (4621 TWh) as the average exposure time to solar irradiation is only 12 hours. A capacity factor (CF_{PV}) can be defined as the ratio of the obtained energy from PV-installations in a year relative to the maximum amount of energy that could have been obtained in a year based on the installed capacity. This factor does not consider the limited exposure time to solar irradiation and hence the maximum capacity factor for PV, CP_{PV}, is only 50%. The effective utilisation of the installed PV-capacity worldwide was only 14.3% in 2022 (see Table 2.8). The capacity factor is a strong function of the location, with countries at a higher latitude experiencing, in general, a lower photon flux (see Figure 1.3), which will thus reduce the power generation. It should be noted that the capacity factor (CF_{PV}) in countries, such as Mexico, United Arabic Emirates, Chile, Argentina and Egypt, is much higher (CF_{PV} of 33.3% in Egypt with an installed capacity of 1.6 GW and a production of 5.0 TWh).

2.6.2 Concentrated Solar Power (CSP)

2.6.2.1 Principle

The energy of solar irradiation can also be captured in the form of heat. Solar light is reflected using a mirror/reflector and concentrated on an absorbing body (see Figure 2.22). The mirror can be arranged in a parabolic concentrator focusing the light on the receiver (parabolic trough) or using flat mirrors focusing the light either on a solar power tower (heliostat) or on a secondary, parabolic concentrator (linear Fresnel collector). Kincaid et al. [103] concluded based on the annualised optical efficiency (defined as the ratio of the energy received at the receiver relative to the energy reflected on the primary reflector) for the parabolic trough is higher than that

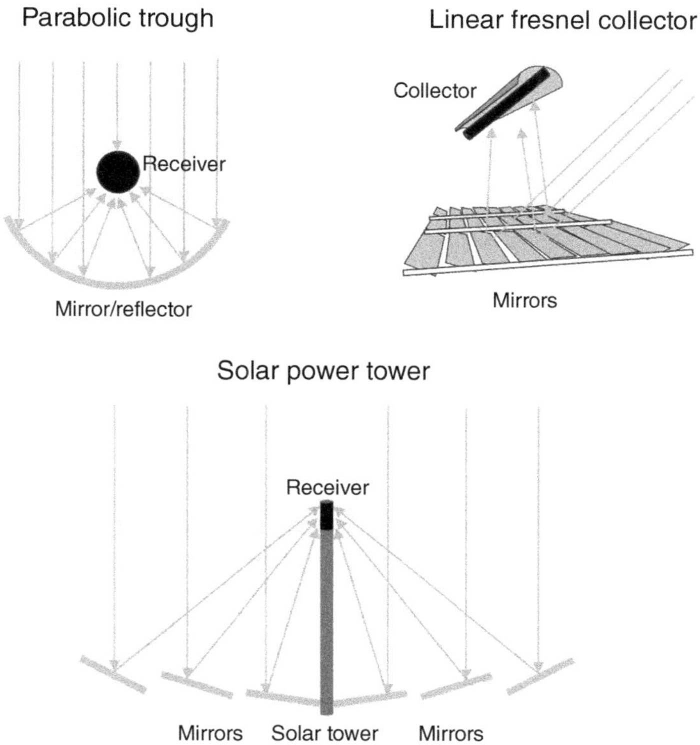

Figure 2.22 Schematic set-up of concentrated solar power.

of the solar power tower, which is higher than that of the linear Fresnel collector. The optical efficiency is dependent on the angle at which radiation is received. The annualised optical efficiency of the parabolic trough at a latitude of 40° was estimated as 60% whilst at a latitude of 21°, it was estimated as 65%.

The mirror should have a high reflectance (ρ) to minimise energy loss. The specular reflectance, i.e. the mirror-like reflection of light in a specific direction, is dependent on the complex refractive index of the material, and thus the wavelength of the incoming radiation, and the angle of the incoming radiation. Diffuse reflectance is caused by imperfections of the surface and thus mirrors/reflectors are typically (highly) polished metals. Table 2.9 shows the reflectance of various materials.

The amount of energy received by a receiver in a parabolic trough of a length L per unit area is related to the amount of energy reflected from the (primary) collector

$$\frac{E_{\text{received}}}{A_{\text{receiver}}} = \alpha_{\text{receiver}} \cdot \frac{E_{\text{reflected}}}{A_{\text{receiver}}} = \eta_{\text{optical}} \cdot \alpha_{\text{absorber}} \cdot \rho_{\text{collector}} \cdot \text{SI} \cdot \text{CF}$$

with SI the annual average solar irradiance, and CF the concentrating factor, i.e. the ratio of the area of the collector relative to the area of the receiver $\left(\frac{A_{\text{reflector}}}{A_{\text{receiver}}}\right)$. The absorbed energy is used to heat up a heat-transfer fluid (thermal oil, molten salt) or

Table 2.9 Absorptivity (α) [105] and reflectance (ρ) [106] of some materials.

Material	A	ρ	Material	α	ρ
Aluminium (highly polished)	0.05	0.95	Red Brick	0.55	0.10–0.15
Copper (carefully polished)	0.02	0.98	Carbon	0.8	0.2
White paint	0.85	0.15	Matt Black paint	0.97	0.03
Glass	0.94	0.06	Water	0.95	0.05

for direct steam generation, whilst some energy is lost due to the surroundings.

$$\frac{E_{\text{received}}}{A_{\text{receiver}}} = \frac{E_{\text{transfer}}}{A_{\text{receiver}}} + \frac{E_{\text{loss}}}{A_{\text{receiver}}}$$

If the temperature of the wall of the receiver ($T_{\text{w,receiver}}$) is uniform, the amount of energy transferred to the heat-transfer fluid flowing at a mass flow rate of $\dot{m}_{\text{htf}}$, which enters the receiver of the parabolic trough at a temperature $T_{\text{htf,inlet}}$, is given by

$$\frac{E_{\text{transfer}}}{A_{\text{receiver}}} = \frac{\dot{m}_{\text{htf}} \cdot C_{\text{P,htf}}}{A_{\text{receiver}}} \cdot (T_{\text{w,receiver}} - T_{\text{htf,inlet}}) \cdot \left(1 - e^{-\frac{h_{\text{htf}} \cdot A_{\text{receiver,w}}}{\dot{m}_{\text{htf}} \cdot C_{\text{P,htf}}}}\right)$$

with h_{htf} the convective heat-transfer coefficient from the wall of the receiver to the heat-transfer fluid, $A_{\text{receiver,w}}$ the wall area of the receiver. The energy loss is in the form of radiative heat loss as well as convective heat loss:

$$\frac{E_{\text{loss}}}{A_{\text{receiver}}} = \varepsilon \cdot \sigma \cdot T^4_{\text{w,receiver}} + h_{\text{air}} \cdot (T_{\text{w,receiver}} - T_{\text{ambient}})$$

with ε the emissivity of the receiver, σ the Stefan-Boltzmann constant ($5.67 \cdot 10^{-8}$ W/m^2/K^4), h_{air} the convective heat-transfer coefficient to the outside air at T_{ambient} (heat loss by convective heat transfer is often low and can be neglected as it contributes only a small amount to the overall heat loss; furthermore, the absorber tube can be encapsulated in glass with an vacuum layer between the glass and the receiver to further minimise the convective heat transfer [104]).

The temperature of the receiver wall can then be estimated by combining the energy absorbed, the energy transferred and the energy lost. For instance, a parabolic trough ($\alpha_{\text{receiver}} = \varepsilon_{\text{receiver}} = 0.97$; $\rho_{\text{reflector}} = 0.95$; $\eta_{\text{optical}} = 0.67$) with a concentrating factor, CF, of 40 and an absorption tube ($L = 4$ m; $d_{\text{in}} = 1''$), where a heat-transfer fluid ($C_{\text{p,fluid}} = 1.5$ kJ/kg/°C, $\mu = 0.0045$ Pa s, $k = 0.4$ W/m/°C) enters at a rate of 0.1 kg/s at 200 °C. The heat-transfer coefficient can then be estimated using the correlation proposed by Haussen cited in [105] to be $h_{\text{htf}} = 80$ W/m^2/K and $h_{\text{air}} = 0.2$ W/m^2/K, respectively, resulting in a wall temperature of 391 °C.

The wall temperature can be controlled by controlling the flow rate of the heat-transfer fluid (see Figure 2.23), and it decreases with increasing mass flow rate of the heat-transfer fluid. The relationship is nonlinear because the heat-transfer coefficient shows a nonlinear relationship of the Nusselt number with the fluid velocity [105]. Similarly, the wall temperature of the receiver can be controlled by changing the mass flow rate of the heating fluid through the receiver.

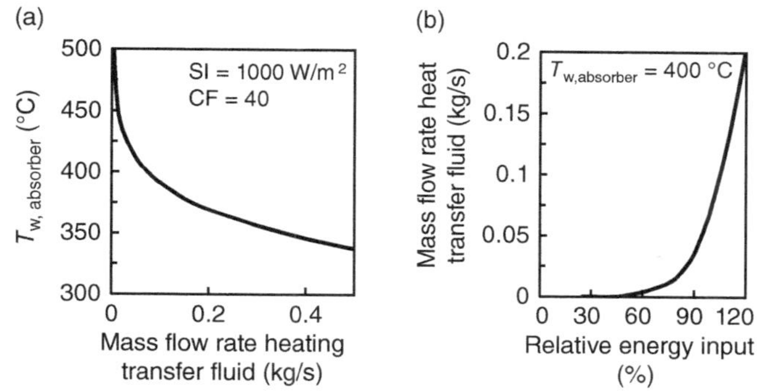

Figure 2.23 Performance of a parabolic trough ($\alpha_{receiver} = \varepsilon_{receiver} = 0.97$; $\rho_{reflector} = 0.95$; $\eta_{optical} = 0.67$; absorption tube with $L = 4$ m and $d_{in} = 1''$) through which a heat-transfer fluid ($C_{p,\,fluid} = 1.5$ kJ/kg/°C, $\mu = 0.0045$ Pa s, $k = 0.4$ W/m/°C) enters with a temperature of 200 °C. (a) The wall temperature of the receiver for a concentrating factor of 40 with a solar irradiation of 1000 W/m^2 as a function of the mass velocity of the heat-transfer fluid. (b) The required mass flow rate to obtain a wall temperature of 400 °C as a function of the energy input relative to the required energy input at a concentrating factor of 40 with a solar irradiation of 1000 W/m^2.

Power is generated from thermal energy using the classical Rankine cycle (see chapter 2.2.1). High-pressure steam is generated upon heat exchange with the hot heat-transfer fluid (the use of water-generating steam directly in the solar light concentrator is possible as well, but the poor thermal conductivity of steam may result in poor heat transfer). The efficient conversion of steam into power requires both a high pressure and a high temperature; the latter is limited by the highest temperature of the heat-transfer fluid, i.e. the temperature at which the heat-transfer fluid leaves the solar concentrator.

The advantage of concentrated solar power (CSP) lies in the possible storage of thermal energy [107], which can be used to power the Rankine cycle if the sun does not shine (see Figure 2.24). During daytime, the heat-transfer fluid coming from the solar concentrator will be distributed over the Rankine cycle and the thermal energy storage, where it exchanges heat with the working fluid of the Rankine cycle and the storage material, respectively. Subsequently, the heat-transfer fluid will be pumped back to the solar plant. During nighttime, the heat-transfer fluid will take up heat from the storage material to give off to the working fluid in the Rankine cycle, after which it will be pumped back over the thermal energy storage, bypassing the solar plant.

The size of the thermal storage capacity is usually dictated by process economics [108]: increasing the size of the thermal storage will increase capital cost of the power plant, whilst at the same time increasing the operational time of the power generation and thus income. Most commonly, thermal energy is stored in the form of sensible heat [109], with molten salts being the material typically used for thermal energy storage, but steam, molten metals and concrete are also used.

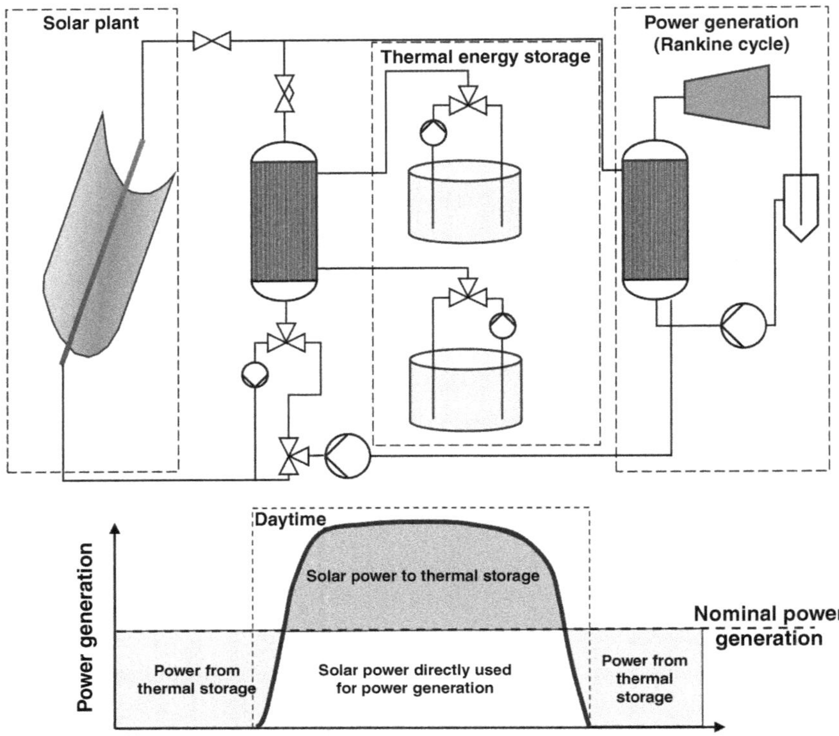

Figure 2.24 CSP with heat storage.

2.6.2.2 Efficiency of Converting Solar Light into Heat

The efficiency of converting solar energy into power using CSP is made up by the two sub-processes, viz. heating up of the heat-transfer fluid and power generation. The maximum efficiency that can be obtained by converting solar light into power using this process can be approximated by the efficiency of the process to transform light into heat and the Carnot efficiency for the conversion of heat into work:

$$\eta_{\text{CSP,max}} = \eta_{\text{solar}\to\text{heat}} \cdot \eta_{\text{Carnot}}$$

The efficiency for the conversion of solar light into heat at a temperature T_H is given by the amount of energy transferred to the heating medium relative to the total energy received:

$$\eta_{\text{solar}\to\text{heat}} = \frac{E_{\text{transfer}}}{E_{\text{received}}} = 1 - \frac{E_{\text{lost}}}{E_{\text{received}}}$$

The total energy received is given by: $E_{\text{received}} = \text{SI} \cdot A_{\text{absorber}} \cdot \text{CF}$, and radiative heat loss is the major form of heat loss. Hence, the efficiency for the conversion of solar energy into heat is given by:

$$\eta_{\text{solar}\to\text{heat}} = 1 - \frac{\varepsilon \cdot \sigma \cdot T_{\text{H}}^4}{\text{SI} \cdot \text{CF}}$$

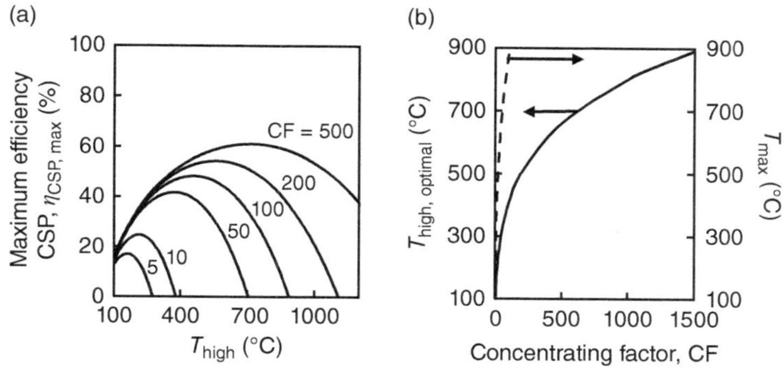

Figure 2.25 Maximum efficiency that can be obtained using CSP as a function of the temperature of the receiver (T_H) for different concentrating factors (CF) of solar irradiation (a) and the temperature, at which the maximum efficiency can be obtained ($T_{H,optimal}$) as a function of the concentrating factor (b).

The Carnot cycle will operate between the highest temperature, T_H, and a temperature at which heat is being given off to the surroundings ($T_{ambient}$) and thus the maximum efficiency of a CSP-plant is

$$\eta_{CSP,max} = \left(1 - \frac{\varepsilon \cdot \sigma \cdot T_H^4}{SI \cdot CF}\right) \cdot \left(1 - \frac{T_{ambient}}{T_H}\right)$$

The maximum efficiency of the CSP-plant is higher when solar light is more concentrated, i.e. with increasing concentrating factors, (see Figure 2.25) and thus a higher efficiency can possibly be achieved with highly concentrated solar power as obtained using heliostats. The maximum efficiency of the CSP-plant passes a maximum as a function of the temperature of the receiver, because the radiative energy loss increases with increasing temperature of the receiver. An optimum temperature for the receiver ($T_{H,optimal}$) can now be defined as the temperature at which the maximum efficiency for the CSP could be obtained.

The actually efficiency of CSP-plants achieved is, however, much lower than theoretical maximum efficiency that can be obtained for a number of reasons: the temperature of the receiver will be lower than the optimum temperature due to heat transfer to the heat-transfer fluid; the Rankine cycle has a lower efficiency than a Carnot cycle; the optical efficiency, as well as the non-ideal reflection and absorption of light. Taking the example of a parabolic trough ($\alpha_{receiver} = \varepsilon_{receiver} = 0.97$; $\rho_{reflector} = 0.95$; $\eta_{optical} = 0.67$) with concentrating factor, CF, of 40: only 61% of the incoming solar light will be absorbed. Energy is being transferred by the flow of a heat-transfer fluid ($C_{p,\,fluid} = 1.5$ kJ/kg/°C, $\mu = 0.0045$ Pa s, $k = 0.4$ W/m/°C, 0.1 kg/s) through the receiver ($L = 4$ m; $d_{in} = 1''$), entering at a rate of 0.1 kg/s at 200 °C. The temperature, at which the fluid will leave the receiver, can be estimated as 361 °C. This implies that only 56% of the absorbed energy is transferred to the fluid with the rest being radiated off. Otherwise stated, the heat-transfer fluid leaving the receiver contains only 35% of the solar energy. The working fluid for the Rankine cycle can be at 350 °C implying an efficiency for the conversion of heat into power of ca. 30%

(see Figure 2.3) and thus the power generation will only be 10% of the solar irradiation, comparable to the reported efficiency of converting solar energy into electric power of 11–16% for the parabolic trough technology, 12–16% for the solar power technology and 8–12% for Fresnel collectors [110].

Thermal storage is the advantage of CSP-techniques. The efficiency of thermal energy storage using molten salt is claimed to be over 99% based on the experience on a 10 MW-plant [111]. Hence, energy loss in storage can be neglected if thermal storage capacity has been designed properly.

Just like PV, land occupation by the installation needs to be considered. The so-called land efficiency is only dependent on the GSR, representing the ratio of the area covered by the reflectors (the generation area) to the total ground area (system area)

$$\eta_{\text{land}} = \text{GSR}$$

because shading has been included in the optical efficiency (η_{optical}) for CSP. The reported land efficiency of the operational units is rather low [110], with a land efficiency for the parabolic trough technology of only 25%, and the solar power tower technology of 19%. However, the locality will strongly influence the land efficiency.

2.6.2.3 Installations

Power generation using CSP is much less common than power generation using PV, with a total operational, installed capacity of CSP for the different technologies worldwide of only 6 GW (see Table 2.10) with another 1.8 GW under construction (mainly in China and the United Arabic Emirates) [112]. Parabolic trough technology is the most commonly used technology making up 72% of the total installed CSP-power (with solar tower power making up 24%, and Fresnel collectors making up 4% of the installed capacity). It is worth noting that 52% of the installations under construction employ the solar power tower technology and only 36% employ the parabolic trough technology (the balance being Fresnel collectors). Parabolic troughs may capture solar energy more efficiently as indicated by a higher annualised solar efficiency [103], but solar power towers are able to achieve higher temperature and thus convert the received energy more efficiently into power.

Table 2.10 lists the countries with the most installed CSP capacity and the type of technology applied. These countries represent 91% of the total installed capacity, with Spain on its own representing 38% of the total installed capacity. Parabolic trough technology is dominant in most countries, with the exception of China where solar power tower technology is the most used technology (the dominance of solar power tower technology in China will become even stronger with 0.8 GW of CSP using this technology currently under construction). The actual power generation using CSP is not well-known [107], and typically only estimated annual production [112] is available, which can be used to estimate a capacity factor. The estimated annual production would indicate a relatively high capacity factor of ca. 32% noting

Table 2.10 The countries with the largest installed capacity of CSP-power, their estimated annual in TWh (1 TWh = 0.0036 EJ) and the capacity factor in 2022 (data from References [112, 113]) for the various technologies.

	Total	Parabolic trough			Solar power tower			Fresnel collector		
	Power (GW)	Power (GW)	Prod.[a)] (TWh)	CF_{CSP}[b)] (%)	Power (GW)	Prod.[a)] (TWh)	CF_{CSP}[b)] (%)	Power (GW)	Prod.[a)] (TWh)	CF_{CSP}[b)] (%)
Spain	*2.31*	2.22	5.90	30.3	0.06	0.18	37.0	0.03	0.05	18.5
USA	*1.35*	0.86	2.28	30.4	0.49	1.58	36.6			
China	*0.60*	0.15	0.55	41.8	0.37	1.24	38.2	0.08	0.35	53.1
South Africa	*0.55*	0.50	1.94	44.3	0.05	0.18	41.1			
Morocco	*0.53*	0.38	1.03	30.6	0.15	0.50	38.1			
India	*0.24*	0.10	0.23	25.8				0.14	0.29	24.8
World	*6.11*	*4.38*	*12.20*	*31.8*	*1.46*	*4.01*	*31.2*	*0.26*	*0.72*	*31.5*

a) Estimated annual production.
b) Estimated capacity factor $CP_{CSP} = \frac{\text{estimated annual production}}{\text{installed capacity}\cdot 24\cdot 365} \cdot 100\%$.

that solar light is only available on average 50% of the time. The actual performance and delivered annualised energy delivery would allow for a better comparison of the various technologies.

The advantage of power generation using CSP is the ability to carry on production using stored thermal energy. The current CSP-plants with a reported storage capacity have, on average, a storage capacity of 7.6 ± 3.6 hours with no statistically significant difference between the various technologies, implying that there will be downtime on the power generation for the installed CSP-installations (the continuous delivery of power would require a storage capacity of over 14 hours with more required at higher latitudes).

2.6.3 Environmental Aspects of Harnessing Solar Energy Using PV or CSP

As with any anthropogenic activity, power generation directly from solar irradiation either through PV or through CSP will have an impact on the environment. The obvious factors to consider are the use of land for capturing solar irradiation, the use of water and chemicals in the conversion of solar energy into power and the manufacture and decommissioning of the required equipment.

Solar energy is received at a rate of 1000 W/m^2, and the area for PV was estimated at 24–50 km^2/GW depending on the location of the installation (see [102, 114]). The estimations for CSP are somewhat less; the average solar field for the parabolic trough and solar power tower can be estimated at 8.5 and 9.1 km^2/GW, respectively (based on solar field and installed power reported by [110]). It should be noted that these values are not statistically different, specifically considering the different locations of these installations (in contrast, the area for the solar field of the Fresnel collectors is estimated at 15.9 km^2/GW). This implies that a significant area will be covered with either PV-cells or reflectors.

The large tracts of land required will be covered with PV-cells/reflectors in utility-scale operations, thus shading a significant area. It is often thought that environmental impact of large-scale solar installations in desert areas will be minimal [114], but shading the area will affect the soil temperature and increase the soil humidity [115], thus affecting flora and fauna below these units even in deserts. The change in insect-life surrounding these installations will affect both bird diversity and other small animals in the solar field, as well as the area surrounding the solar field [116]. Data on altered bird mortality are still scarce, and the consequences of large-scale solar power installations on migratory species will need to be monitored.

It should be noted that the impact of land usage by PV-installations can be reduced by mounting them as a rooftop structure. It has been estimated that 0.2 million km^2 of rooftop area could produce 27 000 TWh of electric power per annum [117], i.e. sufficient to cover the annual, global power consumption in 2022 (see Table 2.1).

Power generation using CSP has a relatively large water footprint due to the cooling required in the Rankine cycle. Water cooled systems are most efficient, but have higher water consumption (with an estimated median water consumption of 3350 l/MWh for CSP-systems [118]), whilst dry systems, which have an estimated median water consumption of 320 l/MWh [118], may need to be implemented in water scarce areas. Water usage during operation is insignificant for PV-installations, where some water is used for cleaning and in some cases for panel cooling. Cleaning of PV-panels and reflectors does not only contribute to water usage but may also result in spillage to the environment [114].

Thermal energy storage is the advantage of CSP over PV, but the storage medium, e.g. molten salt, may also become an environmental risk, if it is spilled, which will need to be addressed. Thermal degradation of the storage medium [119] may also increase the environmental impact of this operation.

The largest environmental impact of solar-generated power lies in the production, construction and decommissioning of the installation [114]. This results in a quoted carbon footprint of 14–73 kg $CO_{2,eq}$/MWh for PV [114] and 15–28 kg $CO_{2,eq}$/MWh for CSP [120] using the current applied technologies. However, the production of PV-panels may have a greater environmental impact due to the processes involved.

2.7 The Grid

The electricity grid transmits power generated at different power stations to the various end users, and as such power generation and consumption must be balanced. The frequency of the grid will vary slightly from its set frequency (e.g. 50 Hz in Europe and 60 Hz in the United States) upon changing the load on the grid, with a drop in the frequency upon an increase in demand and vice versa, and additional generation capacity will be required to counter the increased demand through the so-called automatic generation control (AGC). It should be noted that frequency control for small variations in the load is currently countered by inertia in the synchronous generators in the power stations. PV and wind energy do not provide inertial support to the grid in this form, and alternative strategies must be implemented to ensure grid stability [121].

The total amount of power delivered by the grid depends on the season, as well as the time of the day (see e.g. Figure 2.26). The power demand changes depending on the usage pattern with one or two maxima per day, viz. one located at the start of the evening and the other one sometimes shown at the start of the working day. The variation in the demand on a day can be rather large and, in the example given in Figure 2.26 the variation between the maximum and minimum power demand is as much as 40–50% of the minimum power demand. Furthermore, significant changes in the maximum daily power demand can be noticed. Typically, there is a lower maximum power demand during the weekend, whereas the minimum power demand does not vary as much. Seasonal variations will come on top of the daily demand variations.

Different energy sources can be used to generate the required power. Power generation sources can be categorised as dispatchable power or variable power [123]

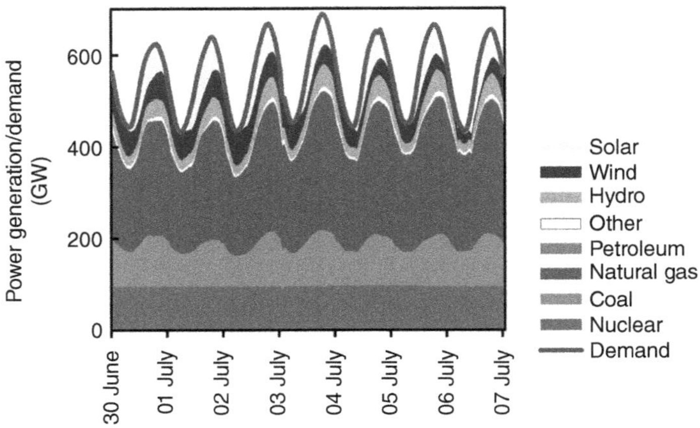

Figure 2.26 Various energy sources making up the demand for power in the United States between 30 June 2024 and 07 July 2024 (data from EIA [122]).

depending on their ability to tune their output to the required demand. Classical power stations, but also renewable energy generated from hydropower, biomass, geothermal energy and CSP-plants with thermal energy storage, are classified as dispatchable power, whereas wind power and PV without intermediate storage are classified as variable power. These variable power sources are thus used when generated (see Figure 2.26). The integration of variable power sources into the grid will need additional measures to stabilise the grid (e.g. with back-up of dispatchable power).

Power generation can be classified into base load, intermediate load and peak load and can be obtained by looking at the annual power demand (see Figure 2.27). The minimum amount of power required at any time in the period considered is called the base load. This needs to be supplied continuously and is preferentially delivered by systems whose power output cannot be adjusted rapidly in response to changing demand. This is supplied in the United States by nuclear energy and coal-based power stations (see Figure 2.26).

Intermediate load makes up a large portion of the generation capacity and needs to cater for seasonal, as well as daily and even hourly variation in the demand. This can be delivered by most dispatchable power sources by ramping up from their minimum load of operation. A logical consequence is the overdesign of the power generation capacity to cater for this (as well as for scheduled and unscheduled maintenance).

Peak load generating units help to meet electricity demand when demand surpasses the maximum load given off by the intermediate load generation units. These units will not need to be operated for long periods of time but will need to start up quickly when the demand arises. Here, natural gas and liquid fuel generators are usually used.

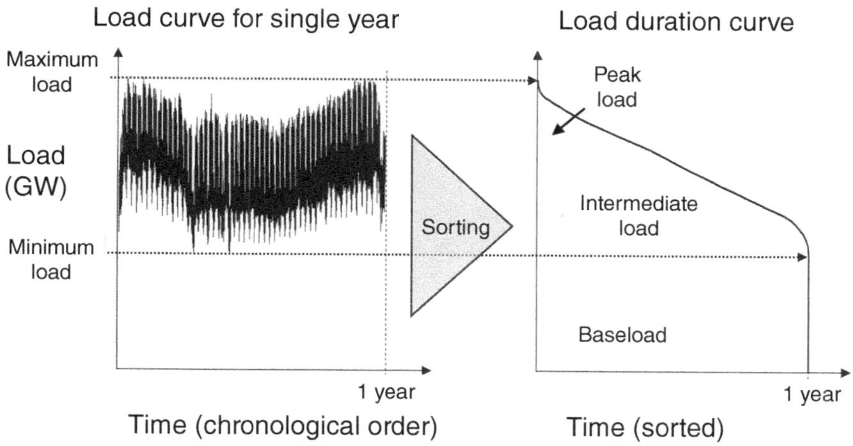

Figure 2.27 Determining base load, intermediate load and peak load in power consumption (adapted from Ueckerdt and Kempener [123]).

Transmission over the grid is associated with some energy loss. On average, the energy loss through transmission and distribution is worldwide ca. 8% of the total power transmitted [124], which is made up by the energy lost in transforming the power to high voltage, energy lost in the transmission lines, energy lost in transforming the power to low voltage and energy lost in the distribution net.

2.8 Outlook

In 2022, the world generated more than 29 000 TWh of electric power [1], which resulted in the emission of 13.4 Gtons of CO_2 (or on average 462 kg CO_2/MWh) contributing 37% of the total annual, anthropogenic CO_2 emission for that year [125]. Power production is expected to increase with the current drive towards a higher degree of electrification of the transportation sector, but also other sectors. Furthermore, it must be anticipated that power usage in currently underdeveloped areas will be expanding in the near future as well, thus requiring more power-generating capacity in these areas (as well as extension of the grid) particularly seeing the anticipated population growth in these areas. The large contribution of power generation towards CO_2 emissions and the anticipated increase in the required power-generating capacity necessitates a rapid decarbonisation of this sector.

Different technologies are available for power generation. These technologies have different inherent efficiencies to convert their energy source into power. The comparison of the efficiencies for different technologies should be avoided, as the different technologies utilise different energy carriers and will have different products (and waste streams). For instance, the efficiency of power generation in coal-based power plants is typically between 30% and 40%, whereas the efficiency for the generation of power using photovoltaics is between 10% and 20%. However, the efficiency for the generation of power from coal does not consider the efficiency of generating coal from solar energy in the first place nor is the emission of CO_2 taken into account. It should be noted that the efficiency of a coal-based based power station is significantly reduced when considering carbon capture utilisation and storage (assuming power generation using a bituminous coal with a heating value of 33 MJ/kg with an efficiency of the Rankine cycle of 40% and a cost of CO_2-capture of 0.6 MWh/tonne CO_2 would reduce the efficiency of power generation from coal to less than 20%). Efficiencies should be used to compare the state of the art within a certain technology and be used as measures for possible improvement in the particular technology.

The large amount of CO_2 emitted in power generation using fossil fuels demands the introduction of more renewable energy into the power generation mix. Table 2.11 lists the mean initial investment cost normalised for their power generation capacity for greenfield operations as well as the range of the investment cost [126]. The range can be large due to differences in the scale of operation, variation in applied technologies (e.g. power generation from pulverised coal power plants, supercritical, pulverised coal power plants and ultra-supercritical, pulverised coal power plants) and the locality. The investment cost for dispatchable power generation using the open cycle gas turbine (natural gas) is the lowest but the investment cost increases substantially upon including a facility for CCUS, because of the additional capital expenditure, as well as the reduced power output when implementing this technology. The same can be observed for the usage of coal for power generation. Nuclear energy has a high investment cost, whilst the investment cost for the generation of variable power using solar or wind power is on par with the investment cost for power generation using natural gas and coal without the obvious benefit of being able to tune the output depending on the demand; their advantage however lies in the strongly reduced CO_2 emissions associated with these technologies.

The investment costs are not the only contributing factor to the costs of power generation, as the cost of the fuel and operational costs will need to be considered as well. Hence, different modes of generating power are often compared based on the so-called Levelised Cost of Electricity (LCOE), which expresses the discounted cost of power generation relative to the discounted economic value of the power generated:

$$\text{LCOE} = \frac{\sum_{t=1}^{t_{\text{final}}} \frac{I_t + \text{O\&M}_t + F_t}{(1+r)^t}}{\sum_{t=1}^{t_{\text{final}}} \frac{E_t}{(1+r)^t}}$$

with t_{final}, lifetime of the project; I_t, the investment cost; O&M$_t$, operations and maintenance costs; F_t, fuel costs; E_t, energy production; r, discount rate.

The LCOE is a useful comparison to obtain an indication of the cost of power production, but the obtained values should be scrutinised as the choice of various

Table 2.11 Mean investment cost[a)] and their range (in brackets) for the various power-generating technologies in 2020 (data from IEA [126]).

	Investment cost[a)] (US$/kW)	Lifetime (yr)
Coal-fired power station	1897 (800–4382)	40
Coal-fired power station + CCUS[a)]	4572 (4490–5991)	40
Lignite-fired power station	2973 (2189–3756)	40
Combined cycle gas turbine (natural gas)	823 (254–1109)	30
Combined cycle gas turbine (natural gas) + CCUS[b)]	2619 (2412–2826)	30
Open cycle gas turbine (natural gas)	630 (325–980)	30
Nuclear	3606 (2157–6920)	60
Hydropower (reservoir >5 MW)	3319 (1899–5819)	80
Wind power (on-shore, >1 MW)	1391 (877–3022)	25
Wind power (off-shore, >1 MW)	2876 (1721–4039)	25
Solar PV (utility scale)	995 (534–2006)	25
Solar PV (residential)	1583 (719–2597)	25
Solar thermal (CSP)	5857 (5238–6475)	25
Geothermal	6647 (3851–3810 959)	25

a) Investment cost excluding interest during construction.
b) Carbon capture utilisation and storage.

variables to determine the LCOE will influence the outcome [127]. For instance, a long (proposed) lifetime of the project will favour projects with a high initial investment cost. Hence, well-established technologies with good estimates on their anticipated lifetime are advantaged here over emerging technologies, whose lifetime is estimated more conservatively as it is not well-known. Projects with a high initial investment cost are also favoured, where the discount rate is low, thus implying that the locality of the project will affect the LCOE.

The LCOE is a good way to track the development of technology over time (see Figure 2.28). For instance, the cost of power from wind energy has come down in the period 2010–2020 from US$ 125/MWh to US$ 42/MWh, and the cost of PV-power has come down from over US$ 250/MWh to US$ 42/MWh, whereas power generated from coal sits at ca. US$ 110/MWh.

The LCOE is only one aspect when considering the energy mix required for power generation. Other factors that need to be considered are the flexibility of power generation to be tuned to demand, as well as the environmental impact of power generation. It has been proposed to capture these in the so-called value-adjusted LCOE (VALCOE) [126].

Renewable energy with its intermittency will play a more prominent role in the global power generation going forward as the carbon emissions will need to be reduced with the actual power generation increasing. The grid will need to be fed

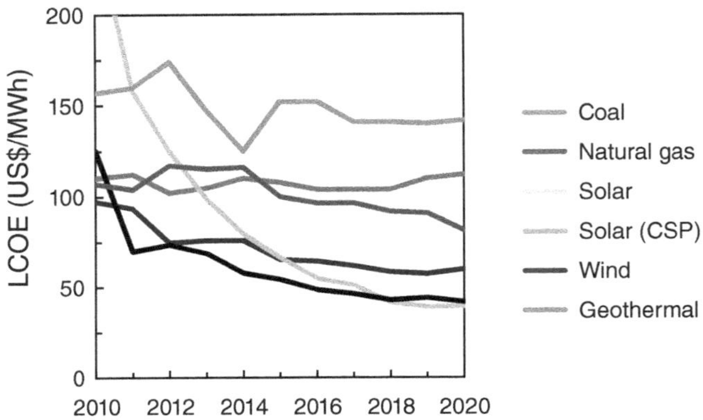

Figure 2.28 Development of LCOE over the years for the various ways to generate power (redrawn from Lazard [128]).

with energy based on the demand. Hence, the introduction of renewable energy into grid may pose a problem, as it cannot be switched on and off on demand. With renewable energy, both short-term and seasonal variability will need to be considered. The short-term variability in wind energy may be balanced by connecting multiple wind turbines at different localities. The short-term variability in solar energy is well documented and well discussed. It can be minimised by introducing storage of energy (e.g. in the form of heat storage for CSP or in other forms – see Chapter 3). Furthermore, the variability can be minimised by combining wind energy with solar energy [129] with larger grids obtaining renewable energy from different geographic location better able to minimise the effect of variability. Furthermore, the variability associated with the renewable energy carriers can be countered by overdesign of the planned installations capturing renewable energy associated with various ways to store energy.

References

1 The Energy Institute (2023). Statistical review of world energy [Online]. Available: https://www.energyinst.org/statistical-review/about (accessed 7 February 2024).

2 EIA - US Energy Information Agency (2023). Electricity explained – electricity in the USA, 30 June 2023 [Online]. Available: https://www.eia.gov/energyexplained/electricity/electricity-in-the-us.php (accessed 2 March 2024).

3 Husaini, D. and Lean, H. (2022). Renewable and non-renewable electricity-growth nexus in Asia: the role of private power plants and oil price threshold effect. *Resources Policy* 78: 102850.

4 Clark, R., Zucker, N., and Urpelainen, J. (2020). The future of coal-fired power generation in Southeast Asia. *Renewable and Sustainable Energy Reviews* 121: 109650.

5 Jaziri, N., Boughamoura, A., Müller, J. et al. (2020). A comprehensive review of thermoelectric generators: technologies and common applications. *Energy Reports* 6 (Supplement 7): 264–287.

6 Kim, H., Liu, W., Che, G., and Chu, C.-W. (2015). Relationship between thermoelectric figure of merit and energy conversion efficiency. *Proceedings of the National Academy of Sciences* 112 (27): 8205–8210.

7 Ahmad, M., Schatz, M., and Casey, M. (2013). Experimental investigation of droplet size influence on low pressure. *Wear* 303: 83–86.

8 Senoo, S. and White, A. (2017). Analysis and design of wet-steam stages. In: *Advances in Steam Turbines for Modern Power Plants* (ed. T. Tanuma), 165–218. Amsterdam: Elsevier.

9 Cebrucean, D., Cebrucean, V., and Ionel, I. (2020). Modeling and performance analysis of subcritical and supercritical coal-fired power plants with biomass co-firing and CO_2 capture. *Clean Technologies and Environmental Policy* 22: 153–169.

10 van der Westhuizen, L. and Gorlach, I. (2021). Performance optimisation of coal-fired boiler control using Flownex® simulation environment and AI. *R & D Journal of the South African Institution of Mechanical Engineering* 37: 9–18.

11 Babatunde, D., Anozie, A., Omoleye, J., and Babatunde, O. (2021). An air-fuel ratio parametric assessment on efficiency and cost of a power plant steam boiler. *Process Integration and Optimization for Sustainability* 5: 561–575.

12 Tumanovskii, A., Shvarts, A., Somova, E. et al. (2017). Review of the coal-fired, over-supercritical and ultra-supercritical steam power plants. *Thermal Engineering* 64: 83–96.

13 Lee, S.-H., Themelis, N., and Castaldi, M. (2007). High-temperature corrosion in waste-to-energy boilers. *Journal of Thermal Spray Technology* 16: 104–110.

14 Viswanathan, R., Coleman, K., and Rao, U. (2006). Materials for ultra-supercritical coal-fired power plant boilers. *International Journal of Pressure Vessels and Piping* 83 (11–12): 778–783.

15 Wikipedia. GE-BWR [Online]. Available: https://en.wikipedia.org/wiki/GE_BWR (accessed 16 March 2024).

16 Rodríguez, M., Rodríguez, J., and Fontes, C.d.H. (2022). Modeling of a combined cycle gas turbine (CCGT) using an adaptive neuro-fuzzy system. *Thermal Engineering* 69: 662–673.

17 Turconi, R., Boldrin, A., and Astrup, T. (2013). Life cycle assessment (LCA) of electricity generation technologies: overview, comparability and limitations. *Renewable and Sustainable Energy Reviews* 28: 555–565.

18 Krall, L., Macfarlane, A., and Ewing, R. (2022). Nuclear waste from small modular reactors. *Proceedings of the National Academy of Sciences* 119 (23): e211833119.

19 Dunbar, W., Moody, S., and Lior, N. (1995). Exergy analysis of an operating boiling-water-reactor nuclear power station. *Energy Conversion and Management* 36 (3): 149–159.

20 Durmayaz, A. and Yavuz, H. (2001). Exergy analysis of a pressurized-water reactor nuclear-power plant. *Applied Energy* 69: 39–57.

21 Harkin, T., Hoadley, A., and Hooper, B. (2009). Process integration analysis of a brown coal-fired power station. *Energy Procedia* 1: 3817–3825.

22 Artanto, Y., Jansen, J., Pearson, P. et al. (2012). Performance of MEA and amine-blends in the CSIRO PCC pilot plant at Loy Yang Power in Australia. *Fuel* 101: 264–275.

23 Krótki, A., Solny, L., Stec, M. et al. (2020). Experimental results of advanced technological modifications for a CO_2 capture process using amine scrubbing. *International Journal of Greenhouse Gas Control* 96: 103014.

24 Pavri, R. and Moore, G.D. Gas turbine emissions and control, General Electric [Online]. Available: https://www.gevernova.com/content/dam/gepower-new/global/en_US/downloads/gas-new-site/resources/reference/ger-4211-gas-turbine-emissions-and-control.pdf (accessed 19 March 2024).

25 Wang, W., Tian, S., Long, J. et al. (2022). Investigation and evaluation of flue gas pollutants emission in waste-to-energy plant with flue gas recirculation. *Atmosphere* 13 (7): 1016.

26 Lee, H., Yi, S.-M., Holsen, T. et al. (2018). Estimation of CO_2 emissions from waste incinerators: comparison of three methods. *Waste Management* 73: 2470255.

27 Aouini, I., Ledoux, A., Estel, L. et al. (2012). Experimental study of carbon dioxide capture from synthetic industrial incinerator flue gas with a pilot and laboratory measurements. *Energy Procedia* 42: 704–720.

28 Silva, J., Teixeira, S., Grilo, E. et al. (2021). Analysis and monitoring of the combustion performance in a biomass power plant. *Cleaner Engineering and Technology* 5: 100334.

29 Glushkov, D., Nyashina, G., Anand, R., and Strizhak, P. Composition of gas produced from the direct combustion and pyrolysis of biomass. *Process Safety & Environmental Protection* 156: 43–56.

30 Husebye, J., Brunsvold, A., Roussanaly, S., and Zhang, X. (2012). Techno economic evaluation of amine based CO_2 capture: impact of CO_2 concentration and steam supply. *Energy Procedia* 23: 381–390.

31 Wu, C., Huang, Q., Xu, Z. et al. (2024). A comprehensive review of carbon capture science and technologies. *Carbon Capture Science & Technology* 11: 100178.

32 Dai, Z. and Deng, L. (2024). Membranes for CO2 capture and separation: progress in research and development for industrial applications. *Separation and Purification Technology* 335: 126022.

33 Hekmatmehr, H., Esmaeili, A., Pourmahdi, M. et al. (2024). Carbon capture technologies: a review on technology readiness level. *Fuel* 363: 130898.

34 Hills, T., Leeson, D., Florin, N., and Fennell, P. (2016). Carbon capture in the cement industry: technologies, progress, and retrofitting. *Environmental Science & Technology* 50: 368–377.

35 Hills, T.P., Sceats, M., Rennie, D., and Fennell, P. (2017). LEILAC: low cost CO_2 capture for the cement and lime industries. *Energy Procedia* 114: 6166–6170.

36 Mangiaracina, A., Zangrilli, L., Robinson, L. et al. (2014). OCTAVIUS: evaluation of flexibility and operability of amine based post combustion CO_2 capture at the Brindisi pilot plant. *Energy Procedia* 63: 1617–1636.

37 Nock, W., Walker, M., Kapoor, R., and Heaven, S. (2014). Modeling the water scrubbing process and energy requirements for CO_2 capture to upgrade biogas to biomethane. *Industrial & Engineering Chemistry Research* 53 (32): 12783–12792.

38 Augustsson, O., Baburao, B., Dube, S. et al. (2017). Chilled ammonia process scale-up and lessons learned. *Energy Procedia* 114: 5593–5615.

39 Ma, Y., Liao, Y., Su, Y. et al. (2021). Comparative investigation of different CO_2 capture technologies for coal to ethylene glycol process. *PRO* 9 (2): 207.

40 Ntiamoah, A., Ling, J., Xiao, P. et al. (2016). CO_2 capture by temperature swing adsorption: use of hot CO_2-rich gas for regeneration. *Industrial & Engineering Chemistry Research* 55 (3): 703–713.

41 Wu, B., Zhang, X., Xu, J. et al. (2015). Assessment of the energy consumption of the biogas upgrading process with pressure swing adsorption using novel adsorbents. *Journal of Cleaner Production* 101: 251–261.

42 Mantripragada, H. and Rubin, E. (2014). Calcium looping cycle for CO_2 capture: performance, cost and feasibility analysis. *Energy Procedia* 63: 2199–2206.

43 Muzio, L., Bogseth, S., Himes, R. et al. (2017). Ammonium bisulfate formation and reduced load SCR operation. *Fuel* 206: 180–189.

44 Schrod, M., Semel, J., and Steiner, R. (1985). Verfahren zur Minderung von NOx-Emissionen in Rauchgasen. *Chemie-Ingenieur-Technik* 57 (9): 717–727.

45 Bowman, F. and Seinfeld, J. (1994). Ozone productivity of atmospheric organics. *Journal of Geophysical Research: Atmospheres* 99 (D3): 5309–5324.

46 Nalbandian, H. (2009). NOx control for coal-fired plant, International Centre for Sustainable Carbon, 1 October 2009 [Online]. Available: https://www.sustainable-carbon.org/wp-content/uploads/dlm_uploads/reports/emmisionscontrol/NOx-control-for-coal-fired-plant-CCC-157.pdf (accessed 25 May 2024).

47 Duffield, W. and Sass, J. (2003). *Geothermal Energy: Clean Power from the Earth's Heat*. Menlo Park, CA: U.S. Geological Survey.

48 Zarrouk, Z. and Moon, H. (2014). Efficiency of geothermal power plants: a worldwide review. *Geothermics* 51: 142–153.

49 Rodríguez, C., Palacio, J., Venturini, O. et al. (2013). Exergetic and economic comparison of ORC and Kalina cycle for low temperature enhanced geothermal system in Brazil. *Applied Thermal Engineering* 52: 109–119.

50 Gutiérrez-Negrín, L. (2024). Evolution of worldwide geothermal power. *Geothermal Energy* 12: 14.

51 Soltani, M., Kashkooli, F., Souri, M. et al. (2021). Environmental, economic, and social impacts of geothermal energy systems. *Reneweable and Sustainable Energy Reviews* 140: 110750.

52 Bošnjaković, M., Stojkov, M., and Jurjević, M. (2019). Environmental impact of geothermal power plants. *Tehnicki vjesnik* 26 (5): 1515–1522.

53 Bassania, A., Previtali, D., Pirola, C. et al. (2018). H_2S in geothermal power plants: from waste to additional resource for energy and environment. *Chemical Engineering Transaction* 70: 127–132.

54 Williamson, S., Stark, B., and Booker, J. (2014). Low head pico hydro turbine selection using a multi-criteria analysis. *Renewable Energy* 61: 43–50.

55 RETscreen® International (2005). Small hydro project analysis. In: *Clean Energy Project Analysis: RETscreen® Engineering & Cases Textbook*, 3e. Minister of Natural Resources Canada.

56 Qin, P., Xu, H., Liu, M. et al. (2020). Climate change impacts on three gorges reservoir impoundment and hydropower generation. *Journal of Hydrology* 580: 123922.

57 Global Infrastructure Hub (2020). Itaipu hydroelectric dam, 30 November 2020 [Online]. Available: https://www.gihub.org/connectivity-across-borders/case-studies/itaipu-hydroelectric-dam (accessed 30 April 2024).

58 GridInfo (2024). GridInfo - Electricity Generation Insight [Online]. Available: https://www.gridinfo.com/plant/grand-coulee/6163 (accessed 1 May 2024).

59 IHA - International Hydropower Association (2021). 2021 Hydropower Status Report.

60 Nautiyal, H. and Varun Goel, V. (2020). Sustainability assessment of hydropower projects. *Journal of Cleaner Production* 265: 121661.

61 Schlömer, S., Bruckner, T., Fulton, L. et al. (2014). Annex III: Technology-specific cost and performance parameters. *Climate Change 2014: Mitigation of Climate Change. Contribution of Working Group III to the 5th Assessment Report of the Intergovernmental Panel on Climate Change.*

62 Ocko, I. and Hamburg, S. (2019). Climate impacts of hydropower: enormous differences among facilities and over time. *Environmental Science & Technology* 53 (23): 14070–14082.

63 Scherer, L. and Pfister, S. Hydropower's biogenic carbon footprint. *PLoS One* 11 (9): e0161947.

64 Ubierna, M., Díez Santos, C., and Mercier-Blais, S. (2022). Water security and climate change: hydropower reservoir greenhouse gas emissions. In: *Water Security under Climate Change. Water Resources Development and Management* (ed. A. Biswas and C. Tortajada), 69–94. Springer.

65 Prairie, Y., Mercier-Blais, S., Harrison, J. et al. (2021). A new modelling framework to assess biogenic GHG emissions from reservoirs: the G-res tool. *Environmental Modelling & Software* 143: 105117.

66 Maguire, G. (2024). German wind-reliant power firms brace for annual 'dunkelflaute', Reuters, 21 February 2024 [Online]. Available: https://www.reuters.com/markets/commodities/german-wind-reliant-power-firms-brace-annual-dunkelflaute-2024-02-21 (accessed 19 October 2025).

67 KNMI (2024). Uurgegevens van het weer in Nederland [Online]. Available: https://www.knmi.nl/nederland-nu/klimatologie/uurgegevens (accessed 29 February 2024).

68 Schubel, P. and Crossley, R. (2012). Wind turbine blade design. *Energies* 5 (9): 3425–3449.

69 Siemens Gemesa Renewable Energy. Wind turbine SG 2.2–122 - on-shore wind turbine [Online]. Available: https://www.siemensgamesa.com/en-int/products-and-services/onshore/wind-turbine-sg-2-2-122 (accessed 29 February 2024).

70 GE Renewable Energy. 2 MW onshore wind turbine platform [Online]. Available: https://www.ge.com/renewableenergy/wind-energy/onshore-wind/2mw-platform#:~:text=GE%E2%80%99s%202%20MW%20wind%20turbine%20platform%20is%20a,generator%20%28DFIG%29%20with%20a%20partial%20power%20converter%20system (accessed 29 February 2024).

71 Ragheb, M. and Ragheb, A. (2011). Wind turbines theory – the Betz equation and optimal rotor tip speed ratio. In: *Fundamental and Advanced Topics in Wind Power* (ed. R. Carriveau), 19–38. IntechOpen.

72 Adeseye, K., Ijumba, N., and Colton, J. (2021). The effect of the number of blades on the efficiency of a wind turbine. *IOP Conference Series Earth and Environmental Science* 801 (1): 012020.

73 Qazi, S. and Mustafa, M. (2016). Review on active filters and its performance with grid connected fixed and variable speed wind turbine generator. *Renewable and Sustainable Energy Reviews* 57: 420–438.

74 Kiranoudis, C. and Maroulis, Z. (1997). Effective short-cut modelling of wind park efficiency. *Renewable Energy* 11 (4): 439–457.

75 Rivero, M., Loera-Palomo, R., and Sellschopp-Sanchez, F. (2018). Wind park electric power estimation based on the Jensen wake model. *2018 IEEE International Autumn Meeting on Power, Electronics and Computing, ROPEC 2018*, Ixtapa, Guerrero, 2018.

76 Lee, J.Z.F., Backwell, B., Clarke, E. et al. (2022). GWEC: Global wind report 2022, Global Wind Energy Council, Brussels.

77 Bonou, A., Laurent, A., and Olsen, S. (2016). Life cycle assessment of onshore and offshore wind energy-from theory to application. *Applied Energy* 180: 327–337.

78 Chowdhury, E., Alam Shakib, M., Xu, F. et al. (2022). Adverse environmental impacts of wind farm installations and alternative research pathways to their mitigation. *Cleaner Engineering and Technology* 7: 100415.

79 Mendecka, B. and Lombardi, L. (2019). Life cycle environmental impacts of wind energy technologies: a review of simplified models and harmonization of the results. *Renewable and Sustainable Energy Reviews* 111: 462–480.

80 Msigwa, G., Ighalo, J., and Yap, P.-S. (2022). Considerations on environmental, economic, and energy impacts of wind energy generation: projections towards sustainability initiatives. *Science of the Total Environment* 849: 157755.

81 Bolin, K., Bluhm, G., Eriksson, G., and Nilsson, M. (2011). Infrasound and low frequency noise from wind turbines: exposure and health effects. *Environmental Research Letters* 6: 035103.

82 Wolf, A., Lutz, T., Würz, W. et al. (2015). Trailing edge noise reduction of wind turbine blades by active flow control. *Wind Energy* 18 (5): 909–923.

83 Zhang, W., Yu, H., Yin, B. et al. (2023). Sustainable transformation of end-of-life wind turbine blades: advancing clean energy solutions in civil engineering through recycling and upcycling. *Journal of Cleaner Production* 426: 139184.

84 Paulsen, E. and Enevoldsen, P. (2021). A multidisciplinary review of recycling methods for end-of-life wind turbine blades. *Energies* 14 (14): 4247.

85 Widen, J. and Munkhammar, J. (2019). Solar radiation theory, Uppsala University [Online]. Available: http://urn.kb.se/resolve?urn=urn:nbn:se:uu:diva-381852 (accessed 30 May 2024).

86 PV Lighthouse (2014). Spectrum AM 1.5g [Online]. Available: www2.pvlighthouse.com.au/resources/optics/spectrum%20library/spectrum%20library.aspx (accessed 30 May 2024).

87 Augustine, J., DeLuisi, J., and Long, C. (2000). SURFRAD—a national surface radiation budget network for atmospheric research. *Bulletin of the American Meteorological Society* 81 (10): 2341–2358.

88 Augustine, J., Hodges, G., Cornwall, C. et al. (2005). An update on SURFRAD—the GCOS surface radiation budget network for the continental United States. *Journal of Atmospheric and Oceanic Technology* 22 (10): 1460–1472.

89 NOAA. Global Monitoring Laboratory [Online]. Available: https://gml.noaa.gov/aftp/data/radiation/surfrad (accessed 10 July 2024).

90 Honsberg, C. and Bowden, S. (2019). Photovoltaics Education Website [Online]. Available: https:// www.pveducation.org (accessed 3 June 2024).

91 Koster, L., Mihailetchi, V., Ramaker, R., and Blom, P. (2005). Light intensity dependence of open-circuit voltage of polymer:fullerene solar cells. *Applied Physics Letters* 86: 123509.

92 Lin, H., Wang, G., Su, Q. et al. (2024). Unveiling the mechanism of attaining high fill factor in silicon solar cells. *Progress in Photovoltaics* 32 (6): 359–371.

93 Landsberg, P. and Baruch, P. (1989). The thermodynamics of the conversion of radiation energy for photovoltaics. *Journal of Physics A: Mathematical and General* 22: 1911–1926.

94 Baruch, P., De Vos, A., Landsberg, P., and Parrott, J. (1995). On some thermodynamic aspects of photovoltaic solar energy conversion. *Solar Energy Materials and Solar Cells* 36 (2): 201–222.

95 Shockley, W. and Queisser, H. (1961). Detailed balance limit of efficiency of p-n junction solar cells. *Journal of Applied Physics* 32 (3): 510–519.

96 Markvart, T. (2022). Shockley: Queisser detailed balance limit after 60 years. *WIREs Energy and Environment* 11 (4): e430.

97 Alves dos Santos, S., Torres, J., Fernandes, C., and Lameirinhas, R. (2021). The impact of aging of solar cells on the performance of photovoltaic panels. *Energy Conversion and Management: X* 10: 100082.

98 Livera, A., Tziolis, G., Theristis, M. et al. (2023). Estimating the performance loss rate of photovoltaic systems using time series change point analysis. *Energies* 16 (9): 3724.

99 de Castro, C., Mediavilla, M., Miguel, L., and Frechoso, F. (2013). Global solar electric potential: a review of their technical and sustainable limits. *Renewable and Sustainable Energy Reviews* 28: 824–835.

100 van deVen, D.-J., Capellan-Peréz, I., Arto, I. et al. (2021). The potential land requirements and related land use change emissions of solar energy. *Scientific Reports* 11: 2907.

101 Martín-Chivelet, N. (2016). Photovoltaic potential and land-use estimation methodology. *Energy* 94: 233–242.

102 Mauro, G. and Lughi, V. (2017). Mapping land use impact of photovoltaic farms via crowdsourcing in the Province of Lecce (Southeastern Italy). *Soalr Energy* 155: 434–444.

103 Kincaid, N., Mungas, G., Kramer, N. et al. (2018). An optical performance comparison of three concentrating solar power collector designs in linear Fresnel, parabolic trough, and central receiver. *Applied Energy* 231: 1109–1121.

104 Zhang, H., Baeyens, J., Degrève, J., and Cacères, G. (2013). Concentrated solar power plants: review and design methodology. *Renewable and Sustainable Energy Reviews* 22: 466–481.

105 Perry, R., Green, D., and Maloney, J. (1997). *Perry's Chemical Engineers' Handbook*, 7the. New York: McGraw-Hill.

106 The Engineering Toolbox (2012). Materials - light reflecting factors [Online]. Available: https://www.engineeringtoolbox.com/light-material-reflecting-factor-d_1842.html (accessed 30 June 2024).

107 Boretti, A. and Castelletto, S. (2021). Techno-economic performances of future concentrating solar power plants in Australia. *Humanities and Social Sciences Communications* 8: 326.

108 Cui, Y., Yang, Z., Zhong, W. et al. (2023). An approach for configuring thermal storage capacity of concentrated solar power considering low-carbon economic dispatch. *2023 International Conference on Power System Technology (PowerCon)*, Jinan (China), 2023.

109 Palacios, A., Barreneche, C., Navarro, M., and Ding, Y. (2020). Thermal energy storage technologies for concentrated solar power – a review from a materials perspective. *Renewable Energy* 156: 1244–1265.

110 Alami, A., Olabi, A., Mdallal, A. et al. (2023). Concentrating solar power (CSP) technologies: status and analysis. *International Journal of Thermofluids* 18: 100340.

111 Roper, R., Harkema, M., Sabharwall, P. et al. (2022). Molten salt for advanced energy applications: a review. *Annals of Nuclear Energy* 108924.
112 Thonig, R., Glimanova, A., and Lilliestam, J. (2023). CSP.guru, 1 July 2023 [Online]. Available: https://csp.guru/about (accessed 4 July 2024).
113 Lilliestam, J., Ollier, L., Labordena, M. et al. (2021). The near- to mid-term outlook for concentrating solar power: mostly cloudy, chance of sun. *Energy Sources, Part B: Economics, Planning, and Policy* 16: 23–41.
114 Tawalbeh, M., Al-Othman, A., Kafiah, F. et al. (2021). Environmental impacts of solar photovoltaic systems: a critical review of recent progress and future outlook. *Science of the Total Environment* 759: 143528.
115 Yue, S., Guo, M., Zou, P. et al. (2021). Effects of photovoltaic panels on soil temperature and moisture in desert areas. *Environmental Science and Pollution Research* 28: 17506–17518.
116 Visser, E., Perold, V., Ralston-Paton, S. et al. (2019). Assessing the impacts of a utility-scale photovoltaic solar energy facility on birds in the Northern Cape, South Africa. *Renewable Energy* 133: 1285–1294.
117 Joshi, S., Mittal, S., Holloway, P. et al. (2021). High resolution global spatiotemporal assessment of rooftop solar photovoltaics potential for renewable electricity generation. *Nature Communications* 12: 5738.
118 Jin, Y., Behrens, P., Tukker, A., and Scherer, L. (2019). Water use of electricity technologies: a global meta-analysis. *Renewable and Sustainable Energy Reviews* 115: 109391.
119 Caraballo, A., Galán-Casado, S., Caballero, Á., and Serena, S. (2021). Molten salts for sensible thermal energy storage: a review and an energy performance analysis. *Energies* 14 (4): 1197.
120 Gamarra, A., Banacloche, S., Lechon, Y., and del Río, P. (2023). Assessing the sustainability impacts of concentrated solar power deployment in Europe in the context of global value chains. *Renewable and Sustainable Energy Reviews* 171: 113004.
121 Tielens, P. and Van Hertem, D. (2016). The relevance of inertia in power systems. *Renewable and Sustainable Energy Reviews* 55: 999–1009.
122 EIA (2024). Real-time Operating Grid - U.S. Energy Information Administration (EIA), [Online]. Available: https://www.eia.gov/electricity/gridmonitor/index.php (accessed 8 September 2024).
123 Ueckerdt, F. and Kempener, R. (2015). From baseload to peak: renewables provide a reliable solution [Online]. Available: https://www.irena.org/-/media/Files/IRENA/Agency/Publication/2015/IRENA_Baseload_to_Peak_2015.pdf?rev=b981f98cd4f145d08c0fe37d7a8ca5aa (accessed 1 July 2024].
124 Worldbank (2018). Electric power transmission and distribution losses (% of output), Worldbank [Online]. Available: https://data.worldbank.org/indicator/EG.ELC.LOSS.ZS (accessed 20 July 2024).
125 Liu, Z., Deng, Z., Davis, S., and Ciais, P. (2023). Monitoring global carbon emissions in 2022. *Nature Reviews Earth & Environment* 4: 205–206.
126 IEA (2020). Projected costs of electricity generation, 2020 Edition [Online]. Available: https://iea.blob.core.windows.net/assets/ae17da3d-e8a5-4163-a3ec-

2e6fb0b5677d/Projected-Costs-of-Generating-Electricity-2020.pdf (accessed 1 July 2024).

127 Branker, K., Pathak, M., and Pearce, J. (2011). A review of solar photovoltaic levelized cost of electricity. *Renewable and Sustainable Energy Reviews* 15 (9): 4470–4482.

128 Lazard (2023). 2023 Levelized Cost Of Energy+, 12 April 2023 [Online]. Available: https://www.lazard.com/research-insights/2023-levelized-cost-of-energyplus (accessed 2 February 2024).

129 Frank, C., Fiedler, S., and Crewell, S. (2021). Balancing potential of natural variability and extremes in photovoltaic and wind energy production for European countries. *Renewable Energy* 163: 674–684.

3

Dealing with Intermittency – Storing Electric Power

3.1 Introduction

Renewable energy in the form of solar energy and wind energy are intermittent and cannot be turned on and off on demand. They are thus typically regarded as variable, non-dispatchable power. Excess power generation capacity could cater for variable demand, but this will increase the capital cost for the required installation. Hence, other options are needed to cater for the intermittency of renewable energy, for instance by storing the excess energy and using it as dispatchable power.

A large variety of technologies have been proposed to store energy based on various physical and chemical principles (a non-exhaustive overview is given in Figure 3.1). Energy generated from renewable systems can be converted to mechanical energy, where it can be stored by increasing the potential energy of the storage material (e.g. pumping of water), by compressing gases, or by increasing the kinetic energy of an object. Renewable energy can also be stored as thermal energy, either directly or by converting generated energy into heat. Furthermore, renewable energy can be converted into chemical energy either in an electrochemical process in batteries or in the form of chemicals such as hydrogen and ammonia (the latter will be dealt with in Chapter 5, as these materials can also be used as a chemical feedstock). Renewable energy in the form of electricity can also be stored directly, for example using supercapacitors.

Important characteristics to evaluate for the different storage technologies are the amount of stored energy per unit mass (the specific energy) or per unit volume (the energy density), as well as the specific power and energy density, which indicates how quickly the stored energy can be discharged per unit mass of storage capacity or per unit volume of storage capacity. The amount of energy lost during storage is indicated by the charge/discharge efficiency, which is also affected by the so-called self-discharge rate. The latter is thus an indication of the duration for which energy can be stored. Another important aspect for the storage is the number of times that it can cycle between charging and discharging (also termed cycle durability).

Renewable Energy in the Process Industry, First Edition.
Eric van Steen, Nicholas S. Featherstone, and Linda H. Callanan.

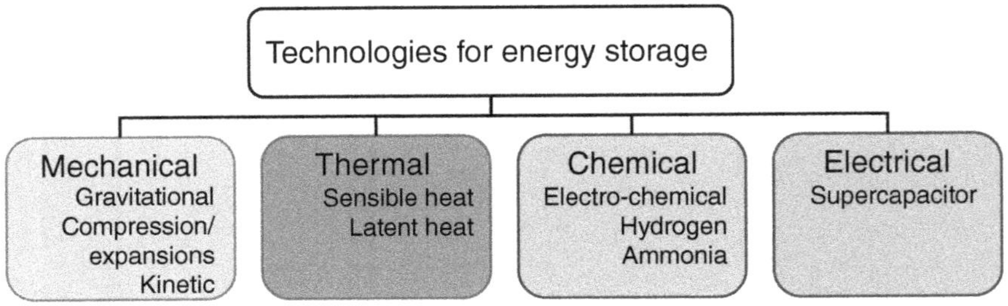

Figure 3.1 Overview of some of the principles on which technologies available for energy storage have been developed.

3.2 Pumped Hydro-Energy Storage (PHES)

Power can be stored in the form of hydro-energy (see Section 2.3), in which two dams at different levels of elevation are used (see Figure 3.2). Excess energy can be used to pump water from the lower level to the upper level. Hence, energy will be stored in the form of an increase in the potential energy of the stored water:

$$\Delta E_\mathrm{p} = m \cdot g \cdot (h_\text{upper level} - h_\text{lower level})$$

The amount of energy stored is 9.8 J/kg of water per metre of height difference between the upper level and the lower level (or a storage capacity of $3.57 \cdot 10^5\ \mathrm{m^3/MWh}$ for a height difference of 1 m). Hence, a storage capacity of 600 MWh (the equivalent storage space for a 50 MW renewable energy generator for a period of 12 hours) would require at least a capacity of $2.2 \cdot 10^6\ \mathrm{m^3}$ (or 2200 Ml) for both the lower and the upper dam if the height difference between the two dams is 100 m (and thus a total capacity of $4.4 \cdot 10^6\ \mathrm{m^3}$, although the dams will not be filled to capacity simultaneously). For an average depth of the reservoir of 10 m, this would imply reservoirs which each will cover an area of at least $0.22\ \mathrm{km^2}$.

Pumped hydro energy storage (PHES) just like hydropower can be classified as either an 'open-loop' system or a 'closed-loop' system [2]. In an open-loop system, the water is connected to a natural, flowing water resource thus reducing the land use,

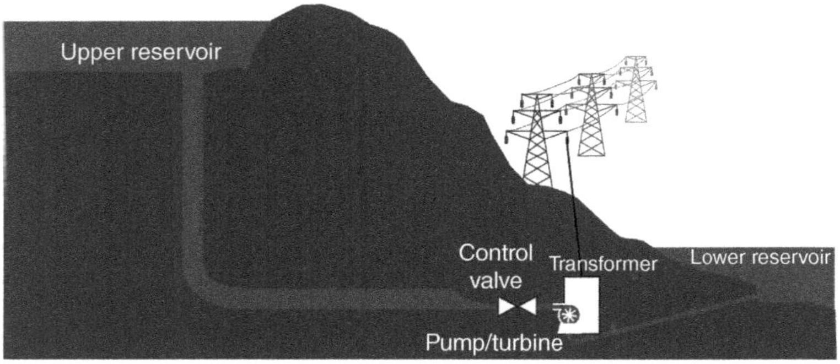

Figure 3.2 Pumped hydro-energy storage (PHES) (adapted from Nikolaos et al. [1]).

but redirecting the water flow may affect aquatic life and the water quality downstream [1]. Furthermore, the ability to store energy in the form of hydropower may be affected by drought conditions. Conversely, a closed-loop system consists of two constructed reservoirs, which has the benefit of a reduction on the impact on the natural aquatic life [2].

Hunt et al. [3] classified PHES according to the storage capacity and the rate of replenishment. The so-called hourly pumped hydro-storage (HPHS) and daily pumped hydro-storage (DPHS), each with a typical storage capacity between 0.001 and 1 km^3, provides small amount of back-up power in case of short-term shortfall in the supply with DPHS built to cater for the variation in power demand between daytime and nighttime. DPHS would be suitable to be used in conjunction with solar power to cater for the day–night variation. Weekly pumped hydro-storage (WPHS) has a larger storage capacity of 0.1–5 km^3 and may cater for longer variation in the power demand. Pumping will then take place over the weekends as well at any time excess power is available, and additional power will be generated during weekdays when the demand rises. Seasonal pumped hydro-storage (SPHS) and pluri-annual pumped hydro-storage (PPHS), with storage capacities of 1–30 km^3 and 5–100 km^3, respectively, may be used in the future to cater for long-term variation in the power supply.

The limitation of PHES lies with the need to have two large water reservoirs at different altitudes, and with only a limited number of natural sites being available. Some aquifers can be used effectively as underground reservoirs of water in an open-loop system with water present in the aquifer being pumped to a reservoir above the ground when excess power is available [2]. Alternatively, old, decommissioned mines can be used to create a closed-loop system with water being pumped from the underground storage facility to a dam on or near the surface. Underground storage may be more effective to store energy than classical dams above ground, because of the high hydraulic head (and thus a higher energy content per unit volume of water). Furthermore, the impact on land use, and flora and fauna are expected to be less compared to the classical PHES systems [4]. However, pumping large amounts of water over prolonged time may affect the geological stability of the underground spaces.

PHES typically utilises a (set of) reversible pump-turbine runner(s) providing for both the pumping action as well as the electricity generation function upon changing the flow of water. The efficiency of PHES is given by the efficiency of the turbine and the pump, which is typically around 90% [5] (although there may be a trade-off between a high pump efficiency and a high turbine efficiency [6]). This would imply that 81% of the stored energy would be retrieved in the form of electric power (and thus 19% of the excess energy supplied to the storage system would be lost to the surroundings). In most operating regimes, a variable speed turbine has a higher efficiency than the cheaper fixed-speed turbine-pump, although this depends on the available/required head [7]. Variable speed turbine-pumps would be more suitable for turning a variable energy supply into dispatchable power [1] because the average operational efficiency will be higher. It should be noted that this can be well below the 90% efficiency at the optimal operation point. Hence, the higher cost for the variable speed turbine-pump will need to be offset by the increased income by suppling dispatchable power. Water loss is usually not considered in the efficiency of PHES, but this will also negatively affect the overall efficiency.

The environmental considerations surrounding PHES are similar to those of hydropower in general (see Section 2.4.4), although the exposed area is typically much smaller than large-scale hydro-plants.

PHES makes up the majority of the installed energy storage schemes with a worldwide capacity in 2020 of over 8300 GWh [8]. Overall PHES is characterised by a long lifetime [1], a high efficiency to return stored power and a possible rapid response to fluctuations in the supply/demand. The downside is the large amount of water involved due to the low energy density, resulting in extensive land use.

3.3 Compressed Air Energy Storage (CAES)

Compressed air energy storage (CAES) has a relatively long history and was developed as an alternative to PHES, when the mountainous features required for PHES were not available [9]. Here, energy is stored in the form of compressed air in other geographical features, such as large, airtight, underground caverns (either salt or rock caverns), disused oil wells or gas fields.

Two large-scale CAES operations are functional, viz. the Huntorf plant in Germany with upgraded capacity of 321 MW connected to a storage facility in a salt cavern of $3.1 \cdot 10^5\ m^3$ (air is stored at a pressure between 46 and 72 bar giving it an energy storage capacity of 642 MWh) and the McIntosh plant in the United States with a capacity of 110 MW connected to a storage facility in a salt cavern of $5.38 \cdot 10^5\ m^3$ (air is stored at a pressure between 46 and 75 bar giving it an energy storage capacity of 2640 MWh) [9]. The initial objective was again to store cheaply produced electricity and dispatch when the need arises.

The principle behind both operations is quite similar (see Figure 3.3); air is adiabatically compressed in 2–3 stages with interstage cooling to a high pressure. This air may then be stored, and subsequently fed to a combustion chamber, where it is mixed with a combustion gas and combusted prior to being fed to a two-stage adiabatic turbine. This may have an intermediate heating step by feeding it again to a combustion chamber. Heat from the hot off-gas of the turbine can be partially recuperated. Essentially this process stores energy in the form of compressed air during a time of surplus energy, which will be used in an open cycle gas turbine plant, rather than consuming part of the plant's power generation in the compression stages. This type of process is named a diabatic process as it utilises heat transfer in the compression stage and prior to the expansion stage.

The need for fuel to feed to the combustion stage implies that CAES is not carbon-neutral and has an overall release of energy into the surroundings. A simplified process flow diagram analysis shows that carbon footprint of this process amounts to ca. 0.12–0.13 kg CO_2/MW, when using methane as a fuel. The overall efficiency of the process is rather low (estimated to be ca. 29–35%), when considering the energy content of the fuel gas. A higher efficiency can be obtained

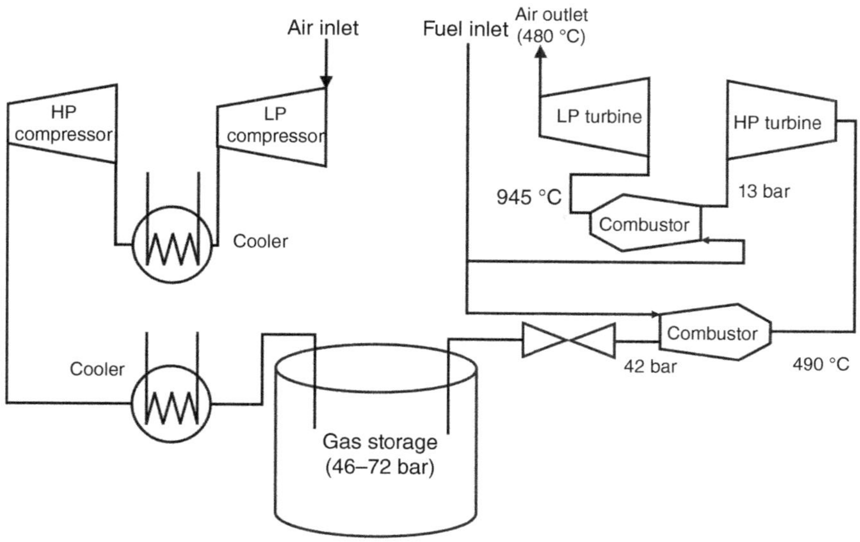

Figure 3.3 Compressed air energy storage (CAES) for a diabatic process without heat recuperation (the indicated temperatures and pressures correspond to the process in Huntorf as given by Budt et al. [9]).

when recuperating the heat from the effluent to pre-heat the stream prior to the first combustor as in the McIntosh plant [9, 10].

The relatively low efficiency of the diabatic process can be attributed to the heat loss in the compression step, the required pre-heating for the turbines in the combustor, and the heat loss in the gas effluent. The efficiency may be improved by creating a fully adiabatic process. We can consider a compressed air energy system using a single compressor and with a fixed compression ratio, *CR*, and a single turbine operating with the same ratio. Hence, the work required by the compressors is

$$\frac{\dot{W}_\mathrm{c}}{\dot{m}} = C_\mathrm{p} \cdot T_\mathrm{initial,c} \cdot \left(\mathrm{CR}^{\frac{R}{C_\mathrm{p}}} - 1\right)$$

and the work given off by the turbines is

$$\frac{\dot{W}_\mathrm{t}}{\dot{m}} = C_\mathrm{p} \cdot T_\mathrm{final,t} \cdot \left(1 - \mathrm{CR}^{\frac{R}{C_\mathrm{p}}}\right)$$

if the gas can be approximated as an ideal gas.

Hence, the maximum efficiency of CAES in an adiabatic system (A-CAES) assuming no loss in compressed air is given by

$$\eta_\mathrm{A-CAES,max} = \frac{\left|\frac{\dot{W}_\mathrm{t}}{\dot{m}}\right|}{\left|\frac{\dot{W}_\mathrm{c}}{\dot{m}}\right|} = \frac{\left|C_\mathrm{p} \cdot T_\mathrm{final,t} \cdot \left(1 - \mathrm{CR}^{\frac{R}{C_\mathrm{p}}}\right)\right|}{\left|C_\mathrm{p} \cdot T_\mathrm{initial,c} \cdot \left(\mathrm{CR}^{\frac{R}{C_\mathrm{p}}} - 1\right)\right|} = \frac{T_\mathrm{final,t}}{T_\mathrm{initial,c}}$$

Hence, the efficiency approaches 1 if the temperature of the compressed gas during storage does not change. The separate storage of the thermal energy from the mechanical energy in the compressed gas would be an alternative approach. This approach is, in practical terms, a necessity, because of thermal stress in the required underground salt caverns [11]. The air passes a thermal energy storage (TES) unit after compression and reverses its flow prior to expansion. The thermal energy can be stored in the form of sensible heat, latent heat or in the form of thermo-chemical energy (with the latter still in the exploratory stage [12]). This could bring the efficiency of energy storage up to 65–75% with a thermal efficiency of energy storage (in a packed bed of crushed rock [11]) ranging from 78% to 90%. The size of the thermal energy storage may become the limiting factor in the adiabatic compressed air storage.

Energy loss in the form of heat can also be minimised in near isothermal operation. This may be achieved/modelled by investigating N adiabatic compressors in series, each working with a compression ratio of $\mathrm{CR}_{\mathrm{individual}}$, with a resulting overall compression ratio of $\mathrm{CR}_{\mathrm{overall}}$, each with interstage cooling (with $\mathrm{CR}_{\mathrm{overall}} = \mathrm{CR}_{\mathrm{individual}}^{N}$), in conjunction with the same number (N) adiabatically operating turbines with interstage heating. The work required by the compressors is now given by

$$\frac{\dot{W}_{\mathrm{c}}}{\dot{m}} = N \cdot C_{\mathrm{p}} \cdot T_{\mathrm{initial,c}} \cdot \left(\mathrm{CR}_{\mathrm{individual}}^{\frac{R}{C_{\mathrm{p}}}} - 1\right) = N \cdot C_{\mathrm{p}} \cdot T_{\mathrm{initial,c}} \cdot \left(\mathrm{CR}_{\mathrm{overall}}^{\frac{R}{N \cdot C_{\mathrm{p}}}} - 1\right)$$

and the work given off by the turbines is given by

$$W_{\mathrm{t}} = N \cdot C_{\mathrm{p}} \cdot T_{\mathrm{final,t}} \cdot \left(1 - \mathrm{CR}_{\mathrm{overall}}^{\frac{R}{N \cdot C_{\mathrm{p}}}}\right)$$

In the pseudo-isothermal compression, the compressed air will be cooled down after each stage, except the last stage. Furthermore, the gas will be heated from the environment upon expansion. Hence, the heat loss will be given by

$$\frac{\dot{Q}_{\mathrm{loss}}}{\dot{m}} = (N-1) \cdot C_{\mathrm{p}} \cdot T_{\mathrm{initial,c}} \cdot \left(\mathrm{CR}_{\mathrm{individual}}^{\frac{R}{C_{\mathrm{p}}}} - 1\right)$$

and the maximum efficiency for the pseudo-isothermal compressed air storage (I-CAES) will be given by:

$$\eta_{\mathrm{I\text{-}CAES,max}} = 1 - \frac{\left|\frac{\dot{Q}_{\mathrm{loss}}}{\dot{m}}\right|}{\left|\frac{\dot{W}_{\mathrm{c}}}{\dot{m}}\right|} = \frac{N-1}{N}$$

This simplified analysis assumes, of course, that the heating requirement for the expansion step is delivered from the surroundings. Near isothermal conditions during compression may also be achieved by compression of air using a liquid piston [13] with enhanced heat transfer by injection of liquid droplets into the compression chamber [14]. This could increase the efficiency of compressed air energy storage to over 90%.

The amount of energy stored depends on the amount of energy that can be retrieved. For instance, in a diabatic process given in Figure 3.3, only energy in compressed air with a pressure greater than 42 bar is recovered (a value set by the pressure reducing valve, which ensures a constant flow rate through the combustors and turbines; this value can be altered when designing for different flow rates), and thus the required storage space is simply given by:

$$V_{\text{storage}} = \frac{E_{\text{stored}}}{p_{\max} - p_{\min}}$$

For instance, for a storage of 600 MWh ($2.16 \cdot 10^{12}$ J) in a facility with a pressure difference of 30 bar would require a volume of 720 000 m^3 (a pressure difference of 30 bar would have a minimum outlet pressure of 42 bar corresponding to a maximum pressure of 72 bar). Otherwise stated, the required storage capacity of CAES under these conditions comes to 1200 m^3/MWh, which is only 16% of the required volume for both reservoirs in PHES (with a height difference between the reservoirs of 100 m). Leakage of air from the storage space will require storing more compressed air than the minimal, necessary amount.

The temperature of the gas will change upon charging/discharging the storage space [11]. Combining the mass and energy balance around the storage space, it can be shown that the temperature in the storage space as a function of time can be described using:

$$T_{\text{storage}} - T_{\text{storage},0} = \left(\frac{\left(\dot{m}_{\text{in}} \cdot \gamma \cdot T_{\text{in}} + \frac{h}{c_{V,\text{air}}} \cdot A \cdot T_{\text{wall}} \right)}{\left(\dot{m}_{\text{in}} + \dot{m}_{\text{out}} \cdot (\gamma - 1) + \frac{h}{c_{V,\text{air}}} \cdot A \right)} - T_{\text{storage},0} \right) \cdot \left(1 - e^{-\frac{\left(\dot{m}_{\text{in}} + \dot{m}_{\text{out}} \cdot (\gamma - 1) + \frac{h}{c_{V,\text{air}}} \cdot A \right) \cdot t}{m_{\text{storage,initial}} + (\dot{m}_{\text{in}} - \dot{m}_{\text{out}}) \cdot t}} \right)$$

from which, using an appropriate equation of state, the pressure as a function of time can be found. Figure 3.4 shows the temperature–time trajectory expected for the filling and emptying of a cavern filled with compressed air at different flow rates. The temperature rises upon filling the cavern with gas at an elevated temperature as it comes from a compressor. The temperature starts dropping slowly, when the cavern is filled and gas is no longer flowing into the reactor. The maximum temperature increases with increasing mass flow rate of air entering the cavern. Discharging air from the cavern will result in a strong decrease in the temperature of the gas. A low heat-transfer coefficient is necessary to keep the thermal energy for a long time in the cavern, but this will also result in a low temperature of the gas upon discharge.

Truly adiabatic operations require not only storage of compressed air, but also storage of thermal energy and heating of the compressed stream prior to the expansion over the turbine is not required. The objective here is to store all heat generated upon compression (as far as possible) and release the heat to pre-heat compressed air prior to expansion. The required volume for the thermal energy storage depends on the type of storage with sensible heat being the simplest system but taking in the largest volume. A thermal energy storage using sensible heat will operate under non-steady-state conditions during charging and discharging with further temperature equilibration taking place in times that no gas is flowing through the system. Numerical simulations have shown that the charging efficiency of the thermal energy storage is affected by the air flow rate and the charging time [15].

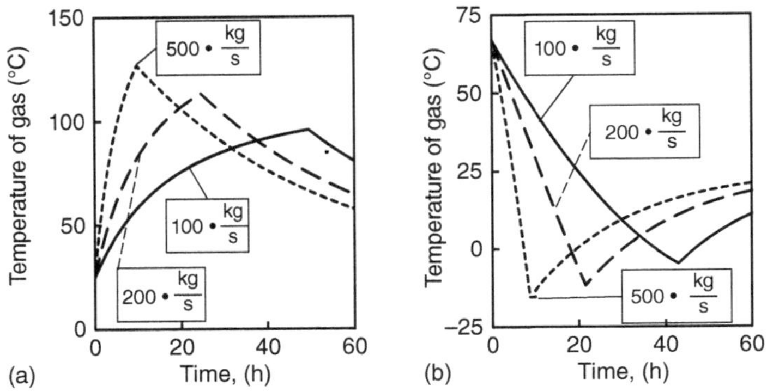

Figure 3.4 Temperature of compressed air during charging of the storage area (a) and discharging of the storage area (b) as a function of time; the discontinuity in the curves represent the time when the storage space has reached the maximum pressure (72 bar; a) or the minimum pressure (42 bar; b). Calculated for air ($C_V = 730$ J/kg/°C; $\gamma = 1.4$) stored in a cavern with an area of $4 \cdot 10^4$ m^2 ($h = 5$ W/m^2/°C [11]) with a minimum pressure of 42 bar and a maximum pressure of 72 bar. Initial temperature at time filling of the cavern: 25 °C; temperature of the gas entering the cavern during charging: 175 °C; initial temperature at time of emptying the cavern 68 °C.

Upon discharging compressed air, the pressure in a fixed volume storage space will decrease. The inlet pressure to the high-pressure turbine can be kept constant by stepping down the pressure over a control valve, as shown in Figure 3.3. This will result in a temperature decrease over the valve (the temperature of the gas is expected to decrease by ca. 12 °C for the adiabatic expansion of air from 72 bar down to 42 bar assuming that compressed air is adequately described using the Peng-Robinson equation of state), which may result in water condensation in the gas lines. Alternative, CAES can be operated with a constant pressure in the storage chamber by using a compensating water column. Compressed air displaces water and pushes it into a surface reservoir, below which the air will sit at a given pressure. This design is claimed to result in the smallest required volume for storage of compressed air [16]. In a similar design, compressed air is stored under the ocean/seabed [17], which is then supplying the required constant pressure for the feed to the high-pressure turbine. In such a design, the compressed air/stream will become saturated with water, thus requiring a higher temperature at the turbines to avoid condensation (a similar problem may be encountered when designing an isothermal compression step using a liquid piston and/or injecting liquid droplets [14]).

CAES typically utilises natural structures (salt caverns, depleted gas reservoirs and aquifers [18]) or purposefully created caverns, which are selected to minimise loss of material. The exposure of these large-scale structures to (large) pressure variations (especially over short periods of time) may be a cause of concern in the long term [19].

Bouman et al. [20] performed a life cycle analysis (LCA) on compressed air energy storage systems to balance a 400 MW wind park using a porous rock formation, a mined cavern in hard rock and a cavern leached out of a salt dome. The storage structure did not appear to have a significant effect in the LCA, though complementing a renewable energy source with CAES will increase the net carbon footprint of the renewable energy source. The adiabatic process will then need to be made operational as it is claimed to be carbon-neutral [21], because it does not require additional heating if adiabatic thermal storage can be achieved. However, the LCA [20] showed that it may have a large environmental impact due to its material requirement.

3.3.1 Liquid Air Energy Storage (LAES)

CAES still requires a large volume under pressure to store energy. This may be circumvented by liquefying air and storing it at close to atmospheric pressure [22]. Figure 3.5 shows a proposed flowsheet for liquid air storage with intermediate thermal energy storage [23]. As in a normal air liquefaction process, air is compressed, and the heat generated in the first compression step(s) can be stored as hot thermal energy. Air is further compressed and cooled down with recycled cold gas before being expanded over a Joule-Thomson valve. Liquid air is then recovered, and the gaseous fraction is recycled and used to cool down compressed air. During energy delivery by this process, liquid air is pumped and heated using the heat stored in the process. The stream can be further heated prior to expansion over the turbine.

Xue et al. [23] investigated the energetic efficiency of LAES as a method of energy storage as a function of the discharge pressure of the compressor and the discharge pressure of the pump, as well as the efficiency of the compressor and the pressure of liquid air. The overall energetic efficiency was determined to be 40–49%. Chino and Araki [22] evaluated the efficiency of such a system as a function of temperature difference over the Joule-Thomson valve and reported efficiencies in the range of 40–80%. Bori et al. [24] reviewed a number of studies on stand-alone LAES units and reported energy efficiencies between 46% and 62%.

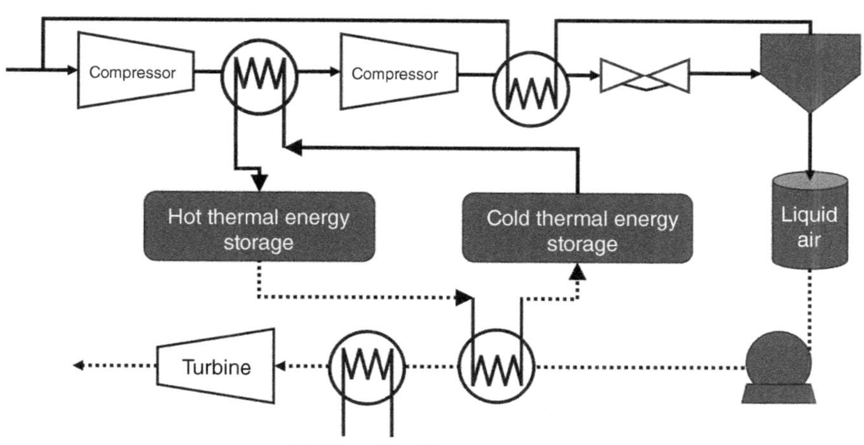

Figure 3.5 Principal flowsheet for liquid air energy storage (LAES); solid line: energy storage; dotted line: energy recovery.

3.4 Flywheel

Energy can also be stored in the form of kinetic energy, and in particular rotational energy as in a flywheel connected to a motor-generator. Storage of energy takes place upon increasing the rotational speed of the flywheel and the release of energy is associated with a decrease in the rotational speed. The amount of energy stored on a flywheel is thus given by the difference in the maximum and minimum allowable rotational speed [25]:

$$E_{\text{flywheel,stored}} = \frac{1}{2} \cdot I \cdot \left(\omega_{\max}^2 - \omega_{\min}^2\right) = \frac{1}{2} \cdot I \cdot \omega_{\max}^2 \cdot \left(1 - \left(\frac{\omega_{\min}}{\omega_{\max}}\right)^2\right)$$

with I the moment of inertia of the flying wheel. The rotational speed can be changed quite rapidly and hence the charging rate and discharging rate can be rather high. The maximum rotational speed is determined by the tensile strength of the rotor ($\sigma_{\max}$), its density (ρ) and the shape factor for the rotor (K). K can theoretically be equal to 1, but is typically less than 0.33. Hence, materials with high tensile strength (σ) and a low density (ρ) can achieve a higher rotational speed and thus a higher mass specific energy content. For a single isotropic material, the specific energy is given by

$$\frac{E_{\text{flywheel,stored}}}{m} = K \cdot \frac{\sigma_{\max}}{\rho}$$

The specific energy is thus a function of the material used. For instance, the maximum specific energy contained in a stainless-steel rotor ($\sigma_{\max} = 1520$ MPa, $\rho = 7700$ kg/m^3) would thus be 200 kJ/kg (assuming $K = 1$), whereas rotors based on less dense carbon-based materials ($\sigma_{\max} = 1650 - 1950$ MPa, $\rho = 1510$ kg/m^3) can have a specific energy of 1100–1300 kJ/kg [26]. The power delivered by a flywheel depends on the motor-generator [27].

The flywheel is supported by either mechanical or magnetic bearings [26], with the latter being preferred as mechanical bearings induce additional energy losses due to friction. The flywheel, the bearings and the motor-generator are placed in a vacuum to minimise frictional losses. Nevertheless, some losses do occur in the various parts of the flywheel set-up because of frictional loss and aerodynamic drag. The loss, also called the flywheel self-discharge rate, depends nonlinearly on the rotational speed and the level of vacuum [25] and can be as high as 25% in a 24 hours period if the system is kept at a pressure of 0.1 Pa. Hence, speed control as well as control of the vacuum level (ensured by an on-site vacuum pump [28]) are essential. Amiryar and Pullen [25] estimated an overall efficiency for energy storage of 83% for a flywheel system operating with a self-discharge rate of 4.7% (10 charging/discharging cycles per day), a motor-generator efficiency of 95% and an efficiency for the power electronics of 98%.

Flywheels are attractive because of both their short response time, as they can deliver power within seconds [27], and the large number of cycles they can go through, and thus their long lifetime. Their performance appears to be independent of the level of discharge, and they can thus be used for both shallow and

Table 3.1 Commercial applications of flywheel technology.

Company	Year	No flywheels	Energy (MWh)	Power (MW)	
Beacon Power					
Stephen Town, USA	2011	200	5	20	[29]
Hazle Township, USA	2014	200	5	20	[29]
Temporal Power					
Wellington County, Canada	2014	10		2	[29]
Clear Creek, Canada	2016	10		10	[30]
Aruba		20	1	5	[31, 32]
Stadtwerke München, Germany		28	0.1	0.6	[33]
Kodiak Island, USA	2016	2		2	[34]
TU Dresden (RKS), Germany	2021		0.5	0.5	[35]
GG&E, Fresno, USA				20	[36]

deep discharge. The relatively high self-discharge rate makes this technology less suitable for storage of power over longer periods of time. The LCA-study on the use of flywheels to store energy by Rahman et al. [28] indicated that the carbon footprint is mainly associated with the carbon footprint of the electric power that is used to charge the flywheels, in addition to the carbon footprint associated with the production of the materials, manufacturing of the flywheel systems, and the decommissioning phase. The intrinsic environmental impact of flywheel technology in the future may be linked to the use of advanced materials for the rotor.

The flywheel technology has been applied in the transportation sector, but also to stabilise power grids and, nowadays, to store renewable energy, in particular wind energy (see Table 3.1). The technology is modular in nature with an energy storage per flywheel of up to 25 kWh.

3.5 Batteries

Energy can also be stored in the form of (electro)chemical energy, a technology which goes back to the old Voltaic-pile (published in 1799), which could achieve a high voltage, albeit over a relatively short time-period. The technology has developed greatly over the last few decades.

3.5.1 Principle

A battery is made up of at least one chemical cell within which a chemical reaction, specifically a redox reaction, is occurring. The spatial separation of the reduction and the oxidation forces an electron flow, which can be used as energy.

Oxidation of the reductant will take place at the anode with the loss of an electron, e.g.

$$R \rightarrow R^+ + e^-$$

and reduction of the oxidant will take up the electron at the cathode, e.g.

$$O^+ + e^- \rightarrow O$$

The cell potential of a battery cell, E, is given by Nernst's equation: $E = -\frac{\Delta G}{n \cdot \mathcal{F}}$, where ΔG is the change in the Gibbs free energy upon the redox reaction, n the number of electrons involved in the reaction and $\mathcal{F}$ the Faraday constant defined based on the elementary charge, e, and Avogadro's number:

$$\mathcal{F} = N_A \cdot e = 6.022 \cdot 10^{23} \cdot \frac{1}{\text{mol}} \cdot 1.602 \cdot 10^{-19} \cdot C = 96\,485 \cdot \frac{C}{\text{mol}} \equiv 96\,485 \cdot \frac{\text{J}}{\text{mol} \cdot \text{V}}$$

The change in the Gibbs free energy upon reaction is given by the change in the Gibbs free energy of reaction under standard conditions and the contribution of the activity of the products relative to the activity of the reactants (also known as the equilibrium quotient, Q_r):

$$\Delta G = \Delta G^0 + RT \cdot \ln Q_r$$

where R is the ideal gas constant (8.314 J/mol/K) and T the temperature in Kelvin. Thus, for an arbitrary half-reaction $M^{n+} + ne^- \rightleftharpoons M$, the change in the Gibbs free energy upon reaction is given by

$$\Delta G = \Delta G^0 + RT \cdot \ln \frac{a_M}{a_{M^{n+}}}$$

where a is the activity coefficient of a species M and its relevant ion. Hence, the cell potential for a half reaction is given by:

$$E = E_0 - \frac{RT}{n \cdot \mathcal{F}} \cdot \ln \frac{a_M}{a_{M^{n+}}}$$

and the cell potential of a battery is given by the cell potential of the half reaction at the cathode and the half reaction at the anode. It should be noted that the voltage of a battery cell is dependent on temperature in the same manner as with the dependency of the change in the Gibbs free energy upon temperature:

$$\left(\frac{\partial \frac{E}{T}}{\partial T}\right)_p = -\left(\frac{\partial \frac{\Delta G}{T}}{\partial T}\right)_p = \frac{\Delta H}{T^2}$$

and thus

$$E(T_2) = E(T_1) \cdot \frac{T_2}{T_1} - \Delta H \cdot \left(\frac{T_2 - T_1}{T_1}\right)$$

if the change in the enthalpy is (almost) independent of temperature.

The thermodynamic cell potential is a property of the materials involved through E_0 and ΔH, the activities of the compounds involved in the two half reactions and temperature.

The current given off by the battery is the rate, at which electrons are given off/taken up at the anode and cathode, respectively, and thus the rate of reaction at the anode and the cathode. The current is controlled by the power drawn by the external circuit and thus the resistance of the external circuit, until the resistance becomes too low and other processes become rate-determining. A higher power demand results in a faster reaction at the anode and cathode and may allow the reaction to go to completion thus limiting the rate of reaction and the power supply.

The rate of reaction of electrochemical reactions is typically described by the Butler-Volmer equation [37], which can be derived from the empirical Tafel expression (alternatively, it can be obtained by considering the electron transfer as the elementary step controlling the rate of reaction). The current per unit surface area, or rate of reaction, is then given by:

$$i = i_0 \cdot \left(e^{\frac{\alpha_{\text{anode}} \cdot \mathcal{F} \cdot \eta}{RT}} - e^{-\frac{\alpha_{\text{cathode}} \cdot \mathcal{F} \cdot \eta}{RT}} \right)$$

with i_0 the exchange current density, which depends on the concentration of the oxidant and the reductant raised to the power of their charge transfer coefficient, α_{anode} and α_{cathode} and the overpotential, η with the overpotential being defined as the difference between the actual potential and the equilibrium potential:

$$\eta = E - E_{\text{eq}}$$

For small values of the overpotential (close to equilibrium), the obtained current density is controlled by the charge transfer resistance.

$$\frac{i}{i_0} = \frac{(\alpha_{\text{anode}} + \alpha_{\text{cathode}}) \cdot \mathcal{F} \cdot \eta}{RT}$$

At high overpotential, the current density is given by:

$$\ln \frac{i}{i_0} = \frac{\alpha_{\text{anode}} \cdot \mathcal{F} \cdot \eta}{RT}$$

The cell potential, E, together with the overpotential and the internal electrical resistance within the battery, will result in a voltage difference over the terminals of the battery. The cell potential E is the maximum voltage, V, over the terminals.

Figure 3.6 shows the principal behaviour of the voltage as a function of current in a battery. The initial region is governed by the dependency of the overpotential on the current density; this drop in voltage is attributed to so-called activation losses, or 'reaction kinetics'. The next region relates to ohmic losses, losses due to resistance.

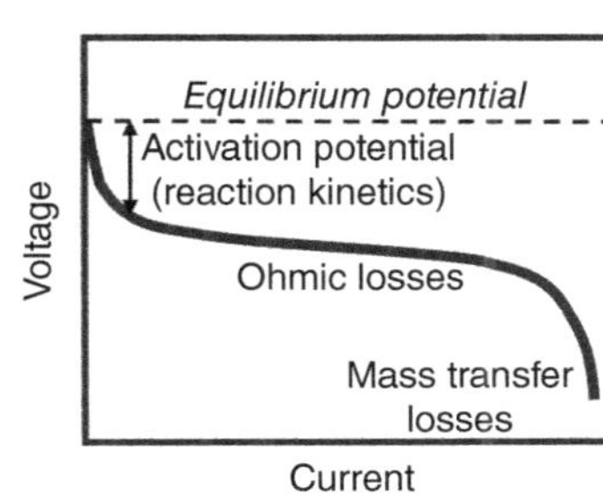

Figure 3.6 Principal behaviour of the voltage of a battery as a function of the current (adapted from Rohman et al. [38]).

Finally, at very high currents there is a large drop off due to mass transfer in the system, essentially the energy is used in support of the mass transfer of the reductant to the anode/the oxidant to the cathode.

The resulting voltage is thus dependent on material constants and the current drawn from the battery. The voltage can be increased by placing a number of electrochemical cells in series. It should be realised that the rate of reaction in each of the cells making up the battery must be consistent for the electron flow to be balanced. Thus, putting cells in series increases the potential or voltage through the battery, i.e. a 9 V battery is made up of six 1.5 V cells, but the maximum current which may be drawn remains the same as when a single cell with a voltage of 1.5 V was used. The current can be increased by putting cells in parallel. In these cases, the driving force across each parallel cell remains the same and thus the voltage remains the same.

The current obtained from a battery is a result of the rate of the electrochemical reaction taking place within the battery. Thus, the current can principally be controlled by the temperature of the battery as per a conventional reaction, although a change in the temperature may affect the stability of the components in the battery as well as the increase in electrical resistance.

A battery stores a certain amount of chemical energy but is commonly defined in terms of the number of electrons stored (typically expressed in Ah), which is related to the time a battery lasts.

$$N_{\text{electrons}}^{\text{stored}} = \int_{t=0}^{t_{\text{final}}} I \cdot \mathrm{d}t$$

Alternatively, the amount of energy stored can be indicated:

$$E_{\text{stored}} = \int_{t=0}^{t_{\text{final}}} P \cdot \mathrm{d}t = \int_{t=0}^{t_{\text{final}}} V \cdot I \cdot \mathrm{d}t$$

For comparison, the energy density in Wh/kg is often indicated.

The voltage given off by a battery is not constant over time (see Figure 3.7). The voltage is typically initially high and drops with time or with depleting capacity of the battery. In a fully charged battery, the concentration at the electrodes is high thus yielding a high current, and the electrochemical reactions are controlled by the activation potential. Increasing the time of discharge will result in depletion of the reductant and the oxidant at the anode and cathode, respectively, and transfer of the reductant and the oxidant to the electrodes will start affecting the rate of reaction and thus the current density. The battery operates at this point in the ohmic regime. After longer time of discharge time, the rate of reaction will drop further, and the battery will operate in the mass transfer regime.

The voltage-time or voltage-capacity behaviour of batteries depends on the rate of discharge. Fast discharging (i.e. relatively high power-draw) will result in a more rapid decrease in the concentration of the reductant and the oxidant at the electrodes and the time at which the power delivered by the battery becomes insufficient for the external circuit will be reached more rapidly, thus reducing the amount of energy that can be extracted from the battery. This can be attributed to an increased overpotential with increasing current density leading to lower efficiency.

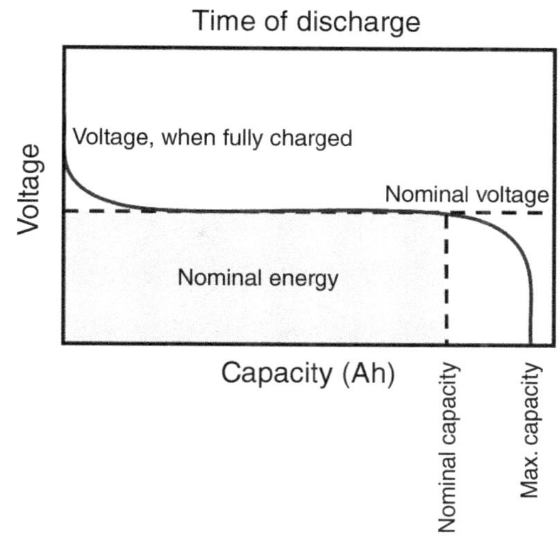

Figure 3.7 Principal behaviour of the voltage of a battery as a function of the capacity at a constant current (adapted from Rohman et al. [38]).

Increasing the temperature at which the battery operates increases the rate of reaction. This will increase the battery capacity at a constant current. However, operating at high temperature may reduce battery lifetime, because not only the rate of the desired reaction increases, but also the rate of possible side reactions, which may damage the battery.

Battery capacity also depends on the age and history of the battery. Batteries degrade with time due to the occurrence of side reactions and how the battery is treated. More side reactions may occur when the battery is discharged to a high extent or when the battery is overcharged (held at 100% capacity through continuous supply for extended periods).

For industrial batteries, UPS's (uninterrupted power supplies) and household inverters, there is a second important term for capacity, that is, the maximum power output of the battery. This is important as a battery needs to be able to provide the power output demanded of it from a system. Hence, not only the nominal voltage of a battery is important, but also its rate of discharge. For example, a battery with a capacity of 500 Ah with a nominal voltage of 36 V that is theoretically discharged to its cut-off voltage in 20 hours will have a discharge rate of 500 Ah/20 h = 25 A and thus a nominal power of 25 A · 36 V = 900 W. This is often compared as a power density, i.e. the nominal power per unit mass of battery (in W/kg).

The rate of discharge of batteries is typically related to the capacity indicated by C/x with x the time it takes to discharge the battery completely. For instance, the current obtained from a battery with a capacity of 1 Ah at C/1 will be 1 A, the current obtained at C/2 will be 0.5 A and the current obtained at C/0.2 will be 5 A. It should be noted that the practical battery capacity typically depends on the rate of discharge. In practical terms, this is summarised in the empirical Peukert's law stating that the time to reach the cut-off voltage (t_i) depends on the rate of discharge (I_i):

$$\frac{t_x}{t_y} = \left(\frac{I_y}{I_x}\right)^{\mathrm{pc}-1}$$

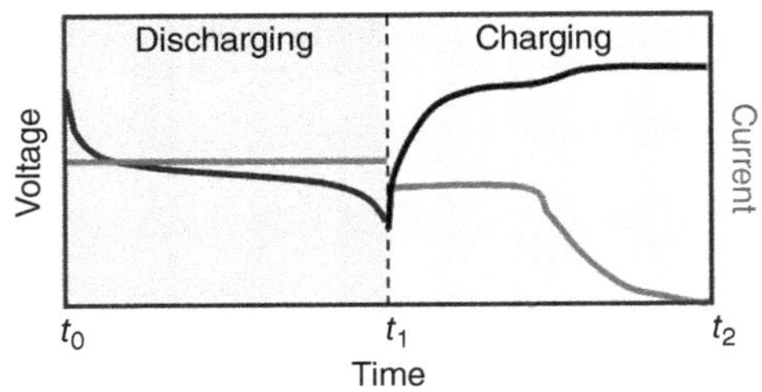

Figure 3.8 Voltage and current during discharging and charging of a battery (adapted from Quan et al. [40]).

with pc being the Peukert's constant [39]. A Peukert's constant of 1 implies that battery capacity and the time till the cut-off voltage is obtained are independent of the discharge rate. The time to reach the cut-off voltage decreases with increasing rate of discharge or current implying a Peukert's constant larger than 1. The reduced capacity at high discharge rate may be ascribed to a more rapid transferring into a more severe mass transfer limitation regime.

Batteries may also discharge themselves with time (known as self-discharge), thus losing their stored charge without any connection between the electrodes or an external circuit, due to other chemical reactions taking place inside the battery.

A battery is further characterised by its state of charge (SOC), which is defined as the fraction of the total energy or battery capacity that has been used, i.e. when 400 Ah of charge is stored in a 500 Ah battery, the SOC would be 80%.

The depth of discharge (DOD) is a similar value but based on to what extent the battery can be discharged from a full charge. This is a limitation imposed on the battery by its chemistry, as batteries cannot be fully discharged without causing serious, and often irreparable damage to the battery. For example, if the DOD of a battery is 25%, then only 25% of the battery capacity can be used by the load. A battery with a capacity of 500 Ah and a DOD of 25% can only provide 125 Ah.

Like all reactions, electrochemical reactions are reversible, and batteries can be charged and discharged. The efficiency of the discharge of electrons is defined as the Coulombic or Faradaic efficiency:

$$\eta^{\text{Coulomb}} = \eta^{\text{Faraday}} = \frac{\int_{t=0}^{t=t_1} I_{\text{discharge}} \cdot dt}{\int_{t=t_1}^{t=t_2} I_{\text{charge}} \cdot dt}$$

This Coulombic efficiency does not account for the difference in voltage between charging and discharging (see Figure 3.8), which contributes further to energy loss. The energy efficiency can be defined as:

$$\eta_{\text{energy}} = \frac{\int_{t=0}^{t=t_1} I_{\text{discharge}} \cdot V_{\text{discharge}} \cdot dt}{\int_{t=t_1}^{t=t_2} I_{\text{charge}} \cdot V_{\text{charge}} \cdot dt}$$

3.5.2 Lead-Acid Battery

The lead-acid battery was developed in the 1800s and is a well-known battery. At the anode, lead is oxidised and at the cathode, lead(IV)oxide is reduced:

Cathode:

$$PbO_2(s) + SO_4^{2-}(aq) + 4H^+(aq) + 2e^- \rightarrow PbSO_4(s) + 2H_2O(l)E^0 = 1.691\ V\ (vs.NHE)$$

Anode:

$$Pb(s) + SO_4^{2-}(aq) \rightarrow PbSO_4(s) + 2e^- E^0 = 0.359\ V\ (vs.NHE)$$

The electrical circuit is closed because of the high conductivity of the degenerate n-type semi-conductor PbO_2 caused by vacancies in the oxygen lattice [41]. The battery can be recharged by reversing the current, although hydrogen may be formed as a side product during the charging process [42].

In lead-acid batteries, the reagents are not present in stoichiometric amounts, and they typically contain 25 wt.% Pb, 35 wt.% lead oxides and 10 wt.% sulphuric acid with the balance being mainly water and the casing (polypropylene) [43]. This implies that sulphuric acid is the limiting reagent in the system and the maximum amount of charge is obtained when all sulphuric acid (0.102 mol/kg) is converted. This corresponds to a capacity of 27.3 Ah/kg.

The theoretical cell potential is 2.050 V and the standard Gibbs free energy for the overall reaction can be determined as:

$$\Delta G^0 = -n \cdot \mathcal{F} \cdot \Delta E^0 = -2 \cdot 96.5 \cdot \frac{kJ}{mol \cdot V} \cdot 2.050\ V = -395 \cdot \frac{kJ}{mol\ Pb} = -198 \cdot \frac{kJ}{mol\ H_2SO_4}$$

The actual cell voltage is dependent on the concentration of the sulphate ion and pH of the solution, because

$$\Delta E = -\frac{\Delta G}{n \cdot \mathcal{F}} = \Delta E^0 - \frac{RT}{2 \cdot \mathcal{F}} \ln \frac{a^2_{PbSO_4(s)} \cdot a^2_{H_2O(l)}}{a_{Pb(s)} \cdot a_{PbO_2(s)} \cdot a^2_{SO_4{}^{2-}(aq)} \cdot a^4_{H^+(aq)}}$$

$$\Delta E = \Delta E^0 - \frac{RT}{\mathcal{F}} \ln \frac{x_{H_2O}}{M_{SO_4{}^{2-}} \cdot M^2_{H^+}} - \frac{RT}{\mathcal{F}} \ln \frac{\gamma_{H_2O}}{\gamma^3_\pm}$$

with a_i the activity of compound i, x_{H_2O} the mole fraction of water, M_i the molality of ion i (mol/kg) and $\mathcal{F}$ the Faraday constant (96 483 J/mol/V).

Figure 3.9 shows the cell potential as a function of the concentration of sulphuric acid solution in the cell [44]. The cell potential increases strongly with increasing concentration of sulphuric acid. This is primarily through the increase in the concentration, although the activity is aided by the decrease in the activity of water and

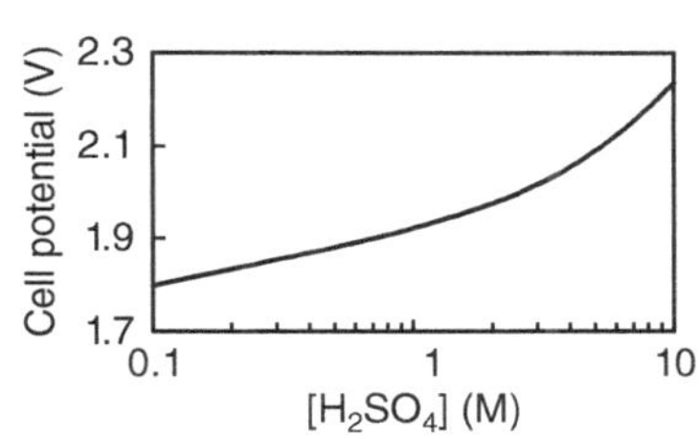

Figure 3.9 Open-circuit cell potential as a function of the sulphuric acid concentration at 25 °C (adapted from Bullock [44]).

the increase in the mean ionic activity coefficient for solutions with a concentration of sulphuric acid larger than 1.5 M. Hence, the cell potential in a solution containing 6 M sulphuric acid is expected to be 2.123 V. A lead-acid battery with six cells, as used in a car battery, and a concentration of 6 M sulphuric acid will thus have a voltage of 12.7 V.

Discharge of the lead-acid cell will thus reduce the cell potential due to the sulphuric acid concentration decreasing upon discharging the cell. For instance, a conversion of 50% of sulphuric acid upon discharge of a lead-acid cell that starts with a 6.00 M H_2SO_4 solution will result in a decrease in the sulphuric acid concentration to 3.26 M, because the number of moles of sulphuric acid and the density of the sulphuric acid solution decreases (the latter from 1.33 to 1.19 kg/l). This change in acid concentration will reduce the open-circuit voltage from 2.123 to 2.026 V. Similarly, a conversion of 80% of sulphuric acid will decrease the open-circuit voltage to 1.933 V.

The open-circuit voltage depends on temperature. The temperature dependency in the lead-acid battery is dominated by the heat of reaction (−159.3 kJ/mol H_2SO_4) and the excess enthalpy of water in the mixture (through the activity coefficient of water) and thus the heat of mixing of sulphuric acid and water. The temperature dependency of the mean ionic activity coefficient is small [44] and can be neglected. Hence, the difference in the cell potential between two temperatures for a given sulphuric acid concentration is given by:

$$\Delta E(T_2) - \Delta E(T_1)$$

$$\approx \left(-\Delta E(T_1) + \frac{159.3 + (1 - x_{H_2O})^2 \cdot (55.17 + 18.23 \cdot (4x_{H_2O} - 1))}{96.5}\right) \cdot \left(1 - \frac{T_2}{T_1}\right).$$

Fitting the mixing enthalpy of water and sulphuric acid with a Redlich-Kister equation gives:

$$\Delta_{mix}H = (1 - x_{H_2O}) \cdot x_{H_2O} \cdot (-55.17 - 18.23 \cdot (x_{H_2O} - x_{SA}))\ \frac{\text{kJ}}{\text{mol}}$$

This implies that the open-circuit voltage for a lead-acid cell in a 6 M H_2SO_4-solution will increase from 2.123 to 2.147 V upon increasing the temperature from 25 to 40 °C. Thus, the lead-acid battery can contain theoretically 2.02 MJ/kg H_2SO_4 or 0.21 MJ/kg of the battery (58 Wh/kg). The quoted specific energy is often lower (see Table 3.2) possibly because of different compositions of the battery.

The Coulombic efficiency of a lead-acid battery is only ca. 85%. The cell operates at a voltage between 1.75 and 2.45 V during discharging and charging, i.e. significantly higher than the standard electrochemical potential for water splitting and the formation of hydrogen and oxygen (specifically during charging) is controlled by the overpotential for water splitting, thus resulting in a low Coulombic efficiency.

Figure 3.10a shows the typical discharge curve for lead-acid batteries. The voltage drops initially after recharging fast (not shown), followed by a slower decrease with time, which may be attributed to the depletion of sulphuric acid near the electrodes.

Table 3.2 Typical characteristics of lead-acid batteries.

Open-circuit cell voltage	2.05 V	
Specific energy	35-40 Wh/kg	[45]
Coulombic efficiency	85%	[46]
Energetic efficiency	70–85%	[45, 46]
Power density	250 W/kg	[45]
Discharge cut-off voltage	1.75 V/cell	[47]
Charge cut-off voltage	2.45 V/cell	[48]
Peukert's constant	1.17–1.19	[39]
Charge time	8–16 h	
Self-discharge	1–5%/month	[48]
Operating temperature	−30–40 °C	[45]

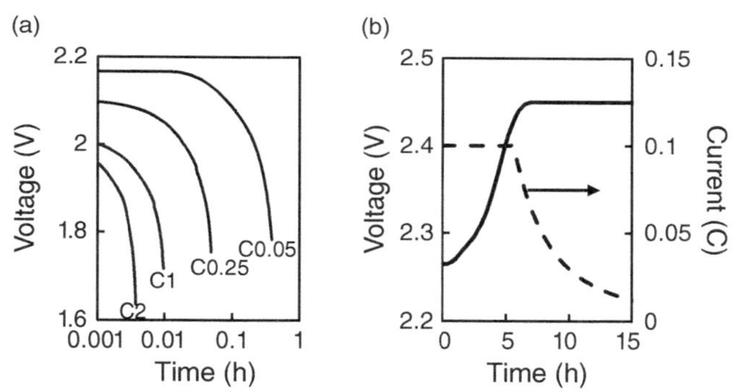

Figure 3.10 Typical dependency of the voltage and current of a lead-acid cell (given as the time to deplete the cell) as a function of time during discharging (a) (redrawn from [47]) and during charging (b) (redrawn from Power Sonic [48]).

A higher rate of discharge will lower the concentration of sulphuric acid near the electrode further thus reducing the voltage with increasing rate of discharge. After some time the voltage starts dropping faster, as the battery starts to operate in the mass transfer limiting regime. The battery will come faster into the region of mass transfer limiting operation when operating at a higher rate of discharge. Interrupting the current and starting it after some time again, will result in some recovery of the voltage [47]. The current is given by the external circuit and for small changes in the voltage is almost constant, and thus the power output of the cell drops with time. The capacity of a cell decreases with increasing current density, as the discharge is stopped when voltage cut-off is obtained (for lead-acid batteries ca. 1.75 V – see Table 3.2).

Charging of the lead-acid battery is relatively slow, typically taking six to eight hours. The required voltage increases with increasing charging time; as the battery gets closer to maximum capacity, a higher voltage is needed to force the reaction.

On a molecular level this is because diffusion becomes controlling of the overall charging mechanism. The typical cut-off voltage for charging the lead-acid battery is 2.4 V.

The maximum capacity of the lead-acid battery is determined by the amount of the limiting reagent in the cell. However, the presence of a cut-off voltage implies that a battery will not be discharged completely, and that the DOD will vary with rate of discharge. Temperature affects the capacity of lead-acid batteries: the cut-off voltage will be reached more readily at a high rate of discharge, which will thus result in a decrease in the DOD.

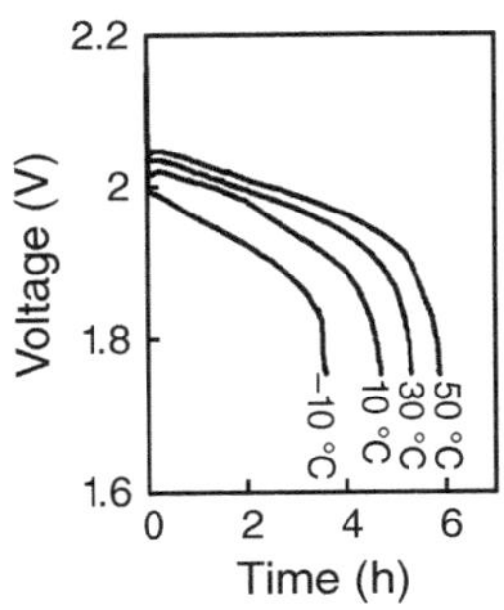

Figure 3.11 Effect of temperature on the voltage of a lead-acid cell as a function of time, for a 1 Ah 6 V battery at a current of C/5 (redrawn from Hariprakash et al. [49]).

Upon increasing temperature, the rate of reaction will increase. This implies that longer discharge times will be possible if the rate of discharge is kept constant (see Figure 3.11). Hence, the capacity that can be recovered, before the cut-off voltage is obtained, will be higher at higher temperatures. This means that a battery which is rated for use at 25 °C, but used at 35 °C, can have a maximum capacity of more than 100%.

A rechargeable battery will have to go through many discharge–recharge cycles. However, the number of cycles a battery can do before seeing significant decrease in performance is affected by the depth of discharge (DOD) and the battery can go through fewer cycles if it is discharged to a higher degree. Lead-acid batteries should not be discharged more than ca. 30% to achieve a good number of cycles (see Figure 3.12). The loss in performance of lead-acid batteries can be attributed to corrosion of metallic lead, active mass degradation, sulfation and short-circuiting [51]. Lead at the anode is unstable and will slowly convert into a variety of different lead oxides [52]. At the cathode, PbO_2 is converted into $PbSO_4$ via a dissolution-precipitation mechanism; the discrepancy in the density of these materials implies that newly formed PbO_2 upon re-oxidation will not adhere well to the cathode [51]. Sulfation is a long-standing problem [53]; upon increasing the number of cycles, the average crystallite size of the formed $PbSO_4$ may increase due to the presence of residual $PbSO_4$-seeds. The larger $PbSO_4$-crystallites will

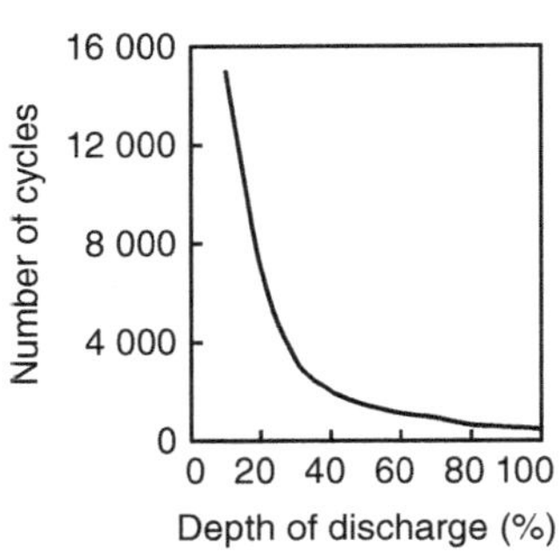

Figure 3.12 Number of cycles with a lead-acid battery as a function of the DOD (redrawn from Fox [50]).

be less readily converted into metallic lead and PbO_2 upon reversing the current. Furthermore, recharging the lead-acid battery may result in the formation of dendritic lead in the separation between the anode and cathode [54], which may result in short-circuiting.

Temperature also affects the undesirable side reactions, reducing the number of cycles of a battery. Thus, whilst higher temperatures can lead to higher capacity, they also reduce the lifetime of the battery.

The rechargeable lead-acid battery is one of the options to mitigate the intermittency associated with the utilisation of renewable energy. The battery requires lead as the anode and lead(IV) oxide as the cathode. The current lead reserves worldwide are estimated at 95 million metric tonnes [55]. This implies that the reserves worked into lead-acid batteries could store 0.09 EJ, assuming 100% efficiency. Otherwise stated, the amount of lead in reserves could store 0.02% of the annual energy consumption in 2022 or 6% of the average daily energy consumption, implying that sufficient lead is in reserve to use lead-acid batteries as a back-up for renewable energy resources, but not enough for long-term storage.

3.5.3 Lithium-Ion Batteries

Lithium-ion batteries typically consist of a lithium metal oxide separated by graphite, which can intercalate lithium (see Figure 3.13). The cathode makes up ca. 25–30% of the weight of the cell [57] and is made up from materials which have a high lithium-ion diffusivity, a high electrical conductivity, as well as strong reversibility, low cost and are safe to use [58]. Various materials fulfil those conditions at least partially, e.g. layered mixed metal oxides, such as $LiCoO_2$ or $LiNi_{1\text{-}x\text{-}y}Co_xMn_yO_2$ (with lithium between MeO_2-sheets), spinel structures, such as $LiMn_2O_4$, and olivine type structures, such as $LiFePO_4$. The anode, making up ca. 15–30% of the weight of the battery [57], can intercalate metallic Li, and utilises materials such as graphite and carbon nanotubes, although other types of materials

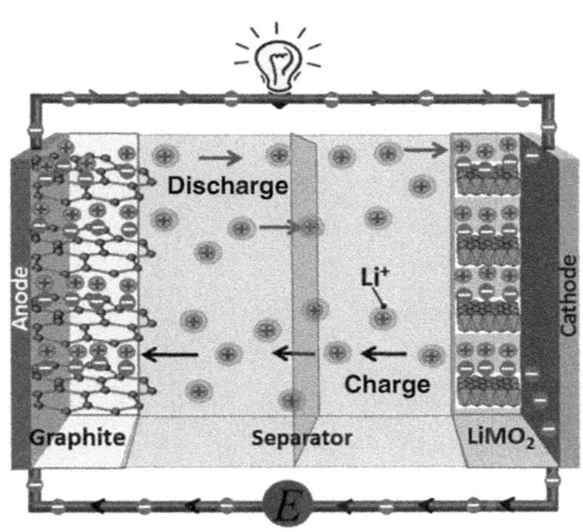

Figure 3.13 Schematic of the operation principles for rechargeable lithium-ion batteries (Madian et al. [56]/CC BY 4.0).

have been tested as well [58]. The cathode and anode are separated from each other typically by a polymer layer to avoid short-circuiting [58]; this layer makes up 3–4% of the mass of the battery [57]. The electrolyte is a medium for the transport of the lithium-ion between the cathode and anode and makes up 10–15% of the battery weight [57]. It is typically comprised of organic carbonates with a dissolved lithium salt (e.g. $LiPF_6$), as aqueous media will react violently with the lithium salt [58].

The half-cell reactions which occur during discharge of a lithium-cobalt cell are

Cathode:

$$Li_xCoO_2(s) + (1 - x)Li^+ + (1 - x)e^- \rightarrow LiCoO_2(s)$$

Anode:

$$(1 - x)\ Li_{in\ carbon}(s) \rightarrow (1 - x)Li^+ + (1 - x)e^-$$

and the nominal cell voltage is ca. 4.0 V. Complete delithiation is not possible [59, 60] and $LiCoO_2$ can be delithiated by ca. 50% (other lithium containing oxides can be delithiated to a larger extent). Furthermore, the typical ratio of Li to carbon in the anode is 6 [61]. This implies that the change in the Gibbs free energy for the overall reaction:

$$Li_{0.5}CoO_2(s) + 0.5LiC_6(s) \rightarrow LiCoO_2(s) + 3C(s)$$

$$\Delta G = -n \cdot F \cdot \Delta E = -0.5 \cdot 96.5 \cdot \frac{kJ}{mol \cdot V} \cdot 4.0\,V = -193 \cdot \frac{kJ}{mol}$$

A typical lithium-cobalt cell contains 17.3 wt.% cobalt and 2.0 wt.% lithium [61]. This corresponds to a capacity of 38.6 Ah/kg. The specific energy is 556 kJ/kg or 155 Wh/kg. Lithium-cobalt batteries are characterised by a high Coulombic efficiency (see Table 3.3) implying that very few side reactions take place within the battery. The energy efficiency is significantly lower (between 78% and 91%) due to the overpotential during charging and discharging the battery [70].

Table 3.3 Typical characteristics of lithium-cobalt batteries.

Open-circuit cell voltage	~3.7 V	[62]
Specific energy	80–280 Wh/kg	[63]
Coulombic efficiency	>99.9%	[64]
Energetic efficiency	78–91%	[65]
Power density	140 W/kg[a)]	
Discharge cut-off voltage	2.2–2.7 V	[65]
Charge cut-off voltage	4.2–4.45 V	[66]
Peukert's constant	1.0–1.28	[67]
Charge time	1–4 h	[68]
Self-discharge	0.4–3.5% per month	Find
Operating temperature	−20–60 °C (optimal: 15–35 °C)	[58, 69]

a) Estimated at a capacity of 37.6 Ah/kg operating at a current of C/1 and a voltage of 3.7 V.

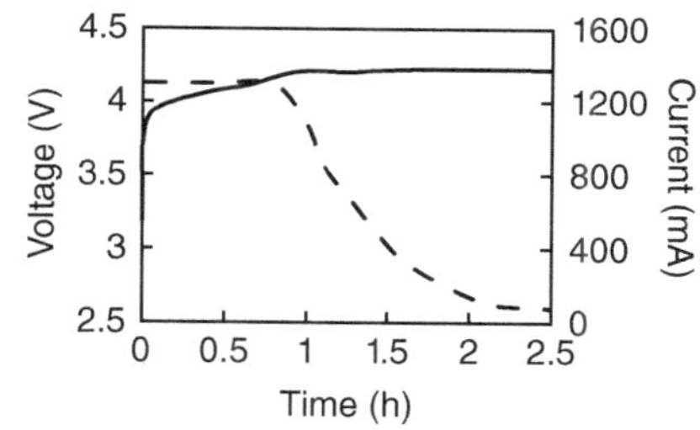

Figure 3.14 Charging a lithium-cobalt battery (redrawn from Gold Peak Industries (Taiwan), Ltd. [68]).

The power density of lithium-cobalt batteries is relatively low and was estimated at 140 W/kg based on the capacity of 38.6 Ah/kg discharged at a rate of 38.6 A (C/1) at a voltage of 3.7 V. Higher discharge rates would generate too much heat, possibly damaging the battery [71].

Some self-discharge takes place within the battery, mainly through the decomposition of the lithium salt in the electrolyte [72]. As this is a chemical reaction, it occurs faster at higher temperatures.

Lithium-cobalt batteries can be charged rapidly (see Figure 3.14). This process typically takes place in two stages, viz. a constant current period, followed by a constant voltage stage to minimise damage to the battery. The change from constant current to constant voltage charging usually occurs at ca. 80% of the total capacity.

Figure 3.15 shows the discharge of a lithium-cobalt battery. Li-Co batteries follow a classical S-shape plot. As was seen previously with other batteries, the voltage drops upon discharging, more so as the current increases. Also, decreasing the temperature yields reduces capacity due to the reaction rate being slowed.

A high degree of discharge reduces the lifetime of the battery. However, in comparison to other batteries, the Li-battery can have a deep cycle or high degree of discharge without power loss and with limited cost in terms of life cycle. Together this gives a distinct advantage, i.e. the lifespan of lithium-cobalt batteries, which can do over 1000 battery cycles if treated properly. Even if treated poorly, the Li batteries last longer than other battery types, although the lifetime of these batteries is affected by the depth of discharge as well.

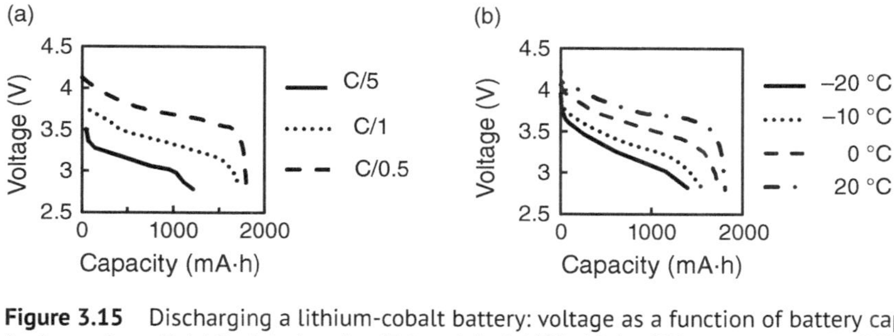

Figure 3.15 Discharging a lithium-cobalt battery: voltage as a function of battery capacity (redrawn from Gold Peak Industries (Taiwan), Ltd. [68]). (a) At different currents and 20 °C. (b) At a current of C/0.5 and different temperatures.

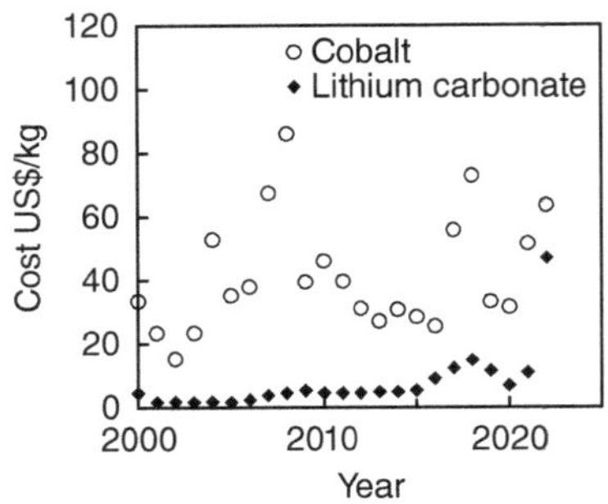

Figure 3.16 Year-on-year prices for cobalt and lithium carbonate (data from Energy Institute Statistical Review of World Energy [73]).

In lithium-cobalt cells, cobalt is the limiting element in terms of availability on earth (world reserve of lithium 28 million tonnes and of cobalt 11 million tonnes [55]). The amount of energy, which can be stored in lithium-cobalt batteries, if all cobalt is used for this purpose only and if the battery operates with an efficiency of 100%, comes to 0.036 EJ or otherwise stated 0.01% of the annual energy demand in the world in 2022 (or 2.4% of the average daily energy consumption). The historical price of cobalt and lithium carbonate are given in Figure 3.16. In the period 2013–2021, lithium was ca. 3.9 times cheaper than cobalt on a molar basis (2.2 times more expensive on a mass basis; lithium carbonate was on average 4.4 times cheaper than cobalt on a mass basis). The prices of cobalt and lithium are subjected to (speculative) market forces, and in 2022 the price ratio of cobalt to lithium on molar basis reduced to 1.2. This sharp increase in the lithium price appears to have been a temporal fluke caused by market forces as the lithium price dropped significantly in 2023 and 2024 [74].

3.5.4 Alternatives to $LiCoO_2$

The lithium-cobalt battery presented a big step forward in the world of batteries, but its performance regarding specific power, lifetime and safety in operation could be improved with a further drive to improve on the cost by replacing the costly cobalt in the cathode. Hence, different materials have been developed to specifically reduce cobalt content in these cathodes. Table 3.4 shows alternatives for the lithium-cobalt cathode together with their specific characteristics.

The lithium-cobalt battery or lithium cobalt oxide (LCO) battery is mainly used in electronic devices. It has a nominal voltage of about 3.7 V but operates during charging and discharging between ca. 3.0 and 4.2 V (with too high a voltage leading to damage of the battery [76]). The specific energy depends on the exact formulation of the battery but is in the range of 150–300 Wh/kg with charge and discharge rates of typically C/0.5 and C/2. The lifetime of these materials is limited, and it can normally cycle between 300 and 500 times, although sometimes even 1000 charge/discharge cycles can be achieved.

Lithium manganese oxide (LMO) is used for power tools. It has a very similar operating voltage to LCO, but it can be discharged at a higher rate (C/0.33) than LCO (C/0.5) over a limited time and can have a longer lifetime. Furthermore, these

Table 3.4 Typical characteristics of lithium-containing cathodes for batteries [75].

Type	$V_{nominal}$ (V)	$V_{operating}$ (V)	$E_{specific}$ (Wh/kg)	Discharge rate	Cycles
$LiCoO_2$ (LCO)	3.7	3.0–4.2	150–200	C/0.5–C/2	300–500
$LiMn_2O_4$ (LMO)	3.7	2.5–4.2	100–150	C/0.33–C/2	500–1000
$LiNi_xMn_yCo_{1-x-y}O_2$ (NMC)	3.6–3.8	3.0–4.2	150–200	C/033–C/2	500–1000
$LiNi_{0.8}Co_{0.15}Al_{0.05}O_2$ (NCA)	3.6–4.0	3.0–4.2	200–250	C/0.2–C2	500–1000
$LiFePO_4$ (LFP)	3.2–3.3	2.5–3.6	90–160	C/0.2–C2	2000–5000
Li-titanate (LTO)	2.4	1.5–2.8	80–120	C/0.1–C2	6000–6010 000

batteries carry less of a safety risk. The disadvantage is the lower specific energy of this system.

A higher specific energy is obtained with the so-called NMC-battery containing nickel, manganese and cobalt (the actual ratio in these materials may change depending on the application). These batteries have a mix of the characteristics of the LCO-battery (high energy density), and the LMO-battery, viz. high discharge rate and lower safety risk. Hence, they are being used in electric vehicles, and energy storage systems, but also in portable electronics.

The NCA-battery, containing nickel, cobalt and aluminium, has a high specific energy and a high discharge rate making it suitable for high power applications, such as in electric vehicles, although its thermal stability is lower than other lithium batteries.

Lithium iron phosphate (LFP) uses no cobalt and has a very good lifetime, due to stability of the iron phosphate structure, and a good thermal stability. Additionally, this material has a low safety risk. However, the energy content is low, and the high discharge rate is not necessarily maintained over long periods of time. These batteries are used in electric buses, for stationary energy storage systems, and back-up power systems.

Lithium titanate (LTO) batteries are different from other lithium batteries where the cathode is a lithium mixed metal oxide, and the anode is graphitic carbon. Here, lithium titanate ($Li_4Ti_5O_{12}$) is used as the anode and either LMO or LFP is used as the cathode. The energy density is thus between the energy density of the LMO and the LFP-battery. The large advantage of this rather costly battery is its long lifetime together with its high discharge rate. Hence, these batteries are targeted for use in electric vehicles, and for energy storage.

It can be useful to compare the characteristics $LiCoO_2$ and the alternative Li-based batteries in so-called spider plots (see Figure 3.17) as a visual representation of the different values. Whilst spider plots can be informative, they should be treated with caution, because a simple visual inspection implies that high values for the various variables are typically desired, although the cost parameter should in reality be as low as possible. Furthermore, categories such as performance and even safety are difficult variables to quantify. More robust techniques, such as multi-criteria

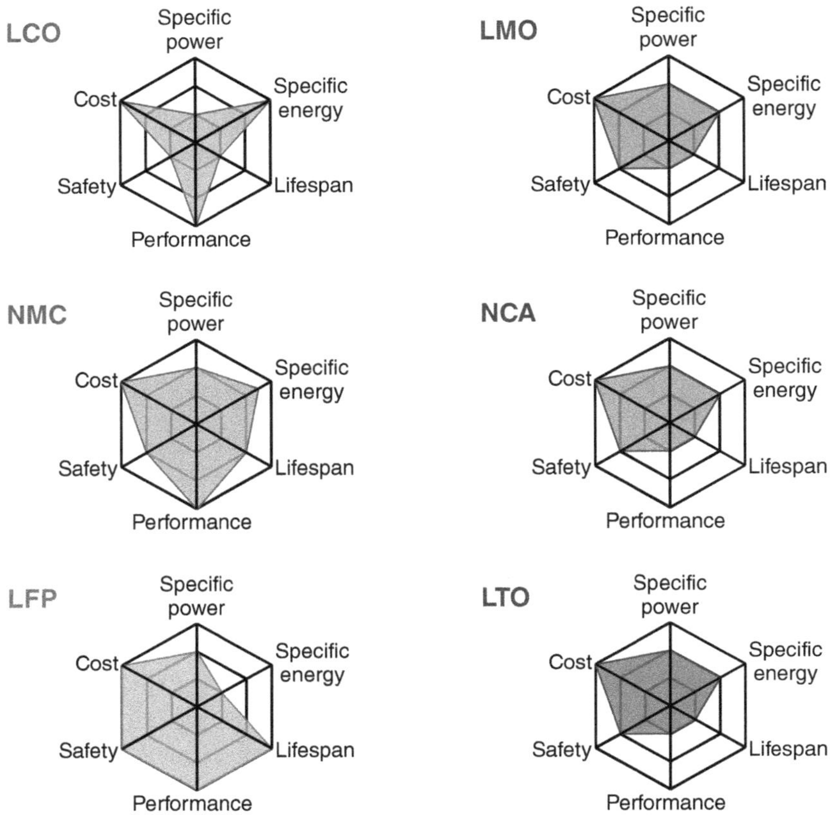

Figure 3.17 Comparison of lithium-ion batteries on spider plots (redrawn from Saldaña et al. [77]).

decision-making [78], should be applied to evaluate the use of different batteries for specific applications.

The alternative lithium-ion batteries have a reduced cobalt content, thus reducing the material cost in the battery. If the whole world-reserve of lithium (28 million tonnes) were converted into LFP (capacity 170 Ah/kg assuming complete delithiation is possible; specific energy 544 Wh/kg assuming a nominal voltage of 3.2 V), the total amount of energy stored could be 1.24 EJ or 0.23% of the total world energy consumption in 2022. On a daily basis, this would amount to 82% of the average daily energy consumption.

There is thus a drive to replace lithium in this type of battery with sodium. The main driving forces here are cost, and supplier security. This type of battery can work on the same principle as lithium-ion batteries, although other systems have been developed as well (see e.g. [79]). Several successes with the sodium-ion batteries have been reported, but the energy density is still low in comparison to lithium-ion batteries [80].

3.5.5 Solid-State Batteries

A major concern around the use of batteries is the presence of liquid electrolytes, in particular in lithium (and sodium) type of batteries when operating at a voltage higher than decomposition voltage of the electrolyte [81]. The elimination of the electrolyte would reduce the hazards associated with batteries, and possibly increase the energy density as the mass of the battery can be reduced, as the electrolyte makes up 10–15 wt.% of the battery [82]. Liquid electrolyte facilitates fast transport of lithium-ions between the electrodes, necessary to avoid side effects caused by retarded transport, such as lithium plating of the graphitic anode and the formation of dendritic structures on the anode [83].

In solid-state batteries, the liquid electrolyte is replaced with a solid electrolyte to facilitate the transport between the electrodes [84] thus requiring a high ionic conductivity in addition to a low electron conductivity. Various materials have been developed as solid electrolytes, viz. oxidic electrolytes such as lithium phosphorus oxynitride (LiPON) and sodium super ion conductors (NASICON), sulphide electrolytes, halide electrolytes and polymer electrolytes. A major challenge for these electrolytes is the contact between the electrolyte and the electrodes [85], as poor contact will increase the internal resistance and thus reduce the power delivered by the battery. Furthermore, the ceramic electrolytes can be brittle and thus pose a challenge in handling.

Solid-state batteries are still under development with a view to develop batteries operating stably at a high voltage for a long time. As such, they may become an important technology going forward.

3.6 Flow Batteries

In batteries, such as the lead-acid battery and lithium-ion batteries, the energy and power density are directly connected and not independently scalable. In flow batteries, the amount of energy stored and the amount of power to be delivered by the battery can be scaled independently of each other.

Flow batteries are comprised of electrochemical cells where instead of the electrolyte being stored within the cell, the oxidant and reductant are stored in holding tanks and the different electrolytic solutions are pumped around (see Figure 3.18). The anode and cathode are separated by an ion-selective membrane ensuring electro-neutrality in each of the solutions. Changing the direction of flow will change the process from charging to discharging. The oxidised and reduced catholyte as well as the oxidised and reduced anolyte can be kept in separate holding tanks to keep the concentration of the ions in the solution as high as possible, although in some configurations the reduced catholyte and oxidised anolyte are mixed with the oxidised catholyte and reduced anolyte, respectively. The dilution of the ions will reduce the potential of the battery and thus the power generation with time. Brée and Mitsos [86] argued that the voltage could be increased by 2% and the energy capacity by 10% by using the split tank design.

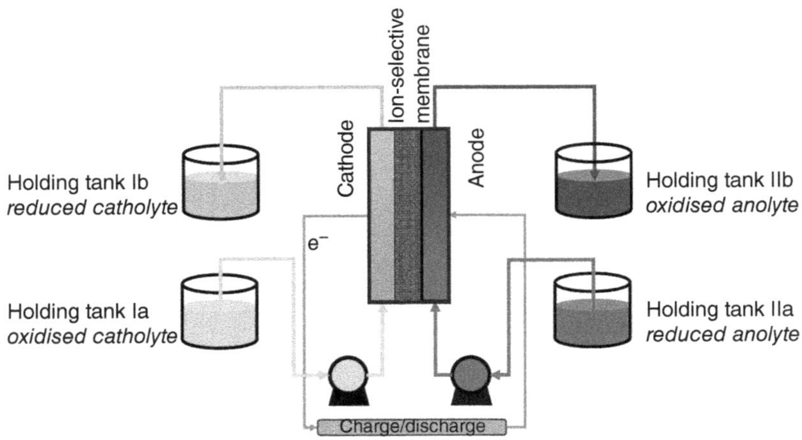

Figure 3.18 Principal set-up of a flow battery (flow direction indicated during discharge).

The amount of energy stored is thus linked to the size of the holding tanks and scales with the concentration of the limiting electrolyte (catholyte/anolyte) and the potential difference, as the stored energy per unit volume in the holding tank is given by:

$$E_{\text{stored}} = -n \cdot \mathcal{F} \cdot E \cdot [\text{electrolyte}]_{\text{limiting}}$$

For instance, an electrolyte solution with an electrolyte concentration of 1 M could store 96.5 MJ/m^3 if the electrochemical potential is 1 V, and the reaction transfers a single electron.

Flow batteries show the typical voltage-current plot (see Figure 3.6), approaching the Nernst potential in the open-circuit mode, dropping the voltage due to so-called activation losses, followed by a transition into a region in which ohmic losses dominate, and lastly a strong drop in the voltage in the mass transfer limited regime, resulting in a limiting current. The latter part depends on the flow rate of the electrolytes through the battery [87] and higher current density can be obtained at higher volumetric flow rate through the cell. However, the maximum conversion of the electrolyte in the cell decreases with increasing flow rate as indicated by the ratio of the limiting current relative to the maximum current [87]. This will thus reduce the amount of energy stored per unit volume.

The voltage-current plot can be transformed into a power-current plot by multiplying the voltage with the corresponding current (see Figure 3.19). The power density for an all-vanadium flow battery as reported by Aaron et al. [87] initially increases almost linearly with the current density until it passes a maximum. The initial, almost linear, increase in the power density with the current density is independent of the flow rate of the solutions through the cell. Hence, the amount of power delivered by a flow battery can be scaled by the pump rate. The maximum power that can be achieved depends on the volumetric flow rate through the cell; it appears to follow a square root law, which might indicate a role of the boundary layer under laminar conditions in the electrochemical conversion.

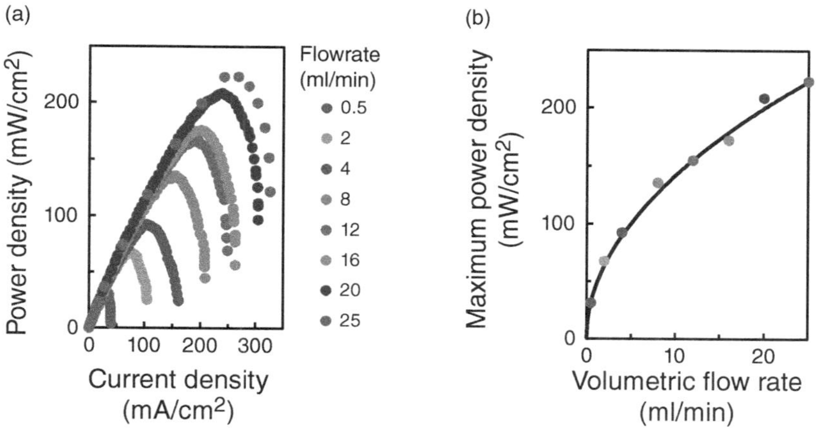

Figure 3.19 Power density as a function of the current density (a) and the maximum power density as a function of the volumetric flow rate (b) in an all-vanadium flow battery ([V] = 1 M; [H_2SO_4] = 5 M) in a 'fuel cell battery' (derived from Aaron et al. [87]).

3.6.1 Electrolytes

The active redox couples determine the cell potential and the energy density. To maximise both parameters, a large difference in the redox potentials of the species involved is required. The oxidised and reduced forms of the electrolytes must both have high solubility so that no solid metals or metal oxides will be formed in the process. They must also show good stability, with minimum degradation of the species.

Table 3.5 shows some of the electrolytes investigated for use in a redox-flow battery. The simplest systems are the iron–chromium system (ΔE^0 = 1.18 V), which may be performed in an aqueous solution [88], as the salts are soluble in water and the potential is below the potential for water splitting. Under discharge, chromium is oxidised, and iron is reduced (upon charging, the reverse reactions occur):

$$Cr^{2+} \rightarrow Cr^{3+} + e^-$$

$$Fe^{3+} + e^- \rightarrow Fe^{2+}$$

The charge in the solutions is balanced by the transport of protons over a membrane. During discharge, protons flow from the side with the chromium-containing solution to the side with the iron-containing solution. The reaction is relatively slow [97], and hence other systems are being investigated as well. The rate of reaction in a zinc–iron battery is higher [98], but the high potential (1.53 V) reduces its ability to be used in an aqueous system with water splitting taking place as a side reaction.

Single metal systems (all-iron or all-vanadium systems) have been investigated extensively for their application in flow batteries. The commonality of the ions used on the anodic and the cathodic side will minimise the effect of contamination of the anolyte and catholyte due to cross-over, and thus improve the durability of these solutions [99]. In aqueous systems, the potential is slightly larger than the potential

Table 3.5 Properties of solvents employed in flow batteries at 20–25 °C.

Redox pairs	Solvent	E^0 (V)	I (mA/cm^2)	$\eta_{energetic}$	$\eta_{Coulombic}$	
Cr^{2+}/Cr^{3+} // Fe^{3+}/Fe^{2+}	Water	1.18	120–200	0.735	0.815	[88]
Zn/Zn^{2+} // Fe^{3+}/Fe^{2+}	Water	1.53	25	0.54–0.78	0.90–0.92	[89]
VO^{2+}/VO_2^+ // Fe^{3+}/Fe^{2+}	Water	1.02	50–150	0.65–0.85	0.95–0.97	[90]
$[Fe(TEOA)OH]^-/[Fe(TEOA)(OH)]^{2-}$ // $Fe(CN)_6^{3-}/Fe(CN)_6^{4-}$	Water	1.34	40	0.70–0.75	0.80–0.90	[91]
VO^{2+}/VO_2^+ // V^{3+}/V^{2+}	Water	1.26		0.75–0.78	0.91–0.93	[92]
VO^{2+}/VO_2^+ // V^{3+}/V^{2+}	Water	1.26	50–150	0.72–0.86	0.94–0.96	[90]
$[V(acac)_3]/[V(acac)_3]^-$ // $[V(acac)_3]^+/[V(acac)_3]$	DMSO[a)]	2.21	0.1	0.895	0.436	[93]
$[V(acac)_3]/[V(acac)_3]^-$ // $[V(acac)_3]^+/[V(acac)_3]$	$[C_2C_1Im][N(Tf_2)]$	~2.0	0.02	0.69–0.75	0.22	[94]
$[V(acac)_3]/[V(acac)_3]^-$ // $[V(acac)_3]^+/[V(acac)_3]$	Choline chloride/ethylene glycol	2.01	0.01	0.49–0.52	0.25–0.31	[95]
$[PW_{12}O_{40}]^{3-}/[PW_{12}O_{40}]^{4-}/[PW_{12}O_{40}]^{5-}/[PW_{12}O_{40}]^{6-}$ // I_3^-/I^-	Water	0.31–0.81	100	0.798	~1	[96]

a) With $TBAPF_6$.

for water splitting, thus necessitating the implementation of non-aqueous systems, when using concentration metal ion solutions. A major concern in flow batteries, is the possible formation of precipitate during the electrochemical conversion [100]. The development of precipitation inhibitors has advanced the development of specifically vanadium flow redox batteries (VFRB) [101].

Heteropolyacids (HPAs), e.g. 12-phosphotungstic acid (PTA, $PW_{12}O_{40}$), a Keggin type HPA, can also be used as the electrolyte, as their electrochemical reactivity is high. Furthermore, the multiple valance states could yield a high theoretical capacity. Hence, the voltage obtained with these batteries will depend on the flow rate through the battery. The relatively large size of these ions will reduce the likelihood of permeation through the membrane, thus promising a long lifetime.

Flow batteries can suffer from self-discharge, when the reduced electrolyte is oxidised within the storage tank. Minimising contact between the reduced electrolyte and air will reduce the extent of self-discharge.

The electrolytes are in the liquid phase, typically dissolved in a solvent. Principally, any liquid can be used as the solvent, provided the electrolytic materials are soluble within it. The properties of some solvents often employed as electrolytes for flow batteries are given in Table 3.5.

Water is an obvious choice, because of the high solubility of a large range of ions, the high conductivity of ions in solution, its low viscosity, its low vapour pressure and its low cost. Furthermore, the environmental impact of spillage of the solvent may be limited. However, water can only be applied in a limited temperature range (0–100 °C at 1 bar). Water may also undergo water splitting if the cell potential is larger than 1.23 V at 25 °C, which will lower the Coulombic and energetic efficiency. The application of water as a solvent will thus limit the maximum voltage and, as a result, the energy density. Some options of solvents are given in Table 3.5, the conductivity of each solvent is greatly dependent on the concentration of the solute.

Organic solvents typically have a much larger electrochemical window of operation (see Table 3.6), and their temperature of operation can be relatively wide. The major disadvantage of organic solvents is the low solubility of ions and relatively low ionic conductivity. Furthermore, they are flammable and may pose an environmental risk.

Ionic liquids will overcome most of the disadvantages of organic solvents, as they have a low vapour pressure, a low flammability and a higher solubility for ions. Their electrochemical window of operation is, however, less than that of typically organic solvents (see Table 3.4; ~2.2–2.7 V [105, 109], compared to ~3.6–5.5 V). Their major disadvantages are the cost, and the high viscosity of these solvents.

Relatively recently, novel solvents have been developed based on mixtures between hydrogen donors (Brønsted acids) and hydrogen acceptors (Brønsted bases) [110]. These have been named deep eutectic solvents as they are liquid far below the melting point of each of the components (although there is some dispute on the deepness of these mixtures – see [107]). Typical mixtures comprise of choline chloride and ethylene chloride, or choline chloride and urea. However, these mixtures can be tuned based on the individual components in the mixture, and they can be designed to have a relatively low volatility. These solvents have

Table 3.6 Properties of solvents employed in flow batteries at 20–25 °C.

	$T_{operating}$ (°C)		Conductivity (mS/cm)	Viscosity (mPa s)	E_{window} (V)	
Electrolytes dissolved in water	~0	~100	0–200	1.0	1.23	[102]
Organic solvents						
Dimethylsulfoxide (DMSO)	19	189		2.0	4.4	[103]
Acetonitrile	−45	82	$\cdot^{-}$	0.38	5.5	[104]
Propylene carbonate	−49	241	$\cdot^{-}$	2.5	3.6	[104]
N,*N*-Dimethylformamide (DMF)	−61	153	$\cdot^{-}$	0.8	4.4	[104]
1,3-Dioxolane	−95	74		0.6	4.0	[103, 104]
Ionic liquids						
[BMIm][Br]			$0.22 \cdot 10^{-3}$	1400	2.7	[105]
Deep eutectic solvents[a)]						
Choline chloride/ethylene glycol	−30[b)]			37	~2[c)]	[106]

a) Properties depend on the ratio of the compounds in the mixture.
b) Agieienko and Buchner [107].
c) Sinclair et al. [108].

a larger solubility for ions than typical organic solvents. The viscosity of these solvents is rather high, and the electrochemical operation can be small depending on the choice of the compounds in the mixture.

3.6.2 Membranes

Membranes are an essential part of flow batteries. They facilitate the required diffusion of some of the ions from one side to the other to maintain charge neutrality in each of the solutions, whilst at the same time retarding the diffusion of other ions, thus allowing for a high Coulombic efficiency. Membranes used in flow batteries must meet certain requirements [104], viz.:

- high ionic conductivity (fast diffusion) of charge balancing ions through the membrane required to maintain electro-neutrality in solution.
- high ionic selectivity with only the desired ions being transported across the membrane to limit unwanted cross-over and maintain a high Coulombic efficiency.
- chemical stability with minimal deterioration of the membrane caused by the high concentrations of electrolytes, which are necessary for a high energy density and a fast discharge.
- mechanical stability to avoid crack formation due to the flow conditions.
- cost-effective.

Protons are often used to balance the charge in the electrolytes. Hence, Nafion, a sulfonated tetrafluoroethylene-based fluoropolymer-copolymer (see Figure 3.20),

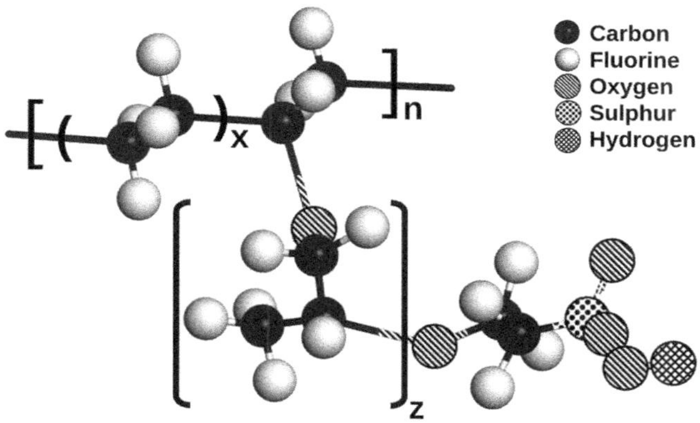

Figure 3.20 Basic structure in Nafion, a sulfonated tetrafluoroethylene-based fluoropolymer-copolymer.

is often used as a membrane [104]. The high density of sulfonated groups allows H^+ ions to 'jump' from one sulphate group to the next, thus allowing conduction of protons (and some other cations) but preventing conduction of electrons and anions. The transport of larger cations can be blocked by covering the Nafion with polytetrafluoroethylene (PTFE) [111]. The thickness of the membrane is typically optimised, with thicker membranes minimising cross-over and having a greater mechanical stability, but a lower rate of proton conduction.

Other membranes [104] targeting the conduction of either generic anions or cations are still under development and may allow the use of a wider range of electrolytes.

3.6.3 Electrodes

The electrochemical oxidation and reduction in flow batteries take place at the electrodes, which need to meet a number of requirements [104], viz.:

- large electro-active area to facilitate the surface oxidation/reduction.
- fast electron-transfer kinetics resulting in a low overpotential.
- high electron conductivity to minimise ohmic losses in the cell.
- ability to operate at high current-density.
- adequate mechanical strength.
- electrochemical stability against corrosion.
- cost-effective.

Carbon is often used as the electrode material as it is cost-effective, can have a high surface area and is corrosion and fire-resistant. Various metal-based electrodes can also be used, but these tend to underperform overall in the above categories compared to carbon.

The electrodes need to be in close contact with the electrolyte solution to ensure that the electrons can be transported from the solution to the greater circuit. In a simple set-up, the electrodes are plates, along which the electrolyte flows. However,

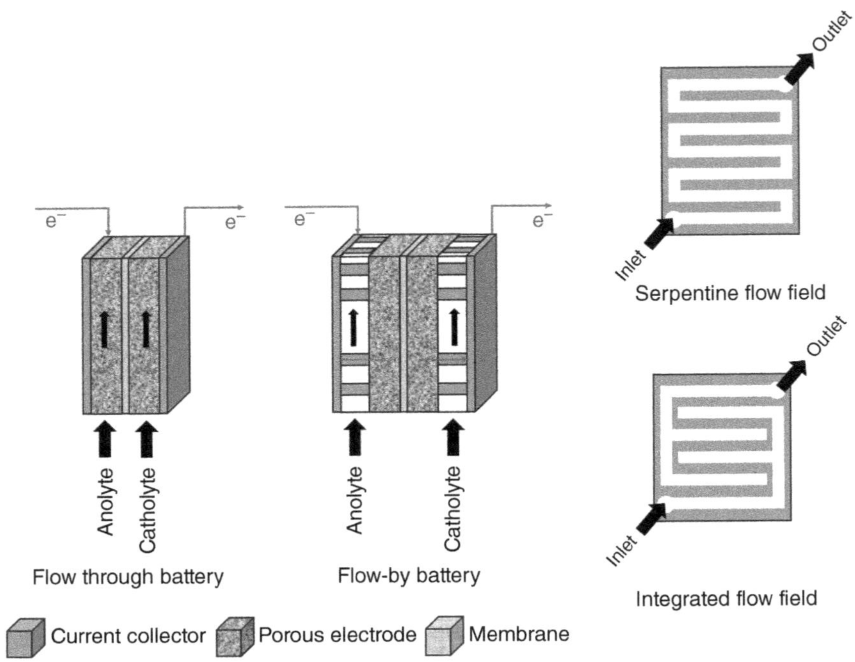

Figure 3.21 Elements in a flow battery and possible flow fields (redrawn from Arevalo et al. [104]).

plates have a low surface area, resulting in a low activity for a given electrode material with a low specific, optimum current density similar as obtained at low flow rates (see Figure 3.19). Furthermore, the flow over flat plates may result in relatively thick boundary layers, thus increasing the likelihood of mass transfer becoming the rate controlling step in the overall process. Hence, porous electrodes are being used in flow batteries. The resulting pressure drop can be negated by creating flow channels between the porous electrodes and the current collector (see Figure 3.21). Flow through the porous electrode layer can be enforced by introducing an integrated flow field [104], where the inlet and the outlet for each of the electrolytes are separated from each other.

3.6.4 Design Considerations for Flow Batteries

The great advantage of flow batteries [104] lies in the independent scaling of energy and power with the former being scaled by scaling the holding tanks, and the latter by scaling the stack. This gives a greater flexibility in the design of batteries for specific applications. The design can be optimised with respect to the cost of stored energy (the levelised cost of energy storage [LCOS]) with the capital costs being separated into costs scaling with the energy capacity (size of holding tanks), costs scaling with the power capacity (stack design, pumps and heat exchange) and capital costs which do not scale well with size (control system and auxiliary equipment) [112].

The electrochemical reactions take place in the liquid phase, thus allowing a rapid response and a greater control of the reaction environment and thus on the resulting electrochemical potential. A critical parameter in the stack design is the required current density, which affects the overall energetic efficiency [112].

The electrolytes can be stored at a distance from the stack thus improving the safety of flow battery operation. The reduced electrolyte can be effectively protected from oxidation, thereby limiting the extent of self-discharge. Elements of the battery, electrolyte and electrode can be replaced independently of each other, which will result in a longer lifetime of the flow battery.

3.7 Supercapacitors

In batteries, energy is stored in a form of chemical energy using electrons directly. Energy can also be stored in the form of potential electro-static energy in capacitors, a device containing two surfaces in close proximity to each other but insulated from each other. The advantage of storing energy in (super)capacitors lies largely in the power density, i.e. the rate at which the energy can be discharged (the power that can be obtained with these devices in comparison to batteries) and in their lifetime [113]. Supercapacitors differentiate themselves from ordinary capacitors in terms of the higher energy density and lower power density.

The stored energy in a capacitor is related to the charge and the potential difference.

$$E_{\text{stored}} = \frac{1}{2} \cdot q_{\text{stored}} \cdot V = \frac{1}{2} \cdot C \cdot V^2$$

with C the capacitance of a capacitor, which depends on the permittivity, ε, (electric polarizability) of the space between the two surfaces, their area, A, and their distance apart (for a capacitor with a single dielectric medium between the plates: $C = \varepsilon \cdot \frac{A}{d}$; for a capacitor with multiple dielectric media layered between the plates: $C = \frac{A}{\sum_i \frac{d_i}{\varepsilon_i}}$ with the distance between the plates being equal to $\sum_i d_i$). Hence, the stored energy and the capacity become larger with increasing area, decreasing distance between the plates (or layers), and with increasing permittivity. Hence, materials with a high permittivity will be desired. The stored energy and capacity can be further increased by utilising interleaved plates (acting as a number of capacitors in parallel). The stored energy can be increased by increasing the voltage, but this is limited by the so-called breakdown voltage, the highest voltage at which the dielectric medium is still an insulator.

Supercapacitors are a type of an electrochemical capacitor utilising liquid electrolytes. The main type are the electrochemical double layer capacitors (EDCL), which utilise a salt, e.g. tetraethylammonium tetrafluoroborate ($TEABF_4$) or a lithium salt dissolved in an aprotic solvent [114] and typically carbon as the electrode [113]. The use of an aqueous electrolyte (e.g. containing NaOH or sulphuric acid) comes at the cost of the achievable voltage. Energy is stored via a

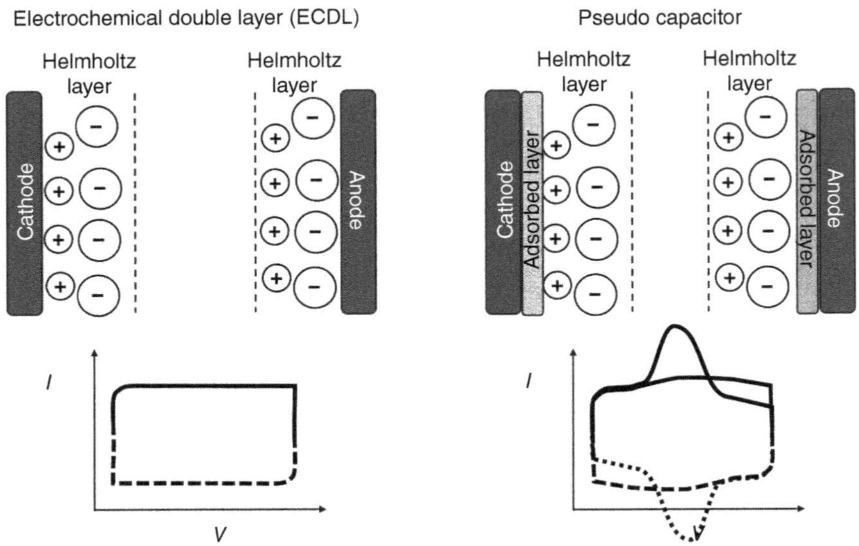

Figure 3.22 Principle of operation of supercapacitors (data from References [114, 115]).

non-Faradaic interaction, i.e. no electron transfer, facilitated by the formation of a Helmholtz-layer near the electrodes (see Figure 3.22).

The disadvantage of the electrochemical double layer capacitors is their relatively low specific energy (see Table 3.7), particularly in comparison to the specific energy of batteries (compare to Table 3.4). This can be enhanced by allowing for the Faradaic electron transfer on and near the electrode surface in the form of electro-deposition and other reversible redox reactions, resulting in a much higher charge accumulation (operating now in a similar manner to batteries).

Table 3.7 Some characteristic properties of supercapacitors [113, 114].

	EDLC	**Pseudo supercapacitor**	**Hybrid supercapacitor**
Energy storage	Non-Faradaic	Faradaic	Faradaic
Cell voltage (V)	1.2–3.3	2.2–3.3	2.2–4.8
Capacitance (F/g)	100–400[a)]		50–350[b)]
Specific energy (Wh/kg)	1.5–70[a)]	3–14	7–14
Specific power (kW/kg)	2–500[a)]	3–14	7–14
Charge time (s)	1–10	1–10	100
Self-discharge (%/month)	60	60	
Cycle life	1 000 000	100 000	500 000

a) Zhang et al. [116].
b) Poonam et al. [117].

The products of the electron-transfer reaction remain in close proximity of the electrodes thus allowing for a more rapid discharge than in batteries, but of course the amount of charge transferred is limited by the size of the electrodes. These so-called pseudo supercapacitors typically use metal oxide electrodes and show the characteristics of a capacitor depending on the mechanism of the Faradaic electron transfer (see Figure 3.22).

Hybrid supercapacitors, or asymmetric supercapacitors, utilise both, having the electrochemical double layer on one of the electrodes and the Faradaic process at the other electrode. This results in both a relatively high energy density as well as a high-power density.

3.8 Outlook

The storage of electric power has been practised for a long time to stabilise the grid, in particular using PHES and to a minor extent CAES to store energy when produced in excess and released on demand. The experience here is mainly with relatively short-term storage, where energy is stored at times of low energy consumption (typically during the night) and released during peak demand.

Energy storage will become more important upon introduction of renewable energy sources, which are intermittent in nature. This will require both short-term and long-term storage of energy as the intermittency is not only related to the short-term day–night pattern but will also fluctuate on an annual basis. The technologies described so far deal mainly with relatively short-term storage solutions as their energy density is insufficient to carry a significant load over a prolonged time. Table 3.8 shows the typical size of the energy storage technologies needed to store 4 GWh, the energy amount delivered by a 400 MW power station operating at full capacity for 10 hours (for comparison, the amount of coal required to supply this energy would be ca. 470 tonne @ 8.5 kWh/kg and the amount of natural gas required would be 260 tonne @ 15.4 kWh/kg). PHES requires large volumes of water at suitable geographical locations, which are not necessarily readily available, and energy stored in these facilities will be best suited for relatively short-term fluctuations balancing out supply with demand at a relatively short notice. This can, in principle, also be obtained using CAES, although the efficiency of this method can be substantially lower than that of PHES, and this also requires relatively specific geology or proximity to appropriate disused facilities. Energy storage using flywheel technology is currently a niche application, and this technology will only be applied for short-term energy storage despite the favourable energy density as the self-discharge rate is relatively high. Batteries can have a similar energy density to flywheels and will be competing with this technology for relatively short-term energy storage. Energy is stored more efficiently in flow batteries than PHES, and the required volume to store 4 GWh ($1.2 \cdot 10^5$ m^3) would correspond to a pool with the dimensions $75 \times 75 \times 20$ m. Supercapacitors could supply a quick response to changes in the energy demand and will thus be applied for very short-term energy storage.

Table 3.8 Characteristic sizes of energy storage facilities to store 4 GWh of energy.

Technology	Size	Comment
Coal	470 tonne	Specific energy 8.5 kWh/kg
Natural gas	260 tonne	Specific energy 15.4 kWh/kg
Pumped hydro-energy storage (PHES)	$1.5 \cdot 10^7$ m^3	Assuming 100 m height difference
Compressed air energy storage (CAES)	$1.9 \cdot 10^6$ m^3	Scaled according to McIntosh plant [9]
Flywheel	$1.2 \cdot 10^4$ tonne	Energy density: 1200 kJ/kg; would require 160 000 flywheels @ 25 kWh each
Li-ion battery	$1.6 \cdot 10^4$ tonne	Assumed energy density: 250 Wh/kg
V-flow battery	$1.2 \cdot 10^5$ m^3	V-battery @ 1.26 V ($[V] = 1$ M)
Supercapacitor	$4.0 \cdot 10^5$ tonne	Assumed capacity: 10 Wh/kg

Energy storage needs to fulfil several criteria, viz. high efficiency of energy storage, high energy density, high power density and low cost of energy storage. The various technologies are often compared based on the so-called Ragone plot, i.e. a plot of the energy density as a function of the power density, which is an appropriate comparison for battery technology. The normalisation per unit mass makes a comparison with technologies, such as PHES, more difficult. An alternative way is to plot the energy density as a function of the typical power of the various technologies (see Figure 3.23). PHES are typically large constructions able to deliver a large amount of power, but the energy density is rather low, implying that a large amount of water will need to be used to store a large amount of energy. CAES has a much higher energy density [118], although the power of the few installations is smaller than that of PHES. Flywheels and batteries can achieve similar energy density, but the typical power obtained by batteries is much smaller. Flow batteries may be able to scale in terms of power density with a constant energy density for a given pair of electrolytes. Supercapacitors have an energy density and a power, which are on the lower end of the batteries.

Cost efficiency will have to be weighed with other factors such as the speed of discharge, the frequency of discharge and the duration of discharge [119] as well as availability, as different technologies have different capabilities and different requirements. Hence, the different technologies are typically evaluated based on their utilisation [119]. For instance, PHES and CAES typically involve large systems (>5 MW), which do not have an immediate response, and are thus more suited for intermediate to long-term energy storage. Flywheels and supercapacitors are characterised by short discharge times, but a rapid response, and are thus well-suited to counter immediate imbalances of the demand-supply on the grid.

Blakers and coworkers [120] calculated the LCOS for a 1 GW PHES using a closed-loop system assuming a discount rate of 5% and a project lifetime of 60 years to be between US$ 30 and 90/MWh depending on the site-specific parameters, such

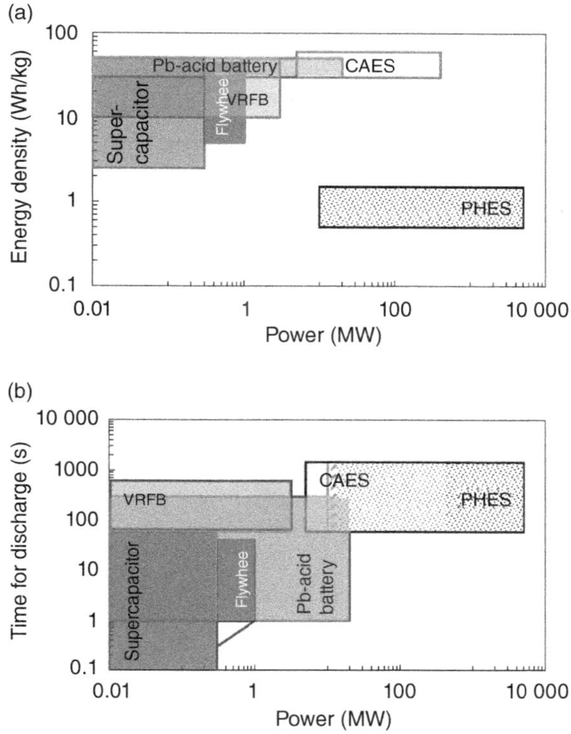

Figure 3.23 Comparison of the various energy storage technologies. (a) Energy density as a function of power (data from Moradi-Shahrbabak and Jadidoleslam [118]). (b) Time of discharge as a function of power (according to May et al. [45]).

as storage capacity, available head, slope, etc. As well as the site-specific capital cost, viz. cost of forming the dam wall, cost of the tunnel to transport water and the construction cost of the powerhouse, dominate the LCOS contributing between 60% and 75% to the overall LCOS. The value is similar to the equivalent LCOS of US$ 110/MWh quoted by Jülch [121] and Viswanathan et al. [122], although Schmidt et al. [119] reported a much larger range: US$ 150–420/MWh (see Table 3.9).

The LCOS for CAES and LAES depends not only on the cost of the purchased power, but also on the cost of the additional fuel, e.g. natural gas [123], although the majority of the cost is still related to the cost of the compressors and turbines [121]. The quoted LCOS ranges from US$ 240 to 400/MWh [121, 123, 124], although Viswanathan et al. [122] quote an optimistic US$ 100/MWh. Schmidt et al. [119] showed a much larger range of US$ 305–865/MWh when used as a secondary response, i.e. addressing imbalances in the power generation system (typically of longer duration).

The investment cost of flywheel energy storage depends on the amount of power output and the amount of energy stored [27], but is dominated by the initial investment cost and the cost of energy to charge the flywheel [119]. The levelised cost of this short-term storage has been estimated to be in the range of

Table 3.9 Typical storage capacity (MW), the investment cost and the levelised cost of energy storage (data from Schmidt et al. [119]).

Technology	Storage capacity (MW)	Investment cost ($US\$_{2015}$/kWh)	LCOS[a)] ($US\$_{2018}$/kWh)
Pumped hydro-energy storage (PHES)	10–5000	80 (±63%)	150–420
Compressed air energy storage (CAES)	5–400	39 (±58%)	305–865
Flywheel	0.01–20	5400 (±67%)	212–1263
Li-ion battery	0.001–35	802 (±24%)	215–535
Lead-acid battery	0.005–100	471 (±38%)	785–2535[b)]
V-flow battery	0.02–50	760 (±17%)	195–400
Supercapacitor	<4	13 560 (±19%)	1000–2700

a) Levelised cost of energy storage predicted for 2020 based on its use as a secondary response (i.e. addressing imbalances in power generation for a duration of 0.25–24 hours).
b) Should be put in perspective with US$ 175–210/MWh quoted by [121].

US$ 120–250/MWh [125]. The lower value quoted by Schmidt et al. [119] (US$ 212/MWh) is within this range, but it indicates a much larger variability in the LCOS.

The levelised cost for batteries varies widely due to the rather large variation in the capital expenditure. Roughly, the cost of lead-acid batteries come at ca. US$ 175–210/MWh, and Li-type of batteries between US$ 250 and 410/MWh [121], thus rendering batteries as the more expensive type of energy storage facilities. Schmidt et al. [119] give similar LCOS for Li-type of batteries of US$ 215–535/MWh, but quote a surprisingly high value for energy storage in lead-acid batteries with a very high investment cost. Vanadium redox flow batteries are still in the experimental stages, but their costs are expected to lie in the higher range of the Li-type of batteries, because of the higher capital expenditure, which is only partly mitigated by the longer lifetime [121]. Schmidt et al. [119] indicate a LCOS for energy storage using vanadium redox flow batteries between US$ 195 and 400/MWh.

Supercapacitors have a high capital cost per kWh of stored energy with quoted capital costs of US$ 10 000/kWh [114] till US$ 14 000/kWh [119], but they have a low operational cost. Nevertheless, the LCOS for supercapacitors is rather high (see Table 3.9).

The cost of energy storage is not insignificant, and there is a major drive towards reducing the cost of energy storage. Schmidt et al. [119] attempted a forecast on the dominance of various energy storage technologies going forward based on a learning or experience curve extrapolation of the price in a limited period as a function of the installed capacity. Care should be taken with this type of extrapolation in the absence of clear technological indicators as an increased installed capacity does not necessarily result in a proportional decrease in the cost. Nevertheless, it is anticipated that the mentioned technologies will be further developed for relatively short-term

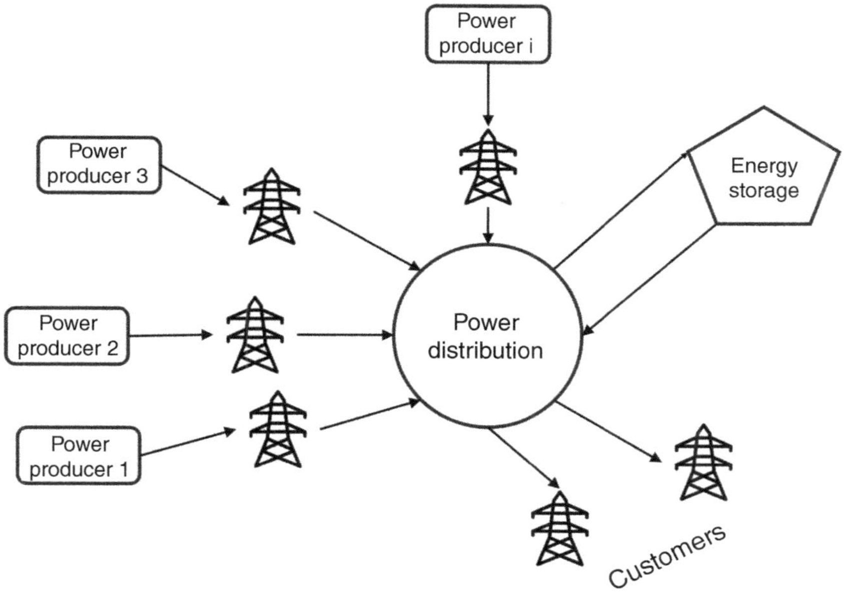

Figure 3.24 Sketch of interplay between power generation, power distribution and transmission and energy storage.

energy storage, whereas long-term (seasonal) energy storage will take the form of higher energy density materials, such as hydrogen.

The cost of energy storage is typically added to the cost of power generation using renewable energy to account for its intermittent nature. This will allow benchmarking the cost of various power generation technologies to deliver power at any time of the day. However, the objective for the introduction of renewable energy is to reduce the carbon footprint of power generation at the most affordable price level. This objective cannot be obtained by only looking at the levelised cost of electricity (LCOE) with the addition of LCOS to the cost of renewable energy, as the cost of CO_2 emissions is not (properly) accounted for in determining the LCOE.

This could be addressed by adding the cost of energy storage to the cost of power distribution (see Figure 3.24). The primary objective of the distribution system is to supply power to customers obtained from various producers at the most cost-effective manner, but at the same time other objectives, such as a target carbon footprint of the distributed power, can be built in from the distribution side. The distribution side has historically played the role of matching power supply with demand. Having access to independent suppliers for energy storage would be a cost to a distribution system but allowing it to obtain its secondary objective to reduce amount of CO_2 emitted per unit of power delivered. This would put the cost of decarbonising the power grid not only on the renewable energy producers, but at the power system at large.

References

1 Nikolaos, P., Fafalakis Marios, F., and Dimitris, K. (2023). A review of pumped hydro storage systems. *Energies* 16 (11): 4516.

2 Office of the Department of Energy (US) (2016). Hydropower Vision. Hydropower Vision Report: Full Report. Available online: https://www.energy.gov/sites/prod/files/2016/10/f33/Hydropower-Vision-Full-Report-10212016.pdf (accessed 20 April 2023).

3 Hunt, J., Zakeri, B., Lopes, R. et al. (2020). Existing and new arrangements of pumped-hydro storage plants. *Renewable and Sustainable Energy Reviews* 129: 109914.

4 Matos, C., Carneiro, J.F., and Silva, P.P. (2019). Overview of large-scale underground energy storage technologies for integration of renewable energies and criteria for reservoir identification. *Journal of Energy Storage* 21: 241–258.

5 Blakers, A., Stocks, M., Lu, B., and Cheng, C. (2021). A review of pumped hydro energy storage. *Progress in Energy* 3: 022003.

6 Zhu, B., Wang, X., Tan, L. et al. (2015). Optimization design of a reversible pump–turbine runner with high efficiency and stability. *Renewable Energy* 81: 366–376.

7 Jukić, D.-K., Kugi, A., and Kemmetmüller, W. (2023). Optimal dynamic operation of pumped storage power plants with variable and fixed speed generators. *Control Engineering Practice* 138: 105601.

8 IEA (2021). Pumped storage hydropower storage capability by countries, 2020-2026, IEA [Online]. Available: https://www.iea.org/data-and-statistics/charts/pumped-storage-hydropower-storage-capability-by-countries-2020-2026 (accessed 10 July 2024).

9 Budt, M., Wolf, D., Span, R., and Yan, J. (2016). A review on compressed air energy storage: basic principles, past milestones and recent developments. *Applied Energy* 170: 250–268.

10 Aruta, G., Ascione, F., Bianco, N., and Mauro, G. (2022). Optimization of a diabatic compressed air energy storage coupled with photovoltaics for buildings: CO_2-eq emissions vs payback time. *Energy Reports* 8: 12686–12698.

11 Geissbühler, L., Becattini, V., Zanganeh, G. et al. (2018). Pilot-scale demonstration of advanced adiabatic compressed air energy storage, part 1: plant description and tests with sensible thermal-energy storage. *Journal of Energy Storage* 17: 129–139.

12 Ali, H., Rehman, T., Arici, M. et al. (2024). Advances in thermal energy storage: fundamentals and applications. *Progress in Energy and Combustion Science* 100: 101109.

13 Van de Ven, J. and Li, P. (2009). Liquid piston gas compression. *Applied Energy* 86 (10): 2183–2191.

14 Qin, C. and Loth, E. (2014). Liquid piston compression efficiency with droplet heat transfer. *Applied Energy* 114: 539–550.

15 Marongiu, F., Soprani, S., and Engelbrecht, K. (2019). Modeling of high temperature thermal energy storage in rock beds – experimental comparison and parametric study. *Applied Thermal Engineering* 163: 114355.

16 Kim, Y., Shin, D., and Favrat, D. (2011). Operating characteristics of constant-pressure compressed air energy storage (CAES) system combined with pumped hydro storage based on energy and exergy analysis. *Energy* 36 (10): 6220–6233.

17 Wang, Z., Carriveau, R., Ting, D.S.-K. et al. (2019). A review of marine renewable energy storage. *International Journal of Energy Research* 43: 6108–6150.

18 Guo, C., Li, C., Zhang, K. et al. (2021). The promise and challenges of utility-scale compressed air energy storage in aquifers. *Applied Energy* 286: 116513.

19 Han, Y., Ma, H., Yang, C. et al. (2021). The mechanical behavior of rock salt under different confining pressure unloading rates during compressed air energy storage (CAES). *Journal of Petroleum Science and Engineering* 196: 107676.

20 Bouman, E., Øberg, M., and Hertwich, E. (2016). Environmental impacts of balancing offshore wind power with compressed air energy storage (CAES). *Energy* 95: 91–98.

21 Ma, L., Zhang, X., Zhang, T. et al. (2024). Design and operation of an adiabatic compressed air energy storage system incorporating a detailed heat exchanger model. *Energy* 304: 132104.

22 Chino, K. and Araki, H. (2000). Evaluation of energy storage method using liquid air. *Heat Transfer - Asian Research* 29 (5): 347–357.

23 Xue, X., Wang, S., Zhang, X. et al. (2015). Thermodynamic analysis of a novel liquid air energy storage system. *Physics Procedia* 67: 733–738.

24 Borri, E., Tafone, A., Ramagnoli, A., and Comodi, G. (2021). A review on liquid air energy storage: history, state of the art and recent developments. *Renewable and Sustainable Energy Reviews* 137: 110572.

25 Amiryar, M. and Pullen, K. (2017). A review of flywheel energy storage system technologies and their applications. *Applied Sciences* 7 (3): 286.

26 Li, X. and Palazzolo, A. (2022). A review of flywheel energy storage systems: state of the art and opportunities. *Journal of Energy Storage* 46: 103576.

27 Choudhury, S. (2021). Flywheel energy storage systems: a critical review on technologies, applications and future prospects. *International Transactions on Electrical Energy Systems* 31 (9): e13024.

28 Rahman, M., Gemechu, E., Oni, A., and Kumar, A. (2021). Energy and environmental footprints of flywheels for utility-scale energy storage applications. *e-Prime Advances in Electrical Engineering, Electronics and Energy* 1: 100020.

29 Beacon Power (2018). Beacon Power - Operating plants [Online]. Available: https://beaconpower.com/operating-plants (accessed 6 August 2024).

30 Power Technology (2021). Clear Creek Flywheel Energy Storage System, Canada, 28 August 2021 [Online]. Available: https://www.power-technology.com/marketdata/clear-creek-flywheel-energy-storage-system-canada/?cf-view (accessed 6 August 2024).

31 Power Technology (2021). WEB Aruba / Temporal Power Phase 1 – Flywheel Energy Storage System, Aruba, 02 September 2021 [Online]. Available: https://www.power-technology.com/marketdata/web-aruba-temporal-power-phase-1-flywheel-energy-storage-system-aruba/?cf-view (accessed 6 August 2024).

32 WEB Aruba (2024). WEB Aruba – Technologies used [Online]. Available: https://webaruba.com/energy/technologies-used (accessed 6 August 2024).

33 Aachener Zeitung (2015). Stadtwerke München vermarkten Jülicher Speicher, 9 November 2015 [Online]. Available: http://www.aachener-zeitung.de/lokales/juelich/stadtwerke-muenchen-vermarkten-juelicher-speicher-1.1220620 (accessed 6 August 2024).

34 Hobson, M. (2016). EnergyWire, *E&E News*, 6 October 2016 [Online]. Available: https://www.eenews.net/articles/on-kodiak-island-flywheels-are-in-and-diesel-is-99-8-out (accessed 6 August 2024).

35 Mitteldeutsche Rundfunk (2021). TU Dresden builds giant flywheel storage facility, 21 September 2021 [Online]. Available: https://www.mdr.de/wissen/naturwissenschaften-technik/tu-dresden-baut-neuartigen-energiespeicher-100.html (accessed 6 August 2024].

36 Theron-Ord, A. (2016). Smart Energy International, 26 February 2016 [Online]. Available: https://www.smart-energy.com/regional-news/north-america/pge-signs-agreement-20mw-flywheel-energy-storage-system (accessed 6 August 2024).

37 Dickinson, E. and Wain, A. (2020). The Butler-Volmer equation in electrochemical theory: origins, value, and practical application. *Journal of Electroanalytical Chemistry* 872: 114145.

38 Rohman, N., Rahman, N., and Zainuri, M. Characteristics of lead-acid and nickel metal hydride batteries in uninterruptible power supply operation. *International Journal of Power Electronics and Drive System (IJPEDS)* 10 (3): 1520–1528.

39 Doerffel, D. and Sharkh, S. (2006). A critical review of using the Peukert equation for determining the remaining capacity of lead-acid and lithium-ion batteries. *Journal of Power Sources* 155 (2): 395–400.

40 Quan, D., Ogliari, E., Grimaccia, F. et al. (2013). Hybrid model for hourly forecast of photovoltaic and wind power. *2013 IEEE International Conference on Fuzzy Systems (FUZZ-IEEE)*, Hyderabad, India.

41 Scanlon, D., Kehoe, A., Watson, G. et al. (2011). Nature of the band gap and origin of the conductivity of PbO_2 revealed by theory and experiment. *Physical Review Letters* 107: 246402.

42 Wang, F., Hu, C., Zhou, M. et al. (2016). Research progresses of cathodic hydrogen evolution in advanced lead–acid batteries. *Science Bulletin* 61 (6): 451–458.

43 Unterreiner, L., Julich, V., and Welter, S. (2016). Recycling battery technologies - ecological impact analysis using lice cycle assessment (LCA). *Energy Procedia* 99: 229–234.

44 Bullock, K. (1991). The electromotive force of the lead-acid cell and its half-cell potentials. *Journal of Power Sources* 35: 197–223.

45 May, G., Davidson, A., and Monahov, B. (2018). Lead batteries for utility energy storage: a review. *Journal of Energy Storage* 15: 145–157.

46 Rand, D. and Moseley, P. (2015). Chapter 13 – Energy storage with lead–acid batteries. In: *Electrochemical Energy Storage for Renewable Sources and Grid Balancing* (ed. P. Moseley and J. Garche), 201–222. Elsevier.

47 http://BatteryGuy.com (2024). Battery glossary – discharge voltage, 3 May 2024 [Online]. Available: https://batteryguy.com/kb/knowledge-base/battery-glossary-discharge-voltage (accessed 8 August 2024).

48 Power Sonic (2011). Technical manual sealed lead acid batteries [Online]. Available: https://www.power-sonic.com/wp-content/uploads/2018/12/Technical-Manual.pdf (accessed 8 August 2024).

49 Hariprakash, B., Mane, A.U., Martha, S.K. et al. (2004). A low-cost, high energy-density lead acid battery. *Electrochemical and Solid-State Letters* 7 (3): A66–A69.

50 Fox, S. (2013). How does depth of discharge factor into grid connected battery systems?, Greentech - Renewables, September 2013 [Online]. Available: https://www.greentechrenewables.com/article/how-does-depth-discharge-factor-grid-connected-battery-systems (accessed 1 August 2024).

51 Ruetschi, P. (2004). Aging mechanisms and service life of lead–acid batteries. *Journal of Power Sources* 127: 33–44.

52 Burbank, J. (1959). The anodic oxides of lead. *Journal of the Electrochemical Society* 106 (5): 369.

53 Catherino, H., Feres, F., and Trinidad, F. (2004). Sulfation in lead–acid batteries. *Journal of Power Sources* 129: 113–120.

54 White, E., Singer, S., Veronica Augustyn, V. et al. (2012). In situ transmission electron microscopy of lead dendrites and lead ions in aqueous solution. *ACS Nano* 6 (7): 6308–6317.

55 US Geological Survey (2024). *Mineral Commodity Summaries 2024*. US Geological Survey.

56 Madian, M., Eychmüller, A., and Giebeler, L. (2018). Current advances in TiO_2-based nanostructure electrodes for high performance lithium ion batteries. *Batteries* 4 (1).

57 Yun, L., Linh, D., Shui, L. et al. (2018). Metallurgical and mechanical methods for recycling of lithium-ion battery pack for electric vehicles. *Resources, Conservation and Recycling* 136: 198–208.

58 Wulandari, T., Fawcett, D., Majumder, S., and Poinern, G. (2023). Lithium-based batteries, history, current status, challenges, and future perspectives. *Battery Energy* 2 (8): 20230030.

59 Basch, A., de Campo, L., Albering, J., and White, J. (2014). Chemical delithiation and exfoliation of LixCoO2. *Journal of Solid State Chemistry* 220: 102–110.

60 Scrosati, B. and Garche, J. (2010). Lithium batteries: status, prospects and future. *Journal of Power Sources* 195 (9): 2419–2430.

61 Wang, X., Gaustad, G., and Babbitt, C. (2016). Targeting high value metals in lithium-ion battery recycling via shredding and size-based separation. *Waste Management* 51: 204–213.

62 Panasonic Corporation of North America (2010). NCR18650B Datasheet Panasonic Specifications [Online]. Available: https://www.orbtronic.com/content/NCR18650B-Datasheet-Panasonic-Specifications.pdf (accessed 20 October 2025).

63 Wang, G., Bi, Z., Zhang, A. et al. (2024). High-voltage and fast-charging lithium cobalt oxide cathodes: from key challenges and strategies to future perspectives. *Engineering* 37: 105–127.

64 Smith, A., Burns, J., Trussler, S., and Dahn, J. (2010). Precision measurement of the coulombic efficiency of lithium-ion batteries and of electrode materials for lithium-ion batteries. *Journal of the Electrochemical Society* 157 (2): A196–A202.

65 Lin, Z., Li, D., and Zou, Y. (2023). Energy efficiency of lithium-ion batteries: influential factors and long-term degradation. *Journal of Energy Storage* 74 (B): 109386.

66 Wu, Q., Zhang, B., and Lu, Y. (2022). Progress and perspective of high voltage lithium cobalt oxide in lithium-ion batteries. *Journal of Energy Chemistry* 74: 283–308.

67 Omar, N., Van den Bossche, P., Coosemans, T., and Van Mierlo, J. (2013). Peukert revisited—critical appraisal and need for modification for lithium-ion batteries. *Energies* 6: 5625–5641.

68 Gold Peak Industries (Taiwan), Ltd. (2007). Lithium ion technical handbook [Online]. Available: https://web.archive.org/web/20071007175038/http:/http://www.gpbatteries.com/html/pdf/Li-ion_handbook.pdf (accessed 28 January 2024).

69 Ji, Y., Zhang, Y., and Wang, C.Y. (2013). Li-ion cell operation at low temperatures. *Journal of the Electrochemical Society* 160: A363–A649.

70 Chen, Z., Danilov, D., Raijmakers, L. et al. (2021). Overpotential analysis of graphite-based Li-ion batteries seen from a porous electrode modeling perspective. *Journal of Power Sources* 509: 230345.

71 Zhang, G., Cao, L., Ge, S. et al. (2014). In situ measurement of radial temperature distributions in cylindrical Li-ion cells. *Journal of the Electrochemical Society* 161 (10): A1499–A1507.

72 Choi, S., Kim, J., and Yoon, Y. (2004). Self-discharge analysis of $LiCoO_2$ for lithium batteries. *Journal of Power Sources* 138: 283–287.

73 Energy Institute Statistical Review of World Energy (2023). Statistical review of world energy, The Energy Institute [Online]. Available: https://www.energyinst.org/statistical-review/about (accessed 7 February 2024).

74 Trading Economics (2024). Lithium [Online]. Available: https://tradingeconomics.com/commodity/lithium (accessed 15 August 2024).

75 Lee, B. (2021). The six main types of lithium-ion batteries, Trek [Online]. Available: https://tritekbattery.com/the-six-main-types-of-lithium-ion-batteries (accessed 15 August 2024).

76 Lin, C., Li, J., Yin, Z.-W. et al. (2024). Structural understanding for high-voltage stabilization of lithium cobalt oxide. *Advanced Materials* 36 (6): 2307404.

77 Saldaña, G., San Martin, J., Zamora, I. et al. (2019). Analysis of the current electric battery models for electric vehicle simulation. *Energies* 12: 2750.

78 Loganathan, M., Mishra, B., Tan, C. et al. (2021). Multi-criteria decision making (MCDM) for the selection of Li-ion batteries used in electric vehicles (EVs). *Materials Today: Proceedings* 41 (5): 1073–1077.

79 Yang, H.-L., Zhang, B.-W., Konstantinov, K. et al. (2021). Progress and challenges for all-solid-state sodium batteries. *Advanced Energy and Sustainability Research* 2 (2): 2000057.

80 Abraham, K. (2020). How comparable are sodium-ion batteries to lithium-ion counterparts? *ACS Energy Letters* 5 (11): 3544–3547.

81 Swiderska-Mocek, A., Jakobczyk, P., Rudnicka, E., and Lewandowski, A. (2020). Flammability parameters of lithium-ion battery electrolytes. *Journal of Molecular Liquids* 318: 113986.

82 Mossali, E., Picone, N., Gentilini, L. et al. (2020). Lithium-ion batteries towards circular economy: a literature review of opportunities and issues of recycling treatments. *Journal of Environmental Management* 264: 110500.

83 Tan, S., Shadike, Z., Cai, X. et al. (2023). Review on low-temperature electrolytes for lithium-ion and lithium metal batteries. *Electrochemical Energy Reviews* 6: 35.

84 Machín, A., Morant, C., and Márquez, F. (2024). Advancements and challenges in solid-state battery technology: an in-depth review of solid electrolytes and anode innovations. *Batteries* 10 (1): 29.

85 Xiao, Y., Wang, Y., Bo, S.-H. et al. (2020). Understanding interface stability in solid-state batteries. *Nature Reviews Materials* 5: 105–126.

86 Brée, L. and Mitsos, A. (2020). Two-tank multi compartment redox flow battery. *Journal of Energy Storage* 29: 101412.

87 Aaron, D., Tang, Z., Papandrew, A., and Zawodzinski, T. (2011). Polarization curve analysis of all-vanadium redox flow batteries. *Journal of Applied Electrochemistry* 41: 1175–1182.

88 Wang, S., Xu, S., Wu, X. et al. (2020). Analyses and optimization of electrolyte concentration on the electrochemical performance of iron-chromium flow battery. *Applied Energy* 271: 115252.

89 Jeena, C., Elsa, P., Moly, P. et al. (2022). A dendrite free Zn-Fe hybrid redox flow battery for renewable energy storage. *Energy Storage* 4: e275.

90 Chen, H., Zhang, X., Zhang, S. et al. (2022). A comparative study of iron-vanadium and all-vanadium flow battery for large scale energy storage. *Chemical Engineering Journal* 429: 132403.

91 Gong, K., Xu, F., Grunewald, J. et al. (2016). All-soluble all-iron aqueous redox-flow battery. *ACS Energy Letters* 1: 89–93.

92 Li, Y., Kienbaum, D., Lüth, T., and Skyllas-Kazacos, M. (2024). Long term performance evaluation of a commercial vanadium flow battery system. *Journal of Energy Storage* 90 (A): 111790.

93 Herr, T., Noack, J., Fischer, P., and Tübke, J. (2013). 1,3-Dioxolane, tetrahydrofuran, acetylacetone and dimethyl sulfoxide as solvents for non-aqueous vanadium acetylacetonate redox-flow-batteries. *Electrochimica Acta* 113: 127–133.

94 Ejigu, A., Greatorex-Davies, P., and Walsh, D. (2015). Room temperature ionic liquid electrolytes for redox flow batteries. *Electrochemistry Communications* 54: 55–59.

95 Bahadori, L.L., Hashim, M., Manan, N. et al. (2016). Investigation of ammonium- and phosphonium-based deep eutectic solvents as electrolytes for a non-aqueous all-vanadium redox cell. *Journal of the Electrochemical Society* 165 (5): A632–A638.

96 Feng, T., Wang, H., Liu, Y. et al. (2019). A redox flow battery with high capacity retention using 12-phosphotungstic acid/iodine mixed solution as electrolytes. *Journal of Power Sources* 436: 226831.

97 Ruan, W., Mao, J., Yang, S. et al. (2020). Designing Cr complexes for a neutral Fe–Cr redox flow battery. *Chemical Communications* 56: 3171–3174.

98 Zhang, Y., Henkensmeier, D., Kim, S. et al. (2019). Enhanced reaction kinetics of an aqueous Zn–Fe hybrid flow battery by optimizing the supporting electrolytes. *Journal of Energy Storage* 25: 100883.

99 Cunha, A., Martins, J., Rodrigues, N., and Brito, F. (2015). Vanadium redox flow batteries: a technology review. *International Journal of Energy Research* 39 (7): 889–918.

100 Lu, W., Li, X., and Zhang, H. (2018). The next generation vanadium flow batteries with high power density - a perspective. *Physical Chemistry Chemical Physics* 20: 23–35.

101 Skyllas-Kazacos, M. (2022). Review — highlights of UNSW all-vanadium redox battery development: 1983 to present. *Journal of the Electrochemical Society* 169 (7): 070513.

102 Haynes, W. (2016). *CRC Handbook of Physics and Chemistry*. Boca Raton: CRC Press.

103 Prem Kumar, T., Prabhu, P., Srivastava, A. et al. (1994). Conductivity and viscosity studies of dimethyl sulfoxide (DMSO)-based electrolyte solutions at 25 °C. *Journal of Power Sources* 50 (3): 283–294.

104 Arevalo-Cid, P., Mendes, P.D.A., and Azevedo, J. (2021). Redox flow batteries: a new frontier on energy. *Sustainable Energy & Fuels* 5: 5366–5419.

105 Kim, K.-S., Shin, B.-K., and Lee, H. (2004). Physical and electrochemical properties of 1-butyl-3-methylimidazolium bromide, 1-butyl-3-methylimidazolium iodide, and 1-butyl-3-methylimidazolium tetrafluoroborate. *Korean Journal of Chemical Engineering* 21 (5): 1010–1014.

106 Mihăilă, E., Aruxande, D., Doncea, S.O.F., and Dincă, C. (2021). Deep eutectic solvents for CO_2-capture in post-combustion processes. *Studia UBB Chemia* 66 (2): 233–246.

107 Agieienko, V. and Buchner, R. (2022). Is ethaline a deep eutectic solvent? *Physical Chemistry Chemical Physics* 24: 5265–5268.

108 Sinclair, N., Shen, X., and Wainright, J. (2021). Electrochemical decomposition of ethylene glycol-choline chloride deep eutectic solvent. ECS Meeting, A01: New Approaches and Advances in Electrochemical Energy Systems, 2021.

109 Costa, R., Pereira, C., and Silva, F. (2010). Double layer in room temperature ionic liquids: influence of temperature and ionic size on the differential capacitance and electrocapillary curves. *Physical Chemistry Chemical Physics* 12: 11125–11132.

110 Hansen, B., Spittle, S., Chen, B. et al. (2021). Deep eutectic solvents: a review of fundamentals and applications. *Chemical Reviews* 121: 1232–1285.

111 Yu, L., Lin, F., Xu, L., and Xi, J. (2017). Structure–property relationship study of Nafion XL membrane for high-rate, long-lifespan, and all-climate vanadium flow batteries. *RSC Advances* 7: 31164–31172.

112 Moore, M., Counce, R., Watson, J. et al. (2012). A step by step design methodology for an all-vanadium redox-flow battery. *Chemical Engineering Education* 46 (4): 239.

113 He, X. and Zhang, X. (2022). A comprehensive review of supercapacitors: properties, electrodes, electrolytes and thermal management systems based on phase change materials. *Journal of Energy Storage* 56 (C): 106023.

114 Libich, J., Máca, J., Vondrák, J. et al. (2018). Supercapacitors: properties and applications. *Journal of Energy Storage* 17: 224–227.

115 Fleischmann, S., Mitchell, J., Wang, R. et al. (2020). Pseudocapacitance: from fundamental understanding to high power energy storage materials. *Chemical Reviews* 120 (14): 6738–6782.

116 Zhang, X., Zhang, H., Li, C. et al. (2014). Recent advances in porous graphene materials for supercapacitor applications. *RSC Advances* 4: 45862–45884.

117 Poonam, K., Sharma, A.A., and Tripathi, S. (2019). Review of supercapacitors: materials and devices. *Journal of Energy Storage* 21: 801–825.

118 Moradi-Shahrbabak, Z. and Jadidoleslam, M. (2023). A new index for techno-economical comparison of storagetechnologies considering effect of self-discharge. *IET Renewable Power Generation* 17: 1699–1712.

119 Schmidt, O., Melchior, S., Hawkes, A., and Staffell, I. (2019). Projecting the future levelized cost of electricity storage technologies. *Joule* 3: 81–100.

120 Stocks, M., Stocks, R., Lu, B. et al. (2021). Global atlas of closed-loop pumped hydro energy storage. *Joule* 5: 270–284.

121 Jülch, V. (2016). Comparison of electricity storage options using levelized cost of storage (LCOS) method. *Applied Energy* 183: 1594–1606.

122 Viswanathan, V., Mongird, K., Franks, R., and Baxter, R. (2022). *2022 Grid Energy Storage Technology Cost and Performance Assessment*. US Department of Energy.

123 Salvini, C. and Giovannelli, A. (2022). Techno-economic comparison of utility-scale compressed air and electro-chemical storage systems. *Energies* 15: 6644.

124 Salvini, C., Giovannelli, A., and Sabatello, D. (2021). Analysis of diabatic compressed air energy storage systems with artificial reservoir using the levelized cost of storage method. *International Journal of Energy Research* 45: 254–268.

125 Rahman, M., Gemechu, E., Oni, A., and Kumar, A. (2021). The development of a techno-economic model for the assessment of the cost of flywheel energy storage systems for utility-scale stationary applications. *Sustainable Energy Technologies and Assessments* 47: 101382.

4 Heating

4.1 Introduction

Only a part of the primary energy is used for electrical power generation (see Figure 4.1). In the United States, the primary energy consumption for the generation of electricity has increased from ca. 10% in 1950 to a maximum of 39% in 2009/2010 (interestingly, the use of primary energy carriers for electricity generation has subsequently declined to ca. 34% of the total energy mix in 2023) [1]. The energy use in the residential sector increased in that period by 250%, mainly due to an increase in the consumption of electricity, which, including the associated energy losses, made up ca. 65% of the energy consumption in the residential sector in 2023. A similar phenomenon can be seen in the commercial sector, whose energy consumption increased by 370% in the period 1950–2023 with electricity making up ca. 72% of the energy consumption in this sector in 2023. Energy usage in the transportation sector increased by 250% in this period; the contribution of electricity in this sector is still marginal commensurate with the estimated 2% of cars in use being electric cars in the United States (the ratio of electric cars to the total cars in use is substantially higher in countries such as Norway, Sweden and China) [2]. The total energy usage by the industry has increased by 120% in the period 1950–2023, and thus, the relative contribution of the industry to the overall energy consumption has decreased from 45% in 1950 to ca. 33% in 2023 in the United States. The use of electricity in this sector, including the losses associated with power generation, has grown from 11% to 28% of the overall energy consumption in this sector. This implies that energy is still used in industry by directly converting the primary energy carrier.

Figure 4.2 shows the energy usage within the process industry in the United States in the year 2014 [3], with electricity making up 22.5% of the total energy consumption. Most of the energy is consumed on-site utilising fuel, although steam import (with steam typically being generated by combusting primary energy carriers) can make up a substantial part of the energy consumption. Most of the energy is used to heat up processes (70% of the total energy consumption) with machinery taking only ca. 20% of the total energy consumption, stressing the importance of heating in the process industry.

Renewable Energy in the Process Industry, First Edition.
Eric van Steen, Nicholas S. Featherstone, and Linda H. Callanan.

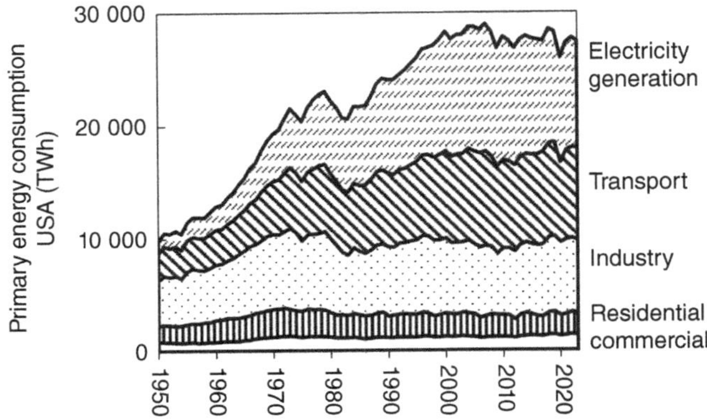

Figure 4.1 Relative primary energy consumption in the United States in the various sectors over the period 1950–2023 (US Energy Information Administration [1] / Public Domain).

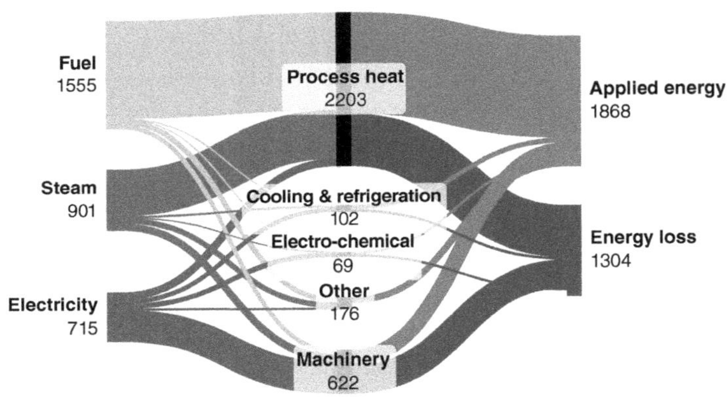

Figure 4.2 Energy usage (in GWh) as process energy in the United States in 2014 (Office of Energy Efficiency & Renewable Energy [3] / Public Domain).

Heating and cooling are used before, in and after different processes, such as distillation, evaporation, drying, extraction, cleaning/sterilisation, curing, heat treatment, melting and providing the heat for endothermic chemical reactions. Heat can be delivered directly or indirectly using a heat-transfer medium (often steam). Direct heating implies contacting the stream to be heated directly with the heat source. This can take place by passing the stream over a resistive heater, exchange heat with a hotter stream (which acts as the source of heat) or passing the stream through a furnace.

Heat, thermal energy, can only be transferred from a hotter source to a colder stream (2nd law of thermodynamics). Hence, the temperature at which the heat will need to be delivered, is important. The temperature at which various processes operate may be roughly classified into a low-temperature regime, a medium-temperature regime and a high-temperature regime (see Table 4.1). It should be noted that there is no formal differentiation between processes based on the temperature range at

Table 4.1 Classification of processes according to their operating temperature and mode of heat transfer.

Temperature range	Mode of heat transfer	Process
Ambient temperature – 250 °C	Hot water	Cooking
	Hot air	Baking
		Drying
	Steam	Laundering
		Sterilisation
		Evaporation
		Distillation
250–600 °C	High-pressure steam	Forced air drying
	Thermal oil	Distillation
	Indirect electrical heating	Moulding
	Direct firing	Heat treatment
>600 °C	Direct electrical	Arc furnace
	Direct firing	Ceramics Reforming natural gas

which they operate but this division may become useful when evaluating alternative sources of energy for heating purposes. In the low-temperature range, heat is needed for typical processes such as cooking, baking and drying, but also for processes, such as sterilisation and evaporation/distillation through the supply of low-pressure steam. In the medium-temperature range, high-pressure steam or thermal oil can be used as heat-transfer media for heating various processes. Resistive heating is currently used in medium-temperature processes such as moulding. High-temperature processes typically utilise furnaces to generate heat, using either a chemical energy source or electrical power in arc furnaces.

Figure 4.3 gives an overview of the amount of energy required in the various industries as well as the temperature range in which heat is required in processes in Europe in 2012 [4]. More than 25% of the energy is required at temperatures less than 100 °C either for space heating and supply of hot water or as process heat. Approximately 21% is required for heating in the range between 100 and 400 °C and ca. 20% for heating between 500 and 1000 °C. Heating up to temperatures larger than 1000 °C takes up almost 34% of the energy supplied. The required temperature range is strongly dependent on the type of industry, with foundries and processing of minerals requiring more than 60% of the heating energy at temperatures greater than 1000 °C. The production of construction materials requires ca. 35% of the energy at these high temperatures, probably attributable to the production of cement, but also requires process heat in the range of 100–1000 °C. The chemical industry utilises a large amount of energy for generating process heat in the range of 500–1000 °C (46%) and 100–500 °C (21%). The wood-processing (excluding

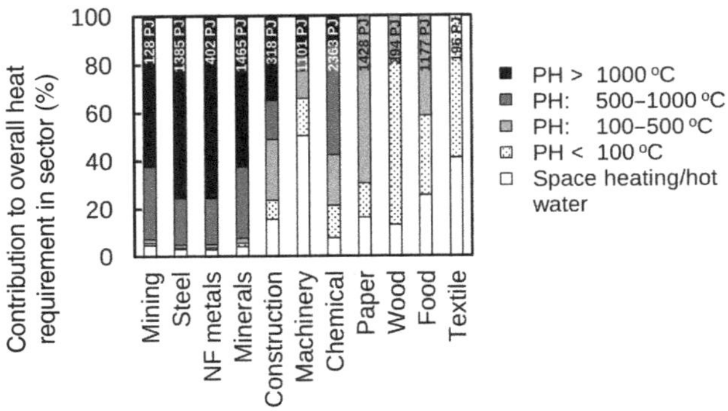

Figure 4.3 Relative amount of the various types of heat required, viz. space heating and hot water less than 100 °C, process heat (PH) at $T < 100$ °C, process heat between 100 and 500 °C, process heat (PH) between 500 and 1000 °C and process heat (PH) at $T > 1000$ °C in different industries in Europe with an indication of the total amount of energy used by these industries. NF, non-ferrous (redrawn from Naegler et al. [4]).

paper and pulp), textile and food industries use process heat at moderate to low temperatures in addition to usage for hot water and space heating.

The choice of the energy source and heat delivery medium for heating process streams is dictated by technical variables, such as required temperature, required rate of heat delivery, product requirements and process safety as well as economic drivers such as fuel availability and cost as well as capital costs.

Hot water can be produced using electric heating, passing water through a furnace, or heat exchange with process fluids (e.g. effluent). Water is a relatively readily available and safe heating medium. Its large heat capacity (see Table 4.2) makes it more useful than other heating media, although liquid water can only be used within a limited temperature range (at atmospheric pressure up to 100 °C), thus limiting the range of activities for which it can be used as a heat-transfer medium. Heat transfer to and from liquid water can be reasonably efficient as the thermal conductivity of liquid water is relatively high (0.598 W/m/°C). Typical problems associated with using hot, liquid water as a heating medium are associated with scaling and fouling of heat-transfer surfaces and corrosion of metal surfaces.

Warm air is typically used for drying activities where it supplies heat, and acts as the carrier gas to remove water (or other solvents) from a product. Warm air can be generated in a similar manner as hot liquid water, i.e. by electric heating, in a furnace or from heat exchange with other streams. In comparison to liquid water, the

Table 4.2 Specific heat (kJ/kg/°C) and thermal conductivity (W/m/°C) of typical liquid heat-transfer media at 20 °C [5–7].

Medium	Air	Water	Ammonia (3.32 bar)	Castor oil	Dowtherm Q™	Ethylene glycol	Ethanol	Kerosene
C_p (kJ/kg/°C)	1.00	4.18	4.74	1.8	1.66	2.36	2.57	2.01
k (W/m/°C)	0.026	0.598	0.521	0.180	0.122	0.258	0.167	0.145

specific heat of air is lower (1.006 kJ/kg/°C), thus requiring a larger mass flow rate and a significantly larger volumetric flow rate to transfer the same amount of heat. Furthermore, its thermal conductivity is much lower than that of liquid water (0.026 vs. 0.598 W/m/°C), thus requiring a larger heat-transfer area for the same amount of heat to be transferred.

Thermal oil (e.g. Dowtherm Q) is usually used where temperatures higher than those which can be readily attained using water are required whilst allowing operation at close to atmospheric pressure. Thermal oil typically has a lower specific heat and a lower thermal conductivity than that of liquid water (see Table 4.2). The system containing thermal oil will need to be leak tight as spillage of thermal oil poses a hazard for the operation due to flammability and environmental concern [6]. The advantage of using thermal oil over water is the lower likelihood of fouling and/or corrosion of the heat exchange system.

Steam is an efficient heat-transfer medium, as it can release a considerable amount of energy through condensation (1500–2250 kJ/kg depending on pressure). Furthermore, heat can be delivered at a fixed temperature by controlling the pressure of the steam, as condensation occurs without a temperature change. It is often produced on purpose by combustion of fuels in boilers, although cooling of reactors may also generate (superheated) steam.

Direct heating, via combustion of fossil fuels (coal, natural gas or fuel oil), is used in many medium- and high-temperature processes. This results in good heat transfer and allows for a quick response to changing conditions in the process stream. However, in these cases the combustion gases will then come into contact with the process stream, potentially resulting in its contamination. Alternatively, burners can fire into a combustion chamber, from which heat is then transferred through the walls to the material to be heated. This method is less efficient than direct heating but prevents contamination of the process stream.

4.2 Heat Pumps

Heat pumps can be used to efficiently convert electric power into heat (see Figure 4.4). They work in an opposite manner to refrigerators, where heat is transferred from the inside of the refrigerator to the surroundings. In a heat pump, heat is taken up from the surroundings by evaporating a working fluid at a low temperature and thus a low pressure. The working fluid, now in the vapour phase, is subsequently compressed adiabatically. Heat is then given off to another medium by condensing the working fluid at high pressure. The stream is subsequently expanded over a Joule-Thomson valve, resulting in the formation of a liquid–vapour mixture at low pressure and allowing for a circular process.

The performance of heat pumps is evaluated using the so-called coefficient of performance (C.O.P.), which is the ratio of the heat given off in the condensation step at the high pressure relative to the amount of work required for the compression:

$$\text{C.O.P.} = \frac{|\dot{Q}_{\text{given off}}|}{|\dot{W}_{\text{compressor}}|}$$

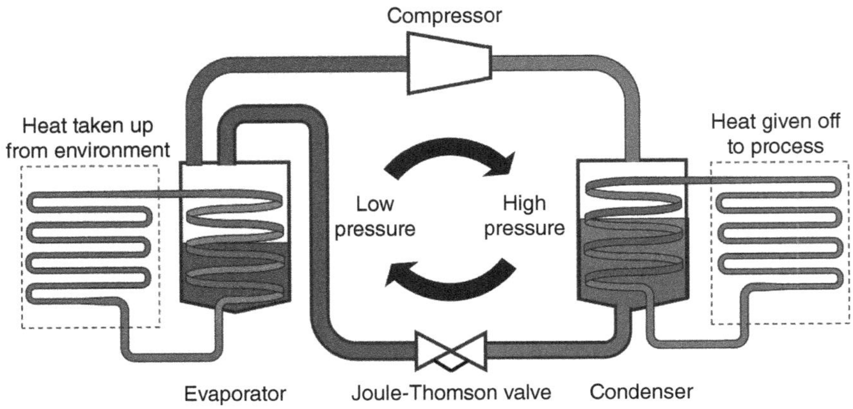

Figure 4.4 Principal operation of a heat pump.

The C.O.P. should always be greater than 1; the overall energy balance over the heat pump gives:

$$\dot{Q}_{\text{given off}} = \dot{W}_{\text{compressor}} + \dot{Q}_{\text{taken-up}}$$

Hence,

$$\text{C.O.P.} = \frac{|\dot{Q}_{\text{given off}}|}{|\dot{Q}_{\text{given off}} - \dot{Q}_{\text{taken-up}}|} = \frac{1}{1 - \left|\frac{\dot{Q}_{\text{taken-up}}}{\dot{Q}_{\text{given off}}}\right|} > 1$$

and more energy in the form of heat is given off than work is being done by the compressor.

The maximum C.O.P. is obtained in a Carnot cycle and can be expressed as:

$$\text{C.O.P.}_{\text{Carnot}} = \frac{T_{\text{high}}}{T_{\text{high}} - T_{\text{low}}}$$

with T_{high} the temperature at which heat is given off, and T_{low} the temperature at which heat is taken up.

A high value for the C.O.P. thus indicates a good utilisation of the work done in the compressor to transfer heat from the surroundings to another medium. The use of heat pumps thus amplifies the electrical energy input for heating.

The process can also be viewed on p-H diagram (see Figure 4.5) or a T-S diagram with the compressor initially operating isentropically, the isenthalpic Joule-Thomson valve, and the condensation and evaporation process operating isobarically (the heat pump is constructed such that the working fluid will leave the condenser as a saturated liquid and the evaporator as a saturated vapour). For instance, the use of ammonia as the working fluid with the evaporator operating at 0 °C and thus a pressure of 4.3 bar (a lower temperature than the temperature of the environment delivering the heat is required), and ammonia being compressed isentropically with a compression ratio of 6 up to a pressure of 25.8 bar, the C.O.P.

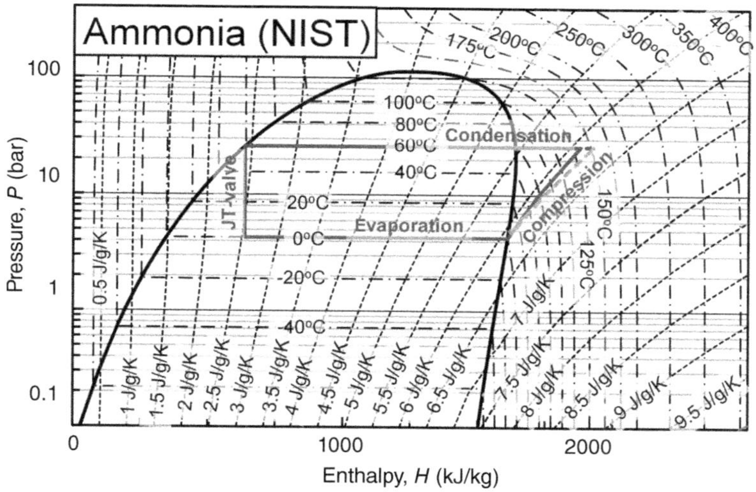

Figure 4.5 Operation of a heat pump on a p-H diagram with ammonia as the working fluid with evaporation at 0 °C and a compression ratio of 6, condensation and a Joule-Thomson expansion (dotted line represents non-ideal process with a compressor efficiency of 85%) (created with data from NIST Chemistry Webbook [8]).

can be determined as:

$$\text{C.O.P.} = \frac{|\dot{Q}_{\text{given off}}|}{|\dot{W}_{\text{compressor}}|} = \frac{|634 - 1878|}{|1878 - 1607|} = 4.60$$

implying that for every kWh of energy inputted in the form of electric work for the compressor, 4.60 kWh of heat is generated. The maximum C.O.P. that could have been achieved is $\text{C.O.P.}_{\text{Carnot}} = 5.4$.

If, for example, the efficiency of the compressor is 85%, the work required by the compressor will increase from 270 to 317 kJ/kg. Hence, the enthalpy of the stream leaving the compressor will increase from 1878 to 1925 kJ/kg, thus increasing the temperature of the stream leaving the compressor from 136 to 153 °C. This reduces the C.O.P. to 4.06.

The heat is given off to another stream upon cooling and condensation of ammonia. This occurs at a temperature ranging from 59 to 136 °C (up to 153 °C for the non-ideal compressor), thus allowing streams to be heated up to temperatures of ca. 121–138 °C in a counter-current heat exchanger assuming an approach temperature of 15 °C with ammonia leaving the condenser at temperatures below its critical point (132.4 °C). Most of the heat transfer will take place due to condensation of ammonia. Higher temperatures may be achieved by utilising higher compression ratios (see Figure 4.6) and, thus, higher pressures in the condenser of the heat pump cycle. However, this comes at the cost of a reduced C.O.P. The compression ratio is limited by the maximum temperature of the compressor. The non-ideality of the compressor will increase the maximum temperature that is achieved at the cost of a reduced C.O.P.

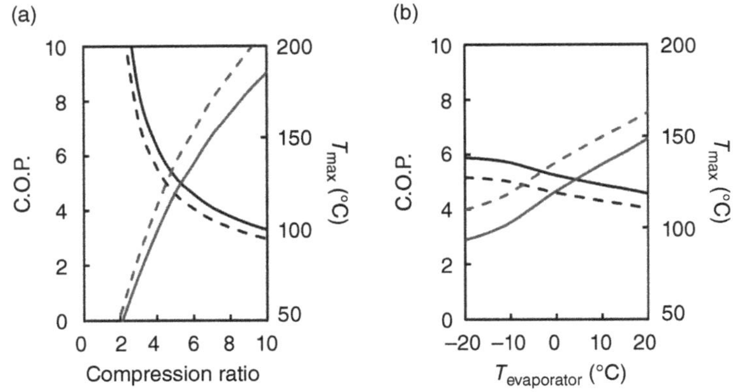

Figure 4.6 Performance of a heat pump using ammonia as the working fluid (black line: coefficient of performance (C.O.P.); grey line: temperature of the fluid leaving the compressor; solid lines: ideal compressor; dashed lines: compressor with efficiency of 85%). (a) Performance as a function of the compression ratio with a heat received in evaporator at 0 °C. (b) Performance as a function of the temperature in the evaporator with a compression ratio of 5.

The temperature in the evaporator is a critical variable determining (together with the compression ratio) the temperature leaving the compressor (and thus the temperature at which heat will be given off). Increasing the temperature in the evaporator will result in a higher temperature of the fluid leaving the compressor, but at the same time result in a lower C.O.P. for a fixed compression ratio.

The heat transferred to the working fluid in the evaporator depends on the temperature difference between the working fluid in the evaporator and the surroundings. A variation in the temperature of the surroundings will result in an increase in the pressure in the evaporator, and thus an increase in the temperature in the evaporator. This will lead to a decrease in the C.O.P. whilst at the same time increasing the temperature in the high-pressure condenser. A drop in the temperature of the surroundings will result in a lowering of the pressure in the evaporator. This is associated with a decrease in the temperature of the high-pressure condenser. The change in the temperature in the high-pressure condenser depends linearly on the temperature in the low-pressure evaporator but the scaling factor is larger than 1, implying that a change in the temperature in the evaporator will result in a larger temperature change in the condenser (for example in Figure 4.6, with ammonia as the working fluid and a compression ratio of 5, a change in the temperature in the evaporator of 10 °C will result in a change in the temperature of the condenser of 14.5 °C).

A constant temperature in the evaporator (and thus in the condenser) would be needed for implementing heat pumps in industrial processes. The use of outside air as the heat source for the heat pump can result in both large daily and large seasonal variations in the temperature of the evaporator. Whilst this is not ideal for industrial purposes, it is classically done for heating homes and for air conditioning purposes, where the daily and seasonal variations are seen as less important, resulting mainly in a change in the time to heat up or cool down an area. Hence, the so-called air source heat pumps are thought to be essential in decarbonising the residential sector upon increasing the renewable energy contribution to the power generation sector [9]. Surface and ground water have a smaller variation in temperature than the outside air and can thus be used as a heat source at a more constant temperature [10], although thermal stratification may occur in surface water [11] and the temperature will, during some periods, be lower than that of the outside air. In a closed-loop configuration, the working fluid exchanges heat with water or water containing anti-freeze which is pumped to heat exchange coils submerged in the basin containing surface water [12]. In contrast, in an open-loop configuration, water from the basin is pumped directly through the heat pump, where it exchanges heat with the working fluid in the low-pressure evaporator. Heat exchange may also take place directly with the ground in a closed-loop configuration, which will have a relatively small seasonal variation, depending on the depth. Depending on the local geography, the heat exchanger can be orientated horizontally at a depth of ca. 2 m [13], but then requires a relatively large area. Alternatively, a vertical configuration like a borehole can be employed, which also takes advantage of the increase in temperature with increasing depth into the lithosphere.

The exit temperature may also be raised by utilising a cascade of two heat pumps. Table 4.3 shows a comparison of the performance of a single heat pump and the overall performance of two heat pumps in series. In both cases, the final exit temperature is 165°, which is suitable to heat up another stream to 145–150 °C. In the example, a maximum temperature of 165 °C may be achieved with a single heat pump using ammonia as the working fluid with the evaporator working at 0 °C, with either an

Table 4.3 Comparison of the performance of a single heat pump to generate a maximum temperature of 165 °C using ammonia as a working fluid with the evaporator operating at 0 °C, the overall performance of 2 heat pumps in series with the compressors working with the same compression ratio using ammonia as the working fluid and a temperature difference between condenser of the 1st heat pump and the condenser of the 2nd heat pump of 20 °C, and the overall performance of 2 heat pumps in series using 2 different working fluids, viz. ammonia and water.

	Single heat pump		**Two heat pumps in series**			
Working fluid	**Ammonia**		**Ammonia**		**Ammonia/water**[a)]	
$\eta_{compressor}$ (%)	100	85	100	85	100	85
C.R	8.16	6.59	3.25/3.25	3.16/3.16	5.01/1.97	4.14/1.83
C.O.P.	3.74	4.26	3.20	3.40	4.25	4.83
$P_{evaporator}$ (bar)	4.3	4.3	4.3/26.5	4.3/24.8	4.3/1.0	4.3/1.0
$P_{condenser}$ (bar)	35.0	28.3	14.1/87.0	13.6/78.4	21.5/2.0	17.8/1.9
$T_{exit,compressor}$ (°C)	165	165	81/165	89/165	120/165	120/165
$T_{exit\ condenser}$ (°C)	72	63	36/117	35/112	52/120	45/118

a) The first value indicates the condition in the ammonia loop and the second indicate in the water loop.

ideal compressor having a compression ratio of 8.16, or a compressor with an efficiency of 85% having a compression ratio of 6.59.

Putting two heat pumps in series results in the required compression ratio being reduced but introduces additional complications to the process. Firstly, the use of 2 heating pumps in series requires heat exchange between the 2 simultaneously operating cycles and, as a result, the evaporator of the 2nd cycle has to operate at a temperature 10–20 °C lower than the temperature of the condenser in the 1st cycle. Secondly, the ratio of the mass flow rates through each of the heat pumps must match the ratio of the specific heat given off by the 1st heat pump to the specific heat taken up for the 2nd heat pump. Hence, the overall C.O.P. will decrease from 3.74 to 3.20 with an ideal compressor when using ammonia as the working fluid with both cycles operating at the same compression ratio noting that the mass flow rate through the 1st heat pump is then only 51% of the mass flow rate through the 2nd heat pump. Similarly, when using a compressor with an efficiency of 85%, the changes will result in a decrease of the C.O.P. from 4.26 when using a single heat pump to 3.40 with 2 heat pumps in series. A disadvantage of utilising two heat pumps, with a lower compression ratio, in series in comparison to a single heat pump, with a higher compression ratio, is the substantially higher pressure required in the 2nd heat pump cycle (see Table 4.3). Putting 2 heat pumps in series will thus result in a higher temperature of the fluid leaving the condenser in the 2nd heat pump and thus a higher-average temperature for heat transfer to the process.

A higher temperature may also be achieved by coupling two or more heat pumps operating with a different working fluid. For instance, as illustrated in Table 4.3, a heat pump operating with ammonia as the working fluid to transfer heat from a low-quality heat source could be coupled with a heat pump utilising water as a working fluid to further upgrade the heat. The 1st heat pump would need to bring up the temperature in the condenser to 120 °C taking into account a 20 °C approach temperature to be able to generate steam at a pressure of in the heat exchanger transferring the heat between the 2 heat pumps to facilitate generating steam at 1 bar.

A dual heat pump system using water as a working fluid in the 2nd heat pump could have various benefits as demonstrated in the example discussed here where low-quality heat at 0 °C is used to heat up a stream to 145–150 °C. The overall C.O.P. is now higher than using either a single heat pump system or a dual heat pump system with ammonia as the sole working fluid, as the required heat transfer between the 2 heat pumps is compensated by the high C.O.P. of the 2nd heat pump using water as the working fluid (assisted by the slightly higher C.O.P. of the 1st heat pump – see Figure 4.6). Furthermore, the required pressure in the heat pumps is significantly lower (especially in comparison with the dual heat pump system using ammonia as the working fluid) and, more importantly, the temperature in the condenser is higher (118–120 °C), and thus increasing the average temperature at which heat is given off to the process stream.

The appropriate choice of the working fluid is essential to obtain efficient heat pump systems [14, 15]. The working fluid should have a low boiling point allowing heat to be taken up from a low-temperature stream (and avoid working at sub-atmospheric pressure), a high critical temperature to allow for a high temperature in the condenser, and be compatible with the materials used, e.g. seals and lubricating oil. As these working fluids may end up in the environment, their environmental impact, in terms of the global warming potential (GWP) and ozone depletion potential (ODP), toxicity and flammability will need to be considered as well.

The highest temperature in the heat pump system is at the exit of the compressor, and thus heat pumps are limited by the maximum temperature, which can be handled by compressors (or, more accurately, the lubricating oil and the seals) [16]. The maximum temperature may be further increased by combining different waste streams.

4.2.1 Chemical Heat Pump

High temperatures are difficult to achieve using traditional heat pumps based on vapour compression. The so-called chemical heat pumps [17] that utilise a reversible chemical reaction may offer an alternative here. In such a system, heat is taken up by an endothermic reaction at low temperature and given off in the reverse exothermic reaction at a higher temperature (see Figure 4.7). This may seem counter-intuitive as, naturally, this system is more readily adaptable to systems requiring cooling as exothermic reactions are more likely performed at low temperatures and the reverse endothermic reactions at higher temperatures. A classic example for the use of reactions/adsorption to cool down systems is the low-temperature adsorption of water in zeolites coupled with the high-temperature desorption of water (see, e.g. [18]). Other systems such as the hydration of MgO [19] and decomposition of metal hydrides [20] have been proposed as well. The challenge a high-temperature 'chemical heat pump' faces is to find reactions where the endothermic reaction can be carried out at low temperatures with the reverse, exothermic reaction not being thermodynamically limited at the high temperature, at which heat will be given off [21]. These heat pumps may become more feasible by incorporating a product separation step to drive

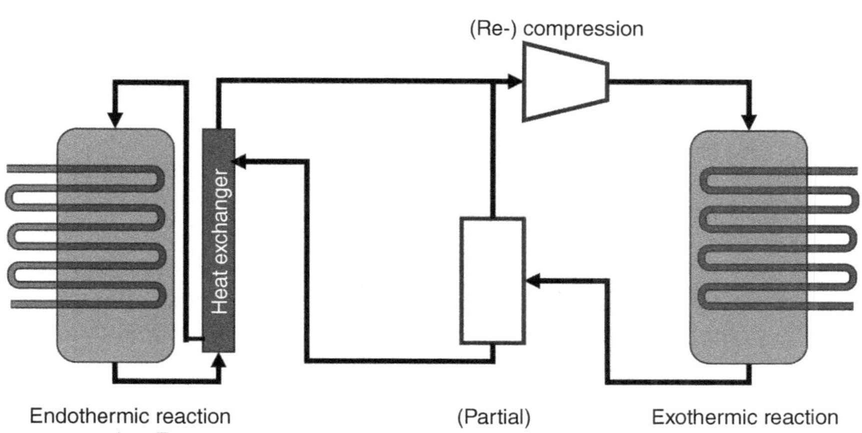

Figure 4.7 Possible configuration for a chemical heat pump.

Table 4.4 Various chemical reactions investigated for use in chemical heat pumps.

Reaction	$\Delta_{rxn}H^{298\,K}$ (kJ/mol)	T_{low} (°C)	T_{high} (°C)	$\eta_{enthalpic}$ [a)]	C.O.P.[b)]	
$(CH_3CHO)_3 \rightleftharpoons 3CH_3CO$	110.3	13	40		6.9	[24]
$NH_2COONH_4 \rightleftharpoons CO_2 + 2NH_3$	159.3	30	70		6.5	[25]
$CH_3CH(OH)CH_3 \rightleftharpoons CH_3COCH_3 + H_2$	54.4	80	200	0.13		[22]
$C_6H_{12} \rightleftharpoons C_6H_6 + 6H_2$	206.2					[25]

a) $\eta_{enthalpic} = \frac{\dot{Q}_{\text{given off at high T}}}{\dot{Q}_{\text{taken up at low T}}}$.

b) $\text{C.O.P.} = \frac{\dot{Q}_{\text{given off at high T}}}{W_{\text{added}}}$.

the equilibrium limited reaction forward using either a thermal separation technique [22] or a membrane separation technique (e.g. in the case of a hydrogenation/dehydrogenation reaction) [23].

Table 4.4 lists some of the possible reactions that can take up heat at a relatively low temperature whilst also being able to give off heat at higher temperatures. The reactions can be classified into condensation/decomposition reactions and hydrogenation/ dehydrogenation reactions. The temperature ranges tested are still in the region of heat pumps based on vapour compression, although the dehydrogenation/hydrogenation of iso-propanol/acetone has been shown to give off heat at 200 °C [22].

The C.O.P. for chemical heat pumps can be defined, in the same manner employed for vapour compression heat pumps, as the ratio of the heat given off relative to the amount of (electric) work added to the cycle. The C.O.P. is then a direct indicator for the amplification of the electric energy input into the system for the generation of heat. Some studies [17, 22] have defined the C.O.P. as the enthalpic efficiency, i.e. the heat given off at high temperature relative to the heat added to the system at low temperature.

4.3 Resistive Heaters

Renewable energy can be readily incorporated into the process industry by using resistive heaters, viz. using electric power directly to heat up a material. With many current processes, electric power is not used for heating as this would involve first generating the electric power by combusting fossil fuels, which is overall less efficient than simply heating the stream by combusting a fossil fuel directly. When it is possible to generate electric power directly from renewable sources, such as photovoltaics (PV) or wind power, the direct heating of the material, using the electric power sourced from renewable resources, may become more efficient. Alternatively, resistive heaters may be incorporated into heat pumps.

In resistive heaters, the supplied power ($P = I_{rms} \cdot V_{rms} = I_{rms}^2 \cdot R = \frac{V_{rms}^2}{R}$ with I_{rms}, V_{rms}: root mean square of the applied AC current and voltage) is converted into heat

(Joule heating), by passing a current through material with a relatively high resistivity (for effective heating), e.g. a heating wire. The maximum efficiency by which electrical power can be converted into heat using resistive heating is 100%, if heat loss to the surroundings can be avoided. Hence, the C.O.P. defined as the amount of heat being given off for heating relative to the amount of electrical energy used for these systems:

$$\text{C.O.P.} = \frac{\dot{Q}}{P} \leq 100\%$$

The material used as a heating wire should have a high resistivity for effective heating, a high melting temperature and a low thermal expansion (for example using Ni-Chrome wires – see Table 4.5). Furthermore, heating wires should not oxidise or be corrosive under the applied conditions. More exotic materials are not often used as they may not be economical.

The choice for a given material as a heating element is often given by the maximum temperature at which it may operate for prolonged periods and its specific resistance ρ_0 (together with the temperature dependency of the specific resistance, α). This may be estimated from a steady-state energy balance around a heating wire (with a diameter d_w and a length L_w), which is used to heat up a stream. The amount of power given off by a heating wire (at T_w):

$$P_\text{given off} = I_\text{rms}^2 \cdot \rho_0 \cdot (1 + \alpha \cdot (T_\text{w} - 298.15)) \cdot \frac{L_\text{w}}{\frac{\pi}{4} \cdot d_\text{w}^2}$$

is equal to the heat transfer through convection and radiation to the surroundings (at T_s), which for a straight heating wire is given by:

$$P_\text{transferred} = h \cdot \pi \cdot d_\text{w} \cdot L_\text{w} \cdot (T_\text{w} - T_\text{s}) + \varepsilon \cdot \sigma \cdot \pi \cdot d_\text{w} \cdot L_\text{w} \cdot \left(T_\text{w}^4 - T_\text{s}^4\right)$$

With ε the emissivity of the wire, and σ the Stefan-Boltzmann constant ($5.67 \cdot 10^{-8}$ W/m^2/K^4). The heat-transfer coefficient can be obtained from the appropriate correlation, e.g. for a straight heating wire with gas flowing across the heating wire [31]:

$$h = Nu \cdot \frac{k}{d_\text{w}} = C \cdot Re^m \cdot Pr^{1/3} \cdot \frac{k}{d_\text{w}}$$

Figure 4.8 shows the steady-state wire temperature as a function of the current through a nickel-chrome heating wire with a diameter of 1 mm with a linear gas flow rate of 0.1 m/s. Radiative heat transfer is the dominant heat-transfer mechanism at high wire temperatures and thus even with a temperature of the surrounding gas of as low as 100 °C, radiative heat transfer is the dominant mechanism at high currents.

The wire temperature cannot go beyond its maximum temperature (as given, e.g. in Table 4.5), thus limiting the current through the wire. The maximum current is strongly dependent on the diameter of the heating wire. The limiting current will determine the minimum length of the heating wire required in the process as

$$R_\text{min} = \frac{V}{I_\text{max}} = \rho \cdot \frac{L_\text{w}}{\frac{\pi}{4} \cdot d_\text{w}^2}$$

Table 4.5 Characteristic of common materials used in resistive heaters at 25 °C.

	Resistive heating materials								Insulator
	Nickel copper	Ni-Chrome	Iron-chrome-aluminium	Molybdenum	Molybdenum disilicide	Silicon carbide	Platinum	Tungsten	Magnesia
ρ (kg/m^3)	8860	7940	7150	10200	5600	3217	21450	19300	3580
C_p (kJ/kg/K)	0.39	0.46	0.46	0.251	0.42	0.67	0.133	0.132	0.922
ρ[a] (μOhm cm)	50	104	139	5.47	544	42–202	10.7	5.39	High
α[b] (10^2/K)	0.006	0.04	0.85	0.47	30	—	0.36	0.48	—
k[c] (W/m/K)	21	13	11	138	15–30	4–21	71.6	174	23
δ[d] (10^6/K)	14.9	15.6	11	4.8	7–8	11	8.8	4.5	12
ε[e]	—	0.88	0.70	0.80[f]	—	—	—	—	0.72
T_{max}[g] (°C)	600	1050	1300	—	1530	1600	1768	2800	—
	[26, 27]	[26]	[28]	[26]	[29]	[30]	[26]	[26]	[26]

a) Specific resistivity $\rho_0 = R_0 \cdot \frac{A}{L}$.
b) Temperature dependency of specific resistivity $\alpha = \frac{d\frac{\rho}{\rho_0}}{dT}$.
c) Thermal conductivity.
d) Thermal expansion ($\delta = \frac{1}{L}\frac{dL}{dT}$).
e) Emissivity.
f) Surface oxidised.
g) Maximum temperature at which heating element can be used for prolonged time.

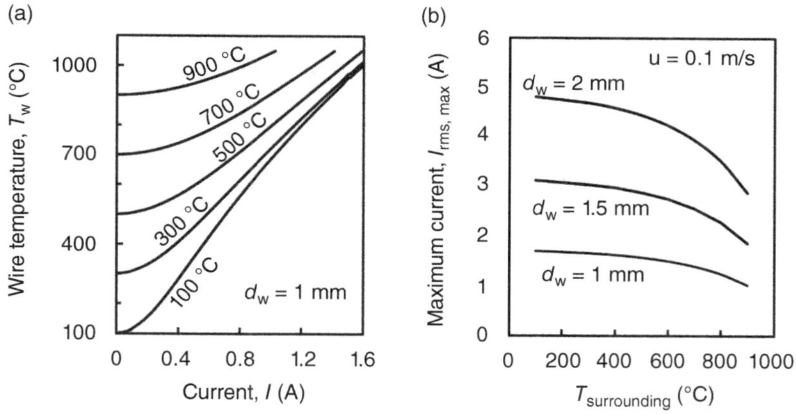

Figure 4.8 Temperature of a Ni-Chrome heating wire as a function of the current and the velocity of the gas surrounding the heating wire (a) and the resulting maximum current through the wire as a function of the diameter of the wire (d_w) and temperature of the surrounding gas ($T_{surrounding}$) (b).

Furthermore, this will determine the maximum power delivered by a single heating wire to the process. A higher power will be obtained by putting multiple heating wires in parallel.

Heating wires degrade with time according to various possible degradation mechanisms. Repetitive heating and cooling may induce thermal stress in the heating wire causing the wire to snap. Furthermore, exposure of heating elements to specific environments may result in structural changes in the heating wire causing degradation of the heating wire. For example, the exposure of SiC to an oxidative atmosphere at high temperature may result in some oxidation yielding silica. The presence of silica, an insulator, on the surface will result in a higher current through the remaining silicon carbide, which will lead to local overheating within the heating wire, thus accelerating the oxidation process, which will ultimately result in failure of the heating wire.

Heating wires are often put into cartridges, where the heating wire is surrounded by an electrical insulator with a high thermal conductivity (e.g. magnesia) encapsulated in a metal sheath (see Figure 4.9). Heating cartridges can then be used to heat up metal blocks acting as heat exchangers to heat up process fluids. The block, heated using heating cartridges containing the heating wire, will be at an approximately constant temperature. The heating block can be designed based on the required heat transferred:

$$P = \frac{V^2}{R} = \dot{m}_{\text{process stream}} \cdot (\underline{H}_{\text{process stream,out}} - \underline{H}_{\text{process stream,in}})$$

with $\underline{H}$ being the enthalpy of the stream per unit mass. The difference in enthalpy will translate into the difference of the temperature, if no phase change takes place during the heating process.

A detailed insight into the temperature distribution throughout the block can be obtained by modelling the combined conductive-radiative problem using discrete

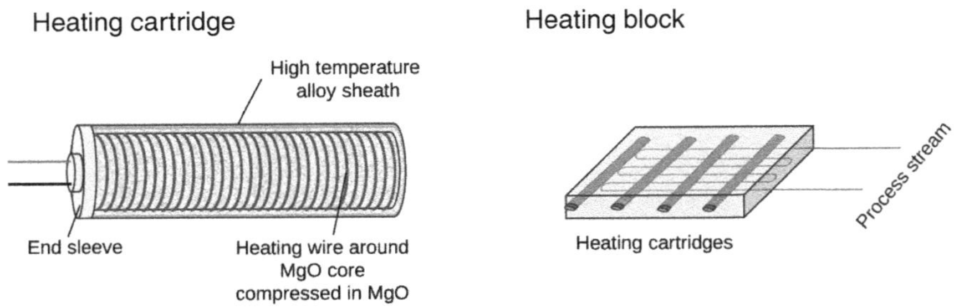

Figure 4.9 Principal sketch of a heating cartridge and a heating block containing multiple heating cartridges.

elements type of techniques [32], but heating cartridges are designed to operate at a given power rating and a given maximum temperature [33]. The number of cartridges is thus given by the power rating of the individual heating cartridges.

The heating block can be considered to be a constant wall temperature heat exchanger and heat is transferred to the fluid flowing through the pipeline from the contact with a constant block temperature. The temperature of the fluid flowing through the pipeline is not constant and its profile can be derived from the steady-state energy balance over a differential length, Δz, of the heated pipe through which the process stream flows, assuming no radial variation.

$$\frac{dH_{\text{fluid}}}{dz} = h \cdot \pi \cdot d_{\text{pipeline}} \cdot (T_{\text{block}} - T_{\text{fluid}})$$

with H_{fluid} being the enthalpy of the fluid, and h as the heat-transfer coefficient. If the heating of the stream results in (partial) evaporation, the heating block will be designed by splitting up the energy balance into a section in which the average temperature of the fluid is raised and a section in which the phase change takes place, ignoring the contribution of axial conduction in the pipeline. For instance, for generating steam from liquid water:

$$\dot{m} \cdot C_{\text{P,water}} \cdot \frac{dT_{\text{fluid}}}{dz} = h_{\text{fc}} \cdot \pi \cdot d_{\text{pipeline}} \cdot (T_{\text{block}} - T_{\text{fluid}})$$

with h_{fc}: heat-transfer coefficient for liquid water flowing through a pipe. Steam is only generated, when the temperature of the fluid is at its boiling point (occurring at a length of the pipeline L_{bp}).

$$\dot{m} \cdot \Delta_{\text{vap}}\underline{H}_{\text{water}} \cdot \frac{dy_{\text{steam}}}{dz} = h_{\text{b}} \cdot \pi \cdot d_{\text{pipeline}} \cdot (T_{\text{block}} - T_{\text{fluid}})$$

with y_{steam} the fraction of water transformed in steam and h_{b} the heat-transfer coefficient for boiling water. The two equations can be integrated:

$$\ln \frac{T_{\text{block}} - T_{\text{saturated water}}}{T_{\text{block}} - T_{\text{fluid,inlet}}} = -\frac{h_{\text{fc}} \cdot \pi \cdot d_{\text{pipeline}}}{\dot{m} \cdot C_{\text{P,water}}} \cdot L_{\text{lb}}$$

$$y_{\text{steam}} = \frac{h_{\text{b}} \cdot \pi \cdot d_{\text{pipeline}}}{\dot{m} \cdot \Delta_{\text{vap}}\underline{H}_{\text{water}}} \cdot (T_{\text{block}} - T_{\text{steam}}) \cdot (L - L_{\text{bp}})$$

Hence, the length of the pipe in the heating block can be sized. For the transformation of liquid water into saturated steam ($y_{steam} = 1$; $T_{saturated\ water} = T_{steam}$):

$$L = \frac{\dot{m}}{\pi \cdot d_{pipeline}} \cdot \left(\frac{C_{P,water}}{h_{fc}} \cdot \ln \frac{T_{block} - T_{fluid,inlet}}{T_{block} - T_{steam}} + \frac{\Delta_{vap}\underline{H}_{water}}{h_b} \cdot \frac{1}{(T_{block} - T_{steam})} \right)$$

with the 1st part of the equation describing heating up liquid water to its saturation point, and the 2nd part the evaporation of water.

For example, generating 10 kg/min saturated steam at 150 °C from liquid water at 25 °C will require 440 kW of energy. The length of pipe, with a diameter of 1 in. (2.54 cm), using a block temperature of 250 °C, with a heat-transfer coefficient in the heating up part of 1000 W/m^2/K and in the evaporation part of 4000 W/m^2/K, can be calculated from:

$$L_{lb} = -\frac{\dot{m} \cdot C_{P,water}}{h \cdot \pi \cdot d_{pipeline}} \cdot \ln \frac{T_{block} - T_{fluid}}{T_{block} - T_{fluid,inlet}}$$

$$= -\frac{10 \cdot \frac{1}{60} \cdot 4184}{1000 \cdot \pi \cdot 2.54 \cdot 10^{-2}} \cdot \ln \frac{100}{225} = 7.1\ \text{m}$$

$$(L - L_{bp}) = \frac{y_{steam} \cdot \dot{m} \cdot \Delta_{vap}\underline{H}_{water}}{h_b \pi d_{pipeline}(T_{block} - T_{steam})}$$

$$= \frac{10 \cdot \frac{1}{60} \cdot 2114.3 \cdot 10^3}{4000 \cdot \pi \cdot 2.54 \cdot 10^{-2} \cdot (250 - 150)} = 11\ \text{m}$$

Hence, generating steam in a single pipeline would require a pipeline of $11 + 7.1 = 18.1$ m. Splitting the water over 10 different pipelines but keeping the same linear velocity through the system in the same heating block would reduce the length of each pipe only by 32%, because the diameter of the pipe has to decrease by $\sqrt{0.1}$ to keep the linear velocity constant.

4.4 Furnaces

A furnace is a chamber, which either holds material to be heated/treated or through which material passes to be heated. As such, furnaces play an important role in the process industry to heat up process streams.

Furnaces can be classified in various ways according to their mode of operation, energy source, method of firing, mode of heat transfer, heat recovery, etc. These classifications are not unique. For example, the blast furnace in the steel industry is directly heated through internal combustion of coal/coke, with heating and reduction taking place in a batch operation, whilst the food industry has boilers producing a continuous flow of steam in tubes indirectly heated within a furnace.

A more useful classification system may be classification according to the energy source. In fuel-fired furnaces, as the name implies, heat is supplied by combustion of fuel either by mixing the fuel and the oxidant in a burner, or by supplying the solid fuel on a grid or in a fluidised bed with the oxidant passing through it. Alternatively, electric power can be used as the energy source. Both methods can be used effectively for heating purposes. However, heating by combusting fuels is more efficient than heating using electric power generated from fuels, as the efficiency of power generation (typically 35–45%) reduces the overall efficiency of heating. Hence, heating by combustion dominates the process industry, although costs associated with the fuel typically make up only a small portion of the total production cost. Heating using electric power is typically used for specific applications, such as arc furnaces. Heating using fossil fuels can have a lower carbon footprint than heating using electric power, if the electricity is generated from fossil fuels.

4.4.1 Combustion Furnace

Furnaces operate as high-temperature heat exchangers in the presence of a flame (see Figure 4.10). The furnace wall is typically made of refractory brick, which can withstand high temperatures, providing insulation and will reflect much of the radiation on to the material to be heated. The burner is often placed at the bottom of the furnace, where the fuel is mixed with incoming air and combusted resulting in the formation of a flame and flue gas, which is rejected from the furnace to the surroundings through the stack (often placed at the top of the furnace). The furnace can be divided into separate sections, viz. a radiative section, where heat transfer takes place mainly via radiation, and a convective section, where heat from flue gas is captured by heat exchange with the material.

The combustion of fuel yields a flame in the radiative section, and the dominant mode of heat transfer in this section is radiation. Here, the material to be heated (e.g. in tubes as depicted in Figure 4.10) is arranged around the flame/burner or is

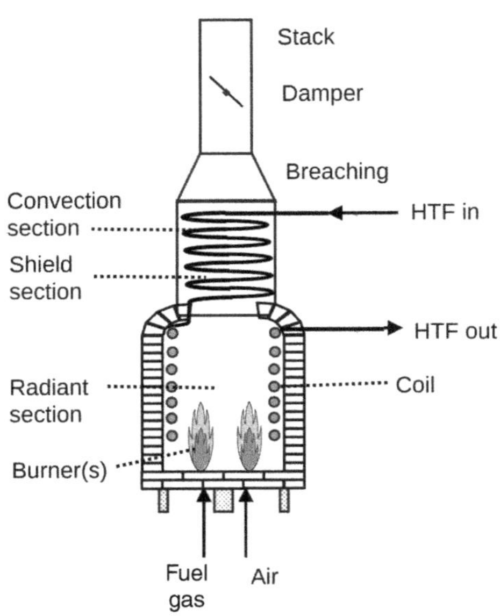

Figure 4.10 Schematic of a furnace.

directly exposed to the flame itself. The rate of heat transfer in this section of the furnace can be approximated by:

$$\frac{\dot{Q}}{A} = \varepsilon \cdot F_{\text{tubes–flame}} \cdot \sigma \cdot \left(T_{\text{flame}}^4 - T_{\text{tube}}^4\right)$$

where ε is the emissivity, $F_{\text{tubes-flame}}$ is the view factor of the tubes area of heat transfer (m^2), $\sigma = 5.67 \cdot 10^{-8}$ W/m^2/K^4 and T_{flame} and T_{tube} are the temperatures of the flame and the tube, respectively. This section can typically be modelled as a well-mixed section, i.e. a section with a constant temperature [31].

Above the radiative section is the convective section, where a bank of tubes is placed above the flame and beyond the upper radiative tubes. These tubes are heated by the high-temperature flue gas coming from the burner/flame. Typically, the tubes are stacked in a bank with cross flow of the flue gas, and the heat transfer may be modelled according to:

$$\frac{\dot{Q}}{A} = h \cdot (T_{\text{flue gas}} - T_{\text{tube}})$$

where h is the convective heat transfer coefficient in $\frac{\text{W}}{\text{m}^2 \cdot \text{K}}$, A is the area of heat exchange, and $T_{\text{flue gas}}$ is the temperature of the flue gas in the convection zone. The convective heat-transfer coefficient is a function of the Reynolds number of the flue gas passing over the material/tubes to be heated and is thus affected by the velocity of the flue gas.

The burner takes the central point in the furnace, with its primary function being to mix the fuel with the oxidant efficiently to obtain complete combustion and thus the highest possible temperature, whilst at the same time reducing the emission of environmental pollutants. Fuel is fed with a slight head pressure through a nozzle to the burner port. In burners utilising liquid fuels, the fuel needs to be atomised to ensure efficient combustion.

In older designs, primary air is added by means of the Venturi effect pulling air in through a gap at the base of the burner, after which it is mixed with the fuel (see Figure 4.11). The amount of air pulled through is typically insufficient for complete combustion and secondary air is pulled into the flame from the air surrounding the flame. A common example is the Bunsen burner, where closing the primary air inlet results in a (relatively) cold yellow flame which produces black smoke because of incomplete combustion, whilst having the primary air inlet open results in a hot, clean burning, blue flame.

More modern burners have moved away from the natural air inlet design and rather use a forced air intake, which may be provided by a fan or compressor. This results in a better burn efficiency due to the ability to provide excess primary air, ensuring complete combustion, but also providing some level of control of the air inlet, meaning that the volume of excess air will be limited.

The energy added to the furnace is controlled by the rate of addition of fuel to the burner and the number of burners in the furnace. The rate of energy addition to the burner is regulated by the (adjustable) fuel nozzle (see Figure 4.11). The flow through the nozzle is given by Bernoulli's equation in conjunction with the

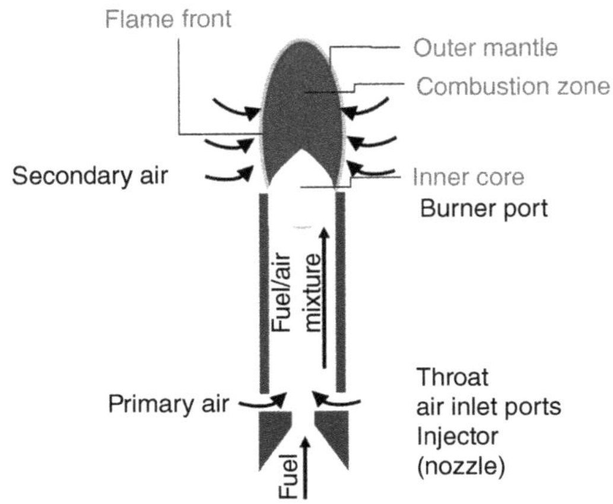

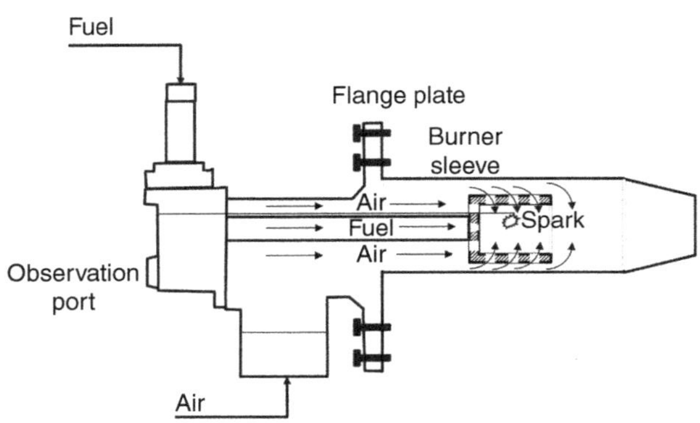

Figure 4.11 Schematic sketch of burner with primary air feed through Venturi effect (a) and an industrial side-burner (b) (redrawn from Liu et al. [34]).

mass balance. The volumetric flow rate of fuel through a nozzle is given by:

$$\dot{V} = C_\mathrm{D} \cdot A_\mathrm{orifice} \cdot Y \cdot \sqrt{\frac{2}{(1-\beta^4)\cdot \rho_\mathrm{fuel}} \cdot \Delta P}$$

with

C_D: drag coefficient (determined by the angle α constricting the flow – see Figure 4.12)

A_orifice: the area of the nozzle also given by $A_0 = \frac{\pi}{4} \cdot d^2_\mathrm{orifice}$ (see Figure 4.12)

Y: expansion coefficient (incompressible liquids $Y = 1$); compressible fluids (ideal gas):

$$Y = \sqrt{r^{\frac{2}{k}} \cdot \frac{k}{k-1} \cdot \frac{1-r^{\frac{k-1}{k}}}{1-r} \cdot \frac{1-\beta^4}{1-\beta^4 \cdot r^{\frac{2}{k}}}}$$

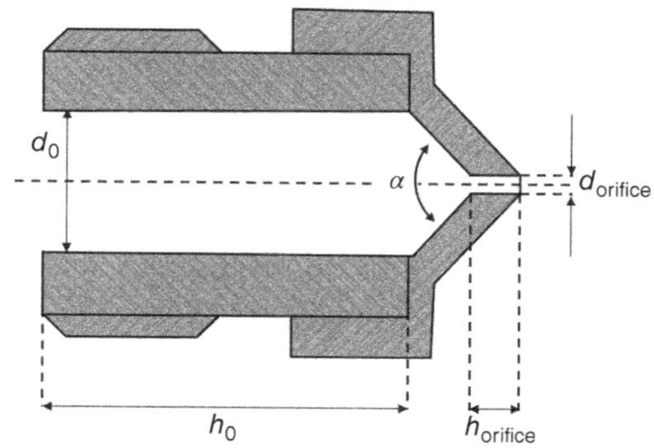

Figure 4.12 Schematic drawing of a fuel nozzle (redrawn from Tian et al. [35]).

with r as the ratio of the pressure after the nozzle (i.e. the pressure at which the fuel is burned) relative to the head pressure (i.e. the pressure before the nozzle), $k = \frac{C_P}{C_V}$

ρ_{fuel}: the density of the fuel at the head pressure,

β: the ratio of the diameter of the orifice relative to d_0 (see Figure 4.12)

ΔP: the pressure difference between the head pressure and the pressure at which the fuel is burned.

The relationship between the volumetric flow rate and the energy feed rate to the burner as a function of pressure difference for different gaseous fuels is given in Figure 4.13. The volumetric flow rate of compressible fluids passes a maximum as a function of the head pressure, as with increasing pressure drop over the nozzle, the expansion factor, Y, decreases and the density of the fluid increases. The difference between the various compressible fluids is mainly determined by the variation in their density and, to a lesser extent, the difference in the expansion coefficient. For incompressible fluids, the volumetric flow rate increases steadily with $\sqrt{\Delta P}$. Of course, the volumetric flow rate of the incompressible fluid is much smaller than the volumetric flow rate of the compressible fluids as its density is much higher.

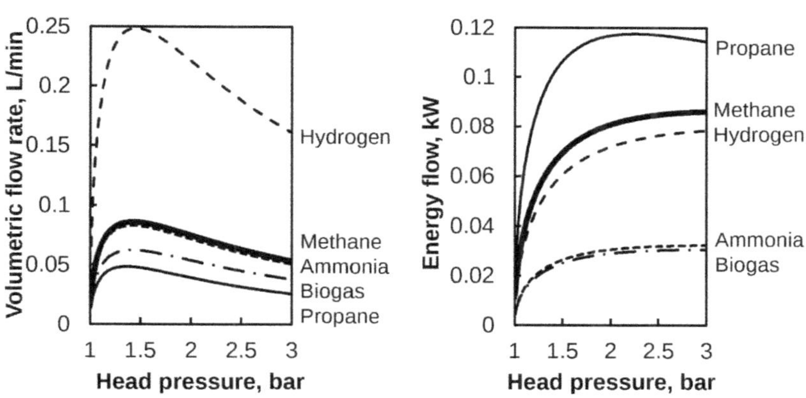

Figure 4.13 Volumetric flow (left) and energy flow (right) of gaseous fuels through a nozzle ($d_{orifice}$ = 0.1 mm, C_D = 0.9; β = 0.1) with pressure beyond the orifice at 1 bar (biogas modelled as 50% CH_4 and 50% CO_2; energy flow calculated using the lower heating value, LHV).

Of more interest is the energy flow to the burner (in Figure 4.13 shown for a flow through a nozzle with an orifice diameter of 0.1 mm and $\beta = 0.1$). The energy flows provided by methane and hydrogen are quite similar as the high volumetric flow rate of hydrogen is compensated by its lower density and its lower heat of combustion. Interestingly, the energy flow with propane as a fuel is high, due to the high heat of combustion and the high density. It can be further noted that there is an optimum head pressure, delivering the highest amount of energy (other fuels show the same characteristic albeit at higher pressures).

Liquid fuels will be pumped into a burner, and the energy flow to the furnace is directly related to the pump rate and the heating value of the fuel.

The classical way to characterise the energy flow to the burner is through the so-called Wobbe index. The Wobbe index is defined as:

$$I_W = \frac{\text{HHV}}{\sqrt{G_S}}$$

with HHV, the higher heating value of the fuel and $G_S = \frac{\rho_{\text{fuel}}(\text{STP})}{\rho_{\text{air}}(\text{STP})} = \frac{M_{\text{fuel}}}{M_{\text{air}}}$. It may be debated as to whether the lower heating value would not be a more appropriate measure as the heat of condensation is not recovered in a furnace. The Wobbe index is typically used to determine the gas composition required to ensure a constant rate of energy delivery to the burner. The Wobbe index is not a complete indicator of the interchangeability of gasses in a given burner/furnace, as the change in the volumetric flow rate between the gases also affects the flame stability. As an example, the Wobbe index for methane and hydrogen show similarity (see Table 4.6), indicating that these gases can be exchanged and mixed to some extent without having to retrofit the nozzles and pressure fittings on the burner.

It should be noted that the Wobbe index (and the energy flow) is not the only criteria to be considered when determining the type of nozzle, as the combustion should result in a stable flame as well.

Flame stability is linked to the (laminar) flame speed [36, 37], i.e. the speed at which the flame moves towards the burner from the main body of the flame. A high flame speed may result in lighting back with the flame reaching the burner head, which may damage the burner. A low flame speed in relation to the gas velocity may

Table 4.6 Wobbe index (I_W) of different fuels.

Fuel		M_{fuel} (g/mol)	LHV (MJ/m³)	HHV (MJ/m³)	I_W (MJ/m³)
Methane	Gas	16.043	35.8	40.8	55.8
Biogas[a)]	Gas	24.433	25.1	28.6	31.6
Biogas[b)]	Gas	30.026	17.9	20.4	20.4
Hydrogen	Gas	2.016	10.8	13.2	51.2
Ammonia	Gas	44.097	14.1	17.9	23.7
Propane	Gas	17.031	99.0	109.0	89.9

a) Biogas comprising of 70% methane and 30% CO_2.
b) Biogas comprising of 50% methane and 50% CO_2.

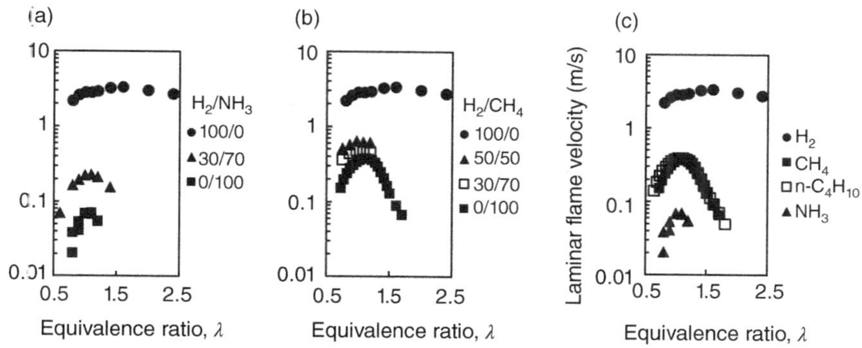

Figure 4.14 Laminar flame speed as a function of the equivalence ratio, λ, for different fuel–air mixtures (a), hydrogen/methane/air mixtures (b) and hydrogen/ammonia/air mixtures (c) at atmospheric pressure (Data: methane, n-butane [38]; hydrogen and methane/hydrogen: [39] ammonia: [40]; ammonia/hydrogen: [41]).

result in flame lift causing detachment of the flame from the burner, which may put the flame out due to a lack of ignition of the main fuel. The flame speed depends on the chemical kinetics of the combustion process and varies for the different fuel types with gaseous hydrocarbon fuels (viz. methane, ethane, propane, butane) showing very similar flame velocities as a function of the equivalence ratio, λ (the ratio between the amount of air present compared to the stoichiometric amount of air) (see Figure 4.14) [37]. The flame speed of hydrogen and synthesis gas are much higher [42], and thus backfire is much more likely when adapting burners designed for hydrocarbon fuels for the combustion of hydrogen and synthesis gas. Ammonia, on the other hand, has a very low flame speed compared to the hydrocarbons (see Figure 4.14), possibly resulting in flame lift in a retrofitted methane burner, though this can be rectified to an extent by co-feeding methane [43]. Co-feeding ammonia with hydrogen can be used to balance the flame speed to approach that of the hydrocarbons [44].

In the burner, an air–fuel mixture will combust, if the mixture is within its flammability region, i.e. if the concentration of the fuel is in the range, in which the combustion reaction sustains itself after ignition using an external source. The flammability region depends on the temperature of the feed gas, the pressure and the heat loss from the flame [45] (the adiabatic flammability region, a value used to assess process safety, can be estimated from thermodynamic properties [46]). Table 4.7 shows the flammability region for air/fuel mixtures at 25 °C. Both the lower flammability level and the upper flammability level depend on temperature [52, 53] and, especially the upper flammability level, on pressure [45, 53]. The flammability range for methane ranges from an air-to-fuel ratio (λ) of 0.54 to 2.28. The flammability ranges of other organic compounds, such as methanol and propane, have similar widths (range of $\lambda = 0.24$–1.95 for methanol and $\lambda = 0.37$–1.94 for propane). Hydrogen has a very wide flammability range ($\lambda = 0.1$–8.0). Ammonia is being discussed as a possible fuel of the future, but its flammability region is comparatively narrow ($\lambda = 0.72$ to 1.59), which could be improved by combusting mixtures of ammonia and hydrogen [54], methane [55], methanol [56], ethers [57] or even diesel [58].

Table 4.7 Combustion characteristics of some fuels.

Fuel		$T_{autoignition}$ (°C)	$T_{adiabatic}$ a) (C)	$y_{stoich.}$ b)	y_{LFL} c)	y_{UFL} d)
Methane	Gas	610 [47]	2053	0.095	0.044	0.164
Biogas	Gas		~1700–1900	0.14–0.21	0.07–0.12 e)	0.12–0.24 e)
Hydrogen	Gas	550 [47]	2523	0.296	0.04	0.75
Ammonia	Gas	630 [48]	2274	0.219	0.15	0.28
Propane	Gas	475 [47]	2281	0.040	0.021	0.101
Methanol	Liquid	464 [49]	2055	0.123	0.067	0.36 f)
Ethanol	Liquid	363 [49]	2018	0.065	0.033	0.19 g)
Toluene	Liquid	480 [49]	2210	0.023	0.011	0.07
Diesel	Liquid	210 [49]	~2250	— h)	— h)	— h)

a) Adiabatic flame temperature for stoichiometric fuel/air mixture at 25 °C.
b) Mole fraction of vapour in air in a stoichiometric mixture.
c) Mole fraction of vapour in air at lower flammability level.
d) Mole fraction of vapour in air at upper flammability level.
e) Pizzuti et al. [50].
f) At 60 °C [45].
g) Upward flame propagation [51].
h) Typically given as mass% of diesel mixed with air.

Radiative heat transfer in the furnace depends on the flame temperature. The highest temperature of the flame is given by the adiabatic flame temperature, which can be estimated knowing the feed ratio of the fuel and the (ideal) composition of the combustion gas from an energy balance:

$$\Delta_{\text{comb}} H(T_{\text{inlet}}) = \int_{T_{\text{inlet}}}^{T_{\text{adiabatic flame}}} \left(\sum_i \theta_i \cdot C_{P,i} \right) \cdot \mathrm{d}T$$

with θ_i the ratio of the components in the effluent relative to the feed of the combustible component. Hence, the adiabatic flame temperature depends on the fuel type, the oxidant and their relative ratio. Figure 4.15 shows the adiabatic flame temperature for the combustion of various fuels in air as a function of the ratio of air to fuel (with $\lambda = 1$ being the stoichiometric air to fuel ratio) at 25 °C, as well as the molar flow of the exhaust gases based on the molar feed rate of the gaseous fuel or the mass flow rate of the solid fuel. The highest temperature is achieved with a stoichiometric air to fuel ratio, as insufficient heat is generated in the combustion when feeding air sub-stoichiometrically and more heat is required to heat up air as well when feeding excess air.

Methane, as a proxy for natural gas, in air can reach an adiabatic temperature of 2053 °C (natural gas taken as a mixture of 10% nitrogen in methane can reach in air an adiabatic flame temperature of 2038 °C). Methane or natural gas does not only yield a high adiabatic flame temperature, an essential characteristic for radiative heat transfer, but also yields a high flow rate of the exhaust gas, thus facilitating

Figure 4.15 Adiabatic flame temperature for the combustion of different fuel types in air within their flammability region as a function of the air-to-fuel ratio (left) and the flow rate of the effluent relative to the fuel feed rate (right) with fuel and air fed at 25 °C (biogas taken as 50% CH_4 and 50% CO_2; diesel taken as $C_{16}H_{31}$ with heat of combustion of 45 MJ/kg, heat of evaporation as 64 kJ/mol and specific heat as 460 J/mol/°C; composition and heating value of anthracite from [59], brown coal from [60], wood pellets from [61] and sludge from [62]; the specific heat of ash was estimated at 0.7 kJ/kg/°C.

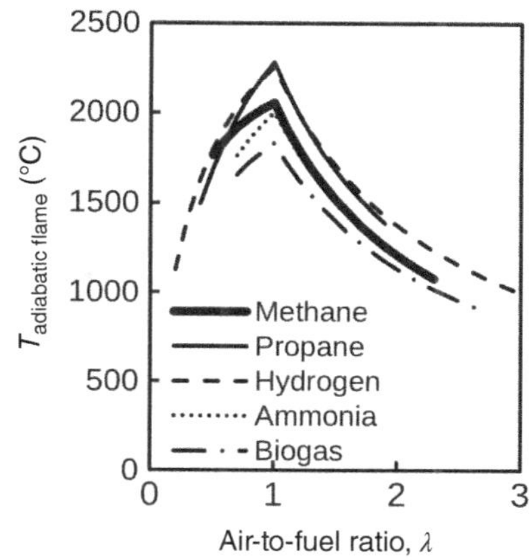

convective heat transfer. Propane as a fuel shows similar characteristics, viz. a high adiabatic flame temperature and a high flow rate of the exhaust gas. The maximum adiabatic flame temperature of alternative fuels, such as hydrogen and ammonia, are similar (H_2: 2250 °C; NH_3: 2001 °C), but the flow rate of the exhaust gas is strongly reduced, which may negatively affect the convective heat transfer in furnaces. Biogas, modelled as a mixture of 50% methane and 50% CO_2, has a lower maximum adiabatic flame temperature than the other fuel types due to the dilution of the combustible fuel.

Liquid fuels can reach similarly high adiabatic flame temperatures, with oxygen-containing fuels reaching a lower flame temperature in the fuel-rich region than fuels containing hydrocarbons. Diesel and toluene, taken as an example of a fuel which is rich in aromatics, reach similar flame temperatures at similar air-to-fuel ratios, but the flow rate of the exhaust gas is higher when diesel is used as the fuel. This may be beneficial for the convective heat transfer in the furnace.

A high adiabatic flame temperature can also be obtained using solid fuels (see Figure 4.15). For instance, anthracite with a lower heating value of 30 MJ/kg containing 6 wt.% ash and, in an ultimate analysis, 92.6% C, 3.2% H and 0.6% S on a dry and ash-free basis [59] will yield a maximum, adiabatic flame temperature of 2048 °C. Taking dry brown coal, with, e.g. a heating value of 23.2 MJ/kg on a dry basis (ash-content: 6 wt.% on a dry basis and according to the ultimate analysis on a dry and ash-free basis 69 wt.% C, 5.0 wt.% H, 0.8 wt.% N, 0.1 wt.% combustible S and 25.1 wt.% O) [60] would result in an adiabatic flame temperature for a stoichiometric feed of 2125 °C, similar to the adiabatic flame temperature of anthracite. The relatively high adiabatic temperature is a consequence of countering the lower heating value with a reduced stoichiometric air requirement because of the high oxygen content in brown coal. Of course, the combustion of (partially) moist brown coal will reduce the adiabatic flame temperature significantly. The lower, stoichiometric requirement for air implies that the combustion of brown coal at the same stochiometric air ratio will result in a lower flow rate of the exhaust gases when operating at the same adiabatic flame temperature. Similarly high adiabatic flame temperatures can be achieved using wood pellets (lower heating value: 18.7 MJ/kg;

ash-content 0.3 wt.%; ultimate analysis 50.9% C, 6.0% H, 42.6% O [61]), as much less air is required due to the higher oxygen content of the wood pellets. Therefore, the flow rate of the exhaust gases is much lower thus affecting the convective heat transfer. Dried sludge, a waste material from water treatment, could be used to fire a furnace with only a slight drop in the adiabatic flame temperature (a dried sludge with a calorific value of 20.4 MJ/kg, ash content of 15.9 wt.% on a dry basis and an ultimate analysis of 58.5% C, 9% H, 5% N, 27.5% O and 0.05% S [62] – the lower heating value was estimated at 20.4 MJ/kg – yields an adiabatic flame temperature of 1800 °C). The flow rate of the exhaust gas obtained in the combustion of sludge drops with the increase in the air flow rate, in a similar fashion to wood pellets.

Pre-heating the air prior to entering the combustion chamber will raise the flame temperature significantly, which is particularly useful when operating with excess air.

Radiative heat transfer also depends on the emissivity of the flame, which is a complex function of the constituents in the flame, pressure and temperature. Water and CO_2 (in the case of a carbon containing fuel) are typically considered the main constituents contributing to the emissivity. The emissivity of water decreases with increasing temperature (see Figure 4.16), whereas the emissivity of CO_2 decreases at temperatures higher than ca. 800 °C and shows a more complex temperature dependency at lower temperatures, depending on the beam length and the partial pressure of CO_2 [63]. The emissivity of a mixture depends on the contribution of the individual components considering the spectral overlap of CO_2 and water (ε_{SO})

$$\varepsilon_{tot} = (C_{CO_2} \cdot \varepsilon_{CO_2} + C_{H_2O} \cdot \varepsilon_{H_2O})(1 - \varepsilon_{SO}) + \sum_{i \neq CO_2, H_2O} \varepsilon_i$$

(the emissivity of CO_2 and water are weighed with the respective pressure correction terms [63]). The presence of soot increases the emissivity, depending on the amount

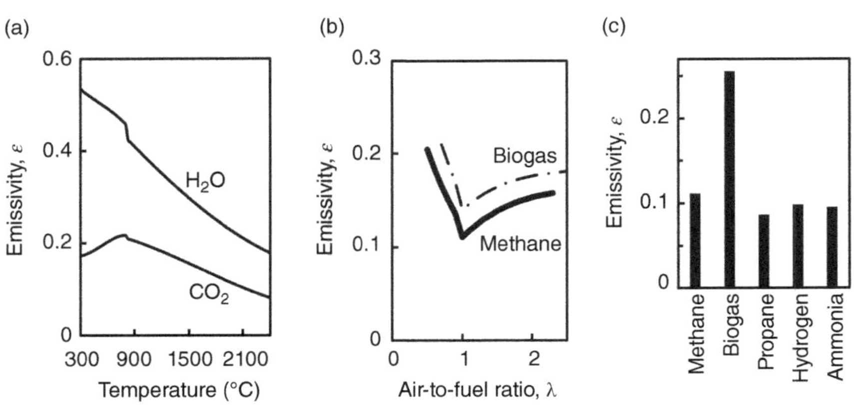

Figure 4.16 Variation in the emissivity of water and CO_2 as a pure component at 100 kPa with temperature (a), the emissivity of an adiabatic flame at 100 kPa generated by combusting methane and biogas (50% CH_4/50% CO_2) as a function of the air-to-fuel ratio yielding CO_2 (b), and the emissivity of a clean-burning, adiabatic flame at 100 kPa generated by various fuels at the stoichiometric air-to-fuel ratio (c) (beam length: 1 m; emissivity calculated using correlations in [63]) (adapted from Mehrotra et al. [63]).

of soot present in the flame [31], and soot is an essential contributor to the total emissivity of flames. Soot is typically formed through radical addition reactions which yield the soot precursor, polycyclic aromatic hydrocarbons [64], whose formation is enhanced at elevated pressure [65].

The emissivity of a flame generated by the adiabatic, sootless combustion of methane decreases initially with increasing air-to-fuel ratio up to the stoichiometric air-to-fuel ratio, which is primarily due to the increase in the flame temperature although the decrease in the emissivity of methane also contributes. The emissivity of this flame increases upon combusting methane using excess air, with again the temperature as the driving force. The emissivity of natural gas is typically low, but increasing the intermediate soot particle concentration in the flame will increase the emissivity significantly [66, 67], and thus improve the radiative heat-transfer properties. The emissivity of an adiabatic, sootless flame by combusting biogas, taken as 50% methane and 50% CO_2, is higher due to the lower flame temperature. The presence of CO_2 should reduce the tendency for soot formation [68], implying that the emissivity will be closer to that of the one obtained for the sootless combustion of biogas, although impurities in this gas may increase the emissivity.

Flames generated by the adiabatic sootless combustion of propane, using a stoichiometric air-to-fuel ratio, will have an emissivity which is slightly lower than the emissivity of a flame generated by the adiabatic sootless combustion of methane ($\lambda = 1$), but the presence of soot may improve the radiative heat transfer from a propane flame. Combustion of hydrogen or ammonia will not result in the formation of soot and thus the emissivity of adiabatic flames of hydrogen and ammonia ($\lambda = 1$) will remain relatively low. This implies that radiative heat transfer from hydrogen/ammonia flames can be as much as 3–4 times less effective than radiative heat transfer from luminous flames, where soot contributes to the emissivity of the flame.

In comparison, the emissivity of a coal flame in the burner zone of a furnace is significantly higher ($\varepsilon_{coal} \sim 0.3$–$0.5$); the emissivity increases with increasing load in the furnace, which may be attributed to a higher soot concentration in the flame [69]), and it decreases with increasing ash content in the coal [70]. Similarly, the emissivity of oil burners will be higher when soot is present (e.g. by increasing the droplet size or reducing the degree of atomisation of the fuel prior to combustion [71]).

Heat transfer by convection in furnaces depends on the flow properties of the gas, and thus the ratio of the rate of the exhaust gas relative to the rate of the energy is delivered through the fuel (which of course depends on the energy content of the fuel and its mass flow rate), as well as the thermal conductivity of the effluent gas. The latter is, in principle, a weighted average of the thermal conductivity of the individual components but is not a strong function of the composition of the gas.

The energetic efficiency of combustion furnaces depends primarily on the heat lost in the stack gas with some of the heat recuperated by, e.g. using the stack gas to pre-heat the air required for the combustion. The efficiency is further affected by both radiative and convective heat loss through openings in the furnace and through the insulated skin of the furnace. Ingress of cold air from the surroundings will also reduce the efficiency of a furnace (this could be mitigated by operating furnaces with a slight overpressure). Si et al. [72] analysed the efficiency of a reheat furnace at a steel mill. At full production capacity, the energy efficiency of the furnace amounted to 60% with 30% of the total energy lost through the stack gas (i.e. 75% of all lost energy was lost through the stack gas). The dominance of heat loss through the stack implies that increasing the amount of excess air fed to the burner may result in a decrease in the overall energy efficiency of the furnace [73].

Furnaces do not typically operate continuously, but typically face some idle time and some times during which the furnace is used only partially. During the idle time, energy is being used to keep the furnace at or close to the operating temperature. For instance, in the reheat furnace analysed by Si et al. [72], the energy consumption during idling amounted to 7.8% of energy consumed at full production capacity. To account for the interchanging periods of full/partial production and idling, it is more useful to define an annualised furnace efficiency (AFE):

$$\mathrm{AFE} = \frac{\eta_{\mathrm{E},100\%} \cdot \mathrm{ff}_{100\%} + \eta_{\mathrm{E,partial}} \cdot \frac{E_{\mathrm{total,partial}}}{E_{\mathrm{total},100\%}} \cdot \mathrm{ff}_{\mathrm{partial}}}{\mathrm{ff}_{100\%} + \frac{E_{\mathrm{total,partial}}}{E_{\mathrm{total},100\%}} \cdot \mathrm{ff}_{\mathrm{partial}} + \frac{E_{\mathrm{total,idle}}}{E_{\mathrm{total},100\%}} \cdot \mathrm{ff}_{\mathrm{idle}}}$$

with $\eta_{\mathrm{E},100\%}$ the energetic efficiency of the furnace operating at 100% capacity, $\eta_{\mathrm{E,partial}}$ the energetic efficiency of the furnace operating at partial capacity and $\mathrm{ff}_{100\%}$, $\mathrm{ff}_{\mathrm{partial}}$ and $\mathrm{ff}_{\mathrm{idle}}$ the time factors at which the furnace operated at 100% capacity, partial capacity and was idle, respectively.

4.4.2 Electric Furnace

There are different types of electric furnaces, which can be classified based on their primary functioning, viz. resistive, microwave, inductive and arc furnaces.

4.4.2.1 Resistive Furnace

Resistive furnaces are like resistive heaters (see Section 4.3), where a current flows through a heating element and heat is generated through Joule heating, i.e. by the resistance encountered in the heating element. The heat generated in the heating element may then be transferred to the material requiring heating via radiation, if the resistor gets hot enough. The maximum wavelength of the radiation emitted from the heating element is given by Wien's displacement law,

$$\lambda_{\mathrm{max}} = \frac{2.897 \cdot 10^{-3}}{T}$$

with λ_{max} the wavelength with the highest intensity in metre and T the temperature in K. It should be realised that the resistor will emit quite a wide spectrum of radiation, as given by Planck's radiation law for black bodies. Heating up an element to 1000 °C will result in radiation with a wavelength at maximum intensity of 2275 nm, thus in the infrared region. Increasing the temperature of the heating element to 2000 °C will decrease the wavelength at maximum intensity to 1274 nm (still in the infrared region).

Heat transfer in resistive furnaces can be improved by (induced) convection, specifically for furnaces operating at lower temperature (see Figure 4.17). Fans can

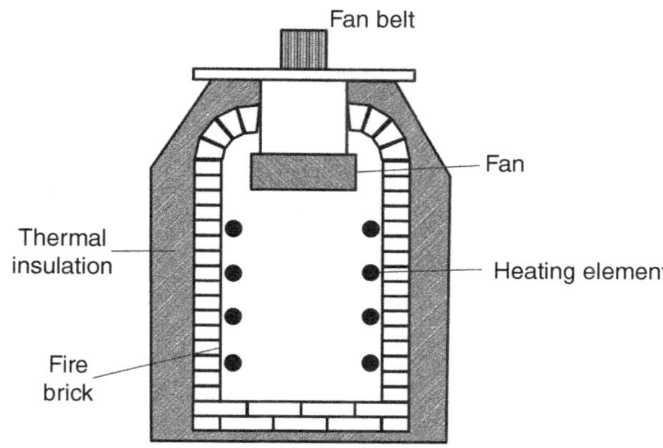

Figure 4.17 Schematic drawing of a resistive furnace.

also be used when radiative heat transfer is the dominant mode of heat transfer to generate a more uniform temperature profile within the furnace.

Resistive furnaces operating at temperatures below 760 °C use chrome–nickel resistors [74], which can come in the form of ribbon, cast alloy, open wire coils or sheathed construction. Alternatively, tungsten resistors encased in glass or quartz to exclude oxygen from the resistors can be used. At high temperatures (approaching 1000 °C), silicon carbide resistors may be required, depending on the design temperature and atmosphere.

Resistive heat can be quite efficient, as all energy is transformed into heat and only heat loss from the furnace reduces its efficiency. Relying on heat transfer through radiation only (e.g. in electric infrared heaters) can result in efficiencies of 70–80% [75].

4.4.2.2 Microwave Furnace

The use of microwaves (i.e. electromagnetic waves with a wavelength of 0.1–100 cm or with a frequency of 0.3–300 GHz) for heating purposes is well-known due to the domestic application of magnetron heating to warm up food (microwave ovens). Not all these frequencies can be used as they are also used for communication purposes. The ITU Radio Regulations released by the International Telecommunications Union has set aside some frequencies for the use of industrial, scientific and medical applications, including domestic applications (see Table 4.8) [76], provided that radiation from the equipment is minimal and that radiation outside the given band results in minimal interference with other communication devices.

The absorption of electromagnetic waves (microwave irradiation) by matter results in heating and the amount of power dissipated into the material per unit volume can be expressed by [77, 78]:

$$P_{\text{dissipated}} = 2\pi f \cdot \left(\varepsilon_0 \cdot \varepsilon''_{\text{r,eff}} \cdot E^2_{\text{rms}} + \mu_0 \cdot \mu''_{\text{r}} \cdot H^2_{\text{rms}}\right)$$

where, f is the frequency of the incoming radiation, ε_0 the electric permittivity in vacuum, $\varepsilon''_{\text{r,eff}}$ the relative effective dielectric loss (the imaginary part of the complex relative permittivity), E_{rms} the strength of the electric field, μ_0 the magnetic

Table 4.8 Frequencies in the microwave range available for industrial, scientific and medical applications) (data from ITU [76]).

Frequency	Frequency range
13.56 MHz	13.553–13.567 MHz
27.12 MHz	26.957–27.283 MHz
40.68 MHz	40.66–40.77 MHz
0.915 GHz	0.902–0.928 GHz
2.450 GHz	2.400–2.500 GHz
5.800 GHz	5.725–5.875 GHz
24.125 GHz	24–24.25 GHz

permeability in vacuum, μ_r'' the relative effective magnetic permeability (the imaginary part of the complex relative permeability) and H_{rms} the strength of the magnetic field. The relative effective dielectric loss has contributions from the dipolar, interfacial and polarisation dielectric loss in the presence of an alternating electric field [77–79], which occur due to the orientation of (induced) dipoles to the electromagnetic field in the presence of a viscous substance. The continuous alignment to an oscillating field results in frictional interaction with the surrounding molecules [77, 80], and thus results in heating.

The response of materials to microwaves depends thus on their absorptive properties and can be classified [78] as

1) Transparent materials: typically, insulators with a low value for the complex permittivity ($\varepsilon''_{r,eff}$) at the frequency of interest, such as Teflon and quartz.
2) Opaque materials: conductor materials reflecting most of the microwaves (such as bulk metals).
3) Absorbing materials: insulators with a high value for the complex permittivity ($\varepsilon''_{r,eff}$), such as water and SiC.

Figure 4.18 shows the loss factor ($\varepsilon''_{r,eff}$) for water and other organics as a function of the microwave frequency. The relative permittivity at 20 °C as a function of the frequency in the microwave range is constant up to ca. 0.1 GHz, after which it drops to zero in the range up to 100 GHz. With increasing temperature, the static relative permittivity decreases, and the frequency at which the relative permittivity decreases strongly as a function of frequency shifts towards higher temperatures. The relative effective dielectric loss factor, $\varepsilon''_{r,eff}$, for water shows a maximum in the microwave region, which shifts with increasing temperature towards higher frequencies (maximum at 20 °C estimated to be at 7.6 GHz and at 100 °C at 32.6 GHz). However, the value of the relative loss factor decreases with increasing temperature (20 °C: $\varepsilon''_{r,eff,max} = 39$; 100 °C: $\varepsilon''_{r,eff,max} = 27.5$) implying a reduced efficiency in microwave heating upon increasing temperature, when operating at the optimum frequency or at a relatively low frequency (e.g. 2.45 GHz as in common household

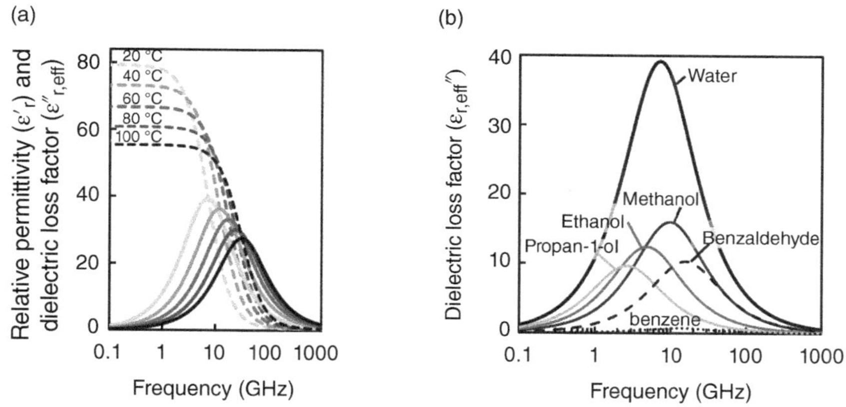

Figure 4.18 Relative permittivity (dashed curve) and loss factor ($\varepsilon''_{r,eff}$; solid curve) of water as a function of temperature (a) and of various organic compounds (b). Permittivity and loss factor calculated using $\varepsilon'_r = \varepsilon_\infty + \frac{\varepsilon_s - \varepsilon_\infty}{1+(2\pi \cdot f \cdot \tau)^2}$ and $\varepsilon''_{r,eff} = \frac{(\varepsilon_s - \varepsilon_\infty)\cdot(2\pi \cdot f \cdot \tau)}{1+(2\pi \cdot f \cdot \tau)^2}$ with the permittivity at infinite frequency (ε_∞) estimated from [81], static permittivity (ε_s) as given in [77] (temperature dependency of static permittivity of water estimated from [82]) and the relaxation time, τ, estimated from [77].

magnetrons). However, the relative loss factor passes through a maximum as a function of temperature when operating at a frequency higher than the frequency at which the relative loss factor is at its maximum (e.g. 24.5 GHz).

Water is a suitable medium to be heated using microwaves. The dissipated power upon microwave irradiation will be transformed into heat and the heating rate for a dielectric can thus be expressed as:

$$\frac{\Delta T}{\Delta t} = \frac{P_{dissipated}}{\rho \cdot C_p} = \frac{2\pi f \cdot \varepsilon_0 \cdot \varepsilon''_{r,eff}}{\rho \cdot C_p} \cdot E_{rms}^2$$

At a frequency of 2.45 GHz, the dielectric loss factor of water decreases with increasing temperature and thus the heating rate will decrease (almost) proportionally.

The static permittivity of low-pressure steam is typically too low [83], resulting in low values for the dielectric loss factor, $\varepsilon''_{r,eff}$ (this is compounded with the relative low relaxation time for low-pressure steam). Thus, magnetron heating is not suitable to heat up steam but can be used to dry solids [84]. Steam can be generated using magnetron heating by heating compressed water followed by flash evaporation.

Microwave heating can also be applied to organic solvents with a sufficiently high dipole moment, as indicated by the low loss factor, e.g. benzene, but may be used for heating of mixtures containing polar molecules around, e.g. distillation units. It should however be realised that the dipolar loss factor, which determines the amount of power dissipated per unit volume is not a linear function of the composition of the mixture [85–87].

Solids, like carbon, can also be heated using microwave irradiation [88]. Here heating is thought to take place due to the mobility of delocalised π-electrons in graphite induced by the interaction with the electromagnetic radiation. This results in interfacial polarisation (Maxwell-Wagner-Sillars polarisation), which in an alternating electromagnetic field results in Joule heating.

Non-reflective magnetic materials, such as hematite, can also be heated using microwaves as the oscillating magnetic field of the electromagnetic wave results in the loss in the magnetic hysteresis, and loss in Eddy currents.

The penetration depth of microwaves in insulators can be expressed as [89, 90]

$$d_{\text{penetration}} = \frac{1}{2\pi f \cdot \sqrt{\begin{array}{l} 0.5 \cdot \mu_0 \cdot \mu_r' \cdot \varepsilon_0 \cdot \varepsilon_r' \\ \cdot \left(\left(\frac{\mu_r''}{\mu_r'} \cdot \frac{\varepsilon_{r,\text{eff}}''}{\varepsilon_r'} \right) - 1 + \sqrt{\left(\frac{\mu_r''}{\mu_r'} \cdot \frac{\varepsilon_{r,\text{eff}}''}{\varepsilon_r'} \right)^2 + \left(\frac{\mu_r''}{\mu_r'} \right)^2 + \left(\frac{\varepsilon_{r,\text{eff}}''}{\varepsilon_r'} \right)^2 + 1} \right) \end{array}}}$$

which for non-magnetic materials ($\mu_r'' = 0$; $\mu_r' = 1$) simplifies to

$$d_{\text{penetration}} = \frac{1}{2\pi f \cdot \sqrt{0.5 \cdot \mu_0 \cdot \varepsilon_0 \cdot \varepsilon_r' \cdot \left\{ \sqrt{1 + \left(\frac{\varepsilon_{r,\text{eff}}''}{\varepsilon_r'} \right)^2} - 1 \right\}}}$$

The penetration depth decreases strongly as a function of the frequency. For example, the penetration depth in water at 20 °C at 0.915 GHz is 9.2 cm, at 2.45 GHz it is 1.4 cm and at 24.5 GHz it is 0.07 cm. Table 4.9 shows the penetration depth of different materials. Materials, such as glass and polyethylene are relatively transparent to microwaves and thus have large penetration depths. Good absorbing materials have penetration depths of up to a few cm, indicating a strong uptake of the electromagnetic radiation by the material.

In metals, the microwave radiation interacts with the electrons in the conductor, and electric charge will accumulate at the surface. The internal electric field decays rapidly, due to the high conductivity of these materials resulting in a small penetration depth. The accumulated electrons at the surface may cause an electric spark or an electric arc, particularly at the edges of metal particles [92].

The efficiency of microwave heating can be defined via the ratio of the difference between the inlet power and the reflected power relative to the inlet power.

Table 4.9 Penetration depth of various materials at 20 °C at 2.45 GHz.

Material	$d_{\text{penetration}}$	
Water	1.4 cm	
Methanol	2.9 cm	
Ethanol	1.9 cm	
Magnetite	2.9 cm	[90]
Carbon	38 μm	[91]
Iron	1.3 μm	[92]
Nickel	2.5 μm	[92]

The experimentally determined efficiency for heating food materials is reported to be between 85% and 90% [93].

4.4.2.3 Induction Furnace

Induction furnaces can be used to heat up metals till their melting point and are used for (surface) hardening, tempering of metals, soldering, smelting of metals and production of alloys. Heating by induction is seen as a relatively clean operation, avoiding direct contact of electrical power with the material to be treated [94]. The process can supply relatively high heating rates with good control of temperature.

An induction furnace works through the generation of current in the material to be heated, induced by an alternating current in a coil surrounding the material. An alternating current in a coil induces an alternating magnetic field, which induces an electric motor force in the second, conductive material, resulting in a flow of Eddy currents opposing the induced magnetic field. The Eddy current will experience internal resistance, thereby generating heat (Joule heating). In magnetic materials, heat will also be developed through hysteresis loss. These materials are then also heated through Joule's law as above.

The currents will be mainly present near the surface of the conductor (skin effect), as the intensity of the electro-magnetic field will decrease with increasing distance through the conductor. Hence, the current density will decrease with increasing distance from the surface, r:

$$J = J_{\text{surface}} \cdot e^{-\frac{r}{\delta}} \text{ with } \delta = \sqrt{\frac{\rho}{\pi \cdot \mu_0 \cdot \mu_r' \cdot f}}$$

At a distance, r, equal to the penetration depth, δ, the current density will have dropped to e^{-1} (37%) of the current at the surface. The penetration depth depends on the resistivity of the material, ρ, the permeability of the material, μ_r', and the frequency of the alternating current. A low frequency (~1–2 kHz) will generate heating throughout most of the material as the penetration depth may start approaching the thickness of the material. A high frequency (2–50 kHz) will generate a thin skin in which Eddy currents are present and, thus, heats a thin skin layer via inductive heating, with the rest of the material being heated via conductive heating. Inductive heating using a very high frequency (>50 kHz) can be used for surface heating and welding.

Figure 4.19 shows different configurations for induction furnaces that are typically used for smelting of metals. The direct core induction furnace comprises a primary winding (typically hollow copper tube through which water flows for cooling) around a magnetic core (typically ferrosilicon), which is supplied with an alternate current. The charge to be heat treated is kept in a crucible around the magnetic core, and acts as a single secondary winding. This results in high currents in the charge, although the magnetic coupling is rather poor [95], which is overcome by utilising low frequencies (~10 Hz). Furthermore, the very high currents may give rise to bubbles in the melt, thereby interrupting the secondary current (so-called pinch effect). This type of furnace is poorly suited for intermittent usage and molten metal is required to be present to start up the operation. The pinch effect may be avoided by having the charge lined with carbon.

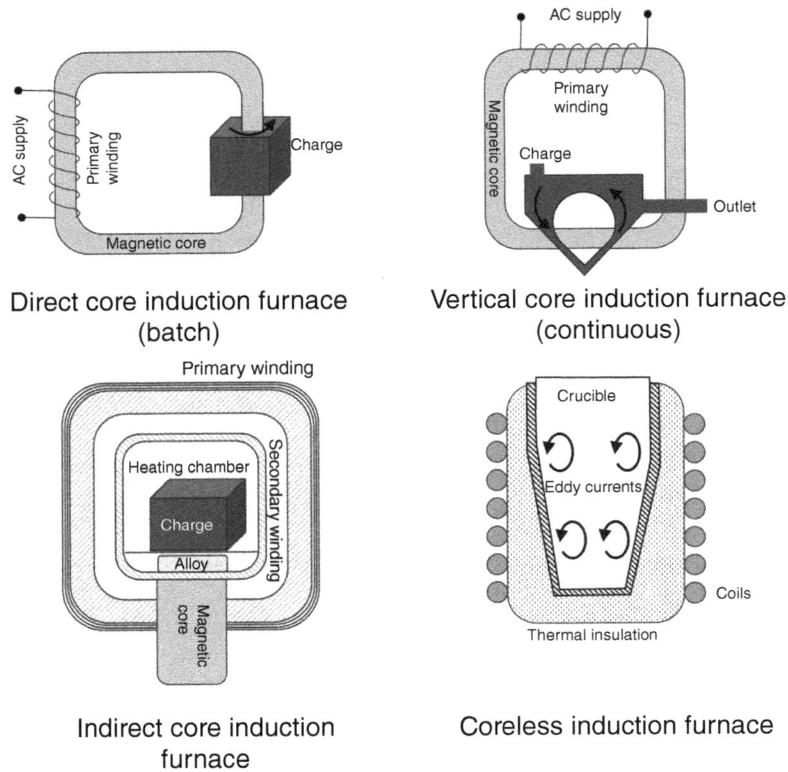

Figure 4.19 Types of induction furnaces (redrawn from EEEguide.com [95]).

The pinch effect, and the associated interruption of the secondary circuit, is minimised by circulating the molten material in a V-shaped channel, enforcing a vertical movement of the liquid in the so-called vertical core induction furnace (the channels may also be U-shaped or even vertically). This type of induction furnace is widely used for melting/refining of non-ferrous metals, such as brass. The power rating of these furnaces ranges from 60 to 300 kW. They operate with a voltage of 600 V and a frequency of ca. 50 Hz at an efficiency of ca. 75% [95].

In the indirect core induction furnace, the alternating current passes through the primary winding and is passed through the magnetic iron core to the wall of the furnace, which acts as the secondary winding. A part of the magnetic circuit is an alloy, which loses its magnetic properties at a given temperature. This will decrease the inductive effect and decrease the current through the secondary winding, thus supplying accurate temperature control to the furnace.

In the coreless induction furnace, the thermally insulated crucible is surrounded by coils supplied with a high-frequency alternating current, which is necessary to induce the required high voltage in the secondary circuit. Most coreless induction furnaces operate with a frequency of 500–1000 Hz, although smaller units can use much higher frequencies [95]. The coils are hollow tubes cooled by water circulation.

The presence of thermal insulation, and in particular the air gap, results in a lower power rating (similar to energy efficiency) which can be improved to near unity by employing static capacitors in parallel to the furnace [95]. The coreless induction unit is mainly used in the steel industry.

4.4.2.4 Electric Arc Furnace (EAF)

Arc furnaces are used to melt solid, electrical conducting, raw materials to a liquid as fast as possible and for further refining in metal processing. Different types of arc furnaces exist, viz. DC single electrode arc furnace, AC arc furnaces and plasma arc furnaces.

A DC arc furnace (see Figure 4.20) consists of a single electrode passed through the roof and electrically connected to the anode at the bottom of the furnace [96] (an alternative design would be having both the anode and the cathode coming through the roof [97]). The furnace is charged with electrically conductive materials, and the electrode is lowered to a point above the material. A voltage difference (600–900 V) will establish an arc between the cathode and the surface of the charge in contact with the anode. The arc transfers most of the heat directly to the charge material by close-proximity radiation, convection and conduction [98]. The resulting high current through the charge will induce Joule heating and melting of the metal. The electrode may also be moved into the charge, resulting in arcing through the charge, in a so-called submerged arc furnace. This type of furnace can have an efficiency of 80–90% [99].

AC arc furnaces have been much more common due to the availability of AC power. Here, three electrodes are used. They can be in direct contact with the material in the furnace (direct EAF). Alternatively, an arc can be established between the electrodes and the material is heated via radiative heat transfer (indirect EAF). The furnace can be operated at reduced pressure to remove dissolved gases from the charge (vacuum arc furnace). Makarov [100] analysed the efficiency of arc furnaces as a function of their load and size with small furnaces showing an efficiency of ca. 55–57% and large furnaces 78–80%.

Higher temperatures can be achieved in a plasma electric arc furnace, with the arc being supported by an ionising gas. The plasma is generated externally using a high-frequency electric field, after which it is fed to the arc at the end of the hollow electrode. The efficiency, defined as the utilisation of electric energy, of a 200 kW plasma arc furnace was determined to be ca. 44% [101].

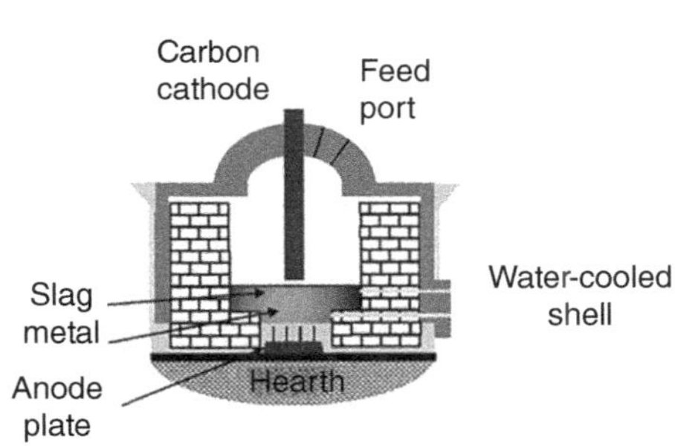

Figure 4.20 DC-type of arc furnace (redrawn from Kotzé [96]).

4.5 Heat Integration

The process industry needs streams at a wide variety of temperatures, depending on the singular processes within the overall process. As a result, process streams are also available at a wide variety of temperatures due to the processes from which they originated. For example, a reactant feed stream may be available at a given temperature and may need to be heated up prior to feeding it to the reactor (see Figure 4.21a). The product stream from the reactor may need to be cooled prior to product separation. Hence, integrating the stream after the reactor with the stream before the reactor to recover as much as possible of the heat will thus reduce the overall energy consumption in the process (a possible configuration is given in Figure 4.21b; note that further heating and further cooling after the integrated heat exchange may be required to accurately meet the target temperatures).

Whilst heat integration can be done via a heat transfer fluid, e.g. steam or molten salts, it is more commonly done by directly exchanging heat between a relatively hot stream needing cooling and a relatively cool stream needing heating (see Figure 4.22a). Heat exchangers are typically designed based on the so-called logarithmic mean temperature difference

$$\Delta T_{\mathrm{LM}} = \frac{(T_{\mathrm{H,out}} - T_{\mathrm{C,in}}) - (T_{\mathrm{H,in}} - T_{\mathrm{C,out}})}{\ln\left(\frac{T_{\mathrm{H,out}} - T_{\mathrm{C,in}}}{T_{\mathrm{H,in}} - T_{\mathrm{C,out}}}\right)}$$

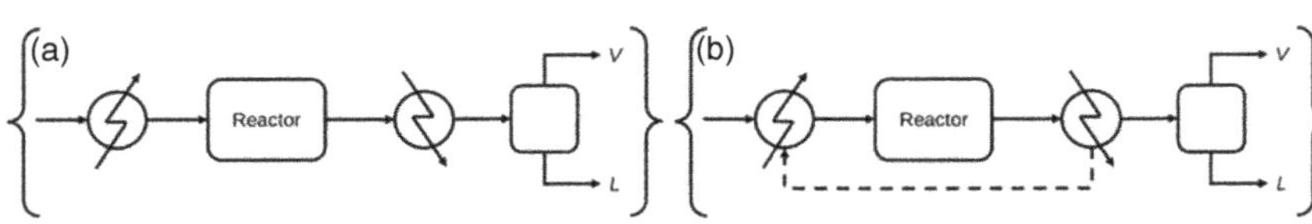

Figure 4.21 Simple heating cooling diagram around a reactor, (a) without and (b) with heat integration about the reactor, solid line = process stream, dashed line = energy stream.

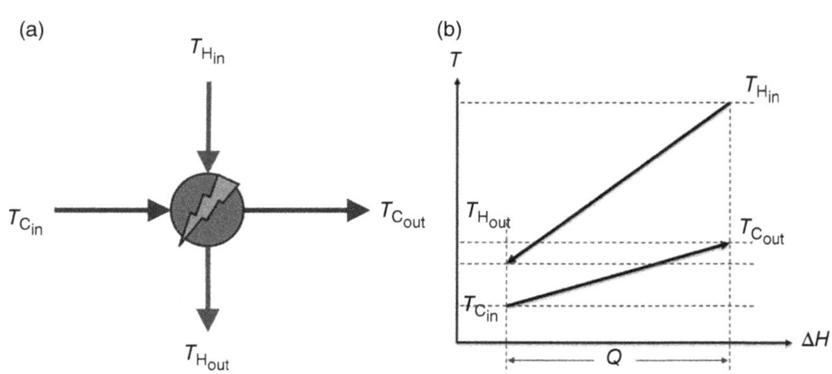

Figure 4.22 (a) Heat exchanger set up between cold and hot streams for a given set of temperatures ($T_{\mathrm{H,in}}$, $T_{\mathrm{H,out}}$, $T_{\mathrm{C,in}}$, $T_{\mathrm{C,out}}$) and (b), relative change in temperature and enthalpy for a hot stream being cooled and a cold stream being heated counter-currently.

(definition of temperatures given in Figure 4.22) and the required heat-transfer area is given by:

$$a = \frac{\dot{Q}}{U \cdot \Delta T_{LM}} = \frac{\dot{m}_c \cdot C_{P,c} \cdot (T_{c,out} - T_{c,in})}{U \cdot \Delta T_{LM}} = \frac{\dot{m}_H \cdot C_{P,H} \cdot (T_{H,out} - T_{H,in})}{U \cdot \Delta T_{LM}}$$

Thus, the heat exchange area required increases when the temperature differences on each side of the heat exchanger (ΔT_{LM}) becomes smaller.

Heat exchangers may be run co-currently or counter-currently. In a co-current operation, the two streams would approach the same temperature at the end of the heat exchanger, which may result in a lower heat recovery than in a counter-current operation. Figure 4.22b shows the heat transfer in a counter-current operation in an enthalpy-temperature diagram with each stream indicated by a vector. The x-axis is given as the change in enthalpy rather than enthalpy and is thus a relative scale, and vectors can thus be shifted along the x-axis. The hot stream is denoted as the stream giving off heat and the cold stream as the stream receiving heat. In counter-current operation, the temperature of the cold stream leaving the heat exchanger can be higher than the temperature of the hot stream leaving the heat exchanger.

The maximum temperature of the cold stream leaving the counter-current heat exchanger may approach the temperature of the hot stream coming into the heat exchanger, whilst the minimum temperature of the hot stream leaving the heat exchanger may approach the temperature of the cold stream entering the heat exchanger. The temperature of the hot stream leaving the heat exchanger will not reach the temperature of the cold stream entering the heat exchanger and vice versa, as this would require an infinitely large heat exchange area to achieve the required heat transfer.

It is unlikely that, as is depicted in Figure 4.22b, the two streams have the same enthalpy difference. More commonly, the required enthalpy difference of one stream is larger than that of the other and additional cooling or heating is then required (see Figure 4.23a,b). Both further heating of the cold stream and further cooling of the hot stream may be required (see Figure 4.23c). For example, an adiabatically operating reactor may require a feed to be heated from 100 °C to an inlet temperature of 800 °C with the effluent at 700 °C, due to the endothermic reaction taking place. The effluent may be required to be cooled down to 50 °C for separation purposes, meaning that further cooling will be needed after the heat exchange. The optimum configuration for this situation will be based on process economics looking at the cost for the heat exchanger and the saving in the energy costs as a function of the approach temperature (the minimum difference in temperature between the stream giving off heat, the hot stream and the stream receiving heat, the cold stream).

4.5.1 Pinch Analysis

Whilst heat integration across a single pair of streams, as shown above, may be intuitive, heat integration across an entire process can be more complex. The most popular methodology for heat integration calculation is the so-called pinch-analysis [102], in which a heat exchange network is developed with the minimum net duty for the hot and cold utility streams.

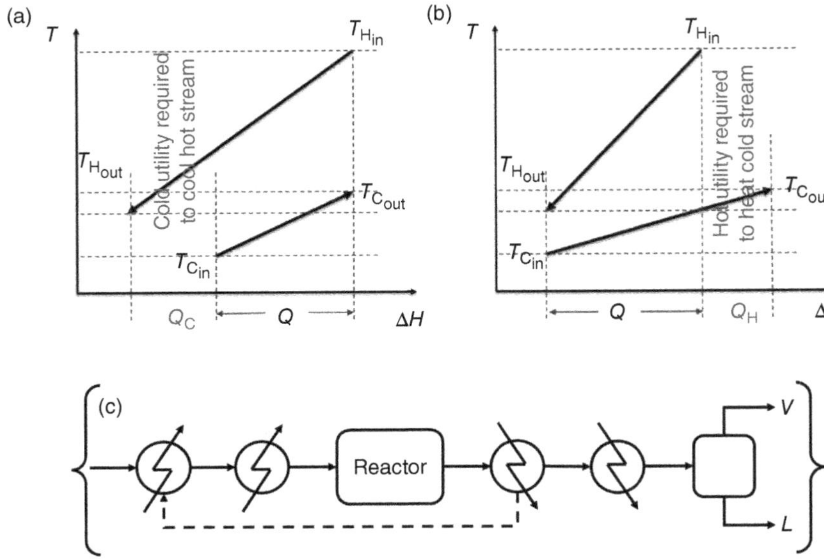

Figure 4.23 (a) Hot stream requiring excess cooling compared to cold stream, requiring cold utility, (b) cold stream requiring excess heating compared to heat utilisation of hot stream and (c) requiring extra heating and cooling after heat exchange.

Graphically, pinch analysis draws so-called composite curves of all the process streams requiring heating and cooling (a simple example of which is shown in Figure 4.24). All streams that are included in the composite cold stream are streams which are receiving heat. The composite cold stream is represented on the enthalpy-temperature graph with the enthalpy of the cold stream at the lowest temperature defined as zero. The slope of each of portion of these composite lines represents the reciprocal value of the mass flow rate of the stream multiplied by the specific heat of the stream. In the overlapping region(s), there are multiple streams, and the slope of the composite is thus given by $\frac{1}{\sum_{i=1}^{k} \dot{m}_i \cdot C_{P,i}}$.

A similar composite curve is generated for the hot streams. Where there is an overlap of the hot and the cold streams, the cold curve can be shifted along the x-axis,

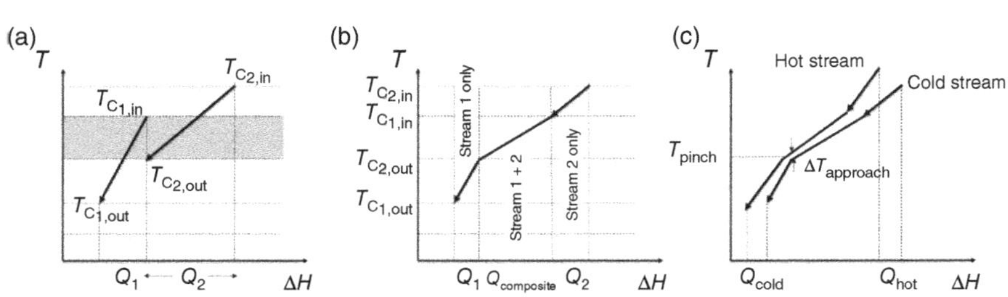

Figure 4.24 Constructing a composite curve (a and b) and graphical representation of 'pinch analysis' showing the pinch temperature, the amount of cold utility and hot utility required (c).

so that there a point is established where the temperature difference between the hot and cold composite curves is at the minimum approach temperature for heat exchange – this is the pinch. In other words, the pinch is defined by the point at which the hot composite curve and the cold composite curve are closest, and the temperature difference corresponds to the minimum approach temperature.

Pinch analysis is a powerful tool which can calculate the minimum duties, both hot and cold, which a process will require, which sets a target for heat integration design. As the heat capacities and temperature changes (and thus heat flows) are known, this can also be used to calculate the required heat exchange area, and thus the capital cost for the required heat exchangers. Further, it can be shown that hot utility is only required above the pinch, and cold below the pinch, further setting targets for heat integration design.

Pinch analysis can set targets for possible heat integration and saving on cold and hot utilities. However, it does not consider process economics. This can be taken into account by developing a heat exchange network whilst minimising the total annualised cost (capital expenditure plus utility cost considering the discount rate and write-off period for the heat exchanger) [103]. Furthermore, location is not considered, which may result in an optimum solution requiring exchange between streams which are far apart from each other, and thus, practically not feasible (also noting that transporting hot streams may result in further heat loss to the environment). This may be aggravated by safety concerns when transporting these streams over prolonged distances. Furthermore, the controllability of a plant may be affected by extensive heat exchange [104], as the interactions induced by heat exchanger networks may resonate throughout the plant. Hence, heat exchange networks should be limited to sections of a plant with well-defined interactions to limit knock-on effects over the whole plant.

Heat exchanger networks do not negate the need for hot and cold utilities, specifically during start-up (and to a lesser extent during shutdown) of a plant. Hence, plants with extensive heat exchange will still need external heat sources (e.g. boilers) during start-up. These boilers may be replaced in future by electrically powered devices, such as resistive heaters, provided sufficient power is delivered to the plant in those periods.

4.5.2 The Future of Heat Integration

Heat integration has been an important part of process design to minimise energy usage within a chemical plant and thus reduce its carbon emissions. This has mainly been achieved by retrofitting existing plants. However, it is an important consideration in the design of new processes and plants, where the energy saving versus the control considerations must be weighed against each other.

Most of heat integration cases investigate the benefits for a single operation. One of the advantages of a so-called Verbund site (a large chemical process site with the production of a variety of different chemicals) is the heat integration between various production units, thus reducing the overall energy consumption and the associated carbon footprint. The disadvantage of a highly integrated operation lies in the dependency of the production units on each other.

Excess heat on a plant can be discarded, converted to power or transferred 'over the fence' by heat integration with neighbouring plants [105]. Heat loss during transport from the plant selling excess heat to the plant/community receiving heat will be a major consideration. Furthermore, the receiving party will not have absolute control over the amount of heat received. Hence, good communication between the receiving and producing sites will need to be maintained. Nevertheless, this will be a feature requiring further attention going forward.

Further, plants may be able to sell heat to a community for district heating [106], where the maximum temperature needed for heating purposes is less than 100 °C. Hence, relatively low-quality (low-temperature) heat will be required, implying that heat loss considerations will play less of a factor compared to higher-temperature heat integration. This may become a reality, especially in towns built around major chemical plants.

4.6 Heat Storage

Intermittency of power supply from renewable energy generation will necessitate some form of energy storage to continue operation when power is not available (see also Chapters 3 and 5). Heat storage is another possible way to store energy, although it cannot be stored indefinitely as some forms may degrade by heat transfer to the surroundings. Heat is not usually stored on chemical plants in the classical sense, but heat storage has become more of a focus area over the last decade with the need to store renewable energy (for heat storage in CSP see Section 2.6.2 and in CAES, see Section 3.3). Heat storage is known in the metallurgical industry, specifically when operating in areas with a less stable grid, or when needing to operate at times with a greater power availability.

The principle behind heat storage is relatively straightforward, with heat being given off from the source generating heat to the storage, with the temperature of the stored heat being lower than the temperature of the heat being generated. Heat is later removed from the storage to the required process stream, with the required temperature of the process stream being lower than the temperature of the stored heat.

Heat can, in principle, be stored as sensible heat, latent heat or chemical heat. The storage of energy as sensible heat requires a material with a high specific heat to store as much heat in a minimal amount of material, a high thermal conductivity so that the heat is readily distributed over the whole medium and the heat is taken up at a relatively high rate, a high melting or boiling point so that heat can be retrieved at a high temperature and a high thermal stability so that the material does not deteriorate with time. Furthermore, the material should preferably be present as a liquid, or even a solid, and should not undergo a phase change over the temperature range of interest. The energy density of the material required to store heat is thus given by

$$\frac{Q_{\text{stored}}}{V} = \int_{T_L}^{T_u} C_p \cdot \rho \cdot dT \sim C_p \cdot \rho \cdot (T_{\text{reservoir,max}} - T_{\text{reservoir,min}})$$

with the storage medium being operated between its maximum and minimum temperature.

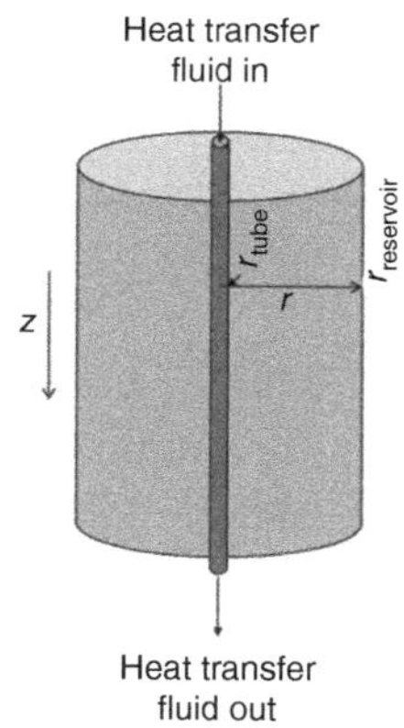

Energy balance

Reservoir (res): $\left(\frac{\rho \cdot Cp}{k}\right)_{res} \cdot \frac{dT_{res}}{dt} = \frac{d^2T_{res}}{dr^2} + \frac{1}{r} \cdot \frac{dT_{res}}{dr}$

Boundary conditions $\quad t = 0 \quad T_{res} = T_{min,res};$

$r = r_{tube} \quad U \cdot \left(T_{res,r=r_{tube}} - T_{htf}\right) = k_{res} \cdot \left(\frac{dT_{res}}{dr}\right)_{r = r_{tube}}$

$r = r_{reservoir} \quad \left(\frac{dT_{res}}{dr}\right)_{r = r_{res}} = 0$ (well insulated)

Heat transfer fluid (htf) (1-dimensional model):

$\frac{(\rho \cdot Cp)htf}{U} \cdot \left(\frac{dT_{htf}}{dt} + v \cdot \frac{dT_{htf}}{dz}\right) = -\frac{2}{r} \cdot \left(T_{htf} - T_{res}\right)$

Boundary conditions $\quad t = 0 \quad T_{htf} = T_{htf,0};$

$z = 0 \quad T_{htf} = T_{htf,inlet};$

Figure 4.25 Modelling the temperature in a heat reservoir with a single heat exchange tube [107] (ρ: density; C_p: specific heat on a mass basis; k: thermal conductivity; U: overall heat-transfer coefficient; v: linear velocity of heat-transfer fluid) (adapted from Barz et al. [107]).

The transport of heat from the source generating heat to the storage and the transport of heat from storage to the process requiring heat can be achieved using a heat-transfer fluid, which is done by means of an embedded pipe system in contact with the heat reservoir. A single pipe immersed in the heat reservoir can be modelled as shown in Figure 4.25 [107]. The rate at which the temperature changes within the reservoir is governed by $\frac{k}{\rho \cdot C_P}$, whereas the transfer of heat is governed by $\frac{U}{(\rho \cdot C_P)_{\text{heat transfer fluid}}}$ and the temperature difference between the heat-transfer fluid and the temperature in the reservoir. Thermal stratification may occur within the reservoir if the rate of heat transfer within the reservoir is slow [108].

Water would be a suitable substance to use as a heat-transfer fluid, given its high specific heat and high thermal conductivity (see Table 4.10) but would need to be operated at a relatively high pressure. For instance, the pressure should be ca. 90 bar to operate between temperatures of 250 and 300 °C. At atmospheric pressure, heating oil can be used to operate in the same temperature regime, but it will require almost twice the volume to store the same amount of energy in the form of heat (albeit now at atmospheric pressure). Furthermore, the low thermal conductivity of heating oil will result in quite a low rate of heat being taken up. Molten salt can operate at higher temperatures, requires less volume than heating oil, but the temperature response to changes is rather slow. Furthermore, molten salts are corrosive, thus requiring special handling. Molten metals can also be used to store heat. For example, molten tin can store heat in a wide temperature range, although its specific heat is not that high, and thus still requires a relatively large volume to store heat,

Table 4.10 Material choice for sensible heat storage.

Material	Conditions	C_p (kJ/kg/°C)	ρ (kg/m^3)	k (mW/m/°C)	Spec. volume (m^3/°C/MWh)	$k/(\rho\ C_p)$ (10^{-6} m^2/s)
Water	$T < 300\,°C$[a]	4.2–5.8	713–1000	553–662[b]	920[b]	0.14[b]
Heating oil[c]	$T = 250–300\,°C$[d]	2.4–2.6[c]	809–848[c]	95–100[c]	1770	0.05
Molten salt[e) f]	$T = 220–540\,°C$[e]	1.55[f]	1773[f]	460[g]	1310	0.002
Molten tin	232–2600 °C[d]	0.24[h]	5821–6982[i]	3000–6000[j]	2150	2.15
Cast iron	<1175 °C	0.52[k]	7200[k]	470[k]	960	0.13
Concrete	<500 °C[l]	0.96[m]	960[m]	550[m]	3900	0.60

a) Up to ~90 bar.
b) At ca. 300 °C and 90 bar.
c) Therminol [109].
d) Atmospheric pressure.
e) 50% $NaNO_3$/50% KNO_3.
f) Engineering Toolbox [110].
g) Zhao et al. [111].
h) Engineering Toolbox [112].
i) Mehdipour et al. [113].
j) Giordanengo et al. [114].
k) Haynes [26].
l) Skinner et al. [115].
m) Barbhuiya et al. [116].

although the rate of energy distribution over the reservoir is high. Solids can also be used as a heat storage medium. For instance, the volume of cast iron required to store heat, and the rate of heat distribution within the reservoir is roughly the same as that of (pressurised) water. Concrete (or rock, sand, soil) is a cheap alternative as a heat storage medium. Its relatively low specific heat results in quite a large volume being required to store heat, but heat dissipation throughout the reservoir is expected to be relatively fast.

Heat can also be stored as latent heat; the reservoir can now operate at a much more constant temperature than when storing sensible heat. The desired phase transition is preferably a liquid-solid phase transition, due to the small volume change between the phases [117], rather than a vapour–liquid equilibrium. Although the latter has a higher enthalpy change, the change in density of the material needs to be dealt with for this to become an effective storage method. The choice of material is dictated by the choice of the phase transition temperature, change in enthalpy upon melting, adequate rate of melting/crystallisation, low vapour pressure, chemical stability and cost [118]. Organics, paraffins and fatty acids, have been proposed as so-called phase-change-materials (PCMs) [118], specifically when incorporated into materials such as concrete [119] as it will enhance the heat storage capacity. However, these materials have a relatively low melting point (<100 °C), thus limiting their application to domestic situations. Metals and metal salts can have melting points in the temperature range of interest. For instance, lithium-based salts and their eutectic mixtures can have a melting temperature between 8 and 850 °C and a change in enthalpy upon melting between 150 and 820 kJ/kg [120]. Hence, these materials are potential candidates to store high-quality heat, although the cyclability of these materials will need to be studied.

Heat can also be stored in chemicals using a reversible reaction. A variety of reactions have been proposed (e.g. redox reactions), but the reaction products should preferably be in the same phase as the reagents, the reaction should be reversible and be highly selective.

4.7 Outlook

Heating is an essential part of the chemical industry, needed to facilitate the desired chemical transformations and separations. Heat is traditionally generated mostly by combusting fossil fuels, after which it is transferred to the required process streams and processes. Fossil fuels have been preferred as the direct transformation of the chemical energy in fossil fuels into heat is more efficient than the use of electric power generated from fossil fuels, due to the low thermodynamic efficiency of the power generation cycle (typically in the order of 30–50%; see Chapter 2). Thus, the current cost of electric heating is greater than fossil fuel-based heating.

Over recent decades, the chemical industry has reduced its energy consumption drastically, primarily through heat integration, improving its carbon footprint. However, there are limits to direct heat integration on process plants due to operability concerns, as variations in sections of a plant may cascade through the plant, rendering it unstable. Here, heat storage will become necessary as an essential tool to minimise the effects of process fluctuation. This may become even more urgent when generating heat in process streams rather than from fuels. Heat may become a commodity which can be traded over the fence, if efficient heat storage facilities and distribution networks can be set up.

The drive to reduce CO_2 emissions requires the industry to reassess its options to heat up streams, with a strong call for electrification of the industry [121]. This may, at first glance, seem like an attractive option with the cost of power generation from renewable energy becoming on par with or even lower than the cost of power generation from fossil fuels. However, direct electric heating will require a significant further drop in the cost of power generation using renewable energy for the cost to become comparable to the cost of direct heating using fossil fuels. At a cost parity for the power generation from fossil fuels and renewable energy, the cost of heating using electric power is then equal to $^1/_{\eta_{\text{power generation}}}$ the cost of heating by combustion fuels (assuming equal efficiency in the furnace) and is thus typically a factor 2.5–3 times more expensive. This factor is, of course, reduced when considering the efficiency of furnaces/boilers versus the efficiency of, e.g. Joule heating. Nevertheless, the cost of electricity from renewable energy will need to be significantly lower than the cost of power generation from fuel combustion to make electric heating economically more advantages (targeting an electricity price from renewable energy of ca. 50% of the electricity price from fuel combustion), not taken into consideration other advantages electric heating may have (e.g. carbon tax/credits).

Heat pumps will lower the barrier to electrify the process industry, provided their C.O.P. will be higher than the cost ratio of electric power relative to the cost of heating using fossil fuels. Otherwise stated, heat pumps will become attractive if their C.O.P. is higher than 2.5–3 if the cost for power generation using fossil fuels and using renewable energy are similar. Heat pumps can be readily implemented into low-temperature processes as they can transform waste heat into higher-quality heat at temperatures of up to 120–150 °C [122, 123].

The electrification of the process industry will need to consider both the required temperature range as well as the material to be heated. For instance, heating of aqueous streams (or streams containing other polar compounds) at low temperatures may be achieved by using magnetron heating, whereas heating of this type of stream using a resistive heater will require additional controls to avoid overheating. The decision concerning the type of furnace is thus not only dictated by the type of heater but also control.

The energy source for high-temperature processes may be more difficult to replace. Resistive heaters are being developed to operate at these temperatures, but durability of heating elements at the targeted operation conditions still needs to be proven. For select applications, primarily metals processing, electric arc or induction furnaces may be used. For some operations, e.g. the cement industry, electrification may not be practically feasible and alternative fuels, such as biomass, biogas, hydrogen or ammonia, will have to be investigated.

Intermittence and fluctuations in the power supply (as may be anticipated relying on renewable energy resources) will affect operability of plants, when relying on electric heating powered by external sources. Various measures can be put into place to mitigate these effects. For instance, heating can be halted/minimised keeping the material hot by minimising heat loss to the surroundings, with start-up of production up restoring power supply with the required minimal power input supplied, e.g. by battery energy storage system (BESS).

References

1. US Energy Information Administration (2024). Total energy [Online]. Available: https://www.eia.gov/totalenergy/ (accessed 5 September 2024).
2. Ritchie, H. (2024). Tracking global data on electric vehicles, OurWorldinData.org [Online]. Available: https://ourworldindata.org/electric-car-sales (accessed 5 September 2024).
3. Office of Energy Efficiency & Renewable Energy (2019). Static Sankey Diagram of Process Energy in U.S. Manufacturing Sector (2014 MECS), US Department of Energy [Online]. Available: https://www.energy.gov/eere/iedo/static-sankey-diagram-process-energy-us-manufacturing-sector-2014-mecs (accessed 5 September 2024).

4 Naegler, T., Simon, S., Klein, M., and Gils, H. (2015). Quantification of the European industrial heat demand by branch and temperature level. *International Journal of Energy Research* 39: 2019–2030.

5 Engineering Toolbox (2003). Specific heats of liquids [Online]. Available: https://www.engineeringtoolbox.com/specific-heat-fluids-d_151.html (accessed 6 September 2024).

6 Dow Chemicals (2024). Technical Data Sheet: Dowtherm Q [Online]. Available: https://www.dow.com/documents/176/176-01467-01-dowtherm-q-tds.pdf?iframe=true (accessed 6 September 2024).

7 Engineering Toolbox (2008). Liquids – thermal conductivities [Online]. Available: https://www.engineeringtoolbox.com/thermal-conductivity-liquids-d_1260.html (accessed 6 September 2024).

8 National Institute of Standards and Technology (2023). NIST Chemistry Webbook, US Departement of Commerce [Online]. Available: https://webbook.nist.gov/cgi/cbook.cgi?Name=Ammonia (accessed 3 March 2024).

9 Chesser, M., Lyons, P., O'Reilly, P., and Carroll, P. (2021). Air source heat pump in-situ performance. *Energy and Buildings* 251: 111365.

10 Wang, Y., Wong, K., Liua, Q. et al. (2012). Improvement of energy efficiency for an open-loop surface water source heat pump system via optimal design of water-intake. *Energy and Buildings* 51: 93–10.

11 Kindaichi, S., Nishina, D., Wen, L., and Kannaka, T. (2015). Potential for using water reservoirs as heat sources in heat pump systems. *Applied Thermal Engineering* 76: 47–53.

12 Chen, X. and Zhang, G. (2014). Study on the application of closed-loop lake water heat pump systems for lakefront buildings in south China climates. *Journal of Renewable and Sustainable Energy* 6: 033125.

13 Olabi, A.-G., Mahmoud, M., Obadein, K. et al. (2023). Ground source heat pumps: recent progress, applications, challenges, barriers, and role in achieving sustainable development goals based on bibliometric analysis. *Thermal Science and Engineering Progress* 41: 101851.

14 McLinden, M., Brown, J., Brignoli, R. et al. (2017). Limited options for low-global-warming-potential refrigerants. *Nature Communications* 8: 14476.

15 Zini, A., Socci, L., Vaccaro, G. et al. (2024). Working fluid selection for high-temperature heat pumps: a comprehensive evaluation. *Energies* 17: 1556.

16 El Samad, T., Żabnieńska-Góra, A., Jouhara, H., and Jouhara, A. (2024). A review of compressors for high temperature heat pumps. *Thermal Science and Engineering Progress* 51: 102603.

17 Wongsuwan, W., Kumar, S., Neveu, P., and Meunier, F. (2001). A review of chemical heat pump technology and applications. *Applied Thermal Engineering* 21 (15): 1489–1519.

18 Ezgi, C. (2021). Design and thermodynamic analysis of waste heat-driven zeolite–water continuous-adsorption refrigeration and heat pump system for ships. *Energies* 14 (3): 699.

19 Kato, Y., Takahashi, F., Watanaba, A., and Yoshizawa, Y. (2001). Thermal analysis of a magnesium oxide/water chemical heat pump for cogeneration. *Applied Thermal Engineering* 21 (10): 1067–1081.

20 Kim, K.J., Feldman, K. Jr., Lloydt, G., and Razanit, A. (1997). Compressor-driven metal-hydride heat pumps. *Applied Thermal Engineering* 17 (6): 551–560.

21 Pardo, P., Deydier, A., Anxionnaz-Minvielle, Z. et al. (2014). A review on high temperature thermochemical heat energy storage. *Renewable and Sustainable Energy Reviews* 32: 591–610.

22 Guo, J. and Huai, X. (2012). The application of entransy theory in optimization design of isopropanol–acetone–hydrogen chemical heat pump. *Energy* 43: 355–360.

23 Aristov, Y., Parmon, V., Cacciola, G., and Giordano, N. (1993). High-temperature chemical heat pump based on reversible catalytic reactions of cyclohexane-dehydrogenation/benzene-hydrogenation: comparison of the potentialities of different flow diagrams. *International Journal of Energy Research* 293–303.

24 Kawasaki, H., Watanabe, T., and Kanzawa, A. (1999). Proposal of a chemical heat pump with paraldehyde. *Applied Thermal Engineering* 19: 133–143.

25 Dai, S., Yin, Y., Li, W., and Zhang, F. (2020). Thermodynamic analysis of a novel chemical heat pump cycle based on the physical-chemical thermal effects of reversible reaction. *Energy Conversion and Management* 225: 113419.

26 Haynes, W. (ed.) (2010). *CRC Handbook of Physics and Chemistry*, 91e, 2670. Boca Raton: Taylor & Francis.

27 Heating Elements Alloys (2024). Materials Used for Heating Elements [Online]. Available: https://www.heating-element-alloy.com/article/heating-elements-materials.html#:~:text=Copper%20Nickel%20%28CuNi%29%20Alloy%20is%20used%20in%20making,continuous%20operation%20at%20a%20temperature%20up%20to%20600%C2%B0C (accessed 30 September 2024).

28 Kanthal (2024). Kanthal® A-1, 5 September 2024 [Online]. Available: https://www.kanthal.com/en/products/material-datasheets/wire/resistance-heating-wire-and-resistance-wire/kanthal-a-1/#:~:text=Kanthal%20%C2%AE%20A-1%20wire%20is%20a%20ferritic (accessed 5 September 2024).

29 Kanthal (2022). Kanthal super electrical heating elements – Handbook, June 2022 [Online]. Available: https://www.kanthal.com/en/products/furnace-products/electric-heating-elements/molybdenum-disilicide-heating-elements/#:~:text=Kanthal%C2%AE%20Super%20-%20electric%20molybdenum (accessed 30 September 2024).

30 AZO Materials (2001). Silicon carbide (SiC) properties and applications [Online]. Available: https://www.azom.com/properties.aspx?ArticleID=42 (accessed 21 March 2025).

31 Knudsen, J., Hottel, H., Sarfim, A. et al. (1997). Heat and mass transfer. In: *Perry's Chemical Engineers' Handbook* (ed. R. Perry, D. Green, and J. Maloney). New York: McGraw-Hill Education.

32 Mishra, S., Poonia, H., Das, A.K. et al. (2014). Analysis of conduction-radiation heat transfer in a 2D enclosure using the lattice Boltzmann technique. *Numerical Heat Transfer, Part A: Applications* 66: 669–688.

33 DwyerOmega (2024). Heavy duty cartridge heaters [Online]. Available: https://assets.omega.com/pdf/process-heaters-and-coolers/cartridge-heaters/C_CTRH_HEATER.pdf (accessed 10 September 2024).

34 Liu, Y., Xue, Q., Zuo, H. et al. (2021). Experimental study of H_2 and/or N_2 addition effects on CO/CO_2-air flames using a combustion diagnostic system. *Journal of Thermal Science* 30 (4): 1268–1277.

35 Tian, X., Jiao, W., and Liu, T. (2022). Intelligent leak detection method for low-pressure gas pipeline inside buildings based on pressure fluctuation identification. *Journal of Civil Structural Health Monitoring* 12: 1191–1208.

36 Vance, F., de Goey, L., and van Oijen, J. (2022). Development of a flashback correlation for burner-stabilized hydrogen-air premixed flames. *Combustion and Flame* 243: 112045.

37 Dirrenberger, P., Glaude, P., Le Gall, H. et al. (2012). Laminar flame velocity of components of natural gas. In: *ASME 2011 Turbo Expo: Turbine Technical Conference and Exposition*. Vancouver.

38 Dirrenberger, P., Le Gall, H., Bounaceur, R. et al. (2011). Measurements of laminar flame velocity for components of natural gas. *Energy and Fuels* 25 (9): 3875–3884.

39 Ilbas, M., Crayford, A.P., Yılmaz, I. et al. (2006). Laminar-burning velocities of hydrogen-air and hydrogen-methane-air mixtures: an experimental study. *International Journal of Hydrogen Energy* 31 (12): 1768–1779.

40 Hayakawa, A., Goto, T., Mimoto, R. et al. (2015). Laminar burning velocity and Markstein length of ammonia/air premixed flames at various pressures. *Fuel* 159: 98–106.

41 Mashruk, S., Zitouni, S.E., Brequigny, P. et al. (2022). Combustion performances of premixed ammonia/hydrogen/air laminar and swirling flames for a wide range of equivalence ratios. *International Journal of Hydrogen Energy* 47 (97): 41170–41182.

42 Han, W., Dai, P., Gou, X., and Chen, Z. (2020). A review of laminar flame speeds of hydrogen and syngas measured from propagating spherical flames. *Applications in Energy and Combustion Science* 1–4: 100008.

43 Zhang, X., Wang, J., Chen, Y., and Li, C. (2021). Effect of CH_4, pressure, and initial temperature on the laminar flame speed of an NH_3–air mixture. *ACS Omega* 6: 11857–11868.

44 Wang, N., Huang, S., Zhang, Z. et al. (2021). Laminar burning characteristics of ammonia/hydrogen/air mixtures with laser ignition. *International Journal of Hydrogen Energy* 46: 31879–31893.

45 Coward, H.F. and Jones, G.W. (1952). Limits of Flammability of Gases and Vapors. Bulletin 503, Bureau of Mines, U.S. Department of the Interior.

46 Di Benedetto, A. (2013). The thermal/thermodynamic theory of flammability: the adiabatic flammability limits. *Chemical Engineering Science* 99: 265–273.

47 Olchewsky, K., Fuller, C., Holton, M., and Gokulakrishnan, P. (1984). Autoignition temperature measurements of hydrogen mixtures. *10th U. S. National Combustion Meeting*, College Park, Maryland.

48 The Engineering ToolBox (2005). Ammonia – NH_3 – Thermodynamic properties [Online]. Available: https://www.engineeringtoolbox.com/ammonia-d_971.html (accessed 9 November 2024).

49 Engineering Toolbox (2003). Fuels and chemicals – autoignition temperatures [Online]. Available: https://www.engineeringtoolbox.com/fuels-ignition-temperatures-d_171.html (accessed 9 November 2024).

50 Pizzuti, L., Martins, C., and Lacava, P. (2016). Laminar burning velocity and flammability limits in biogas: a literature review. *Renewable and Sustainable Energy Reviews* 62: 856–865.

51 Coronado, C., Carvalho, J. Jr., Andrade, J. et al. (2012). Flammability limits: a review with emphasis on ethanol for aeronautical applications and description of the experimental procedure. *Journal of Hazardous Materials* 241–242: 32–54.

52 Mendiburu, A., de Carvalho, J. Jr., Coronado, C., and Roberts, J. (2017). Flammability limits temperature dependence of pure compounds in air at atmospheric pressure. *Energy* 118: 414–424.

53 Mendiburu, A., Carvalho, J. Jr., and Ju, Y. (2023). Flammability limits: a comprehensive review of theory, experiments, and estimation methods. *Energy & Fuels* 37 (6): 4151–4197.

54 Kobayashi, H., Hayakawa, A., Somarathne, K., and Okafor, E. (2019). Science and technology of ammonia combustion. *Proceedings of the Combustion Institute* 37: 109–133.

55 Kohansal, M., Kiani, M., Masoumi, S. et al. (2023). Experimental and numerical investigation of NH_3/CH_4 mixture combustion properties under elevated initial pressure and temperature. *Energy & Fuels* 37 (14): 10681–10696.

56 Lu, M., Dong, D., Wei, F. et al. (2023). Chemical mechanism of ammonia-methanol combustion and chemical reaction kinetics analysis for different methanol blends. *Fuel* 341: 127697.

57 Hussain, B., Fang, J., Zhang, J. et al. (2024). Combustion enhancement of ammonia by cofiring oxymethylene ethers (OMEn, n = 0–2): an experimental and kinetic modeling investigation. *Energy & Fuels* 38 (22): 22516–22526.

58 Deng, X., Zhao, W., Ye, L. et al. (2024). Combustion and emissions characteristics of ammonia/diesel dual-fuel engine with high premixed fractions in different injection modes. *International Journal of Hydrogen Energy* 49 (B): 1232–1244.

59 Bratek, K., Bratek, W., Gerus-Piasecka, I. et al. (2002). Properties and structure of different rank anthracites. *Fuel* 81: 97–108.

60 Reichel, D., Klinger, M., Krzack, S., and Meyer, B. (2013). Effect of ash components on devolatilization behavior of coal in comparison with biomass – product yields, composition, and heating values. *Fuel* 114: 64–70.

61 Bandara, J., Jaiswal, R., Nielsen, H. et al. (2021). Air gasification of wood chips, wood pellets and grass pellets in a bubbling fluidized bed reactor. *Energy* 233: 121149.

62 Jayaraman, K. and Gökalp, I. (2015). Pyrolysis, combustion and gasification characteristics of miscanthus. *Energy Conversion and Management* 89: 83–91.

63 Mehrotra, A., Karan, K., and Behie, L. (1995). Estimate gas emissivities for equipment and process design. *Chemical Engineering Progress* 91 (9): 70–77.

64 Michelsen, H. (2017). Probing soot formation, chemical and physical evolution, and oxidation: a review of in situ diagnostic techniques and needs. *Proceedings of the Combustion Institute* 36: 717–735.

65 Karataş, A. and Gülder, Ö. (2012). Soot formation in high pressure laminar diffusion flames. *Progress in Energy and Combustion Science* 38 (6): 818–845.

66 Pourhoseini, S. (2017). A novel configuration of natural gas diffusion burners to enhance optical, thermal and radiative characteristics of flame and reduce NOx emission. *Energy* 132: 41–48.

67 Lin, J., Zhang, X., Liu, K., and Zhang, W. (2019). Emissivity characteristics of hydrocarbon flame and temperature measurement by color image processing. *Energies* 12 (11): 2185.

68 Hoerlle, C. and Pereira, F.M. (2019). Effects of CO_2 addition on soot formation of ethylene non-premixed flames under oxygen enriched atmospheres. *Combustion and Flame* 203: 407–423.

69 Draper, T., Zeltner, D., Tree, D. et al. (2012). Two-dimensional flame temperature and emissivity measurements of pulverized oxy-coal flames. *Applied Energy* 95: 38–44.

70 Lou, C., Zhou, H.-C., Yu, P.-F., and Jiang, Z.-W. (2007). Measurements of the flame emissivity and radiative properties of particulate medium in pulverized-coal-fired boiler furnaces by image processing of visible radiation. *Proceedings of the Combustion Institute* 31 (2): 2771–2778.

71 Syed, Z. (1975). *The Performance of a Low Pressure Air Atomised Oil Burner*. University of Surrey.

72 Si, M., Thompson, S., and Calder, K. (2011). Energy efficiency assessment by process heating assessment and survey tool (PHAST) and feasibility analysis of waste heat recovery in the reheat furnace at a steel company. *Renewable and Sustainable Energy Reviews* 15 (6): 2904–2908.

73 Karem, S., Al-Obaidi, M., Alsadaie, S. et al. (2022). Significant energy saving in industrial natural draught furnace: a model-based investigation. *Applied Thermal Engineering* 202: 117829.

74 Walton, R. (1994). Furnaces, electric, resistance furnaces. In: *Kirk-Othmer Encyclopedia of Chemical Technology*. John Wiley & Sons, Inc.

75 Wang, T., Xia, L., Ni, M. et al. (2024). Fundamentals of infrared heating and their application in thermosetting polymer curing: a review. *Coatings* 14 (7): 875.

76 ITU (2020). Radio regulations – articles. International Telecommunications Union (ITU).

77 Gabriel, C., Gabriel, S., Grant, E.H. et al. (1998). Dielectric parameters relevant to microwave dielectric heating. *Chemical Society Reviews* 27: 213–224.

78 Mishra, R.S.A. (2016). Microwave-material interaction phenomena: heating mechanisms, challenges and opportunities in material processing. *Composites Part A: Applied Science and Manufacturing* 81: 78–97.

79 Qin, M., Zhang, L., and Wu, H. (2022). Dielectric loss mechanism in electromagnetic wave absorbing materials. *Advanced Science* 9 (10): 2105553.

80 Khaled, D.E., Novas, N., Gazquez, J., and Manzano-Agugliaro, F. (2018). Microwave dielectric heating: applications on metals processing. *Renewable and Sustainable Energy Reviews* 82 (3): 2880–2892.

81 Lyubimov, Y.A. (2006). Permittivity at infinite frequency. *Russian Journal of Physical Chemistry* 80 (12): 2033–2040.

82 Ellison, W.J. (2007). Permittivity of pure water, at standard atmospheric pressure, over the frequency range 0 – 25 THz and the temperature range 0–100°C. *Journal of Physical and Chemical Reference Data* 36 (1): 1–18.

83 Uematsu, M. and Frank, E.U. (1980). Static dielectric constant of water and steam. *Journal of Physical and Chemical Reference Data* 9: 1291.

84 Li, H., Zhao, Z., Xiouras, C. et al. (2019). Fundamentals and applications of microwave heating to chemicals separation processes. *Renewable and Sustainable Energy Reviews* 114: 109316.

85 Yang, L.-J., Yang, X.-Q., Huang, K.-M. et al. (2010). Experimental and theoretic study of the dielectric properties of ethanol + methanol mixture. *Journal of Solution Chemistry* 39: 473–481.

86 Fakhar, Z., Ghanadzadeh, A., and Honarparvar, B. (2015). Dielectric study of molecular association of polar–polar binary mixtures (benzyl alcohol + aliphatic alcohols) at 298.2 K. *Journal of the Iranian Chemical Society* 12: 1363–1368.

87 Reis, J. and Iglesias, T.P. (2011). Kirkwood correlation factors in liquid mixtures from an extended Onsager–Kirkwood–Fröhlich equation. *Physical Chemistry Chemical Physics* 13: 10670–10680.

88 Kim, J.L.T. and Lee, K.-H. (2014). Microwave heating of carbon-based solid materials. *Carbon Letters* 15: 15–24.

89 Sun, J., Wang, W., and Yue, Q. (2016). Review on microwave-matter interaction fundamentals and efficient microwave-associated heating strategies. *Materials* 9 (4): 231.

90 Peng, Z., Hwang, J.-Y., Mouris, J. et al. (2010). Microwave penetration depth in materials with non-zero magnetic susceptibility. *ISIJ International* 50 (11): 1590–1596.

91 Wang, Z., Yu, C., Huang, H. et al. (2021). Carbon-enabled microwave chemistry: from interaction mechanisms to nanomaterial manufacturing. *Nano Energy* 85: 106027.

92 Chen, W., Gutmann, B., and Kappe, C. (2012). Characterization of microwave-induced electric discharge phenomena in metal–solvent mixtures. *ChemistryOpen* 1: 39–48.

93 Xia, Y., Xu, L., Yang, Y., and Zhu, H. (2023). High-efficiency and compact microwave heating system for liquid in a mug. *Case Studies in Thermal Engineering* 51: 103531.
94 Electrical Deck (2022). Induction heating – working principle, types & advantages, 29 May 2022 [Online]. Available: https://www.electricaldeck.com/2022/05/induction-heating.html (accessed 24 March 2025).
95 EEEguide.com (2014). Induction furnace – definition, types, working principle and advantages, EEguid.com [Online]. Available: https://www.eeeguide.com/induction-furnace-definition-types-working-principle-and-advantages/ (accessed 23 March 2025).
96 Kotzé, I. (2002). Pilot plant production of ferronickel from nickel oxide ores and dusts in a DC arc furnace. *Minerals Engineering* 15 (11): 1017–1022.
97 Reynolds, Q. (2011). The dual-electrode DC arc furnace – modelling insights. *Journal of the Southern African Institute of Mining and Metallurgy* 111: 697–703.
98 JonesI, R., Reynolds, Q., Curr, T., and Sager, D. (2011). Some myths about DC arc furnaces. *Journal of the Southern African Institute of Mining and Metallurgy* 111 (10): 665–673.
99 Magnussen, T. (2018). Basic parameters in the operation and design of submerged arc furnaces, with particular reference to production of high-silicon alloys. *Journal of the Southern African Institute of Mining and Metallurgy* 118: 631–636.
100 Makarov, A.N. (2019). Calculation and analysis of the relationship between the efficiency and position of electric arcs and power consumption in electric arc furnaces (EAF) of smaller and larger capacity. Part I. calculation and analysis of the relationship between arc efficiency and power consumption. *Metallurgist* 63: 341–349.
101 Meihack, W., Curr, T., Barcza, N., and Jones, R. (1987). The effect of feed pretreatment on the efficiency of a plasma-arc furnace. In: *International Symposium on Plasma Chemisty, ISPC-8*. Tokyo.
102 Linnhoff, B. and Sahdev, V. (2000). Pinch Technology. In: *Ullmann's Encyclopedia of Industrial Chemistry*. Wiley-VCH Verlag GmbH & Co. KGaA.
103 Short, M., Isafiade, A., Fraser, D., and Kravanja, Z. (2016). Synthesis of heat exchanger networks using mathematical programming and heuristics in a two-step optimisation procedure with detailed exchanger design. *Chemical Engineering Science* 144: 372–385.
104 Escobar, M. and Trierweiler, J. (2010). Operational controllability of heat exchanger networks. *IFAC Proceedings Volumes* 43 (5): 97–102.
105 Wang, Y., Chang, C., and Feng, X. (2015). A systematic framework for multi-plants heat integration combining direct and indirect heat integration methods. *Energy* 90: 56–67.
106 Kapil, A., Bulatov, I., Smith, R., and Kim, J.-K. (2012). Process integration of low grade heat in process industry with district heating networks. *Energy* 44: 11–19.

107 Barz, T., Zauner, C., Lager, D. et al. (2016). Experimental analysis and numerical Modeling of a shell and tube heat storage unit with phase change materials. *Industrial & Engineering Chemistry Research* 55: 8154–8164.

108 Njoku, H., Ekechukwu, O., and Onyegegbu, S. (2014). Analysis of stratified thermal storage systems: an overview. *Heat and Mass Transfer* 50: 1017–1030.

109 Therminol (2025). Therminol 66 (R) – heat transfer fluid [Online]. https://www.therminol.com/sites/therminol/files/documents/TF-8695_Therminol-66_Technical_Bulletin.pdf (accessed 4 March 2025), available in https://www.eastman.com/en/products/brands/therminol/heat-transfer-fluids (accessed 4 March 2025).

110 Engineering Toolbox (2009). Storing thermal heat [Online]. Available: https://www.engineeringtoolbox.com/sensible-heat-storage-d_1217.html (accessed 4 March 2025).

111 Zhao, Q.-G., Hu, C.-X., Liu, S.-J. et al. (2017). The thermal conductivity of molten $NaNO_3$, KNO_3, and their mixtures. *Energy Procedia* 143: 774–779.

112 Engineering Toolbox (2014). Metals – Boiling points and specific heat [Online]. Available: https://www.engineeringtoolbox.com/liquid-metal-boiling-points-specific-heat-d_1893.html (accessed 4 March 2025).

113 Mehdipour, N., Boushehri, A., and Eslami, H. (2005). Prediction of the density of molten metals. *Journal of Non-Crystalline Solids* 351 (16–17): 1333–1337.

114 Giordanengo, B., Benazzi, N., Vinckel, J. et al. (1999). Thermal conductivity of liquid metals and metallic alloys. *Journal of Non-Crystalline Solids* 250–252: 377–383.

115 Skinner, J., Strasser, M., Brown, B., and Selvam, R. (2014). Testing of high-performance concrete as a thermal energy storage medium at high temperatures. *Journal of Solar Energy Engineering* 136 (2): 021004.

116 Barbhuiya, S., Das, B., and Idrees, M. (2024). Thermal energy storage in concrete: a comprehensive review on fundamentals, technology and sustainability. *Journal of Building Engineering* 82: 108302.

117 Fallahi, A., Guldentops, G., Tao, M. et al. (2017). Review on solid-solid phase change materials for thermal energy storage: molecular structure and thermal properties. *Applied Thermal Engineering* 127: 1427–1441.

118 Jouhara, H., Żabnieńska-Góra, A., Khordehgah, N. et al. (2020). Latent thermal energy storage technologies and applications: a review. *International Journal of Thermofluids* 5-6: 100039.

119 Lee, T., Hawes, D., Banu, D., and Feldman, D. (2000). Control aspects of latent heat storage and recovery in concrete. *Solar Energy Materials and Solar Cells* 62 (3): 217–237.

120 Milian, Y., Ushak, S., Cabeza, L., and Grageda, M. (2020). Advances in the development of latent heat storage materials based on inorganic lithium salts. *Solar Energy Materials and Solar Cells* 208: 110344.

121 Sorknæs, P., Johannsen, R., Korberg, A. et al. (2022). Electrification of the industrial sector in 100% renewable energy scenarios. *Energy* 254 (B): 124339.

122 Kosmadakis, G. (2019). Estimating the potential of industrial (high-temperature) heat pumps for exploiting waste heat in EU industries. *Applied Thermal Engineering* 156: 287–298.

123 Arpagaus, C., Bless, F., Uhlmann, M. et al. (2018). High temperature heat pumps: market overview, state of the art, research status, refrigerants, and application potentials. *Energy* 152: 985–1010.

5

Chemical Storage of Energy

The intermittency of renewable energy and in particular long-term intermittency does require storage of relatively large amounts of renewable energy for a prolonged period of time, if this is the sole source of energy. There are multiple ways of storing energy, including pumped hydro, heat and batteries (see Chapter 3). These all achieve good efficiencies of approximately 80%. However, they have very low energy density, thus requiring large amounts of materials/space. Additionally, most of these methods do not necessarily hold the energy indefinitely.

Furthermore, the ideal location of capturing renewable energy may not be at the same location where the energy is to be used and some of these locations will not be useable if efficient long-distance transport of this energy cannot be realised. Direct transport of energy through high-voltage or ultra-high-voltage transmission lines can become cost-limited [1]. Hence, alternative methods for transport of energy, and in particular long-distance transport, will need to be considered where energy can be efficiently captured, stored and transported. The storage of energy in the form of chemicals is one such alternative.

5.1 Power-to-X Concept

The general idea of power-to-X (PtX) is to use renewable energy to produce various chemicals as a way of storing energy. The general train being Power → Hydrogen → Products. The products (X) could be hydrogen, ammonia, methane, methanol, liquid hydrocarbons or even long-chain alcohols. The products can then be used to generate energy, if necessary. A generalised example of a PtX process in which hydrogen, ammonia and other products are formed is given in Figure 5.1.

Both the specific energy and the energy density are important considerations when choosing the chemical to act as an energy storage vehicle. Examples of the energy density vs. specific energy with various PtX options are shown in Figure 5.2.

Whilst hydrogen has the highest specific energy at 120 MJ/kg (lower heating value [LHV]), diesel and petrol have far higher energy densities with the liquid hydrocarbon fuels in general having the highest available energy densities being in the range

Renewable Energy in the Process Industry, First Edition.
Eric van Steen, Nicholas S. Featherstone, and Linda H. Callanan.

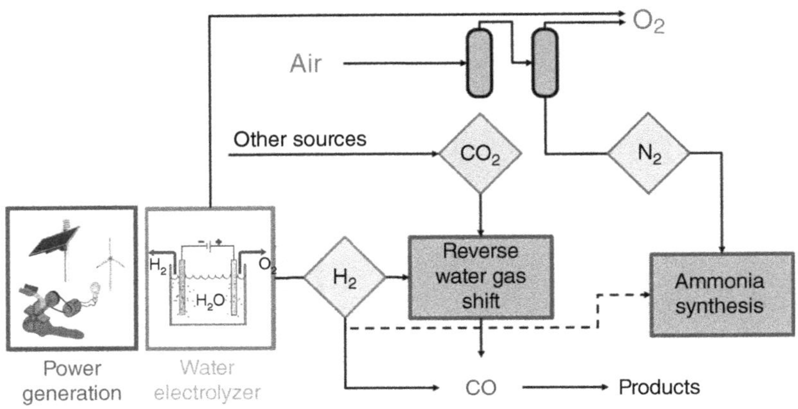

Figure 5.1 Concept of power-to-X process.

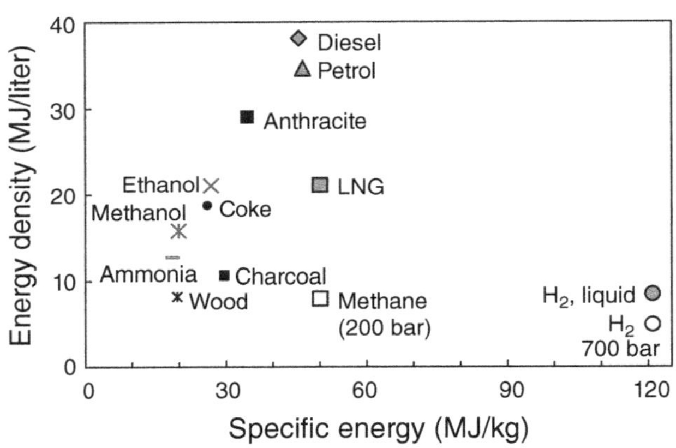

Figure 5.2 Energy density as a function of the specific energy (energy density of solids calculated with a bed porosity of 0.4). Symbols: open symbols: gases; grey symbols: liquids; solid symbols: solids.

of 35–38 MJ/l. This is an important distinction because it is often the volume of fuels that matters more than their mass when they are transported.

5.2 Hydrogen as an Energy Carrier

The most notable option to store energy chemically is hydrogen, because of its high specific energy, and a feasible route to produce of hydrogen using renewable energy. Furthermore, the combustion of hydrogen yields only water and thus no carbon emissions. Hydrogen can be used as a means of storing energy, directly as a fuel, for energy transportation, but is also an important chemical intermediate to produce other compounds such as ammonia and methanol [2, 3].

Table 5.1 Some properties of hydrogen.

Molar mass (H_2)	2.01588 g/mol	T_{melt}	−259.16 °C
H-content	100 wt.%	$T_{normal\ boiling}$	−252.88 °C
Density (0 °C, and 1.01325 bar)	$0.089\frac{kg}{m^3}$	Critical point	−240.04 °C; 1.313 MPa
HHV(H_2(g) + $\frac{1}{2}O_2$ (g) → H_2O(l))	285.6 kJ/mol	Autoignition point[c]	560 °C
LHV (H_2(g) + $\frac{1}{2}O_2$ (g) → H_2O(g))	241.8 kJ/mol	Flash point	−260.14 °C
Specific energy (based on LHV)	119.95 MJ/kg		
Flammability region	LFL: 4 vol.% in air at atmospheric pressure		
	UFL: 75 vol.% in air at atmospheric pressure		

Hydrogen is a permanent gas (see Table 5.1), with a normal boiling point of −259.16 °C. It has quite a large flammability region, although its auto-ignition temperature is rather high (560 °C). The specific energy of hydrogen based on the LHV of hydrogen is 119.95 MJ/kg. The maximum work that can be obtained by oxidising hydrogen is 237.1 kJ/mol (negative value of the Gibbs free energy of formation of liquid water), which corresponds to 117.6 MJ/kg or 0.0327 MWh/kg. Otherwise stated, the generation of 1 MW of power would require the oxidation of at least 30.6 kg H_2/h.

5.2.1 Hydrogen Production

The production of hydrogen is a well-established field, with several processes that have been successfully industrialised. The distribution of global hydrogen manufacturing methods is illustrated in Figure 5.3. At present, most hydrogen in the world is produced through methane steam reforming, followed by oil reforming and coal gasification. In addition to these four processes, a large amount of hydrogen also gets produced from oil on refineries for the internal use in hydro-treating processes. This hydrogen production is not directly associated with CO_2 emissions.

Hydrogen production is often classified by colours, which relate to the origin of hydrogen and the environmental impact of the hydrogen production process. There are principally three colours of hydrogen that are referred to. Grey hydrogen refers to the generation of hydrogen from fossil fuels, resulting in substantial CO_2 release. Sometimes the term brown hydrogen is used to specify the use of coal. Typically, when brown hydrogen is referred to, the term grey hydrogen is used to mean hydrogen from natural gas. Blue hydrogen production refers to the use of fossil fuels, but with carbon capture technologies to prevent the release of the associated CO_2 into the atmosphere. Green hydrogen is formed using renewable energy, e.g. water electrolysis but could also refer to biomass gasification or biomethane reforming. It can be seen (Figure 5.3) that most hydrogen is produced from fossil fuels, with significant carbon output, although, blue hydrogen is in development and starting to replace grey hydrogen. At present, only around 4% of global hydrogen production is

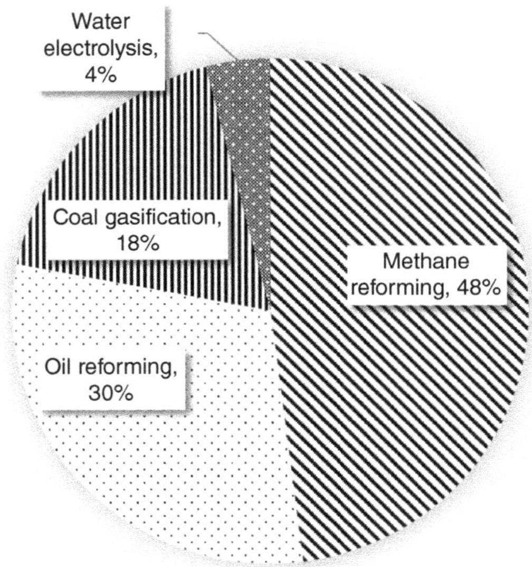

Figure 5.3 Distribution of hydrogen production in 2016 (data from Taibi et al. [4]).

green [4]. The low penetration of green hydrogen in the global hydrogen market is largely driven by the cost of green hydrogen.

Several other colours of hydrogen production are referred to, so one needs to be careful when reading literature in this space, as the colour's exact meaning can be author-dependent. Some examples include purple and yellow which can refer to nuclear energy, aqua where fossil fuels are used but solid carbon rather than CO_2 is produced, and gold which involves electrolysis from non-renewable energy, or may refer to naturally occurring hydrogen, which is also referred to as white hydrogen.

A simplified representation of the block flow diagram for typical hydrogen production processes from fossil sources is shown in Figure 5.4. It is worth noting that the use of gas and coal won't normally happen on the same plant, but representing them together simplifies the diagram and illustrates at which point the processes overlap.

The production of grey hydrogen, i.e. from natural gas, typically requires a pre-treatment (desulphurisation) of the feedstock. In the reformer, hydrogen is produced according to the target reaction:

$$CH_4 + 2H_2O \rightarrow CO_2 + 4H_2$$

Air can be co-fed with natural gas into the reactor to supply heat for the reaction.

The production of brown hydrogen from coal requires milling of the coal to a fine powder before it is fed together with water and air into a gasifier. Air is required to facilitate combusting some of the coal to generate the heat needed. The gas coming out of the gasifier will typically be an equilibrium mixture, which still needs to be desulphurised prior to further processing.

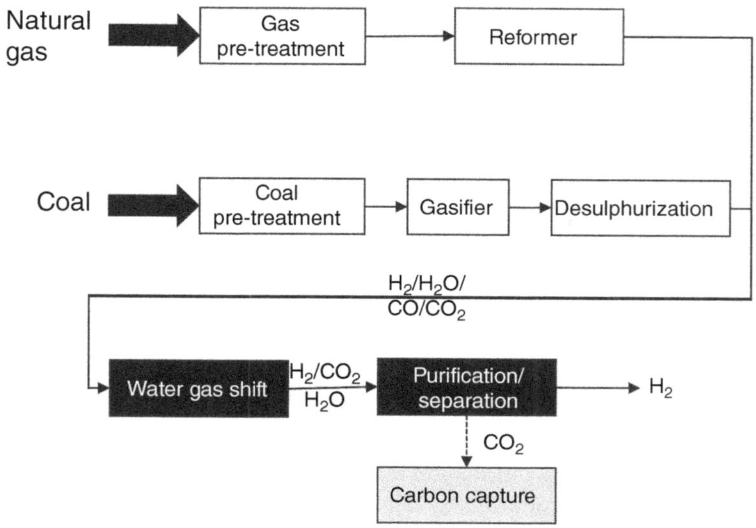

Figure 5.4 Overall block flow diagram to produce hydrogen via from either methane reforming or coal gasification followed by the water gas shift reaction and subsequent hydrogen purification.

Whether using methane or coal, the mixture formed is synthesis gas (also named syngas), which is a mixture of carbon monoxide and hydrogen (together with CO_2 and water). It is named synthesis gas as it is used (ideally with a H_2/CO ratio of 2) for the synthesis of liquid fuels and methanol. When using coal, the synthesis gas is CO-rich, but when using natural gas the resulting syngas is hydrogen-rich. Regardless, with hydrogen synthesis, the target is to make as much hydrogen as possible, hence a water gas shift (WGS) reactor (also called a shift converter) is used to make more hydrogen and consume CO.

$$H_2O + CO \leftrightarrow H_2 + CO_2$$

Purification is represented here by a single block but is typically multiple steps. Residual CO can be either hydrogenated to methane or selectively combusted to CO_2. CO_2 is removed with scrubbing, generating a high-purity CO_2 stream. Water is removed simply through condensation. Further hydrogen purification can possibly be done using a membrane, but in general, components such as methane and nitrogen present at this stage may remain in the product hydrogen stream as inert components.

5.2.1.1 Coal Gasification

Coal gasification may sound like a counterintuitive way to make hydrogen. Coal, typically modelled as pure carbon, is reacted with oxygen and water to make hydrogen. The combustion of coal with oxygen supplies the heat for the endothermic gasification reactions. Inherently, the coal gas comes out of the gasifier at equilibrium, containing some hydrogen, along with CO, CO_2 and water (as well as other impurities). There are a number of reactions that are targeted for different reasons:

Reactions supplying heat:

$$C + 0.5\,O_2 \rightarrow CO \quad \Delta H^0_{rxn} = -110.5\ kJ/mol$$

$$C + O_2 \rightarrow CO_2 \quad \Delta H^0_{rxn} = -393.6\ kJ/mol$$

Reactions yielding hydrogen:

$$C + H_2O \leftrightarrow H_2 + CO \quad \Delta H^0_{rxn} = 131.3\ kJ/mol$$

$$C + 2H_2O \leftrightarrow 2H_2 + CO_2 \quad \Delta H^0_{rxn} = 90\ kJ/mol$$

$$CO + H_2O \leftrightarrow H_2 + CO_2 \quad \Delta H^0_{rxn} = -41.3\ kJ/mol$$

Side reactions:

$$2CO \leftrightarrow C + CO_2 \quad \Delta H^0_{rxn} = -172.4\ kJ/mol$$

$$CO + 2H_2 \leftrightarrow CH_4 + H_2O \quad \Delta H^0_{rxn} = -205.8\ kJ/mol$$

It is worth noting that many other, often undesired, side reactions may occur in the gasifier as well. These include the formation of methane as well as reactions which occur due to an assortment of impurities in the coal. Most notable is sulphur, which can form H_2S (a toxic gas and a notable catalyst poison for downstream processes) at the reactor exist conditions. This occurs particularly at the reactor exit where the reductive conditions are prevalent.

$$S + O_2 \rightarrow SO_2 \quad \Delta H^0_{rxn} = -296.8\ kJ/mol$$

$$S + 1.5O_2 \rightarrow SO_3 \quad \Delta H^0_{rxn} = -396\ kJ/mol$$

$$SO_2 + 3H_2 \rightarrow H_2S + 2H_2O \quad \Delta H^0_{rxn} = -449.2\ kJ/mol$$

NO_x formation is also possible from nitrogen within coal [5]. This will be enhanced when using air in the coal gasification rather than oxygen. Nitrogen in coal will react with H_2 to form NH_3 which, at high temperatures and in the presence of CO, can react further to form HCN.

$$'N'_{coal} + 1.5\,H_2 \rightarrow NH_3$$

$$NH_3 + CO \rightarrow HCN + H_2O \quad \Delta H^0_{rxn} = 49.8\ kJ/mol$$

Several options are available for the construction and operation of coal gasifiers (see Figure 5.5). The gasifier can be operated as a moving-bed gasifier. Here, coal is fed to the top of the gasifier with steam and oxygen, or air, being fed in from the bottom. Different reaction zones can be identified from top to bottom, viz. a pre-heating zone, pyrolysis zone, gasification zone and combustion zone. A fluidised bed is obtained upon increasing the gas velocity ensuring mixing of coal particles throughout the reaction zone (see Figure 5.5b). In an entrained flow gasifier (see Figure 5.5c), gas and coal particles both flow from the top with molten slag being removed from the bottom.

A slightly more detailed view of the internals of the moving-bed coal gasifier is given in Figure 5.6. The highest temperature in the gasifier can be found at the gas inlet side, where some of the coal is combusted to generate the heat required for the endothermic gasification reactions. Gasification occurs above the combustion zone yielding 'water gas' (named due to the addition of water). 'Coal gas' (historically

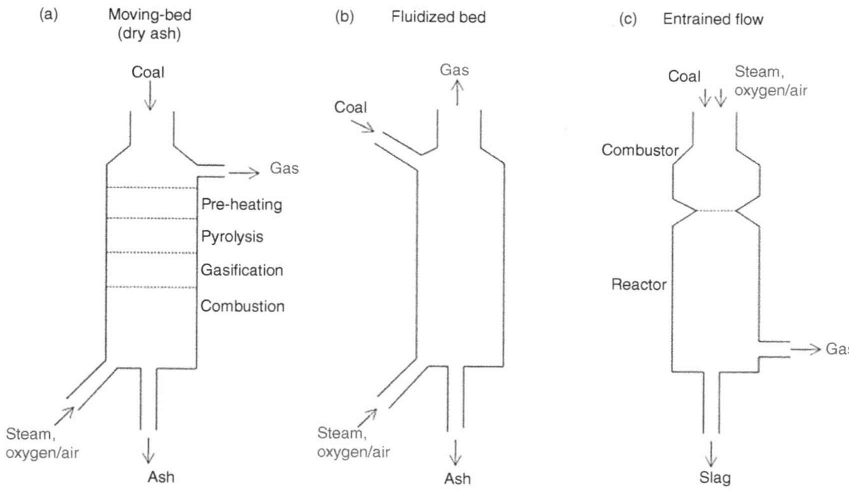

Figure 5.5 Gasifier configurations: (a) moving-bed gasifier, (b) fluidised bed gasifier and (c) entrained flow gasifier (adapted from Kaneko [6]).

Figure 5.6 Internals of a moving-bed coal gasifier (Breault [7] / CC BY 4.0).

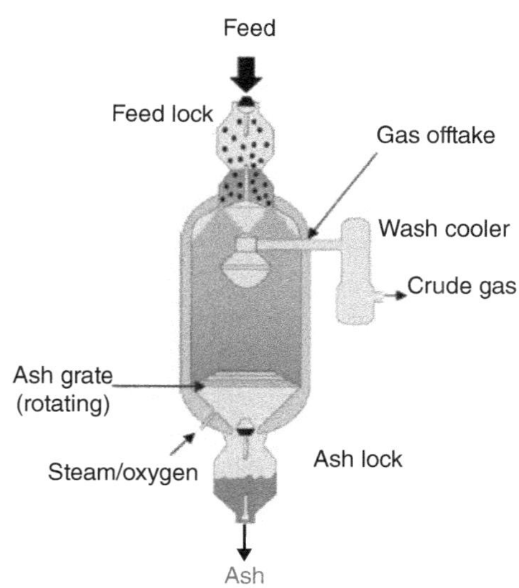

referred to as 'town gas') is formed when coal is heated in the absence of air, though the composition of the gas may be not too dissimilar. Above the gasification zone, pyrolysis occurs which may cause some thermal breakdown of the coal. At the coal inlet side, preheating of the coal occurs before it reaches the other regions. The gas outlet is 'coal gas', a mixture of CO and hydrogen rich in CO, or 'syngas', a mixture of CO and hydrogen. Ash present in coal is removed through the bottom of the gasifier. This normally contains multiple components of the inorganic matter.

When designing a gasifier, several issues need to be considered [6]. Firstly, a fast reaction is needed. Within a moving bed reactor, with normally quite a slow gas flow through the bed, mass transfer limitations may become an issue. Additionally, the temperature in the moving-bed reactor must be limited so that the ash at the exit is not liquefied. In the fluidised bed, there is a higher gas velocity giving lower mass transfer limitations and higher reaction rates. The mixing means that all the steps seen in the moving bed should be occurring throughout the reactor space and that the temperature should be more consistent. However, there is still a need to remove the ash as a solid. It is important here to remember that fluidisation is not easy to control. With entrained flow, the whole feed enters from the top of the reactor. Molten slag is removed from the bottom, and the temperature can be higher. This results in more hydrogen being produced.

The operating pressure of the gasifier is an important design criterion when designing a coal gasifier. One needs to consider what pressure the product gas is required at and whether operating the gasifier at high pressure is more beneficial than compressing the product gas.

The endothermic formation of hydrogen from water requires heat which must be supplied by the (partial) combustion of coal. The amount of heat available is a function of the oxygen/carbon ratio in the feed. This can best be represented by the adiabatic temperature for the combustion of carbon in air as shown in Figure 5.7, noting that this is without the addition of water. The initial rise in adiabatic temperature with the ratio of O_2/C in the feed, up to a value of 0.5, is caused by the formation of primarily CO, whereas the subsequent increase up to an O_2/C feed ratio of 1 is attributed to the formation of CO_2. At higher O_2/C ratios, the adiabatic temperature drops as the energy released by the combustion is simply required to heat the excess

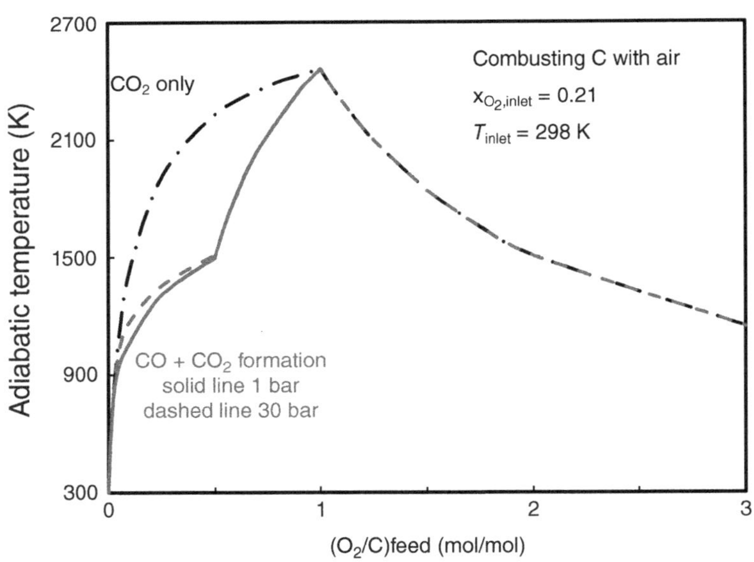

Figure 5.7 Adiabatic temperature upon combustion of coal in air (black line considering CO_2 formation only, grey lines considering formation of both CO and CO_2, dashed lines: $P = 30$ bar, solid line: $P = 1$ bar).

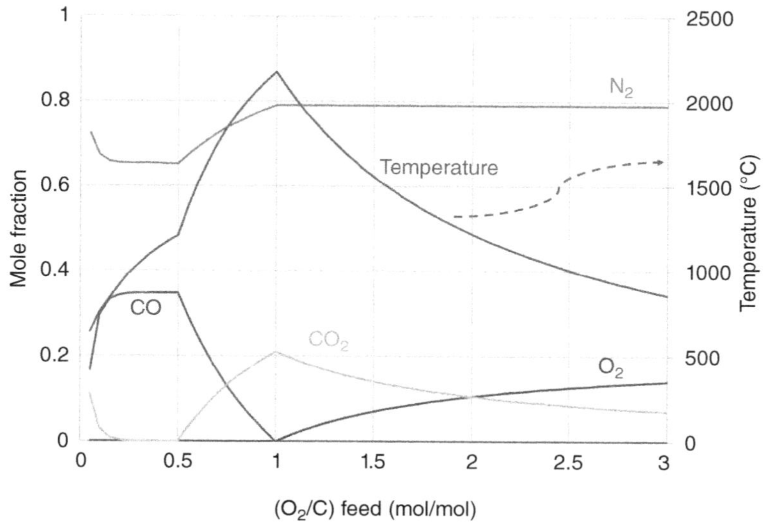

Figure 5.8 Product distribution in the adiabatic combustion of coal with air as a function of O_2/C ratio (T_{inlet} = 25 °C P = 1 bar).

air. Increasing the pressure has almost no effect on the adiabatic temperature trends, with only a slightly earlier onset of the formation of CO_2.

The molar composition of the product compounds in the combustion of coal (modelled as carbon) with air as a function of the O_2/C feed ratio is shown in Figure 5.8. At very low O_2/C feed ratios, all oxygen is consumed, and the temperature remains quite low, thus producing CO_2 and some CO, but the gas phase contains mainly nitrogen. Upon increasing the oxygen to carbon feed ratio, the temperature increases, favouring the formation of CO. At an O_2/C ratio of larger than 0.5, CO_2 appears again in significant amounts in the gas phase. At very high feed ratios, both O_2 and N_2 are in the product stream as the excess air acts as a diluent. Although the combustion of carbon to form CO_2 generates a significant amount of heat, increasing the CO_2 does not contribute to the production of more hydrogen. For this reason, an O_2/C ratio of 0.5 is typically used. It is also worth noting that although coal and water can react directly together, this reaction is kinetically limited [8] meaning that it may be more likely that the coal combusts to form CO which then reacts with the water to form CO_2, however, the exact reaction pathways occurring are complex and depend on numerous parameters [9].

The addition of water is necessary to form the desired hydrogen. The effect of the H_2O/C feed ratio on the adiabatic temperature and hydrogen yield at four different O_2/C ratios is shown in Figure 5.9. An increase in the steam to carbon ratio both lowers the adiabatic temperature and increases the hydrogen yield.

A potential method of overcoming the problem of the low adiabatic temperature seen with high water feed rates is to use isothermal operation using external heating of the gasifier. In the case where no combustion of the coal is used to heat the gasifier,

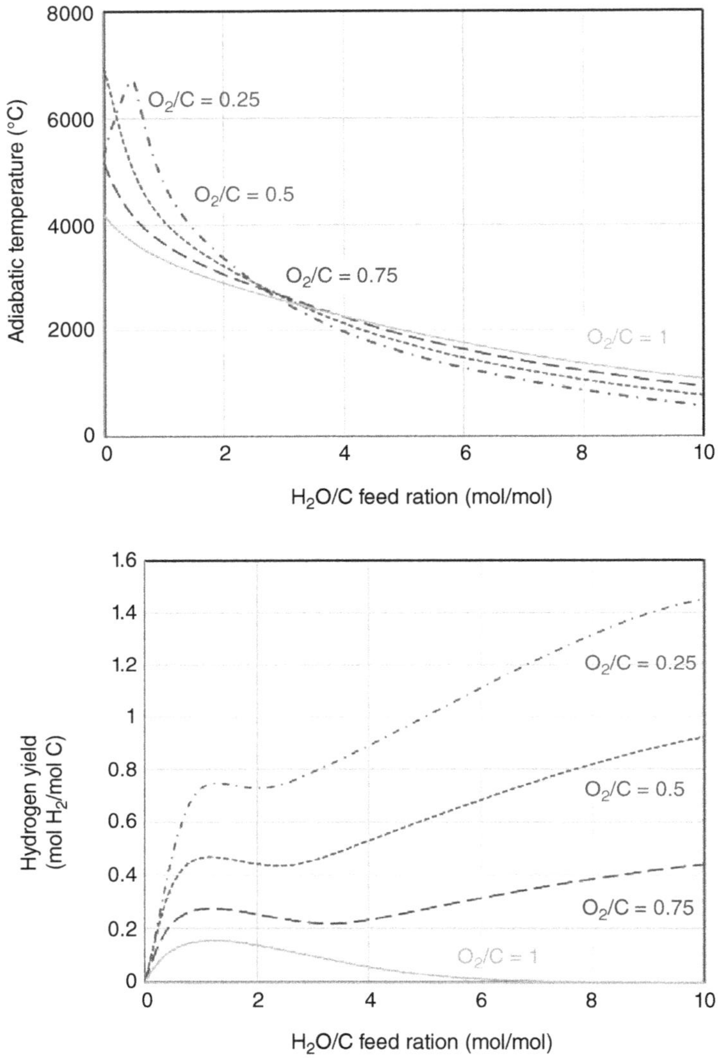

Figure 5.9 Adiabatic gasification of carbon with water and O_2 fed at 25 °C.

only water and coal would need to be added and then the reaction would proceed very readily to the formation of hydrogen as determined by thermodynamics. The hydrogen yield as a function of the reactor temperature is shown in Figure 5.10 at different pressures (a) and H_2O to C feed ratios (b). A higher hydrogen yield is favoured at the lower pressure, which is to be expected given that both reactions between carbon and water lead to an increase in the number of molecules in the gas phase. The hydrogen yield is also favoured by increased levels of water relative to the carbon. For each H_2O to C feed ratio, the hydrogen yield passes through a maximum with temperature.

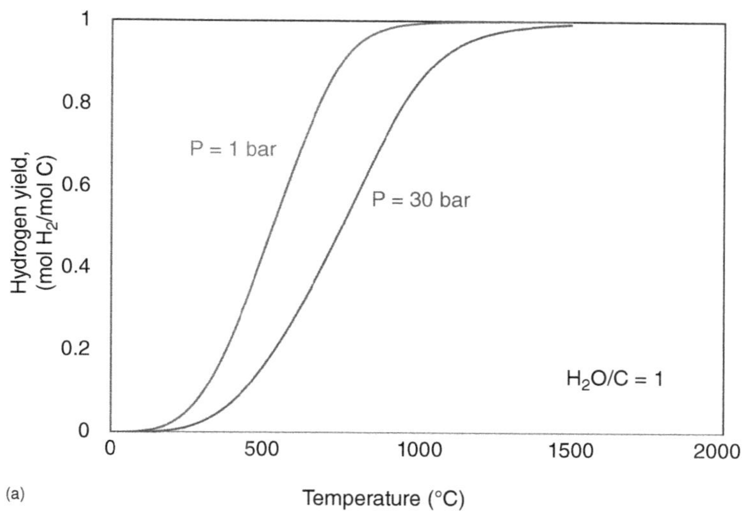

(a)

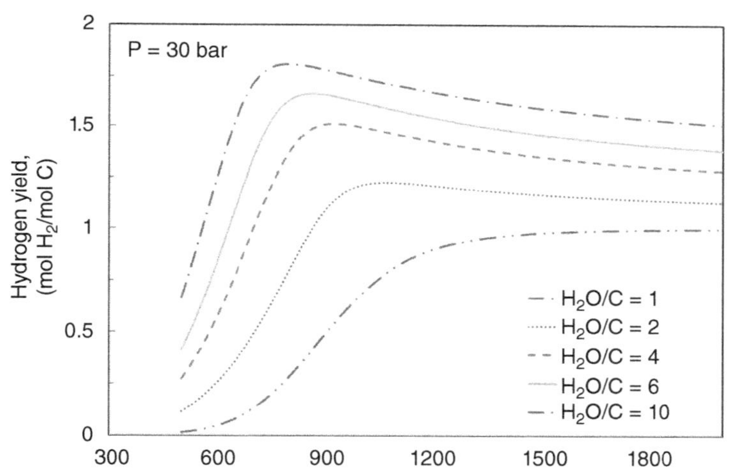

(b)

Figure 5.10 Hydrogen yield in carbon gasification at various (a) pressures and (b) water to carbon feed ratios.

At a given H_2O/C feed ratio (e.g., $H_2O/C = 1$), a hydrogen yield of 0.985 mol H_2/mol C requires a temperature of ca. 900 °C when operating at 1 bar but needs in excess of 1400 °C at 30 bar. At higher pressure, the desired H_2 yield can be gained at a significantly lower temperature by operating at a much higher H_2O/C feed ratio. Whilst adding more water to the inlet gives an increase in hydrogen formation, it will add to the energy required to heat the system; even though a lower temperature may be required, this also adds complexity in the downstream separation of the water. Furthermore, whilst the higher pressure appears beneficial, feeding the coal at high pressure adds to the complexity of the gasifier, as the solid has to be introduced into the high-pressure environment, typically using a screw feed mechanism.

Gasification of coal to generate hydrogen results in the formation of CO_2 with the stoichiometric emission of CO_2 amounting to 11 tonne CO_2/tonne H_2. Heating and partial combustion of coal results in a higher CO_2 emission and the direct CO_2 emission for coal gasification has been estimated at 18 tonne $CO_{2,e}$/tonne H_2 (emissions of the upstream coal extraction process and from the grid power) with the total expected emissions amounting to ca. 20 tonne $CO_{2,e}$/tonne H_2 [10].

5.2.1.2 Methane Reforming

Methane, the primary component of natural gas, may be converted to hydrogen using water in a reforming process. There are several reactions that may occur in this process.

1) Steam reforming: $CH_4 + H_2O \leftrightarrow CO + 3H_2 \quad \Delta H^0_{rxn} = 206\ kJ/mol$
2) Water gas shift: $CO + H_2O \leftrightarrow CO_2 + H_2 \quad \Delta H^0_{rxn} = -41 kJ/mol$
3) Dry reforming: $CH_4 + CO_2 \leftrightarrow 2CO + 2H_2 \quad \Delta H^0_{rxn} = 247\ kJ/mol$
   with the required heat either supplied externally or through the (partial) combustion of methane:
4) Partial oxidation: $CH_4 + \frac{1}{2}O_2 \leftrightarrow CO + 2H_2 \quad \Delta H^0_{rxn} = -36\ kJ/mol$
5) Selective oxidation: $CH_4 + O_2 \leftrightarrow CO_2 + 2H_2 \quad \Delta H^0_{rxn} = -319\ kJ/mol$
6) Combustion: $CH_4 + 2O_2 \leftrightarrow CO_2 + 2H_2O \quad \Delta H^0_{rxn} = -802\ kJ/mol$

There are several processes that may be considered when converting methane to produce hydrogen (Table 5.2), viz. steam reforming, autothermal reforming (ATR), dry reforming or tri-reforming. Of these, only steam reforming and ATR are current, industrial processes.

Methane steam reforming is relatively straightforward. Steam and methane are added to a reactor operating at a sufficiently high temperature to generate hydrogen. The reaction is endothermic but proceeds readily over an appropriate catalyst (typically nickel-based). In the classical set-up (Figure 5.11), methane and water are fed into tubular reactors which are hung inside a furnace. Heat is supplied through combustion of additional methane. Because the reaction is endothermic, the temperature will decrease down the tubes, resulting in the reaction rate slowing and virtually stopping before the bottom of the tubes. Additional heating can be obtained by the combustion of extra methane [12].

Due to the decreasing temperatures down the reactor tubes slowing the reaction kinetics, there will still be some methane in the outlet stream of the reformer. To rectify this and increase the total conversion, a secondary reformer, which is typically an autothermal reformer, is added.

The hydrogen yield in methane steam reforming is equilibrium-limited. As can be seen in Figure 5.12, the maximum hydrogen production is dependent on

Table 5.2 Processes to convert methane into hydrogen (data from Tucker [11]).

	Steam reforming (SMR)	Autothermal reforming (ATR)	Dry reforming	Tri-reforming
Heat addition	External	Internal	External	Internal ("autothermal")
Feed	H_2O/CH_4 (2.4–4)	$O_2/H_2O/CH_4$ (0.6–1/0.5–1.5/1)	CO_2/CH_4 (1:1)	$CO_2/H_2O/O_2/CH_4$
T (°C)	900	950–1050	800–1000	950–1050
P (bar)	3–25	30–50	1–15	30–50
Principal reactions[a)]	(1), (2)	(1), (2), (6)	(4)	(1), (2), (3), (6)
H_2/CO	3–4	2–4	1	1.5–2
Coking	Minimal	Minimal	Likely	Minimal

a) (1) steam reforming; (2) water gas shift; (3) dry reforming; (4) partial oxidation; (5) selective oxidation; (6) combustion.

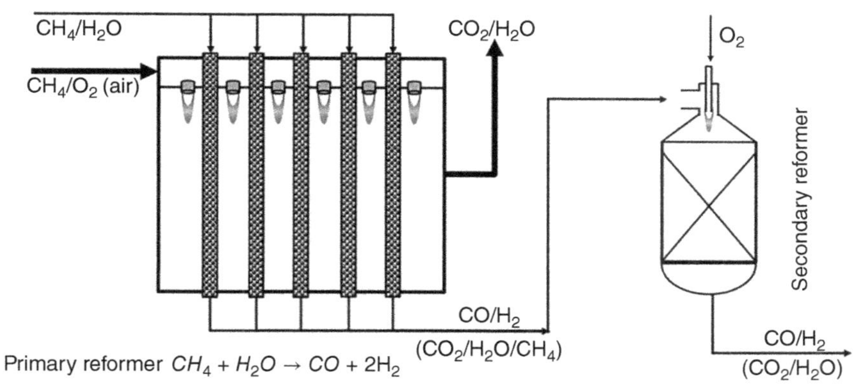

Figure 5.11 Methane steam reforming using a primary and a secondary reformer.

both the temperature and the pressure. At the high temperatures in these reactors, equilibrium can be reached relatively easily. The production of hydrogen is favoured at lower pressures due to the increase in the number of moles during the reaction. However, this must be weighed up with the desired use for the hydrogen, which is typically most useful at high pressures. In terms of energy requirements, pumping water into a high-pressure reactor requires less energy than compressing hydrogen after the reactor.

It is intuitive that the amount of hydrogen produced increases with a higher water/methane ratio (see Figure 5.13a). However, increasing the amount of water in

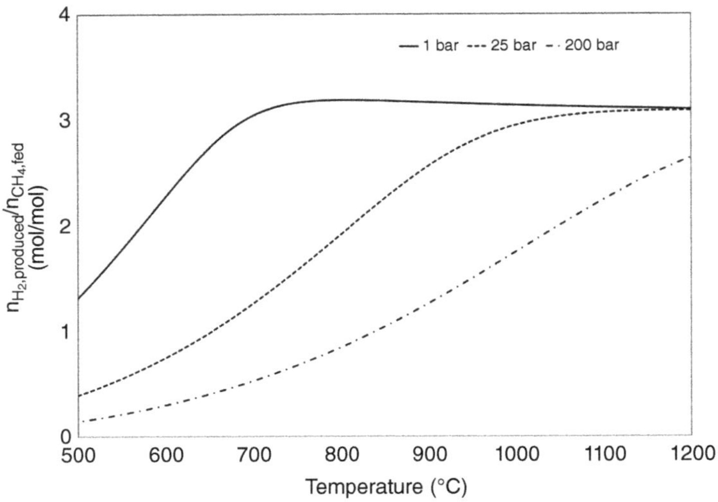

Figure 5.12 Effect of pressure and temperature on hydrogen yield in methane steam reforming ($H_2O/O_2/CH_4$ = 2/0/1; modelled using UNIQUAC-SRK).

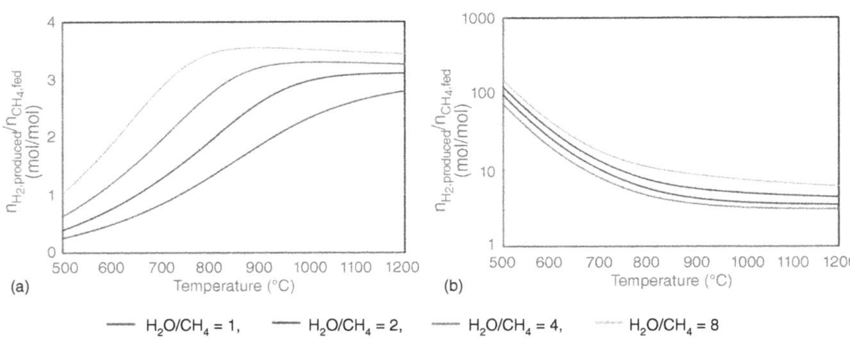

Figure 5.13 Effect of H_2O/CH_4 feed ratio in the methane steam reforming at 25 bar on (a) hydrogen production and (b) H_2/CO ratio in the product (modelled using UNIQUAC-SRK).

the feed also dilutes the product stream. It can also be seen that with external heating, the H_2/CO ratio in the product stream is always higher than 3 (Figure 5.13b). This means that there is a hydrogen-rich gas coming out of the steam reformer and going to the secondary reformer.

Hydrogen production through methane steam reforming is associated with the generation of a large amount of CO_2 (though less than coal gasification). Stoichiometrically, ¼ mol of CO_2 is formed for each mole of H_2 corresponding to a production of 5.5 tonne of CO_2 per tonne of H_2 produced. Heating required for the steam reforming process pushes the amount of CO_2 produced up significantly. A survey of 33 hydrogen production facilities utilising methane steam reforming [13] showed a variation in the CO_2 emissions between 4.7 and 17.3 tonne $CO_{2,eq}$/tonne H_2 with a midpoint of 9.35 tonne $CO_{2,eq}$/tonne H_2 (dominated by the CO_2 emissions with minimal contributions of CO, NO_x and CH_4 emissions), which is in line with the estimated CO_2 emissions of 9.3 tonne CO_2/tonne H_2 quoted by Lewis et al. [10]. This does not include emissions associated with natural gas extraction and power generation, which can bring the emissions up to ca. 10–12 tonne $CO_{2,e}$/tonne H_2 [10]. Carbon capture could reduce direct CO_2 emissions by 80–90% [10, 14, 15], although this would come at a cost of reduced energy efficiency.

ATR (Figure 5.14) is different from steam reforming, in that the feed methane is partially combusted with oxygen inside the reactor rather than using external heating. The overall productivity of hydrogen in this reformer is slightly reduced, if the productivity is based on the amount of methane fed to the reactor, as some methane is combusted in the reactor to supply heat to the reaction. However, the amount of hydrogen based on the amount of methane fed subtracting the amount of methane required to supply heat increases as the presence of additional water and CO_2 in the gas mixture will push the equilibrium towards hydrogen (and CO).

In the autothermal reformer, methane is fed with steam and oxygen in a burner located at the top of the reactor. When used instead of air, the pure oxygen needed is obtained from an air separation unit. It is possible to feed air, but this will give rise to higher gas volumes and, therefore, larger process equipment, such as heat exchangers and compressors [17]. If the nitrogen is needed downstream with the hydrogen, e.g. in the production of ammonia, then air may be the preferred feed. The feed is passed into a hot combustion chamber, generating the heat required for the reforming reactions. After the combustion chamber, the hot gas is passed over a reforming catalyst where hydrogen is produced through both a reaction like that in the methane steam reformer, where methane is converted to hydrogen and CO, and the WGS reaction. The benefit of this set-up is that internal heating will reduce heat loss compared to external heating. A detailed comparison of the reforming methods has been performed by Lewis et al. [10], where it can be seen that the thermal efficiency of steam methane reforming (SMR) and ATR are very similar. This is in

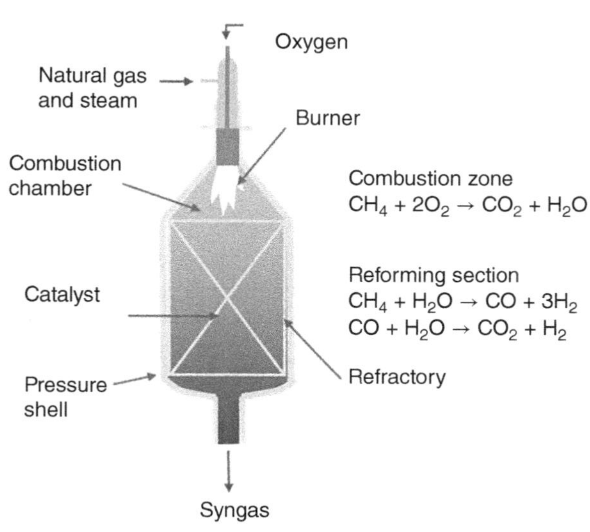

Figure 5.14 Autothermal reformer (redrawn from Faye et al. [16]).

agreement with comparisons of operating autothermal and steam methane reformers, where the performances are very similar overall [18].

ATR produces less CO_2 than steam reforming. The total amount of CO_2 emitted per kg of hydrogen produced has been modelled to amount to ca. 8.2–8.3 tonne CO_2/tonne H_2 [10, 19]. Slightly less than in steam reforming, possibly due to more effective heating. Carbon capture and storage can reduce the CO_2 emissions by 95% resulting in 0.5 tonne CO_2/tonne H_2 [10]. This excludes emissions attributable to natural gas exploration and power generation, which may increase the emissions to 5.7 tonne $CO_{2,e}$/tonne H_2.

In dry reforming, methane is principally reacted with CO_2 to form hydrogen and carbon monoxide. The attractiveness of this idea is that it would both consume CO_2 and produce a viable syngas feed. Unfortunately, with all catalysts at present, this feed will likely lead to coking. The coking can be minimised by co-feeding water [20], but in that situation normal steam reforming occurs, followed by the reverse water gas shift (RWGS) producing the syngas mixture and consuming CO_2. This means that unless a revolutionary new catalyst can be developed, dry reforming is not practical.

Tri-reforming involves the co-feeding of oxygen and CO_2 with the water and methane. Its purpose is to combine all the reactions included in steam reforming, dry reforming and ATR to produce a syngas with the desired H_2/CO-ratio [21]. Tri-reforming produces the heat internally, as with ATR. It should also utilise some CO_2 in the reaction stream. However, at present, it is still challenging to have a consumption rather than generation of CO_2 in the reactor. This is due to the equilibrium conversion of CO_2 being strongly dependent on pressure [22], meaning that low pressures would be needed to effectively convert CO_2. Tri-reforming was developed for the reforming of flue gas from power stations [23].

Nickel is the most common catalyst used in methane reforming [21, 24]. Nickel can get paired with other materials to improve its performance in terms of selectivity, activity, lifespan and strength (typically metal oxides). Furthermore, a very small addition of sulphur can help reduce coking of the catalyst and thus extend catalyst's lifetime. However, the nickel catalyst can easily get poisoned by sulphur, which adsorbs strongly on the catalyst surface and thus blocks the surface from being able to catalyse the desired reaction, so this needs to be carefully balanced. Ultimately between the relevant poisons and other methods of deactivation, the lifetime of the nickel catalyst is typically 1 year. Platinum group metals (PGMs) can also act as a reforming catalyst and can show long-term stability [25]. Unfortunately, the use of platinum is not justifiable due to price as it is far more expensive than nickel. Other materials such as cobalt, copper and molybdenum have been trialled for methane reforming. They typically underperform compared to nickel but can be alloyed with nickel for various overall performance benefits [26].

As with the coal gasification, methane reforming makes use of the WGS reactors following the reformer. The WGS is strongly equilibrium-limited (Figure 5.15). The equilibrium position is independent of pressure due to the equal number of molecules in the forward and reverse reaction. From a strictly equilibrium viewpoint, CO and/or CO_2 methanation can occur, especially at elevated pressures,

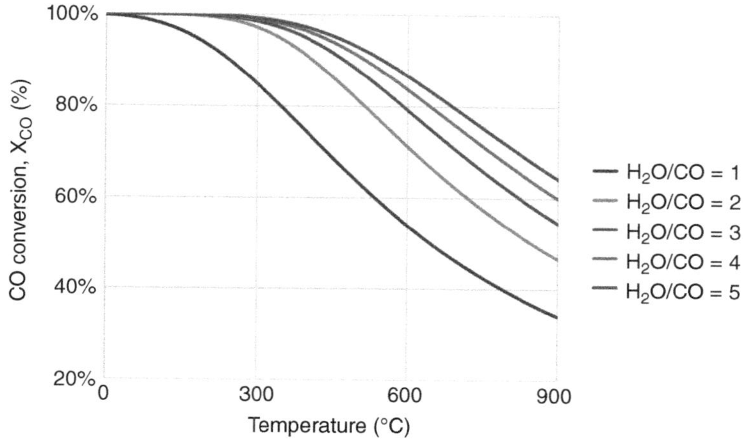

Figure 5.15 Equilibrium in the water gas shift reaction for various feed ratios of water to CO.

though the catalysts of choice for the WGS reaction are naturally poor methanation catalysts.

$$CO + H_2O \leftrightarrow CO_2 + H_2 \quad \Delta H^0_{rxn} = -41.3\frac{kJ}{mol}$$

Because of the equilibrium limitations, the WGS part of the process is typically done with two steps. First, a high-temperature WGS step is used that makes use of an iron/chromium catalyst (Fe_2O_3/Cr_2O_3), operating at 350–400 °C [27]. The catalyst is not very active, but this is countered by the high temperature and the fact that the catalyst is relatively inexpensive. This catalyst is responsible for most of the conversion of CO to hydrogen. The high-temperature step is followed by a low-temperature step. The typically catalyst for the low-temperature WGS reaction, is like that of the methanol synthesis catalyst, $CuO/ZnO/Al_2O_3$, which is far more active than the iron-based catalyst and can thus operate at the lower temperature of 200-250 °C where it converts much of the remaining CO.

The stream leaving the final WGS stage can have a very low CO content (in the ppm range). However, the CO present may still act as a poison to catalysts down the hydrogen value change. This is most notably an issue if the hydrogen is to be used in a hydrogen fuel cell, which uses a platinum catalyst, which is very sensitive to CO poisoning. As such, a CO clean-up stage is often used. This may be done through preferential oxidation of CO to CO_2 (CO-PROX) [28] after which CO_2 is easier to remove and is only a minor poison. Alternatively, selective methanation can be used, although the CO_x hydrogenation will reduce the hydrogen yield.

5.2.1.3 Water Electrolysis

Water electrolysis is a well-studied process that provides exceptionally pure H_2 [29]. Electrolysers were first demonstrated in around 1800 through the very simple process of passing a current through water into which a salt was mixed. The overall electrochemical reaction occurring is:

$$H_2O \rightarrow H_2(g) + \frac{1}{2}O_2(g)$$

Hydrogen at the cathode: $2H^+ + 2e^- \rightarrow H_2 \quad \Delta E^0 = 0\,V$
Oxygen evolution at anode: $H_2O \rightarrow \frac{1}{2}O_2 + 2H^+ + 2e^- \quad \Delta E^0 = -1.23\,V$

The required potential needed for the evolution reaction to occur may be calculated by converting the Gibbs free energy back to voltage using the Nernst equation which was covered in the section on batteries (Chapter 3). For these reactions to occur, a potential of >1.23 V must be applied when using liquid water or > 1.19 V in the vapour phase.

The cell voltage is dependent on temperature, as the Gibbs free energy depends on temperature. Electrolysers can be operated at a wide range of temperatures. Thus, manipulating the Nernst equation, and recognising that the enthalpy is not a strong function of temperature, the required voltage at various temperatures may be calculated:

$$E(T_2) = \frac{\Delta H_{rxn}}{n \cdot \mathcal{F}} \cdot \left(\frac{T_2 - T_1}{T_1}\right) + \frac{T_2}{T_1} \cdot E(T_1)$$

where E is the potential difference, ΔH_{rxn} is the enthalpy of reaction, T is the temperature, n is the number of moles of electrons being transferred in the reaction and $\mathcal{F}$ is the Faraday constant 96 485 A · s/mol.

The total energy required for the electrolysis is the reaction heat (ΔH) which will be supplied in the form of heat (Q) and electrical energy (W_{el}), which in the ideal case corresponds to ΔG. The relationship between these quantities is:

$$\Delta H = \Delta G + T\Delta S = W_{el} + Q$$

where T is the temperature and ΔS is the change in entropy upon the reaction.

The values may be calculated for the electrolysis reaction, allowing for the change temperature. The results of these calculations are shown in Figure 5.16. Although less electrical energy is required at higher temperature, with a notable discontinuity at the vaporisation point (100 °C at a pressure of 1 atm), the heat demand rises with increasing temperature. This means that, overall, there is a slight increase in the total energy required for the electrolysis of water with increasing temperature.

Electrolysers can, in theory, run at any pressure (and even at a different pressure at the cathode and the anode side) meaning that the effect of this pressure change on the Gibbs free energy needs to be considered. To calculate the voltage when the conditions change, the Nernst equation can be adapted to become:

in the presence of liquid water

$$E - E^0 = -\frac{RT}{n \cdot \mathcal{F}} \cdot \ln\left[\left(\frac{f_{H_2}}{p = 1\,\text{bar}}\right) \cdot \left(\frac{f_{O_2}}{p = 1\,\text{bar}}\right)^{0.5}\right]$$

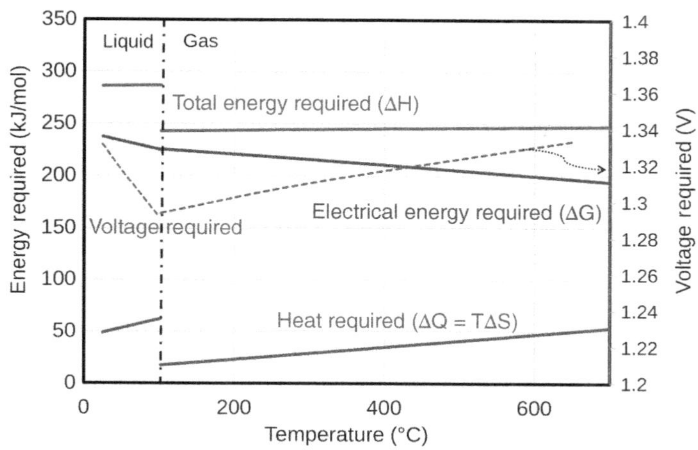

Figure 5.16 Energy required for electrolysis as a function of temperature.

in the presence of steam

$$E - E^0 = -\frac{RT}{n \cdot \mathcal{F}} \cdot \ln\left[\frac{\left(\frac{f_{H_2}}{p = 1\text{ bar}}\right) \cdot \left(\frac{f_{O_2}}{p = 1\text{ bar}}\right)^{0.5}}{\left(\frac{f_{H_2O}}{p = 1\text{ bar}}\right)}\right]$$

where E is the potential difference, R is the ideal gas constant (8.314 kJ/mol K), T is the temperature, n is the number of moles being transferred in the reaction, $\mathcal{F}$ is the Faraday constant and f_i is the fugacity of species i (this becomes the partial pressure of component i if the gas mixture can be approximated as an ideal gas mixture).

Note that here the reference state, E^0, is taken as when the partial pressures of hydrogen and oxygen are each at 1 bar, i.e. each in their reference state.

Operating the electrolyser at high pressure may be desirable for storage of hydrogen, as pumping of water is cheaper (both financially and energetically) than the compression of hydrogen. However, this will require a higher voltage. For example, operating the electrolyser with liquid at 80 °C with both anode and cathode at 200 bar ($E - E^0 = -0.12$ V) will require an applied voltage of at least 1.306 V. Changing only the pressure at the cathode would decrease the required voltage, but increase the likelihood of cross-over, i.e. hydrogen passing through to the anode.

Technologies for Electrolysers The current technologies which exist to produce hydrogen from water include acidic electrolysis with a proton exchange membrane (PEM), alkaline electrolysis (AEL), anion exchange membrane (AEM) electrolysis and solid oxide electrolysis (SOE). A summary of the available technologies is given in Table 5.3. To date, only AEL and PEM have commercial-level implementation. AEL is the most mature technology and is desirable due to its cheaper catalysts compared to PEM [30]. PEM also has some advantages as well, e.g. the significantly

Table 5.3 Available electrolyser technologies (compiled from References [29–36] data as of 2024).

	PEM[a)]	AEL[b)]	AEM[c)]	SOE[d)]
Electrolyte	Ionomer (e.g. Nafion)	Aqueous KOH (20–40 wt.%)	Solid polymer	Y_2O_3-ZrO_2
Anode	RuO_2, IrO_2	Ni/NiCo alloys	Ni/Pt	Lanthanum strontium manganite
Cathode	Pt	Ni/NiMo alloys	Ni/Pt	Ni doped YSZ
Separator	Nafion 117 (180 μm)	Diaphragm Zirfon Perl (500 μm)	*In development*	Y_2O_3-ZrO_2
Cell voltage (V)	1.8–2.4	1.8–2.4	1.6–2.0	0.9–1.3
Current density	0.6–2.0 A/cm^2	0.2–0.4 A/cm^2	*In development*	*In development*
H_2-purity	99.9999	99.5 (99.5–99.9 [37])		99.9
Efficiency (%)	50–83	50–78	57–70	50–89
Stack lifetime (h)	20 000–60 000	>60 000	20 000–60 000	~20 000
Capital cost (US$/kW)	1100–1600	500–1400		2800–5600
Instal. capacity (MW)	126	354		

a) PEM – proton exchange membrane electrolysis (acid electrolysis).
b) AEL – alkaline electrolysis.
c) AEM – alkaline exchange membrane electrolysis.
d) SOE – high-temperature solid oxide electrolysis.

higher current density in comparison to AEL as well as the modularity and the control, which allows for more intermittent use [30].

In the PEM system (Figure 5.17), two half-cells containing liquid water are separated by a proton exchange polymer membrane. At the anode, water is split into oxygen, protons and electrons:

Anode: $H_2O \rightarrow \frac{1}{2}O_2 + 2H^+ + 2e^-$ (oxygen evolution reaction – OER).

The electrons flow from the anode to the cathode via an external circuit, whilst the protons diffuse internally through the membrane to the cathode. At the cathode, the protons react with electrons to form hydrogen:

Cathode: $2H^+ + 2e^- \rightarrow H_2$ (hydrogen evolution reaction – HER).

As the membrane effectively prevents either the oxygen formed at the anode or the hydrogen formed at the cathode from diffusing to the opposite side, both these gases may be removed separately at high purity [29]. These systems show high power

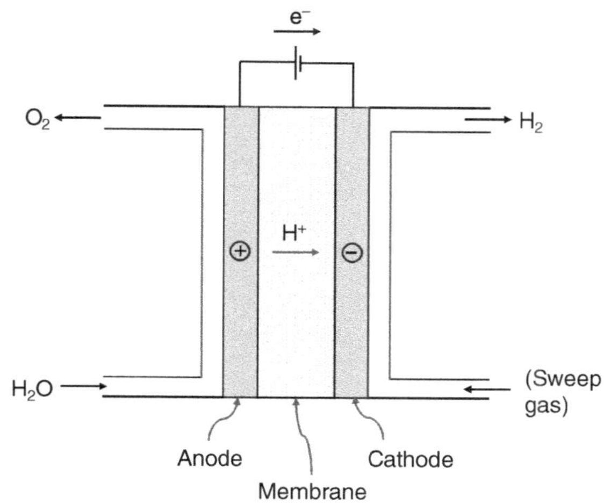

Figure 5.17 Water electrolysis over in a proton exchange membrane (PEM) cell (redrawn from Rego De Vasconcelos and Lavoie [29]).

density (Table 5.3). The use of the membrane additionally allows fast response to changes in input unlike what is seen with dissolved electrolytes. This allows the system to adapt to changing power availability and makes start-up and shutdown relatively simple.

PEM electrolysers are operated at moderate temperatures, normally below 150 °C, and pressures of up to 20 to 50 bar [38]. They may be operated at pressures up to 200 bar [39], but this will add complexity to the system design. At higher pressures, the gases are more able to cross the polymer membrane, thus requiring thicker membranes [40].

A typical polarisation curve for a PEM electrolyser is shown in Figure 5.18. The polarisation curve shows the required overpotential encountered to operate as the current density increases due to various limitations. This means that, with PEM electrolysers, as the voltage increases, the current should increase. This may seem to indicate that fewer electrolysers are needed when the voltage is increased. However, with higher voltage, the efficiency decreases so more power input would be required to achieve the required hydrogen production, meaning a larger power generation section on the plant. In the end, an optimum between the efficiency and the capital cost for the plant needs to be found.

It may also be seen from Figure 5.18 that in a PEM electrolyser, the largest loss is from the anode (OER) kinetics. Due to these losses, research into improving the OER catalyst to make it more active, stable, conductive and selective is of the highest priority. Substantial losses are also seen from ohmic resistance of the PEM membrane [41].

The condition inside the PEM cell is acidic, which means that care must be taken with the construction materials and that the catalysts used need to be corrosion-resistant. For this reason, PGMs are used as the electro-catalyst. Iridium oxides and ruthenium are favoured for the anode OER as the most active catalyst,

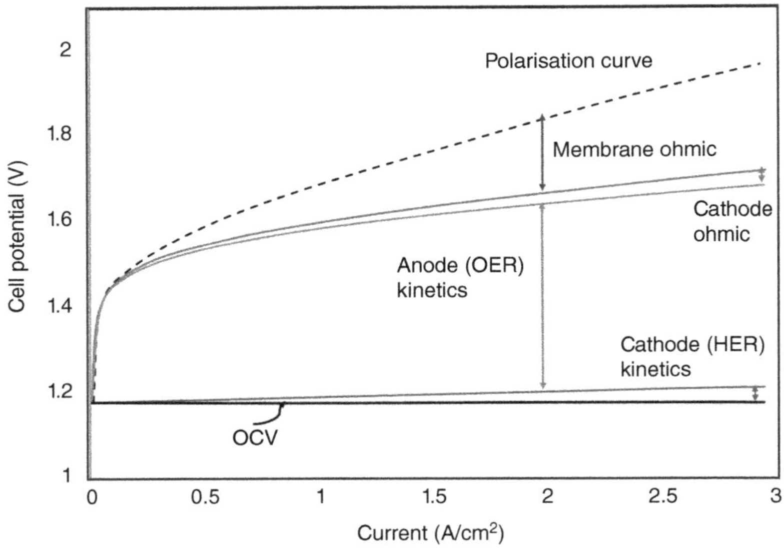

Figure 5.18 Contributing factors to overpotential in PEM electrolysers; 80 °C. HER – hydrogen evolution reaction; OER – oxygen evolution reaction; OCV – open cell voltage (data from Gerhardt et al. [41]).

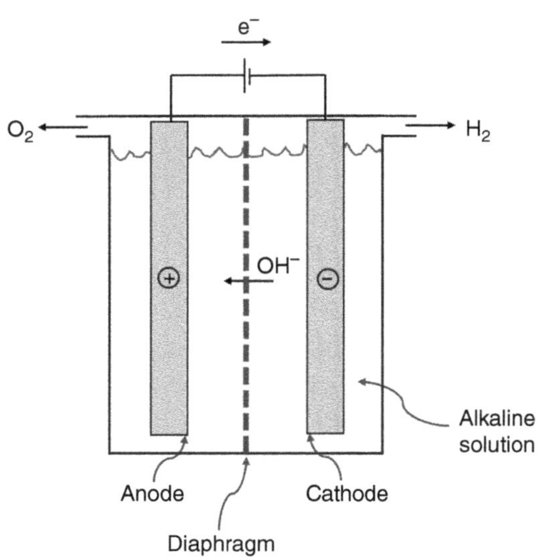

Figure 5.19 Alkaline water electrolysis (AEL) (adapted from References [29, 42]).

and platinum as the cathode catalysing the HER (a comparatively facile reaction) [37, 40]. The disadvantage is the cost and availability of the PGMs.

AEL is a well-established technology that is used on a large scale [37]. In this system (Figure 5.19), two electrodes are immersed in an aqueous alkaline solution. The solution will typically contain KOH or NaOH. There is a diaphragm between the electrodes to separate the gases formed on the two electrodes. At the anode, hydroxyl ions combine to form oxygen and water with the release of electrons:

Anode: $2OH^- \rightarrow \frac{1}{2}O_2 + H_2O + 2e^-$ (oxygen evolution reaction – OER).

The electrons flow from the anode to the cathode via an external circuit, where they react with water to form hydrogen and hydroxyl ions:

Cathode: $2H_2O + 2e^- \rightarrow H_2 + 2OH^-$ (hydrogen evolution reaction – HER).

The hydroxide ions move through the aqueous solution to the anode, completing the circuit, in comparison to the PEM electrolyser which has the cell completed by proton transfer.

As there is water formed on the anode side and consumed on the cathode side, care needs to be taken that the concentration of the alkaline solution remains constant throughout the system. This can be done by mixing the alkaline solutions from the respective side, although it can lead to some contamination of the gas streams obtained. The hydrogen purity obtained using AEL is slightly lower than that obtained with PEM (Table 5.3).

The AEL system may be operated in the temperature range of 40 to 100 °C keeping water in the liquid state. Although these systems are typically operated at pressures below 30 bar, they can also be operated at much higher pressures, reportedly up to 690 bar [29, 43]. The benefit of operating at high pressure is that the hydrogen formed is already pressurised and thus does not need to be compressed before transportation. The disadvantage of high-pressure operation is that it leads to decreased efficiency of the electrolyser and decreased hydrogen purity [44].

The AEL system uses non-noble metal catalysts, giving a significant reduction in the operating cost compared to PEM. The most common material used is nickel, with other metals, e.g. iron added to improve its stability. The electrode performance may be further improved by either adding different transition or noble metals [29].

Notable disadvantages of the AEL systems are the low current density and their sluggish response to changes in input conditions as mixing of the alkaline system introduces a lag time in any response. The system also has a relatively long start-up time making intermittent operation unadvisable [44]. An additional disadvantage is the corrosive nature of the alkaline solution [45].

The third technology researched in earnest is the so-called anion exchange membrane (AEM) water electrolysis. This would combine the benefit of both the PEM cell, viz. the modularity and control, as well as the anionic electrolysis, viz. comparatively cheap catalysts (using the same as those of the AEL). A major problem with this technology at current is the membranes, which need to be capable of transferring hydroxyl (OH^-) species. However, should this technology become available, it may become the dominant player in the future of water electrolysis [46].

SOE is a comparatively recent technology and is still at the prototype stage, with only lab scale systems having been constructed [29]. In this set-up, a solid electrolyte, typically ZrO_2 doped with Y_2O_3, is located between the anode and the cathode (Figure 5.20). The electrolyte is chosen based on its conductivity of the O^{2-} ions

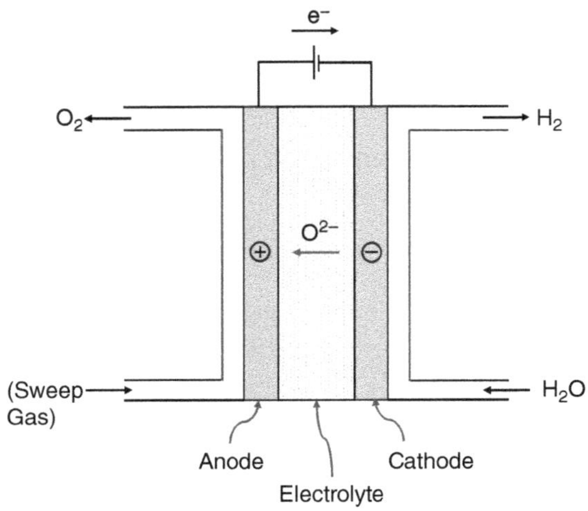

Figure 5.20 High-temperature electrolysis using a solid oxide electrolyser (SOE) (adapted from References [29, 42]).

as well as thermal and chemical stability [44]. At the anode, O^{2-} ions react to form O_2 gas.

Anode: $2O^{2-} \rightarrow O_2 + 4e^-$ (oxygen evolution reaction – OER).

The electrons flow from the anode to the cathode via an external circuit, where they react with water to form hydrogen and O^{2-} ions.

Cathode: $2H_2O + 4e^- \rightarrow H_2 + 2O^{2-}$ (hydrogen evolution reaction – HER).

The O^{2-} ions diffuse through the solid electrolyte back to the anode, completing the circuit.

The SOE process is operated at high temperature (700–1000 °C) with steam as the feed. The high temperature is required for the transport of oxygen anions through the solid oxide, but promotes at the same time the reaction kinetics [42] and favours the endothermic reaction. Together, the high temperature and the endothermic nature of the reaction mean that the equilibrium cell voltage is lower (Table 5.3), further giving lower electricity demands [44]. The high-temperature operation also results in higher efficiencies than either AEL or PEM [37].

The disadvantages of the SOE system are linked to the high temperatures involved. Whilst operating at high temperature has many benefits, it does result in faster material degradation and results in limited long-term stability of the system [37, 44] besides the simple requirement for achieving the high temperatures. The product stream from the SOE electrolyser is a mixture of hydrogen and steam, rather than pure hydrogen as seen with AEL and PEM, and requires additional processing [44]. A secondary benefit which may see an expansion of SOE technology is the ability to co-feed CO_2; this can produce CO through the in situ WGS reaction. Whilst having no current implementation, this is a popular technology in PtX modelling with potential cost savings over other forms of electrolysis followed by WGS.

Intermittent Power Supply and Electrolytic Water Splitting The use of renewable power to produce green hydrogen immediately poses the question on how to deal with the intermittency of the power supply, which may affect the power supply to the electrolyser if not (partially) buffered by a storage facility such as batteries. In the extreme case with no buffer facilities, the power to the electrolyser, and thus the hydrogen production rate, will follow the intermittency pattern of the renewable power. The intermittency in the power can be short-term (on the scale of minutes due to suddenly changing meteorological conditions), diurnal (earth rotation), medium-term (on the scale of days to weeks due to changing long-term meteorological conditions) and long-term (due to seasonal variations).

The effect of intermittent power supply on the performance of electrolyser (stacks) should be viewed from a system's perspective [47], as variable power affects not only the electro-catalytic conversion of water, but also the heat management of the stack, flow and pressure control within the stack and process safety [48].

A drop in the power supply will decrease the electric losses in the cell (see Figure 5.18), and hydrogen will be produced at a reduced rate, but with a higher efficiency. However, a drop in the power supply may also result in a drop in the temperature in the electrolysis cell because of the reduced heat generation in the cell, thus dropping the rate of reaction.

Intermittency may affect product purity due to possible cross-over, which may dictate the lower load capacity that an electrolyser can be operated safely and below which the electrolyser will have to be shut down (permeation of hydrogen into the oxygen-rich stream may result in the formation of an explosive mixture as the lower explosion limit of hydrogen in oxygen at 20 °C is 4%). Hence, alkaline electrolysers must operate at load levels higher than ca. 10–40% of its rated load level [49, 50], where PEM electrolyser and solid oxide electrolyser show a much greater flexibility [37, 51], although this depends on the thickness of the membrane in PEM electrolysers [51].

Regular ramping up and powering down of the electrolysis cell may result in the deterioration of the electrodes and membranes. Operating PEM electrolysers using variable power results in an increase in the ohmic loss due to the degradation of the membrane close to the anode catalyst layer, whereas deactivation due to loss in the electrochemical surface area appears to be modest [52]. Alkaline cells seem to suffer from a similar loss in activity when operating under variable power supply [48]. Shutdown of the electrolysis cells will result in reverse current, which may result in irreversible deactivation of the nickel cathode in the AEL cell [53, 54]. A small protective current through the electrolysis cell will protect the electrode [49].

Design hydrogen production units utilising water electrolysis poses an optimisation problem. The minimum size of the unit and thus capital cost is obtained if continuous power can be drawn either via back-up through batteries or a grid, but this would push the levelised cost of electricity (LCOE) up. Alternatively, the hydrogen production unit operate as an island using solely renewable energy and operates thus under variable load. This requires an overdesign of the hydrogen producing unit, backing up the power supply with some battery storage to always ensure safe

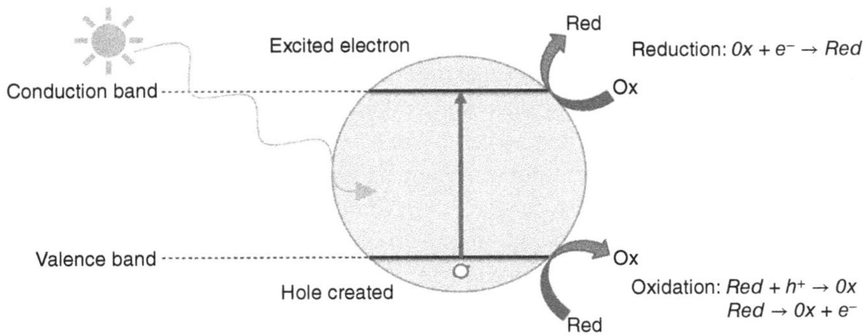

Figure 5.21 Simple schematic of using semi-conductors as a photocatalyst.

operation. The optimum required battery storage should be evaluated over the lifetime of the plant considering the battery cost (including replacement) and the cost of the electrolyser.

5.2.1.4 Hydrogen from Photocatalytic Water Splitting

The major cost associated with obtaining hydrogen from water via electrolysis is the green electricity that is needed to supply energy to the process. It would, therefore, be required to obtain the energy directly from a source that is renewable, which does not result in the release of CO_2 or any other environmental pollutants. One option is to use photocatalytic water splitting. Whilst other catalysts are provided with required activation energy through heat (thermo-catalysis) or from electricity (electro-catalysis), photocatalysts receive energy from light (ideally solar irradiation).

A typical diagram that is seen in the photocatalytic space is shown in Figure 5.21. Firstly, the idea of the band gap gets mapped onto a photocatalytic particle. When light, preferably from the sun, is absorbed by the photocatalytic semi-conductor, it excites an electron from the valence band to the conduction band. This creates an ‘electron site’ or a ‘reduced site’ as it has an extra electron (stylised as e^-). Due to the extra electron, this site may facilitate reduction reactions. The electron leaving the valence band also creates a ‘hole site’ or an ‘oxidised site’ due to the loss of an electron (stylised as h^+). This site may facilitate oxidation reactions.

The reactions occurring on the photocatalyst when being used for hydrogen production using water splitting are as follows:

Charge carrier generation: $\text{light} \rightarrow h^+ + e^-$
Oxidation: $H_2O + 2h^+ \rightarrow 2H^+ + \frac{1}{2}O_2$
Reduction: $2H^+ + 2e^- \rightarrow H_2$

Principally, any semi-conductor can be used as a photocatalyst provided its reduced and oxidised sites can catalyse the desired reaction(s). However, the most popular photocatalyst is titania (specifically P25, a mixture of two different phases of titania: anatase, ~80–90% and rutile, ~10–20%), because it is relatively cheap, has a low toxicity and has a high activity. Titania was shown to be active for photocatalytic water splitting by Fujishima and Honda in 1972 [55]. It has been shown that the mixed phase catalyst is active because of the combination of anatase/rutile with anatase having an indirect band gap (thus more difficult to activate) and rutile having a direct band gap, but a high recombination rate. Hence, activation over rutile may result in holes/electrons spilling over to anatase, where the recombination rate is less thus resulting in a high activity [56]. Titania is often promoted (e.g. with a noble metal) to improve charge carrier separation [57, 58] and thus its performance (although the presence of the metal may also accelerate HER).

Some materials and the associated band gaps and wavelengths are given in Table 5.4. Those with a required wavelength less than that of UV light require UV light to activate the catalyst, whilst the remaining listed may also be activated by visible light.

In systems where only one catalyst is used, said catalyst must meet several criteria, namely thermodynamic potential for water splitting, a sufficiently narrow band gap and corrosion resistance. These criteria mean that there are only limited suitable catalysts [59].

The required wavelength of light is linked to the bandgap of the photo-catalyst. For anatase, which is the dominant photocatalytic phase of titania, this is $E_{\text{bandgap,anatase}} \approx 3.2$ eV, which gives a required wavelength, $\lambda = 388$ nm. This wavelength falls within the UV range and as such is a definite drawback to photocatalytic

Table 5.4 Selected catalytic materials for photocatalysis (dotted line indicates boundary to UV light) (data from Fujishima and Honda [55]).

Semi-conductor	Bandgap (eV)	Wavelength (nm)
ZrO_2 (monoclinic)	4.5	276
ZrO_2 (tetragonal)	4.1	303
ZrO_2 (cubic)	3.8	327
SnO_2	3.5	354
ZnO	3.2	387
TiO_2 (anatase)	3.2	387
TiO_2 (rutile)	3.0	413
CeO_2	2.9	421
WO_3	2.8	442
g-C_3N_4	2.7	458
CdS	2.5	495
V_2O_5	2.4	516
Fe_2O_3	2.3	538
Cu_2O	2.2	563

water splitting. This is especially true using solar light where the availability of sunlight in this range is limited (±5% of incoming light). Other materials, with smaller band gaps (Table 5.4), could potentially use sunlight more effectively.

At present, there are very few systems available that could produce hydrogen efficiently on any feasible scale, with solar to hydrogen efficiencies of only 1.8% reported [57, 60], though it should be noted that solar to hydrogen via electrolysis would only operate at a maximum efficiency of ca. 15%, viz. the efficiency of capturing electric power from solar irradiation (<25%) and the efficiency of electrolysers (~60%; see Table 5.3). Photocatalytic water splitting is thought to deal much better with intermittent solar energy supply than electro-catalytic water splitting (see Section 5.2.1.3.2), which may become advantageous if the efficacy of photocatalytic water splitting can be improved.

5.2.2 Efficiency of Hydrogen Production

It is important to be able to accurately compare various process options, particularly in terms of their relative efficiency. The energy efficiency (η^{energy}) is given by:

$$\eta^{\text{energy}} = \frac{\sum_i (\dot{n} \cdot \underline{H})_{i,\text{target products}} + \dot{Q}_{\text{useful from process}} + \dot{W}_{\text{obtained from process}}}{\sum_j (\dot{n} \cdot \underline{H})_{j,\text{reactants}} + \dot{Q}_{\text{added to the process}} + \dot{W}_{\text{added to the process}}}$$

The enthalpy of the desired, target products and reactants must be taken from a common basis and are typically taken as the LHV of the individual components. Hence, the combustion products in the gas phase are taken as the reference state. Heat, that is not given off to the surroundings and can be used for other purposes, is considered as a useful output of a process ($\dot{Q}_{\text{useful from process}}$). Hence, the energetic efficiency is typically an indication of loss of energy in the form of heat to the surroundings. The drawback of this approach is that not all energy associated with the material flow(s) can be converted into work either, and the process will thus convert some useful energy into entropy, which is lost. Exergy (B) accounts for this formation of entropy by essentially considering the Gibbs free energy with entropy contribution at a reference temperature (T_0, typically 25 °C) rather than process temperature (T).

$$G \equiv H - T \cdot S$$

$$B \equiv H - T_0 \cdot S$$

Exergetic efficiency is thus given by a similar formula to energy efficiency but instead considers exergy rather than enthalpy:

$$\eta^{\text{exergy}} = \frac{\sum_i (\dot{n} \cdot \underline{B})_{i,\text{target products}} + \dot{Q}_{\text{useful from process}} + \dot{W}_{\text{obtained from process}}}{\sum_j (\dot{n} \cdot \underline{B})_{j,\text{reactants}} + \dot{Q}_{\text{added to the process}} + \dot{W}_{\text{added to the process}}}$$

If the process does not generate any entropy, then the exergetic efficiency would be the same as the energy efficiency. Thus, the maximum possible exergetic efficiency for a given process is the same as the energetic efficiency. The use of exergetic efficiency provides a better comparison for individual steps within the overall process by explicitly considering unavoidable entropy generation.

The exact value of the energetic and exergetic efficiency is, of course, dependent on the considered process. In the ideal case, assuming no heat loss and reversible processes, the energetic and exergetic efficiencies are equal to 1, although maximum values can be estimated based on the assumption of the minimum input of work and energy into the system with products and reactants at their reference state, typically taken at 25 °C. For instance, the hydrogen production via coal gasification can stoichiometrically be expressed as:

$$\mathrm{C(s)} + 2\,\mathrm{H_2O(l)} + Q_{\text{required,min}} + W_{\text{required,min}} \rightarrow \mathrm{CO_2(g)} + 2\mathrm{H_2(g)}$$

The minimum, required amount of work for this process feed can be obtained from a combined energy and entropy balance:

$$\begin{aligned}\frac{W_{\text{required,min}}}{n_{\text{C,fed}}} &= \Delta \underline{G}_{\text{rxn}} = \Delta \underline{G}^{25\,°\text{C}}_{\text{f,CO}_2\text{(g)}} + 2 \cdot \Delta \underline{G}^{25\,°\text{C}}_{\text{f,H}_2\text{(g)}} - \Delta \underline{G}^{25\,°\text{C}}_{\text{f,C(s)}} - 2 \cdot \Delta \underline{G}^{25\,°\text{C}}_{\text{f,H}_2\text{O(l)}} \\ &= 80.1 \cdot \frac{\text{kJ}}{\text{mol C}}\end{aligned}$$

and the minimum, required amount of heat for this reaction is given by

$$\begin{aligned}\frac{Q_{\text{required,min}}}{n_{\text{C,fed}}} &= T \cdot \Delta \underline{S}_{\text{rxn}} = T \cdot \left(\Delta \underline{S}^{25\,°\text{C}}_{\text{f,CO}_2\text{(g)}} + \Delta \underline{S}^{25\,°\text{C}}_{\text{H}_2\text{(g)}} - \Delta \underline{S}^{25\,°\text{C}}_{\text{f,C(s)}} - 2\Delta \underline{S}^{25\,°\text{C}}_{\text{f,H}_2\text{O(l)}}\right) \\ &= 98.0 \cdot \frac{\text{kJ}}{\text{mol C}}\end{aligned}$$

Hence, the maximum exergetic efficiency can be obtained from the exergetic input and output of the process (both based on the combustion of the feed):

$$\begin{aligned}\eta^{\text{exergy}}_{\text{max}} &= \frac{2 \cdot B_{\text{H}_2\text{(g)}}}{B_{\text{C(s)}} + 2 \cdot B_{\text{H}_2\text{O(l)}} + \dfrac{Q_{\text{required,min}}}{n_{\text{C,fed}}} + \dfrac{W_{\text{required,min}}}{n_{\text{C,fed}}}} \\ &= \frac{2 \cdot 236.09}{409.87 + 2 \cdot 0.9 + 98.3 + 79.8} = 0.80\end{aligned}$$

(with the maximum energetic efficiency being equal to 100%).

The actual efficiency of hydrogen formation from carbon is much lower. Seyitoglu et al. [61] evaluated the efficiency of hydrogen production via coal gasification (with some minor hydrogen production electrochemically as a way to improve the overall performance of the process – the amount of electrochemically produced hydrogen was less than 5% of the total hydrogen production). They obtained an energetic efficiency of 37–41% and an exergetic efficiency of 34–36.5% depending on the type of coal. The largest inefficiencies in the process were associated with air separation unit required for producing pure oxygen, which is fed to the gasifier to provide heat for the endothermic gasification process without the introduction of nitrogen into the product stream. Younas et al. [24] have summarised the energy and exergy efficiencies for different coal gasification-based processes with different ways to transform as

Table 5.5 Coal gasification-based hydrogen production and coal gasification-based multigeneration (data from Younas et al. [24]).

Coal gasification system	η^{energy} (%)	η^{exergy} (%)	Reference
Coal gasification-based co-generation system	38	27	[62]
Integrated coal-based gasification system	37–41	34–37	[61]
Tri-generation system	57	45	[63]
Combined coal gasification + alkaline electrolysis	58	55	[64]
Coal gasification-based cogeneration system		54	[65]
Gasification + thermochemical water decomposition + H_2 compression systems	51	48	[66]
Integrated underground coal gasification with SOFC fuel cell system	29	26	[67]

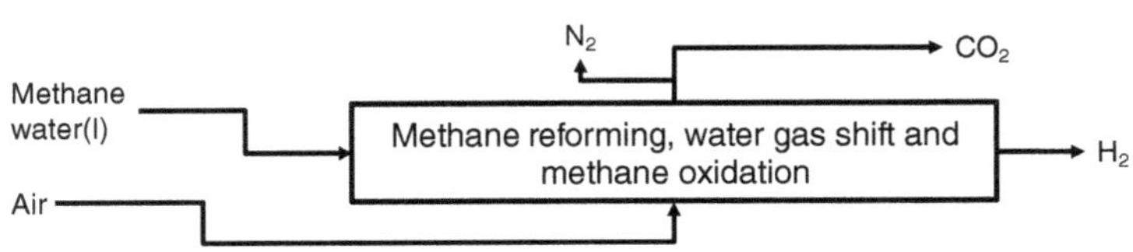

Figure 5.22 Overall process for autothermal methane reforming.

much of the waste energy into hydrogen (see Table 5.5), Integration of various methods to produce hydrogen will improve the overall energetic and exergetic efficiency, of course at the cost of increased capital expenditure.

The maximum energetic and exergetic efficiency for the formation of hydrogen via methane steam reforming can be determined (see Figure 5.22). The overall reaction can be represented by:

$$CH_4(g) + (4x - 2)\, H_2O(l) + (2 - 2x)\, O_2(g) \rightarrow CO_2(g) + 4x\, H_2(g)$$

Here, x is the fraction of methane that is converted to hydrogen in the reforming reactions rather than being combusted. It should be noted that for $x < 0.5$ water is formed in the process, and for $x > 0.5$ water will be consumed in the process.

The minimum, required amount of work for this process feed can be obtained from a combined energy and entropy balance:

$$\begin{aligned}\frac{W_{\text{required,min}}}{n_{CH_4,\text{fed}}} &= \Delta \underline{G}_{\text{rxn}} = \Delta \underline{G}^{25\,°C}_{f,CO_2(g)} + 4x \cdot \Delta \underline{G}^{25\,°C}_{f,H_2(g)} - \Delta \underline{G}^{25\,°C}_{f,CH_4(g)} \\ &\quad - (4x - 2) \cdot \Delta \underline{G}^{25\,°C}_{f,H_2O(l)} - (2 - 2x) \cdot \Delta \underline{G}^{25\,°C}_{f,O_2} \\ &= (-818.1 + 948.4x) \cdot \frac{\text{kJ}}{\text{mol CH}_4}\end{aligned}$$

implying that for $x < 0.86$, $\frac{W_{\text{required,min}}}{n_{CH_4,\text{fed}}} < 0$ and work can be removed from the overall process and for $x > 0.86$ work has to be added to the overall process. The minimum, required amount of heat for this reaction with the temperature at the inlet and outlet at 25 °C is given by

$$\frac{Q_{\text{required,min}}}{n_{\text{CH}_4,\text{fed}}} = T \cdot \Delta \underline{S}_{\text{rxn}} = T \cdot \left(\Delta \underline{S}^{25\,°\text{C}}_{\text{f,CO}_2\text{(g)}} + 4x \cdot \Delta \underline{S}^{25\,°\text{C}}_{\text{f,H}_2\text{(g)}} - \Delta \underline{S}^{25\,°\text{C}}_{\text{f,CH}_4\text{(g)}} \right.$$

$$\left. -(4x-2) \cdot \Delta \underline{S}^{25\,°\text{C}}_{\text{f,H}_2\text{O(l)}} - (2-2x) \cdot \Delta \underline{S}^{25\,°\text{C}}_{\text{f,O}_2} \right)$$

$$= \frac{298.15}{1000} \cdot (-242 + 653x) \cdot \frac{\text{kJ}}{\text{mol C}} = (-72.1 + 194.7x) \cdot \frac{\text{kJ}}{\text{mol CH}_4}$$

Implying that for $x < 0.37$, heat must be removed from the overall process and for $x > 0.37$, heat has to be added to the overall process. If no useful heat or work is removed from the process, the efficiencies can be estimated as:

$x < 0.37$

$$\eta^{\text{energy}} = \frac{4x \cdot \text{LHV}_{\text{H}_2}}{\text{LHV}_{\text{CH}_4}}$$

$$\eta^{\text{exergy}} = \frac{4x \cdot B_{\text{H}_2}}{B_{\text{CH}_4} + (2-2x) \cdot B_{\text{O}_2}}$$

$0.37 < x < 0.50$

$$\eta^{\text{energy}} = \frac{4x \cdot \text{LHV}_{\text{H}_2}}{\text{LHV}_{\text{CH}_4} + (-72.1 + 194.7x)}$$

$$\eta^{\text{exergy}} = \frac{4x \cdot B_{\text{H}_2}}{B_{\text{CH}_4} + (2-2x) \cdot B_{\text{O}_2} + (-72.1 + 194.7x)}$$

$0.50 < x < 0.86$

$$\eta^{\text{energy}} = \frac{4x \cdot \text{LHV}_{\text{H}_2}}{\text{LHV}_{\text{CH}_4} + (4x-2) \cdot (-\Delta H_{\text{vap,H}_2\text{O}}) + (-72.1 + 194.7x)}$$

$$\eta^{\text{exergy}} = \frac{4x \cdot B_{\text{H}_2}}{B_{\text{CH}_4} + (2-2x) \cdot B_{\text{O}_2} + (4x-2) \cdot B_{\text{H}_2\text{O(l)}} + (-72.1 + 194.7x)}$$

$x > 0.86$

$$\eta^{\text{energy}} = \frac{4x \cdot \text{LHV}_{\text{H}_2}}{\left(\begin{array}{l} \text{LHV}_{\text{CH}_4} + (4x-2) \cdot (-\Delta H_{\text{vap,H}_2\text{O}}) \\ \quad + (-72.1 + 194.7x) + (-818.1 + 948.4x) \end{array} \right)}$$

$$\eta^{\text{exergy}} = \frac{4x \cdot B_{\text{H}_2}}{\left(\begin{array}{l} B_{\text{CH}_4} + (2-2x) \cdot B_{\text{O}_2} + (4x-2) \cdot B_{\text{H}_2\text{O(l)}} \\ \quad + (-72.1 + 194.7x) + (-818.1 + 948.4x) \end{array} \right)}$$

The energetic and exergetic efficiency to produce hydrogen using methane steam reforming as a function of the steam to carbon ratio is given in Figure 5.23. The maximum energetic efficiency increases with increasing steam to carbon ratio and reaches 100% at a steam ratio of 1.45 (or $x = 0.86$), because under these circumstances, no energy in the form of work or heat is being generated and all energy input ends up in the product hydrogen. The maximum exergetic efficiency also increases

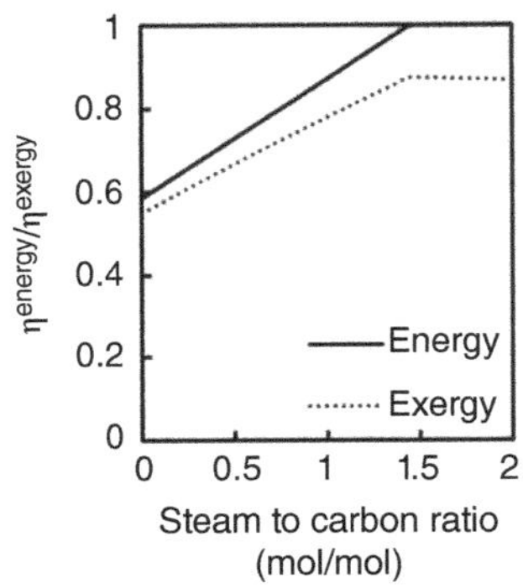

Figure 5.23 Maximum energetic and exergetic efficiency in the steam reforming of methane ensuring complete conversion of methane by co-feeding oxygen as a function of the steam to carbon ratio.

with increasing steam to carbon ratio but is limited to ca. 0.87 at high steam to carbon ratios.

An example of using exergy vs. energy analysis of a methane steam reforming process where heat is added externally is given by Simpson and Lutz [12]. Their specific goal was to use exergy analysis to determine where in the process exergy was being lost. The modelled process showing inlet and outlet streams is given in Figure 5.24. Methane and water are fed to a reformer, after which the resulting syngas is sent to a WGS reactor. Heating of the feed stream is achieved by combusting any retentate from the hydrogen separation membrane together with any necessary extra methane and passing the exhaust through a heat exchanger. Simpson and Lutz [12] concluded that most exergy loss occurred in the reforming stage as a consequence of the irreversibility of combustion and heat transfer. Additionally, they found that substantial exergy was lost with the exhaust stream, a situation that should not occur with the autothermal reactor considered above. They further used their model to investigate the effect of reformer temperature and pressure on the energy and exergy efficiencies (Figure 5.25). Interestingly, energy and exergy efficiencies showed the highest value at the same operating point. However, the efficiencies at this optimum point are quite different, with the exergetic efficiency always being lower than the thermal (energy) efficiency.

At present methane steam reforming is the most efficient method of producing hydrogen. However, it emits CO_2 that will require carbon capture and sequestration (CCS) to mitigate, which adds both to the cost and the inefficiencies to the process. A comparison of hydrogen generation using different methane reforming processes in terms of CO_2 emissions, overall thermal efficiency and cost of hydrogen is given in Table 5.6. The energy efficiency of the hydrogen production is reduced by the inclusion of CO_2 capture in the process, due to the energy cost for CO_2 separation from the product stream (typically an amine wash) and its compression prior to storage. The inclusion of carbon capture technology reduces the so-called Scope 1 emissions, i.e. the direct greenhouse gas emissions from the process, but may increase the Scope 2 emissions, i.e. emissions associated with power generation. Hence, the reduction in CO_2 emissions is limited. For instance, 76% of all greenhouse gas emissions in ATR are so-called Scope 1 emissions; this is reduced to only 16% when including CO_2-capture in the process. Similarly, 81% of all greenhouse

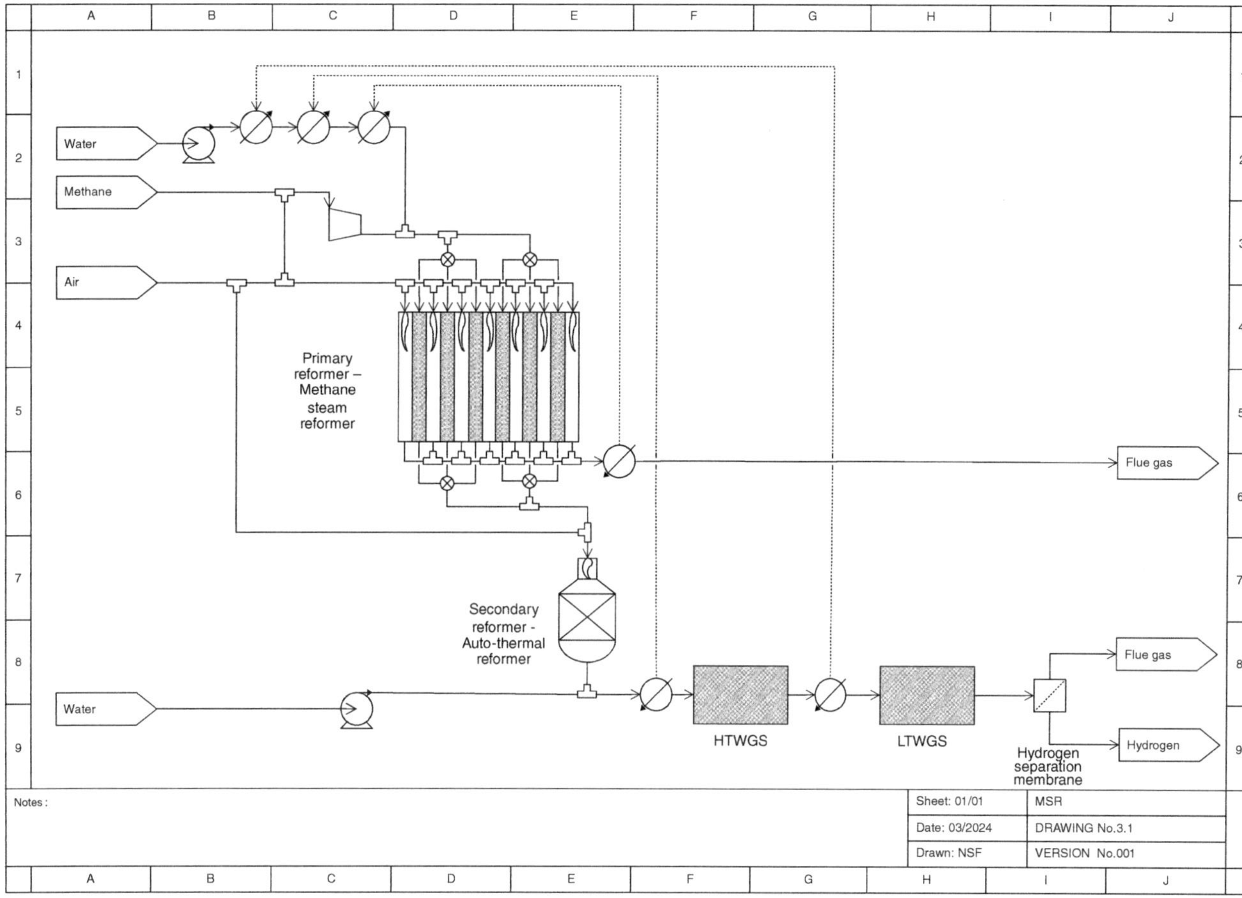

Figure 5.24 Process model for methane steam reforming (redrawn from Simpson and Lutz [12]).

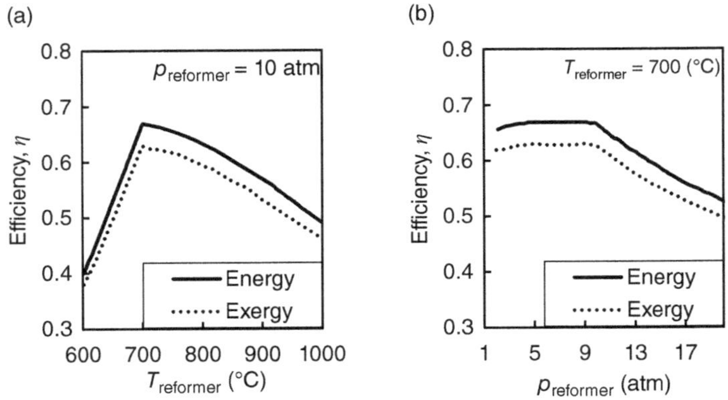

Figure 5.25 Thermal (energetic) and exergetic efficiency in methane steam reforming: (a) as a function of temperature in the reformer and (b) as a function of pressure (redrawn from Simpson and Lutz [12]); steam-to-carbon ratio: 3.2; temperature of shift reactor: 300 °C; hydrogen recovery by membrane separation at 450 °C.

Table 5.6 CO_2 emissions, energetic efficiency and cost in methane reforming (data from Oni et al. [18]).

Process	CO_2-capture[a] (%)	η^{energy} (%)	Total GHG emissions[b] (kg $CO_{2,e}$/kg H_2)	Cost H_2 (US$ 2020[b]/kg)
Steam reforming		63.3	11.35	0.92
	52	51.7	8.20	1.27
	85	43.3	6.66	1.77
Autothermal reforming		73.4	11.01	0.92
	93	65.9	3.91	1.25

a) Amount of CO_2 captured relative to total amount of carbon fed into the system.
b) Originally quoted in CAN$ in 2020 [18]; recalculated using an average exchange rate of 1 CAN$ = 0.75 US$.

gases is directly associated with the process when using steam reforming to generate hydrogen; this is reduced to 30% when capturing 85% of the produced CO_2. The cost of hydrogen comes to ca. US$(2020) 0.90/kg H_2, but CO_2-capture will increase the cost of hydrogen depending on how much CO_2 will need to be captured. This may be offset with increasing carbon taxes going forward.

Hydrogen can also be produced using electrolysis and the maximum exergetic efficiency is then given by

$$\eta_{max}^{exergy} = \frac{B_{H_2(g)}}{B_{H_2O(l)} + \frac{Q_{required,min}}{n_{H_2O(l),fed}} + \frac{W_{required,min}}{n_{H_2O(l),fed}}}$$

With the minimum amount of work required given by the change in the Gibbs free energy upon reaction, and the minimum amount of heat required given by the

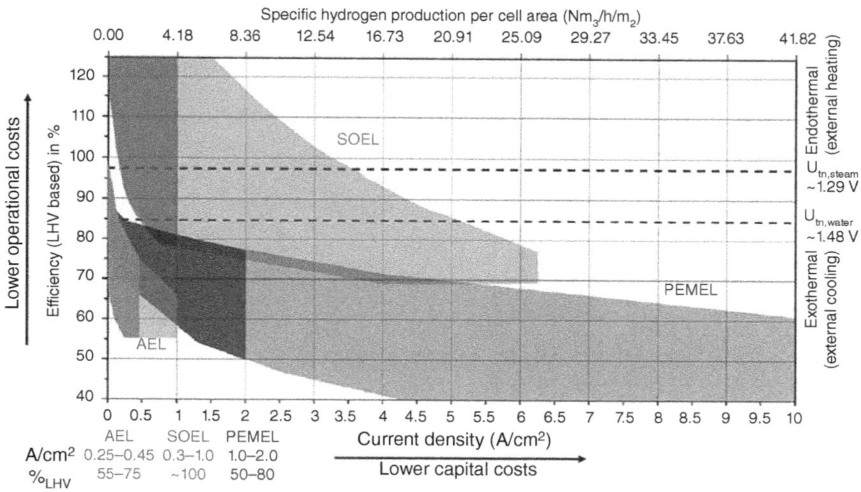

Figure 5.26 Efficiency and operational range of alkaline electrolysis cell (AEL), proton exchange membrane electrolysis cell (PEMEL) and the high-temperature solid oxide electrolysis cell (SOEL) (Buttler and Spliethoff [37], reproduced with permission of Elsevier).

difference in the change in the enthalpy upon reaction and the change in the Gibbs free energy upon reaction:

$$\frac{W_{\text{required,min}}}{n_{\text{H}_2\text{O(l),fed}}} = \Delta G_{\text{rxn}} \quad \frac{Q_{\text{required,min}}}{n_{\text{H}_2\text{O(l),fed}}} = T \cdot \Delta S_{\text{rxn}} = \Delta H_{\text{rxn}} - \Delta G_{\text{rxn}}$$

Hence,

$$\eta_{\max}^{\text{exergy}} = \frac{236.09}{0.9 + 48.7 + 237.1} = 0.823$$

(using steam as the feed would reduce the exergetic efficiency to 0.800). Buttler et al. [37] have given a summary of the efficiency and operational range of various reported AEL, PEM and SOE electrolyser systems (Figure 5.26) basing the efficiency on given cell voltages and assuming a Faraday efficiency of 100%. In this comparison, whilst SOEs have the highest efficiency (between 75% and 100%), proton exchange membrane electrolysis cell (PEMEL) operating at an efficiency between 55% and 80% can reach substantially higher current density. This translates to SOE having potentially lower operational costs whilst PEM will have lower capital cost.

5.2.3 Cost of Hydrogen Production

The levelised cost of hydrogen (LCOH), defined as the discounted lifetime cost of construction, financing, maintenance and operation of the system [68], is the most common way to compare different technologies for hydrogen production. LCOH is

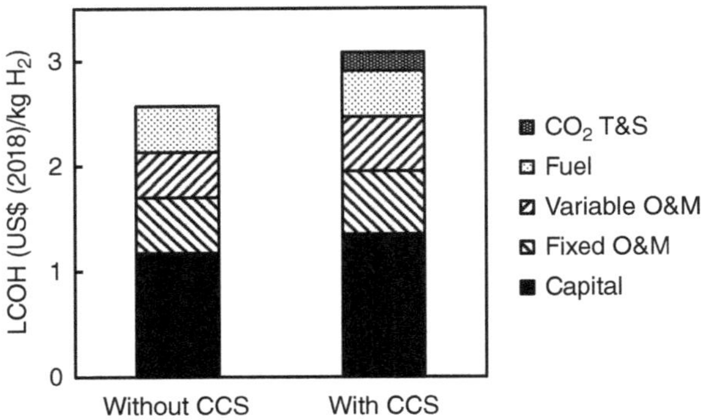

Figure 5.27 LCOH for hydrogen production using coal gasification (Shell gasifier technology: oxygen blown entrained flow; Illinois No. 6 Coal; mass ratio of O_2 to coal = 0.2 kg O_2/kg coal; oxidant 95 vol.% O_2) (redrawn from Lewis et al. [10]).

calculated as:

$$\text{LCOH} = \frac{\sum_{t=1}^{t_{\text{final}}} \frac{I_t + \text{O\&M}_t + \text{F\&E}_t}{(1+r)^t}}{\sum_{t=1}^{t_{\text{final}}} \frac{Q_{H_2,t}}{(1+r)^t}} = \frac{\text{NPV of total costs}}{\text{NPV of hydrogen}}$$

with t_{final}, lifetime of the project; I_t, the investment cost; O&M_t, operations and maintenance costs; F&E_t, costs of feedstock and energy input; $Q_{H_2,t}$, hydrogen production rate; r, discount rate; and NPV, net present value.

It should be noted that there are some limitations associated with using this method of comparison, as the cost of negated emissions is not included, and the values depend strongly on the discount rate. The latter implies a strong location dependency as well as favouring well-established technologies over newer developments. Furthermore, these costs do not include storage or transport of the hydrogen, although the inclusion of these costs may affect the overall cost of hydrogen production [69].

The classical way to generate hydrogen using coal gasification (brown hydrogen) is relatively expensive. Figure 5.27 shows an indication of the LCOH (in 2018 US$) to produce brown hydrogen from Illinois No. 6 Coal using a Shell gasifier (95 vol.% oxygen blown entrained flow; feed rate: feed ratio of 0.2 kg O_2 per kg coal), with and without CO_2 transportation and storage [13]. The LCOH would be US$ 2.6/kg H_2 without CCS with the main cost associated with the capital costs, and the cost of coal only being a minor contributor. It has even been reported that the cost of coal to hydrogen without CCS can be as low as US$ 0.90–1.46/kg H_2 [70]. Adding CCS would increase the LCOH typically by ca.US$ 0.50–0.70/kg H_2 [10, 70, 71] bringing the current estimates for blue hydrogen production from coal gasification with CCS between US$ 1.9 and US$ 3.1 per kg H_2. However, in some regions like China, the cost can be as low as US$ 1.6 per kg [71].

At current, the most cost-effective method of producing hydrogen is grey hydrogen from the reforming of methane. A typical cost breakdown for methane steam reforming and ATR is given in Figure 5.28 [10]. In all cases, the fuel which includes the natural gas as the hydrogen source, has the highest cost. As noted above, including CCS adds significantly to the cost both in terms of CAPEX and OPEX. Without CCS, methane steam reforming comes in at ca. US$1/kg H_2 (in 2018 US$), hence the

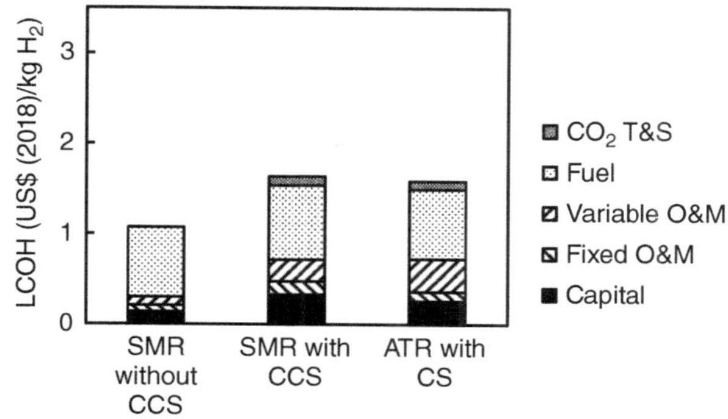

Figure 5.28 LCOH for hydrogen production from methane steam reforming and autothermal reforming (ATR) with and without carbon capture and storage (CCS). CAPEX, capital expenditure; OPEX, operating expenditure; T&S, transport and storage (data from Lewis et al. [10]).

target set by the US Department of Energy (DOE) for US$ 1/kg green hydrogen. The cost of carbon transport and storage comes to ca. US$ 0.5/kg H_2, with most of the cost coming from an increase in CAPEX as well as an increase in variable OPEX. As CCS technologies become more mainstream, it may be expected that these technologies will follow a typical learning cost curve, although expected cost benefit with time is difficult to estimate due to the limited implementation of CCS technologies [72].

It is possible to add biomass to coal in a gasification in order to make it a slightly more renewable process, but such a system will be less efficient and result in a higher hydrogen production cost (as measured by the LCOH) [10].

Hydrogen production from methane pyrolysis yields an attractive side product, viz. carbon black with a global market worldwide of ca. 14.5 million tonnes [73] at a market price (March 2024) between US$ 1.13/kg and US$ 2.06/kg [74]. Figure 5.29 shows an estimate of the cost of hydrogen produced using plasma pyrolysis of methane as a function of the price of carbon black for two different cases, viz. Case I: electricity tariff of US$ 69/MWh, a natural gas price of US$ 7.82/MMBTU (US$ 7.41/GJ) and capital cost of US$ 80 million, and Case II: electricity tariff of

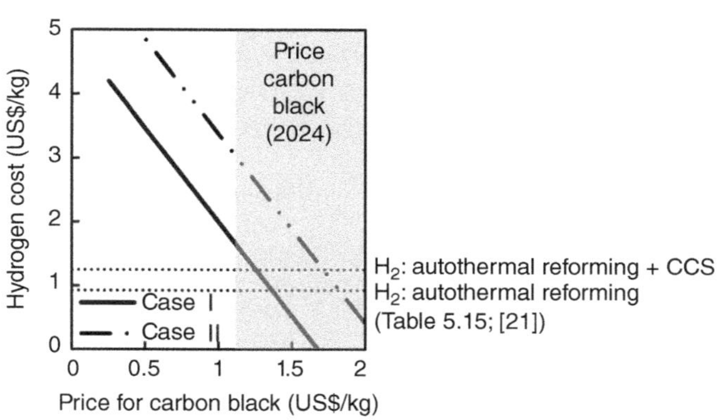

Figure 5.29 H_2 production costs (US$(2020)) by plasma pyrolysis of methane (Case I: electricity tariff of US$ 69/MWh, a natural gas price of US$ 7.82/MMBTU and capital cost of US$ 80 million, and Case II: electricity tariff of US$ 100/MWh, a natural gas price of US$ 8.76/MMBTU and capital cost of US$ 111 million) (redrawn from Da Costa Labanca [75]).

US$ 100/MWh, a natural gas price of US$ 8.76/MMBTU and capital cost of US$ 111 million (lowering the cost of natural gas or the cost of electricity would lower the cost of hydrogen). The obtained cost for hydrogen production using methane pyrolysis is competitive with hydrogen production using methane reforming at the current price of carbon black. The global production of carbon black in 2024 would result in a co-generation of 4.8 Mt of hydrogen, i.e. 5.4% of the 90 Mt global hydrogen demand in 2020 [76]. The carbon black market would be rapidly saturated by producing larger amounts of hydrogen by methane pyrolysis. This would result in a drop in the price for carbon black and push the price for hydrogen up. Hence, economic benefit of hydrogen production by methane pyrolysis should not be linked to the selling price of carbon black (even at a reduced price [77]), but rather linked to avoiding CO_2-capture and sequestration costs (but still taking into account cost associated with disposing of the solid product).

Currently the cost of producing hydrogen with an electrolyser is substantially higher than through the fossil-fuel-based technologies, even when including CCS. The LCOH estimates (construction and equipment costs, operating expenditure and expected production data) are subject to uncertainties and had been estimated in the range between US$ 2 and 10/kg H_2, depending on the exact assumptions that were made in the respective models. The different technologies (e.g. SMR, ATR, PEM, AEL and SOE) showed different susceptibility to scenario change [68]. When looking at the breakdown of the electrolyser cost, it can be seen that although the CAPEX, or the cost of the stacks, is significant (for AEL and PEM in the order of US$ 0.6/kg H_2 and US$ 0.8/kg H_2, respectively), the majority of the expenses are in the operating costs, particularly for electricity usage, which far exceeds that of other methods of hydrogen production (Figure 5.30). This implies that the parity of the LCOE with conventional, fossil-fuel-based electricity generation is not sufficient to make hydrogen competitive. A significant decrease in the cost of renewable energy is required to make the production of hydrogen using electrolytic water splitting competitive to steam reforming including the cost of CCS.

There are (primarily) two ways in which costs for technologies can decrease over time. There is a step change, where a revolutionary change in a process is implemented and thus a sharp drop off in the cost occurs. This is normally a change in the ideas of how the system works. For example, a step change in the cost would occur,

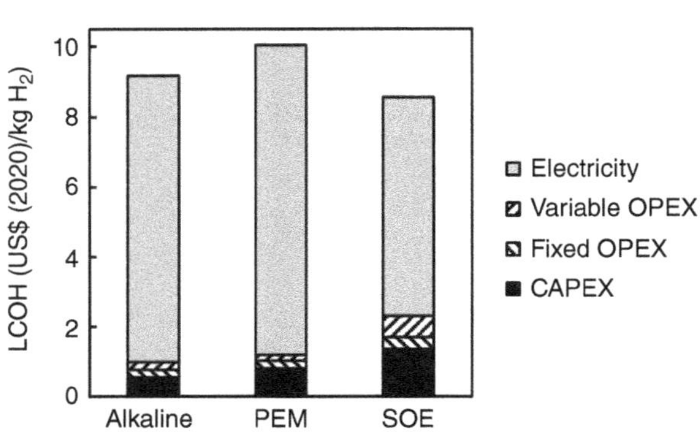

Figure 5.30 Cost of H_2 from water electrolysers based on industrial retail price for electricity (conversion using 2020 average exchange rate 1 US$ = 0.78 £) (redrawn from Department for Business, Energy & Industrial Strategy [68]).

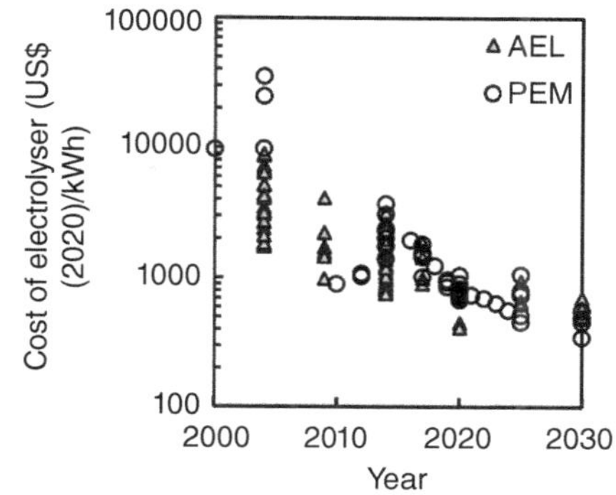

Figure 5.31 Cost learning curve for AEL and PEM electrolysers, with predicted values shown for 2030 (redrawn from Reksten et al. [78]).

if the iridium catalyst for the OER in PEM electrolysers was somehow replaced with something with the cost of iron and had fewer voltage losses than the current OER side.

The alternative, which is the more common way for a cost to decrease is through a learning curve. For instance, the cost of electrolyser plants, C, has been correlated [78] with the size of the electrolyser plant, Q (with a scaling factor α), which decreases with time based on the number of installations, V, relative to the number of installations in a reference year, V_0, scaled with the learning factor, β:

$$C = \left(k_0 + \frac{k}{Q}Q^{\alpha}\right)\left(\frac{V}{V_0}\right)^{\beta}$$

The underpinning idea is that cost of installations will come down due to improvement in the design avoiding potential risks and pitfalls. The learning curve is a simplistic and empirical model looking at past (cost) data to predict via extrapolation future behaviour without pinpointing the driving factor for cost reduction [79]. Figure 5.31 shows the costs for the AEL cells and PEMEL over the years, from which the learning factor and scaling factor for the different technologies can be determined [78]. The scaling factor, α, was similar for both the alkaline and the PEM-type of electrolysers. The learning rate for the AEL cell was estimated to be modest ($\beta = -27$), but the learning rate for the PEM-type electrolysers was possibly overestimated ($\beta = -159$) seeing their recent introduction into the market.

The US DOE has set several targets [80] with the most significant one being the ultimate target cost of green hydrogen being US$1/kg (less expensive than current blue hydrogen, typically estimated at ca. US$1.5–2/kg$H_2$) (Table 5.7). This is likely an essential target value to allow green hydrogen to replace fossil fuels as the servicing chemical for much of the energy generation and chemical synthesis industries.

5.2.4 Hydrogen Storage and Transport

Hydrogen has a LHV of 241.8 kJ/mol, which is equivalent to a high mass specific energy of 120.9 MJ/kg. The useful work, which can be extracted (equal to the Gibbs free energy of formation of water) is 114.3 MJ/kg (95% of the enthalpic energy).

From a transport and storage point of view, mass is not a prominent issue. Lighter objects would be easier to transport and need less fuel, but heavy objects can be

Table 5.7 Technical targets for PEM-stack [80].

	2022 status	2026 target	Ultimate target
Energy efficiency (%, LHV)	61	65	72
Lifetime (h)	40 000	80 000	80 000
Current density (A/cm^2)	2 (@ 1.9 V)	3 (@ 1.8 V)	3 (@ 1.6 V)
Stack capital cost (US$/kW)	450	100	50
System capital cost (US$/kW)	1000	250	150
H_2-production cost (US$/kg)	>3	2	1

transported, especially by boat, quite easily. Volume, on the other hand, is a major consideration, as the size in a transport vessel is given by the tanker size, which cannot be changed, regardless of the content's density. Hence, the energy density, i.e. the product of the density and the specific energy, is important. The density of hydrogen can be obtained from the compressibility factor of hydrogen:

$$Z = \frac{\underline{V}}{\underline{V}_{\text{ideal}}} = \frac{\underline{V}}{\frac{RT}{P}}$$

Z is calculated from an appropriate equation of state. Figure 5.32 shows the compressibility of hydrogen estimated using the Soave-Redlich-Kwong equation of state. The compressibility factor of hydrogen is approximately equal to 1 over a large pressure range (for instance, the compressibility factor at 90 bar is 1.05 implying to take up a volume ca. 5% larger than the volume of an ideal gas) implying that the behaviour of hydrogen in these pressure range can be approximated by the ideal gas law. At pressures larger than 100 bar, the compressibility factor increases significantly beyond 1 implying that its molar volume is larger than the molar volume of an ideal gas at the same temperature and pressure. The storage and

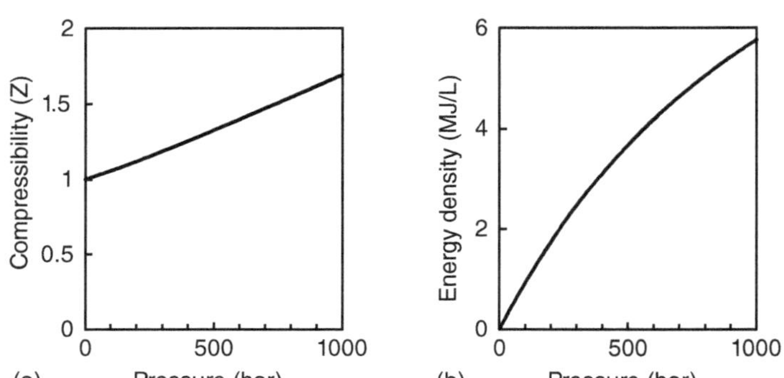

Figure 5.32 Compressibility (a) and energy density (b) of hydrogen at 25 °C (compressibility calculated using the Soave-Redlich-Kwong equation of state).

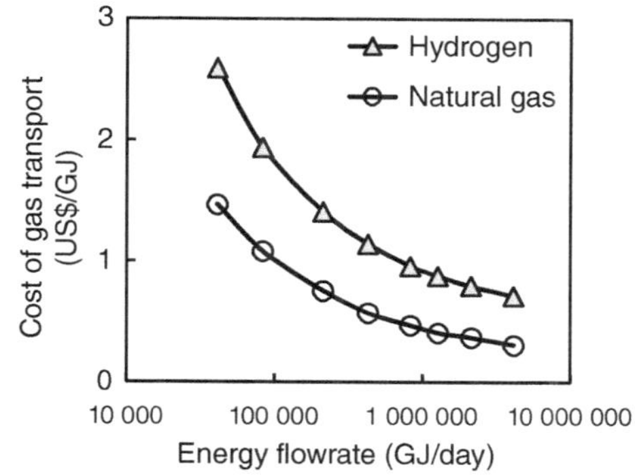

Figure 5.33 Cost of gas transmission (1999) over a length of 1000 km as a function of the energy flow rate for transport of hydrogen (H_2) and natural gas (NG) (pipeline inlet: 69 bar; pipeline outlet: 21 bar; compressor inlet: 14 bar; compressor outlet: 69 bar) (redrawn from Ogden [83]).

transport of energy in hydrogen thus require larger volumes/volumetric flow rates than methane, which is compounded by the difference in the energy density.

5.2.4.1 Hydrogen Transport Through Pipelines

Natural gas is transported extensively with pipelines, so it makes sense to leverage off the infrastructure of these pipelines for hydrogen transport. For natural gas transport, it is cost-effective to use pipelines for distances less than 2500 km, but beyond that it is cheaper to liquefy the gas and transport natural gas as LNG, liquefied natural gas [81]. The shipment of LNG results in some boil-off, with boil-off rates in the order of 0.08–0.16%/day, depending on the shipping vessel, and composition of LNG [82]. The boil-off can be used to power the means of transport of LNG.

Hydrogen pipelines are around 40–50% more expensive than natural gas pipelines [83] (Figure 5.33). This can be attributed to several factors. Firstly, pipe cost due to embrittlement of pipes by H_2: hydrogen can diffuse into the crystal structure of steel over time and weaken it, which is a major concern for long-term projects looking to store or transport hydrogen. ‘Exotic materials’, such as carbon fibre and glass fibre mixtures have been trialled for this use, but this significantly increases the cost. Secondly, the pipeline diameter would need to be 20% larger ([84] cited in [83]) to transport the same amount of energy in the form of H_2 as in the form of natural gas. Next, the compression power per unit energy for H_2 is ca. three times higher than for natural gas. Finally, compressors for H_2 are ca. 20–30% more expensive than natural gas compressors [83].

The flow through a pipeline is associated with a pressure drop, as given by the Bernoulli equation:

$$\frac{1}{2} \cdot v_{\text{in}}^2 + g \cdot z_{\text{in}} + \frac{P_{\text{in}}}{\rho_{\text{in}}} = \text{constant}$$

where v is the fluid velocity, g is the gravitational acceleration, z is the height change and P is the pressure. Figure 5.34 shows the modelled gauge pressure for the adiabatic flow through pipes with different diameters [85]. The pressure drop can be as high as 0.042 MPa/km implying the need to recompress hydrogen at regular intervals with the actual recompression interval depending on the pressure rating of the pipeline, the pipe diameter and the surface roughness.

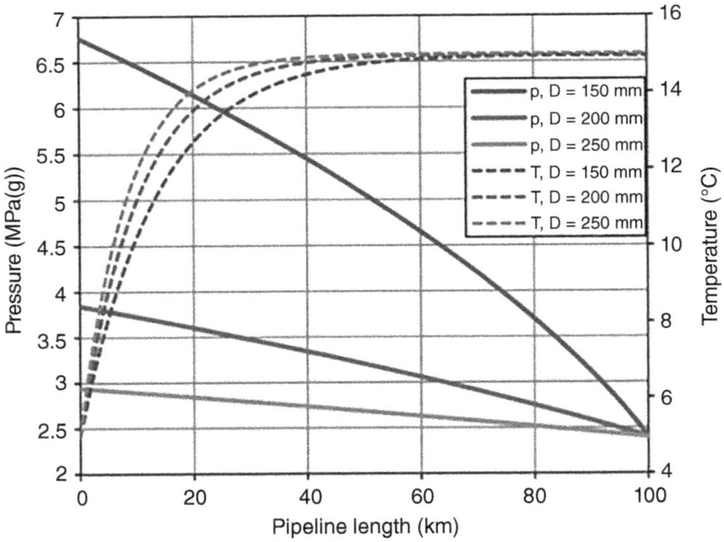

Figure 5.34 Adiabatic hydrogen flow through pipelines Pressure (p) (continuous lines) and temperature (T) (dashed lines) changes for hydrogen mass flow rate of 1.0 kg/s (40 000 Nm^3/h) for different diameters (Kuczyński et al. [85] / MDPI / CC BY 4.0).

5.2.4.2 Liquefying Hydrogen

A higher energy density will be obtained by liquefying hydrogen. Liquid hydrogen requires temperature below its critical temperature of −240 °C (33.2 K) and above its solidification point (triple point −259.3 °C or 13.81 K and 7.042 kPa; see Figure 5.35).

Typically, gasses are liquefied by utilising Joule-Thomson expansion of a compressed gas. This results in the cooling of the gas, if

$$\mu_{JT} = \left(\frac{\partial T}{\partial p}\right)_H = \frac{1}{C_P} \cdot \frac{RT^2}{P}\left(\frac{dZ}{dT}\right)_P > 0$$

For most gases, the Joule-Thomson coefficient, μ_{JT}, is greater than 0 throughout the temperature and pressure range of interest. However, for hydrogen (and helium),

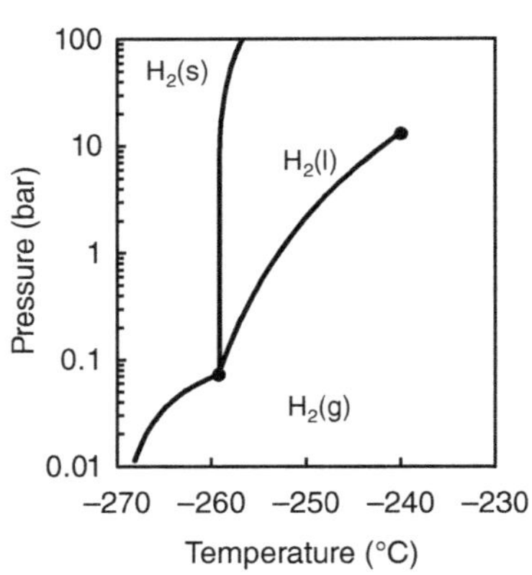

Figure 5.35 Phase diagram for hydrogen (VLE from [86]; VSE calculated using heat of fusion of 0.12 kJ/mol [86]; SLE calculated using a Kechin eq. (data from References [86, 87]).

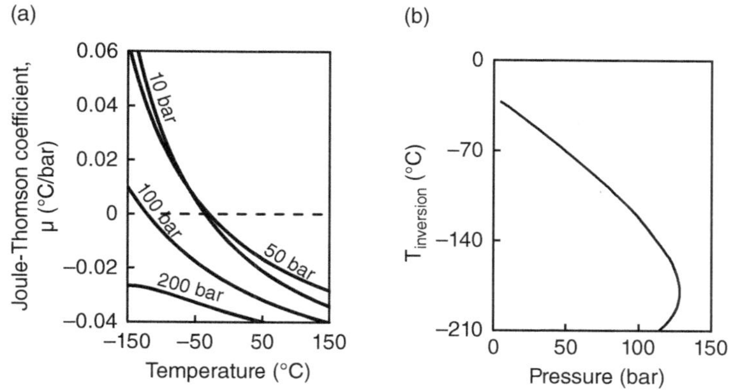

Figure 5.36 Joule-Thompson coefficient for hydrogen as a function of temperature (a) and the inversion temperature for hydrogen as a function of pressure (b) estimated using Soave-Redlich-Kwong equation of state.

this is not the case and at room temperature the Joule-Thomson coefficient for these compounds is less than zero (see Figure 5.36a), implying that expanding hydrogen adiabatically will heat up the gas. The Joule-Thomson coefficient becomes zero at temperatures far below 0 °C (the so-called inversion temperature) and hydrogen will need to be cooled down below this temperature to ensure that gas expansion will result in a further decrease in temperature, which would allow hydrogen liquefaction. It can be further noted that the inversion temperature is dependent on the pressure. Using the Soave-Redlich-Kwong equation of state, it can be estimated that the pressure of hydrogen must be below ca. 130 bar for cooling to take place upon Joule-Thomson expansion (see Figure 5.36b).

Hence, to exploit the Joule-Thomson expansion, in the case of hydrogen at 50 bar, pre-cooling to less than −70 °C is required. There are primarily two processes which may be used, the Claude process or the helium Brayton process. In the Claude process (see Figure 5.37), gaseous hydrogen is compressed to the desired pressure. In the subsequent heat exchanger, hydrogen is cooled to a temperature below the inversion temperature before passing through an expansion engine utilising the Joule-Thompson effect to further cool the gas. This chilled gas is used to cool the

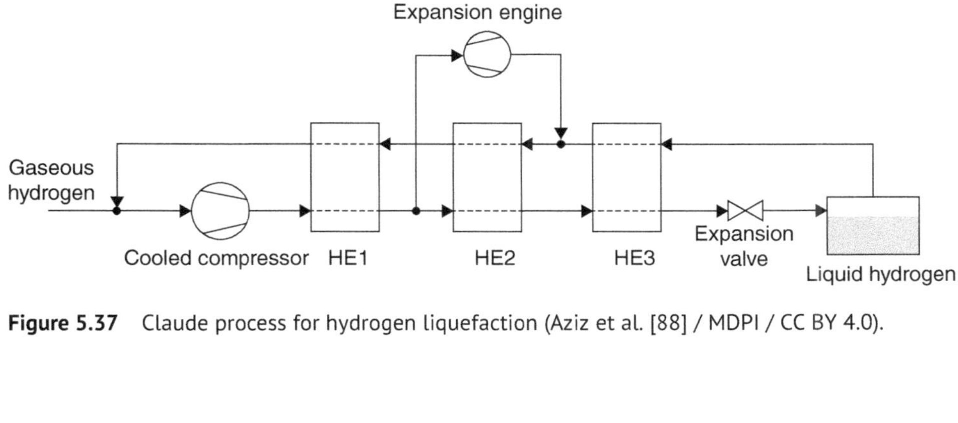

Figure 5.37 Claude process for hydrogen liquefaction (Aziz et al. [88] / MDPI / CC BY 4.0).

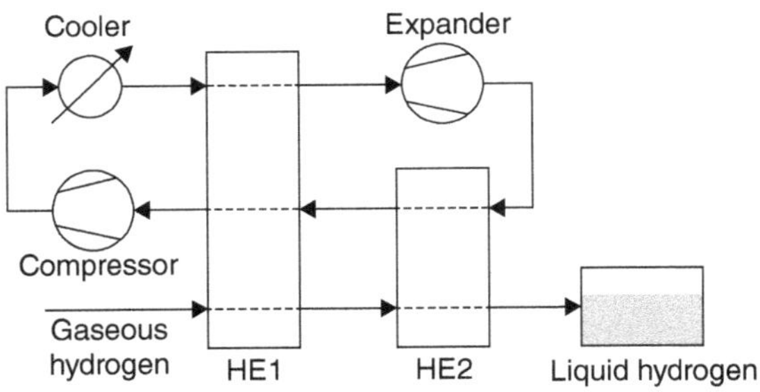

Figure 5.38 Simple helium Brayton process for hydrogen liquefaction (Aziz et al. [88] / MDPI / CC BY 4.0).

recycle stream and is used across the first two heat exchangers. The rest of the inlet stream passes through the second heat exchanger which is cooled from the fluid coming out of the expansion engine. A third heat exchanger is used where the heat is exchanged with gaseous hydrogen evaporating off the liquid hydrogen tank. Finally, an expansion valve is used also making use of the Joule-Thompson effect to liquefy the hydrogen which may be stored.

The helium Brayton process (see Figure 5.38) makes use of a helium loop which follows a similar concept to the classical refrigeration cycle operating with a compressor and a cooler, followed by expansion, and then further heating whilst being used as a cooling fluid. The 'remnant cold' from the helium loop is then used to help further chill the inlet stream to the expander before being returned to the compressor to close the loop. The hydrogen simply passes through a pair of heat exchangers which are cold enough (running with liquid helium) to liquefy the hydrogen.

Whilst storage and transport of hydrogen as a liquid is a mature technology and is suitable for transporting hydrogen over long distances, there are some challenges involved with the process. Firstly, the liquefaction process is energy-intensive, requiring on average for existing installations 13.8 kWh/kg of liquid hydrogen [89, 90], and there can also be significant hydrogen loss due to boil-off. For more recently proposed large-scale plants, the expected energy requirement is 5–8 kWh/kg of liquid hydrogen [90].

5.2.4.3 Need for Hydrogen Carriers

So, whilst hydrogen has a very high specific energy (120.9 MJ/kg), it has a low energy density (6.2 MJ/l @ 1000 bar). This is low in comparison to more classical fuels (see Figure 5.2), with coal (max) and LPG both being ca. 25 MJ/l and liquid hydrocarbons such as diesel and petrol extending above 35 MJ/l. Furthermore, transport and storage extreme high pressures are associated with safety risks as well as high capital costs and compression requirements. Liquefying hydrogen would reduce some of the safety concerns but is energetically quite expensive (with a target of taking up 18–30% of LHV stored in hydrogen).

Alternative hydrogen carriers with a higher energy density such as ammonia, methanol or hydrocarbons could be attractive, provided that their production is associated with both lower capital costs and higher energetic efficiency.

5.2.5 Extracting Energy from Hydrogen

If hydrogen is available, the issue of how to recover the energy stored in hydrogen needs to be considered. The chemical energy stored in hydrogen can be recovered in the form of heat by combustion, in the form of electric power in a fuel cell, or by transferring to another chemical (e.g. by using hydrogen as a reductant).

5.2.5.1 Hydrogen Combustion

Hydrogen can be combusted yielding water:

$$H_2(g) + \frac{1}{2}O_2(g) \rightarrow H_2O(l) \quad \Delta H^0_{\text{combustion}} = \text{HHV} = 286\ \frac{\text{kJ}}{\text{mol}}$$
$$H_2(g) + \frac{1}{2}O_2(g) \rightarrow H_2O(g) \quad \Delta H^0_{\text{combustion}} = \text{LHV} = 241\ \frac{\text{kJ}}{\text{mol}}$$

This option has the benefit that it is a carbon-free combustion, which is its major advantage over the use of fossil fuel in the combustion. The combustion of hydrogen can reach an adiabatic temperature like the one obtained upon combusting propane, and thus the radiative heat transfer from hydrogen flame can be as good as that from a clean-burning propane flame. A higher emissivity, and thus improved radiative heat transfer, is obtained in a propane flame, when forming some soot. Furthermore, the molar flow rate of the combustion gases is much lower in the case of combusting hydrogen and hence the convective heat transfer may be poorer. Hence, furnaces combusting hydrogen to extract its energy will need to be redesigned to enable more effective heat transfer.

The energy obtained by combusting hydrogen can be used to generate power, although this process is immediately limited by the Carnot efficiency, which depending on the steam temperature can be as low as $\eta_{\text{combined cycle}} = 35 - 60\%$. Thus, the combustion of hydrogen to extract its energy may have relatively low efficiency.

5.2.5.2 Fuel Cells

Hydrogen fuel cells are another option to extract energy from hydrogen. From a thermodynamic viewpoint, fuel cells are simply reverse electrolysers, with a hydrogen oxidation reaction (HOR) taking place at the anode and an oxygen reduction reaction (ORR) at the cathode:

Anode: HOR: $H_2(g) \rightarrow 2H^+(aq) + 2e^- \quad \Delta E^0 = 0\ \text{V}$
Cathode: ORR: $O_2(g) + 2H^+(aq) + 2e^- \rightarrow 2H_2O(l) \quad \Delta E^0 = 1.23\ \text{V}$
Overall: $2H_2(g) + O_2(g) \rightarrow 2H_2O(l) \quad \Delta E^0 = 1.23\ \text{V}$

Fuel cells have no moving parts and, thus, have a relatively low maintenance cost.

At standard conditions, a fuel cell has a theoretical voltage of 1.23 V. They can then be stacked in series to generate a higher voltage, or in parallel to allow a higher current or power draw, as with batteries. The difficulty that comes in, in comparison to a battery, is that hydrogen and oxygen need to be supplied to each fuel cell in the fuel cell stack, and similarly the water formed needs to be removed. This removal of the water formed at the cathode is one of the limiting factors in fuel cell operation.

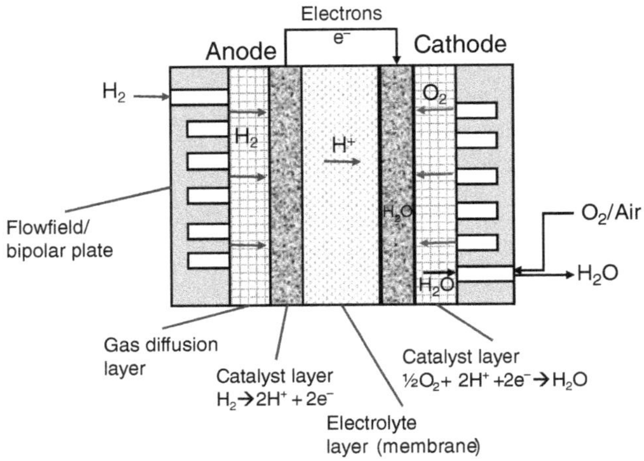

Figure 5.39 Internals of a PEM fuel cell [91].

A schematic representation of a fuel cell is given in Figure 5.39. Hydrogen enters the fuel cell through flow channels in the anode flow field. The name bipolar plate is related to the fact that single cells in a full fuel cell system are stacked against each other so that the plates have an anode on one side and a cathode on the other side, with an insulating layer in-between. The gas is properly distributed across the entire catalyst layer through a gas diffusion layer that contains electrically conductive fibres. In PEM fuel cells, the diffusion layer is often a copolymer of poly(tetrafluoroethylene) and poly-sulphonyl fluoride vinyl ether [91]. Hydrogen reaches the catalytically active material, platinum, where it splits into two protons and two electrons. The protons move from the catalyst to and then through the polymer exchange membrane to the cathode. The electrons move through an external circuit (connection between anode bipolar plate and cathode bipolar plate).

At the cathode, oxygen enters the fuel cell through the flow field (although not shown directly here, the feed is typically air as the cost of pure oxygen is prohibitive). In a similar manner to the anode, oxygen moves through the gas diffusion layer and microporous layer to the catalyst layer. The oxygen adsorbs onto a catalytically active material, platinum, and reacts with protons and electrons to form water. This water, together with oxygen, and any nitrogen leave the cell.

The polarisation curve for a fuel cell is like that for a battery (see Figure 5.40). There are immediate losses (open circuit) due to fuel cross-over followed by activation losses attributed to the sluggish ORR. There is then a linear dependency of the cell voltage on the current due to the so-called ohmic losses, i.e. resistance of the electronic circuit including the ionic transport. At very high current density, the reaction becomes limited by mass transport. The polarisation curve results in an optimum power drawn from the fuel cell typically at a voltage of ca. 0.5–0.6 V.

The maximum achievable energy of a fuel cell equals the change in the Gibbs free energy upon reaction and some energy is lost in the form of heat due to entropy

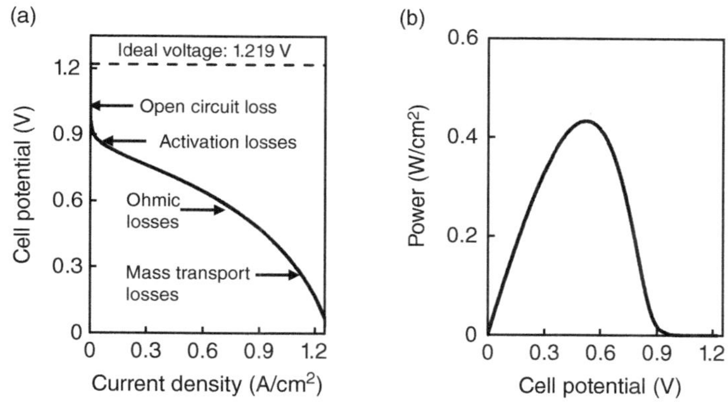

Figure 5.40 Polarisation curve (a) and power-curve (b) for hydrogen-air fuel cell (redrawn from Spiegel [91]).

generation:

$$\Delta G = \Delta H - T\Delta S \text{ at ambient conditions (liquid, 25}^{\circ}\text{C, 1 bar): } \Delta G^0_{\text{rxn}} = 237.34\ \frac{\text{kJ}}{\text{mol}}$$

Hence, the efficiency can be defined by the ratio $\Delta G^0_{\text{rxn}}/\Delta H^0_{\text{rxn}}$:

$$\eta_{\text{HHV}} = \frac{\Delta G}{\text{HHV}} = \frac{237.34}{286.02} = 83\%$$

$$\eta_{\text{LHV}} = \frac{\Delta G}{\text{LHV}} = \frac{237.34}{241.98} = 94.5\%$$

As the fuel cell works with liquid water, the efficiency is typically taken as 83%.

The actual efficiency of a fuel cell can be obtained from the actual voltage relative to the ideal voltage. For an air-hydrogen fuel cell, the ideal voltage is 1.219 V (deviating from the often quoted 1.23 V because of the reduced partial pressure of oxygen). The voltage at the optimum power draw is closer to 0.5–0.6 V (see Figure 5.40). Thus, the actual efficiency of the fuel cell is more in the region of 41–49%.

Fuel cells are not yet widely applied because the hydrogen, which is currently available, is not green. It would be as effective to simply burn the fuel used to make the hydrogen, to generate power. If the hydrogen used can be obtained from a renewable source, e.g. using solar power or wind power to obtain the hydrogen from electrolysis, then the use of hydrogen in a fuel cell to recover stored energy becomes a more viable option.

Fuel cell technology is still in development, as they are expensive, with the electro-catalyst contributing significantly to the total cost [92]. This can be attributed to the use of platinum (supported on carbon) catalyst, with the sluggish ORR necessitating a high platinum loading to obtain the required productivity. Hence, the research on other catalysts as well as other variations of the fuel cell, e.g. an alkali fuel cell, is ongoing [93].

5.3 Ammonia

Ammonia is one of the most discussed hydrogen carriers and is the only one which is not carbon-based, with many others requiring the release of CO_2 upon recovery of the hydrogen/energy.

Ammonia production is an established, large-scale industrial process with a global production of 182 million metric tonnes per annum (2019) [94]. Both the transport and storage of ammonia are very well understood, both of which will assist in the use of ammonia as a hydrogen carrier [95].

It should be noted that replacing the current global ammonia production of 182 million tonnes per annum with a process utilising renewable hydrogen would require the global electrolytically generated hydrogen production to increase by 32.1 million tonne per annum (or 1.02 tonne H_2/s). With an energy cost for the electrolytic production of hydrogen (197.6 MJ/kg, assuming an electrolyser efficiency of 60%), the power requirement for just the ammonia production would amount to 200 GW, which comes to an additional ca. 1.1% of the total, global annual energy consumption in 2023 for just the electrochemical hydrogen production to satisfy the current ammonia market.

5.3.1 Uses for Ammonia

At present, around 80% of the ammonia produced is used for fertiliser production, nitrogen being a key species for the formation of protein and thus the growth of plants. The ammonia is currently used directly as a fertiliser or used as a precursor to a variety of other components (Figure 5.41). Ammonia is also a feedstock for the explosive industry through the conversion of ammonia to nitric acid in the Ostwald process, which in turn is used for the production of the explosives, such as trinitrotoluene (TNT).

At present, ammonia is predominately made with hydrogen from natural gas. The process to produce fertiliser contributes around 1.2% of global CO_2 emissions (based

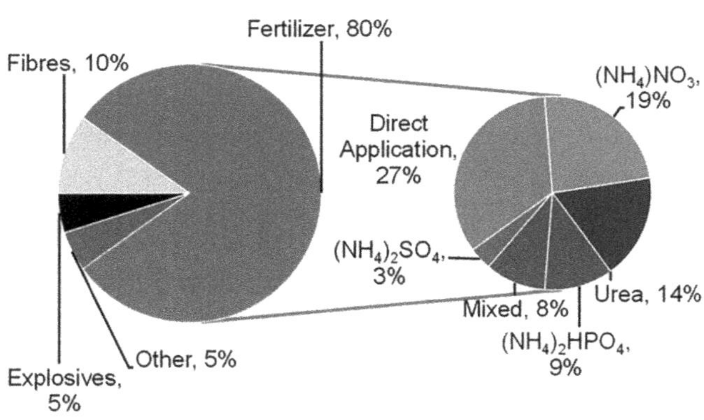

Figure 5.41 Uses of ammonia (2016) (adapted from Afif et al. [96]).

on an average production of 2.4 tonne CO_2/tonne ammonia [94] and a global CO_2 emission of 36.4 Gtonne per annum [97]). Hence, ammonia synthesis using renewable energy is not only looked at as a potential hydrogen carrier, but the production of ammonia from green hydrogen may become an important technology for the fertiliser industry at large.

5.3.2 Ammonia as a Hydrogen Carrier

Ammonia has a high hydrogen content (17.6 wt.%) which is one of the reasons why it is a popular option as a hydrogen carrier. Even though it does have a lower specific energy than hydrogen (18.6 MJ/kg vs. 120.9 MJ/kg for H_2), its energy density makes up for this at 8.8 MJ/l @ 10 bar and 25 °C (liquid ammonia). The energy density of hydrogen at 700 bar and 25 °C is 4.6 MJ/l. It is important to note that the required pressure for ammonia to obtain this energy density is only 10 bar, compared to ca. 700 bar for hydrogen as is used for hydrogen electric vehicles.

Ammonia is a gas at ambient conditions, with a vapour pressure of 9.9 bar at 25 °C (see Table 5.8), showing that ammonia can be readily liquefied.

The difference between ammonia and hydrogen is highlighted via the pressure dependency of the density and thus energy density (Figure 5.42, calculated using the Soave-Redlich-Kwong equation of state). The compressibility of ammonia and hydrogen is similar at low pressure. The compressibility of ammonia drops suddenly at 25 °C and 9.9 bar, as ammonia liquefies at these conditions. The sudden drop in the compressibility results in a jump in the density. The low-pressure phase change of ammonia not only leads to a reduction in the energy required for the densification of ammonia, but ammonia at a lower pressure also presents a much smaller safety risk than hydrogen at 700 bar.

The energy density (Figure 5.42), combined with the compressibility factor (Z), is of most relevance in the context of transporting renewable energy. The difference between ammonia and hydrogen is minimal up to the pressure that is sufficient to liquefy ammonia, where a sharp increase in the energy density to ca. 8.8 MJ/l can be seen. This energy density cannot readily be achieved by hydrogen (requiring pressures over 2500 bar). Hence, ammonia is a desirable chemical compound to store energy.

Table 5.8 Properties of ammonia.

Molar mass	17.031 g/mol	$\Delta_f H$ (25 °C, 1 bar)	−45.9 kJ/mol
H-content	17.6 wt.%	$\Delta_f G$ (25 °C, 1 bar)	−16.4 kJ/mol
Normal boiling point	−33.34 °C	$\Delta_{vap} H$ (25 °C, 1 bar)	19.8 kJ/mol
p^{vap} (25 °C)	9.9 bar	ρ^{liquid} (25 °C)	0.603 g/cm^3
Critical point	111.3 bar, 132.4 °C	Solubility in water	31 wt.% (25 °C)
Auto-ignition point	630 °C		
Flammability	Lower flammability limit: 15%; upper flammability limit: 28%[a]		

a) Flammability region widens significantly with increasing temperature [98].

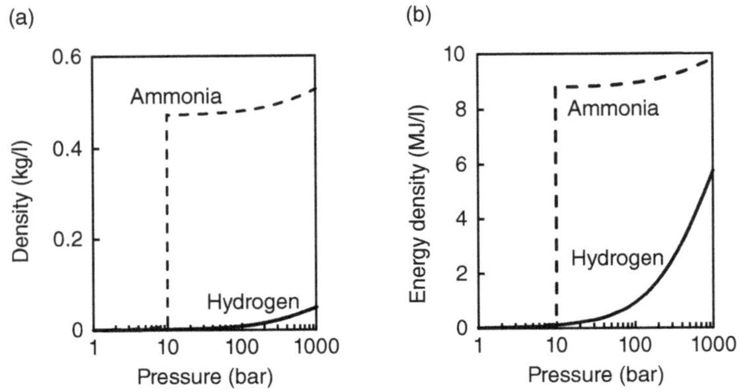

Figure 5.42 Compressibility (a) and energy density (b) of hydrogen (solid line) and ammonia (dotted line) as a function of pressure (25 °C) estimated using the Soave-Redlich Kwong equation of state.

5.3.3 Ammonia Production Process

Haber developed a catalyst which allowed 'for the synthesis of ammonia from its elements' for which he won the Nobel Prize in chemistry in 1918. The catalyst he developed was uranium and thorium based. A major reason for the development of the process was that at the time achieving very high pressure (500–700 bar) was a relatively new technology, and one of the routes they decided to investigate was to liquefy nitrogen.

The catalyst initially used in the development of a reactor used was osmium-based, but by the time reactor development got to the industrial scale an iron catalyst was developed, owing to the cost and limited availability of osmium.

Carl Bosch developed a process to utilise Haber's reaction. His major innovation was the addition of pin holes to the reactor, which allowed the hydrogen to escape, rather than react with the carbon in the steel. For this, he won the 1931 Nobel Prize for his 'contributions to the invention and development of chemical high-pressure methods'. Owing to the contributions from both Fritz Haber and Carl Bosch, the process is now known as the 'Haber-Bosch' process.

5.3.3.1 Thermodynamic Limitations of Ammonia Synthesis

The ammonia synthesis reaction is well known:

$$N_2 + 3H_2 \rightleftharpoons 2NH_3 \quad H^0_{\mathrm{rxn}} = -91.8\frac{\mathrm{kJ}}{\mathrm{mol\ N_2}}$$

The maximum ammonia yield is 0.66 mol NH_3/mol H_2 owing to the stoichiometry of the reaction. An equilibrium calculation shows that close to the maximum yield of ammonia could be attained at room temperature (Figure 5.43). Unfortunately, nitrogen is too stable to react at these temperatures and high temperatures (>400 °C) are required to activate nitrogen. Hence, the process is run at temperatures as high as 500 °C. However, at these temperatures the reaction becomes strongly

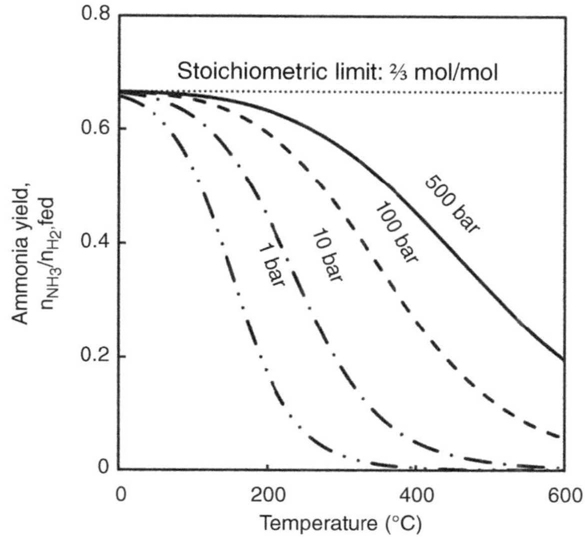

Figure 5.43 Effect of temperature on the equilibrium yield of ammonia for a stoichiometric feed at various pressures.

equilibrium-limited, with the equilibrium conversion being less than 13% at 100 bar and 500 °C.

To achieve higher conversion to ammonia at high temperature, the pressure is increased significantly (see Figure 5.43). Historically, the reaction was run at around 500 bar, though the more modern process uses 100–120 bar. At 1 bar, the temperature required for any activation of nitrogen will lead to virtually no conversion to ammonia. Even increasing the pressure to 10 bar does not result in a significant conversion of nitrogen. At the higher pressures of 100 bar, as shown in Figure 5.43 the yield of ammonia per pass will be limited to about 18% at 450 °C. The historical operation at around 500 bar seems intuitive when it comes to the per pass yield of ammonia, which may increase to almost 40% at 450 °C. However, there is a cost associated with high pressure; in modern processes a higher recycle rate is preferred over operation at higher pressure. This is also coupled with the safety concerns of very high pressure.

5.3.3.2 Catalysts for Ammonia Synthesis

The predominantly used catalyst is iron-based, as it is both cheap and relatively active. Commonly used catalysts are based on iron promoted with K_2O, CaO, SiO_2 and Al_2O_3 [99]. Other alternatives exist such as osmium, ruthenium and CoMo-nitrides [100]. Osmium has been known since 1915, but it is very rare and expensive, and its oxide is very toxic. Ruthenium promoted with barium is also applied but is expensive. CoMo-nitrides are currently under development as a more active material with the hope that they will not be a much more expensive option than iron [100].

The Sabatier, or volcano, plot in Figure 5.44 shows the relative activity of each catalyst given as its turnover frequency (TOF), which indicates the number of molecules consumed or formed on an active site on the catalyst surface per unit time, as a

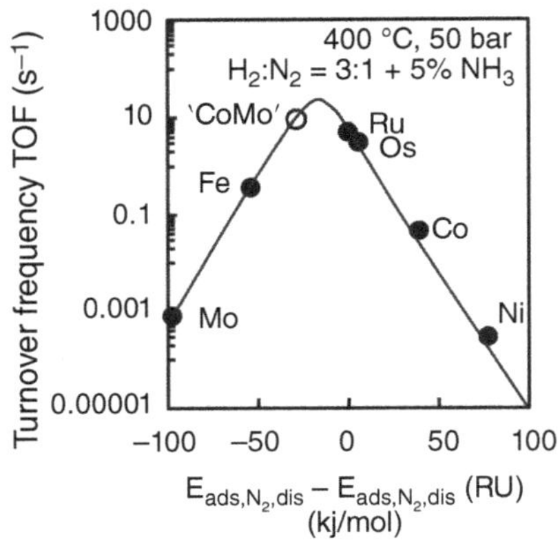

Figure 5.44 Sabatier plot for the activity of different materials as catalysts in the ammonia synthesis (redrawn from Jacobsen et al. [101]).

function of the heat of dissociative chemisorption of nitrogen. In catalysis, this plot is known as a volcano plot due to the shape or a Sabatier plot (named for Paul Sabatier). Weak adsorption of nitrogen will limit the rate of reaction due to lack of reactants on the catalyst surface, whereas strong adsorption of nitrogen will result in self-poisoning of the surface with the adsorbate. Hence, maximum activity is obtained where the adsorption of nitrogen is strong enough for the reaction to take place, but not too strong to inhibit the reaction. The Sabatier plot allows the definition of the catalyst with the highest activity in terms of a characteristic indicator (here the strength of the dissociative chemisorption of nitrogen). Iron and ruthenium are active catalysts for ammonia synthesis, but a more active catalyst could be obtained with a material with an intermediate strength of adsorption for nitrogen, e.g. CoMo-nitrides.

Ammonia synthesis catalysts contain promoters to enhance activity and stability. The iron-based ammonia catalyst is promoted with potassium to improve the dissociation of the N_2 species and speed up the rate of ammonia desorption through electron-donating effects. Alumina is added as a structural promoter, with its primary job to increase the strength of the catalyst and render it more attrition resistant.

5.3.4 Classical Ammonia Synthesis Route

The classical route for synthesising ammonia starts with the preparation of the gas to obtain the correct feed (Figure 5.45). In a primary reformer, steam reforming of methane is performed to make hydrogen. The outlet stream from the primary reformer is mixed with air before being fed to the secondary reformer. Here, the remaining methane is converted, and oxygen from air is removed through the partial combustion of methane in the secondary reformer. It is necessary to feed air as this is the nitrogen source needed in the ammonia synthesis. After the reformers, the exiting stream contains H_2, CO, N_2, H_2O, Ar, some unconverted methane, CO_2, etc. Care is taken that the stream does not contain any residual oxygen as this is a poison for the ammonia synthesis catalyst located further downstream.

WGS reactors are then used to ensure as much hydrogen as possible is produced and to get maximum use out of the reactant by consuming much of the CO, as is done

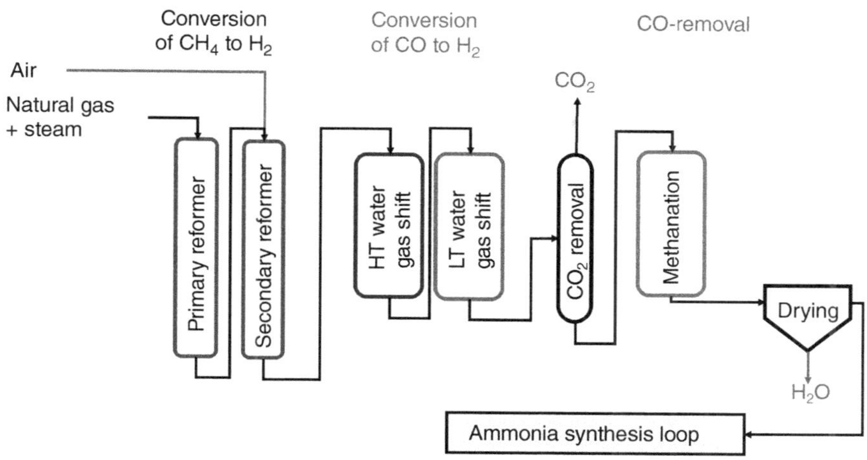

Figure 5.45 Gas preparation section for ammonia synthesis.

in hydrogen production using reforming even when not in conjunction with ammonia synthesis. The stream coming out of the WGS reactors should have a H_2/N_2 ratio of 3. This is controlled by adapting the balance of inlet flow rates to the reformers.

CO_2 is subsequently removed through, e.g. an amine wash, leaving a stream of CO, H_2, N_2 and Ar. This then goes to the CO removal stage, to remove any residual CO another strong poison for the iron-based Haber-Bosch catalyst. CO is removed through methanation, consuming some of the H_2 for the sake of protecting the catalyst, before the stream is dried to remove any water present from the methanation and left over from the reformers and WGS. This then gets sent to the ammonia synthesis loop.

The stoichiometric amount of water that has to be fed to this process to generate hydrogen is 0.653 kg H_2O/kg NH_3 [102], as nitrogen is supplied with oxygen for the combustion of methane and water formed in the combustion together with added is used for the generation of hydrogen:

Combustion of methane with air:

$$0.1505\,CH_4 + 0.3011\,O_2 + 1.1326\,N_2 \rightarrow 0.1505\,CO_2 + 0.3011\,H_2O + 1.1326\,N_2$$

Steam reforming of methane

$$0.8495CH_4 + 1.6989\,H_2O \rightarrow 0.8495\,CO_2 + 3.3978\,H_2$$

Thus, feeding the stoichiometric amount of water will result in combustion of ca. 15% of the fed methane. The required H_2/N_2 mixture may also be produced from coal, viz.

$$0.1505\,C + 0.1505\,O_2 + 0.5663\,N_2 \rightarrow 0.1505\,CO_2 + 0.5663\,N_2$$
$$0.8495\,C + 1.6989\,H_2O \rightarrow 0.8495\,CO_2 + 1.6989\,H_2$$

thus requiring, again, stoichiometrically the combustion of ca. 15% of the fed coal, but now requiring 1.588 kg H_2O/kg NH_3.

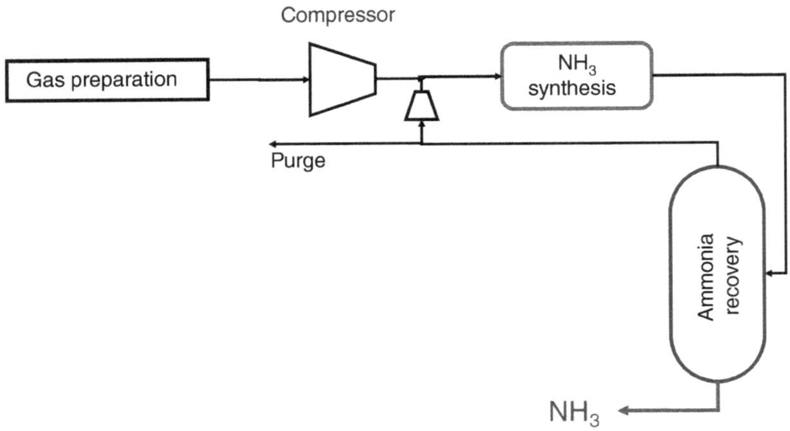

Figure 5.46 Ammonia synthesis loop.

The ammonia synthesis loop (Figure 5.46) is relatively uncomplicated. The incoming gas from the gas preparation section is compressed before the synthesis reactor. The compression costs to reach the required pressure of ca. 100 bar are significant. Ammonia synthesis is carried out at 450–550 °C whilst ammonia recovery takes place at ca. −25 °C. Unconverted reactants are recycled back which is necessary due to the low per pass conversion as dictated by thermodynamics. The temperature difference between the synthesis and separation is one complication in the process. Additionally, as the pressures are in the region of 100 bar, multiple compression steps with interstage cooling are required. A purge stream is used to prevent a build-up of argon, methane and other trace compounds that may have been in the feed air.

The importance of operating the ammonia recovery at low temperatures in the range of the typical operating pressure of the ammonia synthesis reactor (between 100 and 300 bar [103]) can be seen in Figure 5.47. The recovery of ammonia from a product stream containing 11 mol% ammonia at 100 bar and −25 °C is 82%; leaving around 2–3 mol% ammonia in the stream recycled back to the ammonia synthesis reactor. Increasing the pressure, at which ammonia is recovered, improves the recovery, but there is a diminishing return on the effect of increasing pressure; increasing the pressure of the ammonia recovery to 150 or 200 bar, with the ammonia recovery increasing to only 86% and 88%, respectively. Operating at significantly lower pressure, e.g. 50 bar significantly decreases the recovery possible (66% at −25 °C).

The product stream coming out of the ammonia synthesis reactor needs to be cooled down significantly to recover ammonia. This is primarily achieved by heat exchange with fresh feed and recycle streams and to some extent with boiler water to generate steam [104]. Sub-zero temperatures can be achieved by heat exchange with a refrigeration cycle. It should be noted that cooling via Joule-Thomson expansion is not possible as the inversion temperature of the effluent gas mixture is rather high (estimated to be ca. 30 °C for a gas mixture containing 15% ammonia) and the Joule-Thomson coefficient is rather low at the temperatures and pressures of interest implying a large pressure drop to achieve cooling.

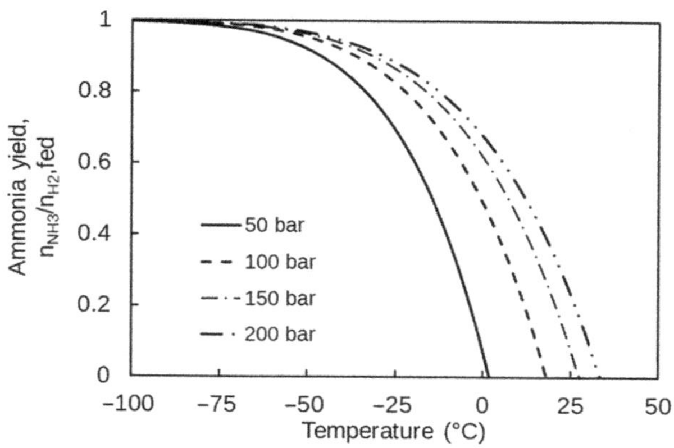

Figure 5.47 Ammonia recovery as a function of temperature at various pressures – modelled using UNIQUAC-SRK, Gibbs reactor fed with $H_2/N_2 = 3$, modelled at 100 bar, 500 °C, achieving a hydrogen/nitrogen conversion of 20%, mole fraction ammonia in $H_2/N_2 = 0.11$ feed to the flash tank (ammonia recovery step).

In the classical process, the partial combustion of the feedstock generates most of the energy required for the process with the total energy requirement amounting to ca. 28 MJ/kg NH_3 when using methane as the feedstock [105]. The total energy requirement amounts to 150% of the heat of combustion of ammonia, which is often equated with the minimum theoretical energy requirement (see e.g. [106]). This negates the energy associated with the feedstock, and the maximum thermal efficiency of the ammonia synthesis defined as the enthalpy of the product and heat generated in the process relative to the enthalpy of all incoming streams.

$$\eta_{\text{thermal,min}} = \frac{\dot{n}_{NH_3} \cdot \Delta_f H^{298.15}_{NH_3(l)} + \dot{Q}_{\text{added}}}{\left(\begin{array}{c}\dot{n}_C \cdot \Delta_f H^{298.15}_{C(s)} + \dot{n}_{CH_4} \cdot \Delta_f H^{298.15}_{CH_4(g)} + \dot{n}_{O_2} \cdot \Delta_f H^{298.15}_{O_2(g)} \\ + \dot{n}_{N_2} \cdot \Delta_f H^{298.15}_{N_2(g)} + \dot{n}_{H_2O} \cdot \Delta_f H^{298.15}_{H_2O(l)}\end{array}\right)}$$

This can be determined from the overall stoichiometric equation and the total energy fed into the process. Using methane as the feedstock yields an overall maximum thermal efficiency of 45.7%. The maximum thermal efficiency using coal as a feedstock would amount to 32.0%. The exergetic efficiency looks at the amount of work which can be generated from the product relative to the amount of work, which could have been obtained from the reactants:

$$\eta_{\text{exergy,min}} = \frac{\dot{n}_{NH_3} \cdot \Delta_f G^{298.15}_{NH_3(l)} + \dot{Q}_{\text{added}}}{\left(\begin{array}{c}\dot{n}_C \cdot \Delta_f G^{298.15}_{C(s)} + \dot{n}_{CH_4} \cdot \Delta_f G^{298.15}_{CH_4(g)} + \dot{n}_{O_2} \cdot \Delta_f G^{298.15}_{O_2(g)} \\ + \dot{n}_{N_2} \cdot \Delta_f G^{298.15}_{N_2(g)} + \dot{n}_{H_2O} \cdot \Delta_f G^{298.15}_{H_2O(l)}\end{array}\right)}$$

Ammonia synthesis utilising methane reforming comes (according to the stoichiometric expression) to a maximum exergetic efficiency of 24.7%, whereas using coal as the feedstock yields a maximum exergetic efficiency of 15.5%.

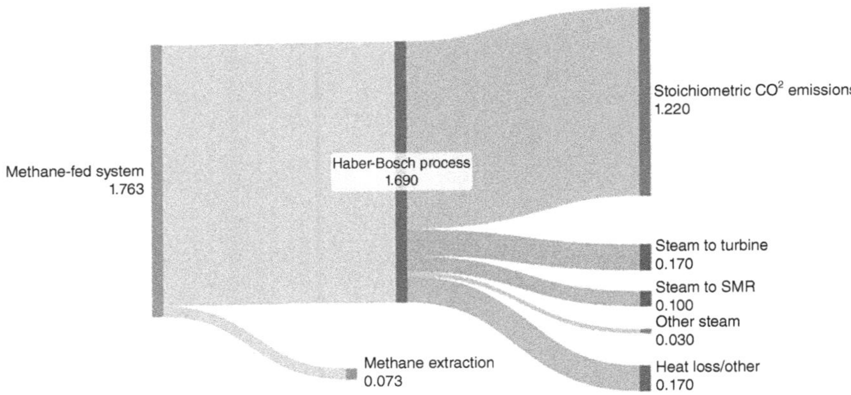

Figure 5.48 CO_2 emission during ammonia synthesis from methane in kg CO_2/kg NH_3 (redrawn from Smith et al. [107], using SankeyMATIC.com).

5.3.4.1 CO_2 Emissions

The overall ammonia synthesis using grey hydrogen can be written as:

$$CH_4 + 1.398\, H_2O + 0.301\, O_2 + 1.133\, N_2 \rightleftharpoons CO_2 + 2.266\, NH_3$$

with thus a minimum, stoichiometric CO_2 emission of 1.143 tonne CO_2/tonne NH_3. Ammonia may also be produced from coal, resulting in twice the amount of CO_2 produced, viz. 2.289 tonne CO_2/tonne NH_3. The actual CO_2 emissions for the Haber-Bosch process are slightly higher due to the inefficiencies in the process (see Figure 5.48); the IEA quotes a higher value of 2.4 tonne CO_2/tonne ammonia [94]. The stoichiometric amount of CO_2 produced is somewhat higher as complete conversion is not achieved. There are additional CO_2 emissions from steam generation, which at present, is primarily provided through either burning additional fossil fuels. This would add to the direct CO_2 emissions from the plant, though steam may also be generated through heat integration with the Haber-Bosch reactor. There is also a requirement for electricity in the process, notably to power the compressors needed for the high-pressure operation. This may, in part, be generated through heat integration with the reactor, forming steam to use to generate electricity/power to the compressors, however additional electricity would still need to be provided from an external source. If the electric power comes from a fossil source, this will also add to the net CO_2 emissions.

5.3.5 Efficiency of Ammonia Synthesis

The overall reaction for the classical Haber-Bosch process using methane steam reforming to produce hydrogen can be represented as:

$$CH_4 + 1.398H_2O + 0.301O_2 + 1.133N_2 \rightleftharpoons CO_2 + 2.266NH_3$$

$$\Delta H^0_{\text{rxn}} = -84.59 \cdot \frac{\text{kJ}}{\text{mol CH}_4}$$

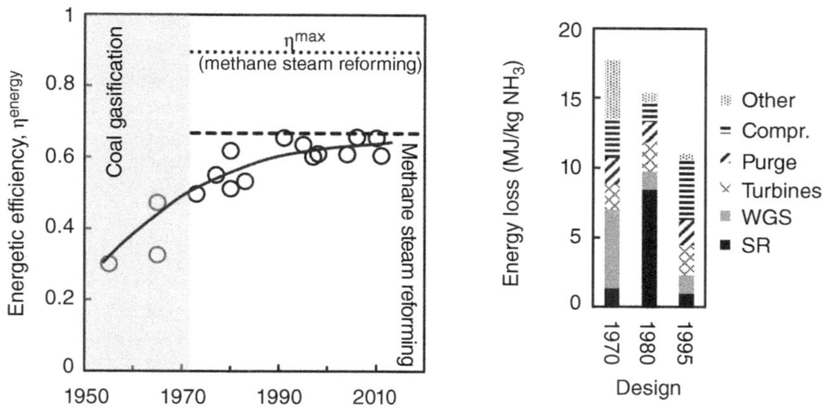

Figure 5.49 Energetic efficiencies and energy losses in the ammonia synthesis (redrawn with data from Smith et al. [107]; 1980 design: ICC pre AMV design; 1995 design: low energy process).

The overall reaction is slightly exothermic. The maximum energetic efficiency that can be obtained can be calculated from the LHV of ammonia relative to the heat of combustion of methane is 89.5%. Smith et al. [107] quote a lower maximum energetic efficiency of 84.1% based on methane steam reforming. The energy efficiency of the ammonia synthesis has improved over the long history of the process (see Figure 5.49). Approaching now an energetic efficiency of 66.7% for the best available technology (BAT). This brings the energy costs for the ammonia synthesis from methane to 28–30 MJ/kg NH_3, i.e. halving from the initial cost, 60 MJ/kg NH_3 (cf. LHV NH_3 316 kJ/mol = 18.6 MJ/kg) [107]. This means that the current process efficiency is around 65%, a significant improvement from 1970s, where the efficiency was only ca. 50%.

The exergy analysis of an ammonia plant [108] showed that ca. 72% of the exergy loss occurs in the generation of the H_2/N_2-mixture, i.e. in the steam reforming step (59%) and 12% in the gas purification step. The exergy loss in the actual ammonia synthesis loop amounts to 18% of the total exergy loss (the balance is lost in process utilities generation and other sections). It is more difficult to define an exergetic efficiency of the overall process [109], as the classical definition (also named the rational exergy efficiency) leads to high values, which are close together, because of the large flow of materials relative to the energy flows.

5.3.6 Alternative Routes for Ammonia Synthesis

When looking to make the ammonia synthesis process renewable, the major consideration will be both getting hydrogen from water using renewable energy, rather than fossil fuels. The process using renewable hydrogen requires a dedicated air separation unit (and will thus produce copious amounts of oxygen from both the air separation and the electrolysers). A comparison of the classical process vs. using renewable hydrogen is given in Table 5.9.

Table 5.9 Classical process vs. using renewable hydrogen for ammonia synthesis.

	Classical process	**Using renewable hydrogen**
Energy source	CH_4/C	Renewable electricity, e.g. PV/wind power
Hydrogen source	H_2 from CH_4 and H_2O	H_2 from H_2O
Nitrogen source	N_2 from air	N_2 from air.
Oxygen removal	O_2-removal by combustion	O_2-removal using air separation
Problematic catalyst poison	CO-poisoning of iron-based Haber-Bosch catalyst	O_2-poisoining of iron-based Haber-Bosch catalyst

5.3.6.1 Production of Ammonia Using Renewable Hydrogen

Ammonia can be produced using renewable resources, i.e. when using hydrogen produced from water electrolysis utilising renewable energy, the overall reaction equation becomes:

$$H_2O + \frac{1}{3}N_2 \rightarrow \frac{1}{2}O_2 + \frac{2}{3}NH_3$$

As such, 1.588 kg H_2O is required per kg of ammonia, identical to the amount used when using coal as the feedstock, but more than double the minimum required amount when obtaining hydrogen from methane reforming (0.656 kg H_2O/kg NH_3). Thus, the process of using renewable hydrogen requires double the amount of water as the current major process, using natural gas, which may be a concern when producing ammonia in water-scarce regions.

When producing hydrogen using renewable energy (Figure 5.50), the actual ammonia synthesis loop will not have to change, only the gas preparation for the feed. This will require nitrogen obtained via air separation as well as production of hydrogen from electrolysis. The obtained nitrogen and hydrogen steams will contain some oxygen, which will need to be removed, before sending the combined stream to the ammonia loop. This can be readily achieved by sacrificing some of the produced hydrogen. As with the classical process, one of the two feeds from the gas preparation will need to be used to ensure the correct stoichiometric feed ratio for the ammonia synthesis reactor.

There are two major methods to obtain nitrogen using air separation, viz. cryogenic separation, or adsorption techniques, the most common of which is pressure swing adsorption. Generally, cryogenic separation is used for large-scale processes [110] whilst pressure swing adsorption is used for small-scale processes [111, 112]. A comparison of the pros and cons of cryogenic separation vs. pressure swing adsorption is shown in Table 5.10. Cryogenic separation can be optimised to produce pure nitrogen and pure oxygen. The temperature of operation implies that the start-up time of this type of operations is relatively long. Air separation using pressure swing adsorption typically yields either a relatively pure nitrogen or a relatively pure oxygen stream, although the purity is typically less than obtained

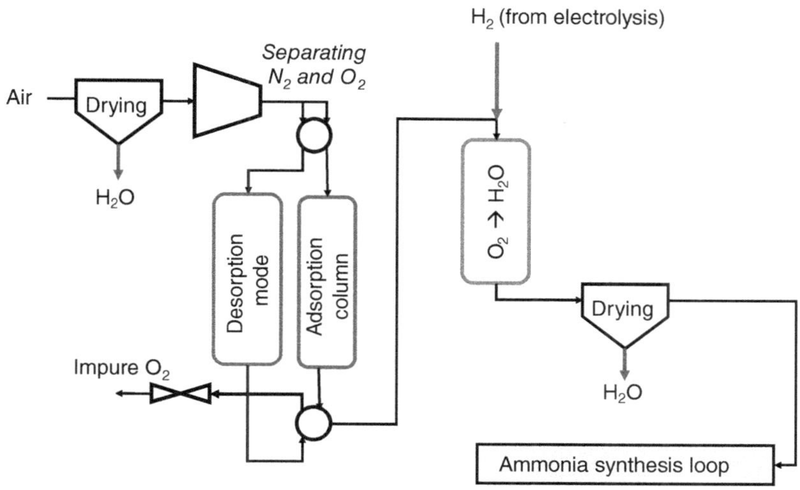

Figure 5.50 Ammonia production using renewable hydrogen with pressure swing adsorption for the purification of nitrogen from air.

Table 5.10 Cryogenic separation vs. pressure swing adsorption.

	Cryogenic separation	Pressure swing adsorption
Optimised for	O_2 and N_2 production	N_2 or O_2 production
Product purity	High	lower purity
Size	Large volume	Small volume
Capital cost	High capital cost	Lower capital cost
Start-up speed	Long start-up	Quick start-up

using cryogenic separation [112]. Pressure swing adsorption responds relatively fast to changes in demand [111].

The imperfect air separation as well as oxygen cross-over in the electrolyser [113] will result in the presence of residual oxygen in the nitrogen/hydrogen stream must be removed prior to entering the ammonia synthesis loop to avoid deactivation of the ammonia synthesis catalyst. This can be achieved by combusting some of the produced hydrogen with residual oxygen, e.g. over a metal catalyst. Hence, the purity of nitrogen achieved in the air separation section will need to be balanced with the cost of hydrogen required to convert the residual oxygen.

The production of renewable ammonia is affected by the intermittency of the power supply, as it is used in the hydrogen generation, but also for compressors, and the air separation unit. Hence, plants producing renewable ammonia will need larger back-up power facilities to keep the plant running at times of low power supply. Furthermore, plant design will need to consider intermediate hydrogen and

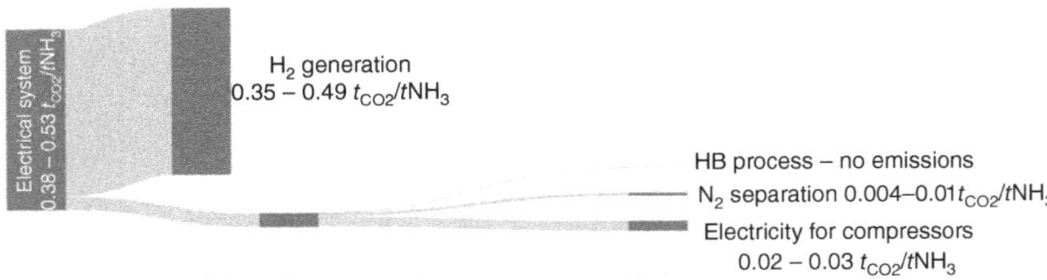

Figure 5.51 CO_2 production in the ammonia synthesis using renewable hydrogen (Smith et al. [107] / Royal Society of Chemistry / CC BY 3.0).

nitrogen storage, as well as operating the Haber-Bosch ammonia synthesis loop at variable loads.

5.3.6.2 CO_2 Emissions from Renewable Ammonia Production

The amount of CO_2 produced in the ammonia synthesis when using renewable hydrogen can be very low. Potentially, there need not be any Scope 1 or Scope 2 emissions, with even the electricity production to power the process coming from renewable sources. The analysis shown in Figure 5.51 was done for an electrically driven Haber-Bosch process with some wind power, supplemented with fossil sources. In this example, the range of CO_2 emissions reported is dependent on the number/size of the wind turbines. As the hydrogen generation takes the most power, it has the most CO_2 emissions, followed by the compressors which are the next greatest power consumer.

5.3.6.3 Energy Requirement for Production of Renewable Ammonia

The energy requirement for the ammonia synthesis using hydrogen obtained from electrochemical water splitting is estimated at 24–36 MJ/kg NH_3 (cf. LHV NH_3 316 kJ/mol → 18.6 MJ/kg NH_3). The efficiency of the process is currently between 52% and 78% but is strongly dependent on H_2 formation and thus the technology readiness level of the method of obtaining green hydrogen, the closest to industrialisation being less efficient than the fossil-fuel based methods, though the processes in development may become far more efficient. As such, efficiency improvement of the electrolysers is needed. Most of the energy loss is associated with the electrochemical water splitting (see Figure 5.52) [107]. The anticipated energy loss of 14.1 and 14.3 MJ/kg NH_3 in an alkaline electrolyser operating at 10 bar, and an PEM electrolyser operating at 80 bar, respectively, would result in an overall energetic efficiency of the process 48–50%. Future improvement, either by using an SOE or an improved PEM electrolyser and minimising the energy loss in the electrolyser to 5–7 MJ/kg NH_3 should improve the overall energy efficiency to 61–63% closer to the best available technology using methane steam reforming.

To summarise, the efficiency of using H_2 from electrochemical water splitting to produce NH_3 can be as low as 50% for the highest level of renewable process, or 62% for the classical synthesis. There is research into the processes where the efficiency is higher [114], although these processes are not all yet ready for implementation. The major benefit of using H_2 from renewable processes is the reduction in CO_2

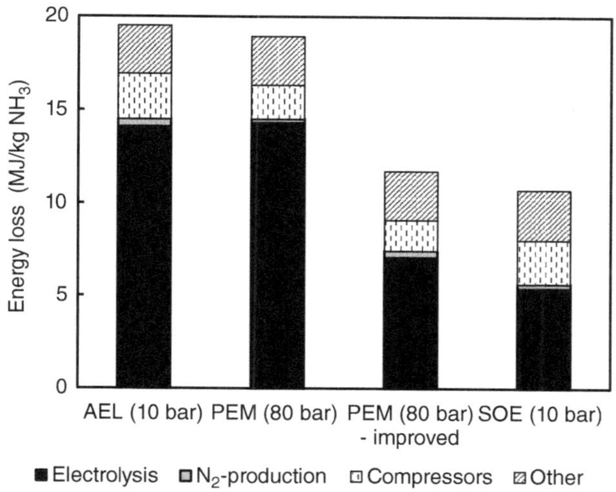

Figure 5.52 Energy loss in the ammonia production using electrochemical water splitting as the hydrogen source. AEL, alkaline electrolysis; PEM, proton exchange membrane electrolysis; SOE, solid oxide electrolysis; pressure of the electrolysis operation indicated; ammonia synthesis at high pressure using the Haber-Bosch process (redrawn with data from Smith et al. [107]).

emissions. This process co-produces 0.24 kg O_2/kg NH_3 in the nitrogen production from air separation which means that there is the opportunity for co-value from O_2.

5.3.7 Power from Ammonia

The use of ammonia as an energy or hydrogen vector, requires the extraction of energy/hydrogen from this source. Ammonia can be combusted (either thermally or electrochemically) or converted back to hydrogen to generate power (Figure 5.53). In the second option, the ammonia will be converted back to hydrogen at the location of end use using a high-temperature decomposition, after which the hydrogen may be used in a combustion fuel to generate power. The direct process will likely have the highest efficiency as there will be fewer options for entropic loss, viz. through both the decomposition and fuel cell, rather than just the ammonia combustion. The direct process will also involve fewer process steps. Conversion of hydrogen may also be performed in a fuel cell to generate power on a smaller scale and at a lower temperature. Using the intermediate generation of hydrogen will allow for some of the hydrogen to be used for purposes other than energy generation, for example as a synthesis gas or as a reducing agent.

5.3.7.1 Energy Recovery Through Ammonia Combustion

Combustion of ammonia is an effective way of using ammonia as an energy source [115]. The ammonia combustion reaction releases a large amount of energy:

$$2NH_3(g) + \frac{3}{2}O_2(g) \rightleftharpoons N_2(g) + 3H_2O(g) \qquad \Delta H^0_{rxn} = -633.2 \cdot \frac{kJ}{mol\ N_2}$$

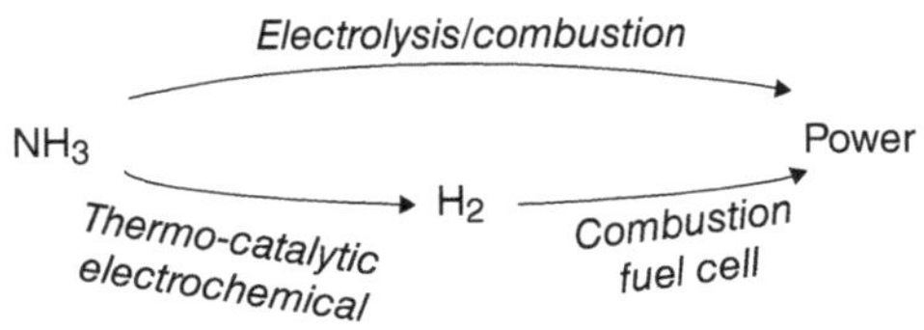

Figure 5.53 Obtaining power from ammonia.

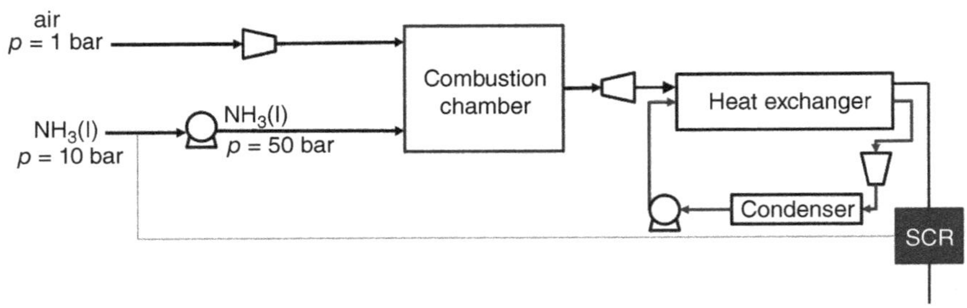

Figure 5.54 Combustion of ammonia in a Combined Open Gas Cycle (COGC).

The adiabatic flame temperature for a stoichiometric feed of air is 2000 °C dropping to ca. 1510 °C with a 50% excess of air (see Figure 4.14). The adiabatic flame temperature with excess air is similar to the adiabatic flame temperature achieved with a methane flame implying that the radiative heat transfer from an ammonia flame is likely to be less than that of methane as less soot will be formed (it should further be noted that the adiabatic flame temperature for ammonia with a sub-stoichiometric air feed is less than the adiabatic flame temperature for methane.). Furthermore, the convective heat transfer from an ammonia flame is expected to be less as well, as the number of moles of flue gas is less.

A typical process for the combustion of ammonia is shown in Figure 5.54. Here, the liquid NH_3 is pumped to the combustion chamber, along with air which has been compressed to the same pressure. It should be noted that the combustion behaviour may be improved by putting an ammonia decomposition reactor prior to the combustion chamber, where ammonia is partially decomposed into hydrogen and nitrogen, to improve the combustion behaviour [116], as ammonia is not very reactive. This is followed by combustion at elevated pressure. The stream exiting the combustion chamber is passed through a turbine, expanding the gas and recovering the work related to the pressure and the slight expansion of the gas in the reaction, a Brayton cycle. A Rankine cycle can be added to recover heat from the effluent; the hot gas is sent to a heat exchanger that uses the excess heat to boil water, giving high-pressure steam forming the so-called combined cycle as discussed in Chapter 2.

The efficiency of the Brayton can be estimated from a system using 50% excess air and assuming a 90% efficiency in all turbines, compressors and pumps. The temperature of effluent after the counter current heat exchanger is 400 °C as required for selective catalytic reduction (SCR).

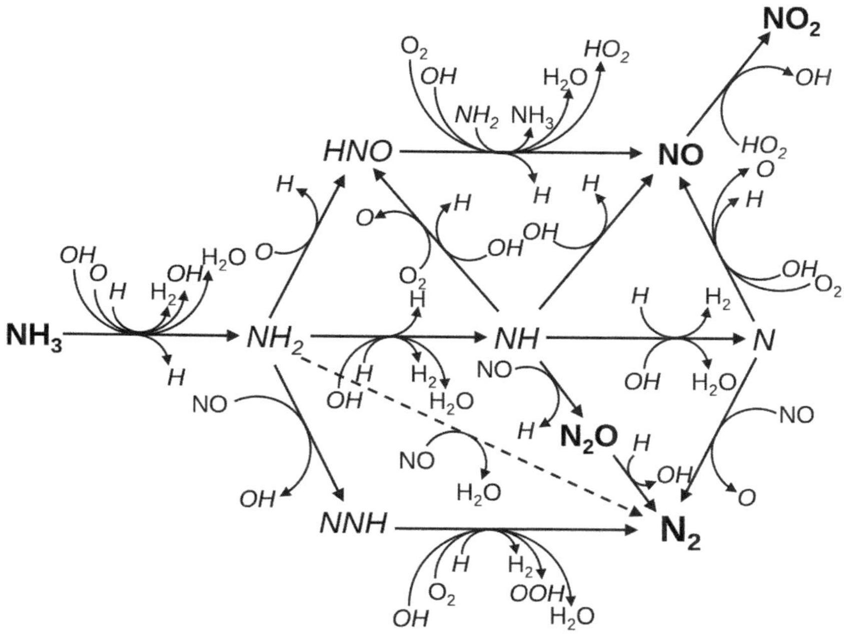

Figure 5.55 Some of the reaction pathways in the high-temperature combustion of ammonia (based on References [116, 118, 119]).

The first step is compressing of the fed air to 50 bar. Multiple compression stages will be needed to increase the feed pressure to 50 bar. As the inefficiency of the compressors may push the temperature above 190–200 °C, a safe option would be to use four stages, each with a compression ratio of $50^{1/4} = 2.7$. With four compressors in series with intermediate cooling down to 40 °C, the final temperature out of the compressors is estimated using the Soave-Redlich-Kwong equation of state to be between 120 and 145 °C. Ammonia is pumped into the combustion chamber (or possibly in a reactor, where ammonia is partially decomposed). The total, required compression work is estimated at 53.5 kJ/mol of ammonia fed. The net efficiency of the Brayton cycle comes to ca. 48–50% (depending on the amount of excess air). A Rankine cycle can be added to improve the overall efficiency, but this is only viable if the temperature of the exhaust gas is significantly higher than 600 °C.

It seems obvious that combusting ammonia, a nitrogen containing compound, will produce NO_x• NO and NO_2 are known pollutants causing the formation of ground-level ozone and in acid rain. Nitrous oxide, N_2O, is itself a strong greenhouse gas, with a greenhouse gas potential 273 times higher than CO_2 based on a 100-year horizon [117]. Figure 5.55 shows the possible pathways for the conversion of ammonia into NO_x at high temperatures via the formation of NH_x•, NNH•, HNO•, N• radicals.

The so-called fuel-NO_x is formed in addition to thermal NO_x, which is produced in a thermal reaction involving oxygen, nitrogen and hydroxyl radicals according to the Zeldovich mechanism.

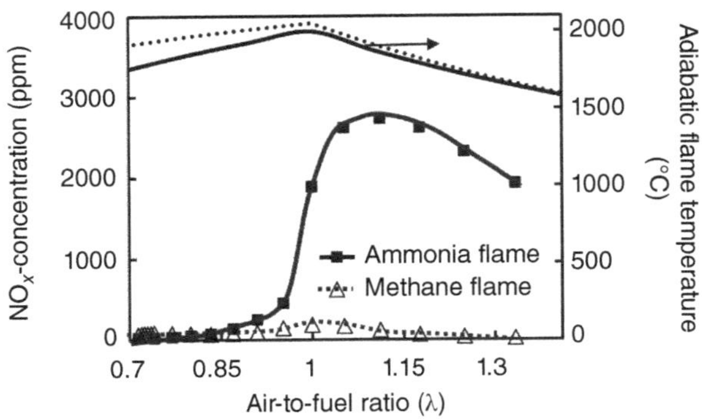

Figure 5.56 Amount of NO_x (NO, NO_2, N_2O) formation at 0.1 MPa and 40 mm downstream the position of maximum heat release as a function of the air-to-fuel ratio (redrawn from Kobayashi et al. [115]; also shown the adiabatic flame temperature).

The contribution of fuel-NO_x and thermal NO_x to the overall NO_x emissions in the combustion of ammonia and methane, respectively, as a function of the air-to-fuel ratio is shown in Figure 5.56 [115]. The adiabatic flame temperature increases until the air/fuel ratio equals one, after which it decreases. At this maximum point, the stoichiometric amount of oxygen is available. Before this point, incomplete combustion occurs whilst after this, there is excess air taking up the heat released. The amount of NO_x formed in the combustion of methane is the highest at the highest temperature, i.e. with a stoichiometric feed ($\lambda = 1$). The amount of NO_x formed is maximum at $\lambda = 1.1$, i.e. with a slight excess of air indicating that the formation of NO_x is favoured by the presence of oxygen (radicals). It can be further noted that the amount of thermal NO_x formed in the combustion of methane is relatively small in comparison to the NO_x formed when using ammonia as a fuel (the maximum concentration of NO_x using ammonia as a fuel is ca. 15 times larger than the maximum concentration of NO_x using methane as a fuel at similar adiabatic flame temperatures).

The formation of NO_x in the combustion of ammonia requires an end-of-pipeline solution to minimise its impact on the environment (modifications to the conditions in the combustion chamber may also reduce the NO_x concentration [115]). Various post-combustion technologies are available for the removal of NO_x from flue gas, viz. oxidation, adsorption and reduction [120]. The most common form currently employed is the SCR reaction/process. This operates over a vanadia supported on titania catalyst and needs a temperature of around 400 °C, limiting the overall process efficiency, again regardless of the fuel [121], and thus reducing the extent energy can be extracted from the flue gas. The reducing agent in the case of the ammonia combustion is ammonia itself, requiring ca. 1–1.3 mol NH_3/mol NO_x (low limit for NO; high limit for NO_2). Slightly excess ammonia is fed to the SCR, which gets trapped in the support, but ensures that all NO is reduced.

NO_x can also be removed in an oxidative process by reacting NO_x with ozone in order to form species, such as N_2O_5 [122], which are easily condensable and reactive meaning that they can readily be removed from the stream. Alternatively, the NO_x species may be adsorbed using, e.g. activated carbon beds, which may then be heated to remove the NO_x species in a near pure form [123].

5.3.7.2 Electrochemical Combustion of Ammonia

It is also possible to electrochemically combust ammonia, using a direct ammonia fuel cell (DAFC). The process is similar to what is used for hydrogen-based fuel cells or direct methanol fuel cells (DMFCs, discussed in Section 5.5.7.2). The energy extracted from the electrochemical combustion is related to the Gibbs free energy of reaction:

$$2NH_3(g) + \frac{3}{2}O_2(g) \rightleftharpoons N_2(g) + 3H_2O(g) \qquad \Delta G^0_{rxn} = -652.8 \cdot \frac{kJ}{mol\ N_2}$$

In the acidic cell, the two electrochemical reactions are:

Anode: $2NH_3 \rightarrow N_2 + 6H^+ + 6e^-$
Cathode: $\frac{3}{2}O_2 + 6H^+ + 6e^- \rightarrow 3H_2O$

$$E^0 = -\frac{\Delta G^0_{rxn}}{n \cdot \mathcal{F}} = -\frac{652.8 \cdot 10^3}{6 \cdot 96458} = 1.128\,V$$

The electrochemical combustion of ammonia can be operated at either high or low temperatures [124], just like the hydrogen fuel cell. The low-temperature cells are similar to PEM-fuel cells or the alkaline cells for conversion of H_2 into water. Here, the ORR is slow (as in PEM-fuel cells) and thus a large overpotential is seen on the cathode. The adsorption of ammonia on the anodic electro-catalyst is much stronger than the adsorption of hydrogen resulting in a more pronounced overpotential on the anode as well. Hence, the efficiency of the DAFC is still a challenge.

Alternatively anionic cells may be used, though the required anionic exchange membranes are still in development for use as a compact cell (as with anionic water electrolysis) [125]. Here, the reactions are:

$$2NH_3(aq) + 6OH^- \rightarrow N_2(g) + 6H_2O(l) + 6e^- \qquad E^0 = -0.77\,V$$

$$\frac{3}{2}O_2(g) + 3H_2O(l) + 6e^- \rightarrow 6OH^-(aq) \qquad E^0 = 0.401\,V$$

This will have an overall voltage of 1.17 V. A benefit of operating the anionic cell (as with the anionic water electrolysis) is the potential for non-noble metal catalysts decreasing cost, though at present this would decrease the operation efficiency.

In high-temperature cells, i.e. solid oxide fuel cells (SOFC), the reaction is likely to follow a reaction path of ammonia decomposition into nitrogen and hydrogen followed by a normal (high-temperature) fuel cell with hydrogen as the fuel.

5.3.7.3 Conversion of Ammonia Back to Hydrogen

It may be preferable to recover (some) hydrogen from the ammonia, rather than to use ammonia as an energy source. For example, the combustion of ammonia is improved by the addition of hydrogen [126], or electric work is to be generated using a hydrogen fuel cell. For recovering hydrogen, there are in principle two options, viz. thermo-catalytic or electrochemical decomposition of ammonia.

The thermo-catalytic conversion of ammonia into nitrogen and hydrogen is simply the reverse of the Haber-reaction for ammonia synthesis with as the key reaction:

$$2NH_3 \rightleftharpoons N_2 + 3H_2$$

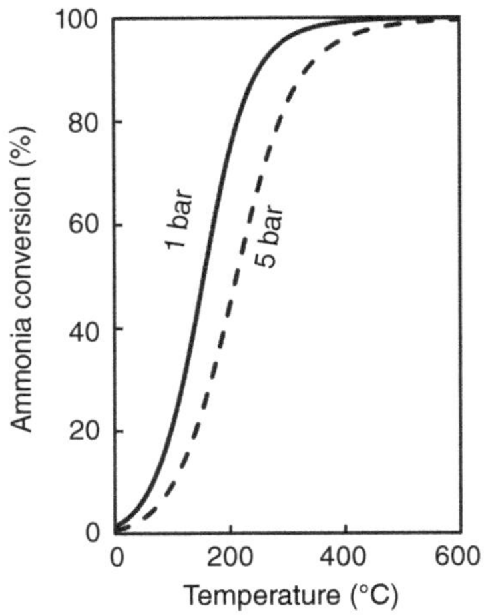

Figure 5.57 Equilibrium for ammonia decomposition into hydrogen and nitrogen.

The catalyst for this reaction could be the same as the ammonia synthesis catalyst, on the principle that if a catalyst can catalyse a forward reaction, it can catalyse the reverse. However, the presence of some oxygen or water in the ammonia stream may deactivate the currently used ammonia synthesis catalyst, and a catalyst which is less sensitive to oxidation, may be needed. From an equilibrium point of view (Figure 5.57, the reverse reaction depicted in Figure 5.43), relatively high temperatures are needed, e.g. at 1 bar, temperatures higher than 400 °C are needed to obtain a conversion of ammonia larger than 99%. This contrasts with the ammonia synthesis that requires the lowest possible temperature, other than it has to be greater than 400 °C to activate the nitrogen, thus meaning high pressures. The increasing number of moles also means that low pressures are beneficial for the ammonia decomposition, in line with the higher pressures previously discussed for the ammonia synthesis.

A very high conversion may be required if the product hydrogen is without further purification to be used in a hydrogen fuel cell. At 1 bar, a high equilibrium conversion of 99.8% requires a temperature larger than 525 °C. The very high required temperature does become limiting as the feed needs to be heated, thus reducing overall efficiency.

An option for achieving high conversion with more moderate temperatures is to use a membrane reactor. Some experimental results in literature have shown that, even at temperatures below 450 °C, ammonia conversions of close to 100% can be achieved [127].

An example of the ammonia decomposition process with hydrogen used in a fuel cell is shown in Figure 5.58. Ammonia is pumped, whilst kept at relatively low pressure, to a furnace to be heated to moderately high temperatures (i.e. >500 °C). The effluent gas from the reactor is heat exchanged with the feed gas to recover some of the heat. Hydrogen is separated from residual ammonia and nitrogen to purify the feed to the fuel cell. This can be achieved using Pd-based membranes. However, the membrane will need a pressure gradient, thus necessitating the operation of

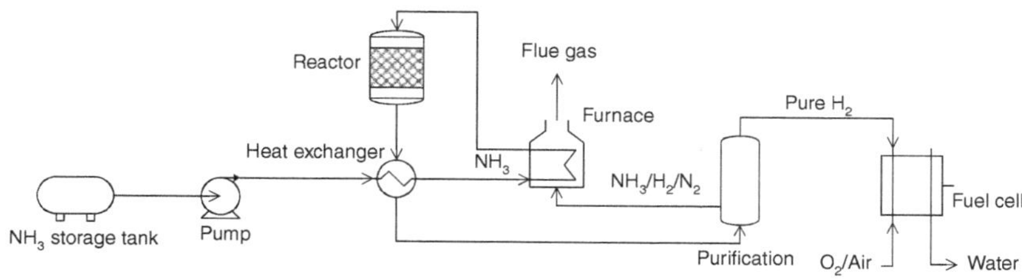

Figure 5.58 Thermochemical conversion of NH_3 into H_2 and its subsequent use in a fuel cell (redrawn from References [115, 128]).

the ammonia decomposition reactor at a slightly elevated pressure (~5 bar). This will reduce the equilibrium conversion of ammonia and the yield of hydrogen, but enhance the flux of hydrogen. For example, Cechetto et al. [127], utilising a packed bed membrane reactor to carry out the ammonia decomposition over a 2% Ru/Al_2O_3-catalyst at 450 °C and 5 bar packed into a Pd-Ag membrane, obtained a conversion of ammonia of 99.7% with a H_2-recovery through the membrane of 90.5% at a purity of 99.98%.

The electrochemical decomposition is an alternative to the same overall reaction but is not an enthalpic process like the thermo-catalytic decomposition above.

$$2NH_3(g) \rightleftharpoons N_2(g) + 3H_2(g) \quad \Delta G^0_{rxn} = +33 \cdot \frac{kJ}{mol\ N_2}$$

This is typically done using an ammonia electrolytic cell (AEC) [128], which makes use of aqueous ammonia solutions with either KOH or NaOH electrolytes. A pictorial representation of the cell is shown in Figure 5.59.

Ammonia oxidation occurs at the anode:

$$2NH_3 + 6OH^- \rightarrow N_2 + 6H_2O + 6e^- \quad E^0 = -0.77\,V$$

Water reduction occurs at the cathode:

$$6H_2O + 6e^- \rightarrow 3H_2 + 2OH^- \quad E^0 = -0.82\,V$$

$$E^0 = -\frac{\Delta G^0_{rxn}}{n \cdot \mathcal{F}} = -\frac{33 \cdot 10^3}{6 \cdot 96458} = -0.057\,V$$

Overall, there is no net water consumption.

5.3.7.4 Efficiency Ammonia as a Hydrogen/Energy Carrier

The utilisation of ammonia as an energy carrier requires the capture of renewable energy (though this is true for any use of renewable energy), the electrolysis of water to produce hydrogen, production of ammonia in the Haber-Bosch process, transport and eventually the conversion of ammonia into power/hydrogen. The overall efficiency of this process is obtained by multiplying the efficiency of each step.

The efficiency of *water electrolysis* to form hydrogen depends highly on the system used but the PEM, which appears to be the most realistic technology presently available, has an efficiency in the range of 55–70%. Ammonia synthesis in general has an efficiency of 65% (see Section 5.3.5).

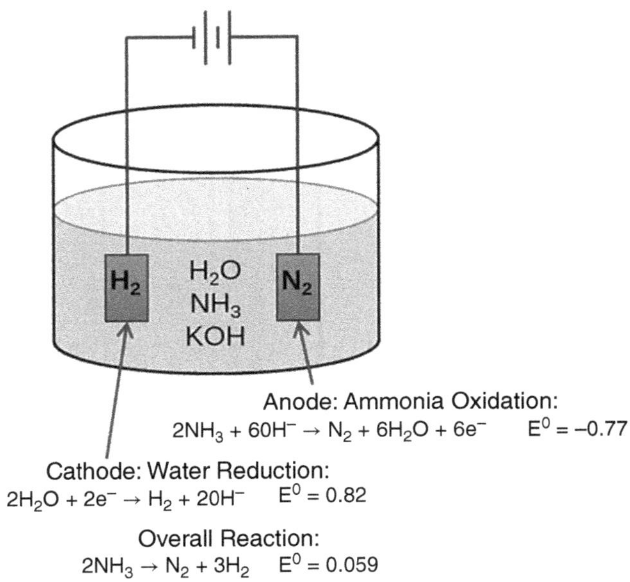

Figure 5.59 Electrochemical conversion of NH_3 into H_2 (redrawn from Abd Ali et al. [128]).

Ammonia transportation can be done via ship with a typical storage volume of 140 000 m^3. At room temperature, this is 84 000 tonne NH_3 (corresponding to 434 GWh NH_3 – LHV). Considering for example transport of ammonia from Cape Town (South Africa) to Rotterdam (the Netherlands), the journey is 7324 km and would take 16 days at 20 knots (ca. 37 km/h). The fuel consumption for the trip would be 0.016 kWh_{el}/km/tonne [129]. This fuel can be provided, at least in part, simply from the boil-off of ammonia from the storage tanks (a boil-off rate of 0.04%/day corresponds to 0.002 kWh_{el}/km/tonne). Considering all the energy required, the transport requires: 3800 tonne NH_3 (assuming engine efficiency of 50%). Thus, the loss in transport due to fuel consumption is 2.3%. The overall efficiency of transport is often quoted as >90% and could be close to, or even exceeding, 95%.

Of the option for converting ammonia to power, combusting in an open gas cycle (Brayton cycle) appears most effective. Here, the efficiencies of 50–65% can be achieved if the gas is not cooled after last compression stage. For increased efficiency, it should be remembered that lower pressures are beneficial and that a large excess of air should be avoided.

Overall Efficiency

Hydrogen production: 55–70%
Conversion of H_2 to NH_3: 65%
Transporting liquid NH_3: 90–95%
Converting ammonia back to power: 50–65%
Overall efficiency: 16–28%

Hence, as an example, the delivery of 1 GW of power in the Netherlands via the Power-Ammonia-Power route would require 3.6–6.2 GW_{el} generated in South Africa (with the conversion of power into ammonia contributing to the largest losses).

5.4 Liquid Organic Hydrogen Carriers - LOHCs

An option that will allow easier transport of hydrogen is to use liquid organic compounds. Liquid organic hydrogen carriers (LOHCs) refer to unsaturated organic molecules, i.e. organic compounds with double bonds that can be hydrogenated. The concept of LOHCs is quite simple, as illustrated in Figure 5.60. At a location where there is abundant renewable energy, hydrogen can be generated through electrolysis, or any of the other methods covered in Section 5.2, and used to hydrogenate the hydrogen carrier in use. This hydrogen carrier can then be shipped or piped to where the hydrogen is needed and dehydrogenated. This will release hydrogen and reform the hydrogen carrier back to the original unsaturated molecule, which can be shipped back to be (re)hydrogenated with the hydrogen used as required at the second location.

There are several requirements for materials to be used as LOHCs. First, the hydrogenation/dehydrogenation must be reversible at practical temperature and pressure ranges and the LOHC in both its hydrogenated and its dehydrogenated form must show high chemical stability. Furthermore, the hydrogenation/dehydrogenation should proceed selectively to minimise its loss due to secondary reactions. To ensure maximum processes efficiency, the hydrogen storage capacity of the carrier chosen should be as high as possible, with most LOHCs being able to carry at least ca. 5.5 wt.%. A higher storage capacity will reduce the mass that needs to be moved to

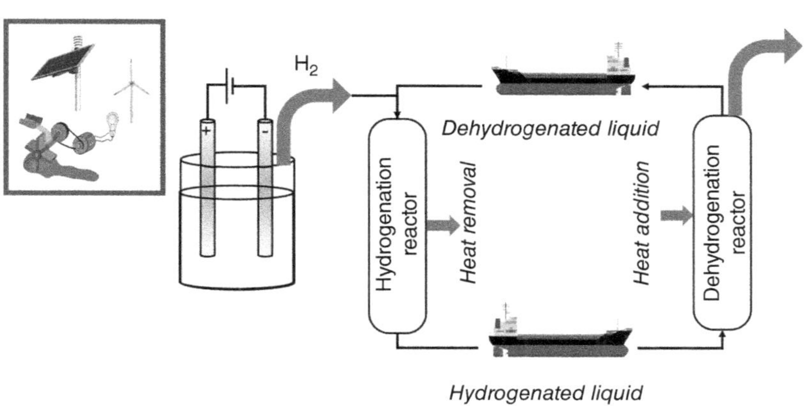

Figure 5.60 Concept for the use of liquid organic hydrogen carriers (LOHCs) to transport hydrogen.

transport a given amount of hydrogen. On a broader level, both the hydrogenated and non-hydrogenated forms of the carrier should have low toxicity and preferably be biodegradable.

The thermodynamic properties of both the hydrogenated and non-hydrogenated forms of the carrier need to be considered. The melting points need to be low to keep the carrier in the liquid form during transport. Likewise, the vapour pressure (often indicated by their normal boiling points) should be low to minimise evaporative losses. The flash points (T_{flash}) and auto-ignition temperature ($T_{auto\text{-}ignition}$) should be high to prevent the LOHC from combusting. Additional factors to consider include density, viscosity and surface tension: higher density means a lower volumetric carrying capacity, and viscosity and surface tension being in appropriate ranges can make the chemicals easier to work with.

There are several compounds that can potentially be used as LOHCs (see e.g. Table 5.11 for some compounds with a hydrogen storage capacity of ca. 5–7 wt.%). It should be noted that hydrogenated forms of CO_2, viz. formic acid or methanol, can also be used as a hydrogen carrier, though are often not referred to as LOHCs. Formic acid has quite a low hydrogen content (only 4.4 wt.%), whereas the conversion of CO_2 into methanol produces water as a co-product and thus consuming more hydrogen than is present in methanol.

Hydrogenation reactions are exothermic, making them thermodynamically favourable at lower temperatures and thus the dehydrogenation reactions are endothermic, meaning that heat must be added for the reaction to occur. This implies some inefficiencies in the system as the hydrogenation and dehydrogenation will take place at different locations and the heat released by the (low-temperature) hydrogenation reaction cannot be used to partially pre-heat the streams for the dehydrogenation reaction. The hydrogenation reaction should preferentially take place at temperatures higher than ca. 200 °C to recover some of the reaction energy, which will otherwise be lost to the surroundings. However, a higher hydrogenation temperature will result in an even higher dehydrogenation temperature. Both reactions require a catalyst, with the preference being heterogeneous catalysts for industrial implementation [130].

Using hydrogen carriers based on the hydrogenation of aromatic rings can lose up to 25–28% of the LHV of the hydrogen that is transported, if no further work is performed in either the hydrogenation or dehydrogenation steps. The loss is reduced to ca. 5% when using CO_2/formic acid as the LOHC; the couple CO_2/methanol loses ca. 10% of the LHV of the transported hydrogen.

5.4.1 Methylcyclohexane-Toluene-Hydrogen

The toluene to methylcyclohexane system is possibly the simplest example of an LOHC that has practical application (principally even ethene to ethane could be used, but the dehydrogenation is too difficult). This system was developed in Japan in the 1980s and is often known as the Methylcyclohexane-Toluene-Hydrogen system

Table 5.11 Properties of potential liquid organic hydrogen carriers in comparison to diesel and gasoline (data from References [130–132]).

	TOL[a)] H0/H6	DBT[b)] H0/H18	NEC[c)] H0/H12	NAP[d)] H0/H10	PHE[e)] H0/H14	FA[e)] H0/H2	MET[f)] H0/H6	MO[g)]	
								D[h)]	G[i)]
T_{melt} (°C)	−95/−126	−34/−45	68/<20	80/−37	172/n.d.	n.d./8	n.d./−98		
T_{boil} (°C)	111/101	398/354	270/280	218/189	360/n.d.	n.d./101	n.d./65	200–350	35–200
T_{flash} (°C)	4/−4	190/n.d.	186/146		160/n.d.	n.d./69	n.d./12	50–80	−65
p^{vap} (40 °C) (bar)	$7.9 \cdot 10^{-2}/1.1 \cdot 10^{-1}$	$7 \cdot 10^{-7}/4 \cdot 10^{-7}$	$1 \cdot 10^{-6}/4.4 \cdot 10^{-5}$	$1.1 \cdot 10^{-3}/4.8 \cdot 10^{-3}$	$1 \cdot 10^{-7}/1.8 \cdot 10^{-6}$	n.d./$1.3 \cdot 10^{-2}$	n.d./$3.5 \cdot 10^{-1}$	$4 \cdot 10^{-4}$	$0.6 \cdot 10^{-1}$
ρ (25 °C) (kg/l)	0.86/0.77	1.04/0.91	1.1/0.94	1.14/0.9	1.3/1	0.71[j)]/1.22	0.71[j)]/0.792	0.82–0.96	0.71–0.77
η (20 °C) (mPa s)	0.6/0.7	44/258	121[k)]/5,9	2/3	3.1/2.2	0.07[j)]/1,8	0.07[j)]/0,55	1.9	5.2
Toxicity (TPI/mg)	19.3/7.3	13.8/n.d.	5.1/n.d.	45.8/1.2	38/n.d.	n.d./9	n.d./29.7	19.1	96.4
H_2-storage (wt.%)	6.2	6.2	5.8	7.3	7.2	4.4	12.6		
Energy density[l)] (kWh/l)	1.6	1.9	1.7	2.2	2.4	1.8	3.3[m)]	9.7–11.4[n)]	8.6–9.3[n)]
$\Delta H_{dehydrogenation}$ (kJ/mol H_2)	68.3	65.4	53.2	66.3	61.3	31.2	16.5	–	–
$T_{dehydrogenation}$ (°C)	250–450	270–310	180–270	210–300	190	50–100	90–420		

a) Toluene/methylcyclohexane.
b) Dibenzyl toluene/perhydro-dibenzyl toluene.
c) N-ethylcarbazol/perhydro N-ethylcarbazol.
d) Naphthalene/decalin; phenazine/tetradecahydrophenazine.
e) CO_2/formic acid.
f) CO_2/methanol.
g) Mineral oils.
h) Diesel.
i) Gasoline.
j) CO_2 as a liquid.
k) Solid.
l) Energy density based on LHV of hydrogen and the density of LOHC in its hydrogenated form.
m) 1 H_2 lost due to water formation (hydrogen can be regained by using methanol reforming as the dehydrogenation reaction).
n) Energy released in combustion.

(MTH). In the hydrogenation reaction, toluene reacts with hydrogen to form methylcyclohexane.

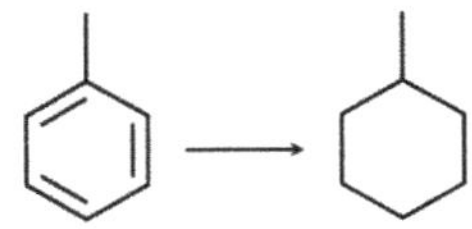

$C_7H_8 + 3H_2 \rightarrow C_7H_{14}$

The hydrogen capacity of a LOHC represents the amount of hydrogen that is available for hydrogen transport per unit mass of the transported molecule. The amount of hydrogen which is transported is not necessarily all the hydrogen atoms that form part of the structure. For example, in methylcyclohexane only 6 of the 14 hydrogen atoms are involved in (de)hydrogenation, the hydrogen capacity is calculated as 6.1 wt.% from

$$H_{\text{capacity}} = \frac{N_{\text{H}_2,\text{transported}} \cdot M_{\text{H}_2}}{M_{\text{hydrogenated molecule}}} = \frac{3 \cdot 2.016 \cdot \frac{\text{g}}{\text{mol}}}{98.189 \cdot \frac{\text{g}}{\text{mol}}} = 6.2\ \text{wt.\%}$$

The conditions for hydrogenation and dehydrogenation are very much controlled by equilibrium limitations as well as Le Chatelier's principle following the change in total moles and the pressure effect thereof. Figure 5.61 shows the distribution between methyl cyclohexane (MCH) and toluene (Tol) as a function of temperature at various total pressures. At low temperatures and/or high pressures, methyl cyclohexane is more stable, and hydrogenation should thus be carried out at those conditions. Conversely, at high temperatures and low pressures, toluene and hydrogen are more stable, and the dehydrogenation should thus be carried out at these conditions. More precisely, the hydrogenation of toluene could be

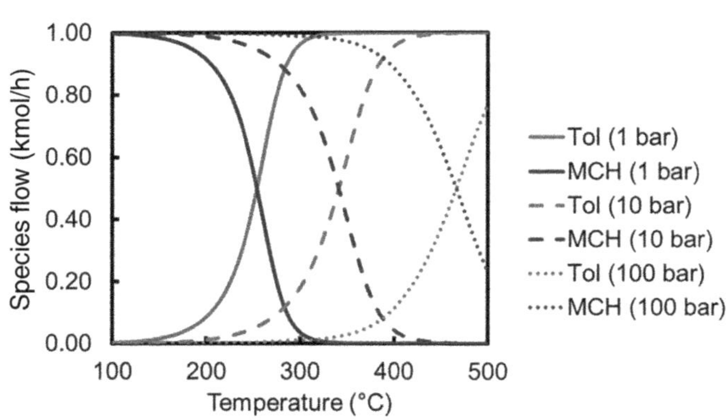

Figure 5.61 Distribution between toluene (grey curves) and methylcyclohexane (black curves) in equilibrium with each other as a function of temperature at various pressures calculated using the SRK model.

carried out at pressures of ca. 10 bar and temperatures below 200 °C, and the dehydrogenation could be carried at atmospheric pressure and temperatures higher than 300 °C.

Although toluene is a good candidate as an LOHC, it does have the disadvantages, such as its toxicity and flammability. Additionally, when using toluene, there is the probability of side reactions, such as disproportionation and dealkylation occurring [133], particularly at the higher temperatures used during the dehydrogenation of methyl cyclohexane. Besides needing to remove the side products, this means that there is systematic loss of the LOHC during the process. The high temperature required for dehydrogenation also means that catalyst deactivation from coking may occur more readily [130].

5.4.2 Dibenzyl Toluene to Perhydro-Dibenzyl Ttoluene

Both dibenzyl toluene and benzyl toluene are heat-transfer oils that can be used as LOHCs. These compounds are sold as isomeric mixtures due to naturally occurring stereoisomers. Dibenzyl toluene is currently one of the most popular molecules for use as an LOHC, because of its lower vapour pressure [134]. The low vapour pressure means that it remains liquid ($T_{\text{boil,DBT}} = 398\,°\text{C}$) at the required operation temperatures, particularly the dehydrogenation. Additionally, the melting point of both compounds is low, so the fluid does not readily solidify. Both compounds also have low toxicity [135]. For simplification, the overall reaction can be written as follows, with the toluene ring being central even though it could be any of the rings:

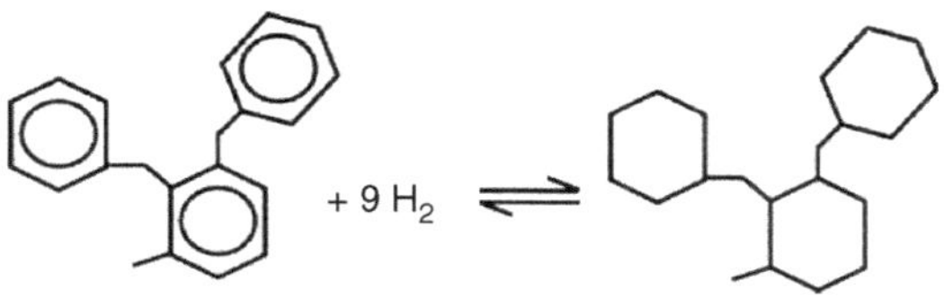

dibenzyl toluene + 9 H_2 → perhydro – dibenzyl toluene

The reaction pathway may be represented as shown in Figure 5.62a. However, each of the above reaction steps shown in (a) is itself a combination of smaller steps, as shown for the first step in Figure 5.62b.

The combined effect of the existing stereo isomers and the total of nine double bonds that can be hydrogenated individually on each molecule means that there are many possible reaction pathways and different intermediate products in the overall hydrogenation reaction. This makes in-depth study of the reaction kinetics challenging. However, the data for the complete reaction may be obtained.

There is also the possibility of a loss in hydrogen capacity due to the degradation of the hydrogen carrier because of C-C cleavage, to simpler aromatics, such as hexane, benzene, toluene and cyclohexane, or carbon deposition (coking).

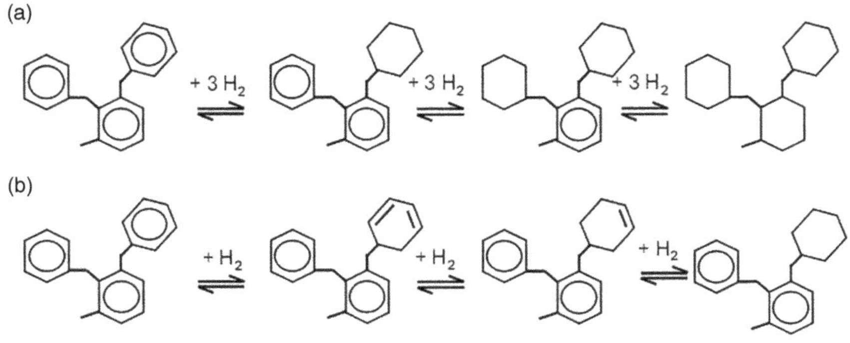

Figure 5.62 Reaction pathway in the hydrogenation and dehydrogenation of 2,3-dibenzyl toluene.

The cost associated with processing and transport of hydrogen in the form of perhydro-dibenzyl toluene is ca. US$ 0.5/kg H_2 as of 2021 [136].

5.4.2.1 Catalysts and Reaction Conditions

The exothermic hydrogenation reaction is relatively facile, and it can be performed over a cheap catalyst, e.g. Raney nickel, although other alternatives are available, such as Ru/Al_2O_3 and Pd/Al_2O_3 [137]. The hydrogenation is typically performed at pressures between 5 and 50 bar and temperatures of 100–200 °C. This relatively low temperature may pose a challenge in recovering the heat of reaction of the hydrogenation reaction.

Ali et al. [137] investigated the effect of various reaction parameters on the hydrogenation of dibenzyl toluene over a Raney nickel catalyst in a stirred reactor (800 rpm). They reported an optimum temperature of 170 °C and an optimum pressure of 70 bar. It was suggested that increased reaction rates at higher temperatures may be hindered by mass transfer limitations. An example of their investigation of the effect of temperature on the hydrogenation of dibenzyl toluene (H0 DBT) is shown in Figure 5.63. The major challenge faced with this reaction is hydrogenation of the final ring , as the uptake of hydrogen is rather quick up to a hydrogen content of ca. 4 wt.%, following which the hydrogen uptake is rather slow. This may be attributed to the hydrogenation of the toluene ring. Prolonged time at elevated temperature may result in the decomposition of the molecules and induce coking (carbon covering of the surface), thus affecting the efficiency of the system.

The endothermic dehydrogenation is a more challenging reaction and requires a platinum catalyst. The reaction is performed at close to atmospheric pressure and temperatures around 250–400 °C. An example of the effect of temperature on the dehydrogenation of perhydro-dibenzyl toluene over a Pt/Al_2O_3 catalyst in the liquid phase is shown in Figure 5.64. The activity for dehydrogenation appears to decrease with increasing reaction time. The maximum hydrogen yield appears to be dependent on the catalyst, the catalyst/LOHC ratio and temperature. The highest yield was obtained at 290 °C [137], although a significant amount of hydrogen was not

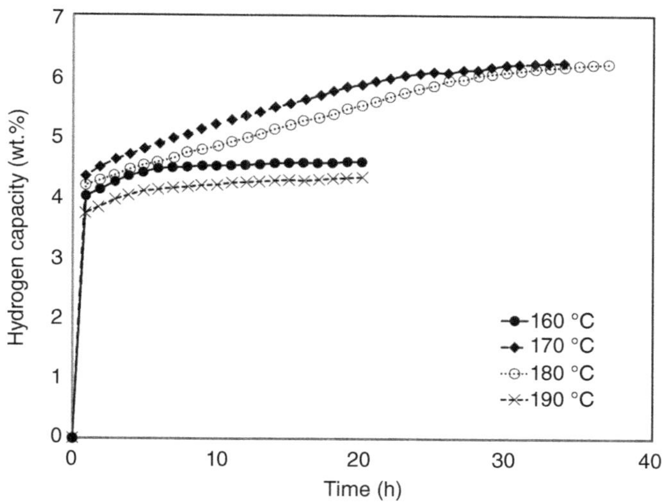

Figure 5.63 Effect of temperature on the hydrogenation of dibenzyl toluene (H0 DBT) at 50 bar (redrawn from Ali et al. [137]).

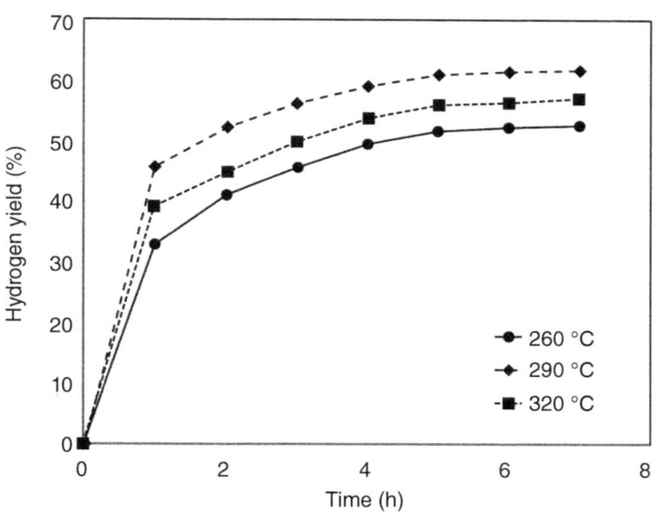

Figure 5.64 Effect of temperature on the dehydrogenation of perhydro-dibenzyl toluene over 3% Pt/Al_2O_3. Y_{H2}: percentage of hydrogen recovered relative to maximum amount of hydrogen recovered [137].

released. In another study [138], it has been seen that although increasing the operating temperature above 290 °C leads to a higher rate of reaction, it also results in an increase in the formation of by-products through decomposing the carrier molecule, thereby decreasing the efficiency of the LOHC.

In summary, dibenzyl toluene can be used as an LOHC with platinum-based catalysts for both the hydrogenation and dehydrogenation step. However, fast hydrogenation requires high-pressure hydrogen and thus the compression of hydrogen will need to be added to the (efficiency) cost of the system. It can be further noted that the hydrogenation of the first two-thirds of all the possible hydrogen is fast (see Figure 5.63), but the subsequent hydrogenation is slow. The dehydrogenation step is also limited, and complete conversion is not readily obtained (see Figure 5.64). Partial hydrogenation/dehydrogenation may become the norm, but this will increase the cost of hydrogen transport significantly.

5.5 Methanol and Dimethyl Ether

Methanol and its derivative, dimethyl ether (DME), are popular choices for hydrogen transport with a hydrogen content of 12.6 wt.% H_2 and 13.1 wt.% H_2, respectively (see Table 5.12). The specific energy of methanol is 19.92 MJ/kg, which is relatively low, as was the case with ammonia with a specific energy of 18.6 MJ/kg. DME has a much larger specific energy of 28.8 MJ/kg.

Methanol is a liquid at ambient conditions with a normal boiling point of 64.7 °C. DME is a gas at room temperature with a vapour pressure of 5.96 bar at 25 °C [139].

Table 5.12 Some properties of methanol and dimethyl ether.

	Methanol	**Dimethyl ether**
Molar mass (g/mol)	32.042	46.069
H-content (wt.%)	12.6	13.1
T_{melt} (°C)	−97.6	−141
$T_{normal\ boiling}$ (°C)	64.7	−26.4
p^{vap} (25 °C) (bar)	0.1696	5.96
$\Delta H_{evaporation}$(25 °C) (kJ/mol)	35.2	21.2 [145]
Critical point	239 °C; 80.84 bar	127 °C; 53.7 bar
Auto-ignition point (°C)	470	350
Flash temperature (°C)	12	−41
Flammability region[a)]	LFL: 6.0%; UFL: 36.5%	LFL: 3.4%; UFL: 27.0%
Solubility in water	Miscible	46 g/l
Density (25 °C)	791.8 g/l	1.9 g/l (as a gas @ 1 bar)[c)] 661 g/l (as a liquid @ 10 bar [145])
Specific energy[b)](MJ/kg)	19.92	28.83
Energy density	15.77 MJ/l	0.05 MJ/l (as a gas @ 1 bar) 18.77 MJ/l (as a liquid at 10 bar)

a) LFL, lower flammability limit; UFL, upper flammability limit (given as vol.% in air).
b) Specific energy based on lower heating value.
c) Estimated based on model using Soave-Redlich-Kwong equation of state (cf. reported density of dimethyl ether at 25 °C).

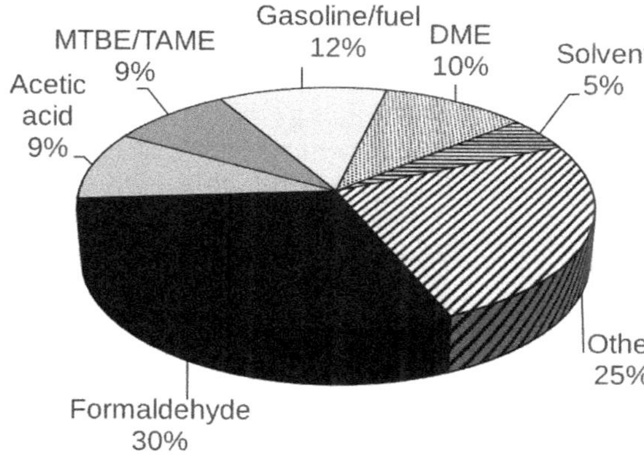

Figure 5.65 Worldwide consumption of methanol in 2016 (redrawn from Ali et al. [146]).

Its normal boiling point is −26.4 °C and if used as a hydrogen carrier will have to be transported as a liquid under pressure.

Methanol is completely miscible in water-making separation of water and methanol non-trivial, though this separation may not be needed depending on the intended use of the methanol. Water and DME are not completely miscible and show a miscibility gap [140].

Methanol, often referred to as wood alcohol, was first synthesised in 1661 by Robert Boyle via the distillation of boxwood, which involves the heating of wood in the absence of air and condensing the vapours formed. In 1926, BASF started an industrial synthesis of methanol using $ZnO\text{-}Cr_2O_3$ catalysts at 320–450 °C and 250–350 bar [141]. In the 1960s, there was further development of a $Cu/ZnO/Al_2O_3$ catalyst by ICI (now Johnson Matthey, JM) in process operating at 200–300 °C and 50–100 bar [142]. Dimethyl ether is produced either via dehydration of methanol [143] or directly in the conversion of synthesis gas [144].

5.5.1 Uses of Methanol and Dimethyl Ether

Methanol is a well-known, large volume chemical with a worldwide in excess of 90 Mio tonnes per annum [146]. It is currently used as a solvent, as a fuel or fuel additive or as a building block for a more complex chemical (an indication of its usage is given in Figure 5.65). Currently, the main usage of methanol is as a feedstock for formaldehyde, which itself is an intermediate to produce, e.g. polymers. Furthermore, methanol was used to produce octane boosters, such as methyl-t-butyl ether (MTBE) and tertiary amyl methyl ether (TAME), but the usage of, specifically, MTBE has dropped due to concerns about seepage into ground water. About 9% of all methanol produced is being carbonylated yielding acetic acid. Methanol is further used as a direct fuel additive, to produce DME, as a solvent, and other uses (primarily as a methylating agent). Green production of methanol will thus also be important for the chemical industry, and not only as a possible hydrogen carrier.

DME is mainly used as a fuel additive, added to LPG or transportation fuel, and as a propellant [122].

5.5.2 Methanol Synthesis Reactions and Equilibrium

At present, methanol is produced from fossil fuel sources, mostly coal and natural gas, which are used to form syngas. The syngas produced contains primarily CO and H_2, with smaller amounts of CO_2. There are two main pathways to synthesise methanol, either from CO or from CO_2:

$$CO(g) + 2H_2(g) \rightleftharpoons CH_3OH(g) \quad \Delta H^0_{rxn} = -90.2\,\frac{kJ}{mol\ CO}$$

$$CO_2(g) + 3H_2(g) \rightleftharpoons CH_3OH(g) + H_2O(g) \quad \Delta H^0_{rxn} = -49.0\,\frac{kJ}{mol\ CO_2}$$

or

$$\begin{cases} CO_2(g) + H_2(g) \rightleftharpoons CO(g) + H_2O(g) \quad \Delta H^0_{rxn} = 41.2\,\dfrac{kJ}{mol\ CO_2} \\ CO(g) + 2H_2(g) \rightleftharpoons CH_3OH(g) \quad \Delta H^0_{rxn} = -90.2\,\dfrac{kJ}{mol\ CO} \end{cases}$$

The pathway starting from carbon monoxide to methanol does not form water, whereas the pathway starting from CO_2 results in the formation of water either directly or indirectly via the RWGS reaction. Hence, one-third of hydrogen fed to the system is converted to water.

Methanol synthesis, whether using either CO or CO_2, is strongly equilibrium-limited, but more favoured with the product(s), methanol and DME + water, in the liquid phase. An example when using CO is shown in Figure 5.66. It is important to note here that the formation of methane had to be suppressed in the model. If the model were allowed to form methane, then methane would form preferentially over methanol. Otherwise stated, catalyst for the methanol synthesis need to suppress the formation of methane to be very selective for the formation of methanol. The equilibrium conversion of the exothermic reaction of CO to methanol decreases with increasing temperature but increases with increasing pressure. The equilibrium yield of methanol yield from CO is Y_{CH_3OH} (100 bar, 250°C) = 71 C% (same value based on hydrogen for a stoichiometric feed). Hence, the reaction is typically carried out at elevated pressures.

Similar limitations are observed when using CO_2 as the carbon source for the methanol synthesis (see Figure 5.67). However, the equilibrium conversion of CO_2

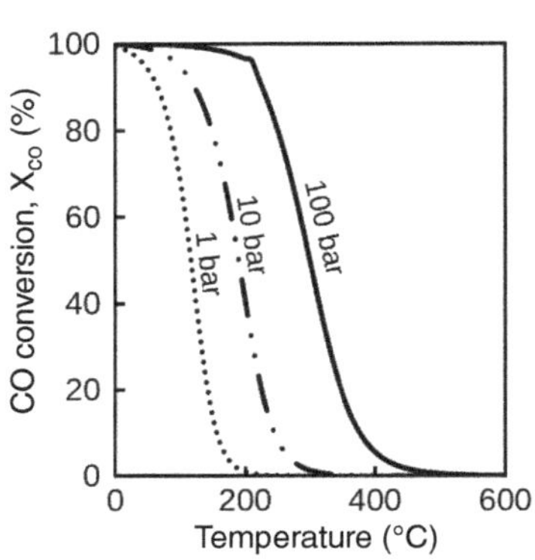

Figure 5.66 Equilibrium conversion in methanol formation from a stoichiometric feed (modelled using UNIQUAC-RK, $[H_2/CO]_{feed} = 2$).

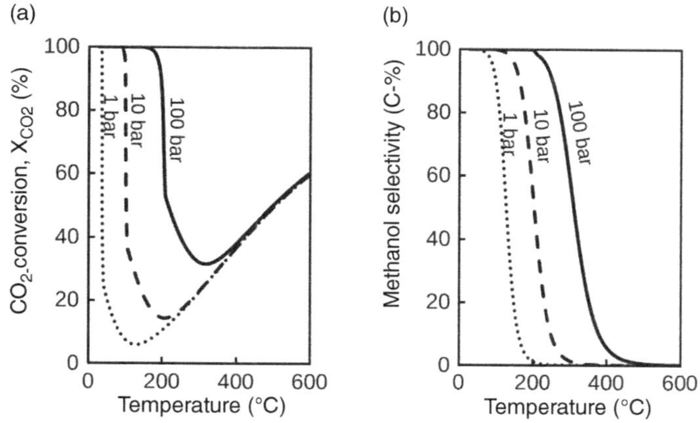

Figure 5.67 Equilibrium limitation in methanol formation from CO_2 using a stoichiometric feed. (a) Equilibrium conversion of CO_2. (b) Equilibrium selectivity for the formation of methanol (modelled using UNIQUAC-RK, $[H_2/CO_2]_{feed} = 3$).

passes now a minimum as a function of temperature. At low temperature, the methanol formation will dominate, whereas at high temperature the endothermic RWGS reaction yielding CO and H_2O starts to dominate (Chapter 5.2.1.2); there are almost two distinct regions on the conversion plot at 100 bar, a methanol synthesis region and a RWGS region. Methanol will form selectively at low temperatures, whereas CO becomes the dominant product at high temperatures. Furthermore, a discontinuity can be seen in the dependency of the CO_2 conversion as a function of temperature, which is caused by the presence of a liquid phase.

The conversion of CO_2 rather than CO limits the yield of methanol (cf. at 100 bar and 250 °C, the yield of methanol in CO_2 hydrogenation is 36 C%, and in CO hydrogenation 71 C%). Furthermore, the CO hydrogenation yielding methanol is stronger affected by the reaction pressure than the hydrogenation of CO_2 as the contraction in the latter reaction is less.

5.5.3 Methanol Synthesis Catalyst

The modern, industrially used catalyst for the methanol synthesis from CO is the one developed by ICI in the 1960s which is a copper, zinc oxide, alumina catalyst [147]. The same catalyst is not as effective for the conversion of CO_2 to methanol as high CO_2 can lead to surface oxygen and a resulting formation of Cu(II), leading to a lower catalyst activity [147]. Furthermore, the use of CO_2 as the carbon feed may lead to the formation of the undesirable methane as a side product [148]. Though modifications to the copper-zinc system are a popular area of research to try and overcome this limitation.

Experimentally it has been seen that a small amount of CO_2 in the syngas feed is beneficial for the high conversion to methanol (Figure 5.68) [148]. This has led to interesting insights into the catalytic mechanism but also provides optimism as to the performance of the catalyst from a pure CO_2/H_2 stream.

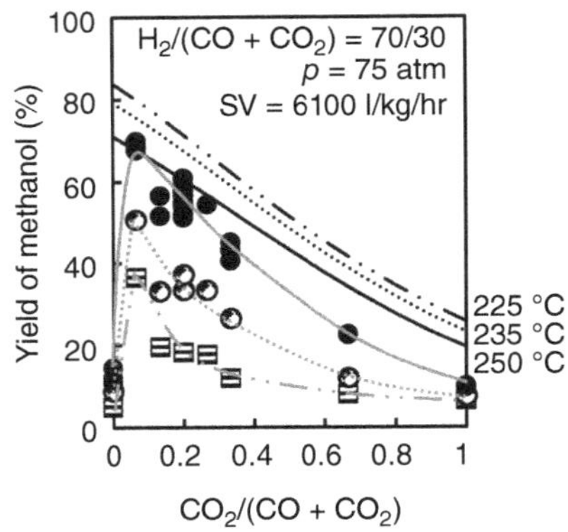

Figure 5.68 Effect of CO_2-content in syngas on the yield of methanol (black curves indicating equilibrium yield of methanol; data points and grey curves indicate the kinetically obtained yield of methanol) [148].

5.5.3.1 Methanol Production

Even with a selective catalyst, the concern is still getting to high conversion, as the exothermic process is bounded by the catalyst activity and the equilibrium limitations. The reaction is often carried out in an adiabatic reactor (see Figure 5.69) [149]. The adiabatic operation results in a temperature increase in the catalyst bed and the reaction is limited by the approach to the adiabatic temperature. Quenching the

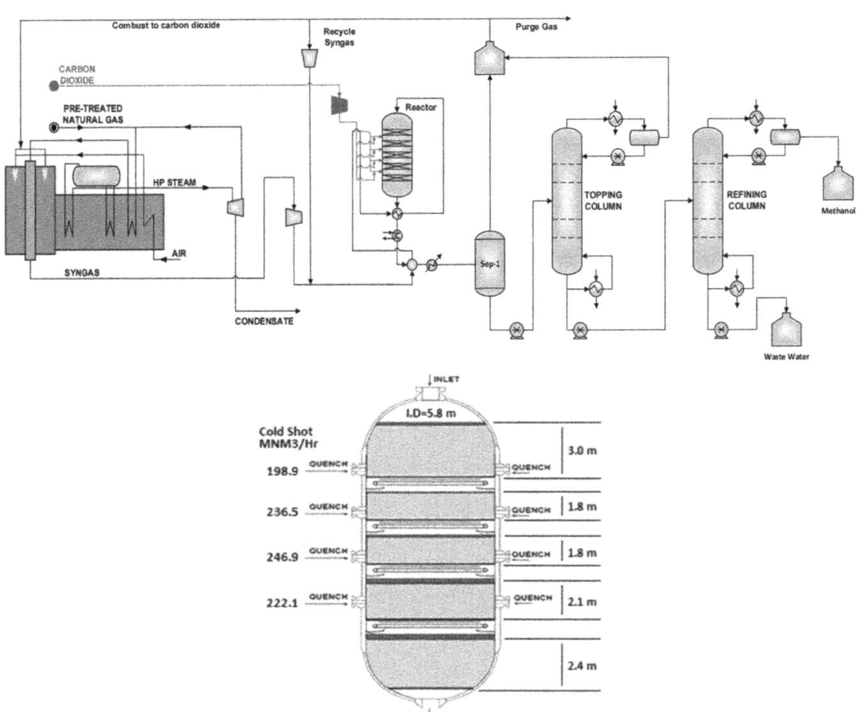

Figure 5.69 ICI process for the production of methanol from CO (Alarifi et al. [149] / with permission of American Chemical Society).

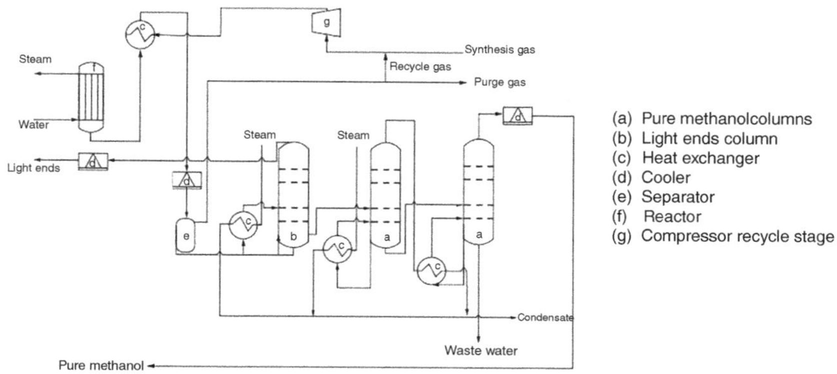

Figure 5.70 Lurgi Ol-Gas-Chemie GmbH process scheme (Palma et al. [150] / with permission of Elsevier).

product stream by adding fresh feed will reduce the temperature allowing for more conversion of the reactant. Quenching can also be achieved by removing heat in a heat exchanger. Full conversion cannot be achieved due to equilibrium limitations. Methanol is recovered via condensation and the remaining gas is recycled.

Higher conversions per pass can also be obtained by removing the heat during the reaction. In the Lurgi Ol-Gas-Chemie GmbH process [150], CO is converted into methanol in cooled reactors (Figure 5.70). The product, methanol, is again separated by condensation with recycle of the unconverted synthesis gas, which is also used as a heat exchange medium in one of the reactors.

5.5.4 Methanol from CO_2

Methanol can also be produced directly from CO_2. However, there is a very important consideration when using CO_2 instead of CO as the carbon source, as now water is a major co-product (one water molecule per molecule of methanol formed: $CO_2 + 3H_2 \rightleftharpoons CH_3OH + H_2O$). The classical methanol synthesis catalyst, $Cu/ZnO/Al_2O_3$, is sensitive to water, due to sintering, oxidation of copper [151–153] and even the formation of copper carbonate. A different catalyst is needed to operate with CO_2, which is much more water-tolerant. There are possibilities based on palladium or indium oxide as well as through modification of the copper-based catalyst (often through replacing the alumina with zirconia), but these are expensive and still in development, respectively.

The generalised process of methanol synthesis from CO_2 is similar to the classical methanol production processes (see Figure 5.71), but with some obvious differences. The feed to the reactor will contain more CO_2 and the adiabatic temperature rise in the reactor will be less. As a consequence, the necessary temperature control within the reactor to minimise catalyst sintering will be easier [154]. However, the reaction produces more water, which could be controlled by minimising the conversion per pass. However, the product stream leaving the condenser will have, besides the desired product methanol, a high concentration of the water as well as

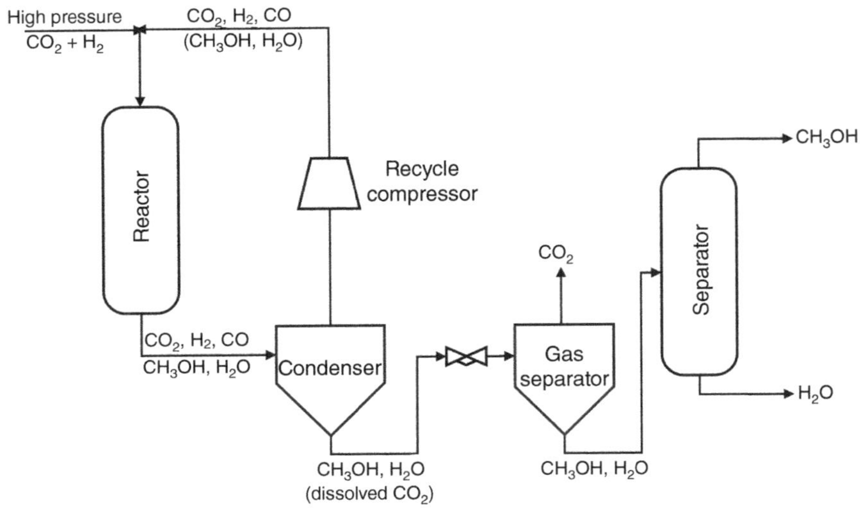

Figure 5.71 Block flow diagram for the formation of methanol from CO_2.

a significant amount of dissolved CO_2. A part of the latter will be released upon decreasing the pressure prior to further product purification (and could be fed back into the overall process). The high concentration of water in the stream prior to product purification implies that more energy is required to obtain high-purity methanol.

5.5.5 Methanol from Renewable Hydrogen

In the future the idea will be to use renewable hydrogen with a carbon source to produce methanol. This will allow for the green production of methanol but also expansion of the methanol market by using it as a hydrogen carrier.

Hydrogen could be formed via any of the processes discussed previously (see chapter 5.2.1). The carbon source could be natural gas that is sent to a reforming process to produce syngas, or preferentially CO_2 that is captured from the flue gas of an existing process or even direct air capture. The separation of the hydrogen source from the carbon source would allow a better control of the feed ratio of CO_x:H_2 and result more readily in a stoichiometric feed required for the methanol synthesis [154].

An example of a modern plant producing methanol from CO_2 and hydrogen is CRI's George Olah Renewable Methanol plant in Svartsengi, Iceland [154]. CO_2 is extracted from a neighbouring geothermal power plant whilst hydrogen is produced via alkaline water electrolysis.

The alternative argument is that this process, which has only CO_2 as the carbon feed and uses a potentially pure H_2 feed, has an overall simpler chemistry leading to an overall simpler purification process [154]. The obvious major advantage is that the process is not only a way to store electrical energy as chemical energy but can also be a way of using CO_2 that would otherwise be released to the atmosphere.

5.5.6 Dimethyl Ether Production

The energy content in methanol can be densified prior to transport by dehydrating it to DME. The specific energy of DME is 45% higher than the specific energy of methanol (28.83 MJ/kg vs. 19.92 MJ/kg). However, energy density of DME (as a liquid at 10 bar) is only ca. 19% higher than the energy density of methanol (see Table 5.12). The relatively modest increase in the energy density is caused by the relatively low density of liquid DME.

The conversion of methanol to DME is an exothermic reaction

$$2CH_3OH(g) \rightleftharpoons CH_3OCH_3(g) + H_2O(g) \quad \Delta H^0_{rxn} = -24.0\,\frac{kJ}{mol}$$

$$2CH_3OH(l) \rightleftharpoons CH_3OCH_3(l) + H_2O(l) \quad \Delta H^0_{rxn} = -14.2\,\frac{kJ}{mol}$$

which is a typical acid catalysed reaction that can be readily performed at ca. 100 °C over an ion-exchange resin [155]. DME can also be formed directly in the methanol synthesis reactor by adding an acid catalyst (e.g. γ-Al_2O_3) to the catalyst formulation [156], especially when converting a CO-rich gas into methanol.

The formation of DME can increase the equilibrium conversion (see Figure 5.72), although this drives at the same time the formation of CO_2. The latter is suppressed at high temperature as the amount of water in the reaction mixture is reduced as the formation of methanol and its dehydration are exothermic. Utilising CO_2 as the carbon source will not result in a large change of the equilibrium conversion; the formation of DME is consuming methanol, but at the same time producing water and thus, the net effect on the CO_2 conversion will become marginal.

5.5.7 Uses of Methanol and Dimethyl Ether

Methanol and DME can be easily transported and stored as a liquid (with storage and transport of DME under pressure with its saturation pressure at 25 °C of 5.96 bar), which makes it attractive as an energy carrier. The energy stored in these energy carriers can be regained by combustion, by electrochemical conversion, or by hydrogen extraction. The conversion of DME back to methanol is in principle possible, although the acid catalysed reaction is thermodynamically strongly limited at low temperatures, which are required to avoid possible side reactions.

5.5.7.1 Methanol/Dimethyl Ether Combustion

The combustion of methanol or DME is one of the methods to recoup the energy stored in this chemical. These are strongly exothermic reactions:

$$CH_3OH(l) + \frac{3}{2}O_2(g) \rightleftharpoons CO_2(g) + 2H_2O(g)$$

$$LHV = -\Delta H^0_{rxn} = 640.5 \cdot \frac{kJ}{mol} = 20.0 \cdot \frac{MJ}{kg} \rightarrow 15.5 \cdot \frac{MJ}{L}$$

$$CH_3OCH_3(l, 10\,bar) + 3O_2(g) \rightleftharpoons CO_2(g) + 2H_2O(g)$$

$$LHV = -\Delta H^0_{rxn} = 1307.2 \cdot \frac{kJ}{mol} = 28.4 \cdot \frac{MJ}{kg} \rightarrow 18.8 \cdot \frac{MJ}{L}$$

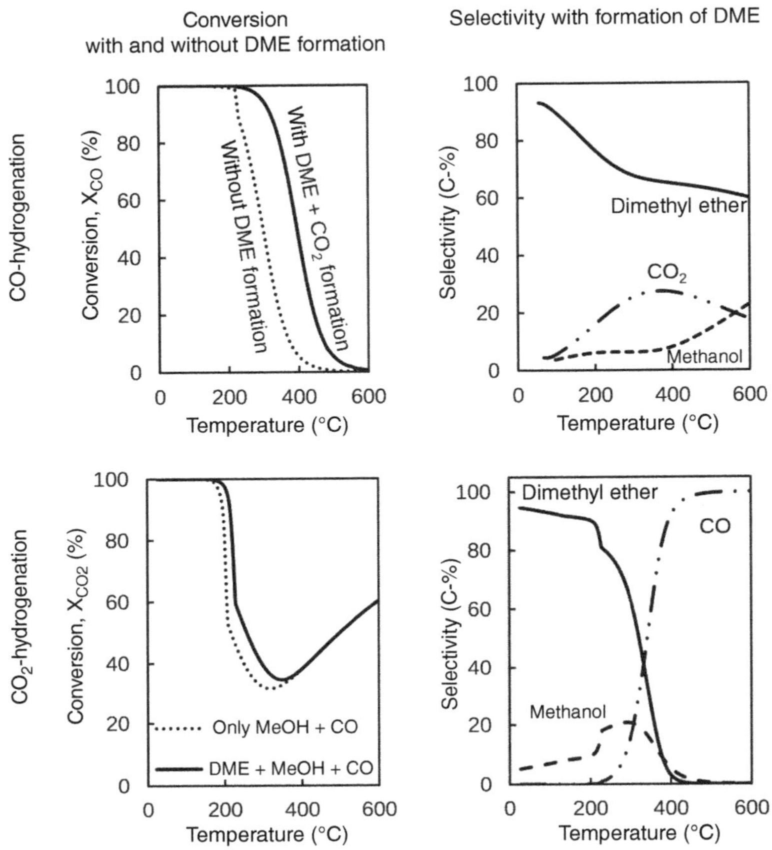

Figure 5.72 Equilibrium conversion (left) and selectivity in the CO_x hydrogenation at 100 bar with and without dimethyl ether (DME) formation (calculated using UNIQAC-SRK model).

The heat generated could be used to heat up process streams or turned into power, through a Brayton/combined cycle, if desired. The maximum temperature that can be achieved in the combustion of methanol with air is 2055 °C. The convective heat transfer in furnaces fired with methanol will be reduced as the amount of flue gas generated is less than, e.g.,with diesel (see Figure 4.14). The maximum adiabatic flame temperature of liquid DME is closer to that of diesel (2267 °C). However, the combustion gases of DME will be less effective in convective heat transfer than the exhaust gas generated by the combustion of diesel as the expansion factor is less (cf. $\delta_{\text{dimethyl ether combustion}} = 1.25$ and $\delta_{\text{diesel}} \sim 1.27$).

5.5.7.2 Direct Methanol Fuel Cell (DMFC)

Energy can be extracted from methanol in the form of electrical power using the DMFC [157]. At present, DMFCs are researched for applications with lower power requirements, such as small vehicles, personal electronic devices or portable power packs [158]. In these fuel cells, methanol is oxidised at the anode, whilst oxygen is reduced at the cathode:

Anode: $CH_3OH(l) + H_2O(l) \rightarrow 6H^+ + 6e^- + CO_2(g)$
Cathode: $\frac{3}{2}O_2(g) + 6H^+ + 6e^- \rightarrow 3H_2O(l)$
Overall: $CH_3OH(l) + \frac{3}{2}O_2 \rightarrow 2H_2O(l) + CO_2(g)$

The reaction is exothermic ($\Delta H^0_{rxn} = -727.7\,\frac{kJ}{mol}$) and a cell potential of 1.21 V ($\Delta G^0_{rxn} = -707.5\,\frac{kJ}{mol}$). Hence, the maximum efficiency of the DMFC is:

$$\eta^{max}_{DMFC} = \frac{-702.5}{-727.7} = 96.7\%$$

The oxidation of methanol is slow compared to the oxidation of hydrogen, and thus a high overpotential will be needed to obtain a certain rate of reaction, i.e. current. CO may be formed in this reaction, which may result in poisoning of the electro-catalysts with CO. Furthermore, the catalyst of the anode side may be prone to coking due to the presence of a carbon-containing feedstock. These disadvantages must be offset against the advantage of transporting energy as a liquid.

5.5.7.3 Methanol/Dimethyl Ether as Feedstock for Fuel Production

Methanol and DME can be upgraded yielding a variety of hydrocarbons. This route can be used as an alternative to the Fischer-Tropsch process for fuel production [159]. This would allow the production of different fuels that may have different energy carrying ability, such as aviation fuel.

Methanol and DME may be readily upgraded to short olefins, long paraffins, or aromatics. The conversion of methanol and DME are equivalent here, as the conversion of methanol in these products proceeds via DME [160]. The latter is converted into initially short-chain olefins, which can be subsequently converted into an olefin-pool. Cyclisation results in the formation of naphthenes, and hydrogen transfer reactions subsequently yield aromatics and paraffins [161]. These types of reactions are typically catalysed using zeolites. Due to the step wise nature of this reaction, it can be possible to produce a given product range selectively, as such the 'methanol to olefins' (MTO) and 'methanol to gasoline' (MTG) processes, which have both been previously industrialised [162]. Due to the 'hard to de-fossilise' nature of the aviation sector it may be expected that 'methanol to jet' (MTJ) may be a process that will be seen in the near future [163].

5.5.7.4 Conversion of Methanol/Dimethyl Ether Back to Hydrogen

Methanol/dimethyl ether can be converted back into hydrogen. There are four primary options for this, which are similar to those used for methane conversion to hydrogen, viz. steam reforming, partial oxidation, ATR and decomposition. The steam reforming route appears the most attractive for the recovery of hydrogen, whilst the other processes may make a more useful syngas feed [164].

With the steam reforming reaction, the endothermic nature of the reactions means that a significant amount of heat is needed to recover the hydrogen. An additional potential problem is that the process may produce CO, which could be a detrimental species depending on the target use of the hydrogen. The reactions are:

$$CH_3OH(l) + H_2O(l) \rightleftharpoons CO_2 + 3H_2 \quad \Delta H^0_{rxn} = 130.7 \cdot \frac{kJ}{mol}$$

$$CH_3OH(l) \rightleftharpoons CO + 2H_2 \quad \Delta H^0_{rxn} = 127.8 \cdot \frac{kJ}{mol}$$

As with the ammonia synthesis and decomposition the equilibrium limitations of the methanol decomposition are the inverse of that of the synthesis (see Figure 5.67). Meaning that higher temperatures and lower pressures favour the decomposition of methanol into hydrogen. It should be noted that a high temperature here implies temperatures higher than ca. 150–350 °C [164], meaning that the classical copper methanol synthesis catalysts might not be usable if operating at the upper limit, but the more historically used metal oxide catalyst may be used.

5.6 Hydrocarbons to Store Energy

It is unlikely that hydrocarbons would be used as a hydrogen carrier, but rather as an energy carrier and feedstock to the chemical industry, where it may replace the fossil fuel sources used at present. Hydrocarbons is a family name of molecules containing carbon and hydrogen, irrespective of internal branch structure, so include alkanes, iso-alkanes, alkenes, aromatics, naphthenes, etc. Fuels based on hydrocarbons are generally classified according to boiling point (see Figure 5.73), although other characteristics do play a role when used in internal combustion engines. Petrol or gasoline has a typical boiling point range between 35 and 200 °C, whereas kerosene and diesel have a boiling point range between 200 and 300 °C and 200 and 350 °C, respectively. The heat of combustion of liquid hydrocarbons is typically in the range 40–45 MJ/kg, with paraffins, olefins and naphthenes in the higher range, and aromatics in the lower range.

In theory, hydrocarbons can be produced from renewable hydrogen using either CO or CO_2 as the carbon source. When producing hydrocarbons from CO, each addition of one carbon atom to the chain results in one molecule of water is being formed using up hydrogen. Overall:

$$CO + \left(2 + \frac{1}{n}\right) H_2 \rightarrow \frac{1}{n} C_nH_{2n+2} + H_2O$$

The reaction of CO with hydrogen to form long-chain hydrocarbons is relatively well established and is called the Fischer-Tropsch (FT) synthesis, or Fischer-Tropsch process.

When producing hydrocarbons from CO_2, two molecules of water are formed for each addition of one carbon atom to the chain. The overall reaction is:

$$CO_2 + \left(3 + \frac{1}{n}\right) H_2 \rightleftharpoons \frac{1}{n} C_nH_{2n+2} + 2H_2O$$

Production from of hydrocarbons form CO is more hydrogen efficient, but it is important to remember that it is likely that CO will be formed from CO_2 and

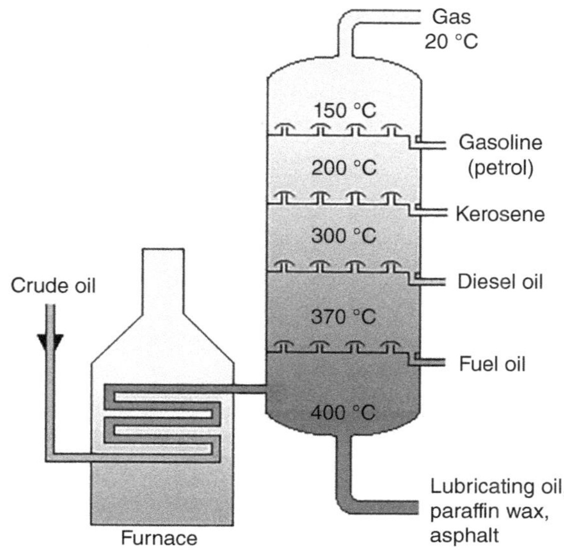

Figure 5.73 Fractionated distillation of crude oil as an indication of the temperature range of various fuels (Knott [165] / Wikimedia Commons / CC BY SA 3.0).

hydrogen in the RWGS reaction net making the overall water requirement the same as that of using the CO_2 directly.

5.6.1 Fischer-Tropsch Synthesis

Formation of hydrocarbons from CO and hydrogen was developed by BASF in the 1910s. It was patented in Germany in 1913, and in the USA in 1916. However, the development got interrupted by World War I. Frans Fischer, looking at the patent, developed the so-called Fischer-Tropsch (FT) process with Hans Tropsch, patented in 1925. They made multiple improvements in both performance and understanding of the process.

The revival of Fischer-Tropsch came with the formation of SASOL (South African Coal (Steenkool), Oil (Olie), and Gas Corporation) in South Africa which was founded in 1950. The first plant was SASOL I, which came online in 1955 in Sasolburg, which still produces up to 6500 barrels per day. At the time, SASOL I was the only working FT plant in the world. This was followed by SASOL II & III in 1980 & 1984 (in Secunda), with a total production of 160 000 barrels per day. This was prompted by the energy crisis of the 1970s which also spurred investment in FT around the world producing synthetic fuels from coal or natural gas. This includes multiple currently operation plants in South Africa, China, Qatar, Nigeria, Uzbekistan and Malaysia.

Due to the agnostic nature of the Fischer-Tropsch synthesis regarding the source of the synthesis gas, the process is ideally suited to use synthesis gas originating from a renewable source, e.g. biomass or CO_2 (either from a point source or direct air capture) combined with renewable hydrogen, which could, after the RWGS reaction yield the required synthesis gas. Especially, CO_2 from direct air capture in conjunction with renewable hydrogen would the introduction of renewable energy in the existing hydrocarbon infrastructure.

5.6.1.1 FT catalysts

Since the early developments of the Fischer-Tropsch (FT) synthesis, it has been generally accepted that there are only four metals which show FT activity, viz. iron, cobalt, nickel and ruthenium (Table 5.13) [166]. Nickel is very active but shows very strong hydrogenation activity leading to it being a prominent methanation catalyst. Nickel can also form carbonyls leading to metal loss from the catalyst surface particularly at conditions where long-chain hydrocarbons are formed. Hence, nickel is not used industrially as a catalyst for the Fischer-Tropsch synthesis but can be used as a methanation catalyst operating at higher temperatures. Ruthenium performs very well for FT, with both high activity and good selectivity, however, it is comparatively very expensive, and the performance gain does not warrant the additional cost. For these reasons, iron and cobalt are the two base materials used for industrial FT.

Cobalt has become popular in recent times for use in the so-called low temperature FT, or LTFT. This is standard for the generation of very long-chain hydrocarbons, such as waxes. Cobalt is more suitable for this than iron as it has a higher activity at elevated water partial pressures and can thus be operated at higher conversion per pass. Furthermore, cobalt catalysts do not produce CO_2 as a side product, in contrast to iron-based catalysts. This significantly decreases the carbon emissions from the process making it more desirable in particular when using hydrogen-rich synthesis gas.

Since the development of cobalt catalysts, the iron catalysts have predominantly been used for the so-called high temperature FT, or HTFT and the so-called medium temperature FT, or MTFT, converting synthesis gas from coal. HTFT is used largely for production of gasoline/petrol type species. The higher temperatures (300–350 °C for HTFT and ca. 280 °C for MTFT) allow a high activity and thus a high throughput, though this will limit the formation of long-chain hydrocarbons.

Table 5.13 Catalysts active for the Fischer-Tropsch synthesis.

	Iron	Cobalt	Nickel	Ruthenium
Active phase(s)	Fe_5C_2 & Fe_3C Fe_3O_4 – (R)WGS	Metallic	Metallic	Metallic
Price[a)]	1	220	120	33 000
Industrially used	Yes	Yes	Methanation	No
Promoters	Alkali (K/Na) Cu/Zn	Pt Mn	—	—
Temperature range (°C)	220–350	180–220	300–400 (methanation)	120–220

a) Price relative to iron.

5.6.1.2 Reactions in the Fischer-Tropsch Synthesis

There are several reactions that take place within an FT reactor that one should be aware of, the first and most important of the formation of hydrocarbons. Carbon monoxide and hydrogen will dissociate on the catalyst surface, recombine and undergo carbon–carbon coupling.

$$\text{Fischer-Tropsch: } CO + 2H_2 \rightarrow (-CH_2-) + H_2O \quad \Delta H_{rxn}^{240\,°C} = -153.7\ \frac{kJ}{mol}$$

However, it is worth noting that the CO can also react with the water formed in the Fischer-Tropsch synthesis and regenerate some hydrogen in the WGS reaction producing CO_2 as a co-product. This is particularly relevant over the iron-based catalyst, with the magnetite phase (Fe_3O_4) thought to be active for the WGS reaction [167], especially at high temperature. Thus, iron-based catalysts are less sensitive to the H_2/CO ratio than cobalt, and can even treat synthesis gas with H_2/CO_2 in theory as low as one half. This comes at the cost of producing CO_2 as a by-product of the reaction. Cobalt is almost inactive for water-gas shift reaction and thus requires an overall ratio of $H_2/CO \sim 2$, the stoichiometric ratio for the long-chain hydrocarbons.

$$\text{Water gas shift: } CO + H_2O \leftrightarrow CO_2 + H_2 \quad \Delta H_{rxn}^{240\,°C} = -39.7\ \frac{kJ}{mol}$$

The formation and consumption of water over an iron-based catalyst as a function of conversion is shown in Figure 5.74.

The Boudouard reaction can also be an important consideration, as it results in the deposition of elemental carbon on surface of the catalyst and is a notable mode of catalyst deactivation in all reactions that have CO as a feedstock or product. This elemental carbon deposition is, however, different from carbonaceous coking. The equilibrium for the Boudouard reaction is shown in Figure 5.75. Lower temperatures favour the forward reaction and hence the formation of CO_2 and carbon. At a higher temperature of operation, less CO_2 is formed, and the less relevant the Boudouard reaction becomes (or otherwise stated less carbon is formed). The rate of carbon deposition can be accelerated by the presence of specific compounds in the feed,

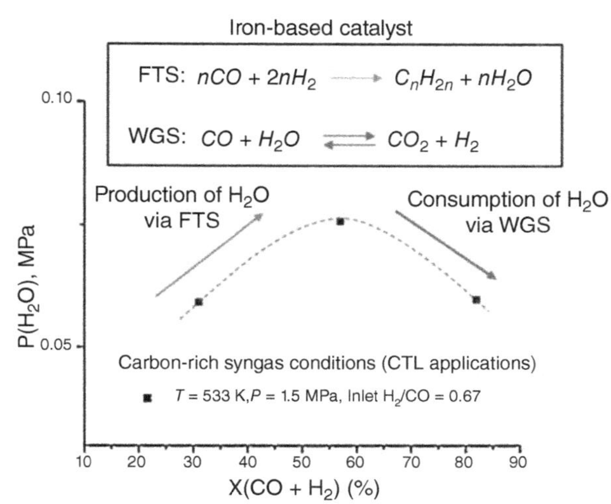

Figure 5.74 Side-reactions to chain growth: Water gas shift reaction (Bukur et al. [168] / with permission of Elsevier).

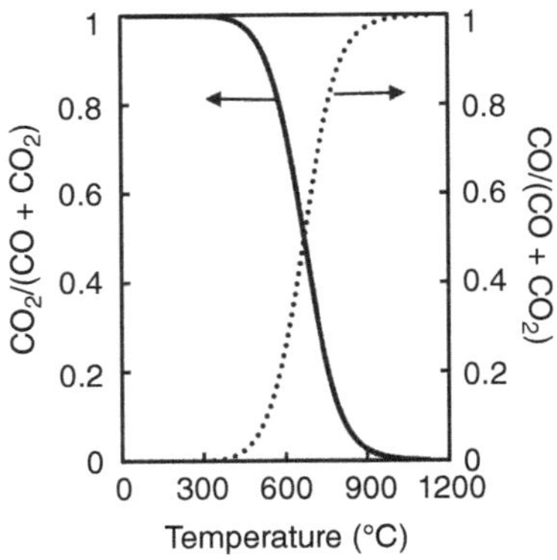

Figure 5.75 Equilibrium of the Boudouard reaction at 1 bar.

For instance, it has been shown that the addition water, hydrogen or ethanol, may increase the deposition rate of carbon whilst the addition of ammonia to the feed will decrease the rate of carbon formation [169].

$$\text{Boudouard: } 2\text{CO} \leftrightarrow \text{CO}_2 + \text{C} \quad \Delta H_{\text{rxn},513\ \text{K}} = -172.9\ \frac{\text{kJ}}{\text{mol}}$$

The overall objective is to form longer chain hydrocarbons; however, the uncertainty is in how this may best be achieved. The exact mechanism for the FT surface reaction is a matter of much debate and is dependent on both the catalyst system used and reaction conditions. The Fischer-Tropsch synthesis is essentially a polymerisation reaction, with the molecules being formed from a combination of hydrogen and CO. Of the proposed mechanisms, there are two that have most commonly been used, namely the alkyl mechanism, also known as the carbide mechanism, and the CO-insertion mechanism. The fundamental difference in these two mechanisms is the point at which the C-O bond is dissociated, viz. prior to insertion or after insertion into a growing chain [170]. In both cases, there are three general steps in the reaction, i.e. formation of the chain initiator, followed by chain growth (propagation) and, finally, growth termination (desorption). The rate of termination is (ideally) inhibited, which will favour the formation of long-chain hydrocarbons.

5.6.1.3 FT Reaction Kinetics

The rate equation for FT over iron-based catalysts may be given as follows [171]:

$$r = \frac{\text{mP}_{\text{H}_2}\text{P}_{\text{CO}}}{(\text{P}_{\text{CO}} + \text{aP}_{\text{H}_2\text{O}})}$$

This accounts for most experimental observations, namely that the rate is strongly dependent on the hydrogen partial pressure but also on the partial pressure of CO. At low conversion levels, this rate equation will simplify to a first order expression in hydrogen partial pressure. At high conversion levels, the increasing partial pressure of water will slow down the rate of reaction. The partial pressure of the products, i.e. hydrocarbons, do not affect the reaction rate.

The WGS reaction, which close to equilibrium at HTFT (>300 °C) operating conditions, will affect the species partial pressures and, therefore the observed reaction rate [172]. At lower temperatures, the WGS will occur concurrently to the FT and may be modelled as first order in CO partial pressure:

$$r_{\mathrm{WGS}} = \mathrm{cP_{CO}}$$

The rate equation for FT over cobalt catalysts is less clear, with various options having been proposed [171]. One example of a typical rate expression is that obtained from macro kinetic studies at Sasol for a Co (20 wt.%) and Pt (0.05 wt.%) supported on γ-alumina [173]:

$$r_{\mathrm{FT-Co}} = A\frac{\mathrm{P_{H_2}^{0.75}P_{CO}^{0.5}}}{\left(1 + k_{\mathrm{CO}}\mathrm{P_{CO}^{0.5}}\right)^2}$$

The key difference between the rate of formation of organic product compounds over iron-based and over cobalt-based catalysts is the inhibition of the rate over iron-based catalysts by water, the major product in the Fischer-Tropsch synthesis. This means that operating iron-based catalysts at elevated conversion is strongly kinetically hindered in comparison to cobalt-based catalysts. When renewable hydrogen is utilised, it may be expected that much of the cost will come from said renewable hydrogen. Thus, it will be imperative to utilise as much hydrogen as possible, meaning high single pass conversions may be of great value, minimising the need for recycle.

5.6.1.4 Product Distribution in the Fischer-Tropsch Synthesis

The Fischer-Tropsch synthesis is a polymerisation reaction [170], and its product distribution can be approximated by the Anderson-Schulz-Flory (ASF) distribution, which is given by the equation:

$$\log\frac{W_{N\mathrm{c}}}{N_{\mathrm{c}}} = N_{\mathrm{c}} \cdot \log\alpha + \log\frac{1-\alpha}{\alpha}$$

where N_{C} is the number of carbon atoms in the chain, W is the weight fraction of species with the chain length N_{C}, α is the chain growth probability defined as the ratio of the rate of chain growth over the rate of chain growth and chain termination. It indicates how likely a chain is to get longer, rather than desorb from the catalyst surface.

The ideal product distribution in the various carbon number fractions as given by the Anderson-Schulz-Flory distribution is shown in Figure 5.76. Looking at the distribution plot, it is important to note that a single product, except for methane and polymethylene, cannot be made selectively, e.g. it would not be possible to produce only C_6. It is, however, possible to shift the product distribution by changing the chain growth probability, e.g. by changing the operating conditions. In the low temperature Fischer-Tropsch synthesis, a chain growth of larger than 0.9 (over iron or cobalt) is typically targeted. This will minimise the formation of the undesired, light products, such as methane, and maximise formation of long-chain products, which can be selectively cracked down to the desired chain length. In the high-temperature Fischer-Tropsch synthesis (HTFT), the chain growth is typically at ca. 0.7 to maximise the yield of products in the gasoline range (the reaction is typically performed

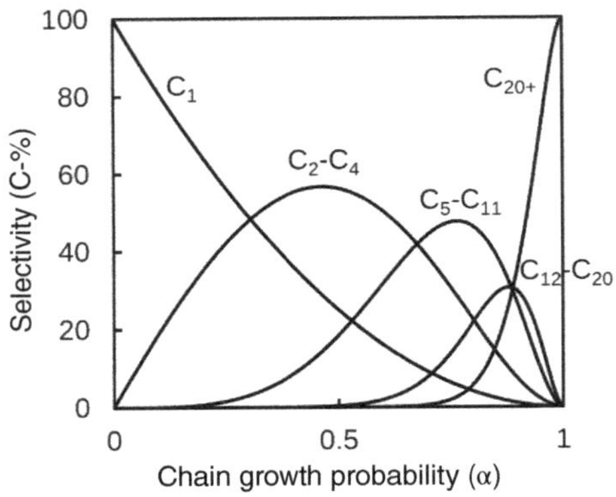

Figure 5.76 Product distribution in the Fischer-Tropsch synthesis according to the ideal Anderson-Schulz-Flory (ASF) distribution.

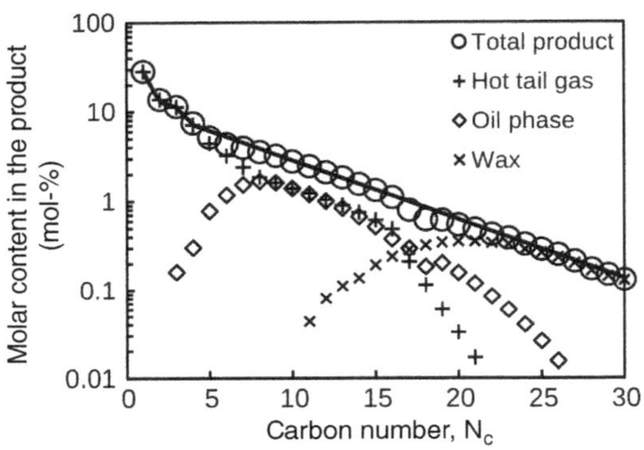

Figure 5.77 Experimental product distribution in the various product fractions in an ASF-plot obtained in the iron-based Fischer-Tropsch synthesis over an ARGE-type iron-based catalyst activated in synthesis gas at 255 °C and 15 bar and operated at 245 °C and 27 bar (adapted from Grobler [175]).

in a fluidised bed reactor and a chain growth probability of slightly less than 0.7 is necessary to avoid formation of liquid products under reaction conditions [174] and thus maintain fluidised bed conditions).

The real product distribution of the Fischer-Tropsch synthesis show some notable deviations from the ideal product distribution described by the Anderson-Schulz-Flory distribution (ASF; see Figure 5.77) Firstly, more methane is typically produced than what is predicted by the ASF model. The formation of methane does not require C–C coupling, and its formation is strictly speaking not a product of the polymerisation reaction, but rather a side reaction of the chain starter, which needs to be minimised. Next to this, there is usually less C_2 than predicted as the major C_2-product, ethene, is very reactive [170], and reacts readily further, e.g. by re-incorporation into the chain growth scheme. Finally, chain growth probability increases with increasing carbon number, which is also attributed to re-incorporation of the reactive product compounds, such as olefins into the chain growth process [170].

The product distribution and the selectivity towards individual classes of product is governed by the principle of hydrogen availability on the surface [176], and thus the product distribution can be shifted by:

- Catalyst choice with more strongly hydrogenating cobalt-based catalyst resulting in a higher methane selectivity, whilst an iron-based catalyst leads to the formation of more C_2-C_4, and more branched products (as well as more CO_2).
- The selectivity for long-chain hydrocarbons is improved by inhibiting desorption, e.g. by reducing hydrogen partial pressure and/or increasing the CO partial pressure, and/or favouring re-introduction of primarily formed olefins into the chain growth scheme, and hence increasing the H_2/CO ratio will increase formation of short chain products, such as methane
- The average chain length is reduced by increasing the temperature, which can be useful, e.g. when olefins or petrol are the desired product.
- Very high conversion levels tend to increase methane and CO_2 formation.

Different fractions of the Fischer-Tropsch product can be used for different purposes (see Table 5.14). The light products can be used as synthetic natural gas or LPG for heating purposes or to generate power. The longer chain products, such as synthetic naphtha [179, 180], synthetic kerosene [181] and synthetic diesel [182],

Table 5.14 Fischer-Tropsch products and some of their applications (data from References [177, 178]).

Product name	Characteristic	Application
Synthetic natural gas (SNG)	- Mixture of C1–C2	- Electricity generation - Heating
Liquefied petroleum gas (LPG)	- Mixture of C3–C4	- Heating - Refrigerant - Propellant
Synthetic naphtha	- Mixture of linear C5–C11 hydrocarbons - Boiling point: 140–205 °C	- Raw material for ethylene and propylene production - Petroleum
Synthetic kerosene	- Linear C10–C14 hydrocarbons - Boiling point: 150–180 °C	- Raw material for the manufacture of surface-active compounds - Jet engine fuel
Synthetic diesel fuel	- Linear C11–C18 hydrocarbons - Boiling point: 180–360 °C - Middle distillate	- Transportation fuel
Lubricant oil	- C18–C44 - Boiling point: ~300 °C	- Lubricating oil for the reduction of friction, heat, and wear in motorised vehicles
Synthetic waxes	- C20–C60 - Boiling point: >360 °C - C21–C30: soft wax - C31–C60: hard wax	- Hot melt adhesives (HMA) - Printing inks and coatings - Bitumen modification - Polymer processing - Polishes and textiles

can replace fossil fuels in their applications. High value can be extracted from the product stream in the form of lubricant oils [183] and synthetic waxes.

5.6.1.5 Process and Reactor Considerations

The Fischer-Tropsch process is a complex process, and even moderate plant size of 34 000 bpd [184]) will need to be fed with a carbon-containing feed (CO or CO_2) of ca. 80–100 m^3(NTP)/s. The required, large amount of raw materials is not readily available over a prolonged time of operation, and certainly not when looking at renewable or unavoidable carbon sources and renewable hydrogen. Hence, modularised, small-scale Fischer-Tropsch units have been developed to reduce capital costs by scale-up of the number of units to be built rather than by scale-up of the size of the plants, and the reduction in the construction and engineering cost should off-set the increased cost associated with the reduced scale of the operation [185, 186]. However, modularised, small-scale Fischer-Tropsch plants do have special requirements, such as operation at a high conversion per pass [187].

The Fischer-Tropsch reactor forms the heart of the process. The central focus in the design of reactors for the Fischer-Tropsch synthesis is the efficient removal of the heat of reaction, as the Fischer-Tropsch synthesis is highly exothermic. The options for this include multi-tubular, fixed bed reactors or moving reactors, such as slurry bubble columns or fluidised beds. The multi-tubular fixed bed reactors are comprised of many small diameter tubes (depending on the catalytic activity ranging from 1″ to 2″ I.D. tubes [188]) filled with catalyst particles. Water is passed through the reactor on the shell side to remove heat. This method of heat removal means that the multi-tubular reactors are restricted by their limited heat transfer capability. In slurry reactors, heat transfer is achieved with internal cooling coils. The reactant gas flow comes in from the bottom through a liquid mixture containing suspended catalyst, the flow is used to mix the liquid phase products and catalyst particles using the movement of the gas bubbles to agitate the fluid.

High-temperature operation uses either fixed fluidised bed reactors or circulating fluid bed reactors [189]. Here, the catalyst particles are fluidised during the reaction by passing a high flow rate of the reagent gas up through the reactor. These reactors tend to be larger, capable of handling higher flow rates. The high flow rate that is needed for fluidisation of the catalyst particles also results in improved rates of heat and mass transfer, allowing for the higher reaction temperature and, in turn, higher production rates. The downside of a moving bed, however, is that the operation puts a large degree of mechanical stress on the catalyst particles, leading to catalyst degradation over time.

5.6.2 CO_2-Hydrogenation

The storage of renewable energy into hydrocarbons requires a (renewable) carbon source, which can then be combined with renewable hydrogen. CO_2 could be a readily available feedstock, e.g. from unavoidable point sources such as cement manufacturers and municipal waste, or even from the atmosphere via direct air capture (DAC). Carbon dioxide can be converted into carbon monoxide in the equilibrium-limited, high-temperature, reverse water gas reaction, but it could also directly be hydrogenated to hydrocarbons. This is sometimes referred to as CO_2-hydrogenation, or CO_2-FT. The hydrogenation of CO_2 can principally follow three pathways. Firstly, there is the so-called direct CO_2 hydrogenation where CO_2 adsorbs on the surface and goes through a series of adsorbed intermediates to then form the long hydrocarbon chains: $CO_2 + 3H_2 \rightarrow (-CH_2-) + 2H_2O$. This would occur over a single catalytic site, likely the iron carbide [190] when using iron-based catalysts, like the classical FT, though may also occur over the metallic cobalt or ruthenium catalysts (though these tend to be stronger methanation catalysts). The second possible pathway is the so-called indirect CO_2 hydrogenation, where CO_2 adsorbs and reacts to form CO in a RWGS step: $CO_2 + H_2 \leftrightarrow CO + H_2O$. The in-situ formed CO then reacts further in the classical Fischer-Tropsch (FT) synthesis. This can theoretically only be done over iron catalysts (magnetite for RWGS and carbides for FT), though a mixed catalyst could be used with a RWGS catalyst mixed with an FT catalyst, operating at the same reaction conditions. The third possible pathway is mediated by methanol. A methanol synthesis catalyst is mixed with an acid catalyst converting methanol into hydrocarbons. For example, mixed metal oxides may be used to convert CO_2 into methanol which is then dehydrated and oligomerised over a zeolite.

5.6.2.1 Limitations in the Direct CO_2-Hydrogenation to Hydrocarbons

The hydrogenation of CO, i.e. the classical Fischer-Tropsch synthesis, is a 'kinetically' controlled reaction rather than equilibrium controlled, as the equilibrium conversion is close to 100%, albeit at equilibrium with a selectivity for the formation of methane of 100%. An example of what occurs during CO_2-FT is given in Figure 5.78. Equilibrium restrictions are not seen at temperatures below 400 °C. At these conditions, all carbon fed to the system is converted to methane. At higher temperatures, the formation of methane becomes less favoured, and the formation of CO becomes more favoured, thus reducing the formation of methane and increasing the formation of CO in the CO_2 hydrogenation. Therefore, the equilibrium conversion of CO_2 passes a minimum as a function of temperature.

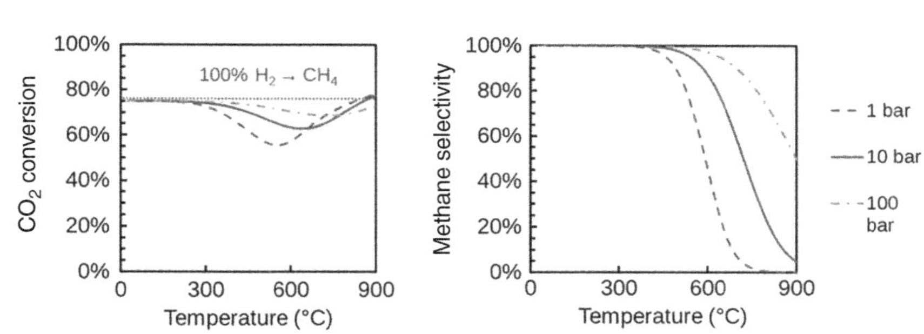

Figure 5.78 Equilibrium limitations in the hydrogenation of CO_2 to methane; modelled using SRK-UNIQUAC with inlet H_2/CO_2 ratio = 3 at 20 bar (adapted from Featherstone and van Steen [191]).

CO_2-hydrogenation for the formation of long-chain hydrocarbons compares unfavourably with the Fischer-Tropsch synthesis, as the selectivity for long-chain hydrocarbons appears to be reduced and the chain growth probability being less than 0.7 with almost all reported catalyst systems [191]. Hence, more work is still needed for the development of suitable catalysts for the direct hydrogenation of CO_2.

5.6.3 FT Process Cost and Efficiency

To conversion of renewable feedstock into hydrocarbons using the Fischer-Tropsch process is costly. It has been estimated, using 2018 economic data, that the production cost for FT fuels will break even at US$ 44 per barrel if natural gas is used as the feedstock, US$ 74 per barrel if using coal, and US$ 92 per barrel if using biomass [184]. This is a notable concern as the price of crude oil is very fluid and varied between $92.50 per barrel and $56.85 per barrel between June and December 2018 [192]. During events such as the 2008 financial crisis or the 2020 COVID-19 pandemic, there was a strong drop in crude price making PtL processes less profitable, though these drops are usually seen by almost equally strong recoveries. Alternatively, events such as the 2022 Russian invasion of Ukraine saw a spike in oil price, thus increasing the operating margin for the PtL market.

There are several parts that contribute to the overall cost of a Fischer-Tropsch plant. One of the more costly is the production and purification (notably the removal of nitrogen in air separation units as well as desulphurisation) of the syngas, highlighting the reason that is important to have the correct ratio of hydrogen to CO in the feed stream, i.e. what will be required in the FT reactor [184, 193]. The high syngas cost also emphasises why it is important that the syngas is used in the best way possible, giving the most desired and valuable FT products.

The initial capital investment for any FT plant is substantial. Table 5.15 shows the initial capital cost of some commercial FT plants normalised with respect to design

Table 5.15 Initial capital cost for some of the commercial Fischer-Tropsch processes (2018 US$) (data from References [184, 194]).

Operation	Capacity (bbl/day)	Initial capital cost (2018, 10^9 US$)	Normalised capital cost (2018, US$/bpd)
Sasol – Secunda (CTL)	160 000	15.63	98 000
MTG-New Zealand	14 750	3.44	238 000
Mossgas (PetroSA – GTL)	47 000	4.12	88 000
Oryx (Qatar)	34 000	1.30	38 000
Pearl (Qater)	260 000 (total liquid fuels, 140 000 from FT)	20.48	79 000
Escavros (Nigeria)	34 000	10.15	239 000
Uzbekistan	38 000		
Synfuels China	>150 000 (est)		

capacity and with the initial capital cost corrected for inflation using headline inflation data to 2018 values.

It is challenging to compare the conclusions of different sources on the efficiency of the Fischer-Tropsch process, primarily because the definition of process efficiency is not standardised across the literature. In particular, the system boundaries used for the calculation of efficiency are varied. The carbon efficiency of Fischer-Tropsch processes is between 25% and 50%, whilst the thermal efficiency is between 50% and 60%, depending on the feed source, i.e. whether coal or gas is used [180, 195]. The overall efficiency is also dependent on how effectively the heat integration on the plant is performed [196].

In the production of syngas, the carbon efficiency depends largely on the feedstock. If methane reforming is used, approximately 20% of the carbon is lost to CO_2. In contrast, 50% of the carbon is lost when coal gasification is employed [193].

When using alternative feedstocks, there are more factors that need to be considered when calculating efficiency. For example, in plants using biomass, the efficiency is also dependent on the moisture content of the feedstock, with biomass to fuel efficiencies of 18–33% [196].

5.7 Outlook

There are many options to store renewable energy, and chemicals are attractive as the specific energy and the energy density are much higher than e.g. batteries, thus allowing for energy transportation, and the loss of the energy content with time can be minimised allowing for energy storage over long periods of time necessary to deal with seasonal fluctuations in the supply of renewable energy.

There are a few chemicals which may be used for the storage and transport of energy. The primary one, and the one that forms the platform for other chemical synthesis is hydrogen. Hydrogen has a high specific energy of 120 MJ/kg and may be synthesised from renewable energy, through water electrolysis, as well as from other carbon-neutral means, through fossil fuel reforming coupled with carbon capture and storage technology. The ability to generate hydrogen from renewable electricity has sparked much interest as a method of having long-term storage. Unlike batteries, which self-discharge over time, the calorific value of hydrogen will not change over any time span (although one could argue some inevitable leakage from storage would result in some loss of energy over time). However, hydrogen faces a similar challenge to those of the other forms of stored renewable energy, that is, difficulty to transport. This is due to the low energy density of hydrogen (Figure 5.2), thus, requiring large volumes of hydrogen required to be transported.

Generation and utilisation of hydrogen is not a new concept by any means, with large existing infrastructure existing in coal gasification, methane (steam and autothermal) reforming, as well as from hydrocarbon cracking. Most of the use of hydrogen at current goes into the synthesis of ammonia, ending up as fertiliser, though a large amount is also used in refinery operations, in synthesis of methanol, and CTL and GTL plants. The problem with the currently produced hydrogen is its use of fossil resources and thus the associated CO_2 release from its production. Whilst CCS technologies can and will have to be implemented in the future to minimise the carbon impact, the generation of hydrogen from fossil fuels only makes sense when hydrogen is utilised on site, as fossil fuels are easier to transport than the hydrogen.

The current major limitation on the production of green hydrogen (and thus green chemicals) is the cost relative to that of the other forms of producing hydrogen. Green hydrogen is currently estimated at ca. US$ 5/kg H_2 whilst grey forms of producing hydrogen can get as low as US$ 1/kg H_2 with the incorporation of CCS to make blue hydrogen still down at ca. US$ 2/kg H_2. The high cost for green hydrogen is mainly attributable to the cost of electricity required to produce green hydrogen. Hence, the price for green hydrogen must be reduced by further reduction of the cost of renewable power generation. This will be assisted by improving the energetic efficiency of the water electrolysis process.

Green hydrogen is an inherently different product from grey hydrogen. Green hydrogen will be a renewable energy carrier, a renewable reductant and a renewable chemical feedstock servicing a new, upcoming, market. The learning curve of renewable hydrogen as well as carbon taxation on the fossil sources will pull the prices together overtime. As such, thinking of the green and grey hydrogen as direct competitors may be unfair to the emerging green hydrogen, and indeed green chemicals market. Green hydrogen should in the first place compete with blue hydrogen, where the cost of rigorous CO_2-capture and long-term storage has been incorporated [197].

Green hydrogen as an energy vector is not readily transportable specifically over long distances, and hence hydrogen carriers are needed, which inherently are then energy carriers as well. There are principally four options for hydrogen/energy carriers which are: ammonia, methanol/DME, liquid organic hydrogen carriers (LOHCs) and liquid hydrocarbon fuels. Each of ammonia, methanol/DME and liquid hydrocarbons, may be converted back to hydrogen, used directly as an energy source/fuel, or used as a green chemical. LOHCs can only operate as hydrogen vectors, due to their cost.

When considering which chemical is the preferred route for PtX options, it is useful to look at the associated transport costs [198]. Figure 5.79 shows the costs that would be involved in the conversion of hydrogen to and the transport of various PtX products as a function of the distance they need to be transported. When travelling short distances, the most economical answer is to simply pipe hydrogen as there is no cost required for conversion of hydrogen into a different carrier. However, hydrogen piping has a sharp gradient when increasing the distance, even when repurposing previously used natural gas pipes, reasserting the difficulty of transporting hydrogen as a gas. Conversion to any other product does not make sense for small distance transport but does for distances over ca. 4000 km, wherein it become viable to liquefy hydrogen or convert it to ammonia. Ammonia being more cost effective than liquid hydrogen throughout the distance range. Conversion to methanol, methane, or through FT does not make financial sense from a purely energy transported point of view, but it is worth remembering that these products, especially methanol and FT

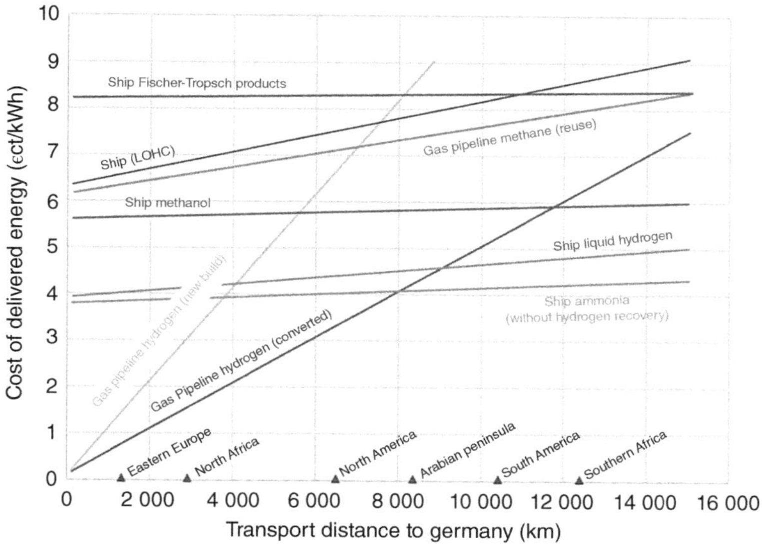

Figure 5.79 Production and transport costs of hydrogen-based energy carriers as a function of the transportation distance (adapted from Staiß et al. [198]).

products have high value use besides just as a hydrogen carrier, either for synthesis of other chemicals or to serve niche industries like aviation.

It is also worth looking at the effect of transporting the products on the hydrogen and power efficiency. Table 5.16 shows the efficiencies of various transport chains in terms of hydrogen or energy available at the point of destination [198]. The latter assumes a fuel cell efficiency of 65%.

The most efficient transport method is the pipeline transport of hydrogen as this incurs very few losses and compression of hydrogen is relatively cheap. By comparison, transporting liquid hydrogen by ship is far less efficient as it involves more losses and requires more energy input for the necessary liquefaction.

Ammonia has the highest round trip energy efficiency of 52% (see Table 5.16) for long-distance energy transport, coupled with this there is a large existing ammonia market. Around 182 million tonnes of ammonia are produced each year to service mainly the fertiliser industry meaning that there is always a 'back-up' market should it be needed in a given location. Along with comparatively cheap reaction and transportation costs, this will likely mean that in the near future green ammonia may become a dominant player in the renewable energy transportation scene.

Methanol has a slightly higher specific energy and energy density than ammonia. The existing market for methanol is also large, although only ca. half the size of the ammonia market, and it is largely used as a platform chemical for synthesis of other higher value products. This could thus provide a 'back-up' market should the production of methanol as an energy vector initially not be viable. The required carbon source in methanol may however become a limiting factor, and favouring ammonia as the future energy carrier, at least in the near future.

Table 5.16 Transport chain efficiencies[a)] (data from Staiß et al. [198]).

Method		Efficiency (%)	
		H_2	Energy
H_2 transported by pipeline (4000 km)	Compression → pipeline	94	61
Liquid H_2 transported by ship	Condenser → ship	75	49
Liquid NH_3 transported by ship	Ammonia synthesis → condenser → ship	80	52
Liquid NH_3 by ship + H_2 regeneration	Ammonia synthesis → condenser → ship → cracker	75	49
LOHC by ship	Hydrogenation → transport	68	44
Methanol	Methanol synthesis → transport	63	41
FT	FT synthesis → transport	56	37

a) Transport distance for all except the H_2 pipeline: 10 000 km.

Ammonia and methanol have an existing market, where it is not used as an energy vector. This presents a major benefit for the introduction of either green ammonia or green methanol as an energy vector as the existing market can take up these products even its initial use as an energy vector may be limited. However, it also poses a limitation. Ammonia and methanol plants scale well, meaning that non-green projects, which have been recently built, will be large, and will be productive for a long time seeing the lifespan of these plants. Hence, there may be a long transition period as grey ammonia and methanol are phased out before green ammonia and methanol can be phased in. Whilst this will likely go through a phase of blue chemicals, the cost of green will need to compete in the end with the cost of grey chemicals.

Green liquid hydrocarbons may come in three forms: methane, 'liquefied fuel gas' (LFG, the successor to LPG, and primarily propane and butane) and liquid fuels, such as diesel and aviation fuel, which are made up of long-chain hydrocarbons. In their liquid form, these energy carriers have a higher energy density than ammonia or methanol (see Figure 5.2), though they lack the 'back-up' market as their fossil equivalents are significantly cheaper. Green methane is a major point of research and may be used as a drop in fuel for systems that currently run on natural gas. Whilst methane has the potential to be transported by pipeline, liquefied methane requires very low temperatures meaning that the relatively high energy density is difficult to capitalise on. Green LFG may also be useful as a drop in fuel for current LPG purposes. Long-chain hydrocarbons will have great potential in terms of long-haul transport as well as servicing the aviation industry, both of which need high energy density fuels to minimise on-board fuel storage space. Synthesis of diesel and kerosene via Fischer-Tropsch (FT) or potentially, e.g. methanol-to-jet (MtJ) processes will therefore be valuable going forward. Methanol will have great value as a general platform chemical and liquid fuels, such as diesel and kerosene (long-chain hydrocarbons), will remain of relatively low value during the phase out period of crude oil. It may be expected that FT pathway will dominate this area (with carbon coming from e.g. biogas and municipal waste), as the MtJ process will convert the currently high-value methanol into a relatively low-value liquid hydrocarbon.

A limitation of using these carbon-based fuels is the release of CO_2 from their combustion. Therefore, their synthesis either needs to be integrated with 'hard-to-mitigate' sectors such as the cement industry (to be discussed in Chapter 6) or through use of 'renewable' carbon via direct air capture (DAC). The amount of carbon available from these sources is relatively small compared to the typical scale of industrial methanol synthesis and Fischer-Tropsch plants meaning that small-scale direct CO_2-hydrogenation or CO_2-FT may further show a valuable pathway for the overall process simplification.

Synthesis of renewable energy vectors will need to compete with the pre-existing market for these chemicals. Building a renewable chemical plant is a high-risk endeavour as it requires an abundant supply of green hydrogen, and thus of green electricity. Furthermore, it requires a levelised market where the increased cost of the green chemical, whether it is to be used as a fertiliser, fuel or as a platform chemical, is offset with the benefit of reduced carbon emissions.

Overall, both renewable generation of hydrogen and chemical storage of hydrogen will prove to be a valuable technology going forward, due to both its ability to function as an energy vector as well as provide the basis for various important downstream methods of further chemical use.

References

1 Qin, B., Wang, H., Liao, Y. et al. (2024). Challenges and opportunities for long-distance renewable energy transmission in China. *Sustainable Energy Technologies and Assessments* 69: 103925. https://doi.org/10.1016/j.seta.2024.103925.

2 Osman, A.I., Mehta, N., Elgarahy, A.M. et al. (2022). Hydrogen production, storage, utilisation and environmental impacts: a review. *Environmental Chemistry Letters* 20 (1): 153–188. https://doi.org/10.1007/s10311-021-01322-8.

3 Ishaq, H., Dincer, I., and Crawford, C. (2022). A review on hydrogen production and utilization: challenges and opportunities. *International Journal of Hydrogen Energy* 47 (62): 26238–26264. https://doi.org/10.1016/j.ijhydene.2021.11.149.

4 Taibi, E., Miranda, R., Vanhoudt, W. et al. (2018). *Hydrogen from Renewable Power: Technology Outlook for the Energy Transition*. Abu Dhabi: International Renewable Energy Agency.

5 Lin, J.-Y., Zhang, S., Zhang, L. et al. (2010). HCN and NH_3 formation during coal/char gasification in the presence of NO. *Environmental Science & Technology* 44 (10): 3719–3723. https://doi.org/10.1021/es1001538.

6 Kaneko, S. (2016). Integrated coal gasification combined cycle: a reality, not a dream. *Journal of Energy Engineering* 142 (2): E4015018. https://doi.org/10.1061/(ASCE)EY.1943-7897.0000312.

7 Breault, R.W. (2010). Gasification processes old and new: a basic review of the major technologies. *Energies* 3 (2): 216–240. https://doi.org/10.3390/en3020216.

8 Kodama, T., Miura, S., Shimizu, T., and Kitayama, Y. (1997). Thermochemical conversion of coal and water to CO and H_2 by a two-step redox cycle of ferrite. *Energy* 22 (11): 1019–1027. https://doi.org/10.1016/S0360-5442(97)00041-8.

9 Moreea-Taha, R. (2000). Modelling and simulation for coal gasification, CCC/42, The International Centre for Sustainable Carbon (ICSC), CCC/42 [Online]. Available: https://www.sustainable-carbon.org/wp-content/uploads/dlm_uploads/reports/gasification/modelling-and-simulation-for-coal-gasification--ccc-42.pdf (accessed 12 February 2025).

10 Lewis, E., McNaul, S., Jamieson, M. et al. (2022). Comparison of commercial, state-of-the-art, fossil-based hydrogen production technologies, DOE/NETL-2022/3241, 1862910, April 2022. doi: 10.2172/1862910.

11 Tucker, C.L. (2020). Waste to fuel: designing a cobalt based catalyst and process for once-through Fischer-Tropsch synthesis operated at high conversion [Online]. Available: http://hdl.handle.net/11427/33059 (accessed 2 March 2024).

12 Simpson, A.P. and Lutz, A.E. (2007). Exergy analysis of hydrogen production via steam methane reforming. *International Journal of Hydrogen Energy* 32 (18): 4811–4820. https://doi.org/10.1016/j.ijhydene.2007.08.025.

13 Cho, H.H., Strezov, V., and Evans, T.J. (2022). Environmental impact assessment of hydrogen production via steam methane reforming based on emissions data. *Energy Reports* 8: 13585–13595. https://doi.org/10.1016/j.egyr.2022.10.053.

14 Antonini, C., Treyer, K., Streb, A. et al. (2020). Hydrogen production from natural gas and biomethane with carbon capture and storage – a techno-environmental analysis. *Sustainable Energy & Fuels* 4 (6): 2967–2986. https://doi.org/10.1039/D0SE00222D.

15 Roussanaly, S., Anantharaman, R., and Fu, C. (2020). Low-carbon footprint hydrogen production from natural gas: a techno-economic analysis of carbon capture and storage from steam-methane reforming. *Chemical Engineering Transactions* 81: 1015–1020. https://doi.org/10.3303/CET2081170.

16 Faye, O., Szpunar, J., and Eduok, U. (2022). A critical review on the current technologies for the generation, storage, and transportation of hydrogen. *International Journal of Hydrogen Energy* 47 (29): 13771–13802. https://doi.org/10.1016/j.ijhydene.2022.02.112.

17 Assabumrungrat, S. and Laosiripojana, N. (2009). Fuels – Hydrogen production | Autothermal reforming. In: *Encyclopedia of Electrochemical Power Sources* (ed. J. Garche), 238–248. Amsterdam: Elsevier. https://doi.org/10.1016/B978-044452745-5.00296-3.

18 Oni, A.O., Anaya, K., Giwa, T. et al. (2022). Comparative assessment of blue hydrogen from steam methane reforming, autothermal reforming, and natural gas decomposition technologies for natural gas-producing regions. *Energy Conversion and Management* 254: 115245. https://doi.org/10.1016/j.enconman.2022.115245.

19 Streb, A. and Mazzotti, M. (2020). Novel adsorption process for co-production of hydrogen and CO_2 from a multicomponent stream—Part 2: Application to steam methane reforming and autothermal reforming gases. *Industrial and Engineering Chemistry Research* 59 (21): 10093–10109. https://doi.org/10.1021/acs.iecr.9b06953.

20 Mtetwa, B.L. (2022). The suitability of Ni-Co catalysts in the dry reforming reaction [Online]. Available: http://hdl.handle.net/11427/36510 (accessed 2 March 2024).

21 Alli, R.D., de Souza, P.A.L., Mohamedali, M. et al. (2023). Tri-reforming of methane for syngas production using Ni catalysts: current status and future outlook. *Catalysis Today* 407: 107–124. https://doi.org/10.1016/j.cattod.2022.02.006.

22 Soleimani, S. and Lehner, M. (2022). Tri-reforming of methane: thermodynamics, operating conditions, reactor technology and efficiency evaluation—a review. *Energies* 15: https://doi.org/10.3390/en15197159.

23 Świrk, K., Grzybek, T., and Motak, M. (2017). Tri-reforming as a process of CO_2 utilization and a novel concept of energy storage in chemical products. *E3S Web of Conferences* 14: 02038. https://doi.org/10.1051/e3sconf/20171402038.

24 Younas, M., Shafique, S., Hafeez, A. et al. (2022). An overview of hydrogen production: current status, potential, and challenges. *Fuel* 316: 123317. https://doi.org/10.1016/j.fuel.2022.123317.

25 Souza, M.M.V.M. and Schmal, M. (2003). Combination of carbon dioxide reforming and partial oxidation of methane over supported platinum catalysts. *Applied Catalysis A: General* 255 (1): 83–92. https://doi.org/10.1016/S0926-860X(03)00646-X.

26 Pham, X.-H., Ashik, U.P.M., Hayashi, J.-I. et al. (2021). Review on the catalytic tri-reforming of methane - Part II: Catalyst development. *Applied Catalysis A: General* 623: 118286. https://doi.org/10.1016/j.apcata.2021.118286.

27 LeValley, T.L., Richard, A.R., and Fan, M. (2014). The progress in water gas shift and steam reforming hydrogen production technologies – a review. *International Journal of Hydrogen Energy* 39 (30): 16983–17000. https://doi.org/10.1016/j.ijhydene.2014.08.041.

28 Fukuoka, A., Kimura, J., Oshio, T. et al. (2007). Preferential oxidation of carbon monoxide catalyzed by platinum nanoparticles in mesoporous silica. *Journal of the American Chemical Society* 129 (33): 10120–10125. https://doi.org/10.1021/ja0703123.

29 De Vasconcelos, B.R. and Lavoie, J.-M. (2019). Recent advances in power-to-X technology for the production of fuels and chemicals. *Frontiers in Chemistry* 7: 392. https://doi.org/10.3389/fchem.2019.00392.

30 Nasser, M., Megahed, T.F., Ookawara, S., and Hassan, H. (2022). A review of water electrolysis–based systems for hydrogen production using hybrid/solar/wind energy systems. *Environmental Science and Pollution Research* 29 (58): 86994–87018. https://doi.org/10.1007/s11356-022-23323-y.

31 Andrew Miller, H., Bouzek, K., Hnat, J. et al. (2020). Green hydrogen from anion exchange membrane water electrolysis: a review of recent developments in critical materials and operating conditions. *Sustainable Energy & Fuels* 4 (5): 2114–2133. https://doi.org/10.1039/C9SE01240K.

32 Nallapaneni, A and Malladi, K. (2021). Green hydrogen through electrolysis: fuelling the future | WRI INDIA [Online]. Available: https://wri-india.org/blog/green-hydrogen-through-electrolysis-fuelling-future (accessed 2 March 2024).

33 IEA (2024). Electrolysers - Energy System – IEA [Online]. Available: https://www.iea.org/energy-system/low-emission-fuels/electrolysers (accessed 2 March 2024).

34 Ursua, A., Gandia, L.M., and Sanchis, P. (2012). Hydrogen production from water electrolysis: current status and future trends. *Proceedings of the IEEE* 100 (2): 410–426. https://doi.org/10.1109/JPROC.2011.2156750.

35 Reytier, M., di Iorio, S., Chatroux, A. et al. (2015). Stack performances in high temperature steam electrolysis and co-electrolysis. *International Journal of Hydrogen Energy* 40 (35): 11370–11377. https://doi.org/10.1016/j.ijhydene.2015.04.085.

36 Wei, X., Sharma, S., Waeber, A. et al. (2024). Comparative life cycle analysis of electrolyzer technologies for hydrogen production: manufacturing and operations. *Joule* 8 (12): 3347–3372. https://doi.org/10.1016/j.joule.2024.09.007.

37 Buttler, A. and Spliethoff, H. (2018). Current status of water electrolysis for energy storage, grid balancing and sector coupling via power-to-gas and power-to-liquids: a review. *Renewable & Sustainable Energy Reviews* 82: 2440–2454. https://doi.org/10.1016/j.rser.2017.09.003.

38 Ahmed, K.W., Jang, M.J., Park, M.G. et al. (2022). Effect of components and operating conditions on the performance of PEM electrolyzers: a review. *Electrochem* 3 (4): 581–612. https://doi.org/10.3390/electrochem3040040.

39 Mo, J., Dehoff, R.R., Peter, W.H. et al. (2016). Additive manufacturing of liquid/gas diffusion layers for low-cost and high-efficiency hydrogen production. *International Journal of Hydrogen Energy* 41 (4): 3128–3135. https://doi.org/10.1016/j.ijhydene.2015.12.111.

40 Carmo, M., Fritz, D.L., Mergel, J., and Stolten, D. (2013). A comprehensive review on PEM water electrolysis. *International Journal of Hydrogen Energy* 38 (12): 4901–4934. https://doi.org/10.1016/j.ijhydene.2013.01.151.

41 Gerhardt, M.R., Pant, L.M., Bui, J.C. et al. (2021). Method—practices and pitfalls in voltage breakdown analysis of electrochemical energy-conversion systems. *Journal of the Electrochemical Society* 168 (7): 074503. https://doi.org/10.1149/1945-7111/abf061.

42 Smolinka, T., Günther, M., and Garche, J. (2011). Stand und Entwicklungspotenzial der Wasserelektrolyse zur Herstellung von Wasserstoff aus regenerativen Energien, Fraunhofer ISE, NOW-Studie [Online]. Available: https://www.now-gmbh.de/wp-content/uploads/2020/09/now-studie-wasserelektrolyse-2011.pdf.

43 Dincer, I. and Zamfirescu, C. (2016). Chapter 3 - Hydrogen production by electrical energy. In: *Sustainable Hydrogen Production* (ed. I. Dincer and C. Zamfirescu), 99–161. Elsevier. https://doi.org/10.1016/B978-0-12-801563-6.00003-0.

44 Götz, M., Lefebvre, J., Mörs, F. et al. (2016). Renewable power-to-gas: a technological and economic review. *Renewable Energy* 85: 1371–1390. https://doi.org/10.1016/j.renene.2015.07.066.

45 Bairachnyi, B.I., Zhelavs'kyi, S.G., Maizelis, A.O., and Voronina, O.V. (2017). Corrosion behavior of electrode materials in the production of hydrogen. *Materials Science* 53: 324–329.

46 Lindquist, G. and Boettcher, S. (2023). Reports from the frontier: overcoming limitations for pure-water anion-exchange-membrane electrolysis. *Electrochemical Society Interface* 32 (2): 32–36. https://doi.org/10.1149/2.F05232IF.

47 Olivier, P., Bourasseau, C., and Bouamama, B. (2017). Low-temperature electrolysis system modelling: a review. *Renewable and Sustainable Energy Reviews* 78: 280–300. https://doi.org/10.1016/j.rser.2017.03.099.

48 Nguyen, E., Olivier, P., Pera, M.-C. et al. (2024). Impacts of intermittency on low-temperature electrolysis technologies: a comprehensive review. *International Journal of Hydrogen Energy* 70: 474–492. https://doi.org/10.1016/j.ijhydene.2024.05.217.

49 Abdel Haleem, A., Huyan, J., Nagasawa, K. et al. (2022). Effects of operation and shutdown parameters and electrode materials on the reverse current phenomenon in alkaline water analyzers. *Journal of Power Sources* 535: 231454. https://doi.org/10.1016/j.jpowsour.2022.231454.

50 Schmidt, O., Gambhir, A., Staffell, I. et al. (2017). Future cost and performance of water electrolysis: an expert elicitation study. *International Journal of Hydrogen Energy* 42 (52): 30470–30492. https://doi.org/10.1016/j.ijhydene.2017.10.045.

51 Sayed-Ahmed, H., Toldy, Á.I., and Santasalo-Aarnio, A. (2024). Dynamic operation of proton exchange membrane electrolyzers—critical review. *Renewable and Sustainable Energy Reviews* 189: 113883. https://doi.org/10.1016/j.rser.2023.113883.

52 Papakonstantinou, G., Algara-Siller, G., Teschner, D. et al. (2020). Degradation study of a proton exchange membrane water electrolyzer under dynamic operation conditions. *Applied Energy* 280: 115911. https://doi.org/10.1016/j.apenergy.2020.115911.

53 Kim, Y., Jung, S.M., Kim, K.S. et al. (2022). Cathodic protection system against a reverse-current after shut-down in zero-gap alkaline water electrolysis. *JACS Au* 2 (11): 2491–2500. https://doi.org/10.1021/jacsau.2c00314.

54 Uchino, Y., Kobayashi, T., Hasegawa, S. et al. (2018). Dependence of the reverse current on the surface of electrode placed on a bipolar plate in an alkaline water electrolyzer. *Electrochemistry* 86 (3): 138–144. https://doi.org/10.5796/electrochemistry.17-00102.

55 Fujishima, A. and Honda, K. (1972). Electrochemical photolysis of water at a semiconductor electrode. *Nature* 238 (5358): 37–38. https://doi.org/10.1038/238037a0.

56 Hurum, D.C., Agrios, A.G., Gray, K.A. et al. (2003). Explaining the enhanced photocatalytic activity of Degussa P25 mixed-phase TiO_2 using EPR. *The*

Journal of Physical Chemistry. B 107 (19): 4545–4549. https://doi.org/10.1021/jp0273934.

57 Eidsvåg, H., Bentouba, S., Vajeeston, P. et al. (2021). TiO_2 as a photocatalyst for water splitting—an experimental and theoretical review. *Molecules* 26: https://doi.org/10.3390/molecules26061687.

58 Harrisankar, N. (2021). Improving hydrogen production from photocatalytic water splitting in an optical fibre monolith reactor by Schottky barrier formation and surface plasmonic resonance in PGM supported TiO_2.

59 Maeda, K. and Domen, K. (2010). Photocatalytic water splitting: recent progress and future challenges. *Journal of Physical Chemistry Letters* 1 (18): 2655–2661. https://doi.org/10.1021/jz1007966.

60 Zhao, Y., Ding, C., Zhu, J. et al. (2020). A hydrogen farm strategy for scalable solar hydrogen production with particulate photocatalysts. *Angewandte Chemie, International Edition* 59 (24): 9653–9658. https://doi.org/10.1002/anie.202001438.

61 Seyitoglu, S.S., Dincer, I., and Kilicarslan, A. (2017). Energy and exergy analyses of hydrogen production by coal gasification. *International Journal of Hydrogen Energy* 42 (4): 2592–2600. https://doi.org/10.1016/j.ijhydene.2016.08.228.

62 El-Emam, R.S., Dincer, I., and Naterer, G.F. (2012). Energy and exergy analyses of an integrated SOFC and coal gasification system. *International Journal of Hydrogen Energy* 37 (2): 1689–1697. https://doi.org/10.1016/j.ijhydene.2011.09.139.

63 Ozcan, H. and Dincer, I. (2014). Thermodynamic analysis of a combined chemical looping-based trigeneration system. *Energy Conversion and Management* 85: 477–487. https://doi.org/10.1016/j.enconman.2014.06.011.

64 Herdem, M.S., Farhad, S., Dincer, I., and Hamdullahpur, F. (2014). Thermodynamic modeling and assessment of a combined coal gasification and alkaline water electrolysis system for hydrogen production. *International Journal of Hydrogen Energy* 39 (7): 3061–3071. https://doi.org/10.1016/j.ijhydene.2013.12.068.

65 Xu, Y., Zang, G., Chen, H. et al. (2012). Co-production system of hydrogen and electricity based on coal partial gasification with CO2 capture. *International Journal of Hydrogen Energy* 37 (16): 11805–11814. https://doi.org/10.1016/j.ijhydene.2012.05.037.

66 Sun, Z., Xiang, W., and Chen, S. (2016). Sorption enhanced coal gasification for hydrogen production using a synthesized CaOMgO-molecular sieve sorbent. *International Journal of Hydrogen Energy* 41 (39): 17323–17333. https://doi.org/10.1016/j.ijhydene.2016.07.145.

67 Bicer, Y. and Dincer, I. (2015). Energy and exergy analyses of an integrated underground coal gasification with SOFC fuel cell system for multigeneration including hydrogen production. *International Journal of Hydrogen Energy* 40: 13323–13337. https://doi.org/10.1016/j.ijhydene.2015.08.023.

68 Department for Business, Energy & Industrial Strategy (2021). Hydrogen production costs 2021 [Online]. Available: www.gov.uk/government/publications/hydrogen-production-costs-2021 (accessed 7 March 2024).

69 Dogliani, P., Ruas Rego Canha, A.N., Elberry, A.M., and Thakur, J. (2024). Multi-option analytical modeling of levelized costs across various hydrogen supply chain nodes. *International Journal of Hydrogen Energy* 70: 737–755. https://doi.org/10.1016/j.ijhydene.2024.05.142.

70 Li, J., Wei, Y.-M., Liu, L. et al. (2022). The carbon footprint and cost of coal-based hydrogen production with and without carbon capture and storage technology in China. *Journal of Cleaner Production* 362: 132514. https://doi.org/10.1016/j.jclepro.2022.132514.

71 Futurecoal (2024). Hydrogen - FutureCoal [Online]. Available: https://www.futurecoal.org/sustainable-coal/hydrogen/ (accessed 9 September 2025).

72 Markusson, N. and Chalmers, H. (2013). Characterising CCS learning: the role of quantitative methods and alternative approaches. *Technological Forecasting and Social Change* 80 (7): 1409–1417. https://doi.org/10.1016/j.techfore.2011.12.010.

73 Chemanalyst (2024). Carbon black market size, growth & forecast, 2035 [Online]. Available: https://www.chemanalyst.com/industry-report/carbon-black-market-440 (accessed 11 March 2024).

74 Intratec (2020). Carbon black prices | Current and Forecast, Intratec.us [Online]. Available: https://www.intratec.us/chemical-markets/carbon-black-price (accessed 10 March 2024).

75 Da Costa Labanca, A.R. (2020). Carbon black and hydrogen production process analysis. *International Journal of Hydrogen Energy* 45 (47): 25698–25707. https://doi.org/10.1016/j.ijhydene.2020.03.081.

76 International Energy Agency (2022). Global Hydrogen Review 2022 [Online]. Available: https://www.iea.org/reports/global-hydrogen-review-2022 (accessed 10 March 2024).

77 Shokrollahi, M., Teymouri, N., Ashrafi, O. et al. (2024). Methane pyrolysis as a potential game changer for hydrogen economy: techno-economic assessment and GHG emissions. *International Journal of Hydrogen Energy* 66: 337–353. https://doi.org/10.1016/j.ijhydene.2024.04.056.

78 Reksten, A.H., Thomassen, M.S., Møller-Holst, S., and Sundseth, K. (2022). Projecting the future cost of PEM and alkaline water electrolysers; a CAPEX model including electrolyser plant size and technology development. *International Journal of Hydrogen Energy* 47 (90): 38106–38113. https://doi.org/10.1016/j.ijhydene.2022.08.306.

79 Pan, H. and Köhler, J. (2007). Technological change in energy systems: learning curves, logistic curves and input–output coefficients. *Ecological Economics* 63 (4): 749–758. https://doi.org/10.1016/j.ecolecon.2007.01.013.

80 Department of Energy (2022). Technical targets for proton exchange membrane electrolysis | Department of Energy [Online]. Available: https://www.energy.gov/eere/fuelcells/technical-targets-proton-exchange-membrane-electrolysis (accessed 2 March 2024).

81 Paltsev, S. (2015). Economics and geopolitics of natural gas: pipelines versus LNG. *2015 12th International Conference on the European Energy Market (EEM)*, May 2015, pp. 1–5. doi: 10.1109/EEM.2015.7216651.

82 Miana, M., Legorburo, R., Díez, D., and Hwang, Y.H. (2016). Calculation of boil-off rate of liquefied natural gas in Mark III tanks of ship carriers by numerical analysis. *Applied Thermal Engineering* 93: 279–296. https://doi.org/10.1016/j.applthermaleng.2015.09.112.

83 Ogden, J.M. (1999). Prospects for building a hydrogen energy infrastructure. *Annual Review of Energy and the Environment* 24 (1): 227–279. https://doi.org/10.1146/annurev.energy.24.1.227.

84 Pottier, J., Blondin, E., Garat, A., and de France, G. (1988). Large scale storage and transmission of gaseous hydrogen. *Proceedings of the 7th World Hydrogen Energy Conference*, Moscow: Pergamon, 1988, pp. 1095–1107.

85 Kuczyński, S., Łaciak, M., Olijnyk, A. et al. (2019). Thermodynamic and technical issues of hydrogen and methane-hydrogen mixtures pipeline transmission. *Energies* 12: https://doi.org/10.3390/en12030569.

86 Rumble, J. (ed.) (2024). *CRC Handbook of Chemistry and Physics*, 105e. Boca Raton London New York: CRC Press.

87 Zha, C., Liu, H., Tse, J.S., and Hemley, R.J. (2017). Melting and high *P–T* transitions of hydrogen up to 300 GPa. *Physical Review Letters* 119 (7): 075302. https://doi.org/10.1103/PhysRevLett.119.075302.

88 Aziz, M. (2021). Liquid hydrogen: a review on liquefaction, storage, transportation, and safety. *Energies* 14 (18): 5917. https://doi.org/10.3390/en14185917.

89 Zhang, T., Uratani, J., Huang, Y. et al. (2023). Hydrogen liquefaction and storage: recent progress and perspectives. *Renewable and Sustainable Energy Reviews* 176: 113204. https://doi.org/10.1016/j.rser.2023.113204.

90 Aasadnia, M. and Mehrpooya, M. (2018). Large-scale liquid hydrogen production methods and approaches: a review. *Applied Energy* 212: 57–83. https://doi.org/10.1016/j.apenergy.2017.12.033.

91 Spiegel, C. (2008). *PEM Fuel Cell Modeling and Simulation Using Matlab*. Amsterdam/Boston: Academic Press/Elsevier.

92 Guerrero Moreno, N., Cisneros Molina, M., Gervasio, D., and Pérez Robles, J.F. (2015). Approaches to polymer electrolyte membrane fuel cells (PEMFCs) and their cost. *Renewable and Sustainable Energy Reviews* 52: 897–906. https://doi.org/10.1016/j.rser.2015.07.157.

93 Wang, Y., Pang, Y., Xu, H. et al. (2022). PEM fuel cell and electrolysis cell technologies and hydrogen infrastructure development – a review. *Energy & Environmental Science* 15 (6): 2288–2328. https://doi.org/10.1039/D2EE00790H.

94 International Energy Agency (2021). *Ammonia Technology Roadmap: Towards More Sustainable Nitrogen Fertiliser Production*. OECD. https://doi.org/10.1787/f6daa4a0-en.

95 David, B., Armstrong, F., Bowen, P. et al. (2020). *Ammonia: Zero-Carbon Fertiliser, Fuel and Energy Store*. London: Royal Society.

96 Afif, A., Radenahmad, N., Cheok, Q. et al. (2016). Ammonia-fed fuel cells: a comprehensive review. *Renewable & Sustainable Energy Reviews* 60: 822–835. https://doi.org/10.1016/j.rser.2016.01.120.

97 Friedlingstein, P., O'Sullivan, M., Jones, M.W. et al. (2023). Global Carbon Budget 2023. *Earth System Science Data* 15 (12): 5301–5369. https://doi.org/10.5194/essd-15-5301-2023.

98 Harris, G.F.P. and MacDermott, P.E. (1977). Flammability and explosibility of ammonia. *IChemE Symposium Series 49*, Rugby, Warwickshire, UK, 1977, pp. 29–36.

99 Ertl, G. (1983). Primary steps in catalytic synthesis of ammonia. *Journal of Vacuum Science and Technology A* 1 (2): 1247–1253. https://doi.org/10.1116/1.572299.

100 Liu, H. (2014). Ammonia synthesis catalyst 100 years: practice, enlightenment and challenge. *Chinese Journal of Catalysis* 35 (10): 1619–1640. https://doi.org/10.1016/S1872-2067(14)60118-2.

101 Jacobsen, C.J.H., Dahl, S., Clausen, B.S. et al. (2001). Catalyst design by interpolation in the periodic table: bimetallic ammonia synthesis catalysts. *Journal of the American Chemical Society* 123 (34): 8404–8405. https://doi.org/10.1021/ja010963d.

102 Ghavam, S., Vahdati, M., Wilson, I.A.G., and Styring, P. (2021). Sustainable ammonia production processes. *Frontiers in Energy Research* 9: 580808. https://doi.org/10.3389/fenrg.2021.580808.

103 Aziz, M., Wijayanta, A.T., and Nandiyanto, A.B.D. (2020). Ammonia as effective hydrogen storage: a review on production, storage and utilization. *Energies* 13 (12): 3062. https://doi.org/10.3390/en13123062.

104 Penkuhn, M. and Tsatsaronis, G. (2017). Comparison of different ammonia synthesis loop configurations with the aid of advanced exergy analysis. *Energy* 137: 854–864. https://doi.org/10.1016/j.energy.2017.02.175.

105 Pattabathula, V. (2019). Ammonia. In: *Kirk-Othmer Encyclopedia of Chemical Technology*, 1–33. John Wiley & Sons, Ltd. https://doi.org/10.1002/0471238961.0113131503262116.a01.pub4.

106 Erfani, N., Baharudin, L., and Watson, M. (2024). Recent advances and intensifications in Haber-Bosch ammonia synthesis process. *Chemical Engineering and Processing* 204: 109962. https://doi.org/10.1016/j.cep.2024.109962.

107 Smith, C., Hill, A.K., and Torrente-Murciano, L. (2020). Current and future role of Haber–Bosch ammonia in a carbon-free energy landscape. *Energy & Environmental Science* 13 (2): 331–344. https://doi.org/10.1039/C9EE02873K.

108 Flórez-Orrego, D. and de Oliveira Junior, S. (2016). On the efficiency, exergy costs and CO_2 emission cost allocation for an integrated syngas and ammonia production plant. *Energy* 117: 341–360. https://doi.org/10.1016/j.energy.2016.05.096.

109 Flórez-Orrego, D. and de Oliveira Junior, S. (2017). Modeling and optimization of an industrial ammonia synthesis unit: An exergy approach. *Energy* 137: 234–250. https://doi.org/10.1016/j.energy.2017.06.157.

110 Smith, A.R. and Klosek, J. (2001). A review of air separation technologies and their integration with energy conversion processes. *Fuel Processing Technology* 70 (2): 115–134. https://doi.org/10.1016/S0378-3820(01)00131-X.

111 Grande, C.A. (2012). Advances in pressure swing adsorption for gas separation. *International Scholarly Research Notices* 2012 (1): 982934. https://doi.org/10.5402/2012/982934.

112 Ahmad, F.N., Sazali, N., and Othman, M.H.D. (2019). Review on development of oxygen/nitrogen separations technologies. *Journal of Advanced Research in Fluid Mechanics and Thermal Sciences* 63 (2): 249–259.

113 Trinke, P., Bensmann, B., and Hanke-Rauschenbach, R. (2017). Experimental evidence of increasing oxygen crossover with increasing current density during PEM water electrolysis. *Electrochemistry Communications* 82: 98–102. https://doi.org/10.1016/j.elecom.2017.07.018.

114 Gharahshiran, V.S. and Zheng, Y. (2024). Sustainable ammonia synthesis: an in-depth review of non-thermal plasma technologies. *Journal of Energy Chemistry* 96: 1–38. https://doi.org/10.1016/j.jechem.2024.04.018.

115 Kobayashi, H., Hayakawa, A., Somarathne, K.D.K.A., and Okafor, E.C. (2019). Science and technology of ammonia combustion. *Proceedings of the Combustion Institute* 37 (1): 109–133. https://doi.org/10.1016/j.proci.2018.09.029.

116 Zhu, Y., Curran, H.J., Girhe, S. et al. (2024). The combustion chemistry of ammonia and ammonia/hydrogen mixtures: a comprehensive chemical kinetic modeling study. *Combustion and Flame* 260: 113239. https://doi.org/10.1016/j.combustflame.2023.113239.

117 Masson-Delmotte, V., Zhai, A., Pirani, S.L. et al. (ed.) (2021). *Climate Change 2021: The Physical Science Basis. Contribution of Working Group I to the Sixth Assessment Report of the Intergovernmental Panel on Climate Change*. Cambridge, UK and New York, NY: Cambridge University Press. https://doi.org/10.1017/9781009157896.

118 Miller, J.A. and Bowman, C.T. (1989). Mechanism and modeling of nitrogen chemistry in combustion. *Progress in Energy and Combustion Science* 15 (4): 287–338. https://doi.org/10.1016/0360-1285(89)90017-8.

119 Purohit, A.L., Nalbandyan, A., Malte, P.C., and Novosselov, I.V. (2021). NNH mechanism in low-NOx hydrogen combustion: experimental and numerical analysis of formation pathways. *Fuel* 292: 120186. https://doi.org/10.1016/j.fuel.2021.120186.

120 Zhu, Z. and Xu, B. (2022). Purification technologies for NOx removal from flue gas: a review. *Separations* 9: https://doi.org/10.3390/separations9100307.

121 Deutschmann, O., Knözinger, H., Kochloefl, K., and Turek, T. (2011). Heterogeneous catalysis and solid catalysts, 3. Industrial applications. In: *Ullmann's Encyclopedia of Industrial Chemistry*. Weinheim, Germany: John Wiley & Sons, Ltd. https://doi.org/10.1002/14356007.o05_o03.

122 Kordylewski, W., Jakubiak, M., and Hardy, T. (2013). Pilot plant studies on NOx removal via NO ozonation and absorption. *Archives of Environmental Protection* 39 (3): 93–106. https://doi.org/10.2478/aep-2013-0025.

123 Lee, Y.-W., Park, J.-W., Jun, S.-J. et al. (2004). NOx adsorption–temperature programmed desorption and surface molecular ions distribution by activated carbon with chemical modification. *Carbon* 42 (1): 59–69. https://doi.org/10.1016/j.carbon.2003.09.019.

124 Valera-Medina, A., Amer-Hatem, F., Azad, A.K. et al. (2021). Review on ammonia as a potential fuel: from synthesis to economics. *Energy & Fuels* 35 (9): 6964–7029. https://doi.org/10.1021/acs.energyfuels.0c03685.

125 Page, M., Achrai, B., Tamir, G. et al. (2019). Alkaline fuel cell with a membrane directly fed with ammonia and a method for operating it, IL260880B, 28 November 2019 [Online]. Available: https://patents.google.com/patent/IL260880B/en?inventor=Ben+ACHRAI (accessed 11 March 2025).

126 Li, J., Huang, H., Kobayashi, N. et al. (2014). Study on using hydrogen and ammonia as fuels: combustion characteristics and NO formation. *International Journal of Energy Research* 38 (9): 1214–1223. https://doi.org/10.1002/er.3141.

127 Cechetto, V., Di Felice, L., Medrano, J.A. et al. (2021). H2 production via ammonia decomposition in a catalytic membrane reactor. *Fuel Processing Technology* 216: 106772. https://doi.org/10.1016/j.fuproc.2021.106772.

128 Abd Ali, Z.D., Allami, S., and Jwad, B.H. (2018). Ammonia as hydrogen storage media, sustainable method to hydrogen evolution. *Journal of Physics Conference Series* 1032: 012062. https://doi.org/10.1088/1742-6596/1032/1/012062.

129 Hank, C., Sternberg, A., Köppel, N. et al. (2020). Energy efficiency and economic assessment of imported energy carriers based on renewable electricity. *Sustainable Energy & Fuels* 4 (5): 2256–2273. https://doi.org/10.1039/D0SE00067A.

130 Rao, P. and Yoon, M. (2020). Potential liquid-organic hydrogen carrier (LOHC) systems: a review on recent progress. *Energies* 13 (22): 6040. https://doi.org/10.3390/en13226040.

131 Niermann, M., Beckendorff, A., Kaltschmitt, M., and Bonhoff, K. (2019). Liquid Organic Hydrogen Carrier (LOHC) – assessment based on chemical and economic properties. *International Journal of Hydrogen Energy* 44 (13): 6631–6654. https://doi.org/10.1016/j.ijhydene.2019.01.199.

132 Haynes, W.M., Lide, D.R., and Bruno, T.J. (ed.) (2014). *CRC Handbook of Chemistry and Physics*, 95e. Boca Raton: CRC Press. https://doi.org/10.1201/b17118.

133 Usman, M.R., Alotaibi, F.M., and Aslam, R. (2015). Dehydrogenation-hydrogenation of methylcyclohexane-toluene system on 1.0 wt% Pt/zeolite beta catalyst. *Progress in Reaction Kinetics and Mechanism* 40 (4): 353–366. https://doi.org/10.3184/146867815X14413752286029.

134 Müller, K., Stark, K., Emel'yanenko, V.N. et al. (2015). Liquid organic hydrogen carriers: thermophysical and thermochemical studies of benzyl- and dibenzyl-toluene derivatives. *Industrial and Engineering Chemistry Research* 54 (32): 7967–7976. https://doi.org/10.1021/acs.iecr.5b01840.

135 Shi, L., Qi, S., Qu, J. et al. (2019). Integration of hydrogenation and dehydrogenation based on dibenzyltoluene as liquid organic hydrogen energy carrier. *International Journal of Hydrogen Energy* 44 (11): 5345–5354. https://doi.org/10.1016/j.ijhydene.2018.09.083.

136 Park, S., Naseem, M., and Lee, S. (2021). Experimental assessment of perhydro-dibenzyltoluene dehydrogenation reaction kinetics in a continuous flow system for stable hydrogen supply. *Materials* 14 (24): 7613. https://doi.org/10.3390/ma14247613.

137 Ali, A., Rohini, A.K., Noh, Y.S. et al. (2022). Hydrogenation of dibenzyltoluene and the catalytic performance of Pt/ AL_2O_3 with various Pt loadings for hydrogen production from perhydro-dibenzyltoluene. *International Journal of Energy Research* 46 (5): 6672–6688. https://doi.org/10.1002/er.7604.

138 Modisha, P., Gqogqa, P., Garidzirai, R. et al. (2019). Evaluation of catalyst activity for release of hydrogen from liquid organic hydrogen carriers. *International Journal of Hydrogen Energy* 44 (39): 21926–21935. https://doi.org/10.1016/j.ijhydene.2019.06.212.

139 Wu, J., Liu, Z., Pan, J., and Zhao, X. (2004). Vapor pressure measurements of dimethyl ether from (233 to 399) K. *Journal of Chemical and Engineering Data* 49 (1): 32–34. https://doi.org/10.1021/je0340046.

140 Pozo, M.E. and Streett, W.B. (1984). Fluid phase equilibria for the system dimethyl ether/water from 50 to 220.degree.C and pressures to 50.9 MPa. *Journal of Chemical & Engineering Data* 29 (3): 324–329. https://doi.org/10.1021/je00037a030.

141 Alwin, M. and Mathias, P. (1926). Synthetic manufacture of methanol, US1569775A, 12 January 1926 [Online]. Available: https://patents.google.com/patent/US1569775A/en (accessed 28 December 2024).

142 Phineas, D. and Forster, S.F. (1967). Production of oxygenated hydrocarbons. US3326956A, 20 June 1967 [Online]. Available: https://patents.google.com/patent/US3326956A/en?oq=US+3326956 (accessed 28 December 2024).

143 Kazuo c/o Toyo Engineering Corporation Shoji and Satoshi c/o Toyo Engineering Corporation Terai (2004). Process for producing dimethyl ether. EP1396483A1, 10 March 2004 [Online]. Available: https://patents.google.com/patent/EP1396483A1/en (accessed 29 December 2024).

144 Fujimoto, K., Shikada, T., Yamaoka, Y., and Sumigama, T. (1995). Method of producing dimethyl ether. US5389689A, 14 February 1995 [Online]. Available: https://patents.google.com/patent/US5389689A/en (accessed 29 December 2024).

145 Wu, J., Zhou, Y., and Lemmon, E.W. (2011). An equation of state for the thermodynamic properties of dimethyl ether. *Journal of Physical and Chemical Reference Data* 40 (2): 023104. https://doi.org/10.1063/1.3582533.

146 Ali, K.A., Abdullah, A.Z., and Mohamed, A.R. (2015). Recent development in catalytic technologies for methanol synthesis from renewable sources: a critical review. *Renewable & Sustainable Energy Reviews* 44: 508–518. https://doi.org/10.1016/j.rser.2015.01.010.

147 Ott, J. (2012). Methanol. In: *Ullmann's Encyclopedia of Industrial Chemistry*, 1e (ed. V. Gronemann, F. Pontzen, et al.). Wiley-VCH. https://doi.org/10.1002/14356007.a16_465.pub3.

148 Klier, K., Chatikavanij, V., Herman, R.G., and Simmons, G.W. (1982). Catalytic synthesis of methanol from CO/H_2. *Journal of Catalysis* 74 (2): 343–360. https://doi.org/10.1016/0021-9517(82)90040-9.

149 Alarifi, A., Alsobhi, S., Elkamel, A., and Croiset, E. (2015). Multiobjective optimization of methanol synthesis loop from synthesis gas via a multibed adiabatic reactor with additional interstage CO_2 quenching. *Energy & Fuels* 29 (2): 530–537. https://doi.org/10.1021/ef502073b.

150 Palma, V., Meloni, E., Ruocco, C. et al. (2018). Chapter 2 - State of the art of conventional reactors for methanol production. In: *Methanol* (ed. A. Basile and F. Dalena), 29–51. Elsevier. https://doi.org/10.1016/B978-0-444-63903-5.00002-9.

151 Kung, H.H. (1992). Deactivation of methanol synthesis catalysts - a review. *Catalysis Today* 11 (4): 443–453. https://doi.org/10.1016/0920-5861(92)80037-N.

152 Xu, C., Yan, Z., Jian, Y. et al. (2021). Development of stable water-resistant Cu-based catalyst for methanol synthesis. *Applied Catalysis A: General* 623: 118299. https://doi.org/10.1016/j.apcata.2021.118299.

153 Prašnikar, A., Pavlišič, A., Ruiz-Zepeda, F. et al. (2019). Mechanisms of copper-based catalyst deactivation during CO_2 reduction to methanol. *Industrial and Engineering Chemistry Research* 58 (29): 13021–13029. https://doi.org/10.1021/acs.iecr.9b01898.

154 Marlin, D.S., Sarron, E., and Sigurbjörnsson, Ó. (2018). Process advantages of direct CO_2 to methanol synthesis. *Frontiers in Chemistry* 6: 446. https://doi.org/10.3389/fchem.2018.00446.

155 Gates, B.C. and Johanson, L.N. (1971). Langmuir-Hinshelwood kinetics of the dehydration of methanol catalyzed by cation exchange resin. *AIChE Journal* 17 (4): 981–983. https://doi.org/10.1002/aic.690170435.

156 Ng, K.L., Chadwick, D., and Toseland, B.A. (1999). Kinetics and modelling of dimethyl ether synthesis from synthesis gas. *Chemical Engineering Science* 54 (15): 3587–3592. https://doi.org/10.1016/S0009-2509(98)00514-4.

157 Giddey, S., Badwal, S.P.S., Kulkarni, A., and Munnings, C. (2012). A comprehensive review of direct carbon fuel cell technology. *Progress in Energy and Combustion Science* 38 (3): 360–399. https://doi.org/10.1016/j.pecs.2012.01.003.

158 Ahmed, A.A., Al Labadidi, M., Hamada, A.T., and Orhan, M.F. (2022). Design and utilization of a direct methanol fuel cell. *Membranes* 12 (12): 1266. https://doi.org/10.3390/membranes12121266.

159 Chakraborty, J.P., Singh, S., and Maity, S.K. (2022). Chapter 6 - Advances in the conversion of methanol to gasoline. In: *Hydrocarbon Biorefinery* (ed. S.K. Maity, K. Gayen, and T.K. Bhowmick), 177–200. Elsevier. https://doi.org/10.1016/B978-0-12-823306-1.00008-X.

160 Chang, C.D. (1983). Hydrocarbons from methanol. *Catalysis Reviews* 25 (1): 1–118. https://doi.org/10.1080/01614948308078874.

161 France, L.J., Edwards, P.P., Kuznetsov, V.L. et al. (2015). The indirect and direct conversion of CO_2 into higher carbon fuels. In: *Carbon Dioxide Utilisation* (ed. P. Styring, E.A. Quadrelli, and K. Armstrong), 161–182. Amsterdam: Elsevier. https://doi.org/10.1016/B978-0-444-62746-9.00010-4.

162 Tian, P., Wei, Y., Ye, M., and Liu, Z. (2015). Methanol to olefins (MTO): from fundamentals to commercialization. *ACS Catalysis* 5 (3): 1922–1938. https://doi.org/10.1021/acscatal.5b00007.

163 Eyberg, V., Dieterich, V., Bastek, S. et al. (2024). Techno-economic assessment and comparison of Fischer–Tropsch and methanol-to-jet processes to produce sustainable aviation fuel via power-to-liquid. *Energy Conversion and Management* 315: 118728. https://doi.org/10.1016/j.enconman.2024.118728.

164 Garcia, G., Arriola, E., Chen, W.-H., and De Luna, M.D. (2021). A comprehensive review of hydrogen production from methanol thermochemical conversion for sustainability. *Energy* 217: 119384. https://doi.org/10.1016/j.energy.2020.119384.

165 Knott, T. (2004). Atmospheric pressure crude oil distillation tower used in petroleum refining (oil refineries) [Online]. Available: https://commons.wikimedia.org/wiki/File:Crude_Oil_Distillation.png (accessed 29 December 2024).

166 van Steen, E. and Claeys, M. (2008). Fischer-Tropsch catalysts for the biomass-to-liquid (BTL)-process. *Chemical Engineering and Technology* 31 (5): 655–666. https://doi.org/10.1002/ceat.200800067.

167 Chou, C.-Y., Loiland, J.A., and Lobo, R.F. (2019). Reverse water-gas shift iron catalyst derived from magnetite. *Catalysts* 9: https://doi.org/10.3390/catal9090773.

168 Bukur, D.B., Todic, B., and Elbashir, N. (2016). Role of water-gas-shift reaction in Fischer–Tropsch synthesis on iron catalysts: a review. *Catalysis Today* 275: 66–75. https://doi.org/10.1016/j.cattod.2015.11.005.

169 Dry, M.E. (1984). *The Fischer-Tropsch Synthesis*, vol. 1 (ed. J.R. Anderson and M. Boudart), 159–255. Berlin: Springer-Verlag.

170 Claeys, M. and van Steen, E. (2004). Chapter 8 - Basic studies. In: *Studies in Surface Science and Catalysis*, Fischer-Tropsch Technology, vol. 152 (ed. A. Steynberg and M. Dry), 601–680. Elsevier. https://doi.org/10.1016/S0167-2991(04)80465-8.

171 Dry, M.E. (2004). FT catalysts. In: *Studies in Surface Science and Catalysis*, vol. 152, 533–600. Elsevier. https://doi.org/10.1016/S0167-2991(04)80464-6.

172 Claeys, M., Steen, E., Botha, T. et al. (2021). Oxidation of Hägg carbide during high-temperature Fischer–Tropsch synthesis: size-dependent thermodynamics and in situ observations. *ACS Catalysis* 11 (22): 13866–13879. https://doi.org/10.1021/acscatal.1c03719.

173 Botes, F.G., Niemantsverdriet, J.W., and Van De Loosdrecht, J. (2013). A comparison of cobalt and iron based slurry phase Fischer–Tropsch synthesis. *Catalysis Today* 215: 112–120. https://doi.org/10.1016/j.cattod.2013.01.013.

174 Biquiza, L.D., Claeys, M., and van Steen, E. (2010). Thermodynamic and experimental aspects of 'supercritical' Fischer–Tropsch synthesis. *Fuel Processing Technology* 91 (10): 1250–1255. https://doi.org/10.1016/j.fuproc.2010.04.005.

175 Grobler, T. (2008). Two-dimensional gas chromatography: a novel technique for Iron Low Temperature Fischer-Tropsch selectivity studies. Thesis. University of Cape Town, Cape Town.

176 Schulz, H., Steen, E.v., and Claeys, M. (1994). Selectivity and mechanism of Fischer-Tropsch synthesis with iron and cobalt catalysts. In: *Studies in Surface Science and Catalysis; Natural Gas Conversion II*, vol. 81 (ed. H.E. Curry-Hyde and R.F. Howe), 455–460. Elsevier. https://doi.org/10.1016/S0167-2991(08)63911-7.

177 Teimouri, Z., Abatzoglou, N., and Dalai, A.K. (2021). Kinetics and selectivity study of Fischer–Tropsch synthesis to C5+ hydrocarbons: a review. *Catalysts* 11 (3): 330. https://doi.org/10.3390/catal11030330.

178 Speight, J.G. (2016). 6 - Production of syngas, synfuel, bio-oils, and biogas from coal, biomass, and opportunity fuels. In: *Fuel Flexible Energy Generation* (ed. J. Oakey), 145–174. Boston: Woodhead Publishing. https://doi.org/10.1016/B978-1-78242-378-2.00006-7.

179 Dancuar, L.P., Mayer, J.F., Tallman, M.J., and Adams, J. (2003). Performance of the Sasol SPD naphtha as steam cracking feedstock. *Performance of the Sasol SPD Naphtha as Steam Cracking Feedstock* 48 (2): 132–138.

180 De Klerk, A. (2011). *Fischer-Tropsch Refining*, 1e. Wiley. https://doi.org/10.1002/9783527635603.

181 Ajam, M. and Viljoen, C. (2011). Synergies between renewable kerosene and Fischer-Tropsch Synthetic Paraffinic Kerosene (FTSPK). In: *Proceedings of the 12th International Conference on Stability, Handling and Use of Liquid Fuels* (ed. R. Morris), 144–185. Sarasota, FL: International Assn for Stability Handling and Use of Liquid Fuels (IASH).

182 Leckel, D. (2009). Diesel production from Fischer–Tropsch: the past, the present, and new concepts. *Energy & Fuels* 23 (5): 2342–2358. https://doi.org/10.1021/ef900064c.

183 Kobayashi, M., Saitoh, M., Ishida, K., and Yachi, H. (2005). Viscosity properties and molecular structure of lube base oil prepared from Fischer-Tropsch waxes. *Journal of the Japan Petroleum Institute* 48 (6): 365–372. https://doi.org/10.1627/jpi.48.365.

184 Van Steen, E., Claeys, M., Möller, K.P., and Nabaho, D. (2018). Comparing a cobalt-based catalyst with iron-based catalysts for the Fischer-Tropsch XTL-process operating at high conversion. *Applied Catalysis A: General* 549: 51–59. https://doi.org/10.1016/j.apcata.2017.09.019.

185 Sievers, S., Seifert, T., Franzen, M. et al. (2017). Fixed capital investment estimation for modular production plants. *Chemical Engineering Science* 158: 395–410. https://doi.org/10.1016/j.ces.2016.09.029.

186 Alhamouri, K.I., O'Connor, J.T., Haapala, K.R., and Paul, B.K. (2024). Conceptual range estimation for total cost of ownership of modular process-intensified chemical plants. *Journal of Advanced Manufacturing and Processing* 6 (2): https://doi.org/10.1002/amp2.10176.

187 Urban, Z. (2016). Modular and scalable Fischer-Tropsch reactor for small-scale gas-to-liquid production. *Chemical Engineering and Technology* 39 (11): 2142–2150. https://doi.org/10.1002/ceat.201600283.

188 Jess, A. and Kern, C. (2009). Modeling of multi-tubular reactors for Fischer-Tropsch synthesis. *Chemical Engineering and Technology* 32 (8): 1164–1175. https://doi.org/10.1002/ceat.200900131.

189 Steynberg, A.P., Espinoza, R.L., Jager, B., and Vosloo, A.C. (1999). High temperature Fischer–Tropsch synthesis in commercial practice. *Applied Catalysis A: General* 186 (1): 41–54. https://doi.org/10.1016/S0926-860X(99)00163-5.

190 Liu, X., Cao, C., Tian, P. et al. (2020). Resolving CO_2 activation and hydrogenation pathways over iron carbides from DFT investigation. *Journal of CO2 Utilization* 38: 10–15. https://doi.org/10.1016/j.jcou.2019.12.014.

191 Featherstone, N.S. and van Steen, E. (2023). Meta-analysis of the thermo-catalytic hydrogenation of CO_2. *Catalysis Today* 423: 113951. https://doi.org/10.1016/j.cattod.2022.11.012.

192 MacroTrends (2024). Crude oil prices - 70 year historical chart [Online]. Available: https://www.macrotrends.net/1369/crude-oil-price-history-chart (accessed 5 January 2025).

193 Dry, M.E. (2002). The Fischer–Tropsch process: 1950–2000. *Catalysis Today* 71: 227–241.

194 Moodley, D., Botha, T., Crous, R. et al. (2024). Catalysis for sustainable aviation fuels: focus on Fischer-Tropsch catalysis. In: *Catalysis for a Sustainable Environment*, 1e (ed. A.J.L. Pombeiro, M. Sutradhar, and E.C.B.A. Alegria), 73–116. Wiley. https://doi.org/10.1002/9781119870647.ch6.

195 Wikipedia (2024). Fischer–Tropsch process, 1 July 2024 [Online]. Available: https://en.wikipedia.org/w/index.php?title=Fischer%E2%80%93Tropsch_process&oldid=1231927963#Process_efficiency (accessed 8 July 2024).

196 Redifuel (2020). State-of-the-art on energy efficiency of various fuels (Jun 2020 - Month 21/40) - REDIFUEL [Online]. Available: https://redifuel.eu/state-of-the-art-jun2020/ (accessed 8 July 2024).

197 Ueckerdt, F., Verpoort, P.C., Anantharaman, R. et al. (2024). On the cost competitiveness of blue and green hydrogen. *Joule* 8 (1): 104–128. https://doi.org/10.1016/j.joule.2023.12.004.

198 Staiß, F., Adolf, J., Austelder, F. et al. (2022). Optionen für den Import grünen Wasserstoffs nach Deutschland bis zum Jahr 2030: Transportwege – Länderbewertungen – Realisierungserfordernisse, acatech - Deutsche Akademie der Technikwissenschaften. doi: https://doi.org/10.48669/ESYS_2022-6.

6

Basic Industries

6.1 Introduction to the Basic Industries

A basic industry is of fundamental economic importance in its provision of raw materials to other industries. Basic industries include, but are not limited to, cement, steel, basic chemicals, fuel, etc. The key feature is that their products feed other industries and are not necessarily used directly by the end consumers. For example, the construction industry uses cement and steel (though these materials may also be sold for domestic use in relatively small quantities). Basic industries tend to operate at very high volumes, manufacturing relatively low-value products. Although these industries are essential, their products are very much down the value chain, before they will eventually reach the end consumer.

This chapter will discuss several basic industries: cement and steel, and three base chemicals: sulphuric acid, sodium hydroxide and hydrochloric acid, with the latter two typically co-produced via the 'chlor-alkali' process. The high-volume production in these industries yields a rather high CO_2 output, even when the CO_2 per unit product is not necessarily high. However, in some of these industries, particularly cement and steel, the amount of CO_2 produced per unit product is high as is the total volume output. Hence, these two industries are responsible for ca. 6% and 8% of global CO_2 emissions, respectively [1].

A list of materials by global production is given in Table 6.1. Every effort is made to include all significantly produced materials with recent values, but some may be missed, and others may be left out in favour of the chemical they are used to produce, e.g. nitrogen being used to produce ammonia. Also listed is a non-exhaustive list of the industries discussed throughout this chapter and in the following two chapters, as well as the hydrogen and ammonia production discussed in Chapter 5.

The sand and gravel industries are by far the largest industries in terms of tonnage produced. These materials feed into the construction industry and into the cement industry. Coal, oil and gas and the associated petroleum industry will not be discussed as it may be expected that in a renewable future, these will be phased out and replaced with the 'green' products manufactured utilising renewable chemicals such as those discussed in Chapter 5. Naturally, coal, oil and gas have a very high CO_2 output, largely through the downstream use of these materials as a fuel used in combustion. It should be noted that the petroleum industry also produces the

Renewable Energy in the Process Industry, First Edition.
Eric van Steen, Nicholas S. Featherstone, and Linda H. Callanan.

Table 6.1 Most produced materials in the world.

Material	Production quantity (Mtonne/ annum)	Reference	Year	Chapter
Sand and gravel	50 000	[2]	2022	—
Coal	8500	[3]	2023	—
Cement	4100	[4]	2023	6
Crude oil	4060	[5]	2023	—
Natural gas	2650	[6]	2023	—
Grain	2200	[7]	2023	8
Iron and steel	1890	[8]	2024	6
Plastic	450	[9]	2023	7
Lime	430	[10]	2023	—
Paper	414	[11]	2022	8
Salt	270	[12]	2024	—
Sulphuric acid	270	[13]	2023	6
Ammonia	183	[14]	2023	5
Hydrogen	120	[15]	2022	5
Glass	100	[16]	2023	7
Chlorine	97	[17]	2022	6
Sodium hydroxide	87	[18]	2022	6
Aluminium	70	[19]	2024	7
Manganese	50	[20]	2021	—
Phosphate pentoxide (phosphate mined)	44 (220)	[21]	2024	—
Chromium	41	[22]	2024	—
Copper	22	[23]	2022	—
Titanium/titania	0.4/10.6	[24]	2024/2023	7
Silicon	9	[25]	2023	7

feedstock for the synthesis of polymers (see Chapter 7), which has a lower carbon output compared to the use of fossil fuels as fuels [26]. During the transitionary period, there will still be a need for fossil fuels and fossil sources for chemical synthesis, meaning that any CO_2 which can be saved within the petroleum industry will have a significant impact on humanity's ability to stay below the 1.5 °C target and within the carbon budget. The same, with respect to the need to phase out, is true of the coal industry, not discussed in this chapter.

Cement (together with lime) is one of the materials with the largest production capacity in the world, as it is an essential component of concrete, being the 'binding agent' to other components in concrete, including sand and gravel/stones, which are mined. Cement is of further importance due to its label as a 'hard-to-mitigate' industry, which cannot reduce its CO_2 emissions readily without a significant overhaul in current composition and production technologies (see Section 6.2).

Iron is, by far, the most mined metal in the world, making up in excess of 93% of all the metals mined by mass [27]. The iron and steel industry is the second largest basic industry. Mitigating CO_2 emissions in the steel industry will require a major overhaul of current production technologies, not just retrofitting. Furthermore, it is not currently well established how the proposed technologies will fare industrially (see Section 6.3).

Sulphuric acid has, with a production of 270 million tonnes per annum, the largest production rate in the entire chemical industry (ca. 50 million tonnes per annum more than second place ammonia, which was discussed in Chapter 5). Comparatively, sodium hydroxide and hydrochloric acid are produced in smaller quantities. However, they are produced from basic, mined/recovered sodium chloride without intermediate steps, and as such, for the purposes of this chapter, they are considered a 'basic industry' [28]. Furthermore, their products feed into other chemical industries. These chemicals are coupled as they are mainly produced together via the 'chlor-alkali process' with the chloride from the salt going to chlorine, and the sodium going to the sodium hydroxide, or 'caustic soda'.

The following chapters do not discuss salt, gypsum and phosphate as these materials are mined. Manganese [29], chromium [30] and nickel [31] (nickel being a far smaller industry than the other two) are used in the production of steel, but the processing of these three materials will also not be discussed further in this chapter or in the following chapters.

6.2 Cement

Cement has two primary uses: one as a 'glue' between building materials – mortar, the material that holds, e.g. bricks together, and second as the 'glue' that holds concrete together. Larger structures and the foundations of almost all buildings are constructed with reinforced concrete, i.e. concrete along with rebar, a hardened form of carbon steel.

6.2.1 Material Specification for Cement and Concrete

Cement is a mixture of finely powdered lime, clay and other minerals and thus contains predominantly CaO (lime) – SiO_2 (silica) – Al_2O_3 (alumina). The lime content is typically in the region of 65 wt.%, but the actual composition is determined by the type of cement. It is not known in detail how the exact composition affects the performance of the cement, which is why the manufacturers are reluctant to change their 'recipe'. The phases that make up Portland cement are shown in Table 6.2 with alite or tricalcium silicate making up the largest fraction, followed by belite (dicalcium silicate) and tricalcium aluminate.

Table 6.2 Indication of the amount of Bogue's compounds in Portland cement (data from Bogues compounds in cement 2024 [32]).

Compound	Formula	Shortform	Content (wt.%)
Tricalcium aluminate	$Ca_3Al_2O_6$	C_3A	10
Tetracalcium aluminoferrite	$Ca_4Al_2Fe_2O_{10}$	C_4AF	8
Dicalcium silicate (belite)	Ca_2SiO_4	C_2S	20
Tricalcium silicate (alite)	Ca_3SiO_5	C_3S	55
Gypsum	$CaSO_4 \cdot 2H_2O$	CSH_2	5
Sodium/potassium oxide	Na_2O/K_2O	N/K	Up to 2

Cement is mixed with water to give a paste or slurry that sets and hardens over time. There are two types of cement which are identified based on how they harden, viz. through the addition of water (called hydraulic cement) or through carbonation by reacting with CO_2 from the air (non-hydraulic cement). Portland cement, one of the most used types of cement, is also a hydraulic cement and reacts to harden via:

$$2(3CaO \cdot SiO_2) + 6H_2O \rightarrow 3CaO \cdot 2SiO_2 \cdot 3H_2O + 3Ca(OH)_2$$

Non-hydraulic cements are named as they do not react with water. Instead, they react with CO_2 in air to form carbonate species. This typically proceeds via two steps: first, the paste is formed where calcium oxide reacts with water to form calcium hydroxide, which subsequently reacts with CO_2 to form calcium carbonate and water. This ultimately reforms the limestone species used to create the CaO, and as it consumes carbon from the air, it is more carbon-neutral than other cement types.

$$CaO + H_2O \rightarrow Ca(OH)_2$$

$$Ca(OH)_2 + CO_2 \rightarrow CaCO_3$$

A major application of cement is concrete, which itself is a mixture of cement, coarse and fine aggregates, water and additives. Cement itself only makes up around 10% of concrete (see Figure 6.1), and the bulk of concrete is aggregates made from materials ranging from stone chips, gravel, pebbles, cinders and fly ash. The prevalence of concrete is also the reason why sand and gravel are such large industries, providing the fine and coarse aggregates, respectively. The usage of concrete (and mortar) globally renders the cement industry as one of the world's largest industries.

6.2.2 Industry Size

A large amount of cement is used throughout the world with every single building using tonnes of it. Most of this cement is made in and used by China (see Figure 6.2; China makes up over 50% of global cement production in 2020, which is sometimes less than their consumption [35]). The demand in China over the last two decades can be attributed to the large-scale building projects expanding their cities in

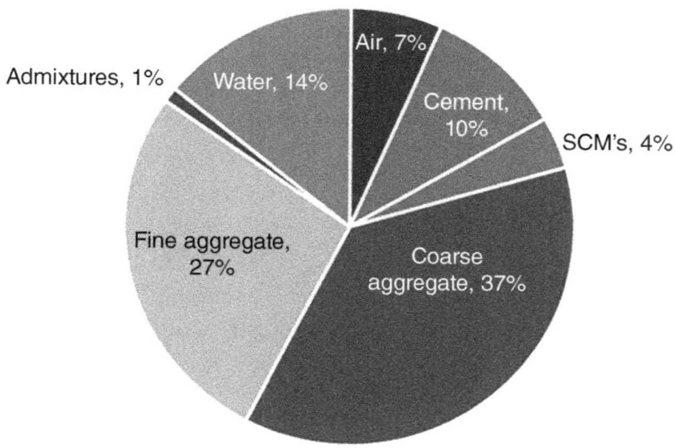

Figure 6.1 Components of concrete (SCM, supplementary cementitious materials) (redrawn from Stanzel and Sembrick [33]).

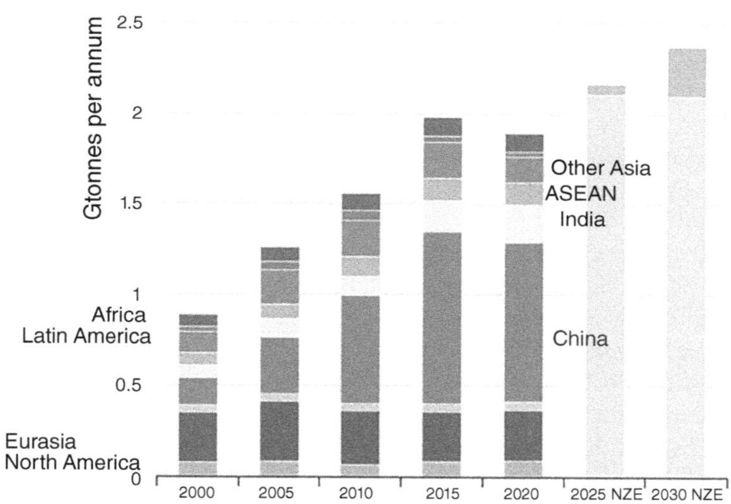

Figure 6.2 Global cement demand for building construction, 2000–2020, and in the Net Zero Scenario (adapted from IEA [34]).

this period. The demand in more developed areas, like North America and Eurasia, has been relatively constant in the time frame 2000–2020, but they make up a relatively small percentage of the total cement demand. In North America, wood is a major building material and is just like Eurasia already largely built up, meaning that less cement is required in these areas. The cement usage in India (ca. 6% of cement use) and the rest of Asia (ca. 20% of cement use) has grown strongly over the last decade and may further increase strongly in these areas over the next decade. Cement usage in Africa is modest based on its population, but a strong increase in demand can be expected due to the ongoing urbanisation in conjunction with the expected growth of their economy in the near future. Hence, the cement market is expected to grow even in a net-zero scenario.

The cement industry is a typical localised industry with less than 4% of its produced value being exported (total value of cement industry 2022: US$ 326.81 billion [36]; export cement in 2022: US$ 15.1 billion [37], p. 2). The anticipated growth of the cement industry in India, Southeast Asia and Africa does not necessarily mean that more cement will be exported or that new cement plants will be built. Plants that are no longer needed (e.g. in China) may be relocated to regions where the cement demand is increasing. Hence, the opening of new markets for cement may continue using relatively old technology without improving energy consumption and reducing CO_2 emissions.

6.2.3 Production of Cement

6.2.3.1 Process Description

A flow diagram of cement production is shown in Figure 6.3. Raw material (limestone and clay) is mined in a quarry, crushed and stored for use as needed. The raw material is transported to the cement plant and loaded into feed bins, which feed a mill for the raw material. This is contacted with water in a conditioning tower for the cleaning of the raw materials. Wet limestone is then dried through a series of cyclones to remove the water before being fed into a rotary kiln, operating at temperatures more than 1500 °C, which is typically fired with coal burners (coal is also milled prior to usage). The rotary kiln is operated in a non-isothermal mode, with the inlet of the kiln on the opposite side of the flame operating at a much lower temperature than near the flame itself. The material in the kiln is largely heated with the flue gas rather than radiation. Out of the kiln comes a material called clinker, which is cooled and stored to feed the next part of the plant. Gypsum and other minerals required are added to the clinker and mixed to make the cement mixture. Finally, the cement is milled into a fine powder, stored in silos, and transported and packaged as needed for specific customers. The most energy-intensive parts of the process are the kiln, with the largest energy demand, requiring thermal energy, and the mills and blowers requiring electrical energy.

The process of forming cement requires a reaction between SiO_2 and Al_2O_3 present in the clay and calcium present in the limestone to form the appropriate phases. The reactions yielding the calcium silicates and calcium aluminate are facilitated by the presence of some molten material in the furnace. To achieve this, the furnace will need to be heated to at least 1370 °C, although this depends on the amount of Al_2O_3 present in the feed [40–42]. Thus, kilns often end up operating such that the temperature at the hottest part is at around 1500 °C. Before reaching these high temperatures, limestone ($CaCO_3$) decomposes into lime (CaO) and CO_2:

$$\mathrm{CaCO_3} \rightarrow \mathrm{CaO} + \mathrm{CO_2} \quad \Delta H^0_{\mathrm{rxn}} = 178.3 \cdot \frac{\mathrm{kJ}}{\mathrm{mol}} (\mathrm{endothermic})$$

This endothermic reaction becomes noticeable at temperatures larger than 750–800 °C and can go to completion at atmospheric pressure at temperatures larger than 868 °C (see Figure 6.4).

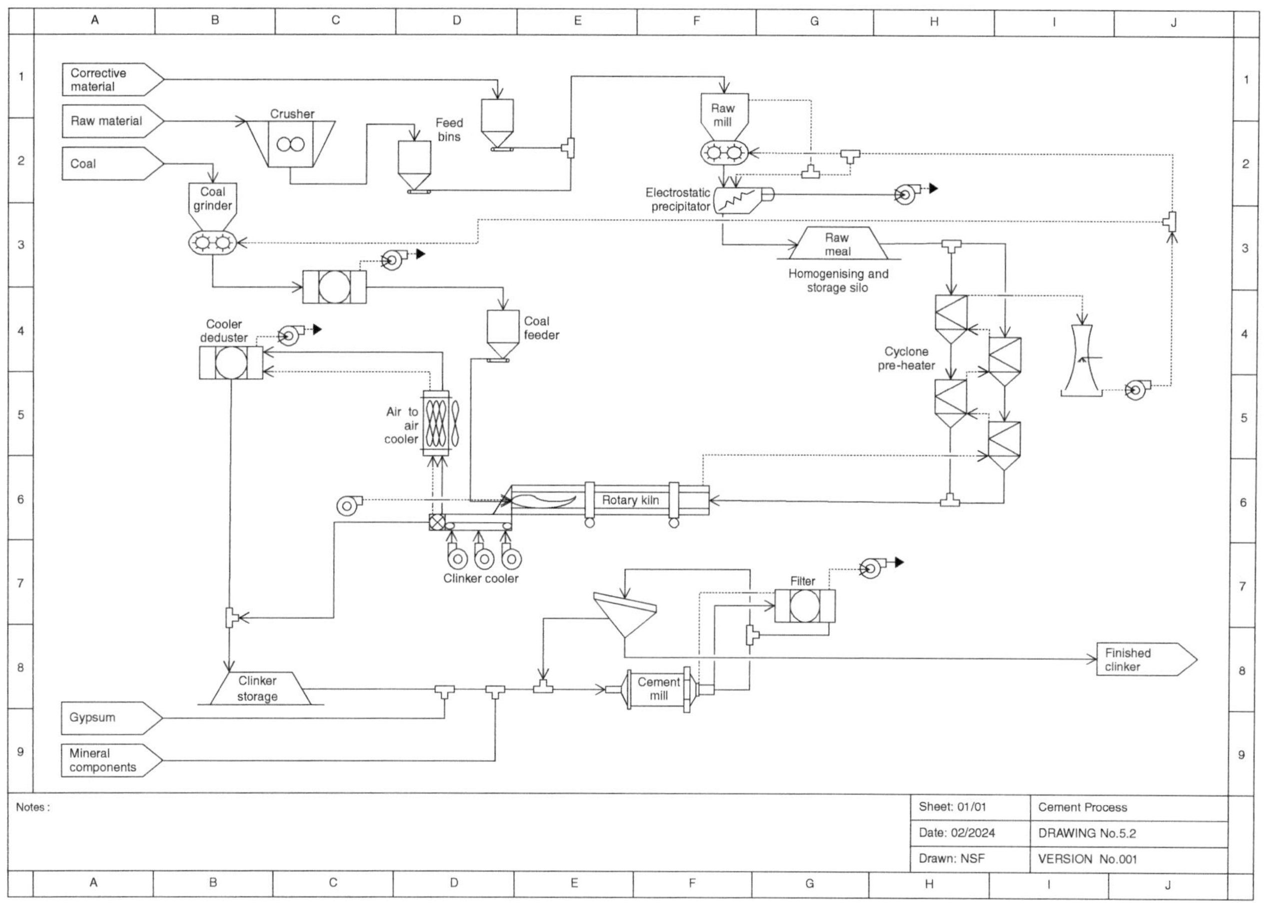

Figure 6.3 Cement production process flowsheet (redrawn from References [38, 39]).

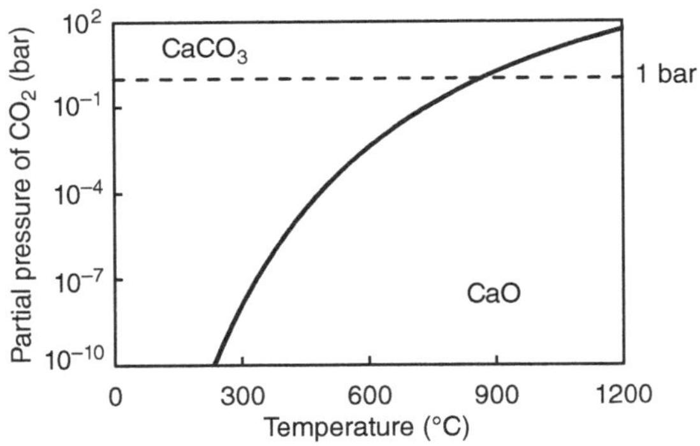

Figure 6.4 Equilibrium between $CaCO_3$ and $CaO + CO_2$, as a function of the partial pressure of CO_2 and temperature.

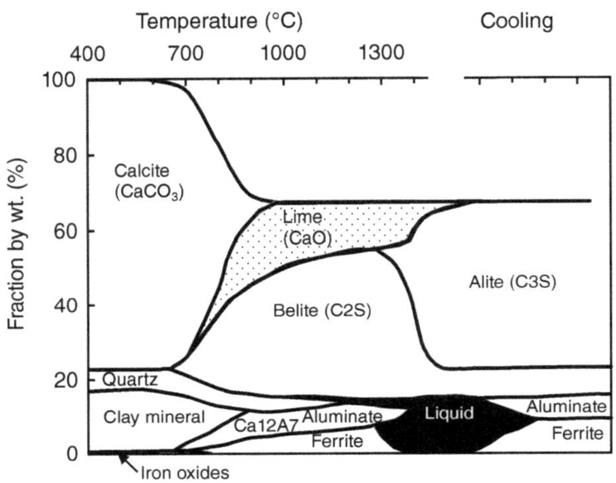

Figure 6.5 Phase development during the clinkering of Portland cement formation in a single-stage operation (redrawn from Hanein et al. [43]).

Figure 6.5 shows the development of the various phases in the kiln during the clinkering of Portland cement as a function of temperature. The desired lime phase starts forming from around 750–800 °C. This is associated with a significant weight loss as calcite (limestone) decomposes releasing CO_2 (the total weight loss in the conversion of calcite to lime is 44.0%). As soon as calcite is formed, it reacts with silica (quartz) yielding belite (Ca_2SiO_4), with alumina yielding dodecacalcium hepta-aluminate ($C12A7 = Ca_{12}Al_{14}O_{33}$) and with iron oxide to ferrite. The eutectic temperature of the system $CaO\text{-}Fe_2O_3$ is 1205 °C [44] and for the system $CaO\text{-}Al_2O_3$ is 1370 °C [45] implying that at temperatures higher than these temperatures, some liquid may be formed. At temperatures between the liquidus and the solidus, a calcium-poorer liquid and solid CaO may form, which can be incorporated into belite, yielding alite (Ca_3SiO_5). At even higher temperatures, calcium aluminate and calcium ferrite will melt completely. It is important to avoid complete liquefaction as this may lead to the kiln blocking. Upon cooling down, the aluminate and ferrite phases are reformed, albeit with a lower calcium content.

The kiln has been set up to maximise the heat integration within the process. Coal is fired in at one end of the kiln, whilst the hot flue gas flows out of the other end before going through several stages in the cyclone pre-heater as shown in Figure 6.3 (blocks H3-I6). In each of these steps, the exhaust gas from the kiln is brought into contact with the solids prior to entering the kiln in a counter-current set-up. Heat transfer from the flue gas to the solids in the cyclones will pre-heat the solids. The last cyclone is called a pre-calciner, as the temperature can be elevated sufficiently for the conversion of the $CaCO_3$ to CaO to begin. The partial pressure of CO_2 in the effluent gas is relatively low as it originates from both burning the carbon-containing fuel and from limestone decomposition.

6.2.3.2 Energy Demand in the Production of Cement

The main energy demand in the cement manufacturing process is for thermal energy owing to the high temperatures required in the kiln with ca. 90% of the total energy requirement needed as thermal energy for the heating in the kiln. This corresponds to approximately 3200 MJ/tonne cement (0.9 MWh/tonne) [46]. The energy demand is a strong function of the moisture content of the material fed to the kiln, as this moisture needs to be boiled off. Some of the moisture is dried off in the cyclones and thus the amount of water in the material fed to the kiln is inversely related to the number of cyclone stages. This means that one of the ways of decreasing the amount of thermal energy required is to increase the number and efficiency of the cyclone stages [47]. Some of the saving in energy with more cyclone stages also comes from the improved heat integration that can be done over multiple steps, increasing the temperature of the stream as well as providing more time for contacting between the hot and cold media. However, there is a trade-off between the operating and capital costs of building more cyclone stages. The inclusion of more stages will increase the feed temperature to the furnace, thus reducing the fuel demand, but increasing the number of stages will also increase the capital cost. Thus, there is a strong variance in the number of cyclone stages required that depends on the desired balance between electrical and thermal energy requirements, as the spinning of the cyclones is electrically controlled. This ultimately amounts to a cost optimisation problem.

The energy demand intuitively goes up with higher production rates for a given plant. In a plant with a given size, a higher feed rate will result in a higher moisture content in the feed to the kiln, as the cyclones will not dry as efficiently. This results in an increase in the thermal energy demand per mass of feed.

Electrical Energy Although most of the energy needed in cement manufacture is in the kiln, some electrical power is also required. Everything, besides the kiln, on the plant uses electrical energy, which amounts to around 0.1 MWh/tonne cement (i.e. ca. 10% of the total plant energy demand). Hence, the total energy cost for cement production is around 1 MWh/tonne cement.

The distribution of electrical energy requirements for cement production are shown in Figure 6.6. Much of this electrical energy is needed by the mills, both for cement and raw material grinding. The cement may be ground finer than the raw material as it needs to meet certain requirements, whilst the raw feed needs only

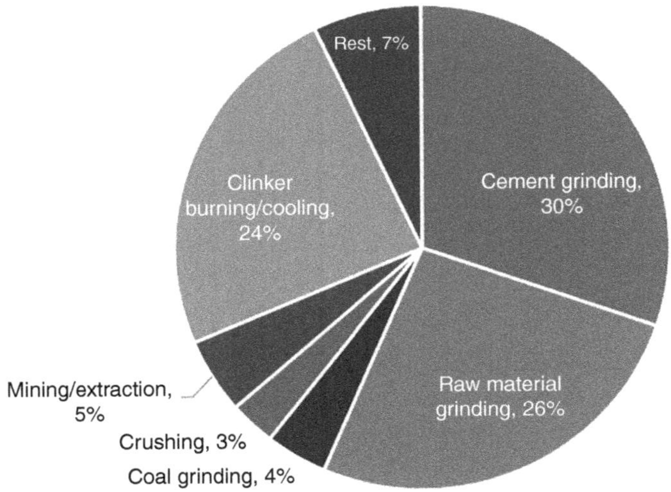

Figure 6.6 Electrical energy requirements for cement production (redrawn from Schneider et al. [46]).

to be fine enough for the rest of the process. Finer grinding requires more energy input; thus, the cement grinding uses slightly more energy than that of the raw material [48]. Some of the clinker treatment is also done with electricity, but this is much lower than what comes from burning, typically coal.

6.2.3.3 CO_2 Production in Cement Manufacture

A large amount of CO_2 is formed in the cement manufacture through the decomposition of $CaCO_3$ yielding CaO and the burning of coal to reach the required temperature of ca. 1400 °C in the furnace. It is reported that the total cement manufacturing process releases ca. 0.88 tonne CO_2/tonne cement [49, 50], with 0.37 tonne CO_2/tonne cement (or 40%) coming from heating.

The minimum amount of CO_2 released due to the chemical conversion of calcite (limestone) into CaO can be estimated based on the reaction stoichiometry and the calcium content in the clinker (e.g. 65 wt.%):

$$\text{minimum CO}_2 - \text{release} = \nu_{\text{CaCO}_3 \rightarrow \text{CO}_2} \cdot \frac{\text{M}_{\text{CO}_2}}{\text{M}_{\text{CaO}}}$$

$$= 1 \cdot \frac{\text{mol CO}_2}{\text{mol CaO}} \cdot \frac{44 \cdot \frac{\text{g}}{\text{mol CO}_2}}{56 \cdot \frac{\text{g}}{\text{mol CaO}}} = 0.786 \cdot \frac{\text{ton CO}_2}{\text{ton CaO}}$$

$$\text{minimum CO}_2 - \text{release} = 0.786 \cdot \frac{\text{ton CO}_2}{\text{ton CaO}} \cdot 0.65 \cdot \frac{\text{ton CaO}}{\text{ton clinker}}$$

$$= 0.51 \cdot \frac{\text{ton CO}_2}{\text{ton clinker}}$$

Alternatively, it can be estimated based on a detailed composition of the clinker. For instance, the amount of CO_2 formed in the production of clinker containing 10 wt.% Ca_3A, 8 wt.% Ca_4AF, 20 wt.% C_2S, 55 wt.% C_3S, 5 wt.% gypsum and 2 wt.%

Table 6.3 Calcium content of cement and CO_2 produced per tonne of a Portland cement.

Component	Formula	M_i a) (g/mol)	w_i b) (wt.%)	n_i c) (kmol)	n_{Ca} d) (kmol)	n_{CO2} e) (kmol)
Ca_3A	$Ca_3Al_2O_6$	270	10	0.370	1.11	1.11
Ca_4AF	$Ca_4Al_2Fe_2O_{10}$	486	8	0.165	0.66	0.66
C_2S	Ca_2SiO_4	172	20	1.163	2.33	2.33
C_3S	Ca_3SiO_5	228	55	2.412	7.24	7.24
Gypsum	$CaSO_4$	136	5	0.367	0.37	—
Total					11.70	11.33

a) Molar weight of compound *i*.
b) Weight fraction of compound *i* in cement.
c) Moles of compound *i*.
d) Moles of calcium in compound *i*.
e) Amount of CO_2 released from having compound *i* assuming calcium originates from $CaCO_3$ (except for gypsum).

Na_2O, can be estimated assuming all calcium, except calcium in gypsum, which is added separately, originates from $CaCO_3$.

Hence, the number of moles of CO_2 coming out of limestone in the production of Portland cement with the composition given in Table 6.3 is $11.33\frac{\text{kmol CO}_2}{\text{ton clinker}}$ or $0.50\frac{\text{ton CO}_2}{\text{ton clinker}}$.

The thermal energy demand can also be estimated from the 0.37 tonne CO_2/tonne cement released in the process for heating purposes. This is typically achieved by combusting coal, although more and more cement plants also utilise carbon-containing waste products [51], and amounts to ca. 3300 MJ/tonne cement. Coal is preferred to maximise radiative heat transfer in these kilns [52] and the presence of soot/unburned coal particles in the flame improves radiative heat transfer (see Section 4.4.1).

This implies that depending on the fuel source (e.g. anthracite: LHV of 30 MJ/kg with carbon content: 85 wt.%; lignite: LHV of 18 MJ/kg with carbon content of 50%), the required amount of fuel is 0.11 and 0.18 tonne coal/tonne cement, for anthracite and lignite, respectively. This will amount to 0.33 tonne CO_2/tonne cement (with this number being dependent on the carbon content of the fuel). Thus, the total amount of CO_2 emitted in the cement production can be expected to be 0.84 tonne CO_2/tonne cement, of which 40% comes from burning coal and 60% comes from the decomposition of calcium carbonate. A production line producing ca. 5000 tonne of clinker per day equivalent to 208.3 tonne/hour, would thus yield 4150 tonne CO_2 per day or 173 tonne CO_2/hour.

The cement production releases a relatively large amount of CO_2 particularly in relation to the profit made (6.9 kg CO_2/\$), compared to chemicals: (0.3 kg CO_2/\$); iron and steel: (1.4 kg CO_2/\$); mining: (0.4 kg CO_2/\$); and oil and gas (0.8 kg CO_2/\$) [53]. This means that the cost margin to decarbonise is relatively small, and a large amount of CO_2 is released in comparison to the value of the product.

Some CO_2 from the atmosphere is taken up by concrete/cement over time in the reverse of the reactions taking place in the kiln, viz. $3CaO \cdot SiO_2 + 3CO_2 \rightarrow 3CaCO_3 + SiO_2$, but the rate of the recarbonation is a strong function of the type of cement, the compressive strength of concrete and the exposure conditions [54]. It has been estimated that between 40% and 50% of the CO_2 emitted in the clinker process (excluding CO_2 emissions from fuel combustion) can be taken up in the recarbonation process [55]. However, this is strongly dependent on the exposure time, and the number quoted here should be taken as an indication on how much of the CO_2 produced in the cement manufacturing process may be taken up by concrete within a reasonable time frame (several decades). Hence, it would be advisable to concentrate on the direct emissions from the cement/concrete production rather than the net CO_2 emissions from this product to evaluate the immediate effect of the use of cement/concrete on the environment and global warming.

6.2.4 Reducing CO_2 Emissions from Cement/Concrete

The production of cement, an essential building material, is responsible for ca. 7% of global CO_2 emissions [53] which illustrates a distinct need to decarbonise this process. CO_2-formation comes from two sources in cement manufacture, i.e.

- Formation of the CaO used in cement that is derived from $CaCO_3$, resulting in around 60% of the CO_2 formed.

$$CaCO_3 \rightarrow CaO + CO_2$$

- Heating up of the reaction mixture to ca. 1400 °C by burning coal.

This means that most of the CO_2 release comes from a chemical requirement (process emissions, ca. 60% or 0.51 kg CO_2/kg cement) rather than a heating purpose (heating emissions, ca. 40% – 0.36 kg CO_2/kg cement). This presents a large difficulty for the decarbonisation of the industry as the only way to remove the process emissions from the decomposition of calcium carbonate is either to replace the feedstock or to change the materials used in the cement industry.

CO_2 emissions from the heating requirements are not insignificant and are necessary to achieve the required high temperature (ca. 1400 °C), which allows for the formation of some liquid in the kiln, which facilitates the formation of the desired solid-state phases. Hence, the implementation of renewable energy in this sector is non-trivial, as e.g. electric power cannot be easily used to create these temperatures in insulators like cement.

Cement is difficult to recycle due to the chemical transition made in the hardening process ($CaO \rightarrow CaCO_3$ or $CaO \rightarrow Ca(OH)_2$). As such, a chemical transition would be required to reform the calcium oxide. Products that use cement, such as concrete, can be partially recycled – as used concrete can be crushed and ground and can then form the basis of new concrete by replacing, in part, the need for sand and gravel in the concrete composition.

Cement and concrete are used in the construction industry, and the more efficient usage of the raw materials as a first step can be used to reduce the carbon footprint of

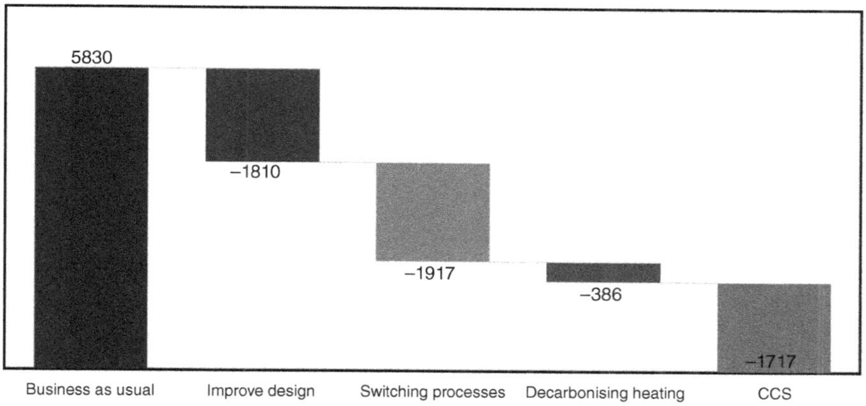

Figure 6.7 CO_2 emissions associated with the materials (cement and steel) used to construct a 30-storey high-rise building skyscraper and the impact of various measures to reduce the emissions (redrawn from Fennell et al. [56]).

the construction energy [56]. For instance, the construction of a 30-storey high-rise building would require ca. 5400 tonnes of cement, but smart design may reduce the cement requirement to 4100 tonnes of cement, thus corresponding to a decrease in the CO_2-footprint associated with the use of cement by 24% without changing the production process (see Figure 6.7). Further reductions can then be achieved by altering the production processes to manufacture cement.

6.2.5 Current Research Around Future Processes

The composition of typical effluent from a classical cement plant is given in Table 6.4. The CO_2-concentration of 17.8 mol% is usually not high enough to benefit from standard carbon utilisation or storage technology. As such, some form of carbon capture integrated with the cement plant is required to make the CO_2 stream more usable and reduce the overall CO_2 emissions.

Many technologies are under development to help decrease the carbon footprint of the cement manufacturing process. These may include the amine scrubbing/methanol looping processes to increase the CO_2-purity (minor components are of importance here, as they may have to be removed before contacting the effluent with an amine-containing solution to remove CO_2 from the stream), though technologies such as calcium looping may be easier to integrate directly into the cement process, as the cement manufacturing process and calcium looping share common reagents ($CaCO_3$, CaO and CO_2). Other technologies have been investigated, such as the oxy-fuel process, which aims to increase the concentration of CO_2 in the effluent gas to make it easier to utilise/store. Furthermore, research aimed at cement replacements is being conducted, which will assist in decarbonising the process by requiring less CaO and thus releasing less CO_2. Besides the process changes mentioned here, many other technologies are examined to further improve the heat integration already shown in the cement industry (which currently uses the heat integration in the form of pre-heaters).

Table 6.4 Typical effluent from a cement plant (wet basis) (data from Schakel et al. [51]).

Condition		Classical process
Temperature (°C)		165
Pressure (bar)		1
Major gas components	CO_2 (mol%)	17.8
	N_2 (mol%)	56.5
	O_2 (mol%)	7.5
	H_2O (mol%)	18.2
Minor gas components	CO (mg/Nm^3)	1470
	NO_x (mg/Nm^3)	250
	SO_2 (mg/Nm^3)	25
	HCl (mg/Nm^3)	10
	dust (mg/Nm^3)	9

6.2.5.1 Making Cement More Sustainable

Implementation of renewable energy in the cement industry is difficult owing to the high-temperature requirement, which makes up ca. 90% of the energy requirement in the process. The remaining 10% of the total energy, which comes from electric power, is used for grinding, milling, etc., and can be replaced comparatively easily by switching to a renewable electricity supply. This could be either self-generated electricity, from e.g. PV or wind, or outsourced to a supplier of green electricity. On the scale of the plant itself, it may be challenging owing to the amount of renewable energy which would need to be generated. For instance, a typical production capacity of 5000 tonne of clinker per day requiring electric power of 0.1 MWh/tonne cement would thus require to be connected to a source delivering continuously 20.8 MW of renewable energy. Intermittency in the power supply can be dealt with by connecting to a larger power source and scheduling the milling of raw materials and the product when power is available, whilst stockpiling the crushed product.

Much of the energy requirement (90%) is to heat the kiln to the required, rather high temperature (~1400–1500 °C). The process is further complicated by the fact that cement is neither electrically conductive nor magnetic and, as a result, some method of combustion seems to be required to reach the required temperatures. Combusting renewable energy carriers, e.g. renewable hydrogen, renewable synthetic fuels (ammonia, methanol, hydrocarbons) or biomass could have a significant impact on making the process energetically more sustainable.

When talking about 'greening' the cement industry, what is often meant is carbon capture. Whilst reducing the CO_2 production would be preferable, capturing CO_2 is the next option and normally more readily achievable. The purpose of some processes that can be introduced in the cement manufacturing process would be to concentrate the CO_2 in the effluent, thereby making CO_2 capture easier. For example, amine scrubbing, calcium looping or a methanol wash could be used to increase the CO_2 concentration in the effluent stream. Another potential way to concentrate the CO_2 is to use O_2 rather than air in the combustion of fuels, with recycling, to get a high-purity CO_2 stream that can be immediately stored or used as required in another process.

A problem faced with carbon capture and storage will be how easily and where it may be stored. This may be a particular problem for the cement industry as mineralisation (i.e. the formation of carbonate rocks) will not be an option with the cement feedstock, limestone, already being a carbonated rock. Further limestone mining does not result in a cavernous mine which could be used for carbon storage as an end-of-life solution.

6.2.5.2 Cement Replacements

A potential method of decarbonising the cement industry is to replace the necessary production of CO_2, viz. the calcination of $CaCO_3$ to CaO. For this reason, a concerted effort is being put into replacing the components in cement. Some examples of potential replacements for traditional cement are given in Table 6.5. Most of these options include a significant increase in the amount of silica, with some options, such as fly ash, also increasing the amount of alumina present in the sample. Slag cement contains less CaO than normal Portland cement, thus decreasing the CO_2 emissions during production. One major problem with cement replacements is that there is currently no industrial-scale implementation of any of these materials for use as cement, and the material strength and longevity may be questioned.

Table 6.5 Example of cement replacement options (data from Neithalath et al. [57]).

Species (wt.%)	Cement (Portland)	Fly ash	Glass powder	Silica fume	Slag cement
CaO	62.2	5.86	5.86	1.91	39.82
SiO_2	20.5	50.24	72.5	93.4	36.01
Al_2O_3	4.7	28.78	0.40	0.42	10.53
Fe_2O_3	2.9	5.72	0.20	0.52	0.67
MgO	2.3	1.74	1.74	—	7.94
Na_2O	0.21	0.06	13.7	0.25	0.28
K_2O	0.83	0.93	0.10	0.79	0.16
SO_3	3.5	0.51	-	0.34	2.11
SSA[a)] (m^2/g)	370	420	420	—	420
S.G.[b)]	3.15	2.38	2.65	—	2.94

a) Specific surface area.
b) Specific gravity.

6.2.5.3 Calcium Looping

Calcium looping (CaL) is technique to capture CO_2 generated on power plants forgoing the energy intensive amine absorption step [49]. Figure 6.8. shows calcium looping being applied to a coal fired power station in tandem with a cement plant. The process is regenerative, allowing for multiple carbonation-calcination cycles. The calcium loop can provide additional calcium oxide to the cement side of the process, with the calcium loop also consuming part of the CO_2 produced by the cement plant.

The calcium looping process is principally infinitely regenerative though it relies on high temperature calcination which is a limiting factor. The coal-fired boiler can be replaced by a methane-fired boiler to decrease the total CO_2 in the process to some extent; the kilns could also be fired with a different fuel. However, this process can be a useful transition process as coal-fired boilers and classical cement manufacturing processes may be gradually replaced by more renewable technologies.

It should be noted that this technology is best implemented with oxy-fuel systems to decrease the nitrogen content of the recovered gas (though the air separation unit will increase the energetic cost of the process [58] by up to 0.06 MWh/tonne clinker [49]). Further, as this process is generating a high purity stream of CO_2, over 90% of the total process CO_2 can be recovered [59].

6.2.5.4 Oxy-fuel

The oxy-fuel combustion is a well-researched method of decreasing the carbon impact in the cement manufacture [49, 60]. Here, pure oxygen rather than air is fed to the burner. It also involves significantly more heat integration than typically used in the current plants. This will thus result in a decrease in the heating demand in the burner, which, naturally, also decreases the amount of fuel required. There are some major changes that are needed compared to the existing plants. Firstly, air separation is required to supply the pure oxygen, secondly the entire plant needs to be rebuilt to allow the heat integration and, thirdly, air ingress into the kiln should be minimised. These required changes necessary for the oxy-fuel combustion would pose a major challenge as an implementation on an existing plant. For this reason, this can only realistically be applied to new processes.

The oxygen feed would be a mixture of O_2 and CO_2. Here, CO_2 acts as a carrier for oxygen, even though it is not really needed, and as a heat-transfer fluid. It is necessary for the heat integration to ensure that there is enough gas for the heat transfer to occur efficiently. The CO_2 'carrier' is obtained by recycling a portion of the exhaust gas, the remainder of which goes to further CO_2 purification, which also allows for some extra heat integration from the exhaust CO_2.

Whilst this system does not capture or purify the CO_2 itself, it is targeted to produce a stream of higher-purity CO_2, thus making the CO_2 capture less energetically intensive (as previously discussed) and thus the storage or utilisation of CO_2 easier.

6.2.5.5 Potential Additional Systems for CO_2 Capture

Different systems have been proposed for potentially improving the CO_2 capture. An example of this is one where a different design for the kiln has been proposed

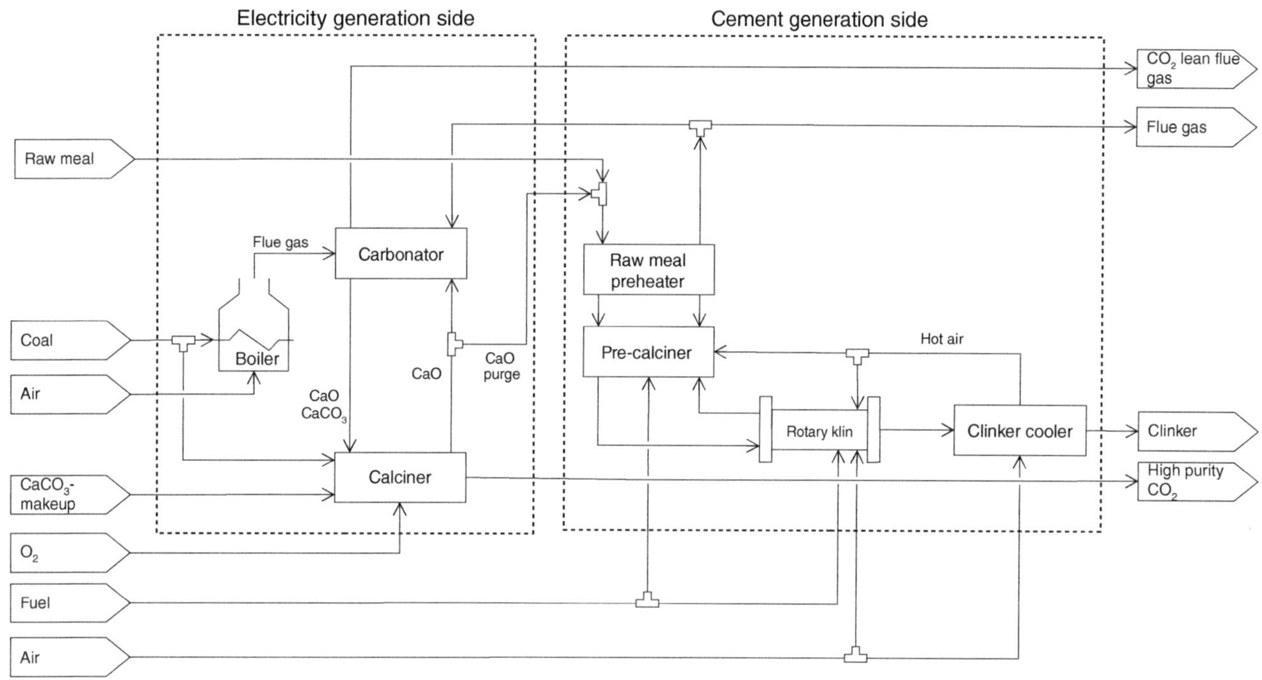

Figure 6.8 Calcium looping: integration with cement plant (redrawn and adapted from Hills et al. [49]).

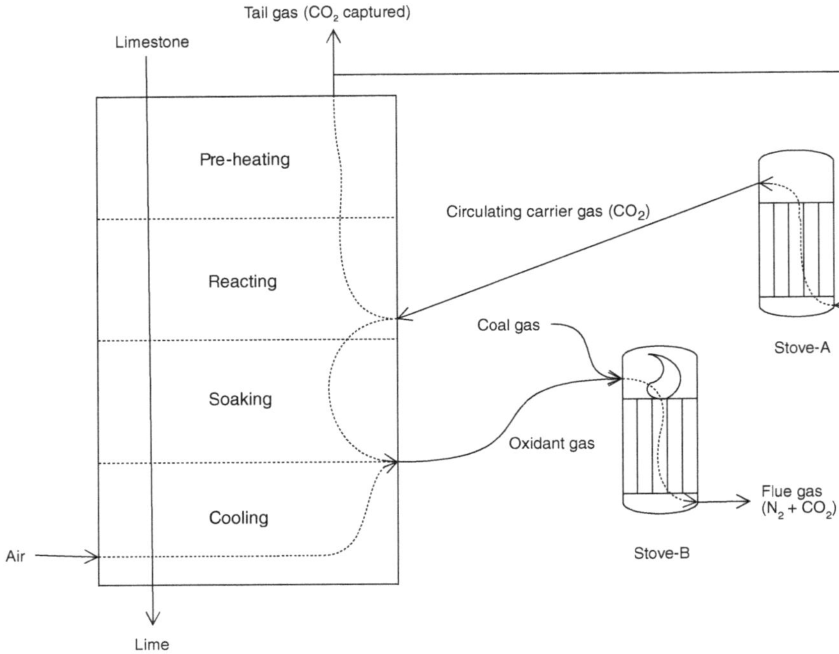

Figure 6.9 Simple diagram for a proposed cement manufacture process for 'high-purity' CO_2 (redrawn from Yang et al. [61]).

using external heating and circulation of the combustion gases as the pre-heating medium (Figure 6.9, [61]). This process includes a 'soaking area' in the kiln which is used to pre-heat the circulating gas. This is said to enable the production of a higher-purity CO_2 stream, which should make the separation of CO_2 from the flue gas easier. The process captures ca. 70% of CO_2 which is said to be of 'high purity' [61]. If this were to be implemented industrially, it would require either the building of a new plant or a large rebuilding of an existing plant, like the oxy-fuel process. Further, the control would be more complex than that of a classical kiln.

6.2.5.6 Using Alternative Fuels

The classical production of cement uses coal as a heat source, requiring ca. 0.1 (up to –0.18) tonne coal/tonne cement (see Section 6.2.3.3). An alternative is to use carbon-containing waste. The cement industry has already used almost any fuel source to heat their processes, not restricting themselves to coal [39, 62–64]. Some examples of fuels which may be used are given in Table 6.6. As kilns are typically designed to take in finely ground coal, it is not necessarily an easy trade to replace the coal with any given fuel source. For example, diesel, being a liquid fuel, requires a different type of injection to that of a solid. Similarly, waste tyres or plastics are not readily ground or cut to the same consistency as powdered coal. For these reasons, design considerations are necessary industrially to allow the usage of mixed fuel in the cement industry. However, the industry is one of the most important consumers

Table 6.6 Potential fuel sources for cement manufacture provision of thermal energy and associated heat of combustion (data from Tokheim and Brevik [39]).

Source	LHV (MJ/kg)
Coal	29.3
Petcoke	33.9
Diesel	42.8
Waste oil	34.0
Plastic	37.7 (32–38[a)])
Solid hazardous waste	14.9
Liquid hazardous waste	15.7
Refuse derived fuels	12.6
CCA waste wood	16.8
Wood	15.7
Dried sewage sludge[b)]	14
Waste tyres[c)]	45

a) Areeprasert et al. [62].
b) Chinyama [63].
c) Rada et al. [64].

of waste products, such as waste rubber (such as tyres) and other industrial waste (including plastic) [65].

Additional problems can come from burning these wastes as they may contain undesired compounds, which may not all end up in flue gas. For example, waste tyres contain a large amount of ash (2–10 wt.%) [66], some of which will ultimately land up in the cement when the tyres are burnt. This may become problematic. For example, zinc oxide is present in tyres added during the vulcanisation process, and may end up in cement retarding the setting of cement through the retardation of the hydration process [67]. Furthermore, sulphur present in tyres will convert calcium oxide to calcium sulphate ($CaSO_4$), increasing the amount of calcium oxide that needs to be produced to meet the same specification of cement, although this is not necessarily bad, provided it is adequately controlled, as gypsum ($CaSO_4 \cdot 2H_2O$) is one of the ingredients of cement. Typically, there is a need to feed more limestone to counter the effects of the ash, as mentioned for the sulphur from tyres above.

Table 6.7 shows the composition of cement compared to the composition of the ash formed from burning various fuels [63]. The ash from coal contains similar compounds to what is required for cement, even though they are not in the desired ratio. The use of dried sewage as a fuel source would require a much larger amount of fuel than when using coal. Furthermore, the fuel is now richer in iron oxide and aluminium oxide, which may affect the performance of cement. Using waste tyres as a fuel reduces the amount of fuel required but will necessitate to increase the limestone feed to the kiln (and reduce the clay) to obtain a similar product.

Table 6.7 Composition of the resulting cement content from using different fuel sources (data from Chinyama [63]).

	Cement	Coal	Dried sewage	Waste tyres
Feed ratio		0.1 t/t	0.23 t/t	0.07 t/t
Ash-content (%)		8–10	26	19
Composition (wt.%)				
CaO	63–67	3	9–22	4
SiO_2	19–23	48	30–49	35
Al_2O_3	3–7	21	8–14	13
Fe_2O_3	1.5–4.5	19	5–23	2
MgO	0.5–2.5	0.8		0.7
K_2O	0.1–1.2	2.9	1–2	1.0
Na_2O	0.07–0.4	0.5		0.9
SO_3	2.5–3.5	4.3		6.7

6.2.6 Outlook

The cement industry is one of the world's largest industries and well has earned its title as a 'hard-to-mitigate sector', due to the so-called process CO_2 emissions, i.e. CO_2 produced by the required chemical transformation in the process. Getting away from these process emissions will be difficult as the formula for cement is quite specific and has been trusted for many years. As it is essential that cement does not fail on a very large scale, it will be non-trivial to replace cement with a new variant containing less calcium. It would need to undergo large-scale testing over decades under extreme conditions before it could be applied on a large scale in the construction industry.

The cement industry requires high-temperature heat, which will, for the foreseeable future, require the burning of materials to provide the required temperature. This is because, whilst it is possible to heat over 1500 °C using electricity; in the case of cement manufacture, it cannot be done as the various oxides involved are both non-conductive and non-magnetic, implying that electric arc and inductive furnaces cannot be used in the cement industry. Furthermore, electrical radiative heating to achieve these temperatures has not been tested close to the required scale or duration the cement industry would require.

The use of alternative fuels as the energy sources, such as dried sewage and other waste products, may be a logical replacement for classical fossil fuels, and indeed these are currently used to some extent. This would not necessarily reduce the direct CO_2 emissions of the cement plant but may have a net decreasing effect on total CO_2 release as the waste materials would naturally decompose and release CO_2 to the atmosphere anyway.

Complete decarbonisation of the cement industry still manufacturing the current product will require carbon capture utilisation or storage (CCUS). This may be done by building in the carbon capture facility, though the most promising approach for the cement industry may be minimising the nitrogen content in the flue gas by carrying out combustion with oxygen-enriched gas. This will make CCUS technologies more financially and energetically feasible, although many of these technologies will require large-scale reconfigurations of existing plants or implementations on new plants, which also have associated costs. It may be preferred to move/reuse equipment rather than build new from a financial viewpoint, noting that the areas with the highest rates of cement industry growth are developing nations.

The use of oxygen-enriched air, or even pure oxygen as the oxygen source, may become more feasible in a future hydrogen economy where pure oxygen may be abundant due to its co-generation alongside the production of green hydrogen via water electrolysis. The carbon savings and relative lack of increased energetic cost associated with the generation of pure oxygen onsite may make this route more cost-competitive [68] when compared to installing air separation units.

6.3 Steel

Iron replaced bronze as a metal of choice due to the abundance of iron and the scarcity of tin, although iron was more difficult to work with. Steel was known at the start of the Iron Age by combining carbon (charcoal) and wrought iron. The large-scale, reliable production of steel only began in the eighteenth century.

6.3.1 Material Specification

Iron is the 26th element on the periodic table, is notably magnetic and is the 4th most abundant element in the Earth's crust. Iron is also known for giving Mars its red colour due to the presence of haematite (Fe_2O_3) or 'rust' on its surface. The red colour comes from iron in the trivalent state, as opposed to divalent iron, which is green. Chemically, iron is a ferro-magnetic metal which is in the body-centred-cubic (BCC) structure (α-Fe) at room temperature, and when heated above 911 °C, it transforms into a paramagnetic face-centred-cubic (FCC) structure (γ-iron; see Figure 6.10).

Iron is mainly used in steel, a mixture or alloy of iron and some other metals and carbon [70]. The exact mixture of metals depends on the type of steel. There are four types of steel:

- Carbon steel is the most affordable and most common steel, which contains between 0.1 and 1 wt.% carbon with the balance being iron (sometimes including other alloying elements). This steel can be heat-treated to provide incredible hardness, if it has an appropriate carbon content. Most of the steel produced is carbon steel, often galvanised or with a protective metal/paint/plastic coating to prevent corrosion.

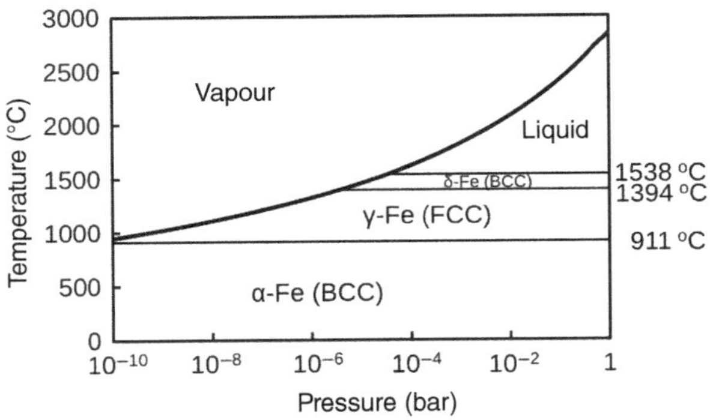

Figure 6.10 Low-pressure phase diagram of iron (drawn with data from Desai [69]).

- Alloy steel, also called low alloy steel, is more durable than carbon steel and can withstand higher temperatures than stainless steel. Although it is comparatively uncommon, it does find some use in kitchen appliances.
- Stainless steel, sometimes called high alloy steel, is known for its resistance to corrosion and staining. Like carbon and low alloy steel, stainless steel can be heat treated to improve hardness and is common for areas where corrosion is highly undesirable, such as food and medical appliances. However, it is far more expensive than carbon steel.
- Tool steel is the toughest type of steel, often including cobalt and/or molybdenum and/or tungsten in the alloy. As the name suggests, this type of steel is used a lot for mechanical tooling, especially for industrial use.

Carbon steel accounts for around 90% of steel produced in the world. It is named for its lack of other alloy materials, rather than the abundance of carbon. The solubility of carbon in α-Fe is very limited (max. 0.02 wt.%) and materials with a higher carbon content contain both α-Fe and cementite (Fe_3C) (see Figure 6.11), although it may also appear as pearlite, a layered structure of α-Fe and cementite, which is the eutectoid formed at 727 °C between α-Fe, γ-Fe and Fe_3C and contains 0.80 wt.% carbon (it is worth noting that the exact location of the eutectic point important to the hardening of carbon steel is inconsistently reported, lying somewhere between 0.68 wt.% C [73] and around 0.83 wt.% C [71]). Carbon has a higher solubility in austenite (γ-Fe) with a maximum solubility of 2.06 wt.%, but again a more limited solubility in δ-Fe.

The most important point on the phase diagram is the eutectic point between ferrite (α-Fe), austenite (γ-Fe) and cementite (Fe_3C), resulting in the formation of pearlite/iron carbide mixture, the ferrite/pearlite/iron carbide mixture and phases containing austenite, as carbon is much more soluble in austenite than in the BCC iron phases (α-Fe, δ-Fe). As such, heating iron with carbon results in a solid solution of carbon dissolved in the FCC iron structure. Upon quenching of the eutectoid at this temperature will result in a strained BCC structure incorporating carbon (see Figure 6.11a) rather than separating into phases [71]. When starting at high temperature with insufficient carbon (hypo-eutectoid), cooling will result in the ferrite

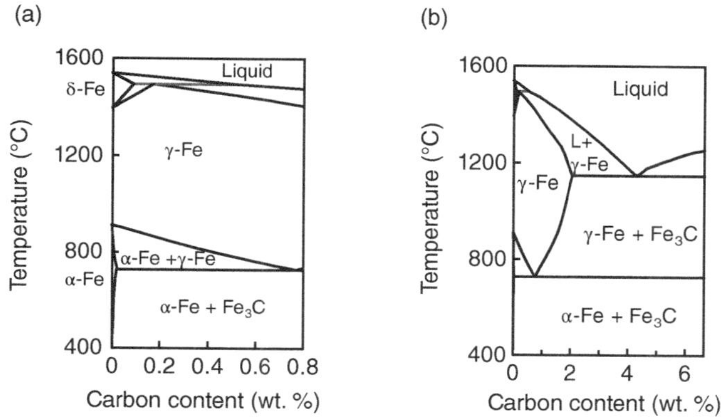

Figure 6.11 Carbon-iron phase diagram at atmospheric pressure (redrawn from References [71, 72]).

(α-Fe) with the remaining material having the composition of the eutectic point. When cooling down a hyper-eutectoid (too much carbon), some cementite will form with the remaining material having the composition of the eutectic point.

Cooling of the iron needs to be rapid (ca. 8.2 °C/s [74, 75]); quenching too fast creates a harder material and causes the material to become brittle due to internal stresses, whilst quenching too slowly creates a more flexible but less hard material. Hence, a quench oil is typically used, rather than water. Rapid cooling is required to minimise the diffusion of carbon out of the structure when iron changes from the FCC to BCC phase (see Figure 6.12). The resulting body-centred tetragonal (BCT) iron phase is super-saturated with carbon and is called martensite, which is very hard but also very brittle [74, 75]. Tempering cycles are used in order to make the steel slightly less hard but also more malleable; this is done by heating the steel to a temperature below that which will form the austenite but high enough to change the solubility of carbon, thus changing the crystal grain structure/crystallite sizes in the material [76, 77]. Properly heat-treated carbon steel is hard and flexible [74, 76].

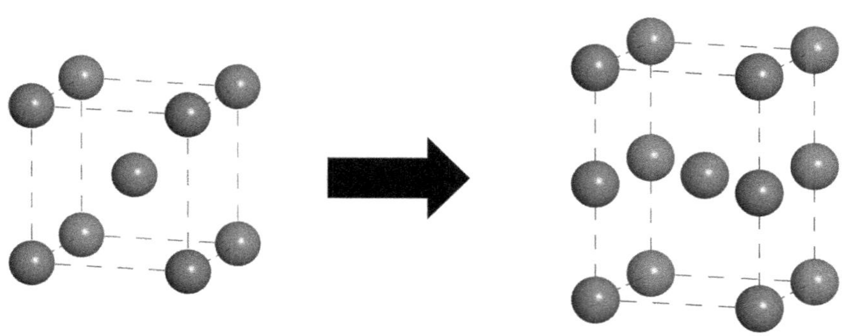

Figure 6.12 Incorporation of carbon into body-centred cubic iron to body-centred tetragonal iron, adding stress to the lattice following carbon incorporation (adapted from Velling et al. [71]).

Carbon steel is named according to AISI/SAE (American Iron and Steel Institute/Society of Automotive Engineers) designation [78]. The general structure of the name is ABXX.

- The A and B represent the other materials present in the steel.
- X values are the carbon content of the steel in wt.% multiplied by 100, e.g. XX = 60 → 0.6 wt.% C
  - 3 X's are required if the carbon content is above 1 wt.%.

There may also be other materials added, for example a B before carbon content would show some boron is present. Some examples of these codes are provided in Table 6.8.

Some steels have 'common names' as they are used extensively in various industries. For example, 'mild steel' has a carbon content between 0.16 and 0.29 wt.% whilst the carbon content in 'high-carbon steel' can be anywhere from 0.6 to 2.0 wt.%. Low, medium and high carbon have maximum carbon contents of 0.25 wt.%, 0.25–0.5 wt.% and 0.5–1.25 wt.%. For tool steel, the carbon grade is preferably as close to 0.8 wt.% carbon as possible to allow the heat treatment to be most effective.

Stainless steel is an iron–chromium alloy containing several other alloying elements to give the material its desired properties. It is known to be highly corrosion-resistant (dependent on the type of steel, see Table 6.9). The corrosion resistance of stainless steel comes from the formation of a thin overlayer of chromium oxide (leached out of the crystal structure of the steel), which protects the steel from further corrosion. Broadly, stainless steel can be grouped into five different types [79]; the properties of the different types of stainless steel are given in Table 6.9.

Table 6.8 Some examples of names of specifications of carbon steel by AISI/SAE convention (data from The Engineering Toolbox [78]).

Structure: ABXX		**XX = Carbon content (100*C-wt.%)**	
A = 1 → Carbon steel		*A = 4 → Molybdenum steel*	
10XX	Plain carbon, Mn <1%	40XX	Mo: 0.2–0.24%
15XX	Mn: 1.0–1.65%	41XX	Mo: 0.12–0.30%; Cr: 0.50–0.95%
A = 2 → Nickel steel		47XX	Mo: 0.25%; Cr: 0.50–0.80%; Ni: 1.82%
23XX	Ni: 3.5%	46XX	Mo: 0.20–0.25%; Ni: 0.85–1.82%
25XX	Ni: 5.0%	*A = 5 → Chromium steel*	
A = 3 → Nickel–chromium steel		50XX	Cr: 0.27–0.65% (50XXX: > 1% C)
31XX	Ni: 1.25%; Cr: 0.65–0.80%	52XX	Cr: 0.60–0.95%; V : 0.10–0.15%
32XX	Ni: 1.75%; Cr: 1.07%	*A = 6 → Chromium–vanadium steel*	
33XX	Ni: 3.50%; Cr: 1.50–1.57%	61XX	Cr: 0.60–0.95%; V : 0.10–0.15%

Table 6.9 Types of stainless steel (data from Unified Alloys [79]).

Alloy group		Austenitic	Duplex	Ferritic	Martensitic	Precipitation hardening
Magnetic		No	Yes	Yes	Yes	Yes
Work hardening		Very high	Medium	Medium	Medium	Medium
Corrosion resistance		High	Very high	Medium	Medium	Medium
Harden		Cold work	No	No	Quench & temper	Age harden
Ductility		Very high	Medium	Medium	Low	Medium
Temperature resistance	*High T*	Very high	Low	High	Low	Low
	Low T	Very high	Medium	Low	Low	Low
Weldability		Very high	High	Low	Low	High

Austenitic steel has an FCC structure supported by high chromium and nickel loadings, with a near-total absence of carbon. A very thin oxidic film is rapidly produced on the material surface, providing very high corrosion resistance. This film is typically only a few atoms thick, so the patina is invisible to the eye. Austenitic steel can be used over a wide range of temperatures and is especially useful for cryogenic storage. It cannot be heat-treated to harden, so it can only be work-hardened [80].

Around two-thirds of stainless steel produced is austenitic as it has the widest range of working conditions and the highest corrosion resistance; as such, it is often the only type of stainless steel, which is called stainless steel outside of metalworking industries. The most prevalent austenitic stainless steel is 304 (according to the SAE grading – 300 series are chromium–nickel alloys). 304 stainless steel is sometimes referred to as 18/8 stainless steel as it contains roughly 18% chromium and 8% nickel (with the balance iron; see Table 6.10). Austenitic stainless steel has a maximum carbon loading, as the inclusion of carbon increases the likelihood of corrosion. Steel's strength comes from its FCC structure, with especially low carbon loadings, receiving special designation. The second most common grade is 316, which has some added molybdenum, further improving its corrosion resistance. It is popular for food and medical uses.

Duplex steel is a mixture of austenitic and ferritic stainless steels, containing a two-phase microstructure with roughly equal amounts of the BCC and FCC structures. It typically contains large amounts of chromium, nickel and molybdenum, whilst having a very low carbon content. Duplex steel has increased strength, toughness, weldability and corrosion resistance, allowing for a wide range of engineering applications [82].

Ferritic steel has a BCC structure, with increased chromium but very low nickel and carbon loadings. This type of steel cannot be hardened. Due to the lack of

Table 6.10 Stainless steel grades and their composition in wt.% (data from AlHazaa and Haneklaus [81]).

	Fe	Cr	Ni	Mo	Mn	Si	N	P	C (max)
304	Balance	18–20	8.0–10.5	—	2	0.75	0.1	0.045	0.08
304l	Balance	18–20	8.0–10.5	—	2	0.75	0.1	0.045	0.03
316	Balance	16–18	10–14	2–3	2	0.75	0.1	0.045	0.08
316l	Balance	16–18	10–14	2–3	2	0.75	0.1	0.045	0.03

nickel, ferritic steel can be used as an alternative to austenitic steel when a relatively low-cost option that provides corrosion resistance is required. It is often used for domestic appliances, car exhaust systems, etc. [83].

Martensitic steel has a BCC structure with a varied chromium loading allowed due to inclusion of materials such as molybdenum and vanadium. It typically contains no nickel and has carbon contents like that of carbon steel. Martensitic steel shows hardening characteristics like those of carbon steel, although it has poor weldability due to brittleness. Its slight corrosion resistance means that it is used as a substitute to carbon steel in engineering applications [84].

Precipitation hardening steel can be heat-treated to achieve high hardness, is weldable and has high corrosion resistance. It has diverse uses and can be customised for each use case.

6.3.2 Iron and Steel Industry

Iron ore makes up approximately 94% of all the metal ores mined in the world, and 98% of the mined iron goes into steel-making [27]. It is also worth noting that most of the chromium, manganese, zinc and nickel that is mined is used to produce steel (largely stainless steel).

Table 6.11 shows the countries with the largest reserves (in 2023) and their annual production of iron ore as in 2021. The largest iron ore deposits can be found in Australia (31.2% of the world reserves) with other large deposits found in Brazil (17.3% of the world reserve), Russia (16.2% of the world reserve) and China (8.0% of the world reserve), though many other countries do have significant iron ore reserves.

Iron ore is principally Fe_2O_3 (haematite) with a weight content of iron of a maximum of 69.9 wt.%. Thus, the highest grade of iron ore is 69.9% iron, though this may be exceeded if the iron ore contains Fe_3O_4 (magnetite). The presence of other elements lowers the iron content; thus, iron ore tends to contain somewhere between 30 and 60 wt.% iron. For example, South Africa has very high-grade ore, containing around 63 wt.% iron, whilst Kazakhstan has much lower-grade ore, with around 36% iron (see Table 6.11).

Australia, with its largest reserves, is also the largest producer of iron ore, mining 37.5% of the total amount of iron ore mined in the world in 2021. It can be noted that most countries mine iron ore with a relatively high iron content (60–70 wt.% Fe),

Table 6.11 Iron ore reserves in million tonnes in 2023 and the amount mined in million tonnes per annum in 2021 (data from References [85–87]).

	Reserves			Mined		
	Crude ore	Fe- ore	% Fe	Crude ore[a)]	Fe-ore	% Fe
Australia	58 000	27 000	46.6	900 (3.3%)	560	62.2
Brazil	34 000	15 000	44.1	380 (2.5%)	240	63.2
Russia	29 000	14 000	44.1	100 (0.7%)	71	71.0
China	20 000	6 900	34.5	360 (5.2%)	220	61.1
Ukraine	6 500	2 300	33.3	81 (3.5%)	51	63.0
Canada	6 000	2 300	38.3	68 (3.0%)	41	60.3
India	5 500	3 400	61.8	240 (7.1%)	150	62.5
Iran	3 300	1 500	45.5			
USA	3 100	1 300	41.9	46 (3.5%)	29	63.0
Peru	2 600	1 200	46.2	16 (1.3%)	11	68.8
Kazakhstan	2 500	900	36.0	64 (7.1%)	13	20.3
Sweden	1 300	600	46.2	40 (6.7%)	28	70.0
South Africa	990	620	62.6	61 (9.8%)	39	63.9
World	191 340	86 620	45.3	2401 (2.8%)	1476	

a) Values in brackets represent the amount of crude ore mined annually relative to the crude ore reserves as an indication of the number of years the production can be sustained at the current level.

much higher than the average iron content of the reserves. This implies that with time the quality of iron ore will decrease, and in the future more effort will be needed to extract iron from the ore.

Figure 6.13 shows the spot price of iron ore (iron content: 62%) delivered, including cost and freight (CFR), at Tianjin port in China over the last few decades. In the period 1990–2003, the price of iron ore was stable and in 2004 US$ 13.07 ± 0.94 per metric tonne, following which the price of iron ore increased sharply reaching US$

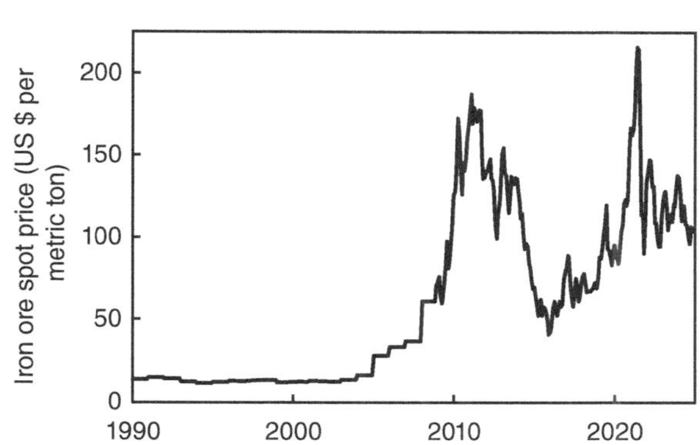

Figure 6.13 Spot price of import of iron ore (Fe-content: 62% Fe) in China (CFR, Tianjin port) in 2024 US$ per metric tonne (data from References [88–90]).

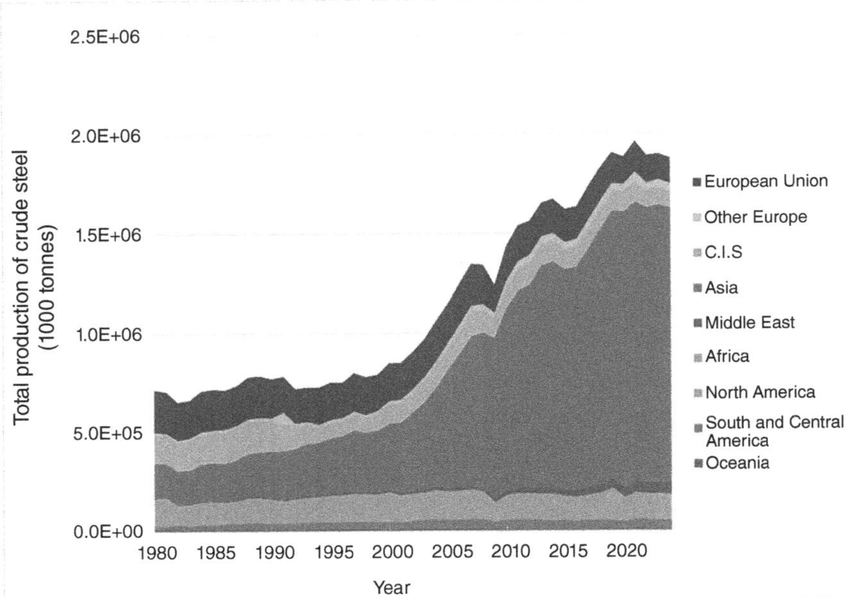

Figure 6.14 Production of steel by country between 1980 and 2024 (data sourced from the World Steel Association [93]).

180 per metric tonne in 2011 because of the growing demand in emerging markets and in particular China (also named the 2000s commodity boom) [91]. Lowering of demand in conjunction with increased production resulted in a decrease in the iron ore price from 2012 to 2016 [92]. From 2016 onwards, the spot price for iron ore has been gradually increasing, albeit with a relatively large variability in its value.

The production of steel by country is shown in Figure 6.14. Europe (including the previous USSR), the United States and Japan were traditionally the largest steel producers, with their large local needs for steel [93]. It should be noted that in Japan, the raw material required for steel production is imported. China is currently the largest producer of steel. It produces this steel to support its rapidly growing infrastructure and exports a large amount of steel. Over the last decade, growth in the steel-making capacity in the rest of Asia has become significant.

Iron and steel mills come in various sizes, depending on both location and the type of process used. The largest plants are located in China, and these tend to make use of blast furnaces in an integrated process that involves both iron and steel manufacture. Plants making use of a blast furnace with an integrated process have capacities ranging from ca. 500 000 to 20 000 000 metric tonnes per annum [94]. Plants making use of electric processing tend to be much smaller, with capacities of ca. 500 000–3 000 000 metric tonnes per annum [94]. Plants employing other technologies tend to range in size. Most notably, direct reduction technologies are the third most used type of iron production plant, representing approximately a quarter of each the blast furnace and electric process. They range in size from 1 000 000 to 10 000 000 million metric tonnes per annum [94]. It is worth noting that whilst there are a similar number of electric and blast furnace integrated plants, the total production of the blast furnace plants is several times more than that of all-electric processes. See the Global Energy Monitor [94] for an interactive map on the size of steel plants around the world.

6.3.3 Current Process(es)

6.3.3.1 Overall Process

There are many sub-processes in the production of steel. Figure 6.15 shows a process flow diagram of a typical steel process. Typically, iron ore is crushed to between 7 and 25 mm before it is added to the blast furnace along with coke, a nearly pure carbon source made from coal, and air. In the blast furnace, iron ore is reduced to metallic iron:

$$Fe_2O_3 + 1.5\,C \rightarrow 2\,Fe + 1.5\,CO_2 \qquad \Delta H^0_{rxn} = 234.5\,\frac{kJ}{mol\ Fe_2O_3}$$

Limestone is added as a flux agent to assist in removing unwanted compounds in the ore.

The resulting material is called pig iron and contains roughly 6 wt.% carbon with the balance iron. The metallic iron then undergoes a decarburisation step, in the 'basic oxygen furnace' (BOF) or in the 'electric arc furnace' (EAF) depending on the technology used, which essentially burns off excess carbon to lower the carbon content to a desired level. From here, the iron/carbon mixture typically goes through a ladle furnace, which may further refine it, after which it can be solidified as carbon steel. Scrap steel is typically added before the BOF or EAF to be recycled with the fresh steel. When other materials are desired in the carbon steel, this is often achieved by adding the ore of that material to the process. This can be done easily as iron is typically a major impurity in the ores of other materials, meaning they do not necessarily bring many undesirable impurities into the mixture. To produce stainless steel, steel is then sent through an 'argon oxygen decarboniser' (AOD) to remove any excess carbon which may be present, with the alloying materials added, typically, in the EAF.

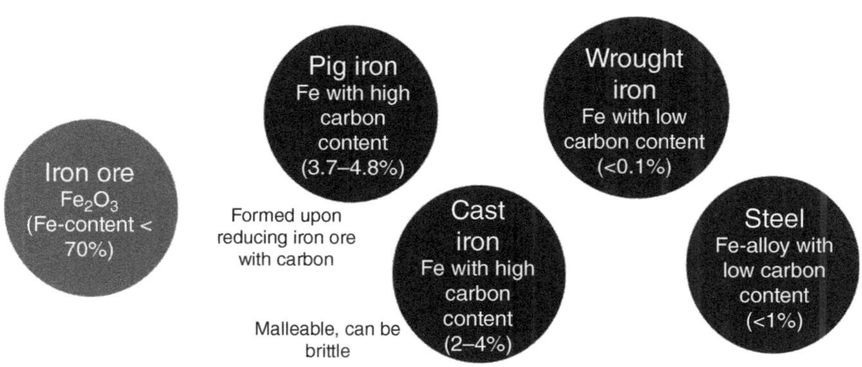

Figure 6.15 Process flow diagram of steel-making process.

An overview of carbon content for various types of steel and iron products is given in Figure 6.16. Note that for stainless steel, the carbon content is nearly zero rather than just 'below 1%' as it is for carbon steel.

6.3.3.2 Coke Production

Classically, coke is used as both the heat source and the reductant in the reduction of iron ore (primarily haematite Fe_2O_3) to pig iron in a blast furnace. To produce coke, coal is pyrolysed, known as the destructive distillation of coal or 'coking'. In the process, coal is heated in the absence of air to around 1100 °C, which removes all the volatile components out of the coal: at 100 °C water is evaporated, tars and oils are formed and evaporated at around 400–600 °C, followed by partial decomposition of coal with hydrogen removal at temperatures around 1100 °C [95]. The remaining material is called coke, an essentially pure carbon material that is highly porous and has a very high surface area.

The coke production is done in a specially designed oven called a coke oven, the design of which has become increasingly complex over the centuries to maximise the energy efficiency of the coke formation process from the coal, where multiple regeneration zones and intricate piping and nozzle designs are used to capture heat from flue gas to pre-heat the entering coal. The primary by-products of the coking process are methane and hydrogen gas, which are typically captured and burnt as a heat source elsewhere on the plant. However, the release of these gases, as well as heat loss in the coking ovens, results in roughly 30% of the enthalpy of the coal coming into the process being lost during the coke formation (see Figure 6.17).

6.3.3.3 Iron Production – The Blast Furnace

The blast furnace (Figure 6.18) is fed with iron ore with iron oxide, Fe_2O_3 (haematite) as the primary source of iron. Lower-purity iron ore, containing less than the 69.9 wt.% iron of pure haematite, may be used, though an increased amount of slag will be produced due to the impurities in the ore. Coke is added as the reducing agent for the iron oxide. The reduction of iron ore requires, stoichiometrically 0.161 tonne C/tonne Fe. However, more carbon is needed than stoichiometrically required as carbon is also used as the source of heat. The reduction of haematite requires 234.5 kJ/mol Fe_2O_3 or 117.8 kJ/mol Fe (2.11 GJ/tonne Fe). Iron leaves the blast furnace typically in the liquid form and the enthalpy change between solid iron at room temperature and liquid iron at its melting point needs to be taken into account. The heat is supplied by combusting some of the coke with air, and thus the change in the enthalpy of the gases emerging from the blast furnace, CO_2 and N_2, will also need to be taken into account. The total amount of coke required relative to the amount of iron fed to blast furnace on a molar basis is thus given by

$$\frac{n_{\mathrm{C,coke}}}{n_{\mathrm{Fe}}} = \frac{n_{\mathrm{C,reductant}}}{n_{\mathrm{Fe}}} + \frac{n_{\mathrm{C,combustion}}}{n_{\mathrm{Fe}}} = \frac{3}{4} + \frac{n_{\mathrm{C,combustion}}}{n_{\mathrm{Fe}}}.$$

CO_2 is formed in the reduction of iron and the combustion of coke $\frac{n_{\mathrm{CO_2}}}{n_{\mathrm{Fe}}} = \frac{n_{\mathrm{C,coke}}}{n_{\mathrm{Fe}}}$.

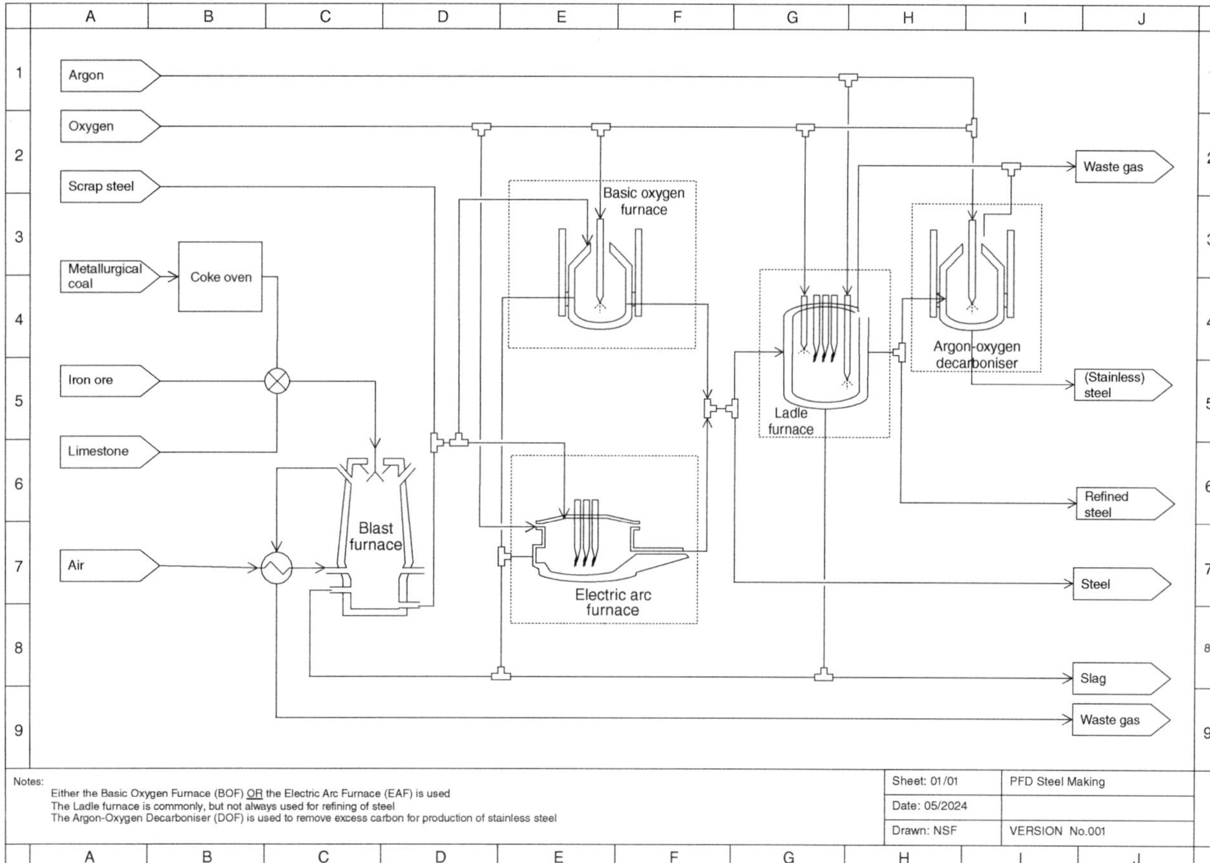

Figure 6.16 Overview of carbon content for various types of steel and iron products.

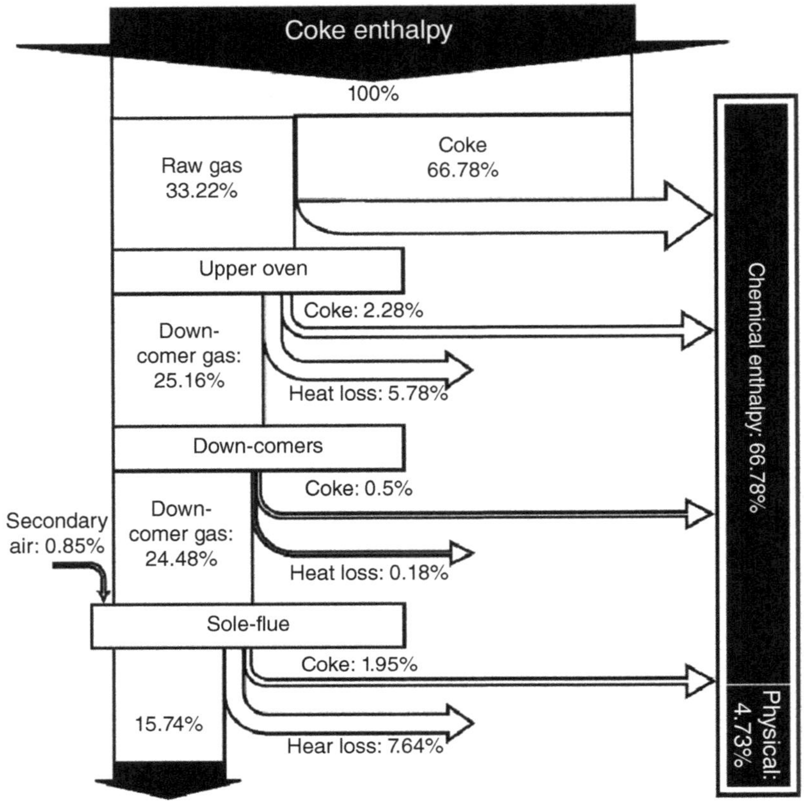

Figure 6.17 Enthalpy loss in coke ovens (redrawn from Buczynski et al. [96]).

The combustion of 1 mol of coke (carbon) requires 1 mol of O_2 in air and thus, the corresponding amount of N_2 equals

$$\frac{n_{N_2}}{n_{Fe}} = \frac{0.79}{0.21} \cdot \frac{n_{O_2}}{n_{Fe}} = \frac{0.79}{0.21} \cdot \frac{n_{C,combustion}}{n_{Fe}} = \frac{0.79}{0.21} \cdot \left(\frac{n_{C,coke}}{n_{Fe}} - \frac{3}{4} \right).$$

The energy balance can now be used to determine the minimum amount of coke required to feed to the blast furnace depending on the temperature of the flue gas leaving the blast furnace ($T_{g,blast\ furnace}$), and assuming air enters the blast furnace at the reference temperature (25 °C):

$$\frac{n_{C,combustion}}{n_{Fe}} \cdot \Delta H_{C,comb.} = \Delta H_{red.:\frac{1}{2}Fe_2O_3+\frac{3}{4}C \to Fe+\frac{3}{4}CO_2} + \Delta H_{fusion,Fe} + \int_{T_{ref}}^{T_{melt}} C_{P,Fe} \cdot dT + \frac{n_{CO_2}}{n_{Fe}} \cdot \int_{T_{ref}}^{T_{g,blast\ furnace}} C_{P,CO_2} \cdot dT + \frac{n_{N_2}}{n_{Fe}} \cdot \int_{T_{ref}}^{T_{g,blast\ furnace}} C_{P,N_2} \cdot dT$$

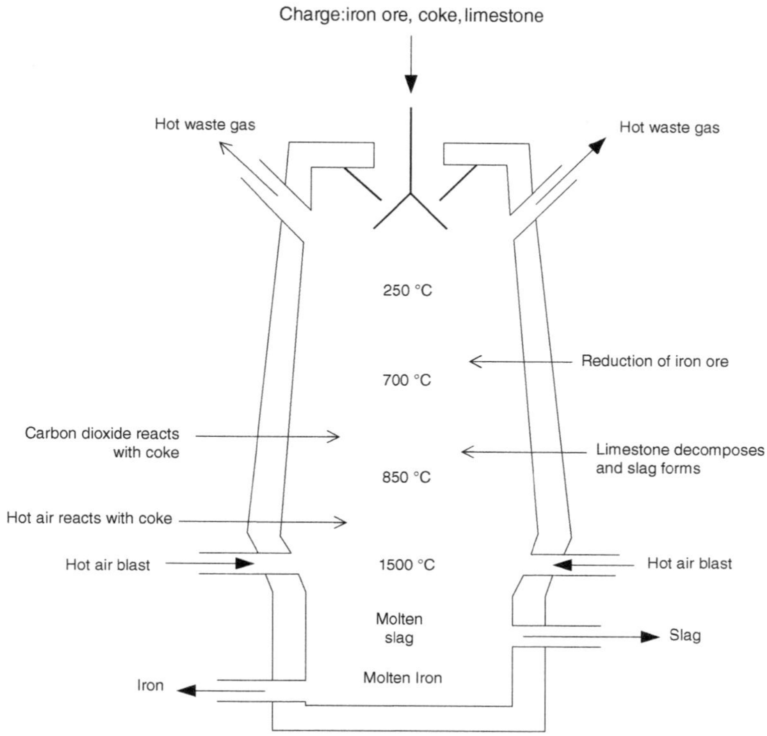

Figure 6.18 Blast furnace (redrawn from Vendatu [97]).

$$\frac{n_{\text{C,coke}}}{n_{\text{Fe}}} = \frac{\Delta H_{\text{red.}:\frac{1}{2}\text{Fe}_2\text{O}_3+\frac{3}{4}\text{C}\rightarrow\text{Fe}+\frac{3}{4}\text{CO}_2} + \Delta H_{\text{fusion,Fe}} + \frac{3}{4}\cdot\Delta H_{\text{C,combustion}} + \int_{T_{\text{ref}}}^{T_{\text{melt}}} C_{\text{P,Fe}}\cdot \text{d}T - \frac{0.79}{0.21}\cdot\frac{3}{4}\cdot\int_{T_{\text{ref}}}^{T_{\text{g,blast furnace}}} C_{\text{P,N}_2}\cdot\text{d}T}{\Delta H_{\text{C,combustion}} - \int_{T_{\text{ref}}}^{T_{\text{g,blast furnace}}}\left(C_{\text{P,CO}_2} + \frac{0.79}{0.21}\cdot C_{\text{P,N}_2}\right)\cdot\text{d}T}$$

Substituting the values ($\Delta H_{\text{red.}:\frac{1}{2}\text{Fe}_2\text{O}_3+\frac{3}{4}\text{C}\rightarrow\text{Fe}+\frac{3}{4}\text{CO}_2} = 117.8\frac{\text{kJ}}{\text{mol}}$; $\Delta H_{\text{fusion,Fe}} = 13.8\frac{\text{kJ}}{\text{mol}}$; sensible heat of Fe: 49.6 kJ/mol) results a total amount of carbon of 2.64 mol C/mol Fe or 0.57 tonne C/tonne Fe, if the flue gas exits the blast furnace at the melting temperature of iron as well. Figure 6.19 shows the amount of coke required as a function of the temperature of the effluent gas, and using the sensible heat contained in the gas will reduce the amount of coke required significantly. In reality, the amount of coke is around 0.4 tonne C/tonne Fe is fed [98], accounting for both the reduced heat loss through the flue gas, as well as the incomplete combustion of carbon.

The actual amount of coke of 0.4 tonne C/tonne Fe fed to the blast furnace [98] implies that the process in the blast furnace will result in ca. 1.47 tonne CO_2/tonne Fe. The total amount of carbon released from the overall iron production is much more, however, as other processes such as the production of coke also contribute to the CO_2 emissions (the production of coke can produce 0.1–0.8 tonne CO_2/tonne Fe [99, 100]).

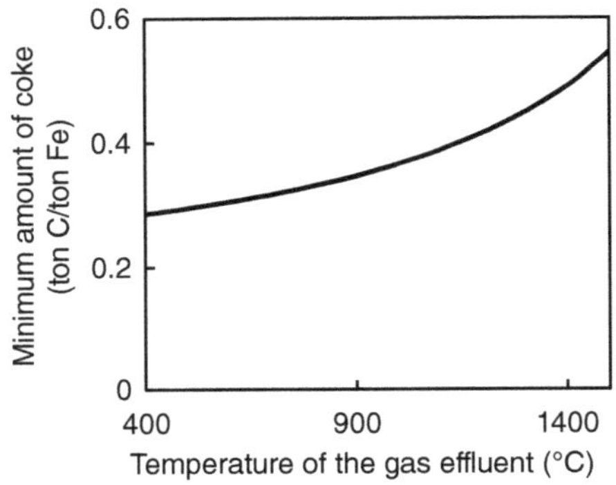

Figure 6.19 The required amount of coke fed to the blast furnace as a function of the temperature of the effluent gas assuming only the formation of CO_2.

The energy requirement for the blast furnace can also be estimated based on the coke requirement and the heat of combustion of carbon (393.5 kJ/mol) to 13.1 GJ/tonne Fe (3.64 MWh/tonne Fe) assuming all carbon is converted to CO_2 (in reality, this is somewhat lower as not all carbon is converted into CO_2).

Multiple reactions occur inside the blast furnace. Starting at the bottom of the furnace, hot air is blasted into the furnace, where it reacts with coke having come down from the top of the furnace. This follows the normal combustion reaction and results in the formation of CO_2 and a large amount of heat. This heat provides the heating for the furnace, raising the temperature to around 1500 °C.

$$C(s) + O_2(g) \rightarrow CO_2(g)$$

When oxygen is depleted, CO_2 formed from the combustion of coke can react with coke coming down from the top of the furnace through the so-called reverse Boudouard reaction yielding CO, a more facile reductant for the reduction of haematite than carbon.

$$CO_2(g) + C(s) \rightarrow 2CO(g)$$

The reduction of iron oxide occurs upon reaction between gaseous carbon monoxide and haematite, forming CO_2 in the process. This reaction produces most of the CO_2 in the blast furnace. If the temperature is high enough (>1184 °C depending on carbon content), molten iron droplets will form dropping to the bottom of the blast furnace to be removed as molten liquid iron, and the overall reaction can be written as:

$$Fe_2O_3(s) + 3CO(g) \rightarrow 2Fe(l) + 3CO_2(g)$$

Not all CO is converted into CO_2 and the effluent gas of the blast furnace is not only rich in CO_2 but also in CO (see Table 6.12). As such, the effluent gas has a relatively high heating value, which can be used in other parts of the plant.

Limestone ($CaCO_3$) is added to the blast furnace together with iron ore as a flux agent to remove impurities in iron ore. At ca. 800 °C, around the middle of the furnace, limestone decomposes to form lime (CaO), releasing more CO_2.

$$CaCO_3(s) \rightarrow CaO(s) + CO_2(g)$$

Table 6.12 Typical composition of flue gas from blast furnace (data from Xi et al. [101]).

Component	Units	Composition
CO_2	vol% (dry)	20%
CO	vol% (dry)	25%
H_2	vol% (dry)	3%
N_2 (may contain O_2)	vol% (dry)	49%
H_2S	mg/Nm^3	10
Particulate matter	mg/Nm^3	5
Mn	mg/Nm^3	0.2
Pb	mg/Nm^3	0.05
Zn	mg/Nm^3	0.05

This is the same reaction as the primary one releasing CO_2 in the cement industry. Here, however, lime is used as a flux rather than as the final product. Lime reacts with sand, which can be modelled as silica (SiO_2), to form calcium silicate, referred to as slag (melting point >1460 °C [102], thus, it does not melt in the blast furnace). In this sense, the blast furnace acts both as a reducer and as a purifier for the iron.

$$CaO(s) + SiO_2(s)(sand) \rightarrow CaSiO_3(s)(slag)$$

The slag drops down the furnace and floats (possibly partially as a liquid depending on the CaO to SiO_2 ratio) on top of the molten iron, as it has a lower density, and can thus be removed from the iron phase. This separation is important for the final purity of the iron as the silica can form inclusions in the crystal structure, significantly weakening the steel product. Hence, the production of liquid iron in the process is necessary to obtain high-purity iron.

The amount of limestone required depends on the purity of the ore as it enters the furnace. Pure ore would not contain any sand and would not need any flux, whilst low-grade ore would need a large amount of flux to purify the iron. Thus, using low-grade ore will increase energy consumption, as the flux and the impurity will need to be heated up to the melting temperature of iron. Furthermore, CO_2 emissions will increase further when using a low-grade ore together with limestone as a flux agent.

The blast furnace produces a stream of liquid reduced iron in the form of iron carbide which contains around 6 wt.% carbon, which is referred to as Pig Iron.

Exergy Analysis of the Blast Furnace An exergy analysis of the blast furnace conducted by Zhang et al. [103], as shown in Figure 6.20, indicated that 64% of the exergy coming into the system via the charge (coke, iron ore, flux) and additional coal used for heating outside the blast furnace is recovered in the form of hot metal – though it will be difficult to recover the heat from the metal in a meaningful way so there will be less exergy recovered once the metal is cooled.

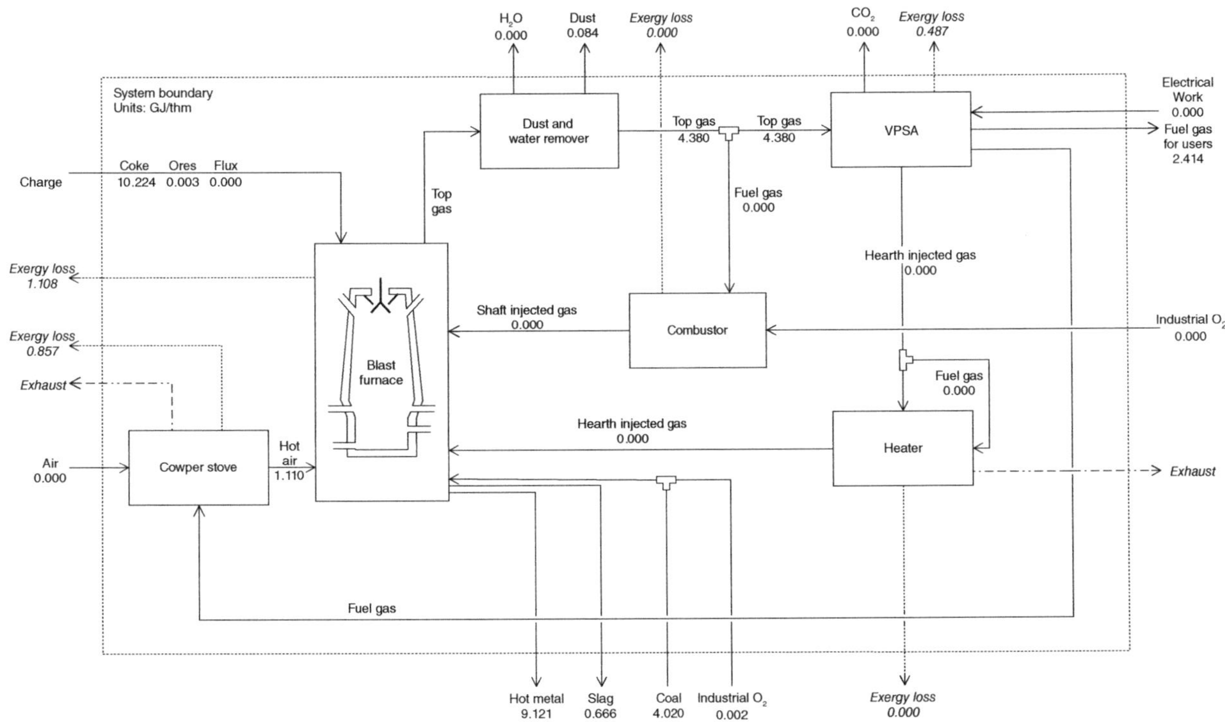

Figure 6.20 Exergy analysis of the blast furnace (redrawn from Zhang et al. [103]).

It is estimated that only around 20% of the exergy is lost entirely, as the gas leaving the top of the blast furnace contains a large amount of CO, which may be recovered by burning, as a fuel, by other users downstream. The slag also carries out some of the exergy, which, though it may not have much energetic value, may have some use downstream. The major source of exergy loss is, intuitively, the blast furnace, as this is the area with the highest temperature and the largest unit and, thus, is the place where most of the exergy is lost as heat.

6.3.3.4 Basic Oxygen Steel-Making

The carbon content out of the blast furnace is typically high, too high for the steel (ca. 5–6 wt.% when ca. 0.8 wt.% is needed for carbon steel). One option to rectify this is to use a basic oxygen steel converter (see Figure 6.21). Here, a supersonic oxygen jet (p = 7–10 bar) is directed onto the surface of a metal bath. Several exothermic reactions take place between the oxygen from the jet and the materials from the blast furnace. Firstly, much of the carbon is removed from the pig iron ('decarburisation'), decreasing the carbon content from around 5–6 wt.% to around 0.5–1 wt.% in the iron melt.

$$C + \frac{1}{2}O_2 \rightarrow CO \quad \Delta H_{\text{rxn},1900\text{K}} = -118.6\ \text{kJ/mol}$$

There is additionally further combustion of the CO in the furnace, which increases the temperature.

$$CO + \frac{1}{2}O_2 \rightarrow CO_2 \quad \Delta H_{\text{rxn},1900\text{K}} = -277.9\ \text{kJ/mol}$$

For instance, the reduction of the carbon content in a material containing 3.67 wt.% carbon (see Table 6.12), to 0.5 wt.% will release 1.06 GJ/tonne steel, if all carbon is converted to CO_2. This amount of energy is sufficient to heat up and melt steel.

Unfortunately, some of the iron is also oxidised to ferrous oxide (FeO):

$$Fe + \frac{1}{2}O_2 \rightarrow FeO \quad \Delta H_{\text{rxn},1900\text{K}} = -242.8\ \text{kJ/mol}$$

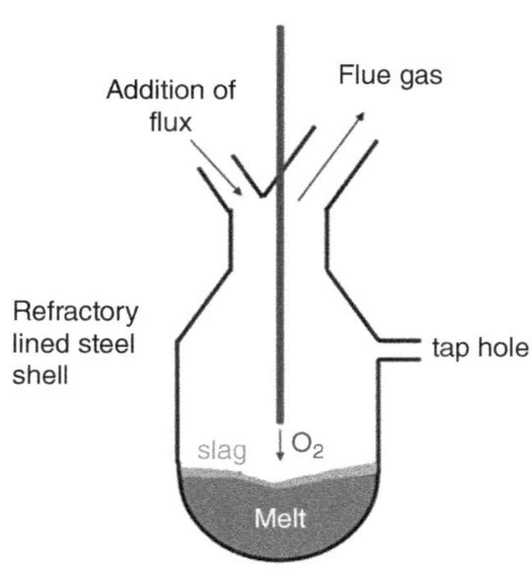

Figure 6.21 Basic oxygen furnace (BOF) steel converter.

Other impurities, such as silicon are also combusted out and will be removed as slag.

$$Si + O_2 \rightarrow SiO_2 \quad \Delta H_{rxn,1900K} = -945.2\,kJ/mol$$

The process utilises oxygen rather than air as nitrogen incorporation in iron may cause embrittlement of the iron [104].

Scrap metal can be added to the process to absorb excess heat, and recycle the scrap metal.

Table 6.13 shows the elements present in the different charges that can be added to the BOF. The masses shown are for a typical feed but may vary substantially, depending on the requirements of the overall process.

Figure 6.22 shows the temperature and carbon content of the steel as a function of the time in the BOF for each batch cycle. The exothermic nature of the combustion reactions occurring raise the temperature substantially over the duration of the 'blow', with the initial decrease in temperature caused by charging some of the flux to the BOF (charging of the flux is performed at discrete times, thus causing a disturbance in the heating profile).

Table 6.13 Approximate amount of charged species and contained elements in a typical basic oxygen furnace feed (data from Madhavan et al. [105]).

Input	Hot charge	Heavy scrap	Light scrap	Pig iron	External scrap
Mass (kg)	170 000	3570	10 000	12 140	4284
C (wt.%)	4.3–5.0	0.08	0.08	4.5	0.05
Si (wt.%)	0.3–0.15	0.05	0.05	0.5	0.001
Mn (wt.%)	0.25–2.2	0.3	0.3	0.5	0.2
P (wt.%)	0.05–0.22	—	—	—	—
S (wt.%)	0.03–0.08	—	—	—	—

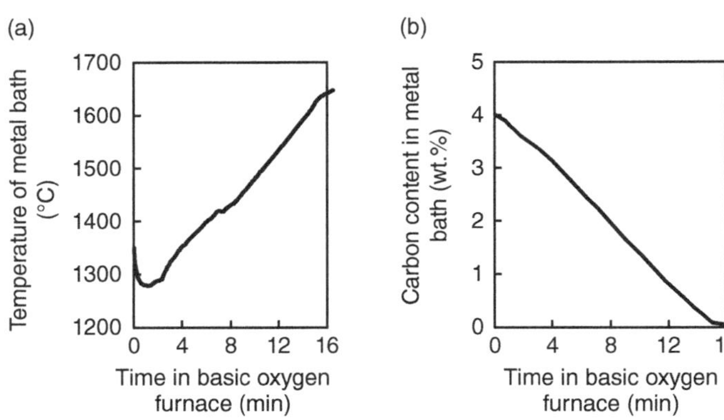

Figure 6.22 Temperature (a) and carbon content (b) in the metal bath in the basic oxygen furnace as a function of time (redrawn from References [106, 107]).

6.3.3.5 Electric Arc Furnace

The alternative and more modern counterpart to the BOF is the EAF [108] making up around 30% of modern steel manufacture (with the remaining steel still being produced using the BOF). A large amount of scrap metal is loaded into the furnace for recycling (recycled steel is predominantly processed in an EAF [109]) and alloys can be produced more easily, with more flexibility, with an EAF in comparison to a BOF.

In the electrical arc furnace (Figure 6.23), heating is done through an electrical arc, hence the name. An electrical arc extends from graphite electrodes towards the bottom of the furnace, with a layer of molten iron with slag on top. Heating is accomplished from the resistance of the air through which the arc extends. As with the BOF, oxygen is fed into the furnace which removes carbon and several of the impurities as slag by oxidation. The energy requirement of the EAF is around 0.5 MWh/tonne steel (1.80 GJ/tonne steel [111]), assuming that all feed is added to the furnace at room temperature (adding hot liquid iron directly to the EAF in an integrated process will reduce the energy cost significantly). This implies that the energy efficiency of the EAF is around 70–75%, as melting the steel starting at room temperature costs ca. 1.14 GJ/tonne steel. The power output from the electric arcs can be varied by raising and lowering the electrodes. As the process runs batch-wise, it has approximately a 37 -minute power-on-time, that is the time in which the metal is melted by the electric arc, and the turnaround time is ca. 40–50 minutes [108]. Hence, a batch containing 300 tonne of steel would thus require 214 MW.

The EAF benefits from using only electricity, meaning that it is relatively easy to adapt the process to renewable energy. Furthermore, scheduling operations will make this process adaptable to the intermittent nature of renewable energy.

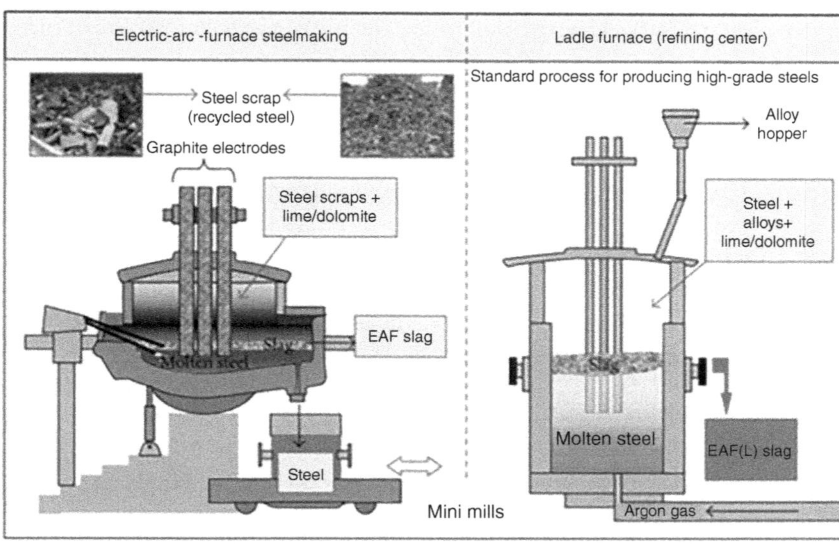

Figure 6.23 Cut-away diagram of an electric arc furnace and ladle furnace (Najm et al. [110] / with permission of Elsevier).

Stainless steel is also produced in an EAF or ladle furnace. The crucial factor here is to get to a sufficiently high temperature so that all the various metals in the alloy will be molten, after which the mixture will solidify in the alloyed phase. It should be noted that there are eutectic points on the solid–solid phase diagram for metallic mixtures, just like with the carbon-iron phase diagram shown previously, and so it is possible to have mixed phases of the various metals in the alloy. After heating, an argon-oxygen decarboniser is used to remove the remaining carbon to meet the specifications required for the steel.

6.3.4 Reducing Emissions in Steel Manufacturing

The total amount of CO_2 emitted in the production of steel can be as high as 2.0–3.1 tonne CO_2/tonne Fe [100, 112]. Emissions from the steel industry originate from the need to reduce iron ore, the high-temperature processes required to perform this reaction and the need to purify iron. Hence, reductions in the CO_2 emissions can be achieved by changing the reductant in the reduction process (or foregoing the reduction process by recycling steel) and changing the method of heating.

6.3.4.1 Potential for Recycle

Recycling of steel is of great importance from an energetic, raw material, and carbon release point of view. The major source of this reduction in the carbon release is the removal of the need to reduce the iron ore into metallic iron, thus not only saving the stoichiometrically required amount of CO_2 (0.59 tonne CO_2/tonne Fe), but all CO_2 emissions associated with the blast furnace. Melting of the (recycled) steel may still be required for its use after recycling, which requires a large amount of energy (ca. 63.4 kJ/mol Fe or 1.14 GJ/tonne steel, which corresponds to 0.32 MWh/tonne steel). However, using electrical energy rather than chemical energy in the form of carbon will reduce the amount of energy lost through the effluent gas.

Steel is, in principle 100% recyclable. However, recycling does have its limitations, when dealing with an expanding market (see Figure 6.15), where the increase in the steel demand will outpace the availability of scrap metal. Hence, the refining of iron ore will remain an important aspect of the steel industry.

6.3.4.2 Alternative Reductants

The most discussed method through which the iron and steel industry can be made more sustainable is using a 'renewable reductant', that is, a reductant that is not fossil-sourced coke. 'Renewable reductants' as a phrase can be ambiguous, as it is sometimes used to refer specifically to charcoal rather than other potential reductants which may be renewable, such as ammonia or hydrogen.

Charcoal may be used as a replacement for the coke, with charcoal essentially being coke derived from wood and other biomass rather than coal through pyrolysis. This can replace coke directly in the current steel refining processes. However, a typical plant size of 3 million tonnes of steel per annum producing 3 tonne CO_2/tonne Fe [112] would thus require 2.45 Mtonne of charcoal per annum. Realising that charcoal is produced from wood (with a yield in the range of 25–50%), assuming a density of wood of 500 kg/m^3 and a tree yielding 1 m^3 of wood to be used for the formation of charcoal, a total of $1–2 \cdot 10^7$ trees per annum would be required. When planting trees at 2000 trees per hectare, this plant would need to harvest each year an area between 49 and 98 km^2.

Direct reduction of iron ore generally refers to a reduction of solid (not molten) iron via a reducing gas, typically a mixture of CO and H_2, rather than the more classical reduction with coke. Reduction with coke is considered 'indirect' as there is first the combustion to CO_2 followed by the reduction of iron ore with the CO formed in situ from coke and CO_2 [113]. Figure 6.24 shows the direct reduction of iron ore using synthesis gas obtained by reforming of natural gas. The reducing gas is then used in a shaft furnace to reduce the iron ore, replacing the blast furnace without melting the iron, as the reaction occurs at ca. 900–1050 °C. The relatively low temperature allows the direct reduction of iron to be carried out as a continuous process with the reduction of iron ore pellets taking place in a shaft furnace, rotary kiln or a moving fluidised bed reactor. The product is typically a sponge iron, containing the same amount of iron as Pig Iron, but which is much more susceptible to re-oxidation and is thus either fed directly into the EAF or briquetted to reduce the rate of re-oxidation. Furthermore, impurities present in the iron ore will remain in the product (and will need to be removed in further processing).

Direct reduction of iron can also take place using hydrogen as the reductant. Hydrogen is a popular option in research [115] as it is discussed as an energy carrier (see Chapter 5) and it can be used directly:

$$Fe_2O_3 + 3H_2 \rightarrow 2Fe + 3H_2O \quad \Delta H^0_{rxn} = 100\frac{kJ}{mol\ Fe_2O_3}$$

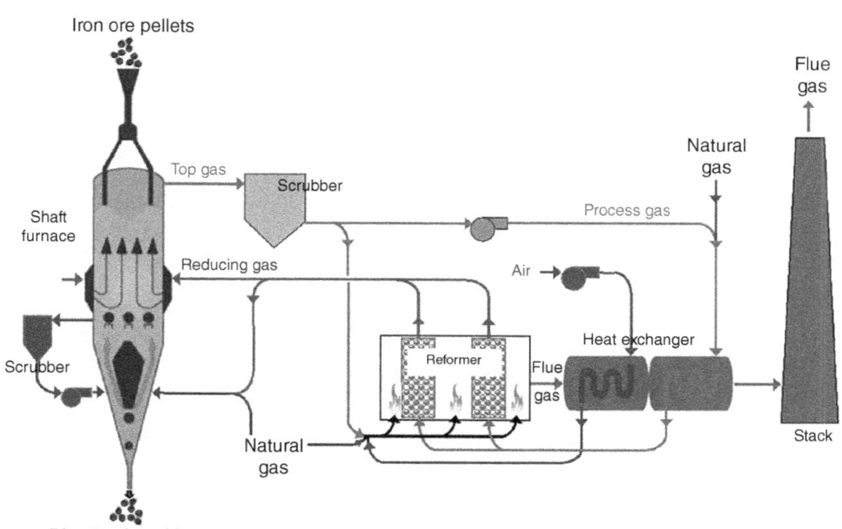

Figure 6.24 Direct reduction of iron using synthesis gas as the reducing gas formed from reforming of methane (Hamadeh et al. [114] / MDPI / CC BY 4.0).

The required amount of hydrogen on a mass basis is rather low owing to the light weight of hydrogen:

$$\frac{3\ \text{mol}H_2}{2\ \text{mol Fe}} = \frac{3\ \text{mol} \cdot \frac{2.016\ \text{g}}{\text{mol}}}{2\ \text{mol} \cdot \frac{55.845\ \text{g}}{\text{mol}}} = 0.054\frac{\text{g}\ H_2}{\text{g Fe}} = 54\frac{\text{g}\ H_2}{\text{kg Fe}}$$

A typical plant producing 3 million tonnes of steel per annum would thus require a steady flow of 5.15 kg H_2/s to reduce the required amount of iron.

The production of the amount of hydrogen required is associated with a large energy requirement. Approximately, 2.4 billion metric tonnes (see Table 6.11) crude iron ore, corresponding to 1.4 billion metric tonnes of iron are produced on an annual basis [116]. If all iron ore were to be reduced using hydrogen reduction, 75.8 million metric tonnes of hydrogen would be required.

Iron can be reduced using any chemical that is a stronger reducing agent than metallic iron. This could principally include multiple different metals, such as aluminium, zinc, magnesium, etc., though it more notably includes ammonia.

$$Fe_2O_3(s) + 2NH_3(g) \rightarrow 2Fe(s) + 3H_2O(g) + N_2(g) \quad \Delta H^0_{\text{rxn}} = 192\frac{\text{kJ}}{\text{mol}\ Fe_2O_3}$$

The use of ammonia as the reducing agent will require the same stoichiometric amount of hydrogen as used when hydrogen is the reductant, but has the advantage that it can be transported and stored readily on site as a liquid. A typical steel plant producing 3 million tonnes of steel per annum would then consume 4170 m^3 daily of liquid ammonia (with a density of 600 kg/m^3), and large storage containers would have to be placed on site. The advantage of using liquid ammonia as the reductant has to be offset with the inefficiencies in the ammonia synthesis, and the overall reduction when using ammonia may become more energy-intensive than using hydrogen itself as the reductant, although now having to deal with the higher physical risk from leaks and spills associated with ammonia. Furthermore, there is also a risk of nitrogen embrittlement when using ammonia.

6.3.4.3 High-Temperature Heat

High-temperature heating is required in the iron and steel industry to facilitate the kinetics of the reduction process, but also for the purification of iron. The reduction of iron ore, as in the direct iron reduction, requires temperatures of ca. 900–1050 °C, but the purification of iron requires molten iron (>1538 °C depending on carbon content).

The required temperature in the direct reduction of iron can be achieved by combusting the reductant in addition to using it as a reductant. This will increase the required amount of the reductant. For instance, when using hydrogen as the reductant, the amount of hydrogen required would increase by 49% when using the combustion of hydrogen with air to provide heat ($54.1\frac{\text{g}\ H_2}{\text{kg Fe}}$ is as the reductant; $7.5\frac{\text{g}\ H_2}{\text{kg Fe}}$ to bring up the heat of reaction – $100\frac{\text{kJ}}{\text{mol}\ Fe_2O_3} = 50\frac{\text{kJ}}{\text{mol Fe}} = 896\frac{\text{kJ}}{\text{kg Fe}}$-; $4.6\frac{\text{g}\ H_2}{\text{kg Fe}}$heating up iron to 1025 °C; enthalpy change: 30.7 kJ/mol; heating up gaseous compounds to 1025 °C: $23.0\frac{\text{g}\ H_2}{\text{kg Fe}}$). This would reduce to an increase of

46%, when using pure oxygen as the combustion gas, but could also be reduced by introducing a counter-current pre-heating system for the ore. Similarly, when using ammonia as the reductant, the required amount of ammonia will increase by 131% ($359\frac{\text{g NH}_3}{\text{kg Fe}}$ is as the reductant, $109\frac{\text{g NH}_3}{\text{kg Fe}}$ to bring up the heat of reaction – $192\frac{\text{kJ}}{\text{mol Fe}_2\text{O}_3} = 96\frac{\text{kJ}}{\text{mol Fe}} = 1717\frac{\text{kJ}}{\text{kg Fe}}$; $35\frac{\text{g NH}_3}{\text{kg Fe}}$ heating up iron to 1025 °C; $326\frac{\text{g NH}_3}{\text{kg Fe}}$ to heat up the gaseous compounds).

Heating the subsequent batch-type of process to purify iron and convert it into steel is more readily adaptable to renewable energy, as the content of the reactor is metallic, and Joule heating or inductive heating is now possible. Both can be readily adapted to utilise renewable energy using either an EAF or an induction furnace, requiring 0.50–0.54 MWh/tonne [117], and by proper scheduling of the operation to deal with its intermittent nature. Connecting the continuous, direct reduction of iron (DRI) with a batch operated EAF will require heat storage in the product of the direct reduction of iron, sponge iron, over a prolonged time. The current EAF design makes use of graphite electrodes. These wear down over about 8–10 hours [118] (equivalent to between every ca. second and fifth batch) and thus release CO_2 in the process. However, these can principally be replaced using bio-based sources or other carbon-neutral graphite sources, enabling the EAF to become carbon-neutral, even though it would release some CO_2.

6.3.4.4 Carbon Capture and Utilisation

The steel industry may present a valuable case for carbon capture and utilisation technologies, due to the relatively high concentration of CO_2 in the effluent gas. Moreover, the effluent gas of the blast furnace is rich not only in CO_2 but also in CO (see Table 6.12), making it a prime candidate to be integrated with modular CO-hydrogenation plants, such as methanol synthesis or the Fischer-Tropsch synthesis (see Chapter 5). Integration with the steel industry may present an opportunity to use the conversion of valuable feedstock of CO_x to organic molecules as a way to store renewable hydrogen in an energy-dense form.

6.3.5 Outlook

In 2021, 1.9 billion tonnes of steel were produced, and this is estimated to increase to 2.2 billion tonnes in 2050 [119] (approaching 250 kg per capita). This would require the use of 27 EJ p.a. of coal, 3 EJ p.a. of natural gas and 5 EJ p.a. of electricity using current technology (and thus consuming almost 6% of the world energy demand) emitting 2 tonne CO_2/tonne steel. As such, the steel industry is one of the major industries which will require a significant amount of decarbonisation going forward.

Most of the CO_2 is released, in the currently used process, from the blast furnace using coke as a reductant to produce metallic iron. This may be replaced by hydrogen as the reductant in 'direct reduction', although, at present, only ca. 7% of steel is produced through the implementation of direct reduction (not necessarily with hydrogen) in conjunction with an EAF. Steel plants currently operating a blast furnace for the reduction of iron would need to reconstruct much of their plant, replacing the blast furnace with a direct reduction unit, thus requiring a significant capital injection.

Carbon is currently introduced in steel in the blast furnace, with the excess being removed in the BOF or in the EAF. The direct reduction process using a mixture of hydrogen with natural gas (or landfill gas) is being developed [120] to introduce carbon into the steel whilst minimising the CO_2 emissions in this process.

Coke is not only used as a reductant but is also combusted to provide heat to the blast furnace. For the direct reduction process, this heat source will need to be replaced by the partial combustion of the renewable reductant, although this will increase the required amount of the renewable reductant significantly. The energetic cost will be lower for the direct reduction of iron in comparison to the blast furnace, as the temperature in the direct reduction of iron is lower (melting of iron is not required in this unit). EAFs are and will be used as the second stage in steel manufacture (ca. 25% of all steel is produced using an EAF [121]), and this is estimated to increase to 50% by 2050. This can be complemented using induction furnaces.

Purification of iron ore is now achieved in the blast furnace, with slag floating on liquid iron that can be easily removed. With the direct reduction of iron taking place below the melting point of iron, the main purification step will have to shift to the second stage of the steel producing part, viz. the EAF/induction furnace, which is a necessary step even when using hydrogen in the direct reduction of iron. These stages will need to deal with the removal of larger amounts of slag, due to the lack of removal in previous steps, and certainly seeing the expected decline in the grade of the iron ore going forward (see Table 6.11). The slag content atop the molten steel may present conductivity issues, further increasing the power requirement.

The electric energy consumption in the combination of a direct reduction unit and an EAF is expected to be higher than the energy consumption in a blast furnace integrated with an electric arc function, even though the product of the direct reduction unit is at a lower temperature. Hence, a shift in the energy consumption is expected to occur from the primary unit to the secondary unit in steel-making.

If electrolysis is to be used to produce the hydrogen (thus requiring 237.1 kJ/mol H_2), the total global, annual energy requirement to produce the hydrogen for iron reduction would be 8.9 EJ, if the electrolyser operated at a 100% efficiency. More realistically, assuming an efficiency of around 60% would imply that 14.9 EJ of energy would be required. By comparison, the total world production of electricity in 2022 was 30.9 PWh [122]or 111.2 EJ, indicating that the total electricity supply would need to increase by 13.4% just to meet the demand of hydrogen reduced steel. It should be noted that the actual energy requirement will be significantly higher as more energy is required to melt iron required for the iron purification.

Green steel production has, expectedly, shown an increased levelised cost compared to the current steel alternatives [123]. This is largely due to three factors: the cost of the direct reduction and EAFs in comparison to the blast and BOFs, and, more importantly, the capital and energy costs associated with the production of the hydrogen. Even so, the estimated prices of grey and green steel are expected to become significantly more comparable in the coming decades as the price of green hydrogen is expected to decrease and carbon taxes are expected to increase [124].

6.4 Sulphuric Acid

Contribution by Sinqobile V.L. Mahlaba.

Humans have a long history with sulphuric acid, which has been known since medieval times under a variety of names, namely oil of vitriol and spirit of vitriol. First produced from sulphate minerals, the systematic production of sulphuric acid, as it is known today, was first achieved by Johann Glauber, who burnt saltpetre (KNO_3) with sulphur in the presence of steam. The saltpetre acts as an oxidising agent upon decomposition, oxidising sulphur to SO_3, which then reacts with water to form sulphuric acid. This process was first implemented on a large scale in the eighteenth century, by Joshua Ward. However, it was not until John Roebuck began using this process in a lead chamber, leading to the invention of the 'lead chamber' process, which truly led to the industrialisation of the production of sulphuric acid. This process achieved a 35–40 vol% sulphuric acid concentration [125, 126].

Further developments, which were developed by Joseph-Louis Gay-Lussac and John Glover, resulted in improvements in the concentration (to achieve 78 vol.%). The present production of highly concentrated sulphuric acid (up to 98 vol.%) is credited to Peregrine Phillips, which involves the oxidation of sulphur dioxide in air, over a platinum catalyst. This was the precursor of the contact process, which received wide-scale implementation. In 1915, the predecessor of BASF developed the vanadium pentoxide catalyst. Around this time, the lead chamber process was phased out in favour of the contact process [125].

6.4.1 Material Specifications

Sulphuric acid is a clear, colourless, viscous, hygroscopic liquid with a boiling point of 280 °C [127]. Sulphuric acid can be produced with concentrations ranging from dilute to 100%. Beyond 100%, the solution is then referred to as oleum, which is made by dissolving sulphur trioxide in sulphuric acid (for example $H_2SO_4{\cdot}0.14SO_3$ is equivalent to $1.14SO_3{\cdot}H_2O$ or net 102.25% strength of sulphuric acid), which forms a fuming, viscous liquid [127]. Sulphuric acid is chemically and thermally stable.

6.4.2 Size and Production Scale in Sulphuric Acid Production

At present, sulphuric acid is the most produced chemical globally by volume (production in 2022: 265 million tonnes per annum [13]), and forms the backbone of human civilisation, being a necessary reactant in crucial processes such as the manufacture of phosphate fertilisers, pigments, textiles, chemicals, polymers, for petroleum refining and for refining of minerals [128]. The use of sulphuric acid by sector is presented in Figure 6.25. Sulphuric acid is predominantly used for the manufacture of fertilisers, followed by use in metal refining for the automotive industry. Sulphuric acid manufacture can be used as a measure of the economic development of a country [130] and the current top three producers are China, the United States of America and India. A typical sulphuric acid plant produces about 4000 tonnes of acid per day [128].

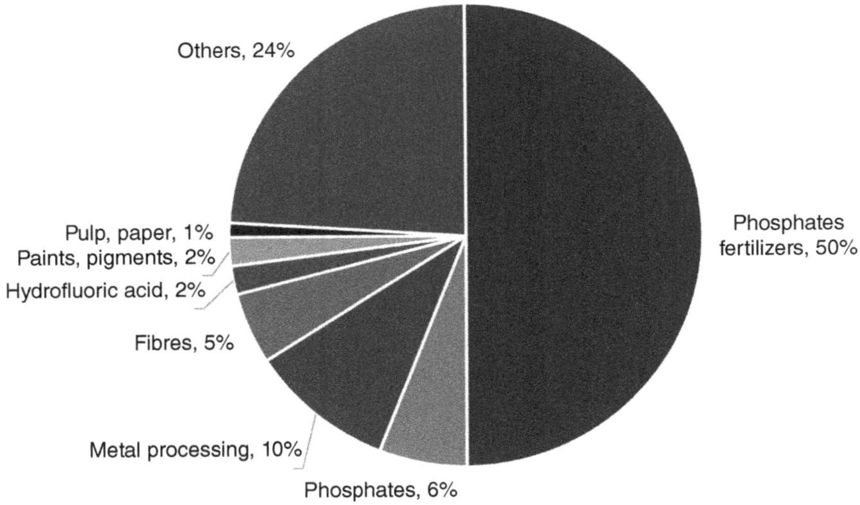

Figure 6.25 Uses of sulphuric acid across different industries in 2012 (adapted from The Essential Chemical Industry [129]).

6.4.3 Sulphuric Acid Production Process

Sulphuric acid is typically produced from elemental sulphur and its production follows four key steps:

A) Gas drying
B) Production of sulphur dioxide (sulphur burning)
C) Sulphur dioxide oxidation to sulphur trioxide (the contact process) and
D) Absorption of sulphur trioxide in water (in sulphuric acid) to form more of the sulphuric acid.

The process then occurs across three main parts of the plant, namely the sulphur burner to produce sulphur dioxide (if elemental sulphur is used), the multi-pass converter which is where the catalytic oxidation of SO_2 with air takes place and the absorber where the SO_3 product reacts with water to form sulphuric acid. Figure 6.26 shows the flowsheet to produce sulphuric acid.

The overall process to produce sulphuric acid is exothermic. Hence, heat recovery is an integral part of the design of modern sulphuric acid plants, with the target of recovering at least 60% of the total energy released during the production of sulphuric acid, mainly in the form of high-pressure steam [127].

Sulphur dioxide (SO_2) can be generated from a number of raw materials, such as solid elemental sulphur (S_8), hydrogen sulphide (H_2S), metal sulphides, e.g. pyrite, and spent sulphuric acid [125]. Elemental sulphur is obtained directly on a large scale from two main sources, viz. sulphur dumps and sulphur mines from geological deposits, which are found next to volcanic regions. Alternatively, sulphur is obtained via the Claus process [132], i.e. the conversion of hydrogen sulphide to elemental sulphur, e.g. from hydrodesulphurisation [133], this results in sulphur dump sites in

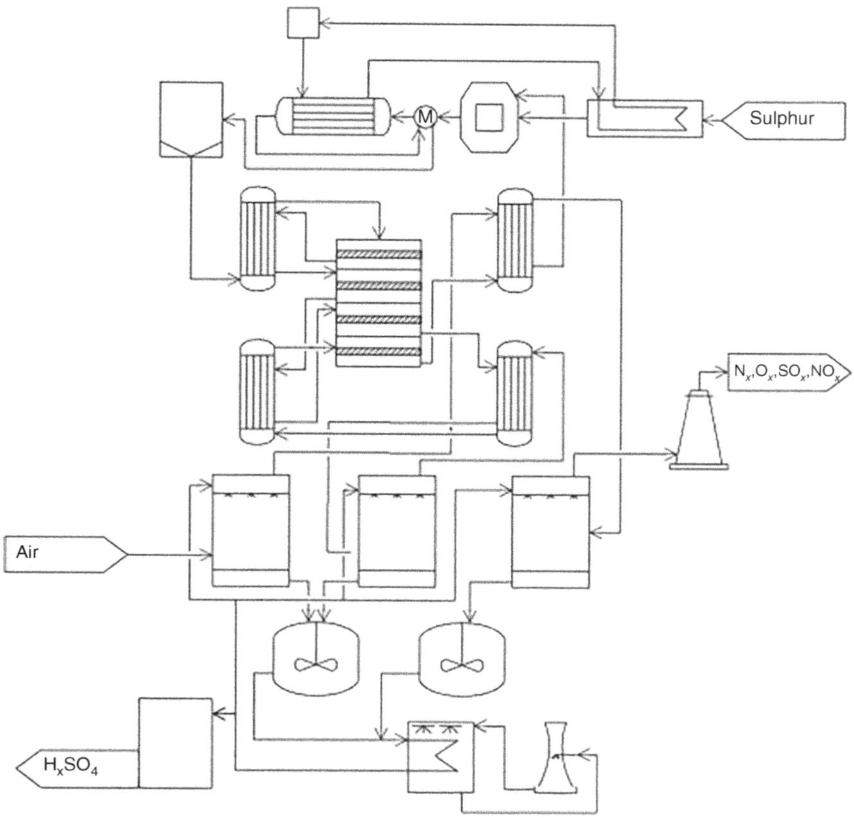

Figure 6.26 Simplified flowsheet of sulphuric acid production plant (redrawn from Civetta et al. [131]).

which the sulphur is stored. At present, most of the sulphur used as a feedstock originates from the hydrodesulphurisation of petroleum and natural gas, (accounting for about 80% of the sulphur used for the production of sulphuric acid) [134, 135]. This sulphur, in the form of hydrogen sulphide is converted to elemental sulphur via the Claus process, after which the elemental sulphur is used to produce sulphuric acid.

About 80% of global sulphur commodities originate from fossil fuels (the rest coming from elemental sulphur and desulphurisation of other process streams). The sulphur is removed in the form of hydrogen sulphide from petroleum using molecular hydrogen, resulting in the formation of hydrogen sulphide (H_2S) via hydrodesulphurisation, or by desulphurisation of natural gas. Hydrogen sulphide may then undergo a sub-stoichiometric oxidation to form elemental, solid sulphur via a process called the Claus process, according to the following overall reaction [136, 137]:

$$H_2S + \tfrac{1}{2}\,O_2 \rightarrow H_2O + \left(\frac{1}{n}\right)\cdot S_n \quad \Delta H^0_{\text{rxn,}} = -221\,\frac{\text{kJ}}{\text{mol}\,H_2S}$$

The reaction may be argued to proceed in two steps, i.e. the combustion of H_2S to form sulphur dioxide, and then the reduction of sulphur dioxide with hydrogen sulphide to form elemental sulphur.

$$H_2S + 3/2\,O_2 \rightarrow H_2O + SO_2 \quad \Delta H^0_{\text{rxn}} = -518\,\frac{\text{kJ}}{\text{mol } H_2S}$$

$$2\,H_2S + SO_2 \rightarrow 2\,H_2O + \left(\frac{3}{n}\right)\cdot S_n \quad \Delta H^0_{\text{rxn}} = -146\,\frac{\text{kJ}}{\text{mol } SO_2}$$

Hydrogen sulphide is fed to a furnace to produce SO_2/H_2S mixture in a ratio of 1:2. Sulphur is formed here as well as in the waste heat recovery unit. The conversion of H_2S is enhanced by passing it over several catalyst beds in series containing either alumina or titania with sulphur removal in its liquid form between the beds. The temperature in each bed is lowered to increase the equilibrium conversion.

In the production of sulphuric acid, sulphur is fed as molten sulphur (melting point 115–120 °C) [138]. Alternatively, hydrogen sulphide may be combusted with larger amounts of air to form sulphur dioxide and water directly, thus avoiding the reduction step to form elemental sulphur. Combustion of sulphur dioxide in the presence of water may yield a highly corrosive sulphuric acid mist. Thus, each option (elemental sulphur or hydrogen sulphide) provides benefits. The use of elemental sulphur allows tighter control over the addition of water to the system. The use of hydrogen sulphide is a simpler process and avoids the need for a reduction step. However, that means the hydrogen sulphide must be available in the same location as the sulphuric acid production plant, as solid sulphur may be more easily stored and transported.

The air supplied to the sulphur burner is dried to remove moisture, which prevents the formation of sulphuric acid mist which might lead to corrosion of downstream components, such as heat exchangers [128]. The air is dried by passing through concentrated sulphuric acid, thus diluting the acid. This acid is reused in the absorption towers for the formation of sulphuric acid from SO_3. Drying is carried out in brick-lined stainless steel vertical cylindrical towers packed with ceramic saddles, which are acid-proof [127, 128, 138]. The dried gas is then compressed by a blower, which introduces the dried air to the sulphur burner [128]. The blower is a steam or electrically driven centrifugal blower.

The actual raw material in the production of sulphuric acid is sulphur dioxide which is obtained from the combustion of sulphur. Solid sulphur (melting point of 115–120 °C) is melted with pressurised steam pipes maintained at ca. 140 °C. Molten sulphur is pumped to the sulphur burners, where it is nebulised in the gas burner. This is achieved by spraying the molten sulphur through nozzles into the sulphur burner, which is typically heated to above 900 °C, at which temperature sulphur is mainly present as S_2 as the sulphur distribution is thermodynamically controlled [138, 139]. Dry air is co-fed into the sulphur burner to provide oxygen for the oxidation of sulphur [140]. The reaction of nebulised sulphur with air thus proceeds according to the following reaction:

$$S(g) + O_2(g) \rightarrow SO_2(g) \quad \Delta H^0_{\text{rxn}} = -723\,\frac{\text{kJ}}{\text{mol}}$$

The reaction is exothermic and leads to a temperature increase above 1100 °C [138, 140]. The gases exiting the burner are then passed through a waste heat recovery boiler, which generates high-pressure steam whilst simultaneously cooling the gases to 390–410 °C, the temperature at which SO_2 can be oxidised to SO_3.

The oxidation of sulphur dioxide to sulphur trioxide forms the heart of this process [125]. Sulphur dioxide and air are passed over a vanadium pentoxide catalyst and oxidised to sulphur trioxide at ca. 400 °C (SO_3) [127, 140]:

$$SO_2 + \tfrac{1}{2}\,O_2 \rightleftharpoons SO_3 \qquad \Delta H^0_{\mathrm{rxn}} = -99\,\frac{\mathrm{kJ}}{\mathrm{mol\,SO_2}}$$

This reaction is also exothermic and generates heat. The oxidation of sulphur dioxide to sulphur trioxide is also equilibrium-limited, and the maximum conversion is dependent on the temperature, the pressure at which the oxidation takes place, as well as the content of oxygen and SO_2 in the feed. The effect of both temperature and SO_2 content in the feed on the maximum equilibrium conversion of SO_2 is presented in Figure 6.27. The conversion of SO_2 in a single, adiabatic reactor is limited, as the temperature increases almost linearly with conversion; a feed temperature of 420 °C with a feed containing 10% SO_2 results in a maximum conversion of ca. 65%, when the temperature is 580 °C. Hence, SO_2 oxidation is performed in a series of reactors with interstage cooling to the minimum operating temperature of each catalyst bed (ca. 400 °C), which enables an increase in the conversion of SO_2 to ca. 98% [125, 127, 140]. The product SO_3 is subsequently absorbed in an absorber that contains sulphuric acid with water dissolved in it. The water reacts with the SO_3 to form more sulphuric acid. Including an additional intermediate absorber removing SO_3, could enhance the conversion further by shifting the equilibrium towards the formation of SO_3 in the intervening reactors. This is referred to as the double contact process and may push the SO_3 conversion to up to 99.5% [125, 127].

The final step in the manufacture of sulphuric acid is the reaction of the formed sulphur trioxide with water dissolved in sulphuric acid (2% H_2O by mass) to yield

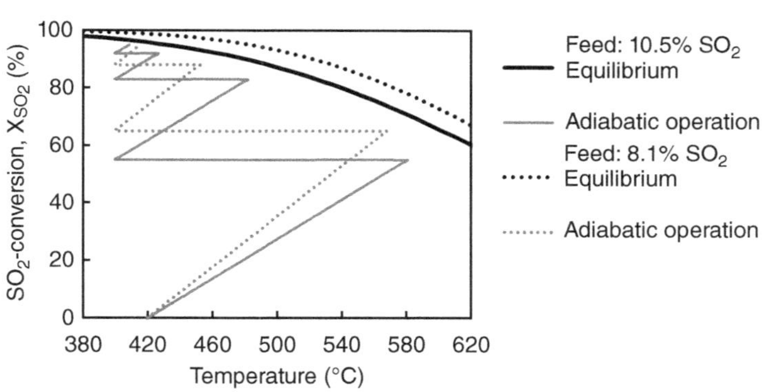

Figure 6.27 Conversion of SO_2 to SO_3 as a function of temperature in four adiabatic reactors in series, showing the effects of interstage cooling on the conversion of SO_2. The lines are shown for a feed containing 8.1 and 10.5 mol% SO_2 at atmospheric pressure obtained by combusting sulphur with air using O_2/S of 2.6 and 2, respectively (redrawn according to References [140, 141]).

highly concentrated acid. This reaction takes place in an absorption tower. The reaction of SO_3 with water is another strongly exothermic reaction; this may result in the formation of sulphuric acid mist, which is difficult to condense [128]. This is a counter-current operation, in which sulphuric acid is pumped into the acid absorption tower and enters from the top, whilst the SO_3 gas enters from the bottom-up [127, 128].

All the steps involved in the production of sulphuric acid, from the combustion of elemental sulphur to the formation of acid via SO_3 absorption in the absorption towers, are exothermic. According to Müller [127], a 1000 tonne/day plant is capable of producing 5.4 GJ of heat per tonne (and thus converting 93% of the reaction enthalpy into heat), resulting in a thermal capacity of about 63 MW. This heat is integrated in the process through the generation of superheated steam, which is used for the generation of electricity. Figure 6.28 shows the Sankey diagram of the energy flow in a double-absorption sulphuric acid plant. The Sankey diagram shows the supply of electricity to air blower for the the sulphur burner. Sulphur melting makes use of heat recovered from the heat released during the process (i.e. in the form of high-pressure or intermediate-pressure steam recovered from the converter) [142].

Most of the heat is generated in the sulphur burner and the converter. This heat is recovered using a waste heat boiler, which generates high-pressure steam [127, 142]. Part of this steam can be re-used in the sulphur burner, as well as to re-heat the gases re-entering part of the converter in a double-absorption contact plant. The rest of the steam is used in turbines to produce electricity, which can be reused in the plant, thereby making the plant self-sustaining [142]. This accounts for a 60% recovery in energy. About 30–40% of the energy is lost in the acid-cooling systems, specifically in the cooling towers [127, 142, 143].

The issue of the generation of waste heat from the absorption part of the plant, where concentrated sulphuric acid is formed, has been an ongoing problem. The temperature in the absorption towers is typically maintained at ca. 100 °C (when using water as a cold utility), which makes recovery of this heat, which is currently being dissipated through cooling towers to cool the acid streams, an important consideration [144]. The heat lost through dissipation accounts for ca. 30–40% of the total energy, of which the majority (24%) is lost in the intermediate absorber circuit in double-absorption plants. The heat loss could potentially be minimised by the generation of low-pressure steam, which will necessitate the operation of the cooling towers at slightly elevated temperatures [143]. Another option that has been investigated is the potential use of heat pumps, which are able to use external energy (i.e. electricity) to absorb heat and upgrade it to useful temperatures. These heat pumps could potentially be interfaced with the cooling towers [145].

6.4.4 Making Production of Sulphuric Acid More Sustainable

The manufacture of sulphuric acid mainly uses sulphur from petroleum resources, and, to a lesser extent, sulphide metal ores for the generation of SO_2. A shift away from petroleum due to the pursuit of renewable, non-petroleum fuels means that ultimately, the supply of elemental sulphur from petroleum resources may decline

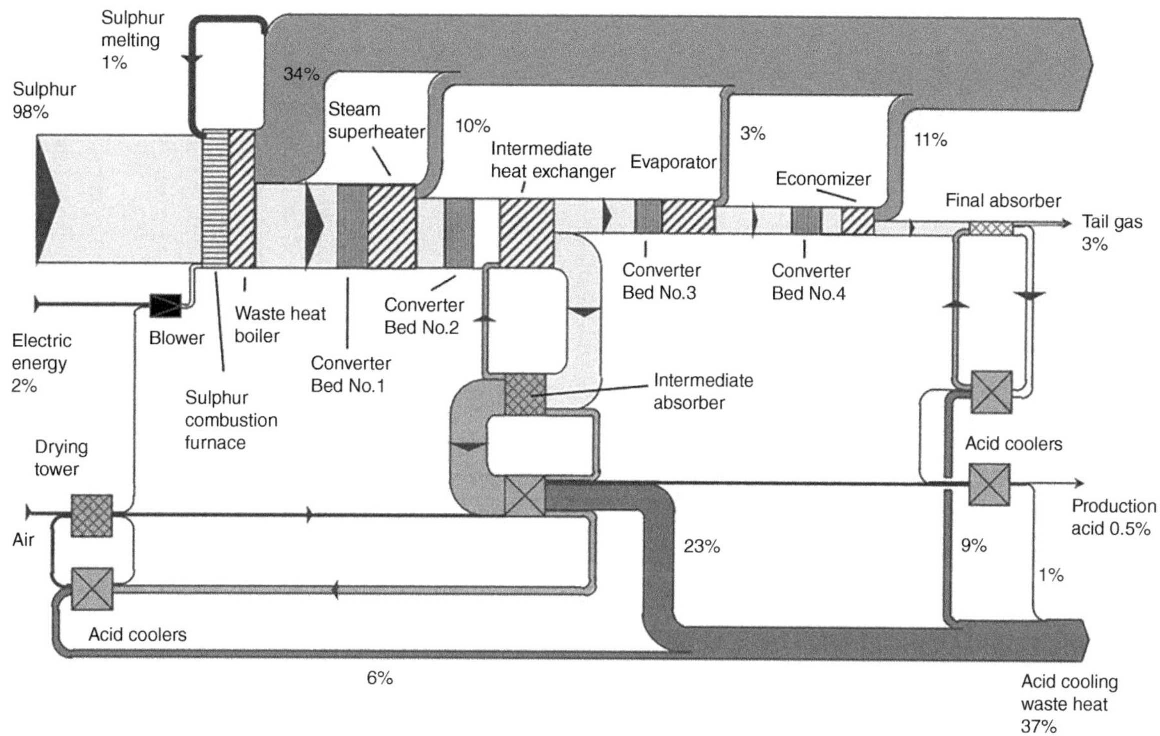

Figure 6.28 Sankey diagram of the energy flow in a double-absorption sulphuric acid plant (Zaker et al. [141] / with permission of Elsevier).

with time [135]. However, at the moment, this is not an urgent worry, due to the abundance of sulphur produced from the Claus process, which has necessitated research into the utilisation of the excess sulphur [146, 147]. Furthermore, whilst sulphuric acid is used to process materials/chemicals, only small amounts end up in the product itself, instead generating dilute and/or spent sulphuric acid. Hence, the reuse and regeneration of this acid waste become more urgent [127]. To that effect, areas for renewable implementation in the manufacture of sulphuric acid encompass alternative energy sources for the provision of heat required during the manufacture of sulphuric acid (and improvements in the recovery of waste heat), and methods for the regeneration of spent acid waste.

Sulphur-burning plants are autothermal in nature and, hence, do not require additional fuel to achieve temperature requirements. External energy sources on the sulphuric acid plant are required during the plant's start-up to generate the heat that is required for the air blower into the sulphur burner/sulphide smelter, as well as to melt the initial feed of solid sulphur. Thereafter, the consecutive processes are all exothermic, generating heat that is typically recovered in the form of high and intermediate-pressure steam, which can be used for the generation of electricity. Furthermore, upon start-up, all units in the plant must be heated to drive off moisture and to bring the catalyst beds up to operating temperature, a process that takes four to six days, thus the process must run continuously after start-up. Heating is also required after the annual shutdown as after the flow of sulphur is stopped, there is no source of heat (this is also true of start-up before the sulphur combustion). Hence, energy provisions must account for the need for extended heating [127].

The external power required during start-up may be provided by an intermittent renewable energy source to melt sulphur.

6.4.4.1 Regeneration of Spent Acid Waste

Spent acid can be used to produce moderate to highly concentrated sulphuric acid, depending on the regeneration method used. Sulphuric acid waste may contain organic impurities in addition to being relatively dilute acid. Several processes have been investigated for the removal of impurities to yield dilute sulphuric acid, which include but are not limited to ion exchange and/or electro-hydrolysis for the removal of inorganic impurities, and solvent extraction or oxidative degradation to remove organic impurities. Vacuum reflux/distillation or rectification has been found to be effective at concentrating 68% sulphuric acid waste to 98% [148]. However, this process may need to be further developed for industrial application.

A more industrially feasible approach to recycling dilute sulphuric acid may be the thermal decomposition of sulphuric acid to yield SO_2 at temperatures above 430 °C. The reaction is endothermic and proceeds via the formation of SO_3 and water, with SO_3 decomposing to SO_2 [149]. Sulphur dioxide can then be converted to SO_3 in the converter, following the traditional contact process. The formation of SO_2 from SO_3 during the decomposition reaction is kinetically limited even at high temperatures and may require the co-feeding of reactants such as hydrocarbons and/or nitrogen-containing species [149]. Catalysts for the decomposition of

SO_3 to SO_2 are being investigated, which have focused on complex oxides of base metals such as iron, nickel, copper, zinc, etc., with iron oxide showing the most promise [150, 151].

6.4.5 Outlook

Sulphuric acid is one of the most important chemicals produced globally and is certainly the one produced most abundantly, with estimates that annual production may increase 60% by 2040 [135]. The manufacture of sulphuric acid using either the contact process or the double-absorption contact process, is a self-sustaining process, with sulphuric acid plants being capable of generating power. This is due to the strongly exothermic combustion of elemental sulphur, oxidation of sulphur dioxide and dissolution of sulphur trioxide in the water/sulphuric acid stream. However, this does not preclude sulphuric acid plants from requiring external power to run blowers and pumps. Furthermore, the process will need external power during start-up. It is reasonable to conclude that sulphuric acid plants may operate well with intermittent power, provided that the external power is sufficient to meet the rather small, but required electrical power to operate the plant. During plant start-up, additional electrical power input may be required (i.e. initial heat requirement to melt sulphur, and to start up the exothermic processes required to sustain the overall process), which can be supplied by connecting it to a heat pump.

There is some energy lost in the process, as the recovery of low-quality heat in the absorption tower is poor. The low-quality heat can, in principle, be upgraded using a heat pump, but current heat pumps do not deliver sufficient heat to generate high- or medium-pressure steam for power generation. Should new heat-transfer mediums for heat pumps be developed to operate at slightly higher temperatures, this may be an ideal opportunity for their implementation and may make sulphuric acid plants a valuable energy source rather than just energy-neutral. This low-quality heat could also be linked to an over-the-fence heat distribution system, where it may be used for systems requiring low-temperature heating, such as domestic heating.

A potential problem facing the sulphuric acid industry is the sulphur feed source. At current, 80% of the sulphur is sourced from petroleum processing and as the demand for sulphuric acid grows (both for the production of fertiliser and light electric vehicles) and the decarbonisation of fuels progresses, it has been estimated, perhaps pessimistically, that there may be a sulphur shortfall between 100 and 320 million tonnes/annum [135] by 2040, depending on the rate of decarbonisation. This means that methods of mining sulphur, e.g. by the Frasch process producing high-grade elemental sulphur [152] may be required. The Frasch process used to be the dominant source of sulphur globally, though was replaced by hydrodesulphurisation as it has a high thermal energy requirement (historically requiring fossil fuels) and produces vast amounts of wastewater [135]. Process improvement in the mining of sulphur may be required in the future to maintain the required feedstock. At present, however, this is not a major issue, as the abundant production of petroleum products has led to an abundance of elemental sulphur formed; this sulphur may typically be stored in waste dumps, and these waste dumps will be a valuable, if finite, source of sulphur during the defossilisation of the world's energy supply.

As well as investigating other sources of sulphur, recycling or reconcentrating spent sulphuric acid to useful sulphuric acid concentrations will need to be developed further. Spent sulphuric acid can be decomposed to yield SO_2, which can then be converted back into concentrated sulphuric acid via the processes described above. This approach is important since it provides a sustainable route for the utilisation of spent sulphuric acid (which is highly corrosive) and may reduce the need for additional sources of sulphur for the generation of SO_2, depending on the rate at which the demand grows.

6.5 Sodium Hydroxide and Hydrochloric Acid – The Chlor-Alkali Process

Contribution by Sinqobile V.L. Mahlaba.

6.5.1 Introduction to Material and Industry

The chloro-alkali process utilises electricity to produce sodium hydroxide and chlorine gas from the electrolysis of brine or sea water and is an example of the large-scale application of electrochemistry. The global production of sodium hydroxide stood at an estimated 82 million tonnes in 2022 [18], with an estimated market value of 42.5 billion US dollars [153]. The global production of chlorine in 2022 was at 97 million tonnes (this is more than the stoichiometric amount from the chlor-alkali process as other processes can also yield chlorine), with a predicted increase in the production to 106 tonnes predicted for 2024 [17] (as of September 2024), and a corresponding market value of 37 billion US dollars [154], and a projected market value of 40 million US dollars in 2024.

Sodium hydroxide finds extensive use in diverse applications, especially in the production of organic and inorganic chemicals. Sodium hydroxide is used directly in the pulp and paper industry, in the manufacture of soaps and detergents (saponification), in petroleum applications for the alkaline flooding of oil fields, which enhances petroleum recovery [155], and in the production of aluminium via the Bayer Process [156]. Sodium hydroxide is also used for the synthesis of organic chemicals in the form of free organic bases, which are then used for neutralising acids and/or wastewater, pH control, off-gas scrubbing and dehydrochlorination. It is also used as a catalyst and a source of sodium, as well as in the production of polymers and resins. The demand for sodium hydroxide by major end user is represented in Figure 6.29a.

Chlorine, on the other hand, is used in the production of materials such as polyvinyl chlorides, isocyanates, epoxy resins, used in coatings and adhesives, as a disinfectant for the treatment of wastewater, as an oxidising agent in the production of pharmaceutical chemicals [159], and as a bleaching agent in the pulp and paper industry. The demand for chlorine by major end user is represented in Figure 6.29b.

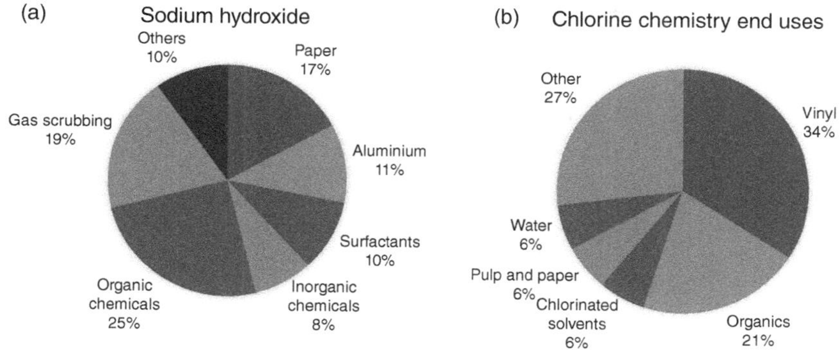

Figure 6.29 Uses of (a) sodium hydroxide (adapted from Kurt and Bittner [157]) and (b) chlorine (adapted from TEADIT [158]) classified according to end user.

6.5.2 The Chlor-Alkali Process

The chlor-alkali process uses brine as a starting material, a solution of sodium chloride, NaCl, or table salt. The brine solution is initially purified through precipitation before being fed into the electrolysis process. The electrolysis production process uses the diaphragm, membrane or mercury cell, with the diaphragm cell being the predominant method (as well as the membrane cell). An additional process using oxygen-depolarised cathodes (referred to as the ODCs) has also been explored in recent times [160]. The mercury cell is expected to increasingly fall out of favour due to the generation of mercury waste [161, 162].

Table 6.14 shows the advantages and disadvantages of the four different methods, their raw material requirements, electrode reactions, power consumption, as well as hydrogen production [161]. The ODC cell does not co-produce hydrogen, unlike the others. Essentially, the electrolysis by direct electric current of sodium chloride converts chloride ions to chlorine gas at the anode, and sodium hydroxide/hydrogen gas at the cathode. The net reaction takes place according to:

$$2\,\text{NaCl(aq)} + 2\,\text{H}_2\text{O(l)} \rightarrow \text{Cl}_2\text{(g)} + 2\,\text{NaOH(aq)} + \text{H}_2\text{(g)} \quad \Delta H^0_{\text{rxn}} = 447.9\,\frac{\text{kJ}}{\text{mol Cl}_2}$$

The change in the Gibbs free energy of reaction (ΔG^0_{rxn}) for the electrolysis of brine with all the compounds in their reference state is 422.2 kJ/mol Cl_2 or 1.65 MWh/tonne Cl_2. The ideal voltage for the electrolysis is 2.188 V. The reaction is endothermic, and heat needs to be supplied to keep the reaction isothermal (at 25 °C : 100 0.1 MWh/tonne Cl_2 or 362 MJ/tonne Cl_2), or the reaction mixture will cool down.

Hydrogen is produced as a by-product and stoichiometrically, a ratio of H_2:Cl_2 in a ratio of 1:1 is expected. In reality, more hydrogen is produced (1 mol of hydrogen gas for every 0.85 mol of chlorine gas, and for every 1.7 mol of sodium hydroxide [163]) implying that water electrolysis takes place as a parasitic side reaction, thus reducing the energy efficiency by ca. 10%.

Table 6.14 Electrolytic chlor-alkali processes as classified along the electrode half reactions, material and energy requirements, product quality, H_2 production and environmental concerns (data from References [160, 161]).

Electrolytic cell	Cell partition/ material	Half-cell reaction	Energy consumption (MWh/tonne Cl_2)	NaOH wt.%	H_2 prod.	Process disadvantages
Mercury cell	None	**Electrolyser** Anode: $2Cl^- \rightarrow Cl_2 + 2e^-$ Cathode: $2Na^+ + 2Hg + 2e^- \rightarrow 2NaHg$ **Decomposer** Anode: $2NaHg \rightarrow 2Na^+ + 2Hg + 2e^-$ Cathode: $2H_2O + 2e^- \rightarrow 2OH^- + H_2$	3.1–3.4[a)]	50	Yes	Hg in cathode material
Diaphragm	Diaphragm, asbestos	Anode: $2Cl^- \rightarrow Cl_2 + 2e^-$ Cathode: $2H_2O + 2e^- \rightarrow H_2 + 2OH^-$	2.3–2.9[a)]	12	Yes	Usage of asbestos High energy for NaOH concentration Low NaOH and Cl_2 quality
Membrane	Cation exchange membrane	Anode: $2Cl^- \rightarrow Cl_2 + 2e^-$ Cathode: $2H_2O + 2e^- \rightarrow H_2 + 2OH^-$	2.1–2.6[a)]	50	Yes	Low Cl_2 quality High brine purity required high energy consumption
ODC cell	Cation exchange membrane	Anode: $2Cl^- \rightarrow Cl_2 + 2e^-$ Cathode: $O_2 + 2H_2O + 4e^- \rightarrow 4OH^-$	1.5–1.6[b)]	50	No	Lack of H_2 produced[c)] High-purity oxygen required High brine purity required

a) Overall reaction: $2\,NaCl(aq) + 2\,H_2O(l) \rightarrow Cl_2(g) + 2\,NaOH(aq) + H_2(g)$ with minimum energy requirement based on ΔG^0 of 1.65 MWh/tonne Cl_2, assuming all reactants at their ideal state and minimum voltage of 2.188 V.
b) Overall reaction: $4NaCl(aq) + 2H_2O(l) + O_2(g) \rightarrow 2Cl_2(g) + 4NaOH(aq)$ with minimum energy requirement based on ΔG^0 of 1.45 MWh/tonne Cl_2 assuming all reactants at their ideal state and minimum voltage of 0.959 V.
c) Lack of hydrogen production may be argued as a disadvantage as it is a valuable side product, especially if it may be considered green due to the use of renewable electricity in the process. However, this could also be seen as an advantage as the electrolysis of water uses extra energy in not forming undesired products of the chlor-alkali process.

6.5.2.1 Raw Materials

The raw material used as the feed is brine; the salt used is either rock salt, solar salt (which is salt produced from the evaporation of seawater), or vacuum-evaporated salt, which is salt purified by a process of evaporative crystallisation [164]. The salt is dissolved in water to form a concentrated brine solution with a concentration of 310–315 g/l [165]. The brine is purified to remove contaminants such as magnesium and calcium ions, which are especially harmful to the membrane cell. These ions are removed via precipitation with sodium bicarbonate or sodium hydroxide, followed by filtration. Hydrochloric acid may also be used for pH adjustment in the electrolysis and dechlorination steps [166]. The purified brine ideally contains Mg^{2+} and Ca^{2+}-ions at concentrations below 1 g/l for the diaphragm process. The concentrations of these contaminants must be much lower when using a membrane cell (<20 ppb), and thus, purification of the brine occurs in two stages, with the second being a fine purification step using polish filtration and ion-exchange resins.

6.5.2.2 Process Flow Diagram

A breakdown of the overall chlor-alkali process is presented in Figure 6.30 [161]. There are four distinct sections to the process, the first two of which, viz. the salt mining and the preparation of brine are common regardless of the chlor-alkali technology selected. Energy input is required in the preparation of brine (i.e. transportation of salt, dissolution of salt via stirring and filtration of brine after precipitation), as this step involves the use of either centrifuges or rotary drum vacuum filters for the separation of precipitated solids from the brine. The brine is then acidified with hydrochloric acid before the introduction to the electrochemical cells, where it is depleted. The depleted brine is typically re-saturated with solid salt, purified and re-introduced in the cell.

The different product streams generated by the different technologies undergo appropriate treatments. The mercury cell produces a sodium amalgam, which is reacted with water to form sodium hydroxide and hydrogen gas. The diaphragm process produces sodium hydroxide dissolved in brine (~12 wt.%), which must be concentrated via evaporation to yield a 50 wt.% NaOH solution and to remove salt. This salt can then be used to re-saturate brine, which can be reused as feed. The membrane cell process produces sodium hydroxide with a concentration of ~35 wt.%, requiring further concentration. The chlorine gas leaving both the diaphragm and membrane cells requires the removal of oxygen, which is achieved by liquefaction and evaporation. Hydrogen from the electrolytic process is of a high purity and is mostly used for organic hydrogenation reactions, or for the generation of heat. The hydrogen is removed through electrically insulated pipework, then transported to where it will be consumed [165], which may include compression for offsite use.

6.5.2.3 The Diaphragm Cell

The diaphragm cell chronically preceded the mercury cell, first being in operation in the middle of the nineteenth century, with several commercial diaphragm plants going into operation towards the end of the century. The early cells underwent several improvements, a notable event being the adoption of asbestos as the

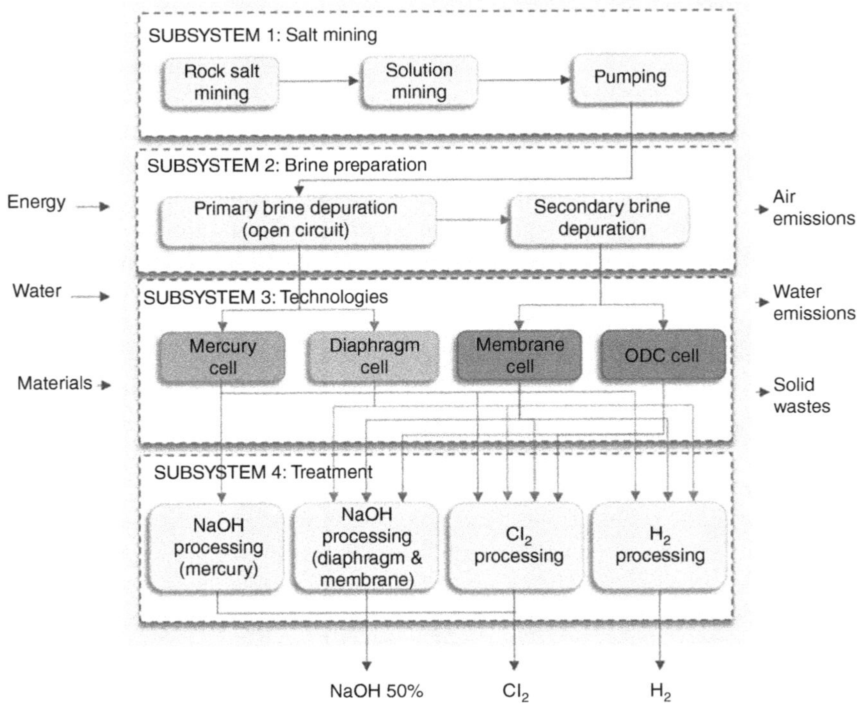

Figure 6.30 Process flow diagram for the electrolysis of brine to produce Cl_2, NaOH and H_2 in the different electrochemical cells (Garcia-Herrero et al. [166] / with permission of Elsevier).

material for the diaphragm, which remained in use for almost a century [167]. The diaphragm cell is constructed of two compartments: the anode compartment, which is the half-cell that produces chlorine gas from brine, and the cathode compartment, which consists of sodium hydroxide and dilute brine. Figure 6.31 shows the set-up of the diaphragm cell. The diaphragm is a current permeable separator that prevents the interaction between the evolved chlorine and sodium hydroxide, which would otherwise react to form sodium hypochlorite.

The anode is constructed of a titanium base coated with a catalytic layer comprised of metal oxides, such as ruthenium or titanium oxides (which are also applicable for membrane cells) [169]. Concentrated brine is fed into this compartment, and contacts the anode, where the half-cell reaction for the oxidation of chloride ions occurs on the surface of the anode. These anodes exhibit low-cell voltage and are quite stable, giving them a long life under operation conditions. They also consume less energy than the preceding magnetite, platinum and graphite anodes (which were expensive, unstable or had current density limitations). The depleted brine and sodium cations diffuse through the diaphragm to the cathode compartment, where water undergoes the reduction half-reaction to yield hydroxide ions in solution. Thus, the cell produces sodium hydroxide in a brine solution at the cathode

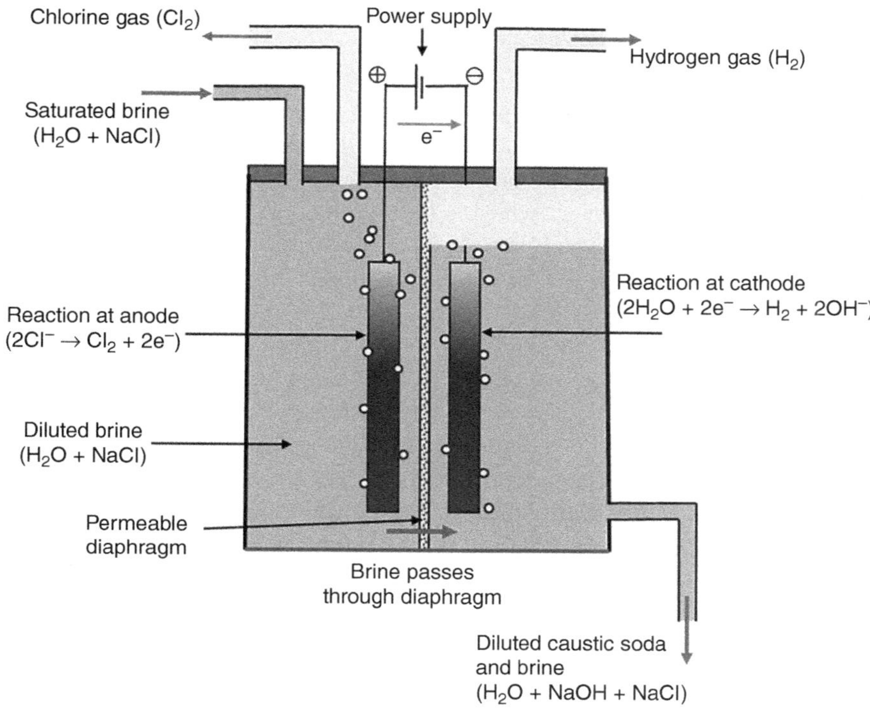

Figure 6.31 Schematic diagram of the diaphragm electrolytic cell showing the evolution of Cl_2 in the anode compartment, H_2 in the cathode compartment and NaOH in the cathode compartment (redrawn from Eurochlor [168]).

with a sodium concentration of 12 wt.%, which is one of the major drawbacks of this process, since the sodium hydroxide must be separated from the brine solution and further concentrated.

Within the diaphragm (as with the membrane) cell, the half-reactions taking place on either side of the diaphragm (or membrane) are:

$$2Cl^- \rightarrow Cl_2(g) + 2e^- \quad E^0 = 1.36\,V$$

and

$$2H_2O + 2e^- \rightarrow H_2 + 2OH^- \quad E^0 = -0.83\,V$$

The overall reaction has a theoretical potential difference of 2.19 V (equivalent to a Gibbs free energy of reaction of 422.1 kJ/mol Cl_2 or 1.65 MWh/tonne Cl_2). However, the cell is typically said to operate at ca. 2.2–3.5 V due to the required overpotential in the system [28].

For the diaphragm cell, the real energy required for the electrolysis process is 2.8–3.0 MWh/tonne Cl_2 and thus, the energetic efficiency of the diaphragm cell is only 55–59%. Furthermore, the generation of steam in caustic evaporators, which is necessary to concentrate the dilute sodium hydroxide product stream, requires an additional 0.6–0.8 MWh/tonne Cl_2, thus reducing the energetic efficiency to ca. 44–49%.

The energy for electrolysis is supplied mainly from the grid, whilst hydrogen from the process can be used to provide heat for the evaporation process [170]. The diaphragm cell is limited to producing low sodium hydroxide concentrations due to the likelihood of higher concentrations to decompose water to oxygen via the oxygen evolution reaction, which is likely to happen in basic conditions (see the alkaline electrolysis section in Chapter 5), and which would thus reduce the current efficiency for the generation of chlorine [167, 171].

6.5.2.4 The Mercury Cell

The mercury cell was developed shortly after the diaphragm cell, and contains an electrolyser, consisting of the anode and cathode (where the half-cell reactions occur) and a decomposer. Unlike the diaphragm cell, which forms hydroxide ions in the cathode, the cathode in the mercury cell forms a sodium amalgam, which then goes to the decomposer to form sodium hydroxide and hydrogen. Figure 6.32 shows the schematic diagram of the process. The anode compartment consists of the brine solution, where chloride ions undergo oxidation on the anode to produce chlorine gas, and sodium ions react with the cathode, composed of flowing mercury, to form the sodium-amalgam. The amalgam then flows to the decomposer, where it comes into contact with deionised water, which reacts with the amalgam to form sodium hydroxide and hydrogen gas. The sodium hydroxide is formed at a high concentration compared to the diaphragm cell, with concentrations of up to 50 wt.%. Mercury can be recovered and recycled to be re-used in

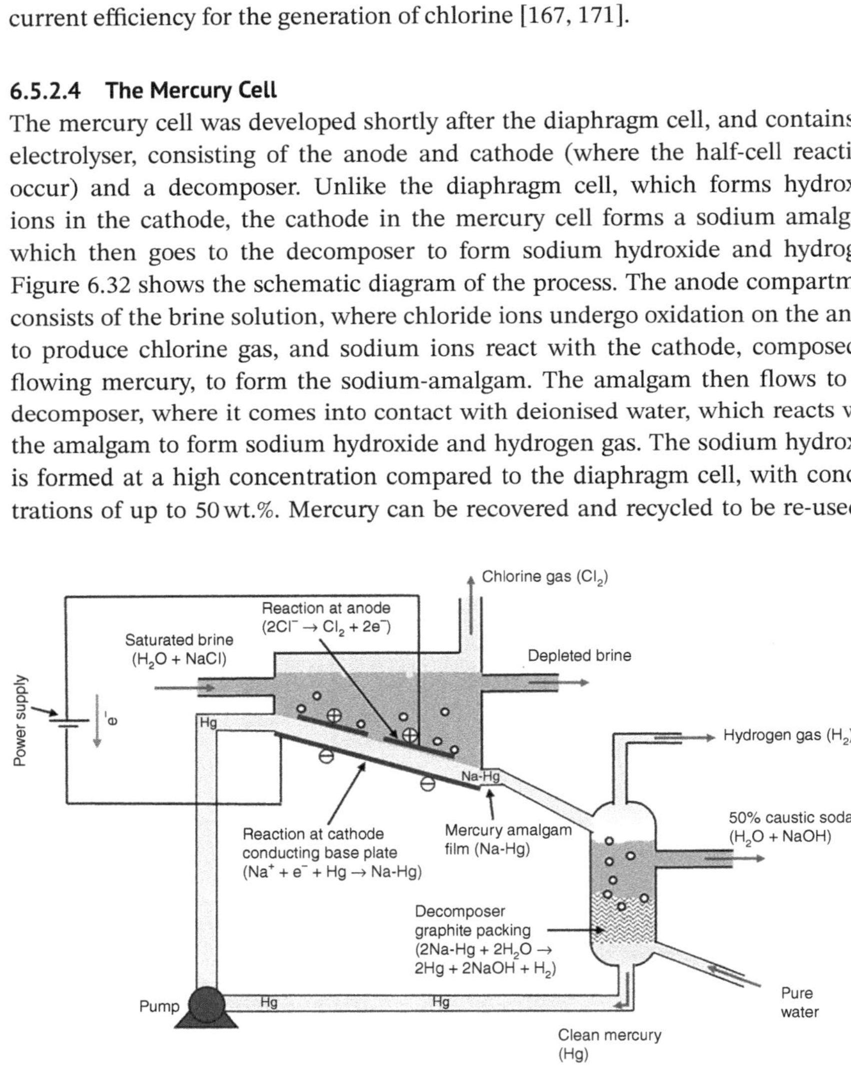

Figure 6.32 Schematic representation of the configuration of the mercury cell, showing the anode, cathode, the reactions that occur at these electrodes and the decomposer which generates sodium hydroxide and hydrogen gas in the presence of water (redrawn from Eurochlor [168]).

the process, whilst the hydrogen and sodium hydroxide are recovered for further processing [167, 170].

The mercury cell is the most energetically intensive of all the electrochemical processes, with an energy requirement for the electrolysis in the range of 3.2–3.6 MWh/tonne Cl_2 and thus an energetic efficiency between 46% and 52%. The added advantage of this process is that the requirements for concentration of sodium hydroxide are not as low as that of the diaphragm cell, since this process produces a fairly concentrated sodium hydroxide stream [170]. A major drawback of this process is the high likelihood of mercury pollution, as mercury recovery is perfect, and some reports of unaccounted mercury exist, which has been found in the environment [167, 172]. Hence, the conversion of mercury cells to less harmful technologies is strongly recommended [173].

6.5.2.5 The Membrane Cell

The membrane cell (Figure 6.33) is currently the most common technology used for the chlor-alkali process and is an alternative to the diaphragm cell. This cell configuration was developed in the mid-twentieth century due to the problems posed by the mercury cell and the use of asbestos in the diaphragm cell [174]. It uses a selectively permeable ion-exchange membrane. This membrane is selective for the diffusion of sodium ions, thereby preventing the diffusion of the chloride ions from the anode compartment into the cathode compartment. Hence, brine is essentially confined to the anode compartment, where it undergoes depletion via the evolution of chlorine gas, and diffusion of sodium cations to the cathode compartment. The membrane cell consumes the least amount of energy in comparison to the mercury

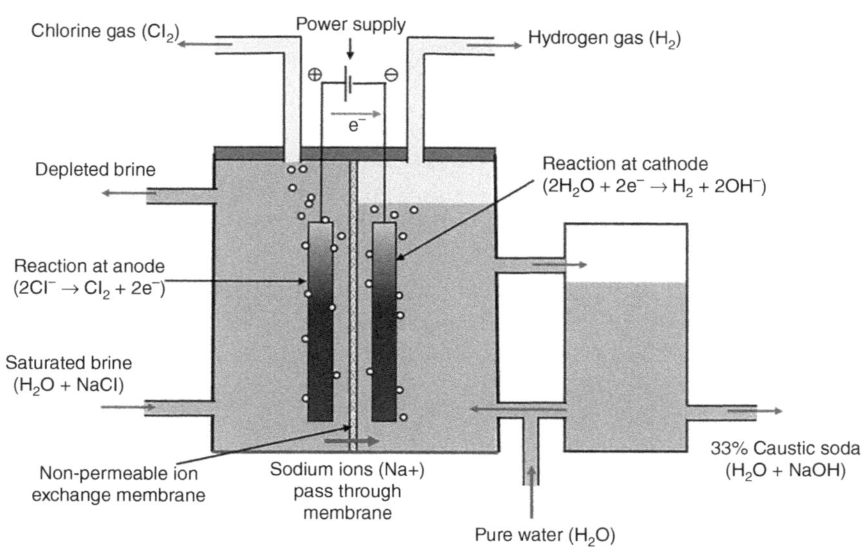

Figure 6.33 Schematic diagram of the membrane cell showing the anode and cathode compartments, partitioned by a selectively permeable membrane preventing the diffusion of chloride anions to the cathode compartment (redrawn from Eurochlor [168]).

and diaphragm cell, requiring between 2.6 and 2.8 MWh/tonne Cl_2 of power for electrolysis, and 0.2–0.3 MWh to generate steam required to concentrate the sodium hydroxide solution [170]. The overall energetic efficiency of the membrane cell including the concentration step comes thus to 53–59%, making it the most energetically efficient option (not considering the energy requirement for the other steps). The membrane is composed of perfluoro compounds to withstand the basic conditions within the cell. Nevertheless, high-purity brine is necessary as feed to ensure the longevity of the membrane.

The membrane cell produces a highly concentrated sodium hydroxide solution, which is almost on par with the mercury cell. However, the process still requires the evaporative concentration of sodium hydroxide solution from 33% to 50%. A diaphragm cell can also be readily converted to a membrane cell, through the replacement of the diaphragm with an ion exchange membrane in an existing cell [171]. Although the energy required by the membrane cell for electrolysis is lower than that of the mercury cell, the total energy required is almost similar, since the mercury cell produces sodium hydroxide solution that may not require further concentration, and the membrane cell will need rather pure brine.

6.5.2.6 Oxygen-Depolarised Cathode (ODC) Cell

The oxygen-depolarised cathode (ODC) cell can be seen as the grandchild of the diaphragm process since it is yet another development of said cell. Instead of hydrogen evolution in the cathode compartment, oxygen is fed, which reacts with water to form hydroxide ions. ODC reduces the electricity demand for the electrochemical process, which can have an energetic efficiency larger than 90%. The minimum required potential difference for the ODC cell is only 0.96 V, compared to 2.19 V for the traditional membrane cell [175, 176]. However, the process requires pure oxygen (0.226 tonne O_2/tonne Cl_2), and air separation will need depending on the required purity and amounts between 0.05 and 0.25 MWh/tonne Cl_2. In a future H_2-economy with hydrogen generation via water electrolysis, pure oxygen may become more readily available, negating the additional energy costs.

The lack of hydrogen production by this process may be seen as a potential disadvantage, since hydrogen could be recovered as a by-product and used as an energy source required, e.g. for steam generation necessary to concentrate the resulting NaOH-solution. However, this process produces an already highly concentrated NaOH-solution. Furthermore, it seems this cell configuration is still under development and optimisation, and is thus not a mature technology, although showing promise [176–178].

6.5.3 Outlook

The chlor-alkali process, as with the other processes discussed here, will remain a process of fundamental importance to human society, likely indefinitely. It is rather energy-intensive and requires a significant amount of energy to drive the process in the form of electricity to drive chemical change. The electrolysis of brine in electrochemical cells makes use of direct current. The exact amount depends on the type of technology applied (see Table 6.14). This process accounts for ca. 10% of the energy used in the entire chemical industry [179], this being electrical energy. At present, this electricity is predominantly supplied by the grid [171] meaning that the environmental impact, in terms of carbon emissions, depends on the type(s) of fuel used to generate grid electricity. The energy efficiency of the electrochemical conversion depends on the applied technology and is somewhere between 42% and 59%. Most of the electricity demand originates from energy losses encountered due to high overpotentials in the transport of sodium ions through the membrane. Furthermore, the accumulation of chlorine and hydrogen gases on the electrodes resulting in electrochemical polarisation is also seen as a possible contributor to losses in efficiency. The use of an ODC cell could improve the energetic efficiency to more than 90%, if the ODC cell can be commercialised successfully.

The phasing out of the mercury cell in favour of the membrane cell could be beneficial both from an environmental and energy standpoint, since the mercury cell is quite energetically intensive, and some mercury, a known neurotoxin, may spill. The utilisation of the membrane cell on a larger scale shows promise.

This process co-produces hydrogen (stoichiometrically, 1 mol of H_2 will be produced from each mole of Cl_2, but in reality, this ratio is higher due to the competitive water electrolysis reaction). Taking the size of the chlor-alkali process being given by the annual production of NaOH (82 million tonne/annum in 2022), the amount of chlorine produced via the chlor-alkali process would be 73 million tonne/annum co-producing at least 2.1 Mio tonne H_2/annum, i.e. 2.1% of the hydrogen produced worldwide in 2022 [15, 180]. As this hydrogen has been produced in the process since its inception there are existing uses for the hydrogen produced within the chlor-alkali process. Though it may be noted that as the industry grows and the market of hydrogen, specifically hydrogen, which is synthesised from non-fossil sources, becomes more apparent. The hydrogen produced from the chlor-alkali process may enter the renewable or carbon-neutral hydrogen market.

Electrolytic processes prefer to be operated with a continuous power draw, as variable load may affect the lifetime of the electrolyser negative [181]. The intermittency of renewable energy may be partially covered by hydrogen generated in the process, using it as a chemical carrier of electrical energy [182]. This may, however, require significant capital investment. It may become economically beneficial to operate electrolysers under variable load [163], when considering the whole lifetime of the electrolyser [181].

6.6 General Outlook for the Basic Industries

The basic industries can encompass many different industries, which take large-scale mined materials and transform them into materials used by other industries. Mined materials are typically at their thermodynamically stable state, and their transformation will require significant energy input. The exception here is the production of sulphuric acid, where the energy is supplied by the exothermic oxidation of sulphur, previously extracted by (hydro)desulphurisation of fossil fuels,

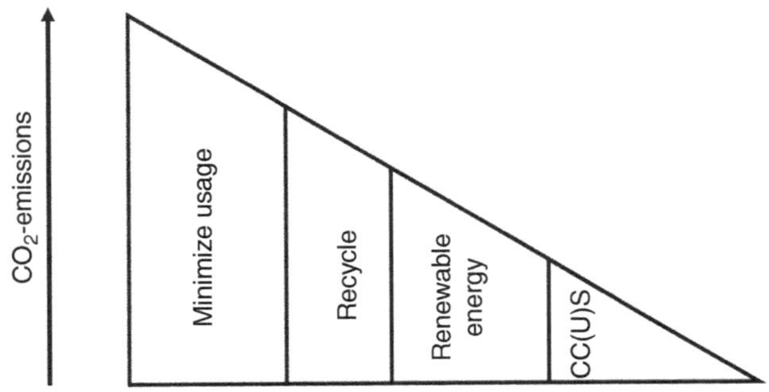

Figure 6.34 Methods to reduce carbon footprint in the basic industries and the possible impact of the various methods.

such as oil and natural gas, or the exothermic roasting of sulphide minerals, and the subsequent oxidation of SO_2 yielding sulphur trioxide.

The resulting CO_2 emissions of a given basic industry are a consequence of the (high) energy input, when using fossil fuels, as well as the use of carbon-based chemicals (specifically $CaCO_3$ in the cement industry and coke as a reducing agent in the steel industry). The reduction in emissions caused by the basic industry can thus be achieved by minimising the manufacture of these products through more efficient usage of the product, recycling of the products, producing the products using renewable energy, and as a last resort, CCUS (see Figure 6.34).

As outlined by Fennell et al. [56], the carbon footprint of skyscrapers can be drastically reduced by using less reinforced concrete. This may be achieved by optimising the design (focusing on light-weight design and reducing overdesign) [183]. Hence, society at large may need to look not only at the energy efficiency of their actions, but also at the efficient usage of materials. Furthermore, the material usage can be minimised by replacing ingredients in the final product. For instance, concrete contains clinker (Portland cement), which is a significant contributor to the carbon footprint of concrete. The clinker content in concrete may be reduced by adding supplementary cementitious materials (SCMs), such as pozzolan, fly ash, slag cement or LC^3 (limestone calcined clay cement). The reduction in the clinker content by adding SCMs with a lower greenhouse gas potential (e.g. fly ash and slag cement are waste products) translates into a reduction of the carbon footprint [184–186]. The substitution of clinker with SCMs can only be within certain ranges to maintain strength and durability of concrete [187]. Minimising the usage of the products does not minimise emissions from the basic industry itself, but it minimises the carbon footprint of society at large through the efficient usage of materials.

Recycling of materials is necessary to obtain a circular economy and will save resources in the long term. However, the lifespan of some products, such as cement and steel, may reach decades or even centuries, and the material will not readily be available for recycle. Recycling steel in EAFs is standard technology, but recycling of other basic chemical products is not readily achieved. For instance, sulphuric acid is used in many different processes, where it leaves rather diluted or as a salt together with other contaminants making centralised recycling problematic.

Renewable energy in its primary form comes as electricity (see Chapter 2), and electrochemical processes, such as the chlor-alkali process, are prime candidates to be connected to a renewable grid. They will need to operate at a variable load, possibly backed up by alternative electricity generation (e.g. by generating on-site power from co-generated hydrogen in the process). Furthermore, electric power can be used directly to heat metallic iron (together with recycled steel) in the steel manufacture in EAFs or induction furnaces. Other high-temperature processes in the cement and iron manufacture will not readily be heated using electricity, but some form of combustible, renewable fuel. Incineration of organic waste may be used here; this may reduce the greenhouse gas potential of waste as the anaerobic conversion of this waste will result in the formation of methane, a gas with a greater greenhouse gas potential (though it may also be used as a fuel).

The cement industry, combined with the steel industry, emits 4.6 billion tonnes of carbon dioxide per annum. This corresponds to the combustion of 2.4 million barrels of oil (sufficient to fill a super tanker) every two hours. CCUS must thus be seen as a last resort. Nevertheless, some processes, like cement manufacture, will need CCUS to reach a net-zero situation, even when using organic waste as a heating source.

Basic industries are large CO_2-emitters per unit production volume operating at relatively low profit margins with production locations spread out over the world. A major dilemma for the basic industries is thus the driving force to decarbonise and become more renewable. Carbon pricing is one of the tools used to drive industry to implement emission-reducing technologies as these technologies may become more economically attractive in the long term. This will drive the cost of production upwards, and carbon pricing may also impact the potential for basic industries to have a financially viable operation. Retrofitting existing plants with the researched technologies that will make the processes 'greener' is difficult (with the exception for the costly carbon capture, though even this is not necessarily trivial to implement). Modern greenfield plants applying 'greener' designs and phasing out of old plants will reduce the carbon emissions for the industry at large, but still a driving force is needed to invest in the 'greener' designs rather than utilising the existing, cheaper technologies. The push will need to come from the downstream users of these products who will be willing to bear a higher price for a greener product, ultimately passing down this cost to the end consumer, e.g. in the building of 'greener' buildings. Alternatively, there may be a need for government subsidy in these industries to support the implementation of carbon capture technologies and benefit the new build projects in adopting the newer technologies. Finally, it may become economically feasible to run the 'greener' technology if, e.g. in the case of the steel industry, the cost of green hydrogen decreases significantly over the interleading years.

References

1 Engel, H., Krishnan, M., Samandari, H. et al. (2023). A sector progress tracker for the net-zero transition | McKinsey, McKinsey Sustainability [Online].

Available: https://www.mckinsey.com/capabilities/sustainability/our-insights/a-sector-progress-tracker-for-the-net-zero-transition (accessed 19 July 2024).

2 Greene, T. (2022). 50bn tonnes of sand and gravel extracted each year, finds UN study, *The Guardian*, 26 April 2022 [Online]. Available: https://www.theguardian.com/environment/2022/apr/26/50bn-tonnes-of-sand-and-gravel-extracted-each-year-finds-un-study (accessed 12 September 2024).

3 IEA (2023). Executive summary - Coal 2023 [Online]. Available: https://www.iea.org/reports/coal-2023/executive-summary (accessed 12 September 2024).

4 Tkachenko, N., Tang, K., McCarten, M. et al. (2023). Global database of cement production assets and upstream suppliers. *Scientific Data* 10 (1): 696. https://doi.org/10.1038/s41597-023-02599-w.

5 U.S. Energy Information Administration(2023). Crude oil including lease condensate monthly, Petroleum and other liquids [Online]. Available: https://www.eia.gov/international/data/world/petroleum-and-other-liquids/annual-petroleum-and-other-liquids-production (accessed 12 September 2024).

6 Statista (2024). Global natural gas production 2023, Statista [Online]. Available: https://www.statista.com/statistics/265344/total-global-natural-gas-production-since-1998 (accessed 12 September 2024).

7 Statista (2024). Global grain production, 2023/2024, Statista [Online]. Available: https://www.statista.com/statistics/271943/total-world-grain-production-since-2008-2009 (accessed 12 September 2024).

8 Worldsteel Association (2024). 2024 world steel in figures [Online]. Available: https://web.archive.org/web/20240621192304/https://worldsteel.org/wp-content/uploads/World-Steel-in-Figures-2024.pdf (accessed 9 December 2024).

9 Ritchie, H., Samborska, V., and Roser, M. (2023). Plastic pollution, *Our World in Data*, December 2023 [Online]. Available: https://ourworldindata.org/plastic-pollution (accessed 12 September 2024).

10 Statista (2023). Global lime production 2023, Statista [Online]. Available: https://www.statista.com/statistics/1006040/production-of-lime-worldwide (accessed 12 September 2024).

11 Statista (2022). Global paper and paperboard production 2022, Statista [Online]. Available: https://www.statista.com/statistics/270314/global-paper-and-cardboard-production (accessed 12 September 2024).

12 Bolen, W.P. (2024). Salt, U.S. Geological Survey, Mineral Commodity Summaries. https://pubs.usgs.gov/periodicals/mcs2024/mcs2024-salt.pdf

13 Statista (2024). Sulfuric acid global market volume 2015-2030, Statista [Online]. Available: https://www.statista.com/statistics/1245226/sulfuric-acid-market-volume-worldwide (accessed 12 September 2024).

14 Saygin, D., Blanco, H., Boshell, F. et al. (2023). Ammonia production from clean hydrogen and the implications for global natural gas demand. *Sustainability* 15: 1623. https://doi.org/10.3390/su15021623.

15 IRENA (2024). Hydrogen [Online]. Available: https://www.irena.org/Energy-Transition/Technology/Hydrogen (accessed 12 September 2024).

16 Hubert, M. (2019). Industrial glass processing and fabrication. In: *Springer Handbook of Glass* (ed. J.D. Musgraves, J. Hu, and L. Calvez), 1195–1231.

Cham: Springer International Publishing https://doi.org/10.1007/978-3-319-93728-1_34.

17 Statista (2024). Chlorine global market volume 2030, Statista [Online]. Available: https://www.statista.com/statistics/1310477/chlorine-market-volume-worldwide (accessed 12 September 2024).

18 Chemanalyst (2023). Decode the future of chlorine[Online]. Available: https://www.chemanalyst.com/industry-report/chlorine-market-2853#:~:text=The%20global%20Chlorine%20market%20stood,the%20forecast%20period%20until%202032 (accessed 12 September 2024).

19 Pistilli, M. (2024). Top 10 aluminum-producing countries [Online]. Available: https://investingnews.com/aluminum-production-by-country (accessed 12 September 2024).

20 Statista (2024). Global manganese mine production 2021, Statista [Online]. Available: https://www.statista.com/statistics/1003518/global-manganese-production-volume (accessed 12 September 2024).

21 Pistilli, M. (2024). Top 10 phosphate countries by production, *Investing News Network* [Online]. Available: https://investingnews.com/daily/resource-investing/agriculture-investing/phosphate-investing/top-phosphate-countries-by-production (accessed 12 September 2024).

22 Pistilli, M. (2024). Top 5 chromium-producing countries, *International News Network* [Online]. Available: https://investingnews.com/daily/resource-investing/industrial-metals-investing/chromium-investing/top-chromium-producing-countries (accessed 12 September 2024).

23 Natural Resources Canada (2024). Copper facts [Online]. Available: https://natural-resources.canada.ca/our-natural-resources/minerals-mining/mining-data-statistics-and-analysis/minerals-metals-facts/copper-facts/20506 (accessed 12 September 2024).

24 U.S. Geological Survey, Mineral Commodity Summaries (2024). Titanium and titanium dioxide, January 2024. Available: https://pubs.usgs.gov/periodicals/mcs2024/mcs2024-titanium.pdf (accessed September 2025).

25 Statista (2024). Global silicon production 2023. Statista [Online]. Available: https://www.statista.com/statistics/573585/global-silicon-production (accessed 12 September 2024).

26 Flores-Granobles, M. and Saeys, M. (2023). Quantitative analysis of CO_2 emissions reduction potential of alternative light olefins production processes. *Green Chemistry* 25 (16): 6459–6471. https://doi.org/10.1039/D3GC01237A.

27 World Economic Forum (2022). Here's what the world mined in 2021 in one infographic. World Economic Forum [Online]. Available: https://www.weforum.org/agenda/2022/10/all-the-metals-we-mined-in-2021-visualized (accessed 28 April 2028).

28 Li, K., Fan, Q., Chuai, H. et al. (2021). Revisiting chlor-alkali electrolyzers: from materials to devices. *Tianjin University* 27 (3): 202–216. https://doi.org/10.1007/s12209-021-00285-9.

29 Britannica (2024). Manganese | uses, facts, & compounds | Britannica [Online]. Available: https://www.britannica.com/science/manganese (accessed 12 September 2024).

30 Britannica (2024). Chromium | uses, properties, & facts | Britannica [Online]. Available: https://www.britannica.com/science/chromium (accessed 12 September 2024).

31 http://Geology.com (2012). Uses of nickel | supply, demand, production, resources [Online]. Available: https://geology.com/usgs/uses-of-nickel (accessed 12 September 2024).

32 PRODYOGI (2018). Bogues compounds in cement | calculation of bogues compound. PRODYOGI [Online]. Available: http://www.prodyogi.com/2018/03/bogue-compounds-cement.html (accessed 7 August 2024).

33 Stanzel, M. and Sembrick, N. (2020). Cement and concrete, still outperforming in the sustainability era. Construction Specifier [Online]. Available: https://www.constructionspecifier.com/cement-and-concrete-still-outperforming-in-the-sustainability-era (accessed 7 August 2024).

34 IEA (2021). Global cement demand for building construction, 2000–2020, and in the Net Zero Scenario, 2025–2030 – Charts – Data & Statistics, IEA [Online]. Available: https://www.iea.org/data-and-statistics/charts/global-cement-demand-for-building-construction-2000-2020-and-in-the-net-zero-scenario-2025-2030 (accessed 7 August 2024).

35 Sanjuán, M., Andrade, C., Mora, P., and Zaragoza, A. (2020). Carbon dioxide uptake by cement-based materials: a Spanish case study. *Applied Sciences* 10: 339. https://doi.org/10.3390/app10010339.

36 SkyQuest (2024). Cement market size, share, growth analysis, by type (ordinary Portland cement (OPC), Portland Pozzolana cement (PPC)), by application (residential, commercial), by region - industry forecast 2025–2032, SQMIG15F2024.

37 The Observatory of Economic Complexity (2023). Cement (HS: 2523) product trade, exporters and importers, The Observatory of Economic Complexity [Online]. Available: https://oec.world/en/profile/hs/cement (accessed 17 January 2025).

38 Arachchige, U., Kawan, D., Tokheim, L.-A., and Melaaen, M. (2014). Waste heat utilization for CO_2 capture in the cement industry. *International Journal of Modeling and Optimization* 4: 438–442. https://doi.org/10.7763/IJMO.2014.V4.414.

39 Tokheim, L.-A. and Brevik, P. (2007). Carbon dioxide reduction by increased utilization of waste derived fuels in the cement industry. *Presented at the International Conference of Sustainable Production and Use of Cement and Concrete* (September 2007).

40 Hermansson, L. (2018). A review of nanostructured Ca-aluminate based biomaterials within odontology and orthopedics. *Journal of the Korean Ceramic Society* 55 (2): 95–107. https://doi.org/10.4191/kcers.2018.55.2.01.

41 Forster, A.M. (2004). Hydraulicity for beginners and the hydraulic lime family. In: *How Hydraulic Lime Binders Work*. Edinbrugh: Masons Mortar.

42 Castro, F., Santos, J., Lacerda, P. et al. (2017). Employment of industrial wastes as agents for inclusion modification in molten steels, p. 394. https://doi.org/10.1201/9781315206172-62.

43 Hanein, T., Glasser, F.P., and Bannerman, M.N. (2020). Thermodynamic data for cement clinkering. *Cement and Concrete Research* 132: 106043. https://doi.org/10.1016/j.cemconres.2020.106043.

44 Murao, R., Harano, T., Kimura, M., and Jung, I.-H. (2018). Thermodynamic modeling of the SFCA phase $Ca_2(Fe,Ca)_6(Fe,Al,Si)_6O_{20}$. *ISIJ International* 58 (2): 259–266. https://doi.org/10.2355/isijinternational.ISIJINT-2017-459.

45 Jerebtsov, D.A. and Mikhailov, G.G. (2001). Phase diagram of $CaO–Al_2O_3$ system. *Ceramics International* 27 (1): 25–28. https://doi.org/10.1016/S0272-8842(00)00037-7.

46 Schneider, M., Romer, M., Tschudin, M., and Bolio, H. (2011). Sustainable cement production - present and future. *Cement and Concrete Research* 41 (7): 642–650. https://doi.org/10.1016/j.cemconres.2011.03.019.

47 Zhang, H., Dewil, R., Degrève, J., and Baeyens, J. (2014). The design of cyclonic pre-heaters in suspension cement kilns. *International Journal of Sustainable Engineering* 7 (4): 307–312. https://doi.org/10.1080/19397038.2013.843605.

48 Kumar, A., Sahu, R., and Tripathy, S.K. (2023). Energy-efficient advanced ultra-fine grinding of particles using stirred mills—a review. *Energies* 16 (14): 5277. https://doi.org/10.3390/en16145277.

49 Hills, T., Leeson, D., Florin, N., and Fennell, P. (2016). Carbon capture in the cement industry: technologies, progress, and retrofitting. *Environmental Science Technology* 50 (1): 368–377. https://doi.org/10.1021/acs.est.5b03508.

50 Hills, T.P., Sceats, M., Rennie, D., and Fennell, P. (2017). LEILAC: low cost CO_2 capture for the cement and lime industries. *Energy Procedia* 114: 6166–6170. https://doi.org/10.1016/j.egypro.2017.03.1753.

51 Schakel, W., Hung, C.R., Tokheim, L.-A. et al. (2018). Impact of fuel selection on the environmental performance of post-combustion calcium looping applied to a cement plant. *Applied Energy* 210: 75–87. https://doi.org/10.1016/j.apenergy.2017.10.123.

52 Mastorakos, E., Massias, A., Tsakiroglou, C.D. et al. (1999). CFD predictions for cement kilns including flame modelling, heat transfer and clinker chemistry. *Applied Mathematical Modelling* 23 (1): 55–76. https://doi.org/10.1016/S0307-904X(98)10053-7.

53 Czigler, T., Reiter, S., Schulze, P., and Somers, K. (2024). Laying the foundation for a zero-carbon cement industry | McKinsey [Online]. Available: https://www.mckinsey.com/industries/chemicals/our-insights/laying-the-foundation-for-zero-carbon-cement (accessed 7 August 2024).

54 Hills, T.P., Gordon, F., Florin, N.H., and Fennell, P.S. (2015). Statistical analysis of the carbonation rate of concrete. *Cement and Concrete Research* 72: 98–107. https://doi.org/10.1016/j.cemconres.2015.02.007.

55 Xi, F., Davis, S.J., Ciais, P. et al. (2016). Substantial global carbon uptake by cement carbonation. *Nature Geoscience* 9 (12): 880–883. https://doi.org/10.1038/ngeo2840.

56 Fennell, P., Driver, J., Bataille, C., and Davis, S.J. (2022). Cement and steel - nine steps to net zero. *Nature* 603 (7902): 574–577. https://doi.org/10.1038/d41586-022-00758-4.

57 Neithalath, N., Persun, J., and Manchiryal, R.K. (2010). Electrical conductivity based microstructure and strength prediction of plain and modified concretes. *International Journal of Advances in Engineering Sciences and Applied Mathematics* 2 (3): 83–94. https://doi.org/10.1007/s12572-011-0023-1.

58 Voldsund, M., Gardarsdottir, S.O., De Lena, E. et al. (2019). Comparison of technologies for CO_2 capture from cement production—part 1: technical evaluation. *Energies* 12 (3): 559. https://doi.org/10.3390/en12030559.

59 Dean, C., Kokot-Blamey, J., Florin, N. et al. (2011). The calcium looping cycle for CO_2 capture from power generation, cement manufacture and hydrogen production. *Chemical Engineering Research & Design* 89: 836–855. https://doi.org/10.1016/j.cherd.2010.10.013.

60 Zeman, F. (2009). Oxygen combustion in cement production. *Energy Procedia* 1 (1): 187–194. https://doi.org/10.1016/j.egypro.2009.01.027.

61 Yang, Y., Wang, L., Xia, D. et al. (2020). Novel lime calcination system for CO_2 capture and its thermal–mass balance analysis. *ACS Omega* 5 (42): 27413–27424. https://doi.org/10.1021/acsomega.0c03850.

62 Areeprasert, C., Asingsamanunt, J., Srisawat, S. et al. (2017). Municipal plastic waste composition study at transfer station of Bangkok and possibility of its energy recovery by pyrolysis. *Energy Procedia* 107: 222–226. https://doi.org/10.1016/j.egypro.2016.12.132.

63 M. Chinyama, "*Alternative Fuels in Cement Manufacturing*," 2011. https://doi.org/10.5772/22319.

64 Rada, E., Ragazzi, M., Maschio, R. et al. (2012). Energy recovery from tyres waste through thermal option. *UPB Scientific Bulletin, Series D: Mechanical Engineering* 74: 201–210.

65 Dreher, P., Faulstich, M., Weber-Blaschke, G. et al. (2004). VI.5 recycling of plastic waste, rubber waste and end-of-life cars in Germany. *Waste Management Series* 4: https://doi.org/10.1016/S0713-2743(04)80035-7.

66 Martínez, J.D., Puy, N., Murillo, R. et al. (2013). Waste tyre pyrolysis – a review. *Renewable and Sustainable Energy Reviews* 23: 179–213. https://doi.org/10.1016/j.rser.2013.02.038.

67 Pandel, B. (2018) Understanding the retardation of cement hydration caused by zinc. MSc. University of Illinois, Illinois.

68 Nhuchhen, D.R., Sit, S.P., and Layzell, D.B. (2022). Decarbonization of cement production in a hydrogen economy. *Applied Energy* 317: 119180. https://doi.org/10.1016/j.apenergy.2022.119180.

69 Desai, P.D. (1986). Thermodynamic properties of iron and silicon. *Journal of Physical and Chemical Reference Data* 15 (3): 967–983. https://doi.org/10.1063/1.555761.

70 Schauwinhold, D., Kalla, U., and Kalwa, G. (2011). Steel, 6. Uses. In: *Ullmann's Encyclopedia of Industrial Chemistry*, 1e. Wiley. https://doi.org/10.1002/14356007.o25_o05.

71 Velling, A. (2020). Iron-carbon phase diagram explained, Fractory [Online]. Available: https://fractory.com/iron-carbon-phase-diagram (accessed 28 April 2024).

72 Nuclear Power. Malleable cast iron, Nuclear Power. (2018) [Online]. Available: https://www.nuclear-power.com/nuclear-engineering/metals-what-are-metals/cast-iron/malleable-cast-iron (accessed 28 April 2024).

73 Ostrovski, O. and Belashchenko, D.K. (2013). Thermophysical properties and structure of liquid Fe-C alloys. *High Temperatures-High Pressures* 42 (2): 137–149. https://doi.org/10.1080/1071441950170102.

74 Jo, H., Kang, M., Park, G.W. et al. (2020). Effects of cooling rate during quenching and tempering conditions on microstructures and mechanical properties of carbon steel flange. *Materials (Basel)* 13 (18): 4186. https://doi.org/10.3390/ma13184186.

75 Larrin. How fast do you have to quench? hardenability of steel, Knife Steel Nerds (2019) [Online]. Available: https://knifesteelnerds.com/2019/02/25/how-fast-do-you-have-to-quench-hardenability-of-steel (accessed 18 January 2025).

76 Zhang, H., Huo, M., Ma, Z. et al. (2022). Effects of quenching and tempering heat treatment processing on the microstructure and properties of high-strength hull steel. *Meta* 12 (6): Art. no. 6: https://doi.org/10.3390/met12060914.

77 Fadare, D.A., Fadara, T.G., and Akanbi, O.Y. (2011). Effect of heat treatment on mechanical properties and microstructure of NST 37-2 steel. *Journal of Minerals and Materials Characterization and Engineering* 10 (03): 299–308. https://doi.org/10.4236/jmmce.2011.103020.

78 The Engineering Toolbox. The AISI/SAE carbon steel and alloys naming conventions,(2009). The Engineering Toolbox [Online]. Available: https://www.engineeringtoolbox.com/aisi-sae-steel-numbering-system-d_1449.html (accessed 28 April 2024).

79 Unified Alloys (2018). Stainless steel grades and families: explained [Online]. Available: https://www.unifiedalloys.com/blog/stainless-grades-families (accessed 28 April 2024).

80 Masteel (2018). Outlining austenitic stainless steel, Masteel [Online]. Available: https://masteel.co.uk/news/outlining-austenitic-stainless-steel (accessed 28 April 2024).

81 AlHazaa, A. and Haneklaus, N. (2020). Diffusion bonding and transient liquid phase (TLP) bonding of type 304 and 316 austenitic stainless steel—a review of similar and dissimilar material joints. *Meta* 10 (5): 613. https://doi.org/10.3390/met10050613.

82 Masteel (2018). Outlining duplex stainless steel, Masteel [Online]. Available: https://masteel.co.uk/news/outlining-duplex-stainless-steel (accessed 28 April 2024).

83 Masteel (2019). Outlining ferritic stainless steel, Masteel [Online]. Available: http://masteel.co.uk/news/outlining-ferritic-stainless (accessed 28 April 2024).

84 Masteel (2018). Outlining martensitic stainless steel, Masteel [Online]. Available: https://masteel.co.uk/news/martensitic-stainless-steel (accessed 28 April 2024).

85 Statista. Iron ore: global reserves by country 2023, Statista (2024)[Online]. Available: https://www.statista.com/statistics/267381/world-reserves-of-iron-ore-by-country (accessed 18 January 2025).

86 Pistilli, M. (2024) Top 10 iron-producing countries, Investing News Network [Online]. Available: https://investingnews.com/daily/resource-investing/base-metals-investing/iron-investing/top-iron-producing-countries (accessed 24 April 2024).

87 World Population Review (2023). Iron ore production by country 2024 [Online]. Available: https://worldpopulationreview.com/country-rankings/iron-ore-production-by-country (accessed 18 January 2025).

88 Statista (2024). Iron ore price 2000-2023, Statista [Online]. Available: https://www.statista.com/statistics/282830/iron-ore-prices-since-2003 (accessed 2 August 2024).

89 Statista (2024). Iron ore mining global market size 2023, Statista [Online]. Available: https://www.statista.com/statistics/1279697/worldwide-iron-ore-mining-market-value (accessed 3 July 2024).

90 International Monetary Fund (2024). IMF primary commodity prices [Online]. Available: https://www.imf.org/en/Research/commodity-prices (accessed 18 January 2025).

91 Wårell, L. (2018). An analysis of iron ore prices during the latest commodity boom. *Mineral Economics* 31 (1–2): 203–216. https://doi.org/10.1007/s13563-018-0150-2.

92 Jégourel, Y. (2020). The global iron ore market: from cyclical developments to potential structural changes. *The Extractive Industries and Society* 7 (3): 1128–1134. https://doi.org/10.1016/j.exis.2020.05.015.

93 World Steel Association (2024). Data - http://worldsteel.org [Online]. Available: https://worldsteel.org/data (accessed 4 March 2025).

94 Global Energy Monitor (2024). Global steel plant tracker, Global Energy Monitor [Online]. Available: https://globalenergymonitor.org/projects/global-steel-plant-tracker (accessed 3 July 2024).

95 Satyendra (2013). Metallurgical coke. ispatguru [Online]. Available: https://www.ispatguru.com/metallurgical-coke (accessed 28 April 2024).

96 Buczynski, R., Weber, R., Kim, R., and Schwöppe, P. (2016). One-dimensional model of heat-recovery, non-recovery coke ovens. *Fuel* 181: 1151–1161. https://doi.org/10.1016/j.fuel.2016.05.033.

97 Vendatu (2024). *Draw a neat diagram of the blast furnace used in the extraction of iron and label the parts* [Online]. Available: https://www.vedantu.com/question-answer/draw-a-neat-diagram-of-the-blast-furnace-used-in-class-11-chemistry-cbse-5f551dfed33d37372b526fae (accessed 28 April 2024).

98 Filatov, S.V., Listopadov, V.S., Sorokin, A.Y. et al. (2020). Technology of blast-furnace smelting of iron using pulverized coal at the NLMK. *Metallurgist* 64 (5–6): 381–387. https://doi.org/10.1007/s11015-020-01006-8.

99 Steelonthenet (2012). Steelmaking CO_2 carbon dioxide emissions by process step [Online]. Available: https://www.steelonthenet.com/kb/co2-emissions.html (accessed 18 January 2025).

100 Fan, Z. and Friedmann, S.J. (2021). Low-carbon production of iron and steel: technology options, economic assessment, and policy. *Joule* 5 (4): 829–862. https://doi.org/10.1016/j.joule.2021.02.018.

101 Xi, L., Lin, Q., Hasan, M. et al. (2019). Assessing the economics of CO_2 capture in China's iron/steel sector: a case study. *Energy Procedia* 158: 3715–3722. https://doi.org/10.1016/j.egypro.2019.01.886.

102 White, J.F., Lee, J., Hessling, O., and Glaser, B. (2017). Reactions between liquid CaO-SiO_2 slags and graphite substrates. *Metallurgical and Materials Transactions B* 48 (1): 506–515. https://doi.org/10.1007/s11663-016-0788-5.

103 Zhang, W., Zhang, J., and Xue, Z. (2017). Exergy analyses of the oxygen blast furnace with top gas recycling process. *Energy* 121: 135–146. https://doi.org/10.1016/j.energy.2016.12.125.

104 Briant, C.L., Banerji, S.K., and Ritter, A.M. (1982). The role of nitrogen in the embrittlement of steel. *Metallurgical Transactions A* 13 (11): 1939–1950. https://doi.org/10.1007/BF02645939.

105 Madhavan, N., Brooks, G.A., Rhamdhani, M.A. et al. (2021). General mass balance for oxygen steelmaking. *Ironmaking & Steelmaking* 48 (1): 40–54. https://doi.org/10.1080/03019233.2020.1731252.

106 Kopeliovich, D. (2023). Steel making (introduction), Substech [Online]. Available: https://substech.com/dokuwiki/doku.php?id=steel_making_introduction (accessed 28 April 2024).

107 Dering, D., Swartz, C., and Dogan, N. (2020). Dynamic modeling and simulation of basic oxygen furnace (BOF) operation. *Processes* 8 (4): 483. https://doi.org/10.3390/pr8040483.

108 Si, M., Thompson, S., and Calder, K. (2011). Energy efficiency assessment by process heating assessment and survey tool (PHAST) and feasibility analysis of waste heat recovery in the reheat furnace at a steel company. *Renewable and Sustainable Energy Reviews* 15: 2904–2908. https://doi.org/10.1016/j.rser.2011.02.035.

109 He, K. and Wang, L. (2017). A review of energy use and energy-efficient technologies for the iron and steel industry. *Renewable and Sustainable Energy Reviews* 70: 1022–1039. https://doi.org/10.1016/j.rser.2016.12.007.

110 Najm, O., El-Hassan, H., and El-Dieb, A. (2021). Ladle slag characteristics and use in mortar and concrete: a comprehensive review. *Journal of Cleaner Production* 288: 125584. https://doi.org/10.1016/j.jclepro.2020.125584.

111 FRC Global (2021). The costs of producing steel with electric arc furnaces | FRC Global [Online]. Available: https://www.frcglobal.com/blog/the-costs-of-producing-steel-with-electric-arc-furnaces (accessed 18 January 2025).

112 Fruehan, R.J., Fortini, O., Paxton, H.W., and Brindle, R. (2000) Theoretical minimum energies to produce steel for selected conditions, United States, May 2000. https://doi.org/10.2172/769470.

113 Cavaliere, P. (2022). Hydrogen direct reduced iron. In: *Hydrogen Assisted Direct Reduction of Iron Oxides*, 233–319. Cham: Springer International Publishing. https://doi.org/10.1007/978-3-030-98056-6_7.

114 Hamadeh, H., Mirgaux, O., and Patisson, F. (2018). Detailed modeling of the direct reduction of Iron ore in a shaft furnace. *Materials* 11 (10): Art. no. 10: https://doi.org/10.3390/ma11101865.

115 Boretti, A. (2023). The perspective of hydrogen direct reduction of iron. *Journal of Cleaner Production* 429: 139585. https://doi.org/10.1016/j.jclepro.2023.139585.

116 Jaganmohan, M. (2023). Iron: usable ore production worldwide 2023, Statista [Online]. Available: https://www.statista.com/statistics/589945/iron-ore-production-gross-weight-worldwide (accessed 11 May 2024).

117 Chaabet, M. and Dötsch, E. (2012). Steelmaking Based on Inductive Melting.

118 Hexagon Resources Limited (2019). Testwork enhances graphite electrode performance, *GlobeNewswire News Room* [Online]. Available: https://www.globenewswire.com/en/news-release/2019/10/16/1930574/0/en/Testwork-Enhances-Graphite-Electrode-Performance.html (accessed 23 September 2024).

119 Wood Mackenzie (2022). Decarbonising global iron ore and steel industry by 2050 necessitates urgent action and US$1.4 trillion of investment | Wood Mackenzie [Online]. Available: https://www.woodmac.com/press-releases/decarbonising-global-iron-ore-and-steel-industry-by-2050-necessitates-urgent-action-and-us$1.4-trillion-of-investment (accessed 18 January 2025).

120 Midrex (2022). MIDREX NG™ with H_2 addition: moving from natural gas to hydrogen in decarbonizing ironmaking, Midrex Technologies, Inc. [Online]. Available: https://www.midrex.com/tech-article/moving-from-natural-gas-to-hydrogen-in-decarbonizing-ironmaking (accessed 18 January 2025).

121 Madias, J. (2014). Electric furnace steelmaking. In: *Treatise on Process Metallurgy*, 271–300. Elsevier. https://doi.org/10.1016/B978-0-08-096988-6.00013-4.

122 Ritchie, H., Rosado, P., and Roser, M. (2024). Energy production and consumption, *Our World in Data*, February 2024 [Online]. Available: https://ourworldindata.org/energy-production-consumption (accessed 11 May 2024).

123 Bhaskar, A., Abhishek, R., Assadi, M., and Somehesaraei, H.N. (2022). Decarbonizing primary steel production: techno-economic assessment of a hydrogen based green steel production plant in Norway. *Journal of Cleaner Production* 350: 131339. https://doi.org/10.1016/j.jclepro.2022.131339.

124 Devlin, A., Kossen, J., Goldie-Jones, H., and Yang, A. (2023). Global green hydrogen-based steel opportunities surrounding high quality renewable energy and iron ore deposits. *Nature Communications* 14 (1): 2578. https://doi.org/10.1038/s41467-023-38123-2.

125 Ashar, N.G. and Golwalkar, K.R. (2013). Processes of manufacture of sulfuric acid. In: *A Practical Guide to the Manufacture of Sulfuric Acid, Oleums, and Sulfonating Agents* (ed. N.G. Ashar and K.R. Golwalkar), 9–30. Cham: Springer International Publishing. https://doi.org/10.1007/978-3-319-02042-6_2.

126 Jones, E.M. (1950). Chamber process manufacture of sulfuric acid. *Industrial & Engineering Chemistry* 42 (11): 2208–2210. https://doi.org/10.1021/ie50491a016.

127 Müller, H. (2000). Sulfuric acid and sulfur trioxide. In: *Ullmann's Encyclopedia of Industrial Chemistry*. John Wiley & Sons, Ltd. https://doi.org/10.1002/14356007.a25_635.

128 Davenport, W.G. and King, M.J. (2006). Chapter 6 - dehydrating air and gases with strong sulfuric acid. In: *Sulfuric Acid Manufacture* (ed. W.G. Davenport and M.J. King), 59–69. Oxford: Elsevier. https://doi.org/10.1016/B978-008044428-4/50006-3.

129 The Essential Chemical Industry (2016). Sulfuric acid, The Essential Chemical Industry [Online]. Available: https://www.essentialchemicalindustry.org/chemicals/sulfuric-acid.html (accessed 5 September 2024).

130 Speight, J.G. (2017). Chapter three - industrial inorganic chemistry. In: *Environmental Inorganic Chemistry for Engineers* (ed. J.G. Speight), 111–169. Butterworth-Heinemann. https://doi.org/10.1016/B978-0-12-849891-0.00003-5.

131 Civetta, W.R., Padilla, R.V., Silvera, A.B., and Quiroga, A.G. (2014). Exergy analysis of an industrial sulphuric acid plant. *IJETAE* 4 (3).

132 Zheng, X., Lei, G., Wang, S. et al. (2023). Advances in resources recovery of H_2S: a review of desulfurization processes and catalysts. *ACS Catalysis* 13 (17): 11723–11752. https://doi.org/10.1021/acscatal.3c02294.

133 Shafi, R. and Hutchings, G.J. (2000). Hydrodesulfurization of hindered dibenzothiophenes: an overview. *Catalysis Today* 59 (3): 423–442. https://doi.org/10.1016/S0920-5861(00)00308-4.

134 Davenport, W.G. and King, M.J. (2006). Chapter 2 - production and consumption. In: *Sulfuric Acid Manufacture* (ed. W.G. Davenport and M.J. King), 11–17. Oxford: Elsevier. https://doi.org/10.1016/B978-008044428-4/50002-6.

135 Maslin, M., Van Heerde, L., and Day, S. (2022). Sulfur: a potential resource crisis that could stifle green technology and threaten food security as the world decarbonises. *Geographical Journal* 188 (4): 498–505. https://doi.org/10.1111/geoj.12475.

136 Piéplu, A., Saur, O., Lavalley, J.-C. et al. (1998). Claus catalysis and H_2S selective oxidation. *Catalysis Reviews* 40 (4): 409–450. https://doi.org/10.1080/01614949808007113.

137 Kent, J.A. (2010). Sulfur and sulfuric acid. In: *Kent and Riegel's Handbook of Industrial Chemistry and Biotechnology*, 997–1016. Springer Science & Business Media.

138 Nehb, W. and Vydra, K. (2000). Sulfur. In: *Ullmann's Encyclopedia of Industrial Chemistry*.

139 García-Labiano, F., de Diego, L.F., Cabello, A. et al. (2016). Sulphuric acid production via chemical looping combustion of elemental sulphur. *Applied Energy* 178: 736–745. https://doi.org/10.1016/j.apenergy.2016.06.110.

140 Bartholomew, C.H.F. and Farrauto, R.J. (2005). Catalytic oxidations of inorganic and organic compounds. Fundamentals of Industrial Catalytic Processes, pp. 560–634.

141 Zaker, M.R., Fauteux-Lefebvre, C., and Thibault, J. (2021). Modelling and multi-objective optimization of the sulphur dioxide oxidation process. *Processes* 9 (6) Art. no. 6: https://doi.org/10.3390/pr9061072.

142 Sampat, S., Kelly, J., de Campos, M., and Mostaghel, S. (2023). Sulphuric acid plants in metallurgical facilities: options for energy optimization. In: *Advances in Pyrometallurgy* (ed. C. Fleuriault, J.D. Steenkamp, D. Gregurek, et al.),

215–224. Cham: Springer Nature Switzerland. https://doi.org/10.1007/978-3-031-22634-2_20.

143 Kemmerich, M. and Storch, H. (2016). Process heat recovery and digitalisation in sulphuric acid plants. *Procedia Engineering* 138: 220–230. https://doi.org/10.1016/j.proeng.2016.02.079.

144 Chouaibi, F., Belghaieb, J., and Hajji, N. (2018). Waste heat recovery in a sulfuric acid production unit. In: *Exergy for A Better Environment and Improved Sustainability 1: Fundamentals* (ed. F. Aloui and I. Dincer), 931–942. Cham: Springer International Publishing. https://doi.org/10.1007/978-3-319-62572-0_60.

145 McLean, S.H., Chenier, J., Muinonen, S. et al. (2020). Low-grade waste heat recovery and repurposing to reduce the load on cooling towers. *Advances in Energy Research* 7 (2): 147–166. https://doi.org/10.12989/eri.2020.7.2.147.

146 Vesborg, P.C.K. and Jaramillo, T.F. (2012). Addressing the terawatt challenge: scalability in the supply of chemical elements for renewable energy. *RSC Advances* 2 (21): 7933–7947. https://doi.org/10.1039/C2RA20839C.

147 Lim, J., Pyun, J., and Char, K. (2015). Recent approaches for the direct use of elemental sulfur in the synthesis and processing of advanced materials. *Angewandte Chemie International Edition* 54 (11): 3249–3258. https://doi.org/10.1002/anie.201409468.

148 Song, K., Meng, Q., Shu, F., and Ye, Z. (2013). Recovery of high purity sulfuric acid from the waste acid in toluene nitration process by rectification. *Chemosphere* 90 (4): 1558–1562. https://doi.org/10.1016/j.chemosphere.2012.09.043.

149 Schwartz, D., Gadiou, R., Brilhac, J.-F. et al. (2000). A kinetic study of the decomposition of spent sulfuric acids at high temperature. *Industrial & Engineering Chemistry Research* 39 (7): 2183–2189. https://doi.org/10.1021/ie990801e.

150 Balla, A., Marcu, C., Axente, D. et al. (2012). Catalytic reduction of sulfuric acid to sulfur dioxide. *Open Chemistry* 10 (6): 1817–1823. https://doi.org/10.2478/s11532-012-0099-x.

151 Sujeesh, S., Ahmed, V.N., Rao, A.S., and Mukhopadhyay, S. (2022). Sulphuric acid decomposition using Cr–Fe_2O_3 catalyst in a tubular packed bed reactor (PBR): modeling and experimental studies. *International Journal of Hydrogen Energy* 47 (23): 11750–11763. https://doi.org/10.1016/j.ijhydene.2022.01.207.

152 Lebowitz, S.H. (1931). A demonstration working model of the Frasch process for mining sulfur. *Journal of Chemical Education* 8 (8): 1630. https://doi.org/10.1021/ed008p1630.

153 Statista (2024). Chlorine global market value 2030, Statista [Online]. Available: https://www.statista.com/statistics/1310482/chlorine-market-value-worldwide (accessed 21 September 2024).

154 FMI (2024). Sodium hydroxide market - analysis, size & forecast 2033 [Online]. Available: https://www.futuremarketinsights.com/reports/sodium-hydroxide-market (accessed 21 September 2024).

155 Li, M., Lin, M., Wu, Z., and Christy, A.A. (2005). The influence of NaOH on the stability of paraffinic crude oil emulsion. *Fuel* 84 (2): 183–187. https://doi.org/10.1016/j.fuel.2004.09.001.

156 Hind, A.R., Bhargava, S.K., and Grocott, S.C. (1999). The surface chemistry of Bayer process solids: a review. *Colloids and Surfaces A: Physicochemical and Engineering Aspects* 146 (1): 359–374. https://doi.org/10.1016/S0927-7757(98)00798-5.

157 Kurt, C. and Bittner, J. (2006). Sodium hydroxide. In: *Ullmann's Encyclopedia of Industrial Chemistry*. John Wiley & Sons, Ltd. https://doi.org/10.1002/14356007.a24_345.pub2.

158 TEADIT (2023). Selecting correct assets for chlorine [Online]. Available: https://teadit.com/us/article/the-importance-of-selecting-the-correct-asset-for-chlorine (accessed 21 September 2024).

159 Evans, R.B. (2005). Chlorine: state of the art. *Lung* 183 (3): 151–167. https://doi.org/10.1007/s00408-004-2530-3.

160 Zhang, L.-N., Li, R., Zang, H.-Y. et al. (2021). Advanced hydrogen evolution electrocatalysts promising sustainable hydrogen and chlor-alkali co-production. *Energy & Environmental Science* 14 (12): 6191–6210. https://doi.org/10.1039/D1EE02798K.

161 Garcia-Herrero, I., Margallo, M., Onandía, R. et al. (2017). Environmental challenges of the chlor-alkali production: seeking answers from a life cycle approach. *Science of the Total Environment* 580: 147–157. https://doi.org/10.1016/j.scitotenv.2016.10.202.

162 Sodeno, R. (2023). Projected global mercury supply, demand, and excess to 2050 based on impacts of the Minamata convention. *Journal of Material Cycles and Waste Management* 25 (6): 3608–3624. https://doi.org/10.1007/s10163-023-01780-y.

163 Wang, X., Teichgraeber, H., Palazoglu, A., and El-Farra, N.H. (2014). An economic receding horizon optimization approach for energy management in the chlor-alkali process with hybrid renewable energy generation. *Journal of Process Control* 24 (8): 1318–1327. https://doi.org/10.1016/j.jprocont.2014.04.017.

164 Bargeman, G., Steensma, M., ten Kate, A. et al. (2009). Nanofiltration as energy-efficient solution for sulfate waste in vacuum salt production. *Desalination* 245 (1): 460–468. https://doi.org/10.1016/j.desal.2009.02.009.

165 Schmittinger, P., Curlin, T.F.L.C., Lüke, B. et al. (2011). Chlorine. In: *Ullmann's Encyclopedia of Industrial Chemistry*. John Wiley & Sons, Ltd. https://doi.org/10.1002/14356007.a06_399.pub3.

166 Garcia-Herrero, I., Margallo, M., Onandía, R. et al. (2017). Life cycle assessment model for the chlor-alkali process: a comprehensive review of resources and available technologies. *Sustainable Production and Consumption* 12: 44–58. https://doi.org/10.1016/j.spc.2017.05.001.

167 Lakshmanan, S. and Thanabalan, M. (2013). The chlor-alkali process: work in progress. *Clean Technologies and Environmental Policy* 16: https://doi.org/10.1007/s10098-013-0630-6.

168 Eurochlor (2022). About Chlor-alkali - Eurochlor [Online]. Available: https://www.eurochlor.org/about-chlor-alkali (accessed 10 February 2025).

169 Bommaraju, T.V., Chen, C.-P., and Birss, V.I. (2001). Deactivation of thermally formed RuO_2 + TiO_2 coatings during chlorine evolution: mechanisms and reactivation measures. In: *Modern Chlor-Alkali Technology*, 57–81. John Wiley & Sons, Ltd. https://doi.org/10.1002/9780470999479.ch5.

170 Stringer, R. and Johnston, P. (2001). Chlorine and the environment: an overview of the chlorine industry. *Environmental Science and Pollution Research* 8: https://doi.org/10.1007/BF02987309.

171 O'Brien, T.F., Bommaraju, T.V., and Hine, F. (2020). *Handbook of Chlor-Alkali Technology, Volume 1: Fundementals*. Springer.

172 Song, Z., Li, P., Ding, L. et al. (2018). Environmental mercury pollution by an abandoned chlor-alkali plant in Southwest China. *Journal of Geochemical Exploration* 194: 81–87. https://doi.org/10.1016/j.gexplo.2018.07.017.

173 Harris, M. (2001). Phase-out issues for mercury cell technology in the chlor-alkali industry. In: *Modern Chlor-Alkali Technology*, 19–43. John Wiley & Sons, Ltd. https://doi.org/10.1002/9780470999479.ch2.

174 Crook, J. and Mousavi, A. (2016). The chlor-alkali process: a review of history and pollution. *Environmental Forensics* 17: 211–217. https://doi.org/10.1080/15275922.2016.1177755.

175 Moussallem, I., Jörissen, J., Kunz, U. et al. (2008). Chlor-alkali electrolysis with oxygen depolarized cathodes: history, present status and future prospects. *Journal of Applied Electrochemistry* 38: 1177–1194. https://doi.org/10.1007/s10800-008-9556-9.

176 Gebhard, M., Tichter, T., Franzen, D. et al. (2020). Improvement of oxygen-depolarized cathodes in highly alkaline media by electrospinning of poly(vinylidene fluoride) barrier layers. *ChemElectroChem* 7: 830–837. https://doi.org/10.1002/celc.201902115.

177 Öhl, D., Franzen, D., Paulisch, M. et al. (2019). Catalytic reactivation of industrial oxygen depolarized cathodes by in situ generation of atomic hydrogen. *ChemSusChem* 12 (12): 2732–2739. https://doi.org/10.1002/cssc.201900628.

178 Hernandez-Aldave, S. and Andreoli, E. (2022). Oxygen depolarised cathode as a learning platform for CO_2 gas diffusion electrodes. *Catalysis Science & Technology* 12 (11): 3412–3420. https://doi.org/10.1039/D2CY00443G.

179 Park, I.K., Ahn, C.-Y., Lee, J.H. et al. (2019). Three-dimensionally interconnected titanium foam anode for an energy-efficient zero gap-type chlor-alkali electrolyzer. *International Journal of Hydrogen Energy* 44 (31): 16079–16086. https://doi.org/10.1016/j.ijhydene.2019.04.230.

180 International Energy Agency (2023). Global hydrogen review 2023.

181 Roh, K., Brée, L.C., Perrey, K. et al. (2019). Flexible operation of switchable chlor-alkali electrolysis for demand side management. *Applied Energy* 255: 113880. https://doi.org/10.1016/j.apenergy.2019.113880.

182 Requena-Leal, I., Carvela, M., Fernández-Marchante, C.M. et al. (2024). On the use of chlor-alkali technology to power environmental electrochemical treatment technologies. *Current Opinion in Electrochemistry* 45: 101461. https://doi.org/10.1016/j.coelec.2024.101461.

183 IEA (2019). Material efficiency in clean energy transitions – analysi. IEA [Online]. Available: https://www.iea.org/reports/material-efficiency-in-clean-energy-transitions (accessed 19 January 2025).

184 Kanagaraj, B., Anand, N., Johnson Alengaram, U. et al. (2024). Limestone calcined clay cement (LC3): a sustainable solution for mitigating environmental impact in the construction sector. *Resources, Conservation & Recycling Advances* 21: 200197. https://doi.org/10.1016/j.rcradv.2023.200197.

185 Gettu, R. and Basavaraj, A.S. (2020). Life cycle assessment of LC3: parameters and prognoses. In: *Calcined Clays for Sustainable Concrete*, RILEM Bookseries, vol. 25 (ed. S. Bishnoi), 277–281. Singapore: Springer Singapore. https://doi.org/10.1007/978-981-15-2806-4_32.

186 Knight, K.A., Cunningham, P.R., and Miller, S.A. (2023). Optimizing supplementary cementitious material replacement to minimize the environmental impacts of concrete. *Cement and Concrete Composites* 139: 105049. https://doi.org/10.1016/j.cemconcomp.2023.105049.

187 Juenger, M.C.G., Snellings, R., and Bernal, S.A. (2019). Supplementary cementitious materials: new sources, characterization, and performance insights. *Cement and Concrete Research* 122: 257–273. https://doi.org/10.1016/j.cemconres.2019.05.008.

7

Manufacture of Materials

7.1 Introduction

Materials are essential for our society, and the production of these materials represents a significant part of the process industry. These materials are typically made from resources extracted from the earth and used by other industries. In principle, the basic industries are also manufacturing materials, which are used by other industries, but they operate at a much larger scale (and hence their impact on worldwide emissions is much larger). A number of these materials are critical for the development of clean energy technology [1] (see Table 7.1). Copper and aluminium play a very important part in the required electrification of the world and its transportation sector. Nickel will have important use in several industries going forward, but will not be considered due to the industry's relatively small size. Similarly, chromium will not be considered due to its nature of primarily (at current) being produced in tandem with steel, which is its primary use. Cobalt and lithium are important for their use in batteries, whereas rare earth elements (REEs) find application in magnets and smart batteries, and this is essential in transforming motion into electricity. Platinum group metals (PGMs) are essential in the transformation of electricity into storable energy in the form of e.g. hydrogen.

These critical elements are mined across the globe, but the known reserves, i.e. deposits which can be recovered economically using current technology, are limited and recycle will become more and more essential going forward. A potential future player in the mining of these critical elements is the so-called deep-sea mining in which metallic nodes may be recovered from the sea floor. This is an idea which has been present since the 1980s, though has recently come back to light with the need for critical minerals [2]. These may include 'cobalt-rich crusts' which contain species such as cobalt and nickel which occur on underwater mountains [3], polymetallic sulphides, which form around hydrothermal vents and contain a variety of minerals such as silver, gold, copper, manganese and zinc [4–6], and polymetallic nodes, typically rich in manganese, nickel, cobalt and copper [7, 8]. These deep-sea minerals, especially the polymetallic nodes [8], may prove an important resource in terms of the future drive for critical minerals as these may have huge deposits of the required critical minerals [3]. However, the so-far unknown environmental impact

Renewable Energy in the Process Industry, First Edition.
Eric van Steen, Nicholas S. Featherstone, and Linda H. Callanan.

Table 7.1 Critical mineral needs for clean energy technologies (data from IEA [1]).

Clean energy technology	Mineral								
	Copper	Aluminium[a)]	Nickel	Chromium	REEs[b)]	Zinc	Cobalt	Lithium	PGMs[c)]
Bioenergy	High	Moderate	Low	Low	Low	Moderate	Low	Low	Low
CSP[d)]	Moderate	High	Moderate	High	Low	Moderate	Low	Low	Low
Electricity networks	High	High	Low	Low	Low	Low	Low	Low	Low
EVs[e)] and batteries	High	High	High	Low	High	Low	High	High	Low
Geothermal	Low	Low	High	High	Low	Low	Low	Low	Low
Hydropower	Moderate	Moderate	Low	Moderate	Low	Moderate	Low	Low	Low
Hydrogen	Low	Moderate	High	Low	Moderate	Low	Low	Low	High
Nuclear	Moderate	Low	Moderate	Moderate	Low	Low	Low	Low	Low
Solar PV[f)]	High	High	Low	Low	Low	Low	Low	Low	Low
Wind	High	Moderate	Moderate	Moderate	High	High	Low	Low	Low

Mineral order based on calculation where dark grey cells are worth 3, light grey worth 2 and white worth 1 point each, ordered on total points accrued per mineral as 'most critical'.

a) Assessed for electricity networks only.
b) Rare earth elements.
c) Platinum group metals.
d) Concentrated solar power.
e) Electric vehicles.
f) Photovoltaic.

of deep-sea mining on marine organisms and biodiversity [9] may limit access to these resources. Further research and governance will be needed to ensure that the environmental impact of deep-sea mining will be assessed before it can take place in full force, which currently and may indefinitely put a halt on deep-sea mining operations [10].

This chapter is not only limited to the manufacture of materials made from these critical minerals. Many mineralogical processes are broadly similar, and there are many other industries of note which have large impacts on renewable energy consumption and fossil fuel release which should also be considered.

Aluminium will be considered as one of the most important elements due to the size of the industry, its carbon output and its status as a critical mineral. Copper is a critical mineral for many of the same reasons as aluminium, viz. its electrical conductivity, which will be essential for a renewable electricity-based future; this will also be considered primarily for its importance going forward. Titanium/titania will be considered as this is an industry which may be expected to grow due to the material properties, if the cost can be reduced. Silicon will be considered an important material for several purposes; most computational chips are silicon-based [11] and silicon is essential to solar photovoltaic cells [12], though it has a multitude of other uses. Furthermore, glass and paper will be considered as important materials used in society. The manufacture of plastics will not be discussed here as a separate industry, as polymerisation reactions are typically exothermic, low-temperature reactions not requiring much heating and further processing of these materials also take place at low temperatures; obtaining a (more) sustainable source for the monomers will be a great challenge for this industry.

There are many materials which need to be manufactured to maintain the current standard of human life and the expansion of many of these processes may be critical to future justice and equality across the planet. These may be used for matters such as access to technology (including renewable energy), sustainable development, as well as health and medical applications. The list of industries dealt with in this chapter are selected based on the types of processes, and principles outlined here could be applied to the manufacture of other materials as well.

7.2 Aluminium

Aluminium has many desirable properties which make it a popular material. It is likely most famous for its low density: aluminium is light compared to other metals, having a density of only 2700 kg/m^3, compared to steel at ca. 7850 kg/m^3, copper: 8960 kg/m^3, nickel: 8900 kg/m^3, titanium: 4506 kg/m^3, etc. Despite the decreased density of aluminium compared to steel, it can have comparable strength characteristics depending on the alloy used. Thus, it can have a very good strength-to-weight ratio, which gives it a wide range of applications.

Aluminium has a high thermal conductivity (247 W/m/K) and is the fourth most thermally conductive metal after silver (429 W/m/K), copper (398 W/m/K) and gold (315 W/m/K) [13]. The thermal conductivity of steel is relatively poor

(45 W/m/K). Aluminium also has a high electrical conductivity ($3.5 \cdot 10^7$ S/m). The order of electrical conductivity is the same as that of thermal: silver ($6.30 \cdot 10^7$ S/m), copper ($5.98 \cdot 10^7$ S/m) and gold ($4.52 \cdot 10^7$ S/m). On the other hand, the electrical conductivity of iron is only $1.04 \cdot 10^7$ S/m [14]. Whilst the thermal conductivity of aluminium is useful, its application is limited by the low melting point (660 °C; cf. copper: 1085 °C). However, it may compete with copper due to its electrical conductivity if increased demand for copper in the future puts a strain on the copper supply.

Aluminium is further known for its corrosion resistance [15], though this is a slight misnomer. Bulk aluminium is corrosion-resistant if exposed to air or neutral water due to the presence of a protective layer of aluminium oxide, which prevents further oxidation. Pure aluminium has poor corrosion resistance in an acid or alkali environment, but alloying with other metals can improve this.

Aluminium finds popular use in a variety of applications. One of the most relevant to everyday life is cooldrink cans, which are typically made of aluminium. Another example of everyday aluminium is aluminium foil, used for its strength and corrosion resistance. A further well-known use of aluminium is 'aircraft grade aluminium', which is popular due to its relatively light weight when considering the strength it provides.

Aluminium is relatively inexpensive, compared to metals with a similar (or slightly higher) thermal and electrical conductivity, viz. copper, silver and gold. In 2024 [16], the price of scrap aluminium is in the range of 2–4 times cheaper than that of copper. Of course, silver (ca. 300 times the price of aluminium [17]) and gold (ca. 30 000 times the price of aluminium [17]) are much more expensive. This relates to the relative abundance of each material with aluminium making up around 2.29% of the Earth's crust [18]. Though, scrap iron is much cheaper than aluminium, with the price of stainless steel being ca. 5 times less than the price of aluminium and carbon steel being ca. 20 times cheaper than aluminium.

7.2.1 Material Specification

As with iron/steel, aluminium is typically used as an alloy each having a composition designed to improve its efficacy for specific applications. The most common alloying elements are copper, magnesium, manganese, silicon, tin, nickel and zinc. Aluminium alloys can be classified into two broad categories, wrought and cast. Wrought alloys are molten and poured to form ingots, after which these are worked into their final form; this removes imperfections in the structure. Cast alloys are cast directly from the molten material into the finalised shape (or close thereto). These do not have the same level of imperfection removal, but the process is overall cheaper.

There are two main nomenclatures for wrought aluminium alloys; the most popular is the 'International Alloy Designation System' for wrought alloys [19]. The naming takes the form: XXXX. The first digit indicates the main alloying material, the second represents the variation of the alloy given by the first number and the last two digits specify the specific alloy in the series. Some of the most relevant groups are:

- 1XXX, essentially pure aluminium, used for aluminium foil
- 2XXX, alloyed with copper, previously used for aerospace applications
- 3XXX, alloyed with manganese, used for aluminium cans
- 5XXX, alloyed with magnesium (and often manganese) to obtain high corrosion resistance and thus used for marine application
- 7XXX, alloyed with zinc (and usually magnesium and copper), used for aerospace applications.

Alloys may further be classified as to how they may be hardened, either through heat treatment, or requiring work hardening [19].

There is a further set of designations (though slightly more descriptive and controlled by different organisations) for cast alloys, which typically have the same alloying material for each X00.0 grade, where the additional numbers provide further specification on the exact alloy composition. There are further definitions based on the temper or heat treatment applied to the alloy to give it its final strength and form.

7.2.2 Aluminium Industry

The primary production of aluminium starts off with bauxite, an aluminium containing ore. Bauxite is mined in several countries around the world with world reserves, i.e. bauxite that can be recovered using current technology under current economic conditions, being estimated at 30 billion tonnes in 2023 [20]. Table 7.2 shows the countries with the largest bauxite reserve, their production and the local work-up to alumina. The largest reserves can be found in Guinea, Vietnam, Australia, Brazil and Jamaica. Guinea, Australia and China (despite having a relatively small total reserve) take an almost equal share of bauxite production, together making up almost 75% of the world's bauxite production. This suggests a strongly unequal depletion rate of the various bauxite deposits, with an expected lifespan of the reserves in Guinea of 76 years, in Australia of 36 years and in China of just 8 years.

The first step in aluminium production is refining bauxite into alumina; about 80–85% of all mined bauxite is used for alumina production (with the remainder being used to produce high alumina cements, non-metallurgical grade alumina, abrasives, etc.). In 2023, 136 million tonnes of alumina was produced out of 400 million tonnes of bauxite. This is not necessarily performed on site, i.e. near the bauxite mining site. China is the largest producer of alumina, taking care of more than 60% of the world's production. The next largest alumina producer is Australia, which produces almost 14% of all the alumina in the world. Although, they export most of the bauxite they mine [21] (just like the other large producers of bauxite), largely to China.

Figure 7.1 shows the primary production of aluminium in the world over the period 1973–2023. Whilst at present, aluminium is a much smaller industry than steel, it has a relatively similar growth rate to the steel industry

Table 7.2 Bauxite reserves, bauxite production and the alumina production in 2023 (data from U.S. Geological Survey [20]).

Country	Bauxite reserves Mio tonne	Bauxite production Mio tonne per annum	Alumina production Mio tonne per annum
Guinea	7400	97	0.33
Vietnam	5800	3.7	1.4
Australia	3500	98	19
Brazil	2700	31	10
Jamaica	2000	6	1.5
Indonesia	1000	20	1.2
China	710	93	82
India	650	23	7.3
Russia	480	5.8	2.4
Saudi Arabia	180	4.6	1.8
Kazakhstan	160	4.3	1.3
Turkey	63	2	0.29
United States	20	a)	0.78
Rest	5100	5.6	8.2

a) Not available.

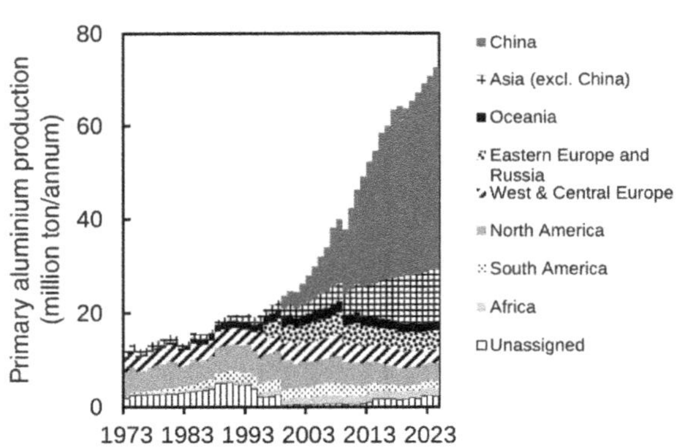

Figure 7.1 Primary aluminium production per annum in various parts across the world (data from International Aluminium Institute [22]; data for 2024 extrapolated).

(ca. 3.6% ± 2.1%/annum, with almost all of the growth coming out of Asia [23]). China has become the largest aluminium producer, accounting for approximately 60% of the global production. The other Asian countries are taking a growing market segment (currently 15%). Typical aluminium production plants, e.g. those operated by Rio Tinto in Australia and Canada, produce in the region of 150 000 to 500 000 tonnes/annum [24].

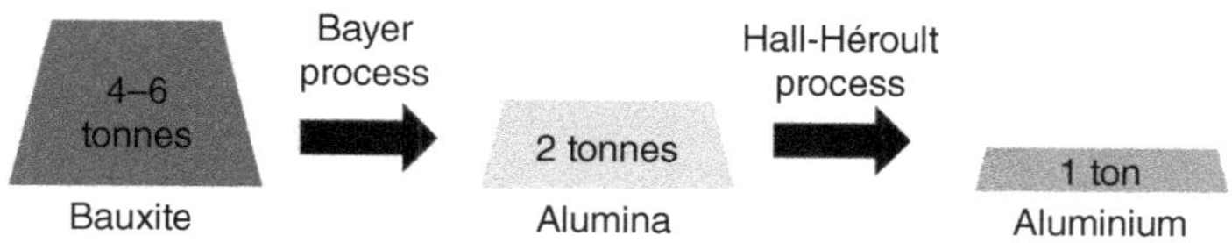

Figure 7.2 Mass efficiency in the refining of bauxite to alumina and aluminium (redrawn from Atlantic [26]).

7.2.3 Current Process

Aluminium is produced through two sequential processes, viz. the Bayer process for the conversion of bauxite into metallurgical grade alumina, and then alumina is converted to aluminium in the Hall-Héroult process [25].

Bauxite consists of various aluminium minerals: gibbsite – $Al(OH)_3$; boehmite – γ-$AlO(OH)$; and diaspore – α-$AlO(OH)$. Mixed in with the aluminium minerals are typically some iron minerals such as haematite, some titanium-containing minerals and silica. Thus, the bauxite has a typical alumina content between 30 and 60 wt.% alumina, depending on the grade of bauxite; alumina itself contains 52.9 wt.% aluminium by mass. Hence, depending on the purity of the starting material 4–6 tonnes of bauxite are required yielding 2 tonnes of alumina, which can be subsequently converted into 1 tonne of aluminium (see Figure 7.2).

7.2.3.1 Bayer Process – Alumina Production

A process flow diagram of the Bayer process is shown in Figure 7.3. Bauxite ore is initially crushed and mixed with a hot solution of caustic soda (ca. 150–300 °C) containing between 100 and 250 g NaOH/l (exact temperature and amount depend on the type of alumina in bauxite) [28]. The solution dissolves the aluminium-containing minerals in bauxite, forming a soluble sodium aluminate (present as $[Al(OH)_4]^-$).

$$Al(OH)_3(s) + OH^-(aq) \rightarrow Al(OH)_4^-(aq) \quad \Delta H^0_{rxn} = 22.4 \frac{kJ}{mol\ Al(OH)_3}$$

$$AlO(OH)(s) + OH^-(aq) + H_2O(l) \rightarrow Al(OH)_4^-(aq) \quad \Delta H^0_{rxn} = 10.6 \frac{kJ}{mol\ AlO(OH)}$$

(heat of reaction based on values given by [29, 30]). The reaction will require stoichiometrically 1.48 tonne NaOH per tonne of aluminium or 0.78 tonne NaOH per tonne of alumina. The reaction is endothermic, requiring 200–430 MJ or 0.06–0.12 MWh per tonne alumina.

Other valuable metals may be extracted from the obtained liquor as well. Bauxite is the world's leading gallium source, with the sodium hydroxide solution accumulating gallium ions. Gallium may then be recovered in part by various techniques, e.g. ion exchange [31].

The sodium aluminate solution is then clarified to remove insoluble impurities, which are known as 'red mud'. Red mud is a mixture of various components present in bauxite ore, which are insoluble in NaOH, and it can have a wide range of compositions depending on the source bauxite (on a mass basis): iron oxide (30–60%),

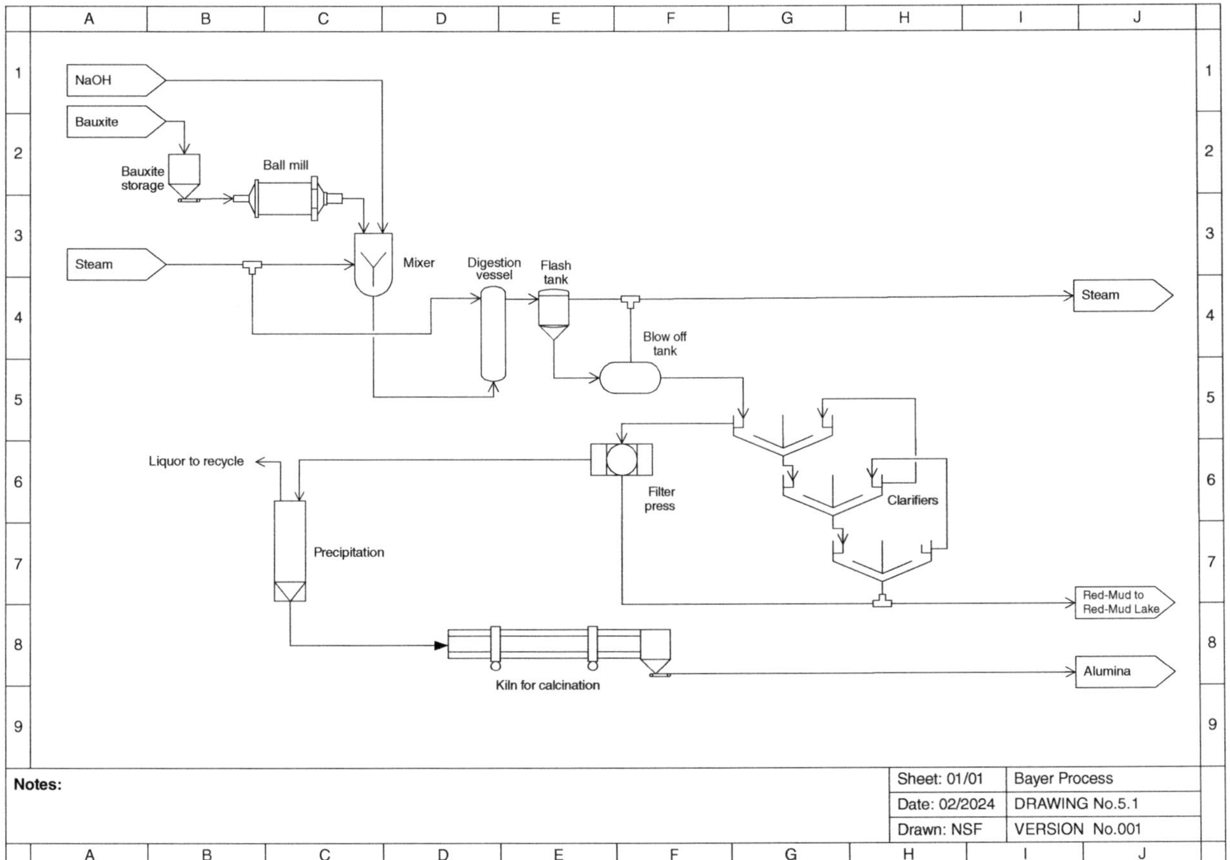

Figure 7.3 Flow diagram for the Bayer process (redrawn from Lancashire [27]).

aluminium oxide (10–20%), silica (3–50%), sodium oxide (2–10%), lime (2–8%) and titania (trace-25%) [32]. Red mud is a known environmental concern largely due to its high alkalinity and the presence of toxic materials. As such, great care is taken to safely store it, in red mud ponds or as a dried powder. There is an ongoing research to recover elements from red mud, such as iron, titanium and heavy metals [33]. Red mud also has the potential to be used as a building material and in correction of the acidity of soil [34].

The sodium aluminate solution is cooled and seeded with aluminium hydroxide (gibbsite) crystals, which causes aluminium hydroxide to precipitate out of the solution. This exothermic reaction is expected to release 430 MJ/tonne alumina (0.12 MWh/tonne). The formed aluminium hydroxide precipitate is calcined in rotary kilns or fluidised bed calciners at ca. 1000 °C releasing water from the structure and yielding anhydrous alumina.

$$2Al(OH)_3(s) \rightarrow Al_2O_3(s) + 3H_2O(g) \quad \Delta H^0_{rxn} = 182.7\frac{kJ}{mol\,Al_2O_3}$$

This calcination is endothermic with a heat of reaction of 1.79 GJ/tonne or 0.50 MWh/tonne alumina; assuming that alumina can be taken as corundum, with a similar amount of energy required to heat up the feed from ambient temperature to 1000 °C. Calcining in a fluidised bed reactor, in comparison to a rotary kiln can reduce the energy requirement for the calcination step by 30% [35].

Figure 7.4 shows the exergy analysis for the Bayer process (based on [36]). The multi-stage gibbsite extraction process from milled bauxite is performed at elevated temperature (up to 300 °C) and pressure, thus requiring utilities to heat up the process stream and to provide the heat of reaction. The reported exergetic efficiency of the gibbsite extraction [36] is rather low at less than 2.8%. In the subsequent step, gibbsite is calcined at ca. 1100 °C yielding metallurgical grade alumina. This high-temperature process requires less than half of the energy input of the gibbsite extraction. The exergetic efficiency of the calcination step is reported as 9.45%.

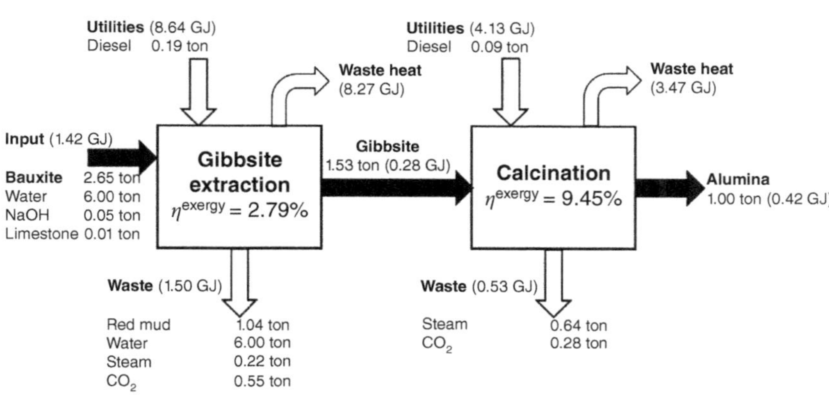

Figure 7.4 Exergy analysis on the production of alumina using the Bayer process (redrawn from data by Balomenos et al. [36]; HHV diesel 45.9 GJ/tonne; carbon content diesel: 78 wt.%).

The overall Bayer process requires an energy input of 12.77 GJ/tonne alumina (corresponding to 3.55 MWh/tonne alumina), and the overall exergetic efficiency amounts to 2.96%. The CO_2 emissions from the process are estimated to be ca. 0.73 tonne CO_2/tonne alumina, when using diesel fuel as a utility.

7.2.3.2 Hall-Héroult Process – Aluminium Production

Aluminium is produced via the high-temperature, electrochemical reduction of alumina in the Hall-Héroult process using carbon electrodes (see Figure 7.5). It produces a liquid stream containing ca. 99% pure aluminium and consumes some of the carbon anode. The overall reaction can be written as:

$$Al_2O_3(s) + \frac{3}{2}C(s) \rightarrow 2Al(l) + \frac{3}{2}CO_2(g) \quad \Delta H^0_{rxn} = 1102\frac{kJ}{mol\ Al_2O_3}$$

if it can be assumed that alumina is present as corundum (carbon can in principle also be converted to CO, but in practice the formation of CO_2 is dominant [38]). The preferential formation of CO_2 over CO is beneficial, as it improves the thermal energetics for the reaction and minimises the required post-process clean-up of the generated flue gas.

The reaction consumes at least 1.89 tonne of alumina per tonne of aluminium (in practice somewhat more due to the impurities present in alumina) and produces stoichiometrically 1.22 tonne CO_2/tonne Al. In reality, the Hall-Héroult process will

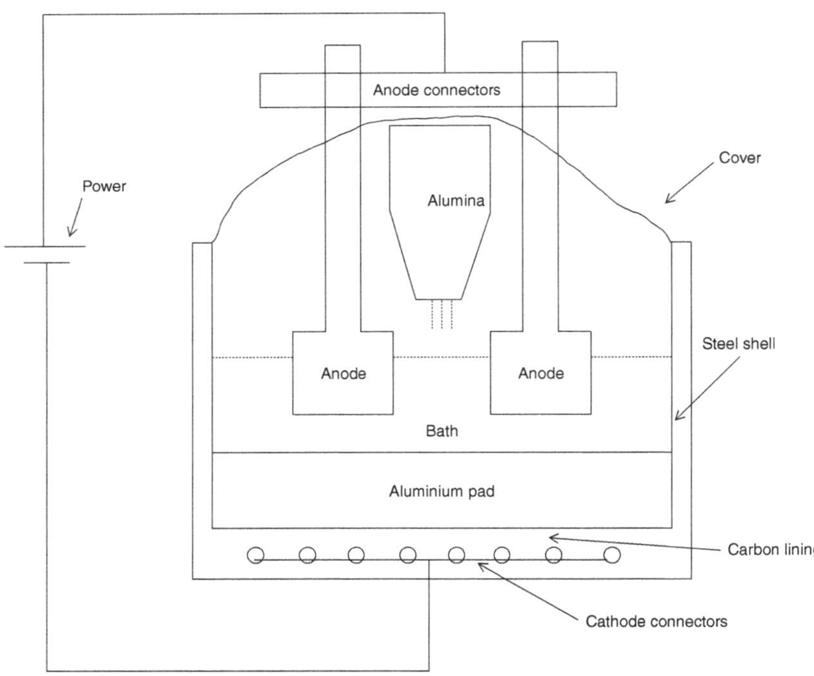

Figure 7.5 Simple diagram of the Hall-Héroult process unit (redrawn from Obaidat et al. [37]).

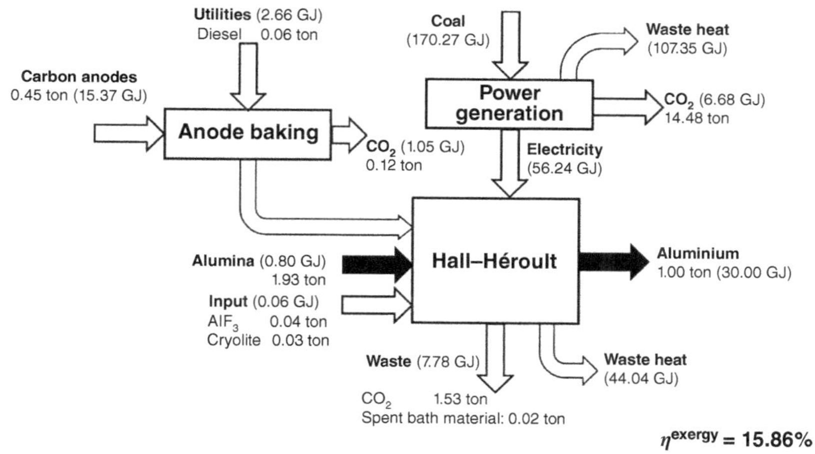

Figure 7.6 Exergy analysis of the Hall-Héroult process for primary aluminium production *using* coal-based electricity (all input and output flows calculated at 25 °C and 1 bar, basis 1 tonne aluminium) (redrawn from Balomenos et al. [36]).

release around 1.53 tonne CO_2/tonne Al [36] (see Figure 7.6). The expected minimum required voltage (E^0) at a typical operating temperature of 1000 °C is 1.84 V with a minimum amount of electrical work of 1.98 GJ/tonne Al or 5.49 MWh/tonne Al. The actual process runs at a significant overpotential, and the applied potential can be as high as 4.6 V [36] corresponding to 13.7 MWh/tonne. The reaction releases some heat at 1000 °C (0.69 GJ/tonne Al or 0.19 MWh/tonne Al), but the heating requirement to bring the reactants from ambient temperature to the reaction temperature of 1000 °C is much higher (1.87 GJ/tonne Al or 0.52 MWh/tonne Al). The process is not very efficient with an energetic efficiency of ca. 35%, as the process typically requires 15.3–15.6 MWh/tonne Al [37].

The electrochemical production of aluminium takes place in the liquid phase. Alumina is dissolved in molten cryolite (Na_3AlF_6). Cryolite is made through the mixture of NaF and AlF_3 salt, where the 'cryolite ratio', NaF/AlF_3 can be controlled to alter the density of the molten mixture as well as the melting point. Pure cryolite melts at ca. 1010 °C but additional aluminium in the mixture (either through dissolved alumina or increasing amounts of AlF_3) will lower the melting point.

The density of the liquid melt needs to be controlled, so that it is below the density of liquid aluminium (at ca. 1000 °C: 2.3 g/ml). This will allow the separation of the product, aluminium, from cryolite, as liquid aluminium can settle to the bottom of the electrolysis unit.

The reactions occurring in the cell are complex [38]. In liquid cryolite, Na_3AlF_6 dissociates into Na^+ and $[AlF_6]^{3-}$, with the latter equilibrating with $[AlF_4]^-$ and fluoride ions. At the cathode, $[AlF_6]^{3-}$ and $[AlF_4]^-$ are reduced yielding aluminium and fluoride ions. The aluminium fluoride complex, $[AlF_6]^{3-}$, may also react with Al_2O_3 yielding $[Al_2O_2F_4]^{2-}$ and $[Al_2OF_6]^{2-}$. At the anode, the oxy fluoride complexes may give off an oxygen anion to the carbon surface, yielding CO/CO_2 whilst generating AlF_3, which reacts further with fluoride ions yielding $[AlF_4]^-$.

The carbon anode is consumed in the process (requiring a minimum of 0.33 tonne of carbon/tonne Al, if CO_2 is produced exclusively). In industry, the typical amount of electrode consumed is higher at ca. 0.40–0.45 tonne of carbon/tonne Al (corresponding to 70–80% of aluminium produced co-generating CO_2 and 20–30% co-generating CO). The carbon anode must be of high purity, as impurities may be incorporated into aluminium. The production of carbon anodes, e.g. from petroleum coke, is associated with further release of CO_2 in the coking process and the high-temperature calcination process.

The Hall-Héroult process also emits, in small amounts, some carbon fluoride compounds with a high global warming potential (CF_4: $GWP^{100\,years} = 7380$; C_2F_6: $GWP^{100\,years} = 12\,400$). Hence, the equivalent CO_2 emissions of the process excluding the electricity generation amounts to 3.82 tonne $CO_{2,eq}$/tonne aluminium, i.e. more than twice the actual CO_2 emissions originating from anode baking and the electrolysis process (1.65 tonne CO_2/tonne Al; see Figure 7.6 [36]). The Hall-Héroult process using electricity from coal has been estimated to produce 16.13 tonne $CO_{2,e}$/tonne Al with 90% of the CO_2 emissions coming from electricity generation (see Figure 7.6).

The exergetic efficiency of the Hall-Héroult process is ca. 16% [36], and the overall exergetic efficiency for the production of aluminium from bauxite is ca. 14%, due to the low exergetic efficiency of gibbsite extraction from bauxite. The relatively small decrease in the exergetic efficiency for the overall process in comparison to the Hall-Héroult process is due to the high energy and exergy demand of the electrochemical process.

7.2.4 Recycling Aluminium

Aluminium is a highly recyclable material, and it has been claimed that 75% of all aluminium ever produced is still being used [39] through recycling. This saves a huge amount of energy as the energy-intensive Bayer and Hall-Héroult process requires ca. 15.5 MWh/tonne (55.8 GJ/tonne) Al electrical power and 1.3 MWh/tonne (4.7 GJ/tonne) Al in fuel (over and above the fuel required to generate electricity). The conventional route to recycle aluminium is by relatively high-temperature melting [40, 41], which would notionally require ca. 0.3 MWh/tonne (1.1 GJ/tonne) Al. Implying a significant energy saving over the primary production of aluminium using the Hall-Héroult process, (it should be noted that further energy savings could be made by changing the recycling process [41]). Some aluminium is lost in the process, as the surface layer of alumina, which forms on aluminium, will be removed as a slag as it will not melt (melting at 2072 °C, compared to 660 °C for pure aluminium). Aluminium alloys are typically recycled as alloys rather than recovering the aluminium.

7.2.5 More Sustainable Aluminium Production

The aluminium production currently emits greenhouse gases at a rate of 5.16 tonne $CO_{2,e}$/tonne Al in addition to the greenhouse emissions from electricity generation [36]. The large electricity demand of 15.3–15.6 MWh/tonne Al [37] may yield a large amount of additional CO_2 emissions, depending on the fuel source used to generate electricity. Replacing the fossil fuel for electricity generation with renewable energy would cut down the emissions to produce aluminium significantly. However, the amount of renewable energy required is significant, and a plant producing 150 000 tonne of aluminium per annum would require on average 325 MW of power. With a capacity factor of renewable energy generation of ca. 33% (see Chapter 2), an aluminium plant of this size would need access to renewable energy generation with a nominal capacity of ca. 1 GW (as well as associated electricity storage).

The Hall-Héroult process may be well suited to deal with variation in the power supply due to intermittency of the renewable energy sources, as a variable supply will affect the productivity, but not necessarily the overall operation.

The Hall-Héroult process requires carbon anodes, which react with the alumina in the production of aluminium yielding CO_2 (1.65 tonne CO_2/tonne aluminium [36]). More importantly, carbon may also react with fluoride present in the liquid reaction mixture yielding compounds with a high global warming potential in small amounts. Hence, novel methods are being developed to replace the carbon anodes.

The use of inert, non-carbon-based electrodes would be a method to reduce the production of the greenhouse gases, but the required potential difference, E^0, will increase to ca. 3.11 V as the reaction is now:

$$Al_2O_3(s) \rightarrow 2Al(s) + \frac{3}{2}O_2(g) \quad \Delta H^0_{rxn} = 1671.9 \frac{kJ}{mol\ Al_2O_3}$$

Possible inert anodes are made from ceramics materials which include tin, nickel-lithium and nickel-iron based oxides [42]. These are reportedly stable during the reaction at normal operating temperatures, which ensures minimal contamination of the aluminium. However, these anodes do have a relatively poor electrical conductivity and mechanical strength [42]. Alternative metal anodes with a higher mechanical strength and conductivity are being studied such as nickel-copper, aluminium-copper and iron-nickel-copper alloys [42], however these may corrode rapidly in the reaction environment. As such various oxide additives may be added to the anode surface, though at increased cost.

7.2.6 Outlook

Aluminium may be a highly valuable metal going forward, not only for its high strength-to-weight ratio and its relative corrosion resistance but also for its electrical conductivity. It is a highly conductive material at a significantly lower cost than those that outperform it. The aluminium industry has several sources of emissions but may be able to reduce its emissions in various steps.

First, there is the obvious requirement for the electricity demand to switch from fossil fuels to renewable sources. Depending on the size of the plant, a relatively large source of renewable energy will be required, and there will need to be detailed strategies on how to operate under variable load conditions or storage of the electricity will be required. This may reduce the CO_2 emissions associated with the Hall-Héroult process significantly, though these may also be taken as Scope 2 emissions, and the industry could try and outsource this method of CO_2-reduction to the grid.

The heating in the Bayer process, i.e. gibbsite extraction and calcination, could be transferred from burning fossil sources to other, renewable heating sources. This currently emits 0.82 tonne CO_2/tonne aluminium which could be saved. This could be done through the use of renewable fuels, e.g. biofuels or hydrogen-based renewable fuels. Furthermore, the use of electric furnaces may assist in at least the extraction process.

The Hall-Héroult process uses carbon electrodes, which results in the direct release of 1.65 tonne CO_2/tonne Al; these may be taken as process emissions, which can be harder to reduce than heating emissions. More importantly, using carbon electrodes reacts with the fluorine in the cryolite that is used to dissolve the alumina. This results in the emission of some perfluoro compounds, which have a very high global warming potential. The formation of these perfluoro compounds limits the value of replacing the carbon electrodes with renewable carbon-based electrodes, as the compounds will still be released. Replacing these carbon electrodes with inert electrodes is still in its infancy but will need to be further developed; this will come with a significantly increased energy cost.

7.3 Copper

Contribution by Chiamaka Amonye.

7.3.1 Material Specification

Copper has a high thermal and electrical conductivity (398 W/m/K and $5.98 \cdot 10^7$ S/m, respectively). These properties make it a popular conductive material for electrical cabling and as well as thermal heating blocks (copper is better than aluminium as it has a notably higher melting point – 1085 °C, vs 660 °C).

It is for its electrical conductivity that copper is regarded as a critical mineral for the renewable energy transition [43]. However, copper is also used as an alloy (e.g. brass), and for copper piping. Furthermore, copper is catalytically active, notably for the low-temperature water gas shift reaction [44] in the synthesis of hydrogen as well as in methanol synthesis [45, 46] as discussed in Chapter 5.

Copper occurs primarily in three different mineral groups (outlined in Table 7.3). Copper is present in different type of ores with ores containing chalcopyrite ($CuFeS_2$) being the most common making up about 70% of the Earth's copper reserves [47]; chalcopyrite contains 34.5 wt.% copper. The next most common ore is chalcocite (Cu_2S), which is a significant contributor to copper production due to its high copper content of 79.8 wt.%. There are also copper oxide or copper carbonate ores; whilst they are available in much smaller quantities than the copper sulphide ores, they are desirable as copper can be extracted much more readily by dissolving the ore in weak acid [48]. Malachite ($Cu_2(OH)_2CO_3$) is the most common copper oxide ore and contains 57.6 wt.% copper by mass. This is typically formed through the oxidation of sulphide ores at the surface of copper ore deposits.

Table 7.3 Elemental composition (wt.%) of the most commonly occurring copper-containing minerals.

Mineral	Cu	Fe	S	CO_2	SiO_2	H_2O
Sulphides						
Chalcopyrite	34.5	30.5	35.0	—	—	—
Bornite	63.3	11.2	25.5	—	—	—
Chalcocite	79.8	—	20.2	—	—	—
Covellite	66.4	—	33.6	—	—	—
Carbonates						
Azurite	55.3	—	—	25.6	—	5.2
Malachite	57.4	—	—	19.9	—	8.2
Silicates						
Chrysocolla	36.1	—	—	—	34.3	20.5

7.3.2 Copper Industry

Copper can be found around the globe, but large reserves can be found in the South American countries: Chile and Peru, as well as Australia, D.R. Congo and Russia (see Table 7.4). The world reserves amount to ca. 1 billion tonnes of copper, which should last for the next 45 years, if mined at the current rate. It should be noted that the rate at which copper is mined depends, to some extent, on location, with countries such as Canada and China having a high depletion rate, but Australia has a low depletion rate.

Compared to the other industries discussed in this chapter and considering the importance of copper to a renewable future, the copper industry is not actually particularly large, with a production of only ca. 21 million tonnes/annum as of 2023 [49]. Almost 50% of all mined copper originates from Chile, Peru and D.R. Congo whilst the largest producer of refined copper is China. Mined copper is not necessarily refined locally and copper concentrate is shipped around the world.

In 2022, the market size of the copper industry stood at 170 billion US dollars globally and is anticipated to grow by a further 90 billion US dollars by 2030 [50]. The copper industry is growing rapidly with a compound annual growth rate of ca. 5% (similar to that of aluminium and steel) [51]. This may be expected to pick up into the future as the need for renewable energy grows and thus the need for copper required to support it is expected to grow as well.

The leading manufacturers of copper globally are Freeport-McMoRan, producing 2.1 million metric tonnes of metal, followed by BHP with a copper output of approximately 1.4 million metric tonnes as of 2023 [52]. Other key players in the copper producing industry include Chilean state-owned Codelco (1.35 million metric tonne), Anglo American (1.1 million metric tonne), Antofagasta and Glencore

Table 7.4 Copper reserves, production and refining worldwide in 2023 (data from National Minerals Information Center [20]).

Country	Copper reserves (Mio tonne)	Copper mined (Mio tonne per annum)	Copper refining (Mio tonne per annum)
Chile	190	5.0	2.00
Peru	120	2.6	0.40
Australia	100	0.8	0.45
D.R. Congo	80	2.5	1.90
Russia	80	0.9	1.00
Mexico	53	0.8	0.48
United States	50	1.1	0.89
China	41	1.7	12.00
Poland	34	0.4	0.59
Zambia	21	0.8	0.38
Indonesia	24	0.2	0.20
Kazakhstan	20	0.6	0.44
Rest	188	3.5	5.94

(each 0.7 million tonne) and KGHM Polska Miedz (0.6 million metric tonne), to name a few [49].

7.3.3 Current Processes

The pyrometallurgical and hydrometallurgical processing routes are the primary routes for copper production, with the former accounting for 67% of total world production of refined copper and the latter 16%, whilst the remaining 17% is obtained from secondary recycled production [53]. The choice of processing routes is dependent on the type of ore present. Copper sulphide ores require a relatively complex, pyrometallurgical process (see Figure 7.7) involving ore concentration through flotation, a high-temperature smelting process to separate copper from iron and sulphur and then refining. By comparison, the extraction process for copper oxide ores (see Figure 7.8) is quite trivial, utilising solvent extraction through leaching and then electrowinning processes to recover the metallic copper.

Both processing routes (hydro- and pyrometallurgy) begin with extracting the mineral from the ground. Typically, lower-grade ores (with a copper content of ~0.5 wt.%) are mined via open-pit operations, whilst higher-grade ores (with a copper content ~1–2 wt.%) are extracted via underground mining [53], with the former being the dominant production method [55]. Blasted rock is loaded and transported to processing plants. The low copper content implies that mainly gangue material is being transported, resulting in significant energy costs being associated with

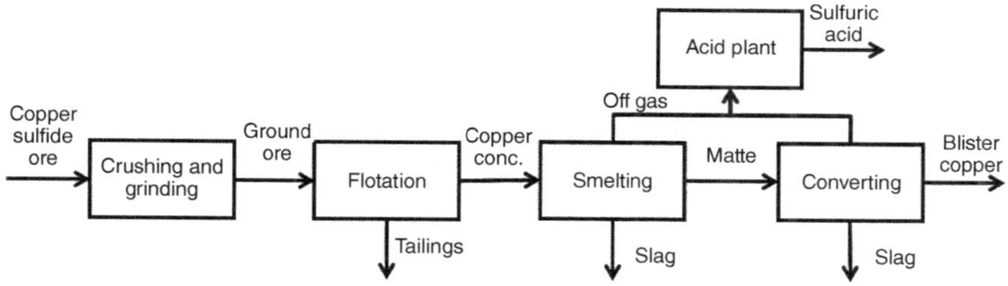

Figure 7.7 Process flow diagram for the pyrometallurgical processing route for the extraction of copper from copper sulphide ores (redrawn from Hawker et al. [54]).

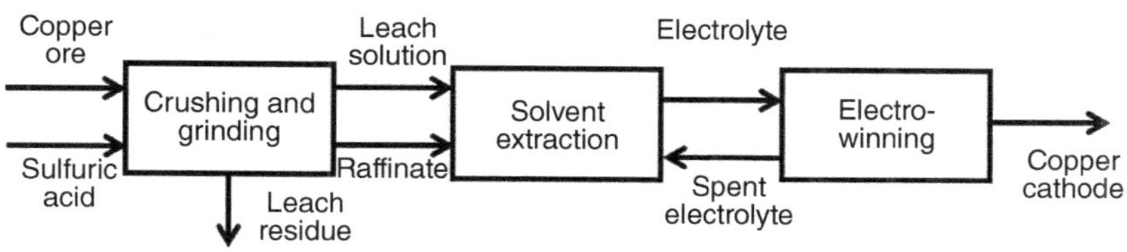

Figure 7.8 Process flow diagram leach/solvent exchange – electrowinning process, typically used for copper oxide ores as well as blister copper purification (redrawn from Hawker et al. [54]).

copper mining depending on the grade of the ore [56] and the distance transported; with a ballpark figure of 2–3 MWh/tonne Cu [57] for transport to the refining plant. Seeing that this is traditionally supplied with diesel (LHV = 42.6 MJ/kg; carbon content [w_C]: 85 wt.%), the associated CO_2 emissions (ECO_2) can be estimated if the efficiency of converting energy in diesel into motion is 40% using:

$$
\begin{aligned}
E_{CO_2} &= 2 \cdot \frac{\text{MWh}}{\text{ton Cu}} \cdot \frac{w_C}{\text{LHV}_{\text{diesel}}} \cdot \frac{M_{CO_2}}{M_C} \cdot \frac{1}{\eta_{\text{diesel}}} \\
&= 2 \cdot \frac{\text{MWh}}{\text{ton Cu}} \cdot \frac{0.85 \cdot \frac{\text{ton C}}{\text{ton diesel}}}{11.83 \cdot \frac{\text{MWh}}{\text{ton diesel}}} \cdot \frac{44 \cdot \frac{\text{ton } CO_2}{\text{tmol C}}}{12 \cdot \frac{\text{ton C}}{\text{tmol C}}} \cdot \frac{1}{40\%} \\
&= 1.32 \cdot \frac{\text{ton } CO_2}{\text{ton Cu}}
\end{aligned}
$$

7.3.3.1 The Pyrometallurgical Route

Figure 7.7 shows an overview of the process involved in the extraction of copper from (typically) copper sulphide ores using the pyrometallurgical route. It begins with crushing and grinding to reduce the size of the ore. To obtain the ideal particle size, various types of crushers and mills are employed (including semi-autogenous mills, ball mills and cyclonic separators [53]). The energy consumption in the comminution circuit is high with reported values of 3.6–9.1 MWh/tonne copper [53, 57, 58] depending on the ore body and the required grind size. The energy needs to be supplied as electrical energy, and the carbon footprint of the milling process depends on the carbon footprint of the source of power (e.g. the CO_2 emissions from the combustion of natural gas is 2.75 tonne CO_2/tonne CH_4 assuming a 45% efficiency).

The milled mineral is concentrated to a copper content of approximately 30 wt.% using froth flotation. In this process, copper-rich particles or copper concentrate material is collected in the foam formed on top of these aerated tanks. Flotation does require the usage of water in often water-scarce regions (0.4–0.6 m^3/tonne depending on the extent of water recycle) [59]. These processes also require electricity for water transportation and/or desalination depending on the source [53].

The recovery of copper from the concentrate is achieved in the high-temperature smelting processes based on density dissimilarities. Silica and limestone are added to aid the separation process. The copper concentrate is smelted at about 1250 °C to produce a copper-rich molten phase, referred to as copper matte, and slag (oxidised impurities). The reaction taking place in the smelter with, e.g. chalcopyrite can be represented by:

$$CuFeS_2(s) + SiO_2(s) + 2O_2(g) \rightarrow \frac{1}{2}Cu_2S(s) + FeSiO_3(s) + 3SO_2(g) \quad \Delta H^0_{rxn} = -585.5\frac{kJ}{mol\ CuFeS_2}$$

The energy required for the smelting process ranges from 1.9 to 2.5 MWh/tonne Cu [53, 57] using mainly fossil fuels to achieve the required smelting temperature [53]. There is also a significant release of sulphur dioxide, 2.4 tonne SO_2/tonne Cu_2S. These are unavoidable emissions when utilising sulphide ores, and the resulting SO_2 will need to be captured.

The resulting copper matte is oxidised by injecting oxygen-enriched air (O_2-content: 30–70%) at ca. 1200 °C yielding 'blister copper' of ca. 98–99% purity. This conversion is autothermal, and the heat required to maintain the operating temperature in this step is generated by the chemical reactions occurring.

$$Cu_2S(s) + \frac{3}{2}O_2(g) \rightarrow Cu_2O(s) + SO_2(g) \quad \Delta H^0_{rxn} = -390.6\frac{kJ}{mol\ Cu_2S}$$

$$\text{followed by } Cu_2S(s) + 2Cu_2O(s) \rightarrow 6Cu(s) + SO_2(g) \quad \Delta H^0_{rxn} = 129.0\frac{kJ}{mol\ Cu_2S}$$

$$\text{and overall } Cu_2S(s) + O_2(g) \rightarrow 2Cu(s) + SO_2(g) \quad \Delta H^0_{rxn} = -217.4\frac{kJ}{mol\ Cu_2S}$$

However, electric power is required for the supply of oxygen-enriched air produced using pressure swing absorption [53, 60].

Blister copper can be refined through a two-step process: fire refining and/or electrolytic refining. Fire refining follows a step of melting and oxidation in which blister copper is melted in a furnace, during which air is blown through the molten copper. This oxidises impurities, which form a slag that can be removed. The melting and oxidation stages are followed by a reduction of the copper from copper oxide, often done using a hydrocarbon such as natural gas as the reducing agent (thus resulting in at least 0.17 tonne CO_2/tonne Cu excluding heating requirements).

Electrolytic refining follows casting of the fire-refined copper into large ingots. These will be used as the anodes in the electrolysis bath. The anodes are placed in an electrolytic cell containing a copper sulphate and sulphuric acid solution (electrolyte). A direct current is passed through the cell, causing copper ions to migrate

Table 7.5 Average elemental composition (wt.%) of copper smelter slag from across the globe as reported in 80 different studies (data from Phiri et al. [62]).

Fe	Si	Ca	Al	Zn	Mg	K + Na	Cu	S
35.7 ± 12.0	15.2 ± 5.6	4.7 ± 3.9	3.1 ± 2.0	2.0 ± 2.3	1.2 ± 0.8	1.7 ± 1.6	1.2 ± 0.7	1.0 ± 1.1
Pb	**Co**	**Mn**	**Ti**	**Ni**	**Mo**	**As**	**Sb**	**P**
0.8 ± 1.9	0.5 ± 0.8	0.3 ± 0.2	0.3 ± 0.1	0.2 ± 0.4	0.2 ± 0.1	0.3 ± 0.7	0.1 ± 0.1	0.1 ± 0.1

from the anode to the cathode, resulting in the build-up of pure copper. Impurities may be removed from the cell as slag or remain in the sulphuric solution. This process is capable of producing copper with ca. 99.99% purity [61].

The waste produced during the pyrometallurgical processing route is called copper slag, a mixture of mainly oxides. The main elements in the slag are iron and silicon, together with aluminium, zinc, calcium, magnesium, sodium and potassium (see Table 7.5). Further, the slag contains copper, sulphur and a host of minor elements. Table 7.5 shows the variation in the slag composition as determined from over 80 studies conducted globally [62]; there is a large variety in copper slag composition as a consequence of the location of the geological deposits.

Gorai et al. estimate that for every tonne of copper produced, approximately 2.2 tonne of slag is generated [63]. The disposal of large amounts of copper slag poses a great waste management challenge as large areas of land are typically required [62–65]. The presence of metals such as lead and arsenic may pollute the underlying soil and migrate into ground and surface water [62].

Extraction of valuable metals from slag (as the slag can still contain a significant amount of copper) can decrease the amount of slag generated. Copper from this type of secondary source may be enriched using flotation [62], where sulphur minerals (Cu_2S and Cu_5FeS_4) are separated from the gangue material aided by specific collectors and frothers [66, 67]. Alternatively, copper (and other metals) may be leached out of the slag treated using hydrometallurgical routes. Slag does have favourable physical and mechanical properties, such as abrasion resistance and high stability [63]. Hence, slag itself can be used to produce blended cements, pavement, abrasives, concrete, tiles, glass, cutting tools and roofing granules [62]. The addition of copper slag has shown improvements in the mechanical performance and durability of concrete [62]. Copper slag is also a promising alternative material for natural aggregates in the construction of road pavements.

7.3.3.2 The Hydrometallurgical Route

Like the pyrometallurgical route, the hydrometallurgical route (see Figure 7.8) does require crushing the mineral rock but does not require grinding it down to a fine powder. Hence, the energy requirement for the comminution process will be reduced. The crushed ore is piled in heaps (typically 7 m high with a top area between 10^4 and 10^6 m^2) that are continuously irrigated with an aqueous solution of sulphuric acid, which leaches the copper from the mineral rock to the solution. The product of the leaching process is a pregnant solution with copper concentrations ranging between 1 and 6 kg Cu/m^3 [53, 68]. At this stage, the copper solution is dilute and impurities are present (1–10 kg Fe/m^3), which is not optimal for electro-deposition [68].

The solution is concentrated using a solvent extraction process, transferring copper from the aqueous solution to a liquid organic phase (e.g. a mixture of oximes). The raffinate with a low copper content (0.3 kg Cu/m^3) is sent back to the leach heap. The Cu-loaded organic phase (4 kg Cu/m^3) is mixed with sulphuric acid from the electrowinning processing, resulting in copper being stripped from the organic phase into the electrolyte. The stripped organic phase (1.5 kg Cu/m^3) is recycled to contact the pregnant solution, whilst the electrolyte (45 kg Cu/m^3) moves on to the next stage of the extraction process.

Finally, metallic copper is recovered from the concentrated copper solution through electrowinning. This step can recover electrical-grade copper without the need for further refining. In the electrowinning process, copper ions are plated on a 99.99% pure cathode [53]. The process operates between 50 and 60 °C [68], and electricity is required to drive the electrochemical reactions whilst heat is needed to maintain the electrolyte temperature for efficient electro-deposition. There are two primary reactions taking place:

$$\text{Cathode: } Cu^{2+}(aq) + 2e^- \rightarrow Cu(s) \qquad E^0 = 0.3419\text{ V}$$

$$\text{Anode: } H_2O(l) \rightarrow \frac{1}{2}O_2(g) + 2H^+(aq) + 2e^- \qquad E^0 = -1.229\text{ V}$$

The sulphuric acid in the solution does not react away and is instead recycled as spent electrolyte as shown in Figure 7.8. The minimum cell voltage is 0.887 V, thus requiring a minimum amount of work of 2.7 GJ/tonne copper (0.75 MWh/tonne copper). The applied voltage is typically in the range of 2.0–2.3 V [69] due to the overpotentials, the resistance of the solution and contact resistance implying an actual amount of work of 1.7–1.9 MWh/tonne copper and an energetic efficiency of ca. 40%. The Faradaic efficiency is between 90% and 95% [70]. This implies that the actual energy consumption in the electrowinning process is in the order of 1.8–2.2 MWh/tonne copper.

CO_2 emissions of primary refined copper, including the mining, shipping, smelting and refining stages, average between 1.6 and 3.4 tonnes CO_2/tonne copper [71] and account for 0.3% of today's global energy consumption [72].

7.3.4 Potential for Recycle

Copper is, in principle, infinitely recyclable, and a large portion of copper is recycled. Figure 7.9 shows the global copper flow as estimated for 2020 [73, 74]. Primary copper (20.66 Mio tonne) is mined and fed together with 3.69 Mio tonne of low-grade copper, or secondary copper, to the smelter/solvent exchange and electro winning circuit, and secondary copper makes ca. 15–16% of the total refined copper stream, an almost constant factor over the last 50 years [75]. Some copper is lost during the

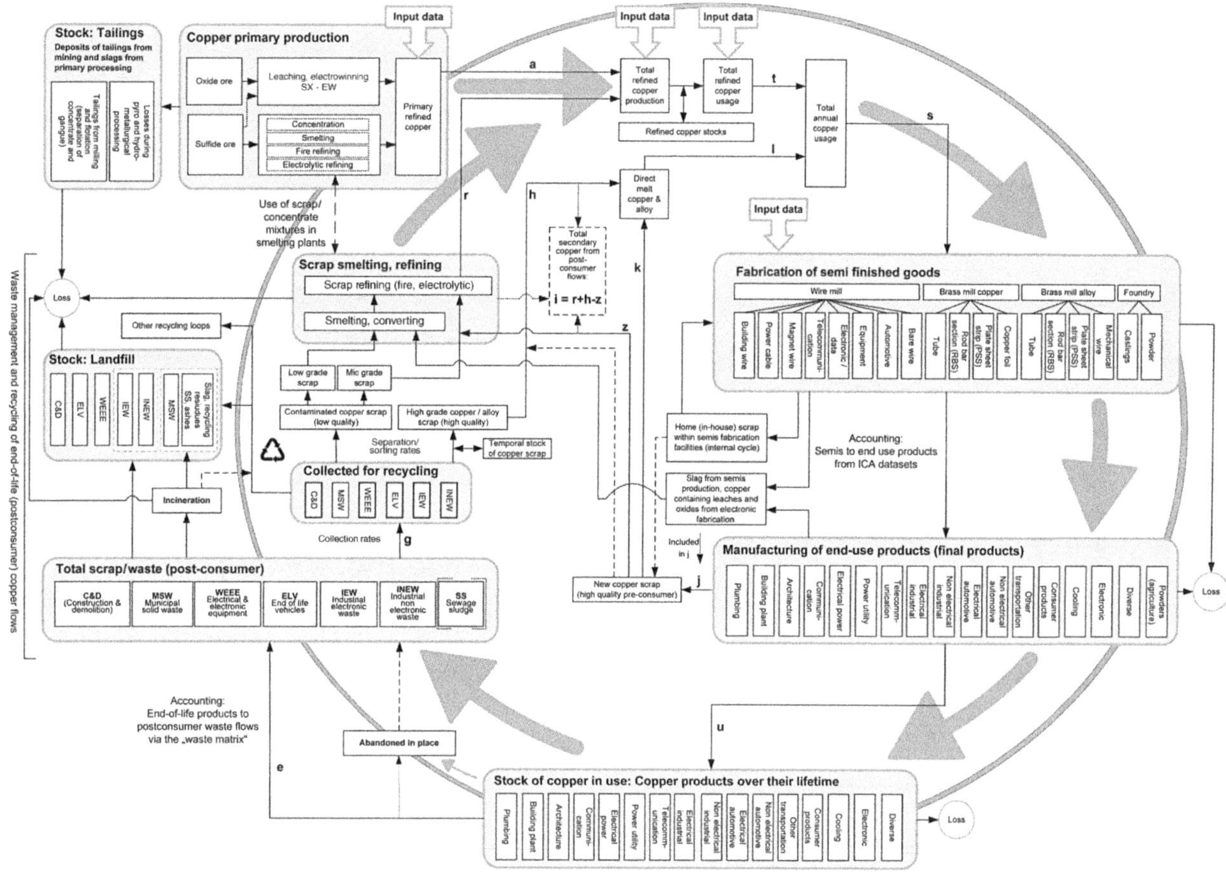

Figure 7.9 Structure of the global copper flow model (Glöser et al. [73] / with permission of American Chemical Society).

refinery stage (ca. 1.7%) and leaves the process together with the slag. High-grade, recycled copper (4.38 Mio tonne in 2020) is used together with refined copper to produce semi-finished products, again with some losses (estimated to be 0.5%). This implies that almost 30% of copper going into semi-finished products is secondary or recycled copper. There are additional losses during further processing of copper and scrap generated during processing is being recycled. Thus, 23.15 Mio tonne of copper was added in 2020 to the 470 Mio tonne of copper in usage. The total amount of copper in usage will need to increase significantly to enable further expansion of electric grid and further electrification worldwide. Copper from end-of-life (Eol) products (13.67 Mio tonne) can be recovered and recycled. However, there are some losses from the copper cycle due to unaccounted scrap (estimated to be 7.76 Mio tonne). Further losses occur in the separation process, and the estimated amount of copper recycled from end-of-life products is only 30%. It should be noted that the OECD estimates a higher recycle rate of copper in end-of-life products of 50% [76].

Recycling copper reduces the world's dependency on the copper reserves necessary to keep production going. This is important as there is an ongoing decrease in the grade of the mined ore. In 2014, the Chilean copper industry experienced a 30% reduction in average ore grade over a decade [77]. As the copper demand increases, more low-grade ore is required to meet the demand. Processing lower ore grades renders the process more energy-demanding compared to current processes [78]. Additionally, the use of secondary copper may reduce the energy consumption in the preparation of semi-finished copper by 85% [79, 80]; lower-grade ores will exacerbate this effect. With the expected strong increase in copper demand, governments have been setting targets for the use of secondary copper.

7.3.5 Carbon Emissions from Copper Production

The production of copper utilises both electrical and thermal energy both currently (primarily) generated from fossil fuels. The carbon footprint associated with copper production based on a Chilean copper production context is shown in Table 7.6; it includes emissions associated with fossil fuel combustion to generate electricity and emissions related to infrastructure [77]. Mining utilises mainly fossil energy and has a carbon footprint of ca. 0.9 tonne $CO_{2,eq}$/tonne Cu. In the pyrometallurgical route, the concentration section with its crushing and grinding utilises large amounts of electrical energy with an estimated carbon footprint of ca. 2.2 tonne $CO_{2,\,eq}$/tonne Cu. The subsequent smelting and electro-refining steps utilise almost equal amounts of fossil fuels and electricity and have a carbon footprint of ca. 2.5 tonne $CO_{2,\,eq}$/tonne Cu. The total carbon footprint for the pyrometallurgical route comes to ca. 6 tonne $CO_{2,\,eq}$/tonne Cu. However, the obtained carbon footprint is dependent on the location, e.g. Norgate et al. [81] determined a carbon footprint for smelting and electro-refining as 3.3 tonne $CO_{2,eq}$/tonne Cu, The hydrometallurgical route in Chile [77] appears to have a smaller carbon footprint of ca. 4.9 tonne $CO_{2,\,eq}$/tonne Cu than the pyrometallurgical route. In contrast, Norgate et al. [81] determined a higher carbon footprint for the hydrometallurgical recovery of copper in Australia. In general, the carbon footprint for copper production appears to be

Table 7.6 Inventory of average specific energy demand in the Chilean copper industry (values from surveys from mining companies representative of 98% of the copper production in Chile) (data from Moreno-Leiva et al. [77]).

	Energy demand (MWh/tonne Cu)		Carbon emissions (tonne CO_2/tonne Cu)	
Process	**Electricity**	**Fossil fuels**[a)]	**Power grid**	**Fossil fuels**[a)]
Pyrometallurgical route (P-Cu)				
Mining	0.23	2.01	0.2	0.7
Concentration	3.00	0.06	2.2	—
Smelting	1.17	1.33	0.9	1.0
Electro-refining	0.36	0.36	0.3	0.3
Water and services	0.25	0.25	0.2	0.2
Total	**5.01**	**4.02**	**3.8**	**2.2**
Hydrometallurgical route (H-Cu)				
Mining	0.23	2.01	0.2	0.7
Water and services	0.25	0.25	0.2	0.2
LX/SX/EW[b)]	3.36	0.80	3.3	0.3
Total	**3.83**	**3.06**	**3.7**	**1.2**

a) Fossil fuel demand assumed as a ratio of natural gas and heavy fuel oil of 3:4.
b) LX, Leaching; SX, Solvent extraction; EW, Electrowinning.

in the range between 1 and 9 tonne $CO_{2,\ eq}$/tonne Cu with an average of 2.6 tonne $CO_{2,eq}$/tonne Cu [82, 83], with a strong dependency on ore grade and the energy intensity of the copper production. The latter depends on:

- Ore grade
- Mineral properties (hardness and size)
- Type of mine and age
- Location of mine
- Chosen processing route
- Process design.

7.3.6 Copper Production Using Renewable Energy

The relationship between copper and renewable energy is symbiotic as copper is needed for renewable energy systems, and renewable energy can improve the carbon footprint of the copper industry [53].

The electricity used in copper production could be replaced by renewable energy. There are already existing implementations of renewable energy to the copper industry. Some copper mines (particularly in Chile) can be found in areas with some of the largest solar irradiance in the world. Therefore, it has been implemented

Table 7.7 Existing solar projects in the copper mining industry (data from Behar et al. [84]).

Mine	Production (tonne Cu/year)	Mining process	Solar technology	Capacity solar (MW)
El Tesoro	85 000	Electrowinning	Solar thermal (parabolic trough)	10.5
Collahuasi	571 000	All-processes	Solar PV	25
El Aquila	66 000	All-processes	Solar PV	2.3
Chuquicamata	855 000	All-processes	Solar PV-Diesel	1
Gabriela Mistral	110 000	Electrowinning	Solar thermal (Flat plate collector)	32
Minera Constanza	118 000	Electrowinning	Solar thermal (Flat plate collector)	0.31

in a few operating copper mines (see Table 7.7). The installed solar capacity ranges from 0.01 to 2.3 MWh/tonne Cu. The smaller solar plants will need to be balanced with grid power to supply sufficient electrical energy. The larger solar plants, viz. at the Gabriela Mistral and El Tesoro mines, with installed solar capacities of 2.3 MWh/tonne Cu and 1.0 MWh/tonne Cu, respectively, could provide more valuable case studies on how to deal with the intermittency of solar power in copper production.

In the electrowinning step in the hydrometallurgical route, the electrolyte is typically slightly heated to temperatures between 50 and 60 °C to maximise copper purity[68]. Water heaters are often integrated into the system to maintain electrolyte temperatures. The operation temperatures of the water heaters are plant-specific but typically operate between 60 and 90 °C [68, 85]. There is a potential for the integration of solar systems to supply both the thermal and electrical energy for the copper production process. Several mines have integrated solar systems to heat the electrolytes [68, 84].

For a total production of 100 000 tonnes Cu/year, Cuevas et al. (2015) proposed several configurations for an enriched electrolyte stream of 300 m^3/hour (with a 45 kg/m^3 Cu concentration) that needs to be warmed to optimum electrowinning temperatures between 50 and 60 °C [68]. The several configurations to warm up the electrolyte are mine-specific. The simulations were assessed for solar field areas from 30 000 to 50 000 m^2, with solar energy storage capacities from 4000 to 6000 m^3. Both technologies (parabolic trough collectors [PTC] and flat plate collectors [FPC]) can supply sufficient energy to heat the electrowinning process. When assessing the same configuration, PTC systems outperformed FPC systems by 7–16% [68]. Future research could focus on the development of a solar-aided flash furnace, taking into account specific metallurgical requirements during the smelting process [86].

Renewable energy can also be captured in the form of wind power as implemented at the Los Pelambres copper mine – one of the largest copper reserves in Chile with an estimated production capacity of 335 Mio tonnes of copper per annum. The El

Arrayán wind power plant will deliver nominally 115 MW to the mine using 50 2.3-MW turbines [87]. The target is to offset around 320 million tonnes of CO_2 in 2020 [88]. The installed power would correspond to ca. 2.4 MWh/tonne copper, which is nominally almost sufficient to power the crushing and grinding section, although considering the intermittency of wind power back-up from the grid is necessary.

7.3.7 Current Research Around Future Processes

There have been various suggestions to optimise the production of copper. For instance the 'synergistic copper process concept' [54] combines the processing of different types/grades of ore by going through a leach process in part as well as a more classical grinding and flotation process. The typical benefit examined for such alternate processes is the minimisation of potentially hazardous waste materials. This may be incorporated with acid recycle as well as a drying stage directly after the precipitation step to minimise the amount of smelting required [54].

The smelting process requires large amounts of thermal energy, and this demand is met by combusting liquid fossil fuels and/or electricity (see Table 7.6). There is a widespread utilisation of solar energy for electricity generation and medium-temperature thermal applications. However, there is minimal commercial use of solar energy for high-temperature applications such as smelting [86]. Solar furnaces present themselves as alternatives to be used in the smelting processes of copper production [77], specifically for remote copper mines which have access to land for solar energy and/or wind energy capture. A few solar furnaces exist worldwide, notably the Odeillo solar furnace in the south of France built in the 1960s, with 63 heliostats fitted onto a parabolic concentrator directed at a receiver. The focal point reaches temperatures up to 3800 °C [86], which is sufficient for the smelting process in addition to providing further heat required in the processing.

7.3.8 Outlook

The copper industry is an interesting one, as it has the potential to benefit greatly from the implementation of renewable electricity and at the same time is an essential mineral in the production and transportation of renewable electricity. Relatively minimal high-temperature heating is required in the process, with large amounts of energy required for the electrowinning process, which is already electrified and so may easily be converted to renewable energy. However, the electrowinning process is run continuously, meaning that there will need to be provision to have access to a constant stream of renewable energy and/or the ability to run the electrolysis cells at a variable load (with the cost of energy storage to be weighed against the increase in the electrolysis capacity).

Key hurdles include the high energy demands of copper production, and the need for reliable and continuous energy supply. Although solar and wind energy projects have been implemented and show significant promise, especially in regions that experience high solar irradiance, achieving 100% integration of renewable energy into copper production processes will require continued advancements, more specifically in energy storage.

Although wider adoption of renewable energy, in the form of renewable fuels, is dependent on additional developments in hydrogen production and infrastructure, the integration of green hydrogen offers an exciting prospect for further decarbonisation of the copper industry. Furthermore, although they are still in their infancy, cutting-edge technologies such as solar furnaces and alternative smelting technologies may provide long-term answers for lowering greenhouse gas emissions.

Recycling of copper is already prominent though this may need to be increased to supplement decreasing copper ore qualities. Copper installed for permanent infrastructure will not be readily recyclable, but shorter-lifetime copper sources will need to be recycled to the fullest possible extent, as the decreasing ore qualities require higher energy requirements to process.

The copper industry will become ever more important in the coming years enhancing the need for sustainability and incorporation of renewable fuels in its production.

7.4 Titanium and Titania

Titanium is the 22nd element of the periodic table and is a popular metal due to its high strength (200–450 MPa; comparable to steel), low density (~4500 kg/m^3; cf. steel ~7700 kg/m^3) and impressive corrosion resistance, including being resistant to corrosion in seawater, aqua regia and chlorine, due to the generation of a titania passive layer. Neat titanium can have the highest strength-density ratio of any metal [89]. At temperatures below 882 °C, titanium will be present in the hexagonal phase (hcp). At higher temperatures (and atmospheric pressure), titanium will be most stable in the bcc-structure.

Titanium alloys are used in the aerospace industry, including in jet engines and spacecraft; this makes up over half of the use of titanium internationally. It is also popular in some industrial and military applications, though its use falls well short of aluminium and steel due to the increased cost. Titanium is also considered a highly biocompatible material, meaning that many prostheses, orthopaedic and dental implants in the medical field are titanium-based [90].

Titania or titanium dioxide (TiO_2) is applied as a strong white pigment in a huge variety of products, viz. in paint, textiles, sunscreen, rubber, plastic, paper, cosmetics, ceramics and the food industry as a colourant [91, 92]. Its use in white paint has dual benefit, as it may also act as a hydrophobic agent preventing ingress of water. It is also one of the most popular choices as a photocatalyst [93], and may be used as a catalyst for several reactions, notably the photocatalytic water splitting, which has the potential to be a low energy, renewable pathway to the generation of green hydrogen.

Titania may exist in different phases, most commonly rutile, anatase and brookite. Both rutile and anatase have tetragonal structures (see Figure 7.10a and b), though have significantly different lattice parameters resulting in quite

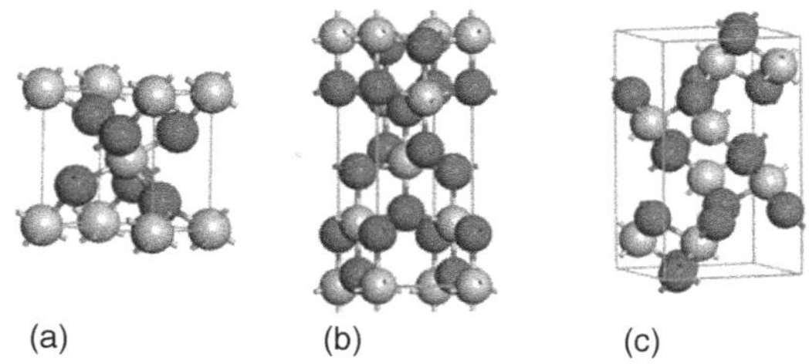

Figure 7.10 Different crystallographic structures of titania: (a) rutile, (b) anatase and (c) brookite. Light grey = titanium, dark grey = oxygen (data from BIOVIA [94]).

different crystallographic structures [95]. Brookite has an orthorhombic structure. Both the density and the hardness of rutile are greater than that of anatase with the density of brookite, the least common phase of titania in the gap between rutile and anatase. Rutile is thermodynamically the most stable phase at relatively low pressures and at temperatures of interest if titania is present as a low surface area material ($S_{BET} < 7\,m^2/g$), but anatase is prevalent in high surface area materials ($S_{BET} > 40\,m^2/g$) due to the difference in the surface energy of the two phases [96], with brookite being the more stable phase for materials with an intermediate surface area.

7.4.1 Titanium Industry

The production of titanium and pigment-grade titania start from titanium-bearing ores, mainly ilmenite ($FeTiO_3$, containing a maximum of 52.6 wt.% TiO_2 or 31.6 wt.% Ti) and to some small extent rutile (TiO_2) with 90% of the titania/titanium coming from ilmenite [97]. Ilmenite may be found in a pure form which is paramagnetic, though may also be found as a solid solution with haematite or, commonly, magnetite, and then is ferromagnetic. The balance of the ore for titania synthesis is naturally occurring rutile, which requires less processing to be utilised as a pigment [98].

Table 7.8 shows the reserves of titanium-bearing ores and the production of these ores. Australia and China have the largest reserves each having ca. 29% of the total world reserves. China is the largest producer of mined titanium ore; its production amounts to 33% of the world production. Other major producers are Mozambique and South Africa, although the depletion rate in these countries is significantly higher than other countries with a large reserve. The typical plant size to produce pigment-grade titania is between 30 000 and 600 000 tonnes TiO_2 per annum.

Titanium ore is mainly converted into pigment-grade titania (with an industry size of ca. 5.4 million tonnes per annum), and only 5% of mined titanium ends up as titanium sponge (used to produce metallic titanium products), hence, there are fewer producers of sponge titanium than titania. The production capacity of titanium sponge has increased from somewhat less than 100 ktonne to over 300 ktonne of titanium sponge per annum from 2001 till 2020 [98]. The increased capacity in China accounts for ca. 70% of the increased world capacity (with increased production capacity in Japan and Russia accounting for most of the other additional capacity). As an indication of the size of a typical sponge titanium producer, Osaka

Table 7.8 Reserves and production of titanium bearing ores in 2022 (in Mio tonnes of TiO_2 per annum) [20] and the production capacity of pigment-grade titania and sponge titanium (in ktonne per annum) [20].

	Titanium ore				
	Reserves (Mio tonne)	**Production (Mio tonne p.a.)**		**TiO_2 pigment-grade (ktonne p.a.)**	**Titanium sponge (ktonne p.a.)**
Australia	215	0.60	China	5500	220
China	210	3.14	United States	1360	a)
India	85	0.21	Mexico	350	—
Canada	52	0.52	Germany	339	
Brazil	43	0.05	Japan	322	60
Norway	37	0.43	United Kingdom	315	—
South Africa	34	1.20	Australia	260	
Madagascar	28	0.32	Saudi Arabia	200	12
Mozambique	23	1.41	Russia	55	20
Rest	16	1.16	Rest	1141	14
World	**743**	**9.42**	**World**	**9842**	**326**

a) Data withheld.

Titanium Technologies Co. Ltd. (OTC) has a production capacity of 40 000 tonnes of sponge titanium per annum.

7.4.2 Current Processes

7.4.2.1 Production of Pigment-Grade Titania

There are two main processes used in the production of titania. The older process is the sulphate process which converts ilmenite into pigment-grade titania using sulphuric acid to leach out the iron present in ilmenite. The sulphate processmay operate intermittently. The chloride process (which is also used to produce titanium) makes use of chlorine. This yields titanium tetrachloride ($TiCl_4$), which may then be converted into titania. The chloride process has steadily grown and taken dominance over the sulphate process around the world over the past decades [99] (see Figure 7.11) with the production of pigment-grade titania using the sulphate process stagnating outside of China. The chloride process is a shorter production process, has a lower energy consumption and yields higher-quality products, although the capital cost is higher [99]. It also requires a higher-grade ore.

Sulphate Process The sulphate process is a simple leaching and precipitation process which follows four primary steps: digestion or leaching, precipitation and centrifuging, hydrolysis and filtration, and calcination [100, 101].

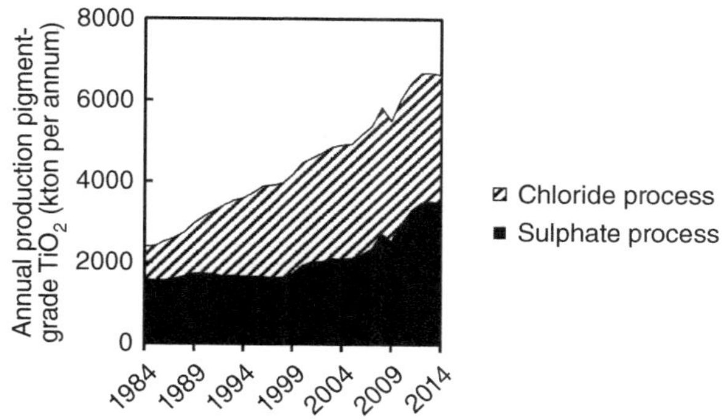

Figure 7.11 Change in process use for titania production over time (data from Völz et al. [99]).

In the digestion stage, ilmenite is digested with sulphuric acid to form ferrous and titanyl ions in solution.

$$FeTiO_3(s) + 4H^+(aq) \rightarrow Fe^{2+}(aq) + TiO^{2+}(aq) + 2H_2O(l) \quad \Delta H^0_{rxn} = -113.8 \frac{kJ}{mol\ FeTiO_3}$$

or

$$FeTiO_3(s) + 2H_2SO_4(aq) \rightarrow FeSO_4(aq) + TiOSO_4(aq) + 2H_2O(l)$$

The exothermic nature of the reaction will increase the reaction temperature to ca. 170–220 °C [102].

Trivalent iron is a major impurity in titanium-containing ore, which has to be reduced to divalent iron by the addition of scrap iron to the digestion process [102]. Iron sulphate, which is less soluble than titanyl sulphate, may then be removed from the solution through precipitation at low temperatures. Upon heating the solution to higher temperatures (ca. 95–110 °C), titanyl sulphate hydrolyses to form hydrated titania as a gel and reform the sulphuric acid.

$$TiOSO_4 + (n + 1)H_2O \rightarrow TiO_2.nH_2O + H_2SO_4$$

Hydrated titania may be recovered through precipitation, acid bleaching (reduction) to remove impurities and washing to remove excess acid, followed by a high-temperature (650–1000 °C) calcination stage. The calcination drives off the water and forms solid titania powder. Depending on the temperature of the calcination and the addition of seeds, either anatase or rutile may be formed.

$$TiO_2.nH_2O \rightarrow TiO_2 + nH_2O$$

The direct energy cost to produce titania from a V-Ti-magnetite ore using this route is estimated at 13 MWh/tonne (46.8 GJ/tonne) TiO_2 [103]. This is besides the energy required to produce the chemicals required for the process. The cumulative energy demand (CED) of the sulphate route, i.e. the direct or indirect total primary energy usage by the process [103], is estimated at 79–117 GJ/tonne TiO_2 (22.5–32 MWh/tonne TiO_2) for a feed of V-Ti magnetite [103]. This is similar to a study from the 1990s, which showed a CED of ca. 22–28 MWh/tonne (79.2–100.8 GJ/tonne) TiO_2 when using ilmenite and slag, respectively [104].

The direct global warming potential of the production of titania using the sulphate route from a V-Ti-magnetite ore amounts to 3.7–5.3 tonne $CO_{2,eq}$/tonne TiO_2 [103].

Becher Process In order to upgrade ilmenite (often required for the chloride process), a synthetic rutile (TiO_2-content of 92–95 wt.%) may be prepared in the Becher process [105]. In this process, ferrous ions are initially oxidised to ferric ions:

$$2FeTiO_3(s) + \frac{1}{2}O_2(g) \rightarrow Fe_2O_3.TiO_2(s) + TiO_2(s) \quad \Delta H^0_{rxn} = -230.4\frac{kJ}{mol\ FeTiO_3}$$

(based on heat of reaction for the formation of Fe_2TiO_5 from Fe_2O_3 and TiO_2 of 8.33 kJ/mol [106]; rest of data from TKB database [107]).

The activated ilmenite is subsequently heated by combusting coal to ca. 1500 °C. Iron in the matrix is then reduced using CO as the reducing agent to the metallic state.

$$Fe_2O_3.TiO_2(s) + 3CO(g) \rightarrow 2Fe(l) + TiO_2(s) + 3CO_2(g) \quad \Delta H^0_{rxn} = -12.5\frac{kJ}{mol\ Fe_2O_3.TiO_2}$$

This process yields 0.82 tonnes of CO_2/tonne TiO_2, from just the stoichiometric requirement. The high temperature in the reduction can be generated by combusting coke, generating the required CO in situ. The formed reduced ilmenite may then be oxidised by aeration in the presence of an aqueous ammonium chloride solution.

$$2Fe(s) + \frac{3}{2}O_2(g) \rightarrow Fe_2O_3(s) \quad \Delta H^0_{rxn} = -825.5\frac{kJ}{mol\ Fe}$$

These two steps successfully separate the iron oxides and titanium oxides from the solid solution, and they may be easily separated as they have different particle sizes as rutile is formed as the titania phase [108]. Finally, any remaining iron may be removed through leaching with dilute sulphuric acid.

Chloride Process The chloride process principally follows four steps, chlorination, condensation, purification and oxidation. However, to minimise chlorine loss, titanium-enriched materials are preferred. For instance, ilmenite smelting in the presence of anthracite at 1200–1600 °C is used to generate a titanium-rich (>85 wt.%) slag together with pig iron [102]. This can be performed in an arc furnace using a carbon electrode [109]. This results stoichiometrically in at least 0.55 tonne CO_2/tonne TiO_2, if all iron is present as ferrous iron (Fe^{2+}). Alternatively, the Becher process described above may be used.

In the chloride process, the dry titanium-containing material is reacted with chlorine gas and coke at ca. 1000 °C in a fluidised bed reactor [110]. For ilmenite, the reaction can be written as:

$$FeTiO_3(s) + \frac{7}{2}Cl_2(g) + 3C(s) \rightarrow TiCl_4(g) + FeCl_3(g) + 3CO(g) \quad \Delta H^0_{rxn} = -112.7\frac{kJ}{mol\ FeTi3}$$

The chlorine consumption can be reduced by reducing the iron content (and other impurities) in the feed. For instance, increasing the titania content to 90 wt.% will reduce the chlorine consumption from 3.11 tonne Cl_2/tonne TiO_2 when using pure ilmenite to 1.94 tonne Cl_2/tonne TiO_2, a reduction by 37.5%.

The reaction, forming CO, which may be further combusted to CO_2, yields stoichiometrically 0.70 tonnes of CO_2 per tonne of titanium tetrachloride, equivalent to 1.65 tonnes of CO_2 per tonne of titania or 2.76 tonnes of CO_2 per tonne of titanium.

The formed titanium tetrachloride may be distilled to achieve very high purities, enabling the formation of high-purity titania or titanium. Titania is produced via the oxidation of the titanium tetrachloride with air, often occurring in an oxygen flame [111] with temperatures in the order of 1500 °C.

$$\mathrm{TiCl_4(g) + O_2(g) \rightarrow TiO_2(s) + 2Cl_2(g)} \quad \Delta H^0_{\mathrm{rxn}} = -180.7 \frac{\mathrm{kJ}}{\mathrm{mol\ TiCl_4}}$$

The direct energy cost to produce titania from a V-Ti-magnetite ore using this route is estimated at 11.7 MWh/tonne (42.1 GJ/tonne) TiO_2 with the CED for the chloride route estimated at 20.4 MWh/tonne (73.4 GJ/tonne) TiO_2 [103]. A study in the 1990s showed a much higher CED for this process of ca. 27–29 MWh/tonne (97.2–104 GJ/tonne) TiO_2 when using rutile and slag, respectively [104]. The global warming potential of the production of titania using the chloride route from a V-Ti-magnetite ore is slightly more than that of the sulphate route at ca. 3.76 tonne CO_2/tonne TiO_2 as opposed to ca. 3.69 tonne CO_2/tonne TiO_2 [103].

A distinct benefit of the chloride process is the comparatively low environmental impact of the slag produced as waste, showing a 43% lower environmental impact. However, this does come with the cost of having a (slightly) higher global warming impact [103].

7.4.2.2 Production of Titanium Using the Kroll Process

Titanium tetrachloride hasuses beyond just that of titania synthesis and is also used in the synthesis of metallic titanium, primarily via the so-called Kroll process. The Kroll process exploits metallic magnesium as a reducing agent to reduce titanium from titanium tetrachloride. The process takes place at 800–850 °C:

$$\mathrm{TiCl_4(g) + 2Mg(l) \rightarrow Ti(s) + 2MgCl_2(l)} \quad \Delta H^0_{\mathrm{rxn}} = -458.0 \frac{\mathrm{kJ}}{\mathrm{mol\ TiCl_4}}$$

forming a titanium sponge covered with magnesium chloride and some magnesium, which is recovered using a high-temperature (1000 °C) vacuum distillation. This is a slow and energy-intensive process, which can take several days and consumes most of the energy in the Kroll process [112].

Magnesium chloride is recovered and electrochemically reduced co-producing chlorine (and some $TiCl_2$ and $TiCl_3$), which can be fed back to chloride process to produce $TiCl_4$. This reaction takes place at ca. 800–850 °C, and both magnesium and magnesium chloride are liquid at these temperatures:

$$\mathrm{MgCl_2(l) \rightarrow 2Mg(l) + Cl_2(g)} \quad \Delta H^0_{\mathrm{rxn}} = -610.6 \frac{\mathrm{kJ}}{\mathrm{mol\ MgCl_2}}$$

The Gibbs free energy for the reaction at 800 °C can be estimated as 468.8 kJ/mol. Hence, the minimum voltage for the electrowinning cell is 2.43 V. With an overpotential of 1.745 V [113], the energetic efficiency of the magnesium electrowinning is ca. 58%. The Faradaic efficiency of the process has been quoted as ca. 90% [114]. Hence, the generation of metallic magnesium adds 10.0 MWh/tonne Ti of electrical energy to the overall energy cost of the production process. In addition, this endothermic reaction requires at least 1.6 MWh/tonne Ti in thermal energy.

The total energy cost for the Kroll process is difficult to estimate, but realising that the reduction of $TiCl_4$ is an exothermic reaction requiring a minimal amount of energy, and more than 50% of the energy for the Kroll process is associated with the vacuum distillation to remove $MgCl_2$ from titanium sponge (some reports even claim 70% [112]), the overall Kroll process would thus require at least 23.2 MWh/tonne (83.5 GJ/tonne) Ti (and 38.8 MWh/tonne [139 GJ/tonne] Ti if vacuum distillation takes up 70% of the total energy demand for the process). The minimum energy requirement for the conversion of $TiCl_4$ into metallic titanium is 763 kJ/mol or 4.4 MWh/tonne (15.8 GJ/tonne) Ti implying a relatively low energy efficiency of this process of 11–19% depending on the energy requirement for the vacuum distillation.

Between the various steps required for the synthesis of titanium, it has been estimated that ca. 10 tonnes of CO_2 is released per tonne of titanium [115] of which around 25% are process emissions with the balance heating emissions.

There are alternatives for the production of metallic of titanium, e.g. the Hunter process may be used [116], which uses sodium as the reducing agent. This process was established before the Kroll process, but it has been replaced due to the higher cost [117].

7.4.3 Recycle

At current, there appears to be very limited recycle of titanium or titania. Waste titanium, often containing a large amount of oxygen and some iron is used as an additive when making titanium alloyed steel [118]. Vacuum arc remelting is the primary technology for the recycling of sufficiently pure titanium waste, as well as being the technology used for the synthesis of titanium-based alloys [119].

Titania is mainly used as a pigment, and as such ends up in a variety of end-products. The dispersion of the end-products renders recycling of titania problematic. There is currently no shortage of titanium-containing reserves, although the work-up of titanium-containing ore for further processing may become more energy-intensive, potentially increasing its carbon footprint.

7.4.4 Implementation of Renewable Energy in the Production of Titania/Titanium

The pre-purification of titanium-bearing ore can be carried out in a direct current arc furnace, which uses carbon as an electrode and coal (anthracite) as a reductant. The arc furnace can readily be powered using renewable energy. This would require scheduling according to availability.

The temperatures required in the calcination of titania produced using either the sulphate or the chloride process are rather high, which may prove difficult to move to non-fossil-based heat sources. Besides this, coke is required in the chloride processes to reduce chlorine in the production titania. This cannot be readily displaced by hydrogen as this would yield the acidic HCl. Hence, the chloride process requires sourcing a renewable form of coke. The Kroll process for the manufacture of titanium sponge requires the production of magnesium, which is performed electrochemically at 800–850 °C. This process requires 9.8 MWh/tonne Mg (35.3 GJ/tonne Mg) on electrical energy in addition to at least 1.6 MWh/tonne Mg (5.8 GJ/tonne Mg) of heat. Intermittent power supply may be negated by sufficient storage for magnesium chloride and intermediate storage.

Vacuum distillation is now the preferred step to remove chlorides from titanium sponge. This is a slow and energy-intensive step. The rate-limiting step appears to be heat transfer to the titanium sponge [120]. Pre-processing the sponge to remove the majority of magnesium and magnesium chloride from the sponge may be a suitable strategy to reduce the distillation time and thus the energy consumption in this step. Furthermore, the long duration of the batch distillation process (several days) makes it less adaptable to a variable energy supply.

7.4.5 Future Processes to Produce Titania and Titanium

Stepwise improvements in the production of titania using either the sulphate or the chloride route are being made. Furthermore, a variety of different leaching processes have been proposed to (pre)-purify titania [121]. For instance, leaching of the titanium-containing ore using hydrochloric acid and reducing the ferric ions in solutions is a possible strategy [97, 122]. Titanium and iron are subsequently separated using solvent extraction, after which titanium chloride may be converted into titania using magnesia. It has also been suggested to react ilmenite with hydrochloric acid which may valuably operate at lower temperatures but does not see industrial implementation currently. The purpose of this process is similar to the ilmenite smelting process and the Becher process, i.e. to increase the titania content of the material, which can then be upgraded to pigment-grade titania.

Hydrogen-assisted magnesio-thermic reduction (HAMR) is a more recently developed technology for the production of metallic titanium rather than the Kroll process [123]. It uses a similar approach to the Kroll process, as magnesium is still used as the primary reducing agent. However, now hydrogen is used as a secondary reductant to destabilise the Ti-O bonds making the reduction with magnesium more favourable. This process is reported to save more than 25% of the energy requirement of the Kroll process, with a more than 65% saving when the magnesium regeneration is considered beyond the system boundary [123].

In order to avoid the complexity, price and energy/carbon intensity of the chloride and subsequent Kroll process, it has been suggested to use electrolysis for the conversion of titania to metallic titanium [121, 124]. This reaction tends to use carbon anodes functioning as the reducing agent and would also require the dissolution of titania in an acidic solution.

$$\mathrm{TiO_2(s) + C(s) \rightarrow Ti(s) + CO_2(g)} \qquad \Delta H^0_{\mathrm{rxn}} = 550.4\frac{\mathrm{kJ}}{\mathrm{mol\ Ti}}$$

The standard Gibbs free energy for this reaction is 494.21 kJ/mol implying that at room temperature, the minimum required voltage would be 1.28 V. The advantage of the direct electrolysis is a reduction in the direct CO_2 emission of 0.92 tonnes of CO_2 per tonne of titanium, which is a marked improvement over the Kroll process as currently used. Furthermore, it could reduce the energy cost significantly as the minimum energy requirement for the electrowinning of titanium is at 2.87 MWh/tonne (10.3 GJ/tonne) Ti. However, the process suffers from a low energy efficiency, low Faradaic efficiency and a low rate of dissolution of ilmenite [121].

7.4.6 Outlook

The titania and titanium industries will be valuable to consider going forward as they are both expected to see significant growth over the next 5 years and beyond with titania expected to have a compound annual growth rate of >4% [125], and titanium of ca. 6.2% [126], respectively. One of the major limitations for the usage of titanium is the high cost associated with its production (aggravated by the loss in machining this material [127]). More cost-effective process technologies, i.e. through the implementation of new technologies may increase this growth rate, as titanium has become popular for alternative uses due to its highly desirable material properties.

There are two primary processes in the titania industry, the chloride and the sulphate, each with distinct pros and cons. The chloride process benefits from the quality of its output and is essential for the synthesis of metallic titanium. However, it requires a feed with a higher titanium content and operates at significantly higher temperatures than the sulphate process. The sulphate process suffers from a longer process time although operating at lower temperatures (and consequently typically produces a mixture of anatase and rutile) and being able to operate in a batch fashion which may add to its appeal in the future. The chloride process has seen more of a rise, outside of China, over the past decades, possibly because of the required purity of pigment titania and the possible production of metallic titanium. Both processes operate with highly hazardous chemicals, viz. chlorine gas and sulphuric acid though both use much of this chemical in a circular manner, which slightly decreases the hazardous waste generated in the process. The sulphate process does produce a more environmentally hazardous waste product meaning that there will need to be an optimisation on new plant design as to whether the clean-up of the waste product, or CO_2 produced will be more cost-effective.

Whilst there are several other options for the manufacture of titanium, there are seemingly one major challenge with all processes: the high temperatures required, which may, for the foreseeable future, be required to be generated with fossil fuels.

7.5 Silicon

Contribution by Chiamaka Amonye.

Silicon is a hard but brittle crystalline metalloid found in group 14 in the periodic table. It is the second most abundant element found in the Earth's crust by mass, making up about 27.7 wt.%. Pure silicon is reactive, so it occurs naturally as silica (SiO_2) or as silicates (minerals containing silicon and oxygen in tetrahedral SiO_4^{4-} units, linked together in various patterns). The oxides occur as sand, quartz, amethyst, etc. Its silicate forms occur as asbestos, granite, clay mica, etc. Silicon-based compounds form a wide variety of rocks and minerals. The common mineral quartz is the most widely used raw material for producing pure silicon. Silicon in its crystalline form was first obtained by Berzelius in 1824 upon heating potassium fluorosilicate with potassium. A silicon product, although contaminated with potassium silicide, was obtained and further purified by stirring in water, removing the silicide impurity and obtaining relatively pure silicon [128].

Silicon is used in various applications in the metallurgical industry, such as the deoxidation and alloying of steel and cast iron (using ferrosilicon, an alloy of iron and silicon) and alloying with other metals, e.g. aluminium. Further, it is used to some extent as a raw material in the chemical industry to produce silanes, but it is much better known for its usage in the semiconductor industry to produce electronic devices (e.g. computers) and photovoltaic cells. Its band gap and absorption spectrum and several other properties [129], as well as its abundance, ensured its use as the predominant semiconductor material.

7.5.1 Material Specifications

Not all types of silicon are similar, and silicon is typically classified into various grades based on purity levels (see Table 7.9). Ferrosilicon is an iron silicon alloy containing between 15 and 70 wt.% silicon, which is mainly used in steel-making. Metallurgical-grade silicon (MG-Si) contains 98–99% silicon, but has quite a high content in other impurities, notably iron, aluminium and calcium. The aluminium

Table 7.9 Typical compositions of existing grades of silicon [130, 131].

Element	Metallurgical-grade silicon (ppm)	Solar-grade silicon (ppm)	Electronic-grade silicon (ppm)
Si	98–99%	6–8 N	9–11 N
Fe	2000–3000	<0.3	<0.01
Al	1500–4000	<0.1	<0.0008
Ca	500–600	<0.1	<0.003
B	40–80	<0.3	<0.0002
P	20–50	<0.1	<0.0008
C	600	<3	<0.5
O	3000	<10	—
Ti	100–200	<0.01	<0.003
Cr	50–200	<0.1	—

Table 7.10 Terminology for various types of crystalline silicon (c-Si) (data from Basore [133]).

Type	Symbol	Grain size	Common growth techniques
Single crystal	Sc-Si	>10 cm	Czochralski (CZ) float zone (FZ)
Multi-crystalline	mc-Si	1 mm–10 cm	Cast, sheet, ribbon
Polycrystalline	pc-Si	1 μm–1 mm	Chemical-vapour deposition
Microcrystalline	μc-Si	<1 μm	Plasma deposition

industry together with the steel industry is a large user of MG-Si, as it is used as an additive to enhance machineability, castability and corrosion resistance of aluminium [130, 132]. Typically, MG-Si is the starting material for refining the other grades of silicon. Solar-grade silicon (SG-Si) is of higher purity and contains at least 99.9999% (6 N) silicon, and it is used to produce photovoltaic cells. Electronic-grade silicon (EG-Si) is required to be of even higher purity, typically 99.9999999% (9 N), and increased purity assists in maintaining computational performance consistency.

Silicon (and other semiconductor materials) can be categorised as single-crystalline, multi-crystalline, polycrystalline or amorphous [133]. The distinguishing factor between these materials is the extent to which their crystal structures are regular and perfectly ordered and, therefore, the achieved crystal sizes. Semiconductor materials may be classified according to the size of the constituent crystals in the material [133], a summary of the different types of crystalline silicon is given in Table 7.10.

Polycrystalline silicon or polysilicon (often grouped with multi-crystalline silicon) is a high-purity form of silicon used as a raw material for the manufacture of solar PV cells and the electronics industry [134, 135]. Polysilicon is typically produced from MG-Si by a chemical process called the Siemens process.

7.5.2 Silicon Industry

The global demand for silicon is driven by rapid industrialisation and infrastructure development across numerous regions. The heavy reliance on silicon metal is a result of industries such as automotive, construction, electronics and photovoltaics. The total silicon production, excluding the United States, amounted to 9.7 Mio tonnes per annum in 2024, almost equally divided between ferrosilicon and silicon metal (see Table 7.11). The silicon production tripled over the last three decades (see Figure 7.12), due to the opening of a novel market (photovoltaics) in addition to the expanding market for ferrosilicon and metallic silicon. The silicon market shows the so-called pork cycle [137], i.e. a high demand induces the increase/creation of supply capacity, generating an oversupply when they come on the market (in the silicon industry, this occurs after a 2–3 year time delay). The oversupply suppresses price forcing from existing suppliers or causes them to reduce production or stop it entirely temporarily. This can lead to another period of relative undersupply,

Table 7.11 Global production[a] of ferrosiliconand silicon metal in 2024 (in ktonne Si per annum) (data from Schnebele [136]).

	Ferrosilicon (ktonne p.a.)	Silicon metal (ktonne p.a.)
China	3500	3900
Russia	470	50
Brazil	200	190
Norway	180	120
Kazakhstan	130	7
Malaysia	130	—
France	20	90
Iceland	70	20
Bhutan	80	—
India	60	—
Rest	260	123
World	**5100**	**4600**

a) Data of the production within the United States withheld.

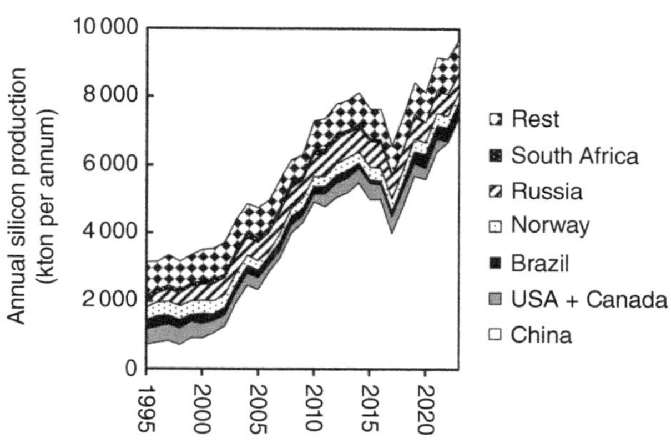

Figure 7.12 Global annual silicon production over the years (data from the US Geological Survey [20]; data from 2022 and 2023 do not include production data from the United States).

restarting the cycle. On top of this behaviour, an unusual decline in silicon production between 2013 and 2017 can be observed. The exact reasons for this are unclear, but it may be ascribed to various factors, such as development of overcapacity in China [137], lowering of feed-in-tariffs in China for PV-generated power [138] and the increased material efficiency in the production of polycrystalline silicon ingots.

Current production is dominated by China, which produces ca. 70% of the world ferrosilicon and almost 85% of the silicon metal demand. Other key players include Russia, Brazil and Norway; Kazakhstan and Malaysia have increased their production significantly over the last few years. The production in the United States did not grow much and the contribution to the global production dropped from ca. 10–13% to ca. 3–5% in the last decade. As of 2023, the total production volume of silicon worldwide is estimated at a little over 9 million metric tonnes [139].

Around 2000, the world market was dominated by the production of ferrosilicon, which contributed to ca. 80% of the global market [20]. In the 1990s, the demand for MG-Si by the semiconductor industry was only a few thousand tonnes per year, making it a niche industry [132]. In recent years, the demand for semiconductors has increased specifically due to the growth of the photovoltaics industry. As of 2010, ca. 60% of the MG-Si is used in the photovoltaic industry and only ca. 30% in the electronics industry [130, 140]. In 2023, the silicon market is no longer dominated by ferrosilicon and is almost equally split between ferrosilicon and silicon metal. A further increase in the photovoltaic industry will switch the silicon market further towards silicon metal.

As of 2019, the market size value stood at US$ 6.3 billion, with a forecasted value of US$ 8.9 billion by 2027 at a compound annual growth rate (CAGR) of 5.5% (like the growth rate of steel). Ferrosilicon is typically priced at US$ 800–1500/tonne. MG-Si is produced at an economic cost of US$ 1900–5000/tonne and an energy cost of 14–16 MWh/tonne Si [141]. Electronic-grade silicon costs more than US$ 50 000/tonne and is produced in much smaller quantities [142].

The global silicon market consists of various small to medium-sized manufacturers. The leading manufacturers of MG-Si globally are Ferroglobe (with a capacity of 275 000 tonnes per year [143]), Baidao Silicon Powder Co., Ltd., Elkem ASA, HPQ Silicon Resources Inc., Silicor Materials, Mississippi Silicon, Canadian Metals, American Elements, Jiuzhou Silicon Industry Co., Ltd., Ardevur, etc.

The average capacity for polysilicon production plants is approximately 44 000 tonnes per annum. Manufacturers of polysilicon include LDK Solar, Tokuyama Corporation, Mitsubishi Polysilicon, Osaka Titanium Technologies.

7.5.3 Current Processes

The raw material for the MG-Si is quartz or silica sand [144]. Sand is found easily; however, high silica content sand sources can be scarce. Quartz mines are present worldwide with deposits of various qualities, namely high-quality (>98%), industrial-grade (95%) and low-quality (65%). The mining of silica from quartz deposits begins with the removal of topsoil to expose quartz deposits/stones, followed by excavation, transportation, sieving, watering and dewatering. Next, depending on the desired purity of the silica, physical processes such as magnetic separation, flotation and gravity separation are used to eliminate impurities [144].

Figure 7.13 shows the distribution of quartz deposits globally. The extracted quartz rock is often transported to silica sand extraction mines if not located in proximity to

Figure 7.13 Global industrial-grade quartz deposits as of 2019 (Heidari and Anctil [144] / with permission of Elsevier).

the deposits. Once the silica sand is extracted, it is transported to its use destination or exported.

Quartz and quartzite (a mineral with more than 97 wt.% SiO_2) are relatively pure forms of silica which are used in metallurgical processes [132]. The impure products of these processes: metallurgical silicon and a variety of grades of silicon carbide, are converted to intermediate chemicals. The production of 1 tonne of polycrystalline silicon requires ca. 4.0 tonne quartzite [145], i.e. significantly more than the stoichiometrically required 2.14 tonne. Quartzite can be found in large surface deposits, and the recovery of quartzite has a minimal energy requirement in comparison to the production of silicon. Silica is initially converted to MG-Si metal, silicon carbide or ferrosilicon, which can subsequently be converted to high-grade silicon and silanes.

7.5.3.1 Production of Metallurgical-Grade Silicon and Ferrosilicon

MG-Si is typically produced industrially by a carbothermic reduction of quartz with carbon in a submerged-electrode arc furnace (SAF) [146, 147], with the overall reaction:

$$SiO_2(s) + 2C(s) \rightarrow Si(l) + 2CO(g) \quad \Delta H^0_{rxn} = 738.4 \frac{kJ}{mol\,SiO_2}$$

The further oxidation of CO to CO_2 will bring the overall heat of reaction down to 565.9 kJ/mol.

This reaction is an idealised form, and the real reaction is more complex [148]. Most of the carbon used comes from a mixture of coal, coke and charcoal. Wood chips are occasionally used to introduce porosity into the reacting masses to inhibit the build-up of gases such as CO and SiO. These raw materials are not pure, and impurities may end up in the final product (see Table 7.9).

The heart of the production is the arc furnace (see Figure 7.14). A typical furnace has a pot with a diameter of roughly 10 metres. The pot is charged with the quartz and carbon source (in the case of ferrosilicon, the iron is added alongside the carbon and quartz). The electrodes are submerged into the pot and supply a current which

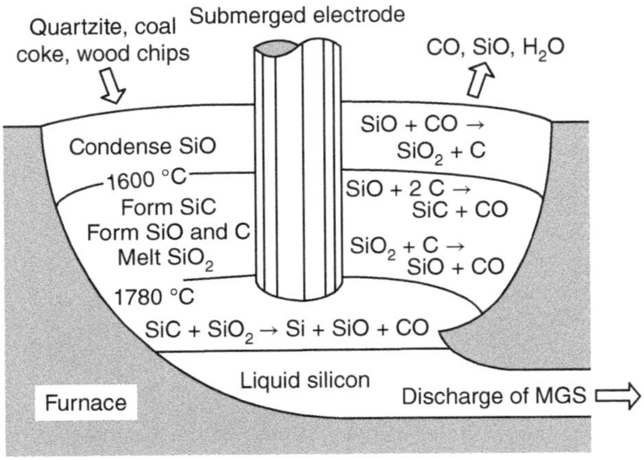

Figure 7.14 Schematic of a submerged arc furnace (SAF) for the production of metallurgical-grade silicon (MG-Si) (Barron [142] / CC BY 3.0).

channels through the contents of the furnace. The provided current heats a portion of the charge to up to 2000 °C in the hottest region. In the colder sections, silicon monoxide may react with carbon in a heterogeneous reaction yielding silicon carbide:

$$\mathrm{SiO(g) + 2C(s) \rightarrow SiC(s) + CO(g)} \quad \Delta H^0_{\mathrm{rxn}} = 73.3\frac{\mathrm{kJ}}{\mathrm{mol\,SiO}}$$

resulting in the condensation of silicon carbide. The high density of silicon carbide (3.2 g/cm^3) will force the downward movement of these particles. The condensation of silica may also occur due to the gas phase reaction between silicon monoxide and CO which will rise in the colder section of the reactor re-forming silica and carbon:

$$\mathrm{SiO(g) + CO(g) \rightarrow SiO_2(s) + C(s)} \quad \Delta H^0_{\mathrm{rxn}} = -697.1\frac{\mathrm{kJ}}{\mathrm{mol\,SiO}}$$

This is a reversible reaction, and at temperatures higher than 1700 °C, the reverse reaction will become noticeable (the reaction of silicon monoxide with carbon yielding silicon carbide is virtually irreversible at the achievable temperatures). At high temperature, silica may react with silicon carbide yielding metallic silicon:

$$\mathrm{SiO_2(s) + SiC(s) \rightarrow Si(l) + SiO(g) + CO(g)} \quad \Delta H^0_{\mathrm{rxn}} = 811.7\frac{\mathrm{kJ}}{\mathrm{mol\,SiO_2}}$$

Silicon monoxide can react further with silicon carbide forming CO and metallic silicon:

$$\mathrm{SiO(g) + SiC(s) \rightarrow 2Si(l) + CO(g)} \quad \Delta H^0_{\mathrm{rxn}} = 158.4\frac{\mathrm{kJ}}{\mathrm{mol\,SiO}}$$

or may dissociate into metallic, liquid silicon and silica:

$$\mathrm{2SiO(g) \rightarrow Si(l) + SiO_2(s)} \quad \Delta H^0_{\mathrm{rxn}} = -330.8\frac{\mathrm{kJ}}{\mathrm{mol\,SiO}}$$

Silicon is tapped from the bottom of the furnace and refined by slag treating or gas purging. During the refinement stage, oxide and carbide particles are removed, resulting in an adjusted composition to a specified value. After refinement, the molten silicon alloy is tapped from the bottom of the furnace at 97–99% purity, cooled in a mould and crushed to a specific size [130]. The off-gas from the reaction in the furnace (SiO + CO) is mixed with excess air under the furnace hood, burned, captured into a gas-cleaning system and filtered [149].

The refining of silica requires ca. 10.5–13 MWh/tonne metallurgical silicon [145], and thus the energy efficiency of the process is defined as:

$$\eta^{energy} = \frac{\Delta H^0_{SiO_2(s)+C(s)\rightarrow Si(l)+CO_2(g)}}{E_{demand}} = 43.1 \text{ to } 53.5\%$$

A more detailed energy analysis for the production of metallurgical silicon [149] on an operating plant in Norway with an SAF using a 48-hour tapping interval period (see Figure 7.15; note data scaled to a 10 MW furnace) showed a lower overall efficiency (34%, energy efficiency being defined as the amount of energy in the product relative to the amount of energy added to the furnace in the form of electrical energy and materials). A large amount of energy, roughly 45%, leaves the system via the off-gas. Thermal losses in the process, including radiative and convective heat loss from the furnace and from the MG-Si product during cooling and solidification are relatively minor, contributing only 8% to the overall energetic losses.

The exergetic losses in the product of metallurgical silicon are high (ca. 46% of the total exergy [149]). Energy recovery from off-gas can improve the thermal efficiency of the process. The off-gas can be used for steam generation in a Rankine cycle to produce electric power; thus, a higher-temperature off-gas increases the exergetic efficiency of the process.

7.5.3.2 Production of Polycrystalline Silicon

MG-Si can be converted into SG-Si using either a metallurgical route or a chemical route [131]. In the metallurgical route: silicon is melted, and impurities in the silicon are removed through various techniques, mainly by slagging, oxidation and volatilisation. Chemical methods are commonly applied as they produce SG-Si of a higher purity than metallurgical methods [130].

In the chemical route, the production of SG-Si occurs through the reaction of MG-Si with hydrochloric acid and subsequent pyrolysis of volatile silanes [130]. Different chemical routes exist such as the Siemens™ Process, Union Carbide™ Monosilane Process and the New Silane Process [132]. The most commonly used Siemens process for the production of high-purity silicon (see Figure 7.16) was developed in the 1960s and contributes globally to roughly 90% of its production [130, 132, 150]. It uses trichlorosilane as the primary precursor to produce bulk polysilicon.

The finely milled MG-Si is hydrochlorinated in a fluidised bed reactor with anhydrous hydrogen chloride at 300 °C and 1 bar in the absence of a catalyst to form trichlorosilane [130, 151, 152]:

$$Si(s) + 3HCl(g) \rightarrow SiHCl_3(g) + H_2(g) \qquad \Delta H^0_{rxn} = -249.9 \frac{kJ}{mol\ Si}$$

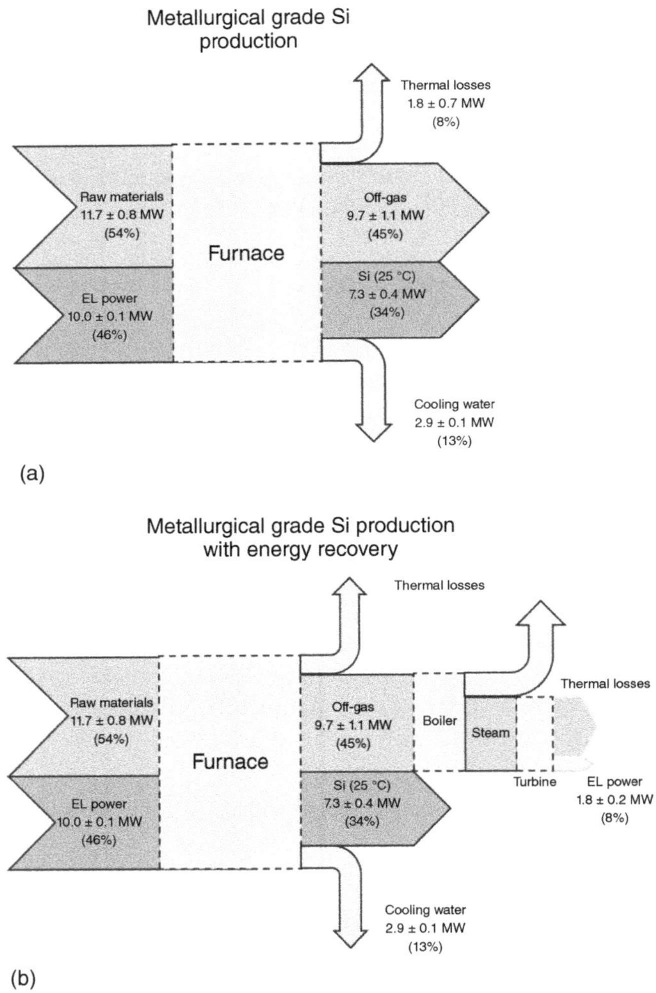

Figure 7.15 Energy flow to produce metallurgical-grade Si (MG-Si) without (a) and with (b) energy recovery (values scaled to a 10 MW furnace) (redrawn from Takla et al. [149]).

This exothermic reaction yields between 80% and 90% $SiHCl_3$ (trichlorosilane) and the rest being $SiCl_4$, which can be described as a consecutive reaction in the hydrochlorination [151]:

$$SiHCl_3(g) + HCl(g) \rightarrow SiCl_4(g) + H_2(g) \quad \Delta H^0_{rxn} = -38.5 \frac{kJ}{mol\ SiHCl_3}$$

As the reactions occur, impurities present within the MG-Si react to form their respective halides such as $AlCl_3$, $FeCl_3$, BCl_3, etc. Trichlorosilane is then purified to a high degree via fractional distillation because of the difference in boiling points of the impurity chlorides present (see Table 7.12). The resulting $SiHCl_3$ contains

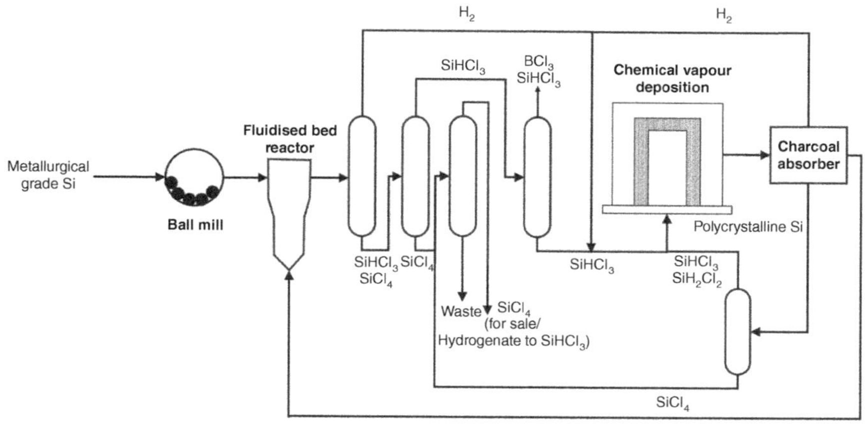

Figure 7.16 Process flow diagram for the production of SG-Si via the Siemens process (redrawn from References [132, 148]).

Table 7.12 Normal boiling point of silicon-based precursors, HCl, and common chloride impurities.

Compound	Boiling point (°C)
HCl	−85
SiH_4	−112
SiH_3Cl	−30
SiH_2Cl_2	8
$SiHCl_3$	32
$SiCl_4$	57
BCl_3	13
PCl_3	74
$AsCl_3$	131
$AlCl_3$	180[a)]
$SbCl_3$	219
$FeCl_3$	~315

a) Sublimation point.

electrically active impurities such as Al, P, B, Fe, Cu or Au, and additional chemical purification can reduce these electrically active impurities to a concentration of ≤1 ppb, for the production of EG-Si [132]. For example, the reduction of boron and phosphorus to concentrations levels below 1 ppb has been achieved by complexing them with partially oxidised $SiHCl_3$ [131, 153].

The resulting high-purity $SiHCl_3$ is vapourised, diluted with high purity hydrogen at temperatures around 1100 °C for ~200–300 hours [132] and passed into an evacuated deposition reactor, forming bulk polycrystalline silicon [154].

$$SiHCl_3(g) + H_2(g) \rightarrow Si(s) + 3HCl(g) \quad \Delta H^0_{rxn} = 249.9 \frac{kJ}{mol\ SiHCl_3}$$

The decomposition of trichlorosilane may also take place at these conditions:

$$3SiHCl_3(g) \rightarrow Si(s) + 2SiCl_4(g) + HCl(g) + H_2(g) \quad \Delta H^0_{rxn} = 173.0 \frac{kJ}{mol\ SiHCl_3}$$

yielding silicon tetrachloride and hydrogen. Many other reactions may take place as well [155]. Hydrochloric acid may leach out deposited silicon; the co-formation of $SiCl_4$ will reduce the HCl concentration, which may prevent the leaching of the formed silicon.

The pure silicon formed is deposited onto polysilicon rods to produce high-purity polysilicon rods – 150–200 mm in diameter. Each batch of polysilicon rods can take up to several days to grow and are broken into smaller chunks after growing to a certain thickness. Depending on the impurities present in trichlorosilane after distillation and on the surface of the polysilicon, chunks/rods are etched off to obtain either SG-Si or EG-Si. The yield of polysilicon and SG-Si is typically ca. 0.7 tonne SG-Si/tonne MG-Si, whereas the yield of EG-Si is ca. 0.44 tonne EG-Si/tonne MG-Si [145].

Purification of MG-Si to SG-Si is quite energy-intensive requiring 41–113 MWh/tonne polysilicon for the Siemens process and a further 41–102 MWh/tonne polysilicon to transform it into SG-Si [145]. Hence, the overall transformation of silica into SG-Si requires 100–233 MWh/tonne SG-Si. A PV-cell with a thickness of the silicon layer of 180 μm will thus contain 1.165 kg of silicon ($\rho_{Si} = 2330\ kg/m^3$) and require 0.117–0.271 MWh to produce. It will thus take between 58 and 136 days to recover this energy during its operation if a PV-cell receives, e.g. on average 200 W/m^2 in solar irradiation and converts it with an efficiency of 15% into electric power. This relatively long time drives the search to minimise the use of silicon in PV-cells and look for more energy-efficient technologies for the conversion of silica into SG-Si.

7.5.3.3 Silicon as a Feedstock for the Chemical Industry

Silicon is a feedstock used in the chemical industry to produce silicones, with the annual production estimated at 2.4 million tonnes as of 2015 [156]. 'Silicone' is a generic term which refers to a class of synthetic polymers that are based on a framework of alternating silicon and oxygen (siloxane) bonds with a least one organic group attached to the Si atom via a direct C-Si bond [157]. These silicone polymers have a host of chemical and physical properties, such as high hydrophobicity, low water solubility, low volatility and significantly low polarisability. Further, they can be synthesised in various forms: solids, liquids, gels, pastes, oils, rubber, etc.

Due to these properties, silicones can be made resistant to moisture, chemicals, heat, cold and numerous other harsh conditions that conventional materials cannot withstand. Hence, silicone polymers are used in thousands of product applications across varying industries, such as construction, consumer products, electronics, transportation and the health care industry.

The production of silicones from MG-Si typically occurs in three stages [156], viz. synthesis of chlorosilanes, hydrolysis of chlorosilanes and condensation polymerisation.

The synthesis of chlorosilanes proceeds in a fluidised bed reactor by passing chloromethane (CH_3Cl) through slightly pressurised, heated powder or small particles of silicon (20–300 μm) at 275 °C in the presence of a copper catalyst containing other metals, e.g. zinc, tin and phosphorus as a promoter [157]:

$$Si(s) + 2CH_3Cl(g) \rightarrow (CH_3)_2SiCl_2(g) \qquad \Delta H^0_{rxn} = -282.8\frac{kJ}{mol\ Si}$$

The reaction is not very selective, and the product mixture contains CH_3SiCl_3, $(CH_3)_2SiCl_2$ and $(CH_3)_3SiCl$, although dimethyl chlorosilane is the main product (70–90% depending on the operational conditions applied). Careful distillation of this mixture produces pure fractions of each chlorosilane [157].

Dichloro silanes, R_2SiCl_2, are hydrolysed to a silanediol

$$R_2SiCl_2 + 2H_2O \rightarrow R_2Si(OH)_2 + 2HCl$$

which may then oligomerise:

$$R_2Si(OH)_2 \rightarrow HO[R_2SiO]_nH + (n-1)H_2O$$

The extent of the oligomerisation depends on the water content in the reaction mixture and temperature. Long-chain polymers can be formed at a relatively low temperature in the presence of an acid catalyst.

$$HO[R_2SiO]_mH + HO[R_2SiO]_nH \rightarrow HO[R_2SiO]_{m+n}H\ (\text{silicone fluids}) + H_2O$$

The values of (m and n) typically range between 2000 and 4000. To form gels, elastomers and resins, the polymer is cross-linked [157].

7.5.4 Direct Greenhouse Gas Emissions from the Production of Silicon

The production of silicon and ferrosilicon has associated emissions, and the management of these emissions is challenging. Mitigation efforts are often hindered due to a lack of sufficient information about emission types and sources. Measurement and mitigation techniques are often scarce, ill-specified and dispersed [158]. The emissions include greenhouse gases, particulate matter in the form of dust, polyaromatic hydrocarbons and heavy metals. These emissions and their origins on the plant are outlined in Table 7.13.

Carbonaceous reductants like coal, coke and bio-carbon are used in the production of silicon and related alloys. The reduction of quartz to MG-Si creates carbon monoxide gas [132], which is converted to carbon dioxide via the Boudouard reaction at the furnace charge top, which is either open or semi-closed [158].

$$2CO \rightarrow CO_2 + C$$

Other gaseous by-products, such as methane and volatile hydrocarbons, are generated during the combustion of these carbonaceous materials and the degradation of the electrodes.

Table 7.13 Typical emission types and their origins on silicon (data from Kero et al. [158]).

Emission type	Source	Emission point
CO_2, CH_4 and other GHGs	Reductants, electrodes, carbonaceous materials	Submerged arc furnace (SAF)
NO_x	Combustion	SAF, tapping
SO_2	Reductants	SAF
Dioxin	Reductants	SAF
Polyaromatic hydrocarbons	Reductants, electrodes, carbonaceous materials	SAF, tapping
Mechanically and thermally generated particulate matter	Solid material handling, liquid alloy in contact with air, SAF charge top	SAF, tapping, refining, casting, conveyor belts, etc.
Heavy metals	All raw materials	SAF, tapping, refining

The extent of greenhouse gas emissions in the production of ferrosilicon is dependent on:

- *Alloy type:* The reduction of quartz will produce stoichiometrically 3.14 tonne CO_2 per tonne Si, whereas the reduction of haematite to metallic iron will produce stoichiometrically 0.6 tonne CO_2 per tonne Fe. Further, the reduction of quartz requires a higher heat input than the reduction of haematite. Thus, ferrosilicon alloys with a higher Si content will have more greenhouse gas emissions [158, 159], e.g. 2.5 tonne CO_2/tonne Si for 45% Si or 4.8 tonne CO_2/tonne Si for 90% Si [158].
- *Reductant material mix:* Different reductants are/can be used in the process, and the amount of carbon and volatile matter (VM) is dependent on the source of carbon. For instance, coke with a fixed carbon content of 84 wt.% and a VM content of 9.5 wt.% (the latter with a carbon content of 80%) will thus result in 3.36 tonne CO_2 per tonne of coke [158].
- *Furnace configuration and operation:* The operation and charging method have a strong influence on process emissions, particularly CH_4 and NO_x emissions. For example, it has been documented that evenly charging furnace feed material reduces emissions compared to batch-wise charging [158].

An important parameter in the production of MG-Si is the portion of incoming silicon in the quartz feed, which is ultimately tapped as Si metal. The reduction of tetra-valent Si in quartz using carbon favours the formation of carbon monoxide rather than carbon dioxide at the required temperatures [160]. Hence, the production of MG-Si, will stoichiometrically yield 3.14 tonne of CO_2 per tonne of tapped silicon. In reality, this is a minimum value as furnace operation has associated inefficiencies and losses which will increase the carbon consumption [160]. The production of MG-Si has been estimated to release 5.0 tonne CO_2/tonne Si [158].

The formation of thermal NO_x is the dominant route of NO_x production. The combustion of gaseous SiO both above the charge surface and in the taphole has the potential to increase the local temperature. Thus, the amount of SiO gas released and combusted affects NO_x formation. NO_x may also be formed during tapping, as an oxygen lance is used to open the tapping channel, increasing metal flow out of the furnace [161, 162]. The high temperatures provided by the oxygen lance (2000–2500 °C) supply the activation energy for thermal NO_x formation via the Zeldovich mechanism [163].

Emissions by the silicon industry can be reduced by changing furnace design and operation [164], in particular by: reducing the combustion temperature via active cooling of the main flame zone, mitigating the ejection of SiO-rich gas through the charge surface, or recycling the flue gas to reduce excess air above the electric charge.

The production of SG-Si is associated with a significant carbon footprint, if the process utilises fossil fuels and power obtained from fossil fuels. Hallam et al. [145] estimated through a lifecycle assessment study, that the total carbon dioxide equivalent emissions for the production of SG-Si, including casting, slicing and wafering ranges between 72.1 and 168 tonne $CO_{2,eq}$/tonne silicon. The production of metallurgical silicon would only contribute 13.1 tonne $CO_{2,eq}$/tonne Si (noting that the production of 1 tonne of SG-Si requires 1.4 tonne of MG-Si).

7.5.5 Potential for Recycle

During the fabrication of PV modules, cutting of large Si crystals occurs. This is done by employing multi-wire slicing technology [165], resulting in the formation of silicon kerf, which can amount to 40% of the polysilicon used. The silicon kerf is currently not recycled as purification of the recovered material is non-trivial. However, recycling could reduce energy consumption and greenhouse gas emissions to produce MG-Si significantly [166]. However, the energy consumption for the formation of MG-Si only contributes 10–24% of the total energy required to make polycrystalline silicon, so this will be a relatively minor saving. Nevertheless, a substantial amount of research has gone into recycling [167] with recycled material being proposed to be used as silicon carbide and ceramics [167].

Similarly, recycling of silicon from end-of-life PV cells has not been developed with only precious metals such as copper and silver being recycled, but the rest typically going to landfill sites [168].

7.5.6 Making Silicon Production More Sustainable

The typical off-gas of a closed, electric, reduction furnace for the production of MG-Si contains 60–90% CO and 5–30% H_2 by volume [169]. This gas can be used as an energy source for combustion and the high temperature of the gas (200–700 °C) can be used to generate electricity via a Rankine cycle [149]. The off-gas CO_2-concentration from the carbothermic process can be up to 4% [170], and the recirculation of the flue gas may increase its concentrations [171]. This means carbon capture and storage (CCS) might be a suitable option as the cost can be significantly reduced if the CO_2-concentration is greater than 10% [170].

Carbon is an essential reagent to produce MG-Si, which may be provided using woodchips. Woodchips contribute to furnace operation by increasing the permeability of the charge, enabling the gaseous SiO to penetrate the charge evenly; this can increase its overall conversion. The replacement of some of the fossil-fuel-derived carbon with woodchips would reduce the carbon footprint of the process [170]. A key aspect of woodchip utilisation is its availability and cost. Unlike fossil carbon sources, the carbon content is lower in woodchips, implying that a larger volume is needed to achieve the same reduction potential. This impacts reactor loading, process efficiency, overall material costs, and may introduce impurities [170].

Traditionally many silicon and ferrosilicon plants have been positioned close to hydropower plants due to the need for the availability of a reliable supply of reasonably priced electricity [170]. Hydropower is the dominant source of renewable energy for newer silicon plants due to its quick start-up procedures. This is the case as silicon smelters are relatively small (compared to other metallurgical industries) and can be shut down without damage to production equipment [170]. Between 2000 and 2021, the production of silicon has risen by a factor of 2.6, primarily due to an increase in capacity in China. A decline from 76% to 50% in the portion of silicon produced by renewable sources has been observed in the same period [170], as the electricity grid in China is fed by fossil fuels.

In polycrystalline silicon production, metallurgical silicon is converted in an exothermic reaction to silanes in a fluidised bed reactor at around 300 °C and 1 bar. Adiabatic operation could maintain this temperature with some pre-heating of the feed (specifically anhydrous HCl). Lower temperatures are needed for the fractional distillation to purify trichlorosilane (see Table 7.12). Renewable implementation may come in the form of direct solar heating or electricity from renewable sources. The Siemens process requires a constant, reliable electricity supply. Hence, the Siemens reactor, which is heated via resistive heating, could be powered primarily using hydropower (which is commonly used where available).

7.5.7 Current Research Around Future Processes

Several initiatives are being developed as alternative processes to produce MG-Si, which are not based on the carbothermic reduction. A viable option is the use of other reducing agents, such as methane, in a plasma furnace. This would reduce, but not eliminate CO_2 process emissions [170].

As in the steel industry, hydrogen can principally be used as a reducing agent in the production of MG-Si for the reduction of SiO_2. However, hydrogen does not thermodynamically favour the reduction of silica, as extremely low ratios of water to hydrogen are required for the formation of either SiO or liquid silicon at temperatures below 1750 °C.

$$\mathrm{SiO_2(s) + H_2(g) \rightarrow SiO(g) + H_2O(g)} \qquad \Delta H^0_{\mathrm{rxn}} = 568.6 \frac{\mathrm{kJ}}{\mathrm{mol\ SiO_2}}$$

$$\mathrm{SiO_2(s) + 2H_2(g) \rightarrow Si(l) + 2H_2O(g)} \qquad \Delta H^0_{\mathrm{rxn}} = 475.7 \frac{\mathrm{kJ}}{\mathrm{mol\ SiO_2}}$$

At temperatures higher than 1750 °C, the formation of gaseous SiO is thermodynamically favoured over the formation of silicon [172]; the extent of reaction is low below this temperature unless performed at very low ratios of water to hydrogen. In the current process, SiO formed in the hotter parts of the arc furnace typically reacts with SiC formed in the cooler portion of the furnace to produce silicon. This route is not an option when using hydrogen as the reductant.

Studies have proposed using methane, as a reducing agent to overcome this barrier as the carbon present in methane reacts with SiO_2 to form SiC, ultimately forming Si metal [173].

$$2SiO_2(s) + CH_4(g) \rightarrow SiO(g) + Si(l) + CO(g) + 2H_2O(g) \quad \Delta H^0_{rxn} = 623.8 \frac{kJ}{mol\ SiO_2}$$

$$SiO_2(s) + CH_4(g) \rightarrow SiC(s) + 2H_2O(g) \quad \Delta H^0_{rxn} = 436.0 \frac{kJ}{mol\ SiO_2}$$

The challenge lies in reactor design, as condensation of SiO into SiO_2 and Si metal droplets leads to uncontrolled deposition that clogs the reactor components and disrupts gas flow [170]. This reduces process efficiency due to material losses and undesired thermal effects, which alters reaction kinetics and product purity. Additionally, the condensation process competes with the desired SiC or Si formation reactions, lowering silicon yield. Therefore, effective reactor design is crucial to maintain temperature control, optimise gas flow and minimise unintended condensation zones to improve process efficiency and scalability [170].

The electrochemical reduction in a molten salt medium or directly from molten oxides has been proposed as an alternative. However, this technology is still in its infancy [174, 175].

SG-Si may also be produced via a metallurgical route (not to be confused with the carbothermal method for MG-Si production), wherein SG-Si is directly produced from MG-Si [176].

The upgrade of MG-Si to SG-Si via pyrometallurgical routes has also been proposed [177]. Numerous pyrometallurgical treatments have been proposed to enhance the removal of boron and phosphorus from MG-Si, which is often a hindrance in traditional MG-Si production methods. Four methods have been proposed: slag treatment, chlorination, vacuum evaporation and solvent refining. Of these four methods, solvent refining proves most promising as an environmentally friendly alternative to produce low-cost SG-Si using MG-Si or Si waste as feed [177].

7.5.8 Outlook

Global silicon production was estimated at 9 million metric tonnes in 2023, and this value is estimated to increase, as the world shifts towards greener energy alternatives. Silicon being a key element in the manufacture of PV cells will see an increase in demand with a forecasted value of US$ 8.9 billion by 2027 [178]. The current carbothermal process to produce MG-Si is associated with large carbon emissions (minimum 3.14 tonne/tonne Si). This is aggravated by the so-called Scope 2 emissions associated with the high electrical energy demand. The silicon industry may experience challenges, such as negative public perception, if strategies are not put in place to mitigate the current Scope 1 and 2 emissions.

Such strategies, which will help reduce emissions, should include the use of reliable renewable energy sources such as hydropower (or even nuclear power), which can significantly reduce emissions and are easy to start up for processes like the production of MG-Si without causing damage to process units. Another strategy involves co-generation by utilising the thermal energy from the off-gas exiting the arc furnace in the MG-Si production process. Additionally, implementing the use of bio-carbon in place of traditional fossil carbon sources and carbon capture technologies may prove useful for emission reduction.

Other processes exist to produce high-purity silicon grades, but they are still in their infancy stage. Significant research still needs to be invested into these processes before commercial application [156].

7.6 Glass

Contribution by Julie-Ann Hoffman.

Glass is defined as a non-crystalline or amorphous solid which exhibits the phenomenon of glass transition [179]. Contrary to other solid materials that are crystalline, glass can be seen as 'a solid with a liquid-like structure' [180]. This is due to the non-periodic arrangement of atoms and molecules in its three-dimensional structure [181] (see Figure 7.17).

The glass transition occurs when a material undergoes cooling that is so rapid that crystallisation does not have sufficient time to occur. The relationship between liquid, crystal and glass can be described by the specific volume of the material as a function of temperature (see Figure 7.18). Crystallisation of a liquid occurs at its melting or freezing temperature (T_{melt}), which is associated with a step change in the specific volume. Glass formation can be described as when the liquid avoids crystallisation and goes to a super-cooled state, where the volume steadily decreases until the

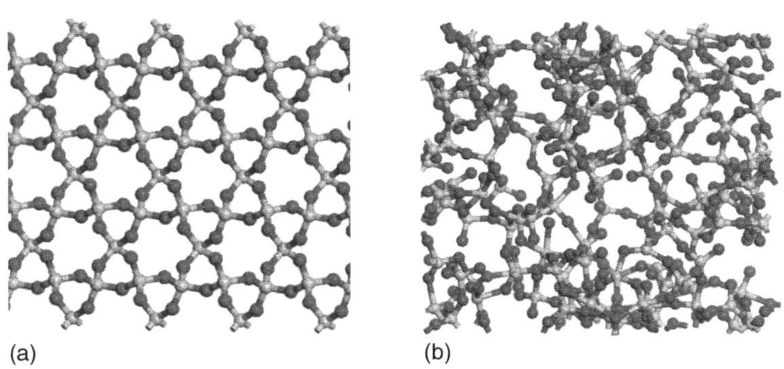

Figure 7.17 (a) Example of periodic arrangement of a crystalline structure of quartz compared to (b) the amorphous structure of glass (data from BIOVIA [94]).

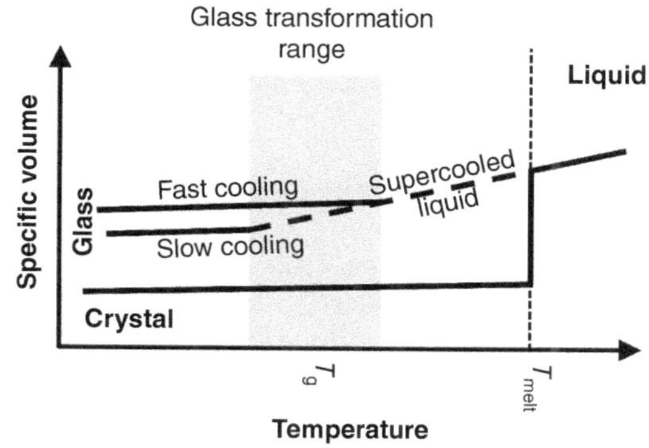

Figure 7.18 Specific volume as a function of temperature depicting the relationship between liquid, crystalline phase and glass (T_s, solidification temperature; T_g, glass transition temperature where the shaded area indicates the glass transformation range) (adapted from Hasanuzzaman et al. [182]).

expansion coefficient reaches a range of temperatures referred to as the glass transformation range. Within this range, the temperature at which the transition occurs is known as the glass transition temperature (T_g). At the glass transition temperature, the derivative of the specific volume with respect to temperature will show a step change. Notably, the glass transition temperature is dependent on the cooling rate. Thus, glass formed with a slow cooling rate results in glass with a lower specific volume, thus a higher density [182].

7.6.1 Introduction to Glass and Its Industry

Glasses are hard, brittle, transparent, chemically inert and are electrical and thermal insulators [183]. Glass is not comprised of a single compound, but rather a combination. Its primary component is silica or silicon dioxide (SiO_2) with alkali oxides (e.g. sodium oxide (Na_2O), potassium oxide (K_2O) and calcium oxide (CaO)) and metal oxides (e.g. lead oxide (PbO), boron trioxide (B_2O_3) and alumina (Al_2O_3)) [183]. The presence, type and quantity of these various oxides play a vital role in determining the characteristics of glass manufactured and its industry application [182]. Hence, glass can be divided into four main types based on its chemical composition [182, 184], viz. commercial glass, lead glass, borosilicate glass and aluminosilicate glass. Commercial glass or soda-lime glass makes up ca. 90% of all manufactured glass and is seen in daily products such as drinking glasses, bottles, food containers and windows. It is relatively inexpensive, has a smooth, non-reactive surface and transmits light well. Lead glass is hard and dense and has a high refractive index; it absorbs X-rays and is an excellent electrical insulator. Hence, it is used for decorative glass as well as electrical and shielding applications. Borosilicate glass is known for its chemical inertness and can be used at relatively high temperatures. It is used for lamps as well as chemical equipment. Aluminosilicate glass can also tolerate high temperatures (and does not soften much with temperature) and has a low thermal expansion coefficient. It is used in furnaces, combustion tubes, and as fibreglass insulation.

Table 7.14 gives an indication of the composition of the various glasses. In addition, various additives or modifying oxides such as zirconia (ZrO_2), alumina

Table 7.14 The chemical compositions of different manufactured glasses (compiled from References [182, 184]).

		Commercial glass	Lead glass	Borosilicate glass	Aluminosilicate glass
Composition (wt.%)	SiO_2	70–75	55–65	70–80	52–58
	Na_2O/K_2O	12–16	13–15	4–8	
	CaO	10–15			4–18
	B_2O_3			13–17	
	Al_2O_3			2–8	15–25
	PbO		18–38		

and titania (TiO_2) are used to improve the mechanical, chemical and thermal properties of glass. For example, alumina has been known to prevent the formation of cracks during porous glass production and improve the mouldability of the glass. Whilst, iron oxide (Fe_2O_3), chromia (Cr_2O_3) and cobalt oxide (CoO) are widely used as colouring additives that result in brown, green and blue hues, respectively [182].

Glass has been used by mankind for a long time. In more recent years, it has also played a key role in various low carbon emissions technologies, such as a main material for the manufacturing of solar panels and wind turbines, but also in the production of high-capacity batteries [185]. It has also been a useful material in the built environment, where it has been used in increasing the energy efficiency of the buildings.

The glass manufacturing industry consists of five main sectors, which is based on the approximate contribution to the total production volume of glass products (see Figure 7.19) [186]. Container glass is the largest sector. This sector includes jars and bottles used for packaging products in various industries such as food, beverage, cosmetics and pharmaceuticals. Flat glass is used in the construction industry (e.g. windows and facades), automotive applications (e.g. windscreens and sunroofs) and solar energy technology (e.g. photovoltaic plates). Fibre glass is used in a variety of industries such as insulation, textiles, composites, automotive and transportation applications. Speciality glass is comprised of products such as laboratory glassware, optical glass and technical glass used in electronic and telecommunication applications. The remainder of the total volume of glass products produced is accounted for by the smaller sectors such as domestic glassware (e.g. tableware, cookware and decorative items) and lighting (e.g. bulbs and tubes) [187].

Glass manufacturing plants often house multiple production lines. The layout typically includes space for raw material storage, furnace areas, forming and annealing lines, quality control laboratories and warehousing facilities for finished products [188]. The size of a glass plant varies by the type/purpose of the glass being produced. An average container glass plant operates continuously and produces about 250 000 tonnes of products per annum [189]. Whilst, in a typical float glass plant,

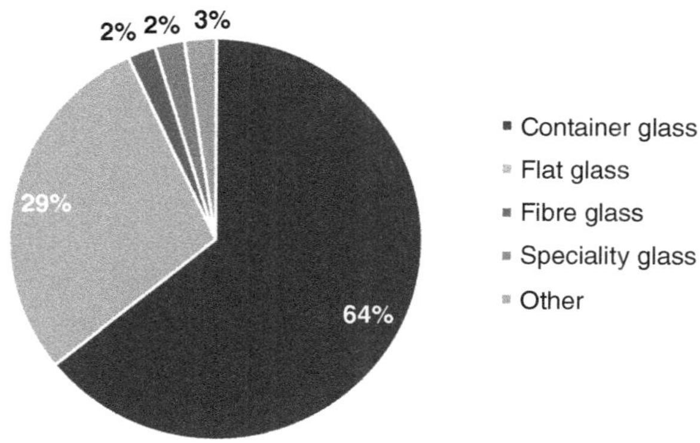

Figure 7.19 The different sectors of the glass manufacturing industry defined by the total volume of production in Europe in 2023 (adapted from Glass Alliance Europe [186]).

the process is fully automated, as such glass can be produced continuously for up to 15 years with approximately 6000 km of glass per annum [190].

The world's top producers of glass are China, North America and Europe. Germany is ranked as the top producer of glass in Europe [186]. The glass industry consists of about 14 700 companies and groups [191]. Globally, about 130 million tonnes of glass is produced per annum [192], with annual market growth of 4–5% per annum and annual sales near $200 billion. This positive market trend can be attributed to increasing urban growth and developments as well as the versatility of glass leading to its utilisation in multiple industries [193].

7.6.2 Current Process

In general, the glass manufacturing process consists of five main stages: raw materials and batch preparation, glass melting, forming, post-treatment and distribution and storage [185, 194]. Due to the size of the individual furnaces as well as the temperatures required in the plants, the process runs continuously and are typically only shut down for a period greater than a month when significant maintenance is required. A block flow diagram of the overall process is given in Figure 7.20.

Different raw materials are used depending on the required chemical composition of the glass. Apart from silica sand, various oxides are introduced into the glass manufacturing process via their raw materials as a carbonate, e.g. soda ash (Na_2CO_3) for Na_2O and limestone ($CaCO_3$) for CaO [194]. These carbonates will decompose at the high temperature used in the glass kiln.

The sourcing of the raw materials is dependent on the location and specific requirements of the glass manufacturer [195], although most of these materials are readily available worldwide. For instance, silica sand can be obtained from beaches, deserts and quarries.

Soda ash is a critical ingredient for commercial and lead glass making. China is the largest producer of soda ash followed by the United States and Turkey; they account for 81% of the soda ash production worldwide [20]. It can be mined in the form of trona, a mixed mineral of soda ash, sodium bicarbonate and water, with a world reserve of ca. 25 billion tonnes (ca. 1000 times the annual production) or produced

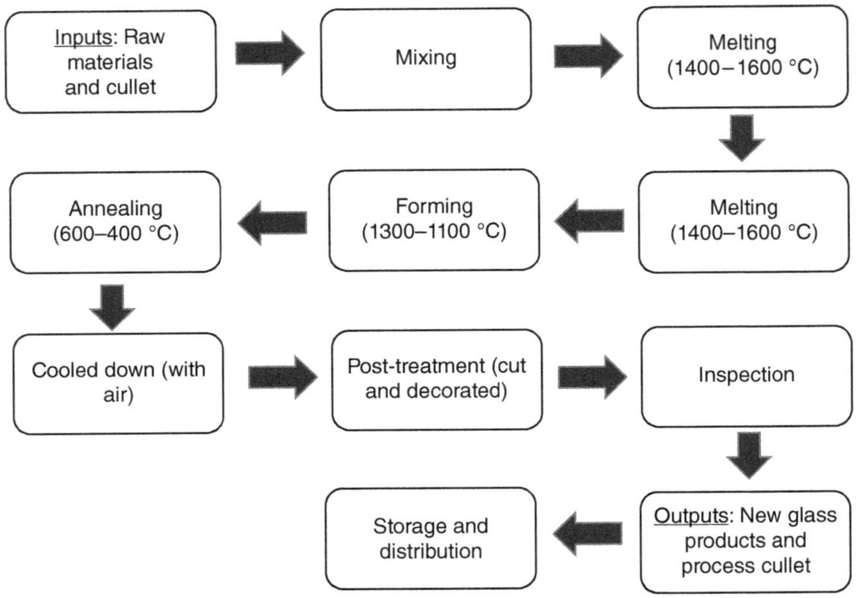

Figure 7.20 Block flow diagram representing the various steps in glass manufacturing.

via the Solvay process. Trona can be converted into soda ash by calcination:

$$2NaHCO_3(s) \rightarrow Na_2CO_3(s) + CO_2(g) + H_2O(g) \qquad \Delta H^0_{rxn} = 66.7\frac{kJ}{mol\ NaHCO_3}$$

Hence, the heat requirement for this process would amount to 1.15 GJ/tonne soda ash if the calcination takes place at 300 °C [196]. Thermodynamically, this conversion could take place at atmospheric pressure at temperatures as low as 125 °C [197], but a reduction in the calcination temperature to 200 °C results in only a slight decrease in the heat requirement to ca. 0.96 GJ/tonne soda ash. The temperature cannot be too high, as otherwise NaOH can be formed [196].

Soda ash can also be readily produced from brine via the Solvay process, if mineral deposits are not available. In 2000, almost 60% of all soda ash was produced using this process,

$$NaCl(aq) + NH_3(aq) + CO_2(g) + H_2O(l) \rightarrow NaHCO_3(s) + NH_4Cl(aq) \quad \Delta H^0_{rxn} = -125.5\frac{kJ}{mol\ NaCl}$$

Sodium bicarbonate is subsequently decomposed into sodium carbonate.

Limestone, to be used in the production of soda ash and as an additive in the production of commercial glass, is present worldwide in sufficient quantities [20] and is readily available. An alternative calcium source is dolomite (calcium–magnesium carbonate). Alumina used in the production of aluminosilicate glass is derived from bauxite ore using the Bayer process (see Section 7.2). Other metallics (e.g. cobalt oxide and iron oxides) are sourced in specific regions known for these minerals. These raw materials are crushed to the desired grain size. The materials are then washed; and in some cases, further chemical and physical processing takes place to reduce the content of the impurities to a specified level [195]. The processed raw materials are mixed together with cullet (broken glass and/or glass waste) [194], stored in silos and transported within the processing plant using conveyors and elevators powered using electrical energy. This is the least energy-intensive part of the glass manufacturing process [198], and can be made renewable through the implementation of renewable electricity. The material is batch pre-heated using the flue or exhaust gas from the furnace required for glass melting and forming, a valuable form of heat integration and energy saving.

The glass melting stage is the most essential part in glass manufacturing, which is also responsible for most of the energy consumption (approximately 70%) in this process. The molten glass must be heated in a furnace to temperatures of approximately 1600 °C to ensure that sufficient fining and uniformity are achieved [194]. During this process, the carbonates present in the raw materials decompose and release large amounts of CO_2.

$$Na_2CO_3 + nSiO_2 \rightarrow Na_2O \cdot nSiO_2 + CO_2$$

The heat of reaction may be estimated from:

$$Na_2CO_3(s) + SiO_2(s) \rightarrow Na_2SiO_3(s) + CO_2(g) \quad \Delta H^0_{rxn} = 80.6 \frac{kJ}{mol\ Na_2CO_3}$$

and similarly for the incorporation of calcium:

$$CaCO_3(s) + SiO_2(s) \rightarrow CaSiO_3(s) + CO_2(g) \quad \Delta H^0_{rxn} = 86.3 \frac{kJ}{mol\ CaCO_3}$$

As the temperature increases, the solubility of gases in the glass melt decreases which could lead to the glass melt becoming saturated with small gas bubbles [199]. To ensure the good quality of the glass, these gases must be removed from the melt. Thus, firing agents, most commonly sodium sulphate (Na_2SO_4), that only decompose at high temperatures, are introduced into the batch. Similarly to the decomposition of soda ash, the decomposition of sodium sulphate may result in the formation of sodium silicate:

$$Na_2SO_4(s) + SiO_2(s) \rightarrow Na_2SiO_3(s) + SO_2(g) + \frac{1}{2}O_2(g)$$
$$\Delta H^0_{rxn} = 437.3 \frac{kJ}{mol\ Na_2SO_4}$$

These produce large amounts of large gas bubbles that absorb the existing small gas bubbles, which are brought to the surface to be released. The sodium sulphate reacts with silica sand above 1200 °C and decomposes at temperatures above 1360 °C and ambient pressure to produce sulphur dioxide and oxygen [200, 201]. This is called refining of the glass.

Currently, 98% of glass is manufactured in continuous glass furnaces due to the size of the furnace as well as the difficulty with start-up – glass being a poor thermal conductor means that once the glass solidifies in the furnace, it is impractical (highly time- and energy-intensive) to restart the furnace. Classically, glass furnaces use natural gas combustion (see Figure 7.21) to supply heat using air, oxygen-enriched

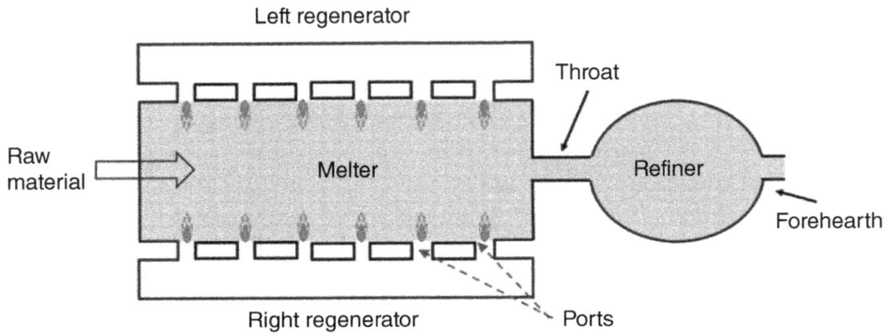

Figure 7.21 Top view of the structure of a fired glass melting furnace (redrawn from Moon and Lee [202]).

air or pure oxygen. The last part of the system is referred to as the refiner working end; its two main functions are to distribute the glass melt to their specific production lines and to cool down the melt to the required working temperature [203].

In the forming stage, the shape that the glass takes is defined, e.g. as container, flat glass or fibre glass. The molten glass is transported to the forming machine through forehearths, ensuring that the condition of the molten glass is at the necessary processing temperature. The glass cools down from about 1300 °C to 400–600 °C. In the container glass industry, the forming machines commonly used are known as individual section machines (IS-machines). Here, molten glass gobs are formed at the forehearth end, which drops into the moulds of the individual section (IS)-machines. The glass containers are formed either by the press-blow process, which is done by a metal plunger, or the blow-blow process using compressed air. The partially moulded glass or parison is then transferred to the finishing mould and blown into its final product using compressed air. A protective layer of tin oxide is sprayed on the finished containers by the hot-end coating unit. Lastly, the containers are transported on conveyor belts [184, 204]. This process takes less than 2% of the total energy required to produce glass containers (see Table 7.15).

Flat glass is processed via the float glass process. The forming machine is a float bath, where the molten glass is poured onto a molten tin bath and forms a continuous

Table 7.15 Indication of energy consumption (in GJ/ton) in glass manufacture (U.S. Department of Energy 2025 [205] / Public Domain).

	Container glass	**Flat glass**	**Glass wool**
Batching	0.72	0.72	0.72
Melting/refining	5.87	6.54	4.42
Forming	0.13	1.59	4.76
Finishing	0.59	2.33	2.11
Total	7.31	11.18	12.01

ribbon. The thickness of the glass ribbon is maintained by water-cooled top rollers. In this process, the molten glass cools from approx. 1100 °C to approx. 600 °C. To prevent the oxidation of the molten tin and ensure the quality of the surface of the flat glass, reducing conditions are employed by purging with a gas mixture of nitrogen and hydrogen. The hot-end coating unit for flat glass sprays on a protective coating of sulphur dioxide [184, 204]. The forming process accounts for ca. 15% of the total energy cost in the production of flat glass (see Table 7.15).

Glass fibres can be divided into two distinct types: continuous and discontinuous. Continuous glass fibres are made by the continuous fibre-drawing process. At the forehearth, they are divided into individual filaments via multiple small nozzles in platinum plates/bushings. The glass filaments are rapidly cooled by water nozzles, high air flows or water-cooled metal plates, and then drawn off by a rotating mechanism. The speed of the drawing determines the filament diameter. Discontinuous fibres can be produced by various processes. One such process is centrifugal fibre drawing, where the molten glass is poured into a hollow drum with multiple small holes in its outer wall. Centrifugal force acts through a rotating motion and numerous filaments are formed. Downward blowers or burners are installed to produce longer fibres. Various coatings can be applied to the glass fibres, depending on the intended use [184, 204]. The forming process is quite energy-intensive taking up ca. 40% of the total energy requirement to produce these fibres (see Table 7.15).

A post-treatment is necessary for all glass products produced. To reduce stresses, container glass and flat glass are cooled in an annealing furnace or 'lehr' under controlled conditions using a combustion air supply, after which, the final glass products are given a cold-end coating to reduce surface damage. Flat glass and glass fibre are cut into the required dimensions. The desired final glass products are then decorated. The final glass products undergo inspection, where all damaged or rejected glass products are returned to the beginning of the process as cullet [194].

7.6.3 Energy Requirement and Carbon Emissions

The thermodynamic minimum amount of energy required in the melting process [205] can be estimated from the composition of glass, the amount of carbonates being decomposed and the material's specific heat. Container glass containing 75 wt.% Si and 15% Na_2O and 10 wt.% CaO and assuming that it is a physical mixture of SiO_2, Na_2SiO_3 and $CaSiO_3$, will require at least 1.51 GJ/tonne (or 0.42 MWh/tonne). The actual energy consumption for the melting process is much higher (see Table 7.15), with glass melting and refining taking 4–6.5 GJ/tonne. This is due largely to the furnaces, which are utilised during the glass melting stage. The other main consumers of energy are the compressors, used for providing air, and the various operating machinery, e.g. conveyors and glass coating equipment. Natural gas is the dominant energy source for the glass industry, contributing 75–85% of the fuel consumed, electricity accounts for about 10–15% and the remaining 5–10% from other fossil fuels [185].

The glass manufacturing industry is responsible for about 1% of global industrial emissions and releases around 95 million tonnes of CO_2 annually [206, 207].

The glass production sector is regarded as an energy-intensive process. 1 tonne of container glass in a gas-fired furnace produces 0.56–0.60 tonne CO_2, where it is comprised of 0.38 tonne from natural gas combustion and 0.18–0.22 tonne from the dissociation of carbonates in the raw materials. The more energy-intensive production of flat glass and glass wool will have a higher carbon footprint, although the contribution due to the decomposition of carbonates remains constant if the glass composition remains unchanged.

Apart from CO_2 emissions released during the glass manufacturing process, toxic gases such as NO_x and SO_x are also released; the former is attributed to the high-temperature combustion during the glass melting stage (this can be reduced by using enriched air or pure oxygen), and the latter is attributed to fuel and fining stemming from sulphur-fined glass. Additionally, other impurities, present in lower quantities, such as potassium hydroxide (KOH) and toxic gases like hydrogen chloride (HCl), can also be released. Moreover, when manufacturing borosilicate glass, borate species, such as metaboric acid (HBO_2) can be produced [185, 194].

7.6.4 Potential for Recycle

Not only is glass up to 99% recyclable (dependent on the composition of glass [198]), but the use of glass waste decreases the energy consumed for the production of the raw materials needed for the glass production process, as well as removing the process CO_2 emissions. There are two strategies involved in the implementation of the glass waste recycling: 'open-loop' and 'close-loop' recycling. 'Open-loop' recycling involves the recovery of glass waste for the use as a raw material, but it specifically entails a slight decrease in quality or value of the glass after reprocessing, also referred to as down-cycling. This process will eventually produce waste that exits the recycling loop. 'Close-loop' recycling involves the utilisation of glass waste as a source for new glass production of the same value/quality [208], e.g. glass bottles being recycled into glass bottles/jars, this process may be repeated infinitely. Close-loop is the more attractive process (never generating waste), and glass's ability to be recycled in a closed-loop system is one of the distinct benefits of using glass over a variety of other materials. Use of 1 metric tonne of recycled glass saves 1.2 tonnes of new raw materials and avoids 60% of CO_2 emissions [185] from the removal of a need for the calcination and production of raw materials. Additionally, the use of cullet instead of new raw materials leads to a reduction in the required furnace temperature for melting by 200–400 °C, as the cullet acts as a catalytic former when new glass is formed – this results in energy savings of about 20% [190]. However, the major pitfall of glass recycling is its low financial value; primarily due to the high cost required for mechanical crushing and separation process as well as for transportation [208]. Hence, the industry has started gravitating towards integration of the reuse of glass products rather than recycling. Producing glass products for reuse would require a redesign of the products (requiring thicker glass products), focussing on multiple uses before its end-of-life. This can lead to further reduction of the energy consumption and related greenhouse gas emissions, though may come at some financial cost to the industry as fewer, if more valuable, products would be produced. For instance,

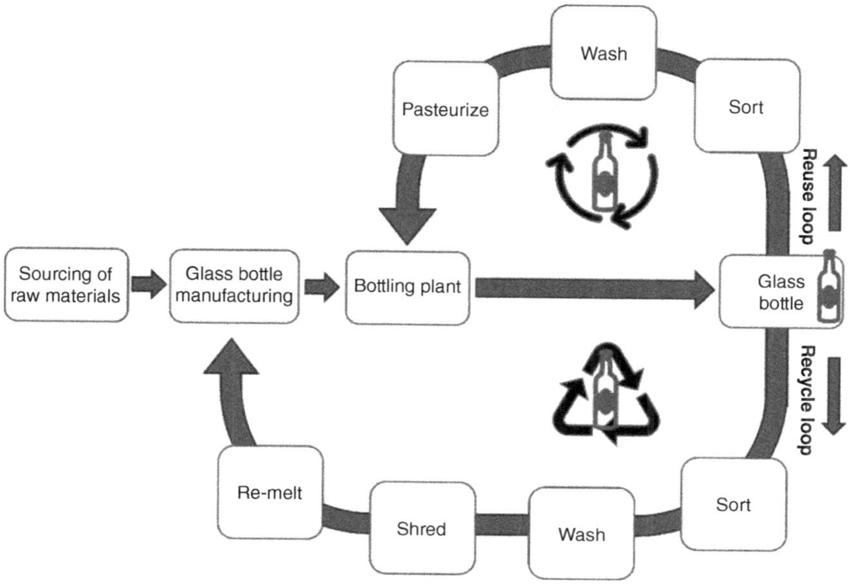

Figure 7.22 The reuse and recycling systems for the glass industry (adapted from Gheewala [209]).

it has been reported that reusing a bottle and then recycling it can save more than 90% of the energy required for glass production from new raw materials and will reduce the amount of glass waste in landfills (see Figure 7.22). However, the reuse of glass comes with drawbacks, such as logistics and contamination. In the food packaging industry, this can restrict the general reuse of products [209] as extensive cleaning may be required which is not always practical.

7.6.5 Making Glass Manufacture More Sustainable

The glass making process has a high carbon footprint due to the required reaction temperature, as well as the release of CO_2 from the raw materials (ca. 0.18–0.22 tonne CO_2/tonne commercial glass depending on its composition).

The furnace's exhaust gases are responsible for 25–30% of furnace energy loss via heat [210]. Whilst there is existing heat recovery in the use of the furnace's exhaust gas to pre-heat the raw materials, fuels and for steam generation on the plant (the so-called internal waste heat uses), there is further potential for the implementation of waste heat recovery. The so-called external uses involve generating electricity, via the Rankine cycle [184], which may be utilised within the plant or sold. This leads to energy savings and enhances thermal energy efficiency by using low- and medium-grade waste heat. Waste heat recovery is very promising; it has been found that the integration of this system on a flat glass line could reduce the energy consumption by 0.84 GJ per tonne of glass [211], i.e. ca. 8% of the total energy requirement for the flat glass production (see Table 7.15). This will not only reduce energy consumption but as a consequence, also reduce emissions of CO_2, NO_x, SO_2 and dust [211].

A large amount of energy is required for glass production (7–12 GJ/tonne), which is typically supplied by combusting natural gas. Hydrogen could replace fossil fuels in the glass manufacturing industry [212]. Principally conventional glass furnaces could be retrofitted to use hydrogen instead of natural gas [213]. Alternatively, renewable methane could be used, as its combustion product has a higher emissivity (see Chapter 4). Hydrogen (or methane) could be utilised in oxy-fuel burners, increasing heating efficiency. This would also prevent the formation of NO_x [214], but would require air separation.

Electric heating to heat glass furnaces is being discussed [215], which may be achieved by inserting electrodes in the glass melt. Joule heating by applying a voltage difference over electrodes can, in glasses, only work in high-temperature melts as the specific resistance of glass at room temperature is high (10^{10}–10^{14} Ω m; cf. resistivity of copper at 20 °C is $1.724 \cdot 10^{-8}$ Ω m), which drops to ca. 400 Ω m in liquid glass at temperatures of ca. 1400 °C [215]. This may be sufficient to keep the melt at a high temperature, but any decrease in temperature could limit the ability to increase the heat input to the melt. This may be circumvented by employing a hybrid design having both burners and electrodes for heating purposes, which could also assist in countering intermittency in the electricity supply. This would also render the process more adaptable to electric power from renewable energy sources. A major concern for electrical heating is the lifetime of the electrodes, as glass furnaces are kept at high temperatures for a long time.

7.6.6 Current Research Around Future Processes

Glass manufacture has a carbon footprint of ca. 0.6 tonne CO_2/tonne glass. Carbon capture technologies may be used [207, 216]; however, the integration of such systems with glass manufacturing has not been extensively studied [217]. However, the glass furnace has a relatively high CO_2-concentration in the flue gas at 7.9% on a dry basis [211].

Currently, the use of alternative furnaces and heating methods is a major focus for the glass industry due to its promising benefits of reducing fossil fuel consumption and thus reducing greenhouse gas emissions, some examples are hybrid furnaces, cross-fired or end-fired glass furnaces. The hot exhaust gas may be passed through regenerators and returned to the combustion air. Generally, these furnaces also have an electric boosting feature, in which electrodes are probed into the melt to provide additional energy [194]. Another alternative is the oxy-fuel furnace, which uses conventional fuel sources with oxygen instead of air for combustion. It mitigates the emitted flue gas by 60–70%, making it a valuable option for future carbon capture, utilisation and storage (CCUS) integration, and reduces energy loss by 25–35% compared to conventional furnaces [214].

Biofuels, such as biodiesel, can be used as an alternative fuel source to achieve the required high temperatures. However, biofuels may generate large amounts of greenhouse gases and are more costly than fossil fuels [218]. As such it has been suggested that biomass and waste should be utilised instead. It has been reported that biomethane could be used for the glass melting process and has the potential to reduce the net CO_2 emissions [219]. Further, the potential of biomass ash as an alternative to sand and limestone raw materials used for glass production has been shown [220].

Alternative heating with microwave or microwave-assisted technology has the potential to quickly and uniformly heat the glass melt whilst reducing the thermal stresses experienced by the glass melt [221]. This technology has been reported to reduce production time by 34% and energy consumption by 25% [221]. Plasma heating is another alternative way of rapidly heating the glass melt whilst reducing fossil fuel consumption and reducing greenhouse gas emissions, however, it requires large amounts of electricity as a power source and involves costly installations [222], meaning it will not be appropriate for retrofitting.

7.6.7 Outlook

Glass production is associated with a large amount of energy in the actual glass production (7.3–12.0 GJ/tonne glass), but also in the production of soda ash, the necessary flux additive in glass production. The latter would add an additional amount of energy of 0.29 GJ/tonne glass, using trona as the starting material to produce soda ash, and 2.49–3.49 GJ/tonne glass with soda ash obtained from the Solvay process, if the glass contains 15 wt.% Na_2O. This would thus bring the total energy consumption to produce glass to 7.6–15.5 GJ/tonne (or 2.11–4.30 MWh/tonne glass). Some parts of these processes could be electrified, and electrifying the main energy consumer, the melting furnace, is a developing technology.

Glass manufacture is associated with not only greenhouse gas emissions from the usage of fossil fuels, but also due to the process CO_2 emissions as a consequence of the high-temperature decomposition of carbonates added to the glass melt. The conversion of trona into soda ash is associated with the stoichiometric emission of 0.14 tonne CO_2/tonne Na_2CO_3, whereas the carbon footprint of the Solvay process is 1.84 tonne CO_2/tonne Na_2CO_3 [223], resulting in a significant increase in the total emissions, if the production of soda ash is taken within the battery limits of the glass manufacture process. In the glass manufacturing process, carbonates are decomposed, resulting in 0.18–0.22 tonne CO_2/tonne of glass depending on its composition. Hence, the overall process emissions of CO_2, i.e. due to the chemical requirements of the process, will amount to 0.20–0.70 tonne CO_2/tonne glass depending on the composition and the origin of soda ash. These emissions can be reduced by using the soda ash obtained from mining of trona rather than using soda ash from the Solvay process. Alternatively, the Solvay process may use CO_2 from an inevitable/renewable CO_2-source rather than producing it from calcium carbonate, thus reducing the CO_2 emissions associated with the Solvay process [224].

However, glass is almost entirely recyclable (and does not need to be downcycled as with, e.g. plastics) and is highly reusable (if the products are designed appropriately). This means that glass as a material may become more popular in the future, replacing currently single-use products, e.g. in food packaging. Glass is an essential material in the production of many forms of renewable energy, including being required to produce photovoltaic cells and wind turbines; this makes glass an important material for a renewable future, increasing the need to make the manufacture of glass as 'green' as possible.

7.7 Paper and Pulp Manufacturing

Paper has been used for thousands of years, and papermaking techniques, specifically papyrus, were evident in ancient Egypt by at least 2900 BCE [225]. In ca. 1 BCE, paper was produced in China, using a simplified version of today's process.

Initially, hemp, flax, cotton and recycled cloth (rags) were used to make paper. However, a global shortage of, particularly, cotton in the mid-1800s inspired the invention of different paper-making techniques. Charles Fenerty is usually credited with developing the wood pulp method of making paper in 1940 (publicised in 1944), but the first patent was filed in Germany by Friedrich Keller for a process that was developed independently [226].

Whilst modern paper production uses primarily wood fibre, fibres obtained from a wide variety of biological sources, including straw or grass, sugar cane bagasse, hemp and linen, amongst others, can also be used. Paper companies produce a wide range of products, including writing and printing paper, newsprint, packaging materials and personal or sanitary items such as tissues and toilet paper.

The paper and pulp industry uses both large amounts of energy and water, which are both potentially problematic.

7.7.1 Introduction to Paper and Pulp and Its Industry

Pulp is a fibrous material that is obtained from either wood or alternative cellulosic material, from which the lignan has been removed. Pulpwood can be obtained by either mechanically grinding and further processing of wood into a fibrous pulp or, alternatively, via chemical separation of the wood fibres. Although most types of wood can be used, the type of wood does affect the quality of the paper that is eventually produced from the pulp. Hardwoods, such as beech, oak and maple, have shorter fibres and produce smoother paper that is best for printing. Softwoods, such as cedar, pinewood, firwood and redwood, have longer fibres and are better for industrial papers. Both woods are typically 40–50% cellulose and ca. 20% hemicellulose. Hardwood typically contains slightly less lignin than softwood (thus sometimes slightly more hemicellulose), and there is a varying amount of extractives (gums, resins, starches, tannins, etc.) in both types, typically 1–10 wt.%, up to 30 wt.% [227].

Board may also be produced for packaging materials, besides paper used for printing and writing. Each paper or board product has its own set of standards that needs to be met, including weight, thickness, colour, brightness and strength. The requirements for paper may be set according to the ISO (International Standards Organisation) or ANSI (American National Standards Institute). Some tests used are given in Table 7.16.

Table 7.16 Paper testing (data from Fogra [228]).

Test	Units	Purpose of test
Thickness	μm	Choice of paper/adjustment of printing machines
Paper density/specific volume	g/cm^3 cm^3/g	Choice of paper to be used
Absolute moisture	mass %	Printing/finishing/physical behaviour affected by moisture content (dimensional change at higher moisture content)
Ash content		Indicative of chemical content of paper
Tensile strength	kN/m	Indicator of the resilience/stability of paper
Bending stiffness	mN m	Influences finishing process/operation of printing press
Splitting resistance		Resistance of paper to splitting perpendicular to sheet
Internal bond strength	J/m^2	Energy needed to split (tear) paper/board
Creasability		Suitability of boxboard for folding
Roughness	μm	Affects feel of paper (haptics)/further processing
Surface energy	mN/m	Indicator adherence of inks to paper
Optical characteristics		Colour, brightness, gloss and whiteness (manufacturers specifications)
Opacity		How well a sheet obscures any printing on a surface below
Printability & writability		Behaviour of ink on paper
Behaviour towards liquids and gases		Adsorption of water or oil by the paper; air permeability

World production of paper and paperboard has continued to increase steadily (see Figure 7.23), reaching 414 Mt. in 2022 [229]. However, the type of paper products being produced has changed substantially this century. Digital media has resulted in a drop in both newsprint (from 12% to 3% of the world's production over 20 years), and printing and writing paper with a drop in the global production by 30% in the last decade [230]. In contrast, paper packaging and board production had increased significantly due to the take-off of e-commerce and represented 65% of the global paper production in 2022 [230].

The world production of pulp (both integrated and market pulp) amounted to 181.8 million tonnes in 2022 with the production relatively equally distributed between North America (30%), Asia (26%), Europe (24%) and Latin America (18%) (see Figure 7.24) produce the highest amounts of pulp, followed by Latin America. Total paper and board production was 425.3 million tonne in 2022. The difference between the paper and board production and the pulp production is due to the paper from recycling and non-fibrous materials used in the paper production.

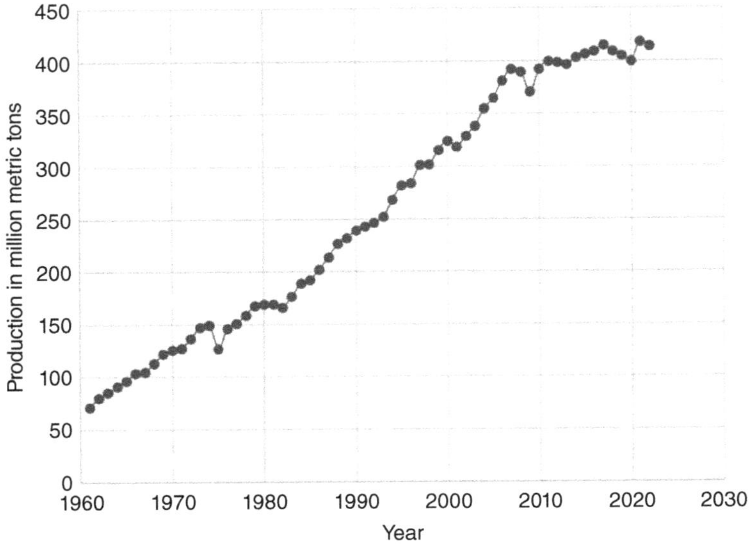

Figure 7.23 Paper and paperboard production from 1961 to 2022 (data from Statista [229]).

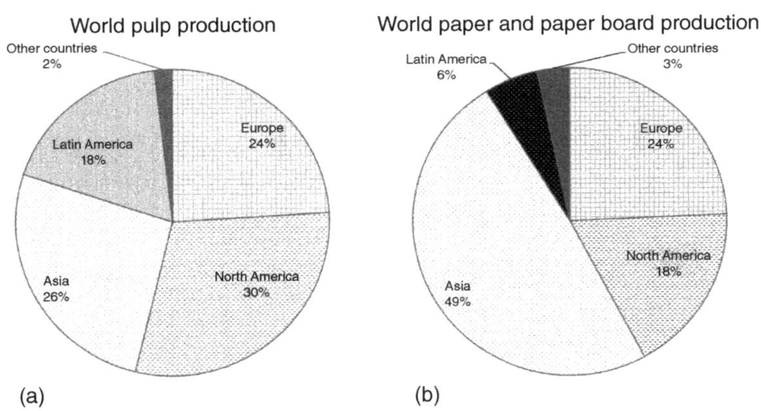

Figure 7.24 Distribution of worldwide pulp (a) and paper and board (b) production in 2022. Total pulp production 181.8 million tonne; total paper and board production 425.3 million tonne (data from Cepi [231]).

Asia had the largest production of paper and paperboard, followed by Europe and North America, with only small amounts produced elsewhere.

The size of the individual operations employed in the paper and pulp industry is extremely varied, with both small to medium-sized and large corporations being present [232]. As early as 2000, more than half of the products in the paper and pulp industry were produced in facilities (mills) with capacities of over 450 tonnes per year [233]. The largest pulp and paper manufacturers (given in Table 7.17) have

Table 7.17 Top pulp and paper manufacturers in 2023 (production in Mio tonnes per year).

Company	Location	Value[a] (2023)	Mills	Pulp	Paper	Packaging	Reference
Smurfit Westrock[b]	Ireland	20[b]	62		24	20	[234]
International Paper	USA	23.3				12.5	[235]
UPM	Finland	11.4	15	5.8	4.6		[236]
Oji Holdings Corporation	Japan	10.9		2.5			[237]
Stora Enso	Finland	10.3	46		0.75	5.3[c]	[238]
Mondi	UK	8		3.6	2.22	2.38	[239]
Nine Dragons Paper Holdings	China	7.8		4	21.7		[237, 240]
Nippon Paper Industries	Japan	7.4	22				[241]
Sappi Limited	South Africa	5.8	19	4.1	5.5		[237, 242]
DS Smith	UK		14			4	[237, 243]

a) In billion US$
b) Smurfit Westrock, formed from the merger of WestRock and Smurfit Kappa (established 2024 and value indicated for 2024).
c) Board and corrugated packaging.

production capacities in the millions of tonnes per year. These total productions are typically spread over several mills and various geographic locations.

Paper and pulp mills require a fibre source as well as access to water and energy supplies. The fibre source is typically obtained from wood, although recycled paper can also serve as a fibre source. Currently, forests (see Figure 7.25) cover approximately 25% of the earth's land area, or one-third of the habitable land area [244]. The density of forests is not even across the globe, with more forests present in Asia, North America and South America. Unsurprisingly, the locations with the highest density of trees are also the areas that traditionally produce the most paper products. In 2020, of the 4.05 billion ha of forest, 3.75 billion ha were naturally regenerating and primary forest, whilst 0.3 billion ha were planted forest [245].

7.7.2 Current Processes

The overall pulp and paper process is multifaceted, as illustrated in Figure 7.26. Most process options in the process are shown with their connections to other parts of the overall system. Note, however, that specific plants will not necessarily contain all the possible units. The two primary sections of the process are the pulping and the papermaking. The raw material, which is typically wood (but can be other materials such as straw or bagasse), must first be pre-treated to make it suitable for pulping. The pre-treatment can involve debarking of the wood if it comes from harvested trees, and chipping to decrease the size. The chips are sent to pulp production, but the bark is removed and potentially used as a fuel source in the process.

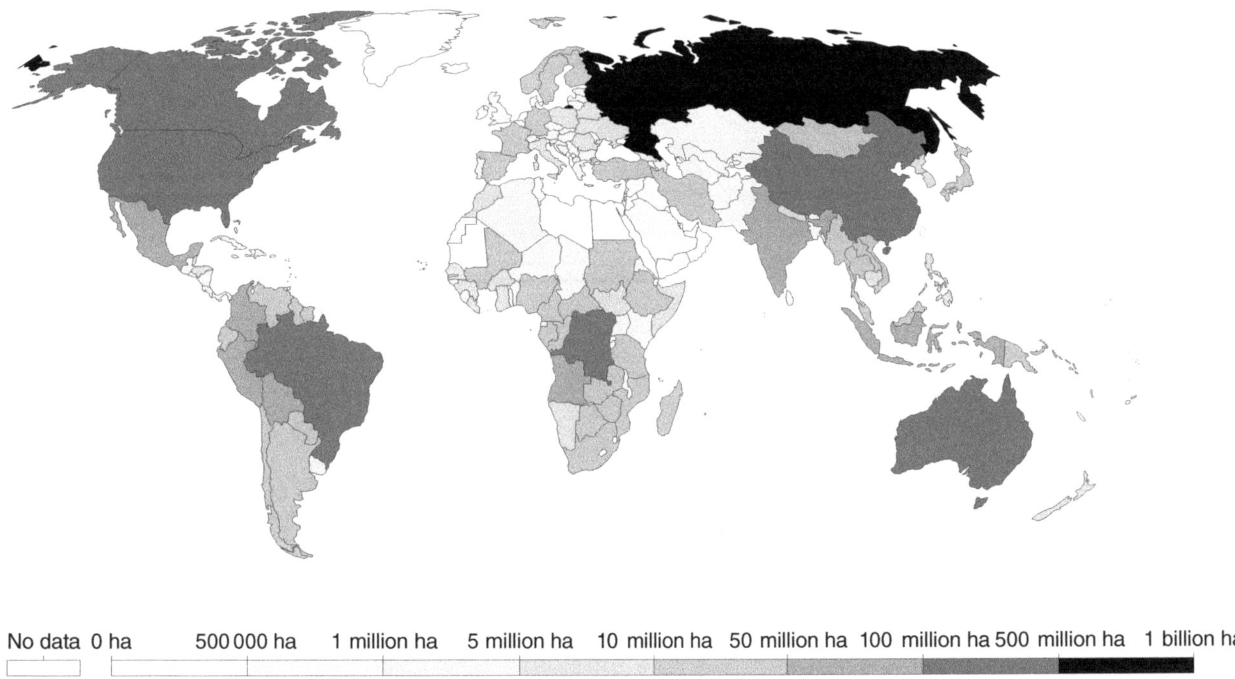

Figure 7.25 Forest area in the world in 2020 (Ritchie [244] using data from FAO [245], ‘creative commons: share of global forest area, 2020’ licensed under CC BY 4.0).

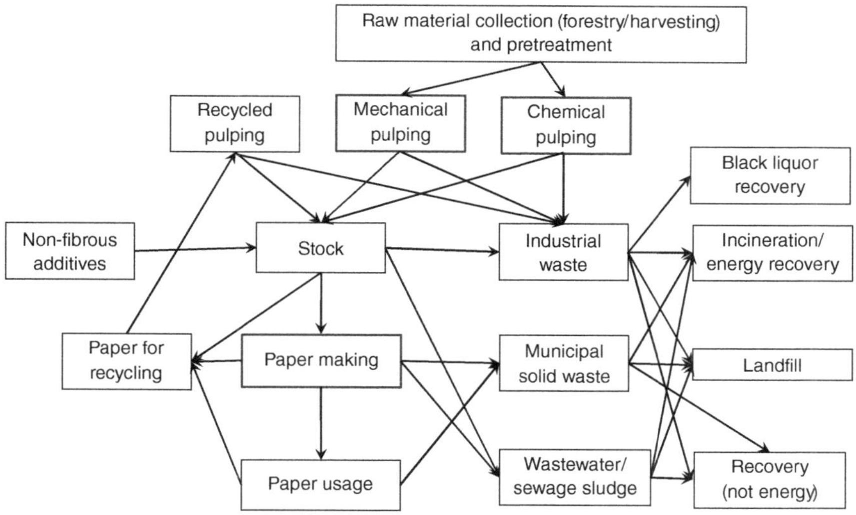

Figure 7.26 Paper and pulp process (compiled from References [246–248]).

The wood chips are sent to the pulping processes where the lignin is removed and the individual fibres separated. There are various options to pulp wood, which can be roughly separated into chemical pulping, mechanical pulping or a combination of the two. Chemical pulping processes, in which the chips are treated at high temperature and pressure in an aqueous solution which digests the pulp, is the more common process (see Figure 7.27). It includes the Kraft process, the sulphite process and the soda process [246]. In mechanical pulping processes, the chips are ground or shredded at elevated temperatures without the addition of chemicals. Only slightly more than 10% of all wood used for papermaking undergoes the mechanical pulping process (see Figure 7.27).

Pulp may also be produced by recycling used paper and packaging materials. Used paper and board must be separated from any unwanted materials, such as metal, plastic, glass or food residues. The paper is mixed with hot water whilst mixing/agitating to separate the fibres and create a suspension. Smaller impurities may be removed from the suspension. This pulp may then be further processed. Pulp from recycled paper accounted for 44% of the papermaking (see Figure 7.27).

The pulp from any of the synthesis routes is then further treated to obtain the grade necessary for the specific paper that will be produced. The pulp treatment includes washing to remove any undesired soluble compounds, screening to remove oversized and unwanted particles and bleaching to allow the production of paper that is bright and soft as well as absorbent [248]. The pulp may be dehydrated for storage or transported as suspensions directly to the papermaking process.

A mixture of pulp and additives, also called stock, is fed to the paper machines. The additives are determined by the quality of the pulp and the required properties of the paper or board that will be produced. During the stock preparation, in addition to washing and separating out the desired fraction, the pulp can be bleached to give the fibres the correct brightness and colour.

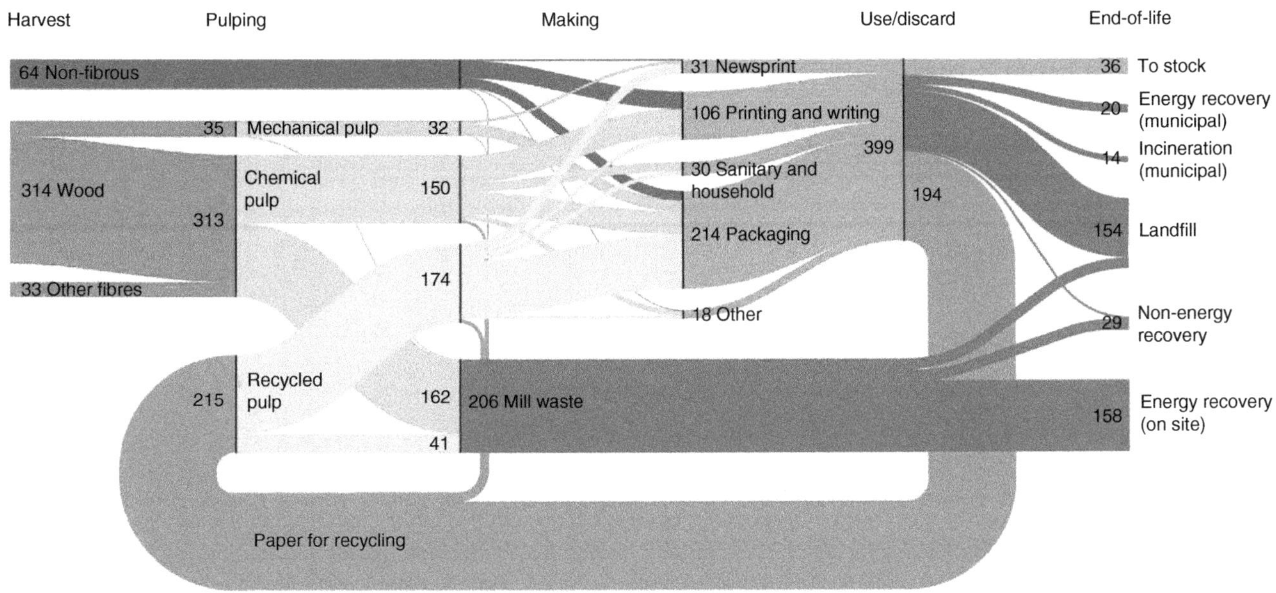

Figure 7.27 Global paper flows in 2012 in million tonnes (Van Ewijk et al. [247] / John Wiley & Sons / CC BY 4.0).

The clean bleached stock is converted into wet paper using a forming device. The wet paper sheets are dried by first passing them through presses to force out as much water as possible. Most of the remaining water is removed by passing the paper sheet through rotating drying cylinders. The machines used to make the paper can have manufacturing capacities of over 300 tonnes/day [249].

The paper industry is associated with large waste streams, with the waste stream being ca. 50% of the papermaking stream (see Figure 7.27). The waste stream is mostly used to recover energy from the process (in 2012, 77% of the mass flow rate of the waste stream was used for energy recovery; see Figure 7.27).

The Kraft process (see Figure 7.28) is the most used chemical pulping process and is used for 75% of the overall pulp production [248]. In the Kraft process, wood chips are treated at high temperature (160–180 °C [233, 251]) and pressure in a concentrated solution of both sodium hydroxide (NaOH) and sodium sulphide (Na_2S), known as white liquor. The process, referred to as either cooking or digesting, results in a breakdown of the bonds in the lignin compounds and can result in up to 95% of the lignin being removed from the wood [252]. Kraft mills convert, on average, 50% of the wood used into pulp [232] (see also Figure 7.27). After digestion, the pulp is separated from the used liquor (black liquor). The Kraft process has advantages in the ease of chemical recovery and in the strength of paper produced.

$$\text{NaOH} + \text{NaS}_2 + \text{wood} \rightarrow \text{Na} - \text{org.} + \text{S} - \text{org.} + \text{NaHS}$$

The black liquor, a by-product of the chemical process, contains wood waste, chemicals and other impurities. Given its high energy content, the liquor is typically burned in recovery boilers (as seen in Kraft mills), thereby producing much, if not all, of the steam needed to drive the steam turbines and produce power. The amount of power generated in this manner is generally more than the energy needed for the chemical pulping process [252]. The additional energy may, be used in paper manufacturing. The black liquor that is separated from the pulp contains a high amount of water, so to enable efficient combustion, the black liquor must first be concentrated by evaporating the excess water.

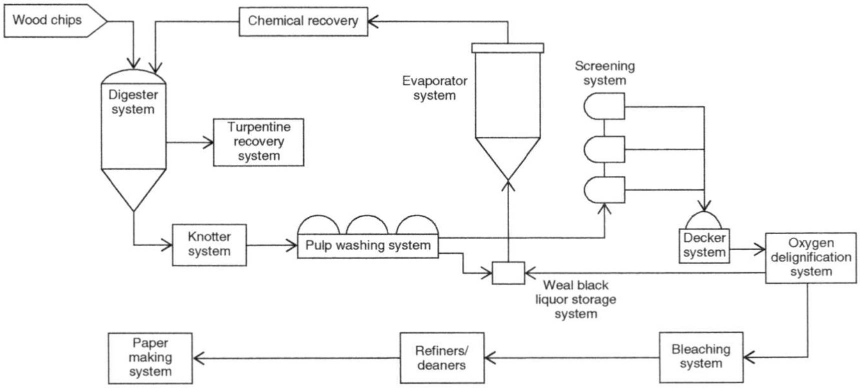

Figure 7.28 Kraft process (redrawn from References [246, 250]).

Instead of burning the black liquor to produce energy, extracting the valuable chemicals contained therein is possible. One of the methods commonly used to extract products from black liquor is membrane separation, which can be tailored by using different membrane cut-offs [253]. The primary objective of membrane separation is to remove the lignin from the liquor. The delignified black liquor may then be used to obtain organic acids, recover the inorganic salts used in the Kraft process, or may be used as a fermentation medium.

7.7.3 Energy Usage and Generation

The paper and pulp industry, as a broad category, is one of the largest industrial consumers of energy, superseded by the chemical and petrochemical, iron and steel, and non-metallic mineral industries [252]. The reported total energy that is required to produce paper varies considerably according to the source, with values between 11 and 35 GJ/tonne of paper [252, 254, 255].

Of the two primary sections of the process, the pulp making requires the majority of the energy use (62%) [256]. The chemical pulping methods, particularly the Kraft process, require less energy than the mechanical ones. Understandably, the energy required for the pulping of recycled paper is much lower than any of the methods using virgin wood/organic matter. The energy required for the individual steps in the production process is indicated in Table 7.18. Most noticeable is the fact that the different process options require significantly different amounts of energy.

Table 7.18 Estimated energy consumption in pulp and paper manufacturing (GJ/tonne) (energy consumption given for different manufacturing steps [252] using information from Bajpai [250]).

Wood preparation		**Paper and paperboard production**	
Debarking	0.11	Paper refining and screening	0.89
Chipping and conveying	0.37	Newsprint FPF[a)]	1.52
Pulping		Newsprint drying	4.4
Kraft (chemical)	2.75	Tissue FPF[a)]	1.92
Sulphite (chemical)	5.68	Tissue drying	8.39
Semi-chemical	4.08	Uncoated paper FPF[a)]	1.9
Mechanical	8.11	Uncoated Paper drying	5.38
Stone ground wood	5.4	Coated paper FPF[a)]	1.9
RMP (refiner mechanical pulping)	6.44	Coated paper drying	5.6
TMP (thermomechanical pulping)	7.49	Line board FPF[a)]	0.97
CTMP (chemi-thermo-mechanical pulping)	8.11	Line board drying	4.28
Recycled paper pulping	1.37	**Kraft chemical recovery**	
Bleaching		Evaporation	4.08
Pulp bleaching	2.43	Recovery boilers	1.19
		Recausticisation	1.08
		Calcination	2.14

a) FPF stands for forming, pressing and finishing.

Table 7.19 Fresh water consumption for paper production (Olejnik [249]; using data from Environmental Technology Best Practice Programme [257]).

Paper grade	Fresh water consumption (m^3/t of paper)
Tissue	5–30
Printing and writing	10–50
Specialty	20–80
Newsprint	15–30
Packaging (virgin and recycled pulps)	6–45
Corrugated boards	6–40

Much of the energy used in the paper and pulp industry is self-generated by burning the black liquor to generate steam. This implies that modelling the exact energy input can be challenging, as the feedstock wood is used both for generating the desired product, pulping and papermaking and, its waste, for the generation of energy required for the production. For instance, in chemical pulping (Kraft Process), half of the wood effectively becomes fuel for the chemical recovery part of the process [254].

7.7.4 Water Usage

The paper and pulp industry requires copious amounts of water for pulping and papermaking. The industry is the third largest industrial user of water, exceeded only by the metal and chemical industries [232].

Water is used at various stages in the overall process, starting with debarking of raw wood, where hydraulic debarking may be used. Next is pulping where a large amount of water is needed to create the needed suspension of fibres in water, with additional washing and bleaching of the pulp. The stock used for papermaking also contains a large amount of water, which may need to be added if the plant uses dry pulp.

The total amount of water that is used to manufacture paper depends not only on the specific process used but also on the type of paper being made. An estimate of the fresh water consumption for different types of paper is shown in Table 7.19. In manufacturing, an additional 100–150 m^3 of process (recycled) water will be needed per tonne of paper produced [249]. The distribution of fresh water consumption for pulp and paper production is shown in Table 7.20. The industry's high water usage subsequently generates high volumes of wastewater, even where there is significant internal recycling of the water on the plant.

7.7.5 Emissions

Emissions from the pulp and paper industry come from different places: during the extraction of raw and natural materials, during product manufacture, during product use and from disposal and/or destruction of product [232].

Table 7.20 Distribution of fresh water consumption in pulp and paper manufacture.

Process stage	Percentage of fresh water usage
Pulp production	40
Paper machine	35
Raw material preparation	15
Stock preparation	5
Utilities	5

Both sections of the process, pulping and papermaking, result in high levels of emissions, with 1 tonne of paper production from wood resulting in between 0.95 and 2.96 tonne $CO_{2,eq}$ greenhouse gases (both CO_2 and methane) [256, 258]. Pulp making generates the higher amount of greenhouse gases with the amount of greenhouse emissions depending on the specific pulping method used. As examples, the Kraft process generates 0.508 tonne $CO_{2,eq}$/tonne paper; the chemi-mechanical method 0.513 tonne $CO_{2,eq}$- /tonne paper; and recycled pulp, 0.408 tonne $CO_{2,eq}$/tonne paper [256]. Besides the CO_2 emissions from the process itself, there are additional CO_2 emissions associated with electricity generation, chemical manufacture and transport [232, 258].

Besides greenhouse gases, such as CO_2, emissions from paper and pulp mills may contain lignin degradation products, such as acrolein, acetaldehyde, o-cresol, carbon tetrachloride, chloroform, cumene, formaldehyde, methanol, methylene chloride, methyl ethyl ketone, phenol, propionaldehyde, 1,2,4-trichlorobenzene and o-xylene [246]. These chemicals all pose public health hazards, and their emissions must, as far as possible, be limited if not eliminated.

Paper and pulp production produces large amounts of wastewater that, if untreated, will pollute the water system. The pollutants can include non-biodegradable organic compounds, adsorbable organic halogens, phenolic compounds, etc. [259]. These compounds can have negative effects on the health of people in the surrounding areas and can negatively affect fresh water and marine ecosystems. When reporting on the level of contamination of water, the two amounts typically given are the AOX (adsorbable organically bound halogens; adsorbable organic halogens in a contaminated water environment: chlorine, bromine and iodine) and the COD (chemical oxygen demand: the amount of oxygen needed for complete oxidation).

Treatment of the wastewater is problematic due to the large variety of contaminants found in the wastewater and the complexity of the treatment methods needed to deal with them [260].

7.7.6 Making the Pulp and Paper Industry More Sustainable

The overall objectives of the paper and pulp industry are to create environmentally responsible, high-quality products [261]. The primary raw material for paper, wood, is intrinsically renewable, but forestation to provide the raw materials for this industry should be carried out in a sustainable manner. The mill should maximise pulp yield, whilst producing an easily recyclable product [248], as the energy consumption can be reduced significantly by utilising recycled paper to produce pulp. Furthermore, using recycled paper reduces water and chemicals usage in the process and produces less pollution than when using virgin wood. Recycled paper is already largely employed in the EU, where more than half of the pulp used is made from it. Paper can be recycled readily and a recycling rate of 78% is achievable [252]. The wood fibres found in reused paper can be recycled five or six times before they are damaged to the extent that can no longer bind together [261]. This recyclability of paper makes paper as a product more sustainable. However, processing recycled paper requires an additional processing step, deinking.

Large amounts of water are used in pulp and paper manufacture, and paper and pulp mills should minimise fresh water intake by recycling process water through adequate treatment methods, although some water losses due to evaporation and spillage will be unavoidable. There are additional benefits to recycling the wastewater, besides the decrease in fresh water consumption, namely that the energy requirements and the cost of effluent treatment will be reduced. There are a number of water conservation methods that may be employed in a pulp mill [248] by using:

- treated effluent for raw material washing
- paper machine backwater in the pulp dilution in the unbleached tower
- backwater in centri-cleaning of pulp and vacuum pump sealing
- improving washing efficiency of pulp washers
- recycling of bleach plant filtrate for pulp dilution to the preceding stage.

The pulp and paper mill do generate a waste stream, but those streams should be minimised. The product should also be safely combusted, as it can only be recycled a limited number of times. Thus, it should be disposed of at the end of its lifetime. Disposal of waste products by incineration is a well-established method within the industry, which offers the opportunity for energy recovery for the overall process. An important aspect of the paper and pulp industry is to maximise energy recovery from waste streams, whilst at the same time minimising the size of these streams.

7.7.7 Current Research Around Future Processes

There are various technologies under investigation, for making the paper and pulp manufacturing process more sustainable. Research is concentrated on minimising energy and fresh water usage as well as decreasing emissions [232, 262].

Current research is focused on a number of different sections of the paper and pulp process, including raw material pre-treatment, pulping technology, papermaking technologies, paper drying technologies and by-product/biomass utilisation [232, 262]. A detailed description of the process innovations for making the pulp and paper manufacturing industry more sustainable is given by Furszyfer Del Rio et al. [232].

For the pre-treatment stage, there are various options that can improve the pulping efficiency. Amongst these, microwave treatment reduces the energy requirement whilst increasing the pulp yield by 40%, although significant capital investment is needed [232]. Both chemical pre-treatment with oxalic acid and biological pre-treatment can decrease the mechanical energy required whilst decreasing resin and pitch, thereby enhancing the paper strength [232]. In pulping, both directed green liquor utilisation and steam cycle washing have been shown to decrease the energy requirement, with steam cycle washing also needing less water (2.5 to 3 tonnes/tonne oven-dry pulp) than conventional washing technology (3 to 4 tonnes/tonne oven-dry pulp) [262].

Many methods have been investigated to improve the efficiency of paper drying, including microwave drying (20% decrease in energy consumption) and using gas-fired drying.

In addition to researching ways of improving the pulp and paper production process, various technologies are available, which can be integrated into paper and pulp mills to enhance their performance and minimise the environmental impact (Table 7.21).

7.7.8 Outlook for the Paper and Pulp Industry

Paper has been in existence for millennia and has seen many changes over this time. Most recently, the introduction of digital media has meant that the use of paper for books and writing has decreased. However, at the same time, the focus on environmentally sustainable processes together with the introduction of e-commerce has meant that the use of paper-based packaging materials, which are biodegradable and recyclable (multiple times), has seen a substantial increase. Overall, this means that the demand for the products as a whole continues to grow, with current global production at over 400 million metric tonnes per annum.

Although the industry traditionally had high energy usage, specifically to produce pulp, this is now largely balanced with secondary product usage (black liquor combustion) to produce heat, although only half of the wood is used in the final paper product (e.g. with the Kraft process). This raises the question of whether the process is using more raw materials than alternative processes, as the energy generated from the black liquor replaces other fuel sources (e.g. wood) to such an extent that the overall process has become less resource-intensive.

Modern mills have the capability for recycling water and extraction of chemicals from the production cycle, and they can expand their portfolio by including export of heat, generating electricity and extracting chemicals for use outside the paper mill. Careful recycling of the chemicals also means that the overall discharge to the environment is reduced.

Table 7.21 Technologies that can be integrated into pulp and paper mills (Kong et al. [232] compiled from References [262–265]).

Technology	Product	Benefits
Thermal gasification and synthesis	Heat, electricity and/or synthesis gas can be turned into fuels and chemicals such as ethanol, methane, hydrogen methanol, diesel and dimethyl ether	Gasification is flexible in terms of final product and raw materials. The benefits of employing BLG consists in reducing electricity and steam production and increases the capacity of pulp production
Production of ethanol through enzymatic hydrolysis of cellulose or acid followed by fermentation	By-products and ethanol (e.g. lignin and acetic acid)	Reduces electricity and steam production
Lignin extraction and refining	Solid fuels, lignin oil, dispersants additives for food production, phenols, carbon fibres, sorbents, binders, etc.	Reduces electricity and steam production and increases the capacity in pulp production
Separation and refining of hemicelluloses from solid biomass or black liquor	Solid fuels, lignin oil, dispersants additives for food production, phenols, carbon fibres, sorbents, binders, etc.	Reduces electricity and steam production and increases the capacity of pulp production
Refining and separation of extractives from bark and wood	Triglycerides, tall crude oil, turpentine, rosin acids and medical substances (e.g. antioxidants)	Given that in mechanical pulp process, fibres are more intact. This approach enables to recover extractives found in low concentrations
Converting cellulose to alternative products	Dissolving ethanol, pulp, biocomposites and textile fibres	Displaces pulp for paper production

The use of virgin wood in the industry does need careful management. Of particular importance is the sourcing of the wood. There are various options for ethically sourced wood where deforestation is mitigated. These include cultivating fast-growing trees, reforestation and the use of less invasive harvesting techniques that preserve the forests' biodiversity.

7.8 Outlook for Industrial Manufacture of Materials

Humankind will need a variety of materials going forward in an environment of increased population with the associated increase in the demand for these products,

Table 7.22 Specific energy consumption in various materials manufacturing processes.

Industry	Production (Mio tonne p.a.)	$E_{specific}$ (MWh/tonne)	Process CO_2 emission[a)] (tonne CO_2/tonne)
Cement	4100	1.0	0.51
Steel	1890	3.6	1.47[b)]
Sulphuric acid	270	~0	—
Chlorine/sodium hydroxide	87	2.8–3.0	—
Paper and pulp	414	3.1–9.7[c)]	—
Glass	130	2.1–4.3	0.18[d)]
Soda ash	73	0.3–3.8	0.20–0.41
Aluminium	70	19.2	1.53
Copper	22	6.9–9.0	—
Metallurgical grade Si (MG-Si)	9	4.6	3.15
Titanium	0.3	45–70	2.76[e)]
Solar grade Si (SG-Si)		100–233	

a) CO_2 emissions from the process without considering CO_2 emissions from fuel consumption.
b) Carbon for the reduction of haematite.
c) Mainly sourced from waste generated with the process from wood feedstock.
d) Emissions without considering emissions associated with production of soda ash.
e) Emissions associated with chloride process.

and decreased availability of the raw materials from mineral resources. The specific energy consumption of these industries lies between being self-sufficient in the case of the pulp and paper industry, where the required energy (3.1-9.7 MWh/tonne of paper is extracted from the waste generated in the process, and 100–233 MWh/tonne for SG-Si as an example of a product where a high purity is required but is often significantly higher than the specific energy consumption for the basic industry (see Table 7.22). However, the overall energy consumption in these industries is smaller than in the basic industries due to the smaller volumes involved. However, the industries are sufficiently large and as such there is still an urgent need for decarbonisation in the industry, as they have both high volumetric outputs as well as relatively high energy requirements and thus a significant CO_2 output.

All material synthesis industries will have varying extents of importance going forward, but all may be expected to grow at significant rates. This means that there will need to be some implementation of renewable energy and renewable processes to keep these industries sustainable and utilise them to aid the renewable energy transition.

Defossilisation is different with each industry. Industries utilising electrochemical processes, such as the aluminium and copper industry, may be more readily adaptable to renewable electricity as a primary energy input than industries requiring high-temperature operation. A complication that arises from the material synthesis industries is that they tend to operate as continuous processes, which means that the implementation of renewable electricity will need to be available on a continuous basis. Fluctuations in renewable energy supply to electrochemical processes can, to some extent, be tolerated and mitigated by a temporary variation in the production capacity. Mitigation by battery storage may be impractical for many of these processes, especially from a capital point of view, although some systems will need battery back-up for safe operation.

A part of the emission profile of the materials manufacturing industry originates from transport, as most of the transportation modes are powered using fossil fuels. This is specifically relevant for operations processing low-grade ore as large amount of gangue material will be transported as well, and the amount of energy required per unit mass of product is high. Transport of large amounts of ore requires exceptionally large vehicles for the transportation, which are powered using fossil fuels but may be fed with sustainable diesel. Alternatively, the use of diesel can be replaced by batteries, fuel cells or an overhead power supply. However, this will come with substantial capital cost and may not be practical in all circumstances. To this end, several mining companies have expressed interest in green hydrogen. Anglo American commissioned the first green hydrogen station in Chile in 2021 [266]. This pilot is located at the Las Tórtolas flotation plant in Colina, Chile and has a production capacity of 2 kg of H_2 per day. This can be used to power a forklift crane and a stationary fuel cell which reinjects energy to the local electricity grid [266].

The polymers/plastics industry has not been discussed in this chapter largely as the temperature requirements in the industry are low (up to ca. 250 °C depending on the melting point of the plastic, with mould temperatures being <ca. 140 °C; polymerisation reactions may operate at temperatures between room temperature up to 300 °C, but these reactions are exothermic and can be performed without heat input). The major problem facing the plastics industry is associated with the supply of monomers going forward, i.e. ethylene, propylene, butadiene, styrene, terephthalic acid, adipic acid and ethylene glycol. At present, these monomers come from petrochemical sources. Not only is the source of these materials a fossil source, but e.g. ethene (the most common monomer), propene and butadiene originate from the steam cracking of naphtha (medium chain length hydrocarbons), a process, which operates at temperatures up to ca. 800–900 °C. This may be drawn outside the boundary of the plastics industry itself, and may fall under the petrochemical industry, but is a large source of Scope 3 carbon emissions. Ethylene and propylene may be generated through other processes from renewable carbon sources using the methanol to olefins (MTO) or Fischer-Tropsch targeting short-chain olefins (FTO) process, although these processes are not yet widely industrialised, and the synthesis of more complex monomers may prove difficult.

Quite a few of the processes in the discussed material synthesis industries require high temperatures, which are currently obtained by combusting fossil fuels. This could also be supplied by alternative fuels, such as hydrogen. This may, however, require careful redesign of the furnace if radiative heat transfer is dominant in the heat transfer as the emissivity of a hydrogen flame is much lower than typically observed with either natural gas or coal flames. Alternatively, furnaces may be heated using renewable, sustainable fuels. Achieving heating at the required temperatures is currently being investigated for various industries, but these furnaces typically operate continuously, and the lifetime of the heating elements needs to be extended significantly.

The question will always remain what the driving force will be to incorporate renewable energy in the manufacture of materials. Governmental pressure, to some extent, will naturally be required if there is no specific need for a change in the currently used industrial processes.

A highly important factor for the materials synthesis industries going forward will be the reuse of products as well as recycle. The third of the famous 'three R's', reduce, will likely not be possible in the context of the material synthesis industries. Within the six industries discussed above, all of aluminium, copper and silicon are essential to the renewable energy transition – aluminium and copper for the electrical conductivity and silicon for the product of solar panels (as well as computer chips, another fast-growing industry). Glass and paper are growing industries in part to remove dependency on fossil-based plastic/polymer products. Glass being infinitely reusable and recyclable and paper being far more recyclable than plastics, which may only be downcycled, i.e. made into less valuable products, such as recycled cooldrink plastic bottles being made into plastic park benches.

All metals may be recycled infinitely, including those essential to the renewable energy transition. Whilst metal recycling is something currently implemented, the scale of operation will need to be upgraded significantly with a distinct shift in mentality not only from the industry but also from the public. Similar to the need to choose to 'go green' with electricity will be the choice to 'go green' with the products supporting said industry. Further, as paper and glass are further rolled out to replace plastic packaging, there will be a need for the public to pay attention to the potential for the recycle/reuse of the product packaging. Naturally, the need for recycle/(reuse) is something that may need government regulation and provides an additional consideration for recycle in these industries.

Recycle will also be an important consideration due to the diminishing reserves of raw materials with the associated diminishing grade of many of those materials (and thus increased energy requirement for processing of that type of feed). With all metal ores, there is a tendency to pick the high-grade locations first and avoid the use of lower-grade ores; this, too, will start to present a problem as the higher-grade ore stock runs lower. A way to curb the higher energy intensity, lower-grade stock processing will be to supplement the low-grade feed with recycled materials. Deforestation (in the case of papermaking) cannot continue indefinitely; besides just the massive environmental damage, at some point there will be no more forest. Thus, either specific plantations for papermaking trees or mass recycle of paper products will be required to fuel the paper industry.

All industries which may be cast under the blanket of 'material synthesis' will require their own forms of innovation and invention to become more renewable. These industries will likely show large growth in the future, and this may be expected to come with interesting manufacturing process modifications.

References

1 IEA (2021). Mineral requirements for clean energy transitions – The Role of Critical Minerals in Clean Energy Transitions – Analysis [Online]. Available: https://www.iea.org/reports/the-role-of-critical-minerals-in-clean-energy-transitions/mineral-requirements-for-clean-energy-transitions (accessed 16 September 2024).

2 Koschinsky, A., Heinrich, L., Boehnke, K. et al. (2018). Deep-sea mining: interdisciplinary research on potential environmental, legal, economic, and societal implications. *Integrated Environmental Assessment and Management* 14 (6): 672–691. https://doi.org/10.1002/ieam.4071.

3 Du, K., Xi, W., Huang, S., and Zhou, J. (2024). Deep-sea mineral resource mining: a historical review, developmental progress, and insights. *Mining, Metallurgy & Exploration* 41 (1): 173–192. https://doi.org/10.1007/s42461-023-00909-9.

4 Kowalczuk, P., Bouzahzah, H., Kleiv, R., and Aasly, K. (2019). Simultaneous leaching of seafloor massive sulfides and polymetallic nodules. *Minerals* 9 (8): 482. https://doi.org/10.3390/min9080482.

5 Fuchs, S., Hannington, M.D., and Petersen, S. (2019). Divining gold in seafloor polymetallic massive sulfide systems. *Mineralium Deposita* 54 (6): 789–820. https://doi.org/10.1007/s00126-019-00895-3.

6 Wang, S., Li, H., Zhai, S. et al. (2017). Mineralogical characteristics of polymetallic sulfides from the Deyin-1 hydrothermal field near 15°S, southern Mid-Atlantic Ridge. *Acta Oceanologica Sinica* 36 (2): 22–34. https://doi.org/10.1007/s13131-016-0961-3.

7 Cheng, Y., Dai, Y., Zhang, Y. et al. (2023). Status and prospects of the development of deep-sea polymetallic nodule-collecting technology. *Sustainability* 15 (5): 4572. https://doi.org/10.3390/su15054572.

8 Hein, J.R., Koschinsky, A., and Kuhn, T. (2020). Deep-ocean polymetallic nodules as a resource for critical materials. *Nature Reviews Earth and Environment* 1 (3): 158–169. https://doi.org/10.1038/s43017-020-0027-0.

9 Niner, H.J., Ardron, J.A., Escobar, E.G. et al. (2018). Deep-sea mining with no net loss of biodiversity—an impossible aim. *Frontiers in Marine Science* 5: https://doi.org/10.3389/fmars.2018.00053.

10 Durden, J.M., Lallier, L.E., Murphy, K. et al. (2018). Environmental Impact Assessment process for deep-sea mining in 'the Area. *Marine Policy* 87: 194–202. https://doi.org/10.1016/j.marpol.2017.10.013.

11 Uddin, W., Khan, B., Dewan, S., and Das, S. (2022). Silicon-based qubit technology: progress and future prospects. *Bulletin of Materials Science* 45 (1): 46. https://doi.org/10.1007/s12034-021-02621-0.

12 Kim, S., Hoang, V.Q., and Bark, C.W. (2021). Silicon-based technologies for flexible photovoltaic (PV) devices: from basic mechanism to manufacturing technologies. *Nanomaterials* 11 (11): 2944. https://doi.org/10.3390/nano11112944.

13 Wilson, K. (2024). Top 10 Thermally Conductive Materials - Thermtest, Themtest Instruments [Online]. Available: https://thermtest.com/thermal-resources/top-10-resources/top-10-thermally-conductive-materials (accessed 8 May 2024).

14 Bell, T. (2024). Electrical conductivity of metals, ThoughtCo. [Online]. Available: https://www.thoughtco.com/electrical-conductivity-in-metals-2340117 (accessed 5 August 2024).

15 Yasakau, K.A., Zheludkevich, M.L., and Ferreira, M.G.S. (2018). Corrosion and corrosion protection of aluminum alloys. In: *Encyclopedia of Interfacial Chemistry* (ed. K. Wandelt), 115–127. Oxford: Elsevier. https://doi.org/10.1016/B978-0-12-409547-2.13870-3.

16 Scrap Metal & E-Waste Collection (2024). Scrap Metal Prices Near Me: The Full Scrap Metal Guide: 2024, Scrap Metal & E-Waste Collection [Online]. Available: https://scrapmetaljoburg.co.za/scrap-metal-prices/ (accessed 5 August 2024).

17 Business Insider (2024). Commodities, Markets Insider [Online]. Available: https://markets.businessinsider.com/commodities (accessed 12 August 2024).

18 Wang, H.S., Lineweaver, C.H., and Ireland, T.R. (2018). The elemental abundances (with uncertainties) of the most Earth-like planet. *Icarus* 299: 460–474. https://doi.org/10.1016/j.icarus.2017.08.024.

19 European Aluminium Association (2022). Materials – Designation system, *The Aluminium Automative Manual* [Online]. Available: https://european-aluminium.eu/wp-content/uploads/2022/11/aam-materials-3-designation-system.pdf (accessed 12 August 2024).

20 U.S. Geological Survey (2023). *Mineral commodity summaries 2023*, USGS Publications Warehouse. https://doi.org/10.3133/mcs2023.

21 Ober, J.A. (2018). Mineral commodity summaries 2018, U.S. Geological Survey. https://doi.org/10.3133/70194932.

22 International Aluminium Institute (2025). Primary aluminium production, International Aluminium Institute [Online]. Available: https://international-aluminium.org/statistics/primary-aluminium-production/ (accessed 4 March 2025).

23 Home, A. and Blair, E. (2019). Global aluminium production growth brakes sharply in 2018, Bloomberg Opinion [Online]. Available: https://www.mining.com/web/global-aluminium-production-growth-brakes-sharply-2018/ (accessed 5 July 2024).

24 RioTinto (2024). Aluminium [Online]. Available: https://www.riotinto.com/en/products/aluminium (accessed 10 September 2024).

25 Brough, D. and Jouhara, H. (2020). The aluminium industry: a review on state-of-the-art technologies, environmental impacts and possibilities for waste heat recovery. *International Journal of Thermofluids* 1–2: 100007. https://doi.org/10.1016/j.ijft.2019.100007.

26 Atlantic (2024). Atlantic Pty Ltd » Development Projects » Bauxite, Atlantic [Online]. Available: https://atlanticptyltd.com.au/?id=314 (accessed 8 May 2024).

27 Lancashire, R.J. (2014). The chemistry and processing of Jamaican Bauxite[Online]. Available: http://wwwchem.uwimona.edu.jm/lectures/bauxite.html (accessed 8 May 2024).

28 Evans, K., Hudson, L.K., Misra, C. et al. (2024). Aluminum oxide. In: *Ullmann's Encyclopedia of Industrial Chemistry*, 1e, 1–59. Weinheim: Wiley. https://doi.org/10.1002/14356007.a01_557.pub2.

29 Bénézeth, P., Hilic, S., and Palmer, D.A. (2016). The solubilities of gibbsite and bayerite below 100 °C in near neutral to basic solutions. *Journal of Solution Chemistry* 45 (9): 1288–1302. https://doi.org/10.1007/s10953-016-0470-1.

30 Königsberger, E., May, P.M., and Hefter, G. (2006). Comprehensive model of synthetic Bayer liquors. Part 3. Sodium aluminate solutions and the solubility of gibbsite and boehmite. *Monatshefte fuer Chemie* 137 (9): 1139–1149. https://doi.org/10.1007/s00706-006-0526-9.

31 Frenzel, M., Ketris, M.P., Seifert, T., and Gutzmer, J. (2016). On the current and future availability of gallium. *Resources Policy* 47: 38–50. https://doi.org/10.1016/j.resourpol.2015.11.005.

32 Pontikes, Y. (2005). Characteristics – Red Mud Project [Online]. Available: https://redmud.org/red-mud/characteristics/ (accessed 12 September 2024).

33 Rao Borra, C. (2024). Iron extracted from hazardous waste of aluminium production. *Nature* 625: 665–667. https://doi.org/10.1038/d41586-024-00071-2.

34 Silveira, N.C.G., Martins, M.L.F., Bezerra, A.C.S., and Araújo, F.G.S. (2021). Red mud from the aluminium industry: production, characteristics, and alternative applications in construction materials—a review. *Sustainability* 13 (22) Art. no. 22: https://doi.org/10.3390/su132212741.

35 Sanders, R.E. and Updated by Staff (2012). Aluminum and aluminum alloys. In: *Kirk-Othmer Encyclopedia of Chemical Technology*, 1e, 1–64. John Wiley & Sons. https://doi.org/10.1002/0471238961.0112211319200112.a01.pub3.

36 Balomenos, E., Panias, D., and Paspaliaris, I. (2011). Energy and exergy analysis of the primary aluminum production processes: a review on current and future sustainability. *Mineral Processing and Extractive Metallurgy Review* 32 (2): 69–89. https://doi.org/10.1080/08827508.2010.530721.

37 Obaidat, M., Al-Ghandoor, A., Phelan, P. et al. (2018). Energy and exergy analyses of different aluminum reduction technologies. *Sustainability* 10 (4): 1216. https://doi.org/10.3390/su10041216.

38 Frank, W.B., Haupin, W.E., Vogt, H. et al. (2000). Aluminum. In: *Ullmann's Encyclopedia of Industrial Chemistry*, a01_459. Weinheim: Wiley-VCH Verlag GmbH & Co. KGaA. https://doi.org/10.1002/14356007.a01_459.

39 International-Aluminium (2021). 75% of all aluminium ever produced is still in use today, International Aluminium Institute [Online]. Available: https://international-aluminium.org/landing/75-of-all-aluminium-ever-produced-is-still-in-use-today/ (accessed 4 March 2025).

40 Bleam, W., McKinney, D., Tempest, R., and Jacobsen, E.K. (2012). Aluminum Recycling, ChemMatters – April 2012 Teacher's Guide. American Chemical Society [Online]. Available: https://www.acs.org/content/dam/acsorg/education/resources/highschool/chemmatters/archive/chemmatters-teachers-guide-april-2012-pdf.pdf (accessed 13 September 2024).

41 Al-Alimi, S., Yusuf, N.K., Ghaleb, A.M. et al. (2024). Recycling aluminium for sustainable development: a review of different processing technologies in green manufacturing. *Results in Engineering* 23: 102566. https://doi.org/10.1016/j.rineng.2024.102566.

42 Padamata, S.K., Yasinskiy, A.S., and Polyakov, P.V. (2018). Progress of inert anodes in aluminium industry: review. *Journal of Siberian Federal University Chemistry* 11 (1): 18–30. https://doi.org/10.17516/1998-2836-0055.

43 UN Panel on Critical Energy Transition Minerals (2024). Resourcing the energy transition: principles to guide critical energy transition minerals towards equity and justice, New York: United Nations Digital Library.

44 Ebrahimi, P., Kumar, A., and Khraisheh, M. (2020). A review of recent advances in water-gas shift catalysis for hydrogen production. *Emergent Materials* 3 (6): 881–917. https://doi.org/10.1007/s42247-020-00116-y.

45 Tursunov, O., Kustov, L., and Kustov, A. (2017). A brief review of carbon dioxide hydrogenation to methanol over copper and iron based catalysts. *Oil and Gas Science and Technology* 72 (5): https://doi.org/10.2516/ogst/2017027.

46 Niu, J., Liu, H., Jin, Y. et al. (2022). Comprehensive review of Cu-based CO_2 hydrogenation to CH_3OH: insights from experimental work and theoretical analysis. *International Journal of Hydrogen Energy* 47 (15): 9183–9200. https://doi.org/10.1016/j.ijhydene.2022.01.021.

47 Mejía, E.R., Ospina, J.D., Osorno, L. et al. (2015). Mineralogical characterization of chalcopyrite bioleaching. In: *Fourier Transform - Signal Processing and Physical Sciences*. IntechOpen. https://doi.org/10.5772/59489.

48 Feng, Q., Yang, W., Wen, S. et al. (2022). Flotation of copper oxide minerals: a review. *International Journal of Mining Science and Technology* 32 (6): 1351–1364. https://doi.org/10.1016/j.ijmst.2022.09.011.

49 Pistilli, M. (2024). Top 10 copper producers by country (updated 2024) | Nasdaq [Online]. Available: https://investingnews.com/daily/resource-investing/base-metals-investing/copper-investing/copper-production-country/?utm_source=nasdaq&utm_medium=syndication (accessed 4 December 2024).

50 Statista (2024). Copper market size worldwide 2030 [Online]. Available: https://www.statista.com/statistics/1470350/global-copper-market-size/ (accessed 4 December 2024).

51 Precendence Research (2024). Copper market size, share, and trends 2024 to 2034 [Online]. Available: https://www.precedenceresearch.com/copper-market (accessed 13 September 2024).

52 Statista (2024). Copper producers ranking by production output 2023 [Online]. Available: https://www.statista.com/statistics/281023/leading-copper-producers-worldwide-by-output/ (accessed 4 December 2024).

53 Moreno-Leiva, S., Haas, J., Junne, T. et al. (2020). Renewable energy in copper production: a review on systems design and methodological approaches. *Journal of Cleaner Production* 246: 118978. https://doi.org/10.1016/j.jclepro.2019.118978.

54 Hawker, W., Vaughan, J., Jak, E., and Hayes, P. (2017). The synergistic copper process concept. *Mineral Processing and Extractive Metallurgy* 127: 1–11. https://doi.org/10.1080/03719553.2017.1375768.

55 Schlesinger, M.E., King, M.J., Sole, K.C., and Davenport, W.G. (2011). *Extractive Metallurgy of Copper*, 5e. Amsterdam: Elsevier.

56 Koppelaar, R.H.E.M. and Koppelaar, H. (2016). The ore grade and depth influence on copper energy inputs. *Biophysical Economics and Resource Quality* 1 (2): 11:1–11:16. https://doi.org/10.1007/s41247-016-0012-x.

57 Kuckshinrichs, W., Zapp, P., and Poganietz, W.-R. (2007). CO_2 emissions of global metal-industries: the case of copper. *Applied Energy* 84 (7–8): 842–852. https://doi.org/10.1016/j.apenergy.2007.01.014.

58 Allen, M. (2021). Mining energy consumption 2021 [Online]. Available: https://www.ceecthefuture.org/resources/mining-energy-consumption-2021 (accessed 1 March 2025).

59 Donoso, M., Mellado, M.E., Gálvez, E.D., and Cisternas, L.A. (2012). Water and energy use in mineral processing. In: *Computer Aided Chemical Engineering*, vol. 30, 627–631. Elsevier. https://doi.org/10.1016/B978-0-444-59519-5.50126-X.

60 Ferguson, W.C. (1910). The determination of copper in blister and refining copper. *Journal of Industrial and Engineering Chemistry* 2 (5): 187–195. https://doi.org/10.1021/ie50017a003.

61 Chen, T.T. and Dutrizac, J.E. (1990). The mineralogy of copper electrorefining. *JOM* 42 (8): 39–44. https://doi.org/10.1007/BF03221053.

62 Phiri, T.C., Singh, P., and Nikoloski, A.N. (2021). The potential for copper slag waste as a resource for a circular economy: a review – Part II. *Minerals Engineering* 172: 107150. https://doi.org/10.1016/j.mineng.2021.107150.

63 Gorai, B., Jana, R.K., and Premchand (2003). Characteristics and utilisation of copper slag—a review. *Resources, Conservation and Recycling* 39 (4): 299–313. https://doi.org/10.1016/S0921-3449(02)00171-4.

64 Alp, İ., Deveci, H., and Süngün, H. (2008). Utilization of flotation wastes of copper slag as raw material in cement production. *Journal of Hazardous Materials* 159 (2–3): 390–395. https://doi.org/10.1016/j.jhazmat.2008.02.056.

65 Al-Jabri, K.S., Al-Saidy, A.H., and Taha, R. (2011). Effect of copper slag as a fine aggregate on the properties of cement mortars and concrete. *Construction and Building Materials* 25 (2): 933–938. https://doi.org/10.1016/j.conbuildmat.2010.06.090.

66 Osborn, G. (1990). Pilot-Plant study for recovery of copper metal from secondary copper slag by froth flotation. *Recycling of Metalliferous Materials* 173–179.

67 Bulut, G. (2006). Recovery of copper and cobalt from ancient slag. *Waste Management & Research* 24 (2): 118–124. https://doi.org/10.1177/0734242X06063350.

68 Cuevas, F., Murray, C., Platzer, W., and Heimsath, A. (2015). Large scale solar plants integration in electro-winning copper recuperation process. *Energy Procedia* 70: 605–614. https://doi.org/10.1016/j.egypro.2015.02.167.

69 Forner, E.L., Scheepers, J., Du Toit, A.J., and MillerII, G.M. (2018). Copper electrowinning circuit design: optimized costing as a function of cell arrangement, productivity, rectiformer size, and throughput. *Journal of the Southern African Institute of Mining and Metallurgy* 118 (11): 1141–1148. https://doi.org/10.17159/2411-9717/2018/v118n11a3.

70 Moongo, T.E. and Michael, S. (2021). A continuous quality improvement framework for electrowinning current efficiency. *Journal of the Southern African Institute of Mining and Metallurgy* 121 (1): 21–28. https://doi.org/10.17159/2411-9717/1243/2021.

71 Klose, S. and Pauliuk, S. (2021). Quantifying longevity and circularity of copper for different resource efficiency policies at the material and product levels. *Journal of Industrial Ecology* 25 (4): 979–993. https://doi.org/10.1111/jiec.13092.

72 Fizaine, F. and Court, V. (2015). Renewable electricity producing technologies and metal depletion: a sensitivity analysis using the EROI. *Ecological Economics* 110: 106–118. https://doi.org/10.1016/j.ecolecon.2014.12.001.

73 Glöser, S., Soulier, M., and Tercero Espinoza, L.A. (2013). Dynamic analysis of global copper flows. Global stocks, postconsumer material flows, recycling indicators, and uncertainty evaluation. *Environmental Science & Technology* 47 (12): 6564–6572. https://doi.org/10.1021/es400069b.

74 International Copper Association (2025). Stocks and Flows - International Copper Association [Online]. Available: https://internationalcopper.org/sustainable-copper/about-copper/cu-demand-long-term-availability/stocks-flows/ (accessed 7 March 2025).

75 International Copper Study Group (2025). Copper Factbook 2024, International Copper Study Group [Online]. Available: https://icsg.org/copper-factbook/ (accessed 7 March 2025).

76 OECD (2019). *Global Material Resources Outlook to 2060: Economic Drivers and Environmental Consequences*. OECD. https://doi.org/10.1787/9789264307452-en.

77 Moreno-Leiva, S., Díaz-Ferrán, G., Haas, J. et al. (2017). Towards solar power supply for copper production in Chile: assessment of global warming potential using a life-cycle approach. *Journal of Cleaner Production* 164: 242–249. https://doi.org/10.1016/j.jclepro.2017.06.038.

78 Harmsen, J.H.M., Roes, A.L., and Patel, M.K. (2013). The impact of copper scarcity on the efficiency of 2050 global renewable energy scenarios. *Energy* 50: 62–73. https://doi.org/10.1016/j.energy.2012.12.006.

79 Kölking, M., Flamme, S., Heinrichs, S. et al. (2024). More resource efficient recycling of copper and copper alloys by using X-ray fluorescence sorting systems: an investigation on the metallic fraction of mixed foundry residues. *Waste Management & Research* 42 (9): 814–822. https://doi.org/10.1177/0734242X241241601.

80 Rankin, W. (2011). *Minerals, Metals and Sustainability: Meeting Future Material Needs*. CSIRO Publishing. https://doi.org/10.1071/9780643097278.

81 Norgate, T.E., Jahanshahi, S., and Rankin, W.J. (2007). Assessing the environmental impact of metal production processes. *Journal of Cleaner Production* 15 (8–9): 838–848. https://doi.org/10.1016/j.jclepro.2006.06.018.

82 Northey, S., Haque, N., and Mudd, G. (2013). Using sustainability reporting to assess the environmental footprint of copper mining. *Journal of Cleaner Production* 40: 118–128. https://doi.org/10.1016/j.jclepro.2012.09.027.

83 Castro-Molinare, J. (2014). Sustainability analysis of copper extraction and processing using LCA methods. PhD. Imperial College of Science, Technology and Medicine.

84 Behar, O., Peña, R., Kouro, S. et al. (2021). The use of solar energy in the copper mining processes: a comprehensive review. *Cleaner Engineering and Technology* 4: 100259. https://doi.org/10.1016/j.clet.2021.100259.

85 Schlesinger, M.E., King, M.J., Sole, K.C., and Davenport, W.G. (2011). Chapter 1 - Overview. In: *Extractive Metallurgy of Copper*, 5e (ed. M.E. Schlesinger, M.J. King, K.C. Sole, and W.G. Davenport), 1–12. Oxford: Elsevier. https://doi.org/10.1016/B978-0-08-096789-9.10001-0.

86 Chudinzow, D., Switon, D., and Eltrop, L. (2017). A solar furnace for copper smelting in Chile: assessment of economic benefits and reductions in greenhouse gas emissions. In: *Proceedings of SWC2017/SHC2017*, 1–8. Abu Dhabi: International Solar Energy Society. https://doi.org/10.18086/swc.2017.26.03.

87 Choi, Y. and Song, J. (2017). Review of photovoltaic and wind power systems utilized in the mining industry. *Renewable and Sustainable Energy Reviews* 75: 1386–1391. https://doi.org/10.1016/j.rser.2016.11.127.

88 Power Technology (2014). El Arrayán Wind Farm - Power Technology [Online]. Available: https://www.power-technology.com/projects/el-arrayn-wind-farm/ (accessed 4 December 2024).

89 Hampel, C.A. (1968). *The Encyclopedia of the Chemical Elements*. New York: Reinhold Book Corp [Online]. Available: http://archive.org/details/encyclopediaofch00hamp (accessed 6 September 2024).

90 Kim, S.C., Hanawa, T., Manaka, T. et al. (2022). Band structures of passive films on titanium in simulated bioliquids determined by photelectrochemical response: principle governing the biocompatibility. *Science and Technology of Advanced Materials* 23: 322–331. https://doi.org/10.1080/14686996.2022.2066960.

91 Singh, H.B. and Bharati, K.A. (2014). 6 - Enumeration of dyes. In: *Handbook of Natural Dyes and Pigments* (ed. H.B. Singh and K.A. Bharati), 33–260. Woodhead Publishing India, https://doi.org/10.1016/B978-93-80308-54-8.50006-X.

92 Britannica (2024). Titanium dioxide [Online]. Available: https://www.britannica.com/science/titanium-dioxide (accessed 6 September 2024).

93 Molinari, R., Caruso, A., and Palmisano, L. (2010). 3.07 - Photocatalytic processes in membrane reactors. In: *Comprehensive Membrane Science and Engineering* (ed. E. Drioli and L. Giorno), 165–193. Oxford: Elsevier, https://doi.org/10.1016/B978-0-08-093250-7.00039-6.

94 BIOVA, Dassault Systèmes, Materials Studio, San Diego (2020) (licensed through NICIS-CHPC).

95 Oi, L.E., Yee, C., Lee, H. et al. (2016). Recent advances of titanium dioxide (TiO_2) for green organic synthesis. *RSC Advances* 6: 108741–108754. https://doi.org/10.1039/C6RA22894A.

96 Ranade, M.R., Navrotsky, A., Zhang, H.Z. et al. (2002). Energetics of nanocrystalline TiO_2. *Proceedings of the National Academy of Sciences* 99 (suppl_2): 6476–6481. https://doi.org/10.1073/pnas.251534898.

97 Haverkamp, R.G., Kruger, D., and Rajashekar, R. (2016). The digestion of New Zealand ilmenite by hydrochloric acid. *Hydrometallurgy* 163: 198–203. https://doi.org/10.1016/j.hydromet.2016.04.015.

98 Wang, W. and Wu, F. (2023). Dataset of annual metal scrap circularity of titanium industry in China from 2005 to 2020. *Scientific Data* 10 (1): 435. https://doi.org/10.1038/s41597-023-02351-4.

99 Völz, H.G., Kischkewitz, J., Woditsch, P. et al. (2006). Pigments, inorganic. In: *Ullmann's Encyclopedia of Industrial Chemistry*. John Wiley & Sons, Ltd, https://doi.org/10.1002/14356007.a20_243.pub2.

100 Sadeghi, M.H. and Esfahany, M.N. (2022). Development of a safe and environmentally friendly Sulfate process for the production of titanium oxide. *Industrial and Engineering Chemistry Research* 61 (4): 1786–1796. https://doi.org/10.1021/acs.iecr.1c03364.

101 Habashi, F., Kamaleddine, F., and Bourricaudy, E. (2015). A new process to upgrade ilmenite to synthetic rutile. *Metall* 69: 27–30.

102 Auer, G., Woditsch, P., Westerhaus, A. et al. (2017). Pigments, inorganic, 2. White pigments. In: *Ullmann's Encyclopedia of Industrial Chemistry*, 1–36. John Wiley & Sons, Ltd, https://doi.org/10.1002/14356007.n20_n01.pub2.

103 Dai, Y., Dong, H., Sun, L. et al. (2024). Life cycle environmental impact assessment of titanium dioxide production in China. *Environmental Impact Assessment Review* 105: 107412. https://doi.org/10.1016/j.eiar.2023.107412.

104 Middlemas, S., Fang, Z.Z., and Fan, P. (2015). Life cycle assessment comparison of emerging and traditional titanium dioxide manufacturing processes. *Journal of Cleaner Production* 89: 137–147. https://doi.org/10.1016/j.jclepro.2014.11.019.

105 Welham, N.J. (1996). A parametric study of the mechanically activated carbothermic reduction of ilmenite. *Minerals Engineering* 9 (12): 1189–1200. https://doi.org/10.1016/S0892-6875(96)00115-X.

106 Navrotsky, A. (1975). Thermodynamics of formation of some compounds with the pseudobrookite structure and of the $FeTi_2O_5$-Ti_3O_5 solid solution series. *American Mineralogist* 60 (3–4): 249–256.

107 Iorisch, V.S. and Yungman, V.S. (2008). TCH Database, Institute of Thermophysics of Extreme States of the Russian Academy of Sciences, Joint Institute for High Temperatures of the Russian Academy of Sciences, Faculty of Chemistry of Lomonosov Moscow State University, https://www.chem.msu.ru/cgi-bin/tkv.pl?show=welcome.html (accessed July 2024).

108 Xiang, J., Pei, G., Lv, W. et al. (2020). Preparation of synthetic rutile from reduced ilmenite through the aeration leaching process. *Chemical Engineering and Processing - Process Intensification* 147: 107774. https://doi.org/10.1016/j.cep.2019.107774.

109 Mabentsela, A. and Mainza, A. (2023). Numerical investigation into slag solidification inside an ilmenite DC arc furnace using a finite element method approach. *Journal of the Southern African Institute of Mining and Metallurgy* 123 (3): 113–122. https://doi.org/10.17159/2411-9717/1247/2023.

110 Greenwood (1996). *Chemistry of the Elements*. Elsevier Science & Technology Books.

111 Jones, T. and Egerton, T.A. (2012). Titanium compounds, inorganic. In: *Kirk-Othmer Encyclopedia of Chemical Technology*, 1–62. John Wiley & Sons, Ltd. https://doi.org/10.1002/0471238961.0914151805070518.a01.pub3.

112 Fang, Z.Z., Paramore, J.D., Sun, P. et al. (2018). Powder metallurgy of titanium – past, present, and future. *International Materials Reviews* 63 (7): 407–459. https://doi.org/10.1080/09506608.2017.1366003.

113 Liu, C., Sun, Z., Lu, G. et al. (2014). Scale-up design of a 300 kA magnesium electrolysis cell based on thermo-electric mathematical models. *Canadian Journal of Chemical Engineering* 92 (7): 1197–1206. https://doi.org/10.1002/cjce.21940.

114 Demirci, G. (2006). Electrolytic magnesium production using coaxial electrodes. PhD, Middle East Technical University, Ankara (Turkey).

115 Toho Titanium Co. Ltd. (2021). Titanium takes on a greener hue [Online]. Available: https://www.nature.com/articles/d42473-021-00166-8 (accessed 5 September 2024).

116 Hunter, M.A. (1910). Metallic titanium. *Journal of the American Chemical Society* 32 (3): 330–336. https://doi.org/10.1021/ja01921a006.

117 Roza, G. (2008). *Titanium*. The Rosen Publishing Group, Inc. ISBN: 978-1-4042-1412-5.

118 Takeda, O., Ouchi, T., and Okabe, T.H. (2020). Recent progress in titanium extraction and recycling. *Metallurgical and Materials Transactions B* 51 (4): 1315–1328. https://doi.org/10.1007/s11663-020-01898-6.

119 Jing, Z., Liu, R., Geng, N. et al. (2024). Simulation of solidification structure in the vacuum arc remelting process of titanium alloy TC4 based on 3D CAFE method. *PRO* 12 (4) Art. no. 4: https://doi.org/10.3390/pr12040802.

120 Shibata, K., Yamaguchi, M., Katayama, H. et al. (1996). Temperature measurement in a titanium sponge and mathematical modeling for the vacuum distillation in Kroll process. *Shigen-to-Sozai (Journal of the Mining and Materials Processing Institute of Japan)* 112 (4): 257–262. https://doi.org/10.2473/shigentosozai.112.257.

121 Zhang, W., Zhu, Z., and Cheng, C.Y. (2011). A literature review of titanium metallurgical processes. *Hydrometallurgy* 108 (3): 177–188. https://doi.org/10.1016/j.hydromet.2011.04.005.

122 Gireesh, V.S., Vinod, V.P., Krishnan Nair, S., and Ninan, G. (2015). Catalytic leaching of ilmenite using hydrochloric acid: a kinetic approach. *International Journal of Mineral Processing* 134: 36–40. https://doi.org/10.1016/j.minpro.2014.11.004.

123 Xia, Y., Lefler, H.D., Fang, Z.Z. et al. (2020). Chapter 17 - Energy consumption of the Kroll and HAMR processes for titanium production. In: *Extractive Metallurgy of Titanium* (ed. Z.Z. Fang, F.H. Froes, and Y. Zhang), 389–410. Elsevier, https://doi.org/10.1016/B978-0-12-817200-1.00017-X.

124 Zhu, H., Jiao, S., Xiao, J., and Zhu, J. (2020). Chapter 13 - Titanium production through electrolysis of titanium oxycarbide consumable anode—the USTB process. In: *Extractive Metallurgy of Titanium* (ed. Z.Z. Fang, F.H. Froes, and Y. Zhang), 315–329. Elsevier, https://doi.org/10.1016/B978-0-12-817200-1.00013-2.

125 MarketsandMarkets (2024). Titanium dioxide market - report, size - industry share [Online]. Available: https://www.mordorintelligence.com/industry-reports/titanium-dioxide-market (accessed 10 September 2024).

126 Grand View Research (2023). Titanium market size, share & growth analysis report, 2030 [Online]. Available: https://www.grandviewresearch.com/industry-analysis/titanium-market-report (accessed 10 September 2024).

127 Childerhouse, T. and Jackson, M. (2019). Near net shape manufacture of titanium alloy components from powder and wire: a review of state-of-the-art process routes. *Metals* 9 (6): 689. https://doi.org/10.3390/met9060689.

128 Royal Society of Chemistry (2017). Silicon - element information, properties and uses | Periodic Table [Online]. Available: https://www.rsc.org/periodic-table/element/14/silicon (accessed 27 September 2024).

129 Nabil, M., Elnouby, M., Al-Askar, A.A. et al. (2024). Porous silicon nanostructures: synthesis, characterization, and their antifungal activity. *Open Chemistry* 22 (1): 20230169. https://doi.org/10.1515/chem-2023-0169.

130 Chigondo, F. (2018). From metallurgical-grade to solar-grade silicon: an overview. *Silicon* 10 (3): 789–798. https://doi.org/10.1007/s12633-016-9532-7.

131 Jarkin, V.N., Kisarin, O.A., and Kritskaya, T.V. (2021). Methods of trichlorosilane synthesis for polycrystalline silicon production. Part 1: Direct synthesis. *MoEM* 7 (1): 1–10. https://doi.org/10.3897/j.moem.7.1.64953.

132 O'Mara, W.C., Herring, R.B., and Hunt, L.P. (1990). *Handbook of Semiconductor Silicon Technology*. Noyes Publications.

133 Basore, P.A. (1994). Defining terms for crystalline silicon solar cells. *Progress in Photovoltaics: Research and Applications* 2: 177–179.

134 Streetman, B.G. and Banerjee, S. (2000). *Solid State Electronic Devices*. Prentice Hall.

135 Ghosh, A.K., Fishman, C., and Feng, T. (1980). Theory of the electrical and photovoltaic properties of polycrystalline silicon. *Journal of Applied Physics* 51 (1): 446–454. https://doi.org/10.1063/1.327342.

136 Schnebele, E.K. (2025). *Mineral Commodity Summaries: Silicon*. U.S. Geological Survey.

137 Fu, R., James, T.L., and Woodhouse, M. (2015). Economic measurements of polysilicon for the photovoltaic industry: market competition and manufacturing competitiveness. *IEEE Journal of Photovoltaics* 5 (2): 515–524. https://doi.org/10.1109/JPHOTOV.2014.2388076.

138 Song, X., Huang, Y., Zhang, Y. et al. (2023). An appraisal on China's feed-in tariff policies for PV and wind power: implementation effects and optimization. *Sustainability* 15 (6): 5137. https://doi.org/10.3390/su15065137.

139 Statista (2023). Global silicon mining - statistics & facts | Statista [Online]. Available: https://www.statista.com/topics/1959/silicon/ (accessed 12 September 2024).

140 Saga, T. (2010). Advances in crystalline silicon solar cell technology for industrial mass production. *NPG Asia Materials* 2 (3): 96–102. https://doi.org/10.1038/asiamat.2010.82.

141 Honsburg, C. and Bowden, S. (2019). Refining silicon, Photovoltaics Education Website. https://www.pveducation.org/pvcdrom/manufacturing-si-cells/refining-silicon (accessed 20 September 2024).

142 Barron, A.R. (2020). Chemistry of the main group elements, Chemistry LibreTexts [Online]. Available: https://chem.libretexts.org/Bookshelves/Inorganic_Chemistry/Chemistry_of_the_Main_Group_Elements_(Barron) (accessed 27 September 2024).

143 Shabalala, Z. (2022). Ferroglobe aims to restart South African silicon plant, raise U.S. output | Reuters [Online]. Available: https://www.reuters.com/world/africa/ferroglobe-aims-restart-safrican-silicon-plant-raise-us-output-2022-04-09/https://www.reuters.com/world/africa/ferroglobe-aims-restart-safrican-silicon-plant-raise-us-output-2022-04-09/ (accessed 27 September 2024).

144 Heidari, S.M. and Anctil, A. (2022). Country-specific carbon footprint and cumulative energy demand of metallurgical grade silicon production for silicon photovoltaics. *Resources, Conservation and Recycling* 180: 106171. https://doi.org/10.1016/j.resconrec.2022.106171.

145 Hallam, B., Kim, M., Underwood, R. et al. (2022). A polysilicon learning curve and the material requirements for broad electrification with photovoltaics by 2050. *Solar RRL* 6 (10): 2200458. https://doi.org/10.1002/solr.202200458.

146 Fairchild, W.T. (1970). Electric furnace manufacture of silicon metal. *Journal of Metals* 22: 55–58. https://doi.org/10.1007/BF03355657.

147 Dubrous, F.Y. and Septier, L.G. (1971). Silicon furnace performance related to the carbon in the load. *Transactions of the Metallurgical Society of AIME* 635–649.

148 Schei, A., Tuset, J., and Tveit, H. (1998). *Production of High Silicon Alloys*. Tapir. ISBN: 9788251913171.

149 Takla, M., Kamfjord, N.E., Tveit, H., and Kjelstrup, S. (2013). Energy and exergy analysis of the silicon production process. *Energy* 58: 138–146. https://doi.org/10.1016/j.energy.2013.04.051.

150 Siffert, P. and Krimmel, E.F. (ed.) (2004). *Silicon*. Berlin: Springer, https://doi.org/10.1007/978-3-662-09897-4.

151 Luque, A. and Hegedu, S. (ed.) (2010). *Handbook of Photovoltaic Science and Engineering*, 1e. John Wiley & Sons, https://doi.org/10.1002/9780470974704.

152 Böer, K.W. and Pohl, U.W. (2018). *Semiconductor Physics*. Cham: Springer International Publishing, https://doi.org/10.1007/978-3-319-69150-3.

153 Ingle, W.M. and Darnell, R.D. (1985). Oxidative purification of chlorosilane silicon source materials. *Journal of the Electrochemical Society* 132 (5): 1240. https://doi.org/10.1149/1.2114079.

154 Filtvedt, W.O., Javidi, M., Holt, A. et al. (2010). Development of fluidized bed reactors for silicon production. *Solar Energy Materials and Solar Cells* 94 (12): 1980–1995. https://doi.org/10.1016/j.solmat.2010.07.027.

155 Ravasio, S., Masi, M., and Cavallotti, C. (2013). Analysis of the gas phase reactivity of chlorosilanes. *The Journal of Physical Chemistry. A* 117 (25): 5221–5231. https://doi.org/10.1021/jp403529x.

156 The Essential Chemical Industry (2019). Silicones [Online]. Available: https://www.essentialchemicalindustry.org/polymers/silicones.html (accessed 27 September 2024).

157 Andriot, M., DeGroot, J.V., Meeks, R. et al. (2007). Silicones in industrial applications. In: *Inorganic Polymers* (ed. R. De Jaeger and M. Gleria), 61–161. Nova Science Publishers ISBN: 1600216560, 9781600216565.

158 Kero, I., Grådahl, S., and Tranell, G. (2017). Airborne emissions from Si/FeSi production. *Journal of Metals* 69 (2): 365–380. https://doi.org/10.1007/s11837-016-2149-x.

159 Ali, H.H.M., El-Sadek, M.H., Morsi, M.B. et al. (2018). Production of metallurgical-grade silicon from Egyptian quartz. *Journal of the Southern African Institute of Mining and Metallurgy* 118 (2): 143–148. https://doi.org/10.17159/2411-9717/2018/v118n2a7.

160 Sævarsdottir, G., Kvande, H., and Magnusson, T. (2021). Greenhouse gas emissions from silicon production -development of carbon footprint with changing energy systems. *SSRN Journal*, https://doi.org/10.2139/ssrn.3926088.

161 Kadkhodabeigi, M., Tveit, H., and Johansen, S.T. (2011). Modelling the tapping process in submerged arc furnaces used in high silicon alloys production. *ISIJ International* 51 (2): 193–202. https://doi.org/10.2355/isijinternational.51.193.

162 Kamfjord, N.E. (2012). Mass and energy balances of the silicon process: -improved emission standards. Doctoral thesis. Norges teknisk-naturvitenskapelige universitet, Fakultet for naturvitenskap og teknologi, Institutt for materialteknologi [Online]. Available: https://ntnuopen.ntnu.no/ntnu-xmlui/handle/11250/249043 (accessed 27 September 2024).

163 Lavoie, G.A., Heywood, J.B., and Keck, J.C. (1970). Experimental and theoretical study of nitric oxide formation in internal combustion engines. *Combustion Science and Technology* 1 (4): 313–326. https://doi.org/10.1080/00102206908952211.

164 Myrhaug, E., Tveit, H., Kamfjord., N. et al. (2012). NOx emissions from silicon production. In: *Conference on Silicon for the Chemical and Solar Industry XI* (ed. H.A. Øye, H. Brekken, L. Nygaard, et al.), 95–106. Bergen-Ulvik, Norway (25–29 June 2012).

165 Koch, E. and Clément, D. (2007). Special materials in pyrotechnics: VI. Silicon – an old fuel with new perspectives. *Propellants Explosives Pyrotechnics* 32 (3): 205–212. https://doi.org/10.1002/prep.200700021.

166 Blömeke, S., Arafat, R., Yang, J. et al. (2023). Environmental assessment of silicon kerf recycling and its benefits for applications in solar cells and Li-ion batteries. *Procedia CIRP* 116: 179–184. https://doi.org/10.1016/j.procir.2023.02.031.

167 Ciftja, A., Engh, T., and Tangstad, M. (2008). *Refining and Recycling of Silicon: A Review*. Trondheim: NTNU Trykk. ISBN: 978-82-997357-2-8.

168 Cao, J., Sim, Y., Tan, X.Y. et al. (2022). Upcycling silicon photovoltaic waste into thermoelectrics. *Advanced Materials* 34 (19): 2110518. https://doi.org/10.1002/adma.202110518.

169 Schubert, E.S. and Gottschling, R. (2011). Co-generation: a challenge for furnace off-gas cleaning systems. In: *Southern African Pyrometallurgy* (ed. R.T. Jones and P. den Hoed). Johannesburg: Southern African Institute of Mining and Metallurgy.

170 Saevarsdottir, G., Kvande, H., and Magnusson, T. (2023). The path to zero carbon dioxide emissions in silicon production. In: *Advances in Pyrometallurgy*, The Minerals, Metals & Materials Series (ed. C. Fleuriault, J.D. Steenkamp, D. Gregurek, et al.), 123–135. Cham, Switzerland: Springer Nature, https://doi.org/10.1007/978-3-031-22634-2_12.

171 Andersen, V., Solheim, I., Gaertner, H. et al. (2023). Pilot-scale test of flue gas recirculation for the silicon process. *Journal of Sustainable Metallurgy* 9 (1): 81–92. https://doi.org/10.1007/s40831-022-00639-0.

172 Han, G. and Sohn, H.Y. (2005). Kinetics of the hydrogen reduction of silica incorporating the effect of gas-volume change upon reaction. *Journal of the American Ceramic Society* 88 (4): 882–888. https://doi.org/10.1111/j.1551-2916.2005.00144.x.

173 Monsen, B., Kolbeinsen, L., Prytz, S. et al. (2013). Possible use of natural gas for silicon or ferrosilicon production. *Proceedings 13th International Ferroalloys Congress: Efficient Technologies in Ferroalloy Industry*, Vol. 1, pp. 467–478. Almaty, Kazakhstan (9–13 June 2013).

174 Jiang, T., Xu, X., and Chen, G.Z. (2020). Silicon prepared by electro-reduction in molten salts as new energy materials. *Journal of Energy Chemistry* 47: 46–61. https://doi.org/10.1016/j.jechem.2019.11.005.

175 Tian, F., Pang, Z., Hu, S. et al. (2023). Recent advances in electrochemical-based silicon production technologies with reduced carbon emission. *Research* 6: 0142. https://doi.org/10.34133/research.0142.

176 Braga, A.F.B., Moreira, S.P., Zampieri, P.R. et al. (2008). New processes for the production of solar-grade polycrystalline silicon: a review. *Solar Energy Materials and Solar Cells* 92 (4): 418–424. https://doi.org/10.1016/j.solmat.2007.10.003.

177 Lei, Y., Ma, X., Wang, Y. et al. (2022). Recent progress in upgrading metallurgical-grade silicon to solar-grade silicon via pyrometallurgical routes. *International Journal of Minerals, Metallurgy, and Materials* 29 (4): 767–782. https://doi.org/10.1007/s12613-022-2418-3.

178 Wood, L. (2021). Outlook on the Silicon Metal Global Market to 2027 - Opportunity analysis and industry forecasts, *GlobeNewswire News Room* [Online]. Available: https://www.globenewswire.com/news-release/2021/03/26/2199987/28124/en/Outlook-on-the-Silicon-Metal-Global-Market-to-2027-Opportunity-Analysis-and-Industry-Forecasts.html (accessed 12 March 2025).

179 Gupta, P.K. (1996). Non-crystalline solids: glasses and amorphous solids. *Journal of Non-Crystalline Solids* 195 (1–2): 158–164. https://doi.org/10.1016/0022-3093(95)00502-1.

180 Varshneya, A.K. (1994). Fundamentals of the glassy state. In: *Fundamentals of Inorganic Glasses*, 14. New York: Academic Press.

181 Axinte, E. (2011). Glasses as engineering materials: a review. *Materials & Design* 32 (4): 1717–1732. https://doi.org/10.1016/j.matdes.2010.11.057.

182 Hasanuzzaman, M., Rafferty, A., Sajjia, M., and Olabi, A.-G. (2016). Properties of glass materials. In: *Reference Module in Materials Science and Materials Engineering*. Amsterdam: Elsevier, https://doi.org/10.1016/B978-0-12-803581-8.03998-9.

183 Vedrtnam, A. and Pawar, S.J. (2025). Laminated glass: classification, characterization, and future perspectives. *Journal of Materials Education* 42: 51–61.

184 Musgraves, J.D., Hu, J., and Calvez, L. (ed.) (2019). *Springer Handbook of Glass*, Springer Handbooks. Cham: Springer International Publishing, https://doi.org/10.1007/978-3-319-93728-1.

185 Furszyfer Del Rio, D.D., Sovacool, B.K., Foley, A.M. et al. (2022). Decarbonizing the glass industry: a critical and systematic review of developments, sociotechnical systems and policy options. *Renewable and Sustainable Energy Reviews* 155: 111885. https://doi.org/10.1016/j.rser.2021.111885.

186 Glass Alliance Europe (2024). Statistical Report 2023–2024 [Online]. Available: https://www.wko.at/oe/industrie/glasindustrie/annual-report-of-gae (accessed 07 June 2024).

187 Wintour, N. and International Labour Organization. Sectoral Policies Department (2015). The glass industry: Recent trends and changes in working conditions and employment relations. Geneva: ILO. EISBN: 9789221301189.

188 Environment Agency (2001). Integrated Pollution Prevention and Control (IPPC): Guidance for the Glass Manufacturing Sector (A1 processes), Bristol.

189 Lubitz, G. (2023). Success factors for efficient container glass production. *Glass International* 23–24.

190 Kafizas, A. and Parkin, I.P. (2012). Chapter 20: Glass and new technologies. In: *Materials for a Sustainable Future* (ed. T.M. Letcher and J.L. Scott), 644–667. Cambridge: The Royal Society of Chemistry, https://doi.org/10.1039/9781839168949.

191 OGIS GmbH. Glass Global - The global glass competence (2025). [Online]. Available: https://www.glassglobal.com/ (accessed 03 March 2025).

192 Business Waste (2024). Glass waste facts and statistics, Business Waste [Online]. Available: https://www.businesswaste.co.uk/your-waste/glass-recycling/glass-waste-facts-and-statistics/ (accessed 12 March 2025).

193 Colangelo, S. (2024). Reducing the environmental footprint of glass manufacturing. *International Journal of Applied Glass Science* 15 (04): 350–366. https://doi.org/10.1111/ijag.16674.

194 Berger, G., Raonic, Z., Forthuber, D. et al. (2022). Energy and material flow evaluation with CO_2 emissions in the glass production process. *Advances in Material Physics and Chemistry* 12 (05): 82–105. https://doi.org/10.4236/ampc.2022.125007.

195 Di Pierro, S. (2021). Raw materials for glassmaking: properties and constraints. In: *Encyclopedia of Glass Science, Technology, History, and Culture*, vol. 1 (ed.

P. Richet, R. Conradt, A. Takada, and J. Dyon), 39–51. John Wiley & Sons, https://doi.org/10.1002/9781118801017.ch1.2.

196 Thieme, C. (2000). Sodium carbonates. In: *Ullmann's Encyclopedia of Industrial Chemistry*, 1e. Weinheim: Wiley-VCH, https://doi.org/10.1002/14356007.a24_299.

197 Vanderzee, C.E. (1982). Thermodynamic relations and equilibria in (Na_2CO_3 + $NaHCO_3$ + H_2O): standard Gibbs energies of formation and other properties of sodium hydrogen carbonate, sodium carbonate heptahydrate, sodium carbonate decahydrate, trona: ($Na_2CO_3 \cdot NaHCO_3 \cdot 2H_2O$), and Wegscheider's salt: ($Na_2CO_3 \cdot 3NaHCO_3$). *The Journal of Chemical Thermodynamics* 14 (3): 219–238. https://doi.org/10.1016/0021-9614(82)90013-1.

198 Zier, M., Stenzel, P., Kotzur, L., and Stolten, D. (2021). A review of decarbonization options for the glass industry. *Energy Conversion and Management: X* 10: 100083. https://doi.org/10.1016/j.ecmx.2021.100083.

199 Hujova, M. (2017). Influence of fining agents on glass melting: a review, Part 2. *Ceramics - Silikaty* 202–208. https://doi.org/10.13168/cs.2017.0017.

200 Hujova, M. (2017). Influence of fining agents on glass melting: a review, Part 1. *Ceramics - Silikaty* 119–126. https://doi.org/10.13168/cs.2017.0006.

201 Muller-Simon, H. (2011). Fining of glass melts. *Reviews in Mineralogy and Geochemistry* 73 (1): 337–361. https://doi.org/10.2138/rmg.2011.73.12.

202 Moon, U.-C. and Lee, K.Y. (2003). Hybrid algorithm with fuzzy system and conventional PI control of TV glass furnace. *Control Systems Technology, IEEE Transactions on* 11: 548–554. https://doi.org/10.1109/TCST.2003.813385.

203 Beerkens, R. (2008). Analysis of elementary process steps in industrial glass melting tanks - some ideas on innovations in industrial glass melting. *Ceramics-Silikaty* 52: 206–217.

204 Pfaender, H.G. (1996). *Schott Guide to Glass*. Dordrecht: Springer Netherlands, https://doi.org/10.1007/978-94-011-0517-0.

205 U.S. Department of Energy (2017). Bandwidth Study U.S. Glass Manufacturing, Energy.gov. [Online]. Available: https://www.energy.gov/eere/iedo/articles/bandwidth-study-us-glass-manufacturing (accessed 12 March 2025).

206 Wilkinson, C. (2024). Glass and a carbon-free United States: what is glass's role in the upcoming green revolution? *Journal of the American Ceramic Society* 107 (03): 1533–1542. https://doi.org/10.1111/jace.19360.

207 Caudle, B., Taniguchi, S., Nguyen, T.T.H., and Kataoka, S. (2023). Integrating carbon capture and utilization into the glass industry: economic analysis of emissions reduction through CO2 mineralization. *Journal of Cleaner Production* 416: 137846. https://doi.org/10.1016/j.jclepro.2023.137846.

208 Yuan, X., Wang, J., Song, Q., and Xu, Z. (2024). Integrated assessment of economic benefits and environmental impact in waste glass closed-loop recycling for promoting glass circularity. *Journal of Cleaner Production* 444: 141155. https://doi.org/10.1016/j.jclepro.2024.141155.

209 Gheewala, S.H. (2024). Material recycling in a circular economy—a systems view. *WIREs Energy & Environment* 13 (1): e503. https://doi.org/10.1002/wene.503.

210 Saha, B.K., Chakraborty, B., and Dutta, R. (2020). Estimation of waste heat and its recovery potential from energy-intensive industries. *Clean Technologies and Environmental Policy* 22: 1795–1814. https://doi.org/10.1007/s10098-020-01919-7.

211 Li, Z., He, X., Wang, Y. et al. (2014). Design of a flat glass furnace waste heat power generation system. *Applied Thermal Engineering* 63 (1): 290–296. https://doi.org/10.1016/j.applthermaleng.2013.10.038.

212 Griffiths, S., Sovacool, B.K., Kim, J. et al. (2021). Industrial decarbonization via hydrogen: a critical and systematic review of developments, socio-technical systems and policy options. *Energy Research & Social Science* 80: 102208. https://doi.org/10.1016/j.erss.2021.102208.

213 Choudhary, M.K., Venuturumilli, R., and Hyre, M.R. (2010). Mathematical modeling of flow and heat transfer phenomena in glass melting, delivery, and forming processes. *International Journal of Applied Glass Science* 1 (2): 188–214. https://doi.org/10.1111/j.2041-1294.2010.00018.x.

214 Krijgsman, R. and Marsidi, M. (2019). *Decarbonisation Options for the Dutch Glass Wool Industry*. The Hague: PBL Netherlands Environmental Assessment Agency and ECN part of TNO.

215 Fluegel, A., Earl, D.A., and Varshneya, A.K. (2007). Electrical resistivity of silicate glass melts calculation based on the SciGlass Database [Online]. Available: https://www.glassproperties.com/resistivity/ (accessed: 17 June 2024).

216 Glass for Europe (2020). 2050: Flat-glass in climate neutral Europe: Triggering a virtuous cycle of decarbonisation". [Online]. Available: https://glassforeurope.com/wp-content/uploads/2020/01/PR-2050-Flat-Glass-Climate-Neutral-Europe.pdf (accessed 17 June 2024).

217 Barón, C., Perpiñán, J., Bailera, M., and Peña, B. (2023). Techno-economic assessment of glassmaking decarbonization through integration of calcium looping carbon capture and power-to-gas technologies. *Sustainable Production and Consumption* 41: 121–133. https://doi.org/10.1016/j.spc.2023.07.029.

218 Sutherland, B.R. (2020). Sustainably heating heavy industry. *Joule* 4 (1): 14–16. https://doi.org/10.1016/j.joule.2019.12.020.

219 Chan, Y., Petithuguenin, L., Fleiter, T. et al. (2019). *Industrial Innovation: Pathways to Deep Decarbonisation of Industry. Part 1: Technology Analysis*. London, ICF.

220 Deng, W., Backhouse, D.J., Kazi, F.K. et al. (2019). An ancient technology could help deliver decarbonisation. *Glass International* 42 (07): 47–49.

221 Kogut, K., Kasprzyk, K., Zboromirska-Wnukiewicz, B., and Ruziewicz, T. (2016). The research of ceramic materials for applications in the glass industry including microwave heating techniques. *IOP Conference Series: Materials Science and Engineering* 113: 012014. https://doi.org/10.1088/1757-899X/113/1/012014.

222 Sikarwar, V.S., Hrabovský, M., Van Oost, G. et al. (2020). Progress in waste utilization via thermal plasma. *Progress in Energy and Combustion Science* 81: 100873. https://doi.org/10.1016/j.pecs.2020.100873.

223 Prashantsinh, M.R., Bhagchandani, C.G., and Gupta, A. (2015). Environmental impact of soda ash using LCA tool. *International Journal for Innovative Research in Science & Technology* 1 (12).

224 Rahimpour, H., Fahmi, A., and Zinatloo-Ajabshir, S. (2024). Toward sustainable soda ash production: a critical review on eco-impacts, modifications, and innovative approaches. *Results in Engineering* 23: 102399. https://doi.org/10.1016/j.rineng.2024.102399.

225 Capua, R. (2015). Papyrus-Making in Egypt. In: *Heilbrunn Timeline of Art History*. New York: The Metropolitan Museum of Art [Online]. Available: https://www.metmuseum.org/essays/papyrus-making-in-egypt (accessed 21 September 2024).

226 Black, H. (2008). *Canadian Scientists and Inventors: Biographies of People who Shaped Our World*. Pembroke Publishers Limited.

227 Tsoumis, G.T. (2024). Wood. Encyclopedia Britannica, Wood | Properties, Production, Uses, & Facts | Britannica [Online]. Available: https://www.britannica.com/science/wood-plant-tissue (accessed 28 September 2024).

228 Fogra Forschungsinstitut für Medientechnologien e.V. (2024). Paper testing - paper specifications [Online]. Available: https://fogra.org/en/testing/materials-environment/paper-testing (accessed 30 September 2024).

229 Statista (2024). Global paper and paperboard production 2022 [Online]. Available: https://www.statista.com/statistics/270314/global-paper-and-cardboard-production/ (accessed 21 September 2024).

230 Statista (2024). Statista - The Statistics Portal [Online]. Available: https://www.statista.com/markets/410/topic/961/pulp-paper/ (accessed 17 September 2024).

231 Cepi (2024). Key Statistics 2023 European Pulp & Paper Industry. Cepi statistics [Online]. Available: https://www.cepi.org/wp-content/uploads/2024/09/Key-Statistics-2023-FINAL-2.pdf.

232 Furszyfer Del Rio, D.D., Sovacool, B.K., Griffiths, S. et al. (2022). Decarbonizing the pulp and paper industry: a critical and systematic review of sociotechnical developments and policy options. *Renewable and Sustainable Energy Reviews* 167: 112706. https://doi.org/10.1016/j.rser.2022.112706.

233 Martin, N., Anglani, N., Einstein, D. et al. (2000). *Opportunities to Improve Energy Efficiency and Reduce Greenhouse Gas Emissions in the U.S. Pulp and Paper Industry*, LBNL-46141. Ernest Orlando Lawrence Berkeley National Laboratory.

234 Smurfit Westrock (2024). A global leader in sustainable packaging [Online]. Available: https://www.smurfitwestrock.com (accessed 2 October 2024).

235 International Paper (2024). International Paper 2023 Annual Report, International Paper, Annual Report, March 2024 [Online]. Available: https://www.internationalpaper.com/resources/reports/annual-report.

236 UPM (2024). A material solutions company [Online]. Available: https://www.upm.com/ (accessed 2 October 2024).

237 Greycon (2024). Top 10 pulp & paper manufacturers in 2023 [Online]. Available: https://www.greycon.com/us/paper-us/top-10-pulp-paper-manufacturers/ (accessed 30 September 2024).

238 Stora Enso (2023). Stora Enso Annual Report 2023 [Online]. Available: https://www.storaenso.com/-/media/documents/download-center/documents/annual-reports/2023/storaenso_annual_report_2023.pdf (accessed 3 October 2024).

239 Statista Research Department (2023). Mondi Group's production volumes worldwide from 2018 to 2022, by product [Online]. Available: https://www.statista.com/statistics/1294077/mondi-production-volumes-by-product/ (accessed 2 October 2024).

240 Nine Dragons Paper (2024). Financial Reports | Nine Dragons Paper (Holdings) Limited [Online]. Available: http://www.ndpaper.com/en/ir/reports.php (accessed 3 October 2024).

241 Nippon Paper Group (2024). Domestic Mills | Network | Nippon Paper Group [Online]. Available: https://www.nipponpapergroup.com/english/about/branch/factory/ (accessed 3 October 2024).

242 Sappi (2023). Sappi Annual Integrated Report 2023 [Online]. Available: https://sappi-ir-reports.co.za/online-reports/air-2023/downloads.php (accessed 3 October 2024).

243 DS Smith (2024). DS Smith Paper Mills [Online]. Available: https://www.dssmith.com/company/locations/our-paper-mills (accessed 3 October 2024).

244 Ritchie, H. (2024). Forest area - Our World in Data [Online]. Available: https://ourworldindata.org/forest-area (accessed 7 October 2024).

245 FAO (2020). Global Forest Resources Assessment [Online]. Available: https://fra-data.fao.org/assessments/fra/2020 (accessed 7 October 2024).

246 Environmental Protection Agency. (1998). Pulp and paper NESHAP: a plain english description, Environmental Protection Agency, Office of Air Quality Planning and Standards [Online]. Available: https://ntrl.ntis.gov/NTRL/dashboard/searchResults/titleDetail/PB99124067.xhtml (accessed 23 September 2024).

247 Van Ewijk, S., Stegemann, J.A., and Ekins, P. (2018). Global life cycle paper flows, recycling metrics, and material efficiency. *Journal of Industrial Ecology* 22 (4): 686–693. https://doi.org/10.1111/jiec.12613.

248 Bajpai, P. (2015). *Green Chemistry and Sustainability in Pulp and Paper Industry*. Cham: Springer International Publishing, https://doi.org/10.1007/978-3-319-18744-0.

249 Olejnik, K. (2011). Water consumption in paper industry – reduction capabilities and the consequences. In: *Security of Industrial Water Supply and Management*, NATO Science for Peace and Security Series C: Environmental Security (ed. A.T. Atimtay and S.K. Sikdar), 113–129. Dordrecht: Springer Netherlands, https://doi.org/10.1007/978-94-007-1805-0_8.

250 Bajpai, P. (2016). *Pulp and Paper Industry: Energy Conservation*. Elsevier.

251 Rullifank, K.F., Roefinal, M.E., Kostanti, M. et al. (2020). Pulp and paper industry: an overview on pulping technologies, factors, and challenges. *IOP Conference Series: Materials Science and Engineering* 845 (1) Art. no. 1: https://doi.org/10.1088/1757-899X/845/1/012005.

252 Joint Research Centre (European Commission), Pavel, C.C., and Moya, J.A. (2018). *Energy efficiency and GHG Emissions: Prospective Scenarios for the*

Pulp and Paper Industry. Publications Office of the European Union [Online]. Available: https://data.europa.eu/doi/10.2760/035301 (accessed 25 September 2024).

253 Pola, L., Collado, S., Oulego, P., and Díaz, M. (2022). Kraft black liquor as a renewable source of value-added chemicals. *Chemical Engineering Journal* 448: 137728. https://doi.org/10.1016/j.cej.2022.137728.

254 Szabó, L., Soria, A., Forsström, J. et al. (2009). A world model of the pulp and paper industry: demand, energy consumption and emission scenarios to 2030. *Environmental Science & Policy* 12 (3) Art. no. 3: https://doi.org/10.1016/j.envsci.2009.01.011.

255 Sun, Z., Chen, X., Lu, F. et al. (2022). Effect of Rb promoter on Fe_3O_4 microsphere catalyst for CO_2 hydrogenation to light olefins. *Catalysis Communications* 162: 106387. https://doi.org/10.1016/j.catcom.2021.106387.

256 Sun, M., Wang, Y., Shi, L., and Klemeš, J.J. (2018). Uncovering energy use, carbon emissions and environmental burdens of pulp and paper industry: a systematic review and meta-analysis. *Renewable and Sustainable Energy Reviews* 92: 823–833. https://doi.org/10.1016/j.rser.2018.04.036.

257 Environmental Technology Best Practice Programme (1998). *Practical Water Management in Paper and Board Mills*. Environmental Technology Best Practice Programme.

258 Man, Y., Li, J., Hong, M., and Han, Y. (2020). Energy transition for the low-carbon pulp and paper industry in China. *Renewable and Sustainable Energy Reviews* 131: 109998. https://doi.org/10.1016/j.rser.2020.109998.

259 Kamali, M. and Khodaparast, Z. (2015). Review on recent developments on pulp and paper mill wastewater treatment. *Ecotoxicology and Environmental Safety* 114: 326–342. https://doi.org/10.1016/j.ecoenv.2014.05.005.

260 Gholami, M., Ghaneian, M.T., Fatemi, S.S. et al. (2022). Investigating the efficiency of coagulation and flocculation process in wastewater treatment of paper and cardboard recycling industry. *International Journal of Environmental Analytical Chemistry* 102 (17): 5860–5872. https://doi.org/10.1080/03067319.2020.1804891.

261 WWF International (2007). The WWF Guide to Buying Paper, WWF International [Online]. Available: https://assets.wwf.org.uk/downloads/wwf_paper_guide.pdf.

262 Kong, L., Hasanbeigi, A., and Price, L. (2016). Assessment of emerging energy-efficiency technologies for the pulp and paper industry: a technical review. *Journal of Cleaner Production* 122: 5–28. https://doi.org/10.1016/j.jclepro.2015.12.116.

263 Ericsson, K. and Nilsson, L.J. (2018). Climate innovations in the paper industry: Prospects for decarbonisation, Vol. IMES/EESS report 110 in IMES/EESS report series. Miljö- och energisystem, LTH, Lunds universitet.

264 Karltorp, K. and Sandén, B.A. (2012). Explaining regime destabilisation in the pulp and paper industry. *Environmental Innovation and Societal Transitions* 2: 66–81. https://doi.org/10.1016/j.eist.2011.12.001.

265 Ericsson, K. (2017). Biogenic carbon dioxide as feedstock for production of chemicals and fuels : A techno-economic assessment with a European perspective. Miljö- och energisystem, LTH, Lunds universitet [Online]. Available: http://lup.lub.lu.se/record/67d3a737-cf7c-4109-bc4f-a6346956d6a2 (accessed 15 October 2024).

266 Anglo American (2021). , Anglo American commissions first hydrogen station for electromobility in Chile [Online]. Available: http://www.angloamerican.com/our-stories/environment/anglo-american-commissions-first-hydrogen-for-electromobility-in-chile (accessed 4 December 2024).

8

Consumer Goods Industries

8.1 Overview of Consumer Goods Industries

Consumer goods, for the purposes of this chapter, may include any product that is targeted for use by the end user. Whilst the 'basic industries' were defined as producers of large volumes of products feeding into other major industries, e.g. the use of sulphuric acid feeding into copper manufacture, the 'manufacture of materials' chapter looked at the production of materials used on a large scale. Consumer goods will include areas such as the food industry, which makes food with the direct target market of the individual consumer. The textile industry may be argued as either material manufacture or consumer goods, but certainly the clothing industry would be a consumer good. Packaging and domestic appliances may be considered consumer goods, as can some forms of construction. For the purposes of this chapter, three essential areas of consumer goods will be considered, viz. the food industry, the pharmaceutical industry and waste management, specifically wastewater treatment, as water is a fundamental necessity required both on an individual basis as well as an industrial scale.

The food industry is responsible for a staggeringly large amount of global CO_2 emissions, accounting for around 28% of global emissions or 13.7 billion metric tonnes of $CO_{2,eq}$ [1]. Though, greater estimates have been made, up to a third of the global equivalent $CO_{2,eq}$ emissions [2]. The vast majority of this comes from the agricultural sector rather than food processing itself; it is estimated that in the United States 9.9% of their $CO_{2,eq}$ emissions in 2020 originated from agriculture, whilst only 0.6% came from food processing [3]. Within the agricultural sector, the majority of CO_2 release comes from cattle, with red meat making up over half of the amount of CO_2 released in the food production for an average American diet, with an additional 20% being attributed to dairy. An example of this can be seen in the equivalent amount of CO_2 emissions released per serving (see Figure 8.1), where the emissions associated with the consumption of beef is more than double the emissions associated with cheese, which is the next largest greenhouse gas emitter. The major CO_2 release from cattle comes from two locations, firstly the cattle themselves, but also the food grown as cattle feed.

There are three main reasons why the agricultural industry contributes to such a large extent to the overall equivalent CO_2 emissions [4, 5]:

Renewable Energy in the Process Industry, First Edition.
Eric van Steen, Nicholas S. Featherstone, and Linda H. Callanan.

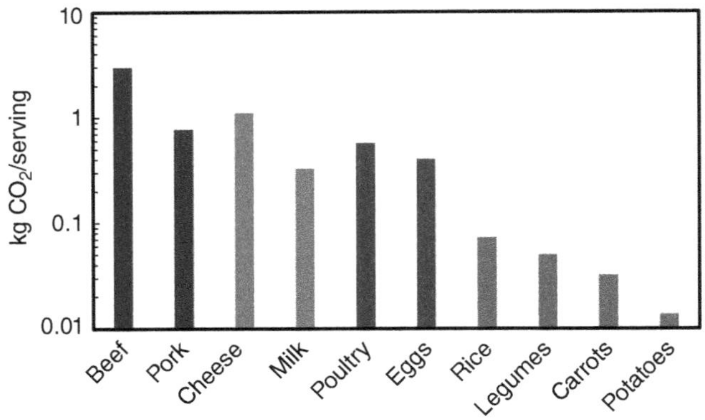

Figure 8.1 The equivalent CO_2 release per serving for various foods (note the logarithmic scale; the unit per serving is an objective measure, so this is primarily for illustrative purposes) (data from University of Michigan [3]).

- *Land-use*: Raising animals, especially cattle, as well as domesticated grains require large tracts of land, with much of this land being cleared before use through burning. This has the dual negative effect of deforestation and CO_2 release (besides the additional CO_2 that may have been taken up by the pre-existing plants).
- *Fermentation*: Cattle produce methane upon digestion of food in their guts, acting as anaerobic digestors producing methane and CO_2. Indeed, manure is a major source of global methane release. To put this in perspective, around 40% of America's methane release comes from livestock [3] (and ca. 40% from petrochemical processes and 20% from various waste streams). The high global warming potential of methane thus results in a large negative impact.
- *Fertiliser*: The increased need for food globally requires fertilisation of crops with artificial fertilisers, typically produced using ammonia synthesised using the Haber-Bosch process. However, not all of the fertiliser distributed over the field is taken up by the crop, but some is broken down into NO_x, a species with a high global warming potential. To put this in perspective, around 90% of the NO_x emissions by the United States comes from the agricultural sector, contributing to ca. 7% of their total equivalent CO_2 emissions [3].

Growing crop and raising cattle have a strong effect on the emission of greenhouse gases. However, here we will focus on what is done with the crop post farming, i.e. in food processing, and how renewable energy may be implemented in these areas. Processing is an important part of the overall food production chain as it takes food that has calorific value but is not necessarily in a format easy to eat/digest and makes it into food that may be readily consumed and/or has some manner of preservation increasing the shelf life of the product. Processes that will be looked at include industrial baking and canning, though there are many other food production processes which may also see meaningful implementation of renewable energy.

Pharmaceuticals cover a wide range of things, including drugs, which will be discussed here. It is worth noting that other areas, such as vaccines, could also be examined, but these have broadly speaking similar energy needs between the various pharmaceutical industries, viz. sterilisation, but otherwise low temperatures.

Pharmaceutical drugs comprise a set of essential medications, which assist in the priority health care of the general population. The pharmaceutical industry is relatively small in terms of total amount produced; however, the CO_2-footprint per kilogram of synthesised material can be large, and as such, it may have a significant impact on global warming.

A major aspect of the pharmaceutical industry will be the so-called Scope 3 type of emissions, specifically those associated with the transport of the drugs to the final consumer. This will be especially true of drugs (and vaccines) which require cold-chain transport.

Waste and waste generation are inherent to human existence, and thus waste treatment is essential for a healthy human society. There are many kinds of waste and all of them are treated differently depending on their composition. Waste types may be broken down into multiple categories depending on the definition and target in terms of treatment. For the purposes of this introduction, six categories will be discussed (expanded from [6]), of which wastewater will be a focus in this chapter.

Municipal waste is the most common form of waste in a person's everyday life. It includes things such as food scraps, paper, plastics, glass, metals, other biological matter (e.g. from garden plants), textiles, etc. Typically, the countries producing the most municipal waste per capita are the first-world countries. Most of this waste will end up as landfill, though many components of this waste may have a potentially better use. As has been discussed in the previous chapters, paper, plastics, glass and metal all have the potential to be recycled. Further food waste and much biological waste have the potential to be made into compost. Through a societal shift away from waste generation, there is the potential to significantly decrease the amount of waste generated each year by everyone. There is also a requirement for the use of 'waste as a resource' which is an area of much research and rapid growth [7–11].

Industrial waste may include sludge materials, such as red mud produced in the synthesis of alumina. Sludges are often treated by being dumped into sludge ponds, through evaporation of the water involved; this leaves only a solid sludge which may be disposed of in a landfill. Industrial waste may also include heavy metals, e.g. lead, mercury and cadmium, which are especially commonly found in the manufacturing by-products of batteries and electronics [12], though many other heavy metals may be present as well depending on the industrial source of the waste. Asbestos may also be present, though its use is being phased out due to its associated health risks.

Chemical waste may include things such as solvents, e.g. acetone and benzene; acids, e.g. sulphuric acid; pesticides, e.g. hazardous pesticides like DDT (dichloro-diphenyl-trichloroethane) and organophosphates; and fertilisers, specifically NO_x species which may be released from the use of fertilisers. Many of these chemicals do come from the industrial sector, meaning the lines between industrial and chemical waste are blurred. These may also include the remnants of batteries, which may also be classified as hazardous waste.

Medical waste primarily includes two things: contaminated instruments and pathological waste. Contaminated instruments may include used needles and scalpels; these can spread infectious diseases as they are contaminated and can easily pierce the skin. Pathological waste, such as human tissues from hospitals and laboratories, may also be present, which can spread diseases. The most robust method of dealing with the contamination effect of medical waste is incineration [13].

Nuclear waste deserves its own category here due to the specific ways in which it is processed, viz. in a cement bunker far away from human/animal/plant contact due to its high toxicity from radiation. Large amounts of research go into making nuclear waste treatment more sustainable [14, 15]. Nuclear energy is often seen as an essential backbone/base load provider in a fossil-free energy future and thus improved waste disposal techniques will be required.

Wastewater treatment can be broken down into two subsections, water treatment and sludge treatment. In treating the wastewater, water of a given purity may be recovered as well as sludge. Sludge is the sediment removed from the water. This sludge may be treated in an aptly named 'sludge pond' in which anaerobic digestion may be used, which uses up nutrients in the waste (i.e. making it less polluting) as well as reducing the total waste volume [16]. Anaerobic digestion may also be done in dedicated units in order to recover the methane formed as a fuel. This form of waste treatment will be discussed later in this chapter due to the reusability of water, as well as its essentiality to life on Earth.

8.2 Industrial Baking

Contribution by Noor Hassan.

8.2.1 Introduction

Throughout history, people from many cultures have incorporated some form of baked food into their diets. Baked goods include grain-based foods, such as bread, cakes, pies and pastries, with bread being considered a staple food around the world.

Food processing can be considered in the same way as any other chemical process – reactants (ingredients) are combined, undergo a transformation through chemical reactions under certain process conditions and exit the process as the desired product. Unlike traditional chemical processes, the 'specification' of bread is not in the form of product purity. Rather, the most important consideration is the characteristics of the final product, which include its taste, texture, appearance, shelf life and safety. These properties are ensured by careful control of the process variables in the production steps [17, 18].

Bread comes in many forms (see Figure 8.2) – from the baguettes of France to Indian roti to sourdough, which saw a resurgence during the COVID-19 pandemic. Bread that caters to specific dietary requirements, such as gluten-free bread, can be made by changing the flour. The scale of the bakery often influences the specification of the bread produced. Small artisan bakeries produce loaves that are of a higher quality (and therefore more expensive) in batches, with parts of the process

Figure 8.2 Different types of bread (source article distributed under the terms and conditions of the Creative Commons Attribution (CC BY) license (https://creativecommons.org/licenses/by/4.0/)) (Rahman et al. [19] / MDPI / CC BY 4.0).

performed manually. Bread baked on an industrial scale tends to be simpler and more uniform, with machinery doing all the work. Here, we will focus on industrial bread production, particularly in the form of sliced loaves commonly found in supermarkets [19].

8.2.1.1 Industry Size

A bakery can be classified as either commercial or industrial. Commercial bakeries are typically on a relatively smaller scale. One such bakery is the Anthony and Sons Bakery, in New Jersey which produces 10 000 loaves per day [20]. Industrial factories run by Premier Foods, a South African company, produce on average about 177 000 loaves per day [21]. Similarly, the UK brand Hovis produces 162 500 loaves per day at each of their 8 factories (on average) [22].

The total global revenue generated from food production in 2024 was valued at $9.12 trillion per year [23]. As of 2024, the grain farming industry was valued at $1585 billion with a predicted compound annual growth rate (CAGR) of 8.8% [24]. This makes it one of the largest industries, which is comparable in size to the iron and steel market [25]. The most commonly produced grains are corn, wheat and rice (see Figure 8.3) with wheat and rice production being ca. 36% and ca. 68% less than the production of corn, respectively [26]. China with its large population is the largest overall consumer of wheat (see Figure 8.4) and consumes per capita ca. 105 kg of wheat per person per year, generally in the form of steamed buns [27]. The consumption of wheat per capita varies strongly regionally possibly linked to the prevalent food culture, and income levels.

The revenue from bread production industry was valued at $500 billion, as of 2024, with 192 million tonnes being produced per year. This revenue is expected to grow by 6.34% per annum [28]. Figure 8.5 shows the revenue and production volumes of bread from 2018 to 2024 and the predicted values until 2029. The rapid growth in the

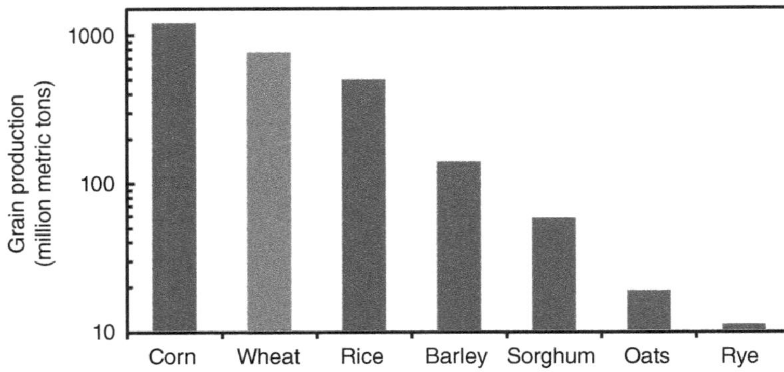

Figure 8.3 Grain production 2023/2024 by type (note logarithmic scale) (data from Shahbandeh [26]).

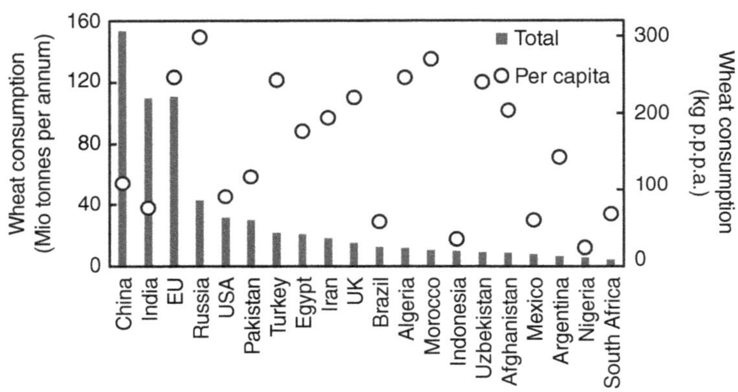

Figure 8.4 Wheat consumption 2023/2024 by country (data from Shahbandeh [27]).

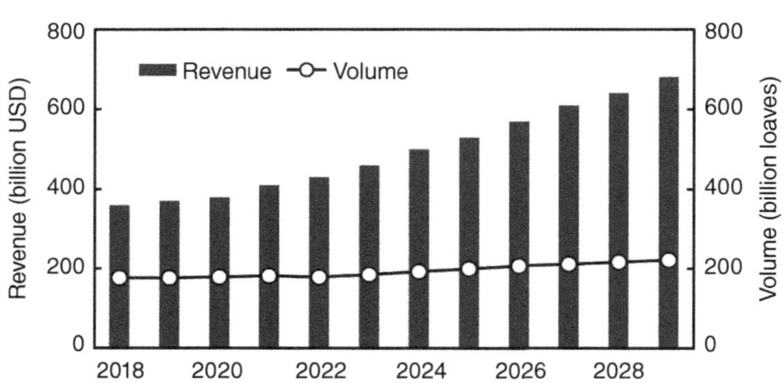

Figure 8.5 Global bread revenue and production volumes (redrawn from Statista Market Insights [28]; extrapolated from 2025 onwards).

predicted revenue is accompanied by the more gradual increase in the production volume over the same period, which indicates that the price of bread will increase (in line with inflation). The bakery packaging industry had a global market value of $3.54 billion as of 2023, with an expected CAGR of 4.8% [29]. Note that this includes packaging for all baked goods, not just bread.

8.2.2 Raw Materials and Process Steps

The main ingredient in bread production is flour, which comes from wheat. Milling of the flour is an important process step and widely varying amounts of energy consumption have been reported for the conversion of wheat into flour ranging 28–280 kWh/tonne of wheat, with milling accounting for ca. 70% of these energy cost [30–33]. This process requires mainly electrical energy and will be kept outside the scope of the process boundary for the generation of bread. The quality of the wheat that is used to make the flour greatly affects the volume and structure of the product. For bread production, softer wheat with a protein content of 8–11% is used [17].

Water is used to bind the dough, with salt used to control the dough's taste and texture. Other ingredients, such as vegetable fat and acetic acid, can be added to improve the softness and shelf life of the bread, respectively [34].

Lastly, yeast is a key ingredient. Yeast is a single-cell fungus, which consumes sugars in a process called fermentation. Ethanol and gaseous CO_2 are released during fermentation, with the trapped gas bubbles expanding the dough volume, and causing the dough to rise [35].

The major steps in the overall breadmaking process (see Figure 8.6) are mixing of the ingredients (dough formation), fermentation, proofing, baking and post-processing (cooling and packaging).

Upon mixing flour with water, gluten molecules, a type of protein present in flour, interact with each other to result in a gluten network via hydrogen bonds and disulphide cross-links [18]. This process is aided by kneading, which uncoils the gluten molecules, resulting in a stronger network. The application of mechanical energy via a mixer (i.e. the kneading) is thus an essential contributor to the gluten network formation. There are various configurations of the mixer, including vertical spindle,

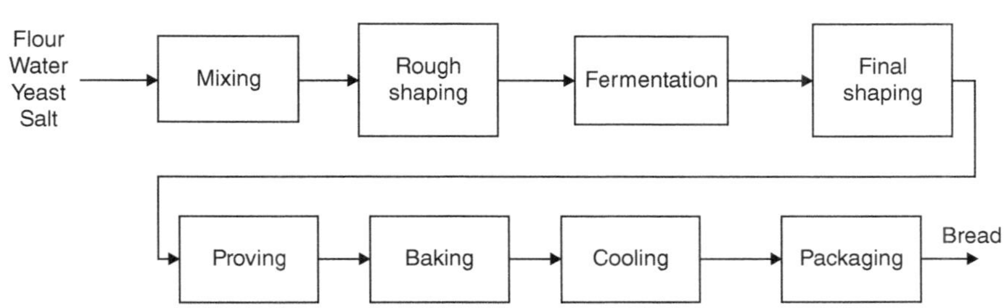

Figure 8.6 Block flow diagram of bread production process.

high-speed horizontal, double-arm and planetary mixers. For bread production, continuous mixers are preferred as they allow for homogeneous dough formation [18]. Furthermore, the addition of mechanical energy results in frictional energy losses in the form of heat, causing a temperature increase. At higher temperatures, the gluten network forms more quickly, however the gluten is at risk of denaturing. Therefore, controlling the mixing temperature – recommended at 17 °C – is important for the formation of a stable gluten network in the dough [18]. The Chorleywood Bread Process (CBP) deliberately uses a large amount of mechanical energy (44.6 kJ/kg) to mix the dough, which is much higher than traditional methods [36]. This causes the dough to heat up rapidly and the gluten network is formed more readily, allowing for an increased production rate. This is the primary method of bread production in the United Kingdom and produces a softer loaf, however there are concerns about the nutritional losses [36].

This gluten network forms the basis for the crumb, such that when flour with a low-protein content is used, the result is a low-volume bread – like baguettes [37]. In addition to adding flavour, salt plays an important role in slowing down the fermentation process, forming a more elastic dough. Vegetable fat can be used to soften the bread by stabilising the gas bubbles in the dough, increasing the volume [34].

During fermentation, carbohydrates in wheat, present as starch, which are long chains of sugar molecules, are broken down in the presence of amylase into simple sugars – maltose, glucose, sucrose and fructose [38]. Yeast consumes the glucose, which releases CO_2 as a gas, as well as ethanol [35]:

$$C_6H_{12}O_6 \rightarrow 2C_2H_5OH + 2CO_2$$

The expansion of the gas in the dough causes it to rise. This process is carried out in a warm, humid environment, such that the dough doubles in size [36]. The ethanol formed has a high relative volatility and evaporates during baking at a temperature above its boiling point of 78 °C. A skin forms on the outside of the dough, which will become the crust. It is important that the dough is handled gently after this stage in order to keep the skin intact and the CO_2 inside the dough [36].

In the Bulk Fermentation Process (BFP), the mixed ingredients are fermented together for about 1–20 hours [34]. When using the Chorleywood Bread Process (CBP), this bulk fermentation step is not required. After the fermented dough has been shaped into individual loaves, these are fermented again in a step called proofing, which occurs between 30 and 40 °C and a relative humidity (RH) of around 65% [36]. Provers can be batch, or on a conveyor belt for a continuous process.

Baking is the stage where the dough is transformed into bread. The bread must be baked at a high enough temperature to deactivate the yeast, but not so high that moisture is lost rapidly. A higher moisture retention is desirable as it allows for less raw material to be used to make a loaf of a certain weight and uses less energy for the evaporation of water. Furthermore, it has been linked to a softer crumb and a longer shelf life [36]. The gluten structure is considered to be set once the dough reaches 74 °C, with the CO_2 molecules trapped in this structure to form the porous crumb. A core temperature of 96–98 °C should be reached in order for the bread to be considered cooked [36].

During baking, heat transfer occurring via conduction, convection and radiation is important. Conductive heat transfer is used in industrial baking to produce biscuits or crackers, since the rapid heat transfer results in the desired crisp texture [39]. Convective heat transfer from the moving air or steam within the oven allows for control of temperature, pressure and humidity, all affecting the bread properties. Whilst this movement can occur naturally due to density and temperature differences in the air, it is more common to force the flow by using fans to circulate the hot air. Radiative transfer of heat via the electromagnetic spectrum, usually in the form of infrared radiation (IR), occurs when the dough faces the heating element. The low-energy waves do not penetrate far below the surface of the bread, allowing for browning of the crust to a desired colour [39].

The temperature in the ovens used to bake the bread is around 180–250 °C [36]. Direct ovens involve the burning of fuel inside the chamber where the dough is, resulting in direct contact between the dough and hot fumes [40] (see Figure 8.7). The obvious downside is that there should be no toxic impurities in the fuel source to ensure a safe product. Indirect ovens involve the burning of the fuel outside the main chamber. The resulting hot air circulates through a pipe that travels through the oven, heating the air via convection, which in turn cooks the bread. Direct-fired ovens have a high burner efficiency, with 92% of the fuel turned into heat in the oven, whilst this is only 65–80% for indirect-fired ovens, since heat is lost in the flue gas and heat exchanger efficiency [41]. Electricity can also be used to heat an element, which would work in the same way as pipes in the indirect oven.

In addition to the method of heat transfer, the way the dough is loaded into the oven is also considered. Continuous ovens are used in industrial-scale bread baking, allowing for a steady-state process with a very large production volume and uniform product. These so-called tunnel ovens are usually 30–40 m long and are split into zones such that the conditions can be controlled differently in different sections [36]. Batch ovens are used in commercial applications, where the bread is baked on trays that are inserted and removed manually.

The bread needs to be cooled down before it is sliced and bagged. The cooling is performed at atmospheric temperature with an RH of greater than 75%, to minimise

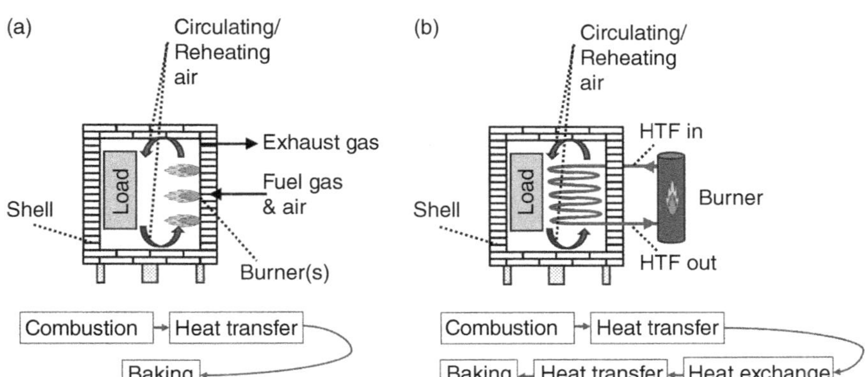

Figure 8.7 Diagram of direct (a) and indirect (b) ovens.

moisture losses. Steam is produced by natural gas burners [36]. The cooling, like the proofing and fermenting, can be done as a batch or continuously on a spiral conveyor belt. Vacuum cooling can be used to speed up this process, or alternatively, the air can be chilled to below ambient temperature to increase the temperature gradient, resulting in more rapid cooling [36]. However, these methods are not widely adopted in industrial baking.

The bread is sliced mechanically before it is packaged [36]. The material that the bread is transported and stored in is vital in maintaining health standards and the longevity of the bread. Fresh bread from a commercial or artisan bakery is usually served in paper packaging, which does not allow for the bread to last a long time. Sliced loaves usually come in plastic packaging that is secured with a tag or tape, but is not airtight, which keeps the bread fresh for longer [42].

8.2.3 Carbon Footprint of Breadmaking

Espinoza-Orias et al. [43] calculated that the carbon footprint for a loaf of bread (standardised to 800 g) is between 997 and 1244 g $CO_{2,eq}$ emissions for the entire life cycle. Various factors affect the emissions, such as wheat type – with wholewheat bread having lower emissions due to the decreased milling requirements – and the thickness of the slice – thicker slices mean fewer slices and therefore a decrease in the energy requirements for toasting. This study includes a cradle-to-grave analysis, meaning that the scope includes the emissions from the fertiliser used to grow the wheat to the emissions released by uneaten bread in a landfill. For this chapter, the scope is defined as the breadmaking process only, as illustrated in Figure 8.6. Espinoza-Orias et al. [43] determine that 16% of the total emissions come from the bread manufacturing itself (the rest originating mainly from growing wheat), resulting in an estimated 160–200 g $CO_{2,eq}$ per loaf. The packaging accounts for 1% of the total emissions.

Looking at the energy requirements for the individual processing steps allows for the identification of where green energy integration would be most useful:

For mixing, the maximum energy requirement occurs when using the Chorleywood Process (44.6 kJ/kg bread [36]), and results in 2.1 g$CO_{2,eq}$ per loaf, using the 2023 UK electricity emission factor of 0.207 kg $CO_{2,eq}$/kWh [44]. Standard mixing uses 15 kJ/kg [45], which would produce 0.7 g$CO_{2,eq}$ per loaf.

Paton [36] investigated the energy requirements of a standard industrial prover, which has contributions from the electrical air handling unit (AHU) to control air distribution, the gas burner to heat the air and the steam injected to control the RH. The prover has a dough flow rate of 8000 kg/h (10 000 loaves/h). Average weekly electricity and gas consumption rates were calculated as 3747 kWh and 633.4 m^3 (CV_{gas} = 40 000 kJ/m^3), respectively. Using the 2023 UK electricity emission factor, as above, and the natural gas factor of 0.183 kg $CO_{2,eq}$/kWh [44], the yearly emissions were calculated as 40.3 and 67.1 t $CO_{2,eq}$ for the electricity and gas consumption, respectively. An hourly saturated steam (8 bar) injection rate of 100 kg/h was also determined. Using the specific evaporation enthalpy of 2048 kJ/kg [46] and a natural gas boiler efficiency of 75% [36], the annual emissions were calculated as 115 t $CO_{2,eq}$. Combined, the prover has a footprint of 2.7 g $CO_{2,eq}$ per loaf.

Additionally, the fermentation process has inherent and hard-to-abate CO_2 emissions from the reaction itself. Istudor et al. [47] found that an increase in fermentation temperature resulted in an increase in the amount of CO_2 released. For this study, 9 portions of dough at 360 g each (targeting a final loaf weight of 300 g) were fermented in a 0.36 m^3 chamber at temperatures from 30 to 39 °C, resulting in an increase in CO_2 air concentration of 1900–2892 ppm, respectively. Based on the proofed dough dimensions, a temperature of 36 °C was recommended, which corresponds to an increase in the CO_2-concentration increase of 2624 ppm. Standardising this for an 800 g loaf results in 49 g $CO_{2,eq}$ emissions per loaf. It is clear that the emissions released by the fermentation reaction far outweigh the emissions from the energy input.

Baking is the most energy-intensive step in the bread production process. Energy can be lost to heating the baking tins/lids, through the oven walls, to heating the conveyor, and exiting with the flue gas, resulting in an energy efficiency specific to each oven [36]. Paton [36] suggests that the theoretical minimum energy required to bake white bread is 476 kJ/kg, with the actual energy requirements for a direct-fired tunnel oven being 804 and 920 kJ/kg for ovens with bread flow rates of 6500 and 9000 kg/h, respectively. Using natural gas as the fuel, this corresponds to 33 and 37 g $CO_{2,eq}$ per loaf, respectively. Similarly, in a direct gas-fired batch oven, the energy consumption was calculated as 796 kJ/kg bread, which releases 33 g $CO_{2,eq}$ per loaf [48]. Indirect ovens have a lower efficiency due to the use of heat exchanger pipes, so the CO_2 emissions would be slightly higher. For an electrical oven, Zisopoulos et al. [45] calculate a requirement of 1931 kJ/kg, which would give off 89 g $CO_{2,eq}$ per loaf.

The process modelled by Zisopoulos et al. [45] uses 3 × 400 W fans for 30 minutes to cool 40 kg of bread to 20 °C. This corresponds to an energy demand of 54 kJ/kg bread and 2.5 g $CO_{2,eq}$ per loaf. Additional electrical power is required for steps like shaping, slicing and packaging, but these are negligible compared to the calculated steps.

A summary of the carbon emissions for each main step is shown in Table 8.1, with emissions per 800 g loaf and total annual emissions for a process with 8000 kg

Table 8.1 Carbon dioxide equivalent emissions for various process steps to produce one 800 g loaf from flour, the annualised emissions of an industrial bakery assuming 10 000 loaves per hour operating 159 hours/week and the relative contribution of the individual process steps to the overall carbon footprint.

Process step	**Emissions**		
	g $CO_{2,eq}$/loaf	**t $CO_{2,eq}$/annum**	
Mixing	0.7–2.1	57–170	~1%
Proofing (reaction)	49	4015	34–56%
Proofing (operation)	2.7	222	2–3%
Baking	33–89	2702–7347	38–61%
Cooling	2.5	925	2–3%
Total	**89–145**	**7389–11 960**	

bread/h operating for 159 h/week. Carbon dioxide generated in the fermentation reaction contributes 1/3–1/2 to the total carbon footprint. The largest energy consumption occurs in baking, which then contributes between 38% and 61% to the overall carbon footprint. The other processes are minor contributors.

Like all processes, industrial baking generates waste. Waste from food processes can be anaerobically digested to generate electricity and heat [49]. Bread waste can be used as a feedstock for a biorefinery, where it is made into chemicals such as ethanol, lactic acid, succinic acid and 2–3-butanediol [50]. However, Vandermeersch et al. [49] reported that valorising food waste with over 60% dry solids content (like bread waste) into animal feed has 30% lower environmental impact than the anaerobic digestion.

8.2.4 Renewable Energy in Breadmaking

The electrified parts of the process are usually powered by the grid. On-site renewable power generation via solar panels can eliminate the emissions associated with these process steps. The mixing in a CBP producing 10 000 loaves per hour, proofing and cooling require 99.1 and 23.6 and 1.2 kW, respectively. For the mixing alone, using a solar irradiance of 1000 W/m^2, a solar panel efficiency of 10% and a field efficiency of 40%, the land area required would be approximately 2500 m^2. This is an area larger than what the roof of the factory would be, meaning that unless there is available land, the cost of integrating solar power into the process would be high. Solar panels could be built onto the roof, with the remaining power requirement offset by the grid. This will not eliminate the associated emissions but reduce them. There is also the issue with solar energy that the power supply would only be available during the day. Therefore, either battery storage is required, or the extra power requirements would also come from the grid.

The heating requirement for baking is the major contributor to the energy demand (see Table 8.1). Heat integration is an important part of the design of the baking plant and allows for the reduction of the overall energy demand. Often, the heat from the oven flue gas is used to heat the air going into the prover [36].

The natural gas that is used to generate the steam and heat the air in the direct oven can be replaced with hydrogen, which is a clean-burning fuel. The oven with an energy requirement of 920 kJ/kg bread has a power requirement of 2100 kW. Using the LHV of H_2 of 241.8 kJ/mol, it can be determined that 503 tonne/annum of H_2 is needed to fulfil this requirement. The hydrogen can be generated with an electrolyser. Using the liquid water Gibbs free energy of 237.1 kJ/mol to determine the work required, and an electrolyser efficiency of 58% (typical of PEM electrolysers), the power to the electrolyser should be 3625 kW. The land area of solar panels that can supply this energy is 90 600 m^2. A similar calculation can be done for the hydrogen required to make steam in the boiler for the prover – 11 tonne H_2 per annum, which would require approximately 2000 m^2 of solar panels.

However, replacing natural gas with hydrogen is not straightforward, since hydrogen behaves differently during combustion. The natural gas burners can only tolerate hydrogen in a mixture with a maximum of 20 vol% [51]. Therefore, in the long term, if hydrogen is chosen as the burner fuel, the burners will have to be replaced with ones designed for hydrogen. Furthermore, hydrogen burns approximately 100 °C hotter in air [52]. Therefore, the oven temperature will be affected, and gas flow rates may have to be adjusted to produce the desired product.

8.2.5 Current Research Around Future Processes

Research is being done into how solar energy can be used to bake bread directly. Forsythe and Madyira [53] investigated a solar-powered oven, which uses evacuated tube collectors (ETC) to heat air in pipes before it is sent into the oven. It was also found that using parabolic collectors improved the efficiency from 72% to 78%. This design was for a domestic scale oven, with plans for scaling up.

Electrification is a promising way to make the baking process green. Researchers at the Danish Technological Institute investigated the potential of using heat pumps as an alternative to gas burners for supplying heated air to a tunnel oven, using water as the working fluid [54]. These can be fitted retroactively onto existing tunnel ovens. Since temperatures above 200 °C are required for baking, the heat pump would be in the form of a compressor system with a compression ratio of about 20. This design showed promising results mechanically with a COP between 2 and 3, but further development of the compressor spindle and water injection designs was suggested, since the required spindles are difficult to manufacture and there were issues with the water film seal.

8.2.6 Outlook

The global food industry is very large and as the global population continues to grow, so will the demand for food, specifically bread. However, due to the relatively low temperatures in the system, compared to other industrial processes, the energy demand is not as high. Even though it is not possible to eliminate all carbon emissions, as seen by the CO_2 released by the fermentation reaction, an effort should still be made to reduce the environmental impact. Decarbonisation of the power supplied by the grid, will reduce the carbon emissions of the process. In countries where there is an effort being made towards adding renewable power to the grid, further electrification of the breadmaking process utilising heat pumps and/or resistive heaters may be the way forward. Alternatively, the fuel for the oven could be changed to green hydrogen or biogas, potentially replacing natural gas.

Within the bread-processing system boundary, the baking step is the most energy-intensive, but if that boundary is extended to the entire life cycle, it is actually the growing of the wheat that contributes 35–40% of the overall bread manufacturing emissions [43, 55]. Therefore, by using green ammonia to make the fertiliser, a significant portion of the carbon emissions can be eliminated. There is the issue that presently, green ammonia is more expensive to produce than grey (conventional) ammonia. However, due to predicted technological developments that will bring renewable electricity and electrolyser costs down, as well as the introduction of a carbon tax on fossil fuels, it is predicted that the green ammonia cost will become economically competitive by 2030 [56].

8.3 Canning and Related Processes

Canning has been an important technique in food preservation since its first implementation by Nicolas Appert, a French chef, in 1804. His initial observation came from storing syrups, which had been heated and stored in stoppered glass bottles, which appeared to last indefinitely. He expanded this work to include other food types and was, in 1810, able to claim a prize offered by the French government in 1795 to combat malnutrition and scurvy amongst soldiers. The technique was poorly understood, although various people kept on working on the idea of food preservation. In 1859, when Louis Pasteur demonstrated that microorganisms influenced food spoilage, the reasoning behind thermal processing was understood. Pasteur went on to develop a method for heat treating milk to preserve it and thus had the technique of pasteurisation named after him. Since then, a large amount of scientific research has gone into understanding microbiology and food spoilage, and food preservation has very much become a well-understood science [57].

Canning is a somewhat ambiguous term in the sense that what is often meant by canning is 'thermal processing', despite the initial development in glass jars. In 1910, Peter Durand of England patented preserving food in tin-coated iron cans, and in 1813 Bryan Donkin and John Hall set up a factory producing 'canned' food for the British army, and the name stuck. In terms of the process used (to heat treat foods in a sealed container with the intention of killing microorganisms to preserve the food), multiple different types of packaging may be employed, including glass jars, cans, glass bottles, laminated (plastic and/or metal) pouches, plastic tubs – essentially any packaging which can withstand the required temperatures [57].

Nowadays, the most important factor in the food industry is the safety of the food; this cannot be compromised, regardless of the potential for the processes to become more renewable and implement renewable energy and other renewable technologies. There are several considerations that come into the preservation and thus the shelf life and sterility of the food: these are referred to as 'hurdles' where more hurdles increase the safety of the food. Hurdles are factors in the food and/or the process which decreases the likelihood of the food spoiling or becoming dangerous [58]. The basis of hurdle technology is that the more hurdles there are, the less microorganism growth there can be, thus improving the safety of the food. Some of the hurdles are:

- the pH: strongly acidic or basic foods prevent bacterial growth (products may be acidified to assist with this hurdle);
- preservatives: added chemicals to prevent the growth of microorganisms;
- the water activity: water content is required for the growth of bacteria, this may be controlled through, e.g. sugar content (or ^{o}Bx), with too high a sugar content decreasing the availability of other nutrients required for growth (e.g. honey is known to have near infinite shelf life);
- the process: higher temperature kills more of the bacteria;
- aseptic operation: preventing the potential for bacteria to enter the product, where the feed may be sterilised;

- anaerobic sealing: to remove oxygen (using nitrogen or steam, for example) or to pull a vacuum on sealing, thus removing oxygen required for microbial growth;
- low-temperature storage: lower temperature slows down growth rate, and spoilage reaction rate (It should be noted that freezing is a very effective method to ensure a long food shelf life, although requiring a high energy consumption [59] due to the long-term requirement of maintaining the low temperature).

The most important factor (hurdle) that determines the process required to sterilise food is the pH [58]. The so-called high-acid foods require lower processing temperatures due to the preservation effects presented by the acidity of the product, viz. the bacterium which causes pathogens cannot grow in a high-acid environment (only bacteria which would cause spoilage), and thus a lower temperature process is required. 'Low-acid food' requires higher temperature sterilisation as bacteria which form pathogens can live in a more basic environment. Thus, high temperatures are required in a process called retorting to ensure a sufficient reduction in the number of bacteria present to preserve the food.

The most infamous of these pathogen-forming bacteria is Clostridium botulinum (or C. botulinum) which forms the botulinum toxins under low oxygen environments (i.e. the type of environment typically present in a sealed can). These toxins are some of the most lethal substances known; it prevents the release of the neurotransmitter acetylcholine, causing flaccid paralysis and eventual death. The disease is known as botulism. To get an idea for the toxicity of this, the median lethal dose of the botulinum toxin is ca. $2\,\text{ng/kg}_{\text{body mass}}$ [60], assuming an average human body mass of 70 kg – approximately 1 kg of the toxin could kill the entire human population.

The cut-off pH, and the pH at which C. botulinum can no longer grow is 4.6, as such any food more acidic than this is classified as high acid, and any food more basic than this is classified as low acid [57]. Most fruit-based products are high acid, whilst meat, vegetables, milk and coffee are low-acid foods. Table 8.2 gives a summary of the pH of some common foods which may need to be thermally processed.

Owing to the two different temperature ranges required, this chapter will look at two processes, retorting and pasteurisation, as there is a significant temperature difference between them. Typically, retorting occurs in the region of 121 °C (250 °F),

Table 8.2 Examples of some high-acid and low-acid foods (data from Featherstone [61]).

High acid (pH < 4.6)		Low acid (pH > 4.6)	
Apple juice	**3.5–3.7**	**Milk**	**6.5–6.7**
Peaches	3.3–4.0	Canned meat	5.9–6.6
Grapefruit	3.0–3.7	Baked beans	5.1–5.8
Ketchup	3.9	Coffee	4.8–5.1
Lemon juice	2.0–2.6	Corn	5.9–7.5
Pineapple	3.2–4.0	Tomatoes	4.3–4.9

thus, in high-pressure steam, whilst pasteurisation occurs in hot water, typically in the region of 90 °C.

The canned food market has a value of ca. 89 billion US$ in 2023 and is expected to reach ca. 109 billion US$ by 2029, showing a CAGR of 4.19%, very much in line with the other industries discussed in the previous two chapters despite the major difference in product type.

Production plants vary greatly in size depending on the type of product being produced, with some factories producing niche products like abalone, producing only a few hundred cans per week. Whilst larger factories for more common foodstuffs, such as baked beans, may produce over a billion cans per annum [62].

8.3.1 Current Processes

8.3.1.1 Overall Process

The process of canning follows several steps which vary slightly depending on the product in question. Broadly speaking, there are five primary steps:

- Receiving and storage
- Preparation and washing (including the can)
- Canning and sealing
- Sterilising
- Storage and dispatch.

The raw food will typically come from a farm (fruit, vegetables, meat, etc.), though it may also come from a fishing boat. Depending on the food in question and the final product intended, this will go through quality checks and may be cold-stored before being put into the preparation lines.

Preparation will also take different forms depending on the product in question. For example, canned peaches will be peeled, depitted and sliced, whilst pilchards will be headed and gutted. Sizing of the feedstock may also be done.

Canning and sealing involve adding the product into a sterilised can, with any liquids that may go in with the product, i.e. peaches with syrup or pilchards with tomato sauce, after which the can may be sealed. This will be done whilst weighing the product to ensure that the specified amount of both the base product and any appropriate sauce has been met.

- Jars are sealed using a metal lid with a sealing compound on the inside; as they are sealed a blast of steam is added to the top of the jar in the 'headspace' (the gap between the top of the product and the lid) which draws a vacuum upon condensing (hence releasing this vacuum gives the 'pop' that comes when the lid is removed from a sealed jar).
- Pouches are sealed through melting two of the edges together, typically with a sealing compound between them. These may first have a vacuum drawn inside the pouch.
- Plastic tubs are sealed by melting a laminate to the top of the tub once filled with the product.
- Cans are sealed through what is called a double seam (see Figure 8.8). The can lid has a lip which overhangs a lip on the top of the can body, the outer lip

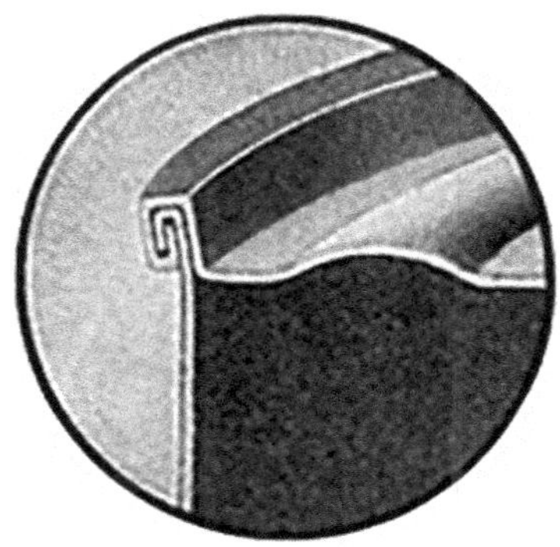

Figure 8.8 Double seam (image in public domain due to age) (Wikimedia [63] / Seattle Public Library / Wikimedia Commons / Public Domain).

(from the lid) then wraps around the lip from the body. After which, the can body lip (now surrounded by the lid lip) is flattened onto the body of the can. This too may be done with a blast of steam to draw a vacuum or with a blast of nitrogen to displace any air present in the headspace.

Sterilisation is the most important part of the process. Here, the food is cooked, killing microorganisms and rendering the food safe. This is also the part of the process where most of the work is done for what will make the consumer most satisfied, i.e. the product is not overcooked, the correct texture is achieved, etc. The product is subsequently cooled, typically using recirculating cooling water, which is then dried with jets of air to get the product ready for labelling, storage and dispatch. It has been estimated that, for fish canning, ca. >85% of the energy used on the canning plant is in the form of steam which, depending on the process, is almost all used for sterilisation [64].

Storage and dispatch take place after the product has been sterilised. For a sterilised product, the storage is relatively trivial, provided the storage temperatures are moderate (ca. 21 °C). Retorted canned food typically has a prescribed shelf life of around three years (though this is not true of pasteurised foods). Higher storage temperatures (e.g. 37 °C) would reduce this shelf life to ca. one year as the various spoilage reactions are accelerated at higher temperatures; hence, storage in a relatively cool warehouse is typically sufficient to support a long shelf-life, meaning refrigeration during storage is dependent on the location and not always a necessity.

Dispatch poses an interesting problem; trucking is still done using diesel power and thus is non-renewable and may be so for the foreseeable future depending on the development of renewably powered trucks. The scope of the effect of dispatch can be taken as beyond the boundaries of the food factory as it would form 'Scope 3' emissions and thus energy requirements.

8.3.1.2 Retorting – A Batch Process

Retorting occurs in a steam retort, using saturated steam to reach the target processing temperature, typically 121 °C (250 °F). Thus, retorting requires above atmospheric steam, with further increases in pressure required if there is air in the steam mixture, and a pressure vessel is required resulting in retorting being a batch process (see Figure 8.9). The temperature is measured throughout the retort,

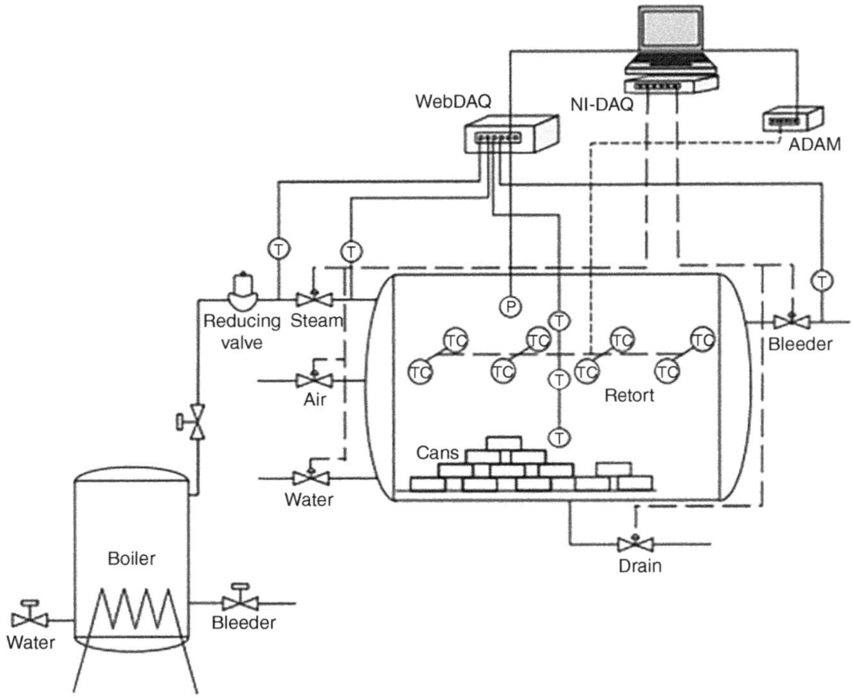

Figure 8.9 Schematic diagram of a horizontal retort (Escaño et al. [65] / with permission of Elsevier).

and there is a continuous feed of steam into the retort; part of the steam is ‘bled’ out whilst most condenses and leaves via a drain. The water may then principally be recycled. Modern retorts are computer-controlled, but human operators are still required should there be a process disruption.

The steam principally can come from any source but is largely produced in coal- or gas-fired boilers (both being well-understood feedstocks with relatively easy storage so that the plants can comfortably run through the day, especially important during the harvest season).

Calculating the amount of steam can be a complex problem due to heat losses in the retort which is a function of process time and temperature. Process time is highly dependent on the type of product being processed and the method of heat transfer therein. For example, water may be assumed to be a perfect convection pack (i.e. may be modelled assuming the only mode of heat transfer is convection and the contribution from conduction is negligible), thus requiring a short heating time to reach a given temperature. A product such as canned meat can be assumed to be a perfect conduction pack (assuming the contribution from convection is negligible) with far slower heat conduction to the cold spot (geometric centre of the can) when compared to water, and thus much longer processing times.

The minimum amount of energy required to heat the product to, e.g. 101 °C (the minimum temperature required to obtain a stable, storable product [66]) can be easily estimated from the heat capacity of water, 4.2 kJ/kg/°C, and the heat capacity of canned meat ca. 3.4 kJ/kg/°C [67]:

$$Q = m \cdot C_\text{p} \cdot \Delta T$$

Assuming a typical can mass of ca. 400 g starting at 25 °C, heating the water would require ca. 167 kJ/can and the canned meat would be ca. 136 kJ/can; notably, as most foodstuffs are water-based, the heat capacity is not a strong function of the product type. However, to heat the water, the process would require only a few minutes, whilst the meat product may be processed for over an hour [68]. Thus, the major concern with longer processes from an energy efficiency viewpoint is incomplete use of the steam, i.e. large amounts of waste steam leaving the retort, and heat loss from the walls of the retort.

The energy footprint of retorted products varies with the type of product and process used. Energy intensity has been estimated at: 2.7 MJ/kg for spinach, 3.9 MJ/kg for pork, 4.7 MJ/kg for baked beans and 6.6 MJ/kg for hominy (a maize-based product) [69]. Most of required energy is needed in the form of electrical energy [70] in areas such as cold storage (when achieved through refrigeration loops), product preparation, packaging and labelling. The existing electrical energy requirement can be readily replaced using various sources of green electricity with appropriate storage depending on the plant type and location. The primary non-electrical energy consumption comes from sterilising or cooking of the food.

8.3.1.3 Pasteurising – A Continuous Process

Pasteurising is a process taking place at a lower temperature than retorting using heated water rather than steam. This means that instead of needing a high-pressure retort, a continuous process can be used, in which hot water is sprayed over the packaged products, or the packaged products are conveyed through a bath of hot water. Temperatures used are typically up to ca. 90 °C and are more commonly in the range of 70–80 °C.

Pasteurising may be used on a range of products which do not need retorting, viz. high-acid products, alcoholic products, products with a high sugar content, etc. The energy consumption for pasteurisation depends on the product type, which is significantly lower than that of retorted products due to the lower temperatures and easier heat integration. For example, pasteurisation of beer has been estimated to require 19–23 kWh per 1000 bottles, approximately 0.2 MJ/kg [71].

However, at present, it is most common for the energy for the hot water to be provided by heat exchange with steam generated from coal/diesel-fired boilers simply due to the existing infrastructure of the plants. However, low temperatures are required and the heat of vaporisation of steam does not need to be provided. This makes the implementation of renewable energy in the form of heat pumps or resistive heaters relatively trivial. However, as mentioned above, pasteurisers may run continuously, which means that the intermittency of renewable energy will need to be overcome.

8.3.2 Implementation of Renewable Energy in Thermal Food Processing

Owing to the temperature requirements of the steam in the retort process (with a maximum required temperature of ca. 127 °C), coal-fired (or natural gas-fired) boilers could be relatively easily replaced using a heat pump or resistive heaters (see Chapter 4) powered using renewable electricity, either from the grid or generated on site. The use of heat pumps may be more readily applicable for pasteurisation processes where lower temperatures are required, and notably, the significant heat of vaporisation of water will not need to be overcome. However, due to their continuous nature, pasteurisation processes will need constant access to renewable electricity.

The batch processing nature of retorted products makes them prime candidates for using renewable energy, as they may be run when renewable energy is available rather than at any given point in the day. This may often be done regardless, depending on the product intake in a day, though it should be noted that this may not be an option during harvest season when the factory may need to run at full capacity, i.e. 24 hours a day. Should the factory be one that has such times, viz. a fruit or vegetable factory, provision would need to be made to be able to provide power 24/7 which could add significant capital cost to the investment in storage of renewable energy, though, conveniently harvest season often tends to coincide with the most solar availability, i.e. summer. Alternatively, at least in the short term, it would be possible to keep boilers (e.g. fed with biomass) operational during harvest season, whilst any other processing done may be operated by renewable sources. It should be noted as well that many larger factories operate multiple retorts, meaning that the process runs continuously even if each retort runs batch.

Continuous processes, such as pasteurisation or aseptic lines (processes that take product in one side and do not expose it to the environment again before it ends up in aseptic packaging) may be more challenging to switch to renewable energy, if it is operated 24/7. However, it should be noted that it is not uncommon for pasteurisation processes to run in a batch manner, due to the nature of requiring hands-on workers and the length of the workday.

Steam is of vital importance in the canning industry being used as the heating fluid directly in retorting processes and as an intermediate to heat water to be used in the pasteurising processes. This steam is required at a given temperature to be useful in food processing, though it will leave the process as a heated water stream. It is thus logical to include some heat integration to improve the energy efficiency of steam utilisation (see Chapter 4) (though heat integration is already implemented in some cases). This could lead to utilising hot water as the feed to the boiler, with the distinct advantage of minimising water usage on the plant, another consideration which will become more urgent in the future. Residual hot water from retorting processes may also be used in pasteurisation processes, but this requires the relevant plant to run both processes in sufficient proximity and with sufficient temperature requirements in order to make this viable.

Various improvements can be made in the steam system, including increasing boiler efficiency through increasing combustion efficiency (e.g. through feeding oxygen enriched air), and thermal efficiency (through minimising scaling on the water pipe side). The distribution network may also be optimised to minimise the length of steam travel, allow for heat integration as mentioned above [72].

Only part of the food feedstock is appropriate to be canned and made available for human consumption. For example, peach pits and skins are not canned and are discarded. These wastes are high in nutrients and so may commonly be used in purposes such as animal feed or fertilisers through composting. However, this potential also can lead to contamination, having negative environmental impact, therefore waste management is an important part of food processing [73].

The processing of waste is dependent on the type of waste generated, for example the non-edible parts of the sugar cane plant may be most appropriate for a process such as anaerobic digestion, to recover valuable renewable methane through biogas generation, which could be used to fire boilers.

8.3.3 Recycling of Packaging Materials

The canning industry primarily uses metal cans; the modern-day cans are so-called tinplate cans, which are steel with a tin coating, whereby a very thin tin-oxide layer provides very good corrosion resistance. Both the inside and the outside of the can are coated, with a thicker tin layer on the side exposed to the product, as the product, especially high-acid products, are more corrosive than the outside environment. For many products, the can has a further layer, known as a lacquer, which is a polymer coating providing extra protection from the can as well as serving other purposes, such as preventing discolouration from leaching of tin, or aesthetic reasons, e.g. a red tomato product on a white lacquer.

Beverage cans are not unlike food cans in terms of having a lacquer, though they are typically made from aluminium. They are more likely to have the label printed on the can rather than a paper label glued on the can (this is because beverage cans are plain/smooth-walled, whilst food cans are beaded/ribbed for extra strength).

Metal from cans is highly recyclable (especially aluminium cans) [74–76]. It is highly important that end users separate cans from general waste for them to be recycled. Municipalities/companies will need to make recycling facilities available (metal cans can take 100 years to decompose [77]). However, the lacquer is non-recyclable and needs to be burned off, resulting in carbon emissions in the recycling process. Further, a new lacquer coating will be required.

A more modern approach, especially for pasteurised foods is to use plastic laminates (multiple sheets of plastic laminated together as each sheet serves a different purpose, e.g. strength, etc.) such as those used for baby food pouches or plastic pots of fruits. These are typically chosen for convenience to the end user (and sometimes for aesthetic purposes) though they are much harder to recycle. The development of packaging for such products towards recyclability would be beneficial.

Glass may also be used as the packaging material, with a metal cap. A major benefit of glass is reuse, rather than recycling (which is energy-intensive due to the need to melt the glass or steel). Though there are still cleaning (requiring water) and sterilising requirements, so reuse is not an energy-free process.

Recycling food packaging is a major area through which individuals can make their lives more sustainable, and canned foods present an easy opportunity to do so. However, at current recycling is not always easy (e.g. through access to recycling facilities), and is seldom incentivised. Thus, in the future, it will be important to encourage individuals to recycle as well as companies to collect recyclable/reusable products and make greater use of them (which would likely also improve how easy it is for the end user to recycle).

8.3.4 Outlook

The canning industry has one primary prerogative, produce safe food, and any environmental or energetic concerns would need to be implemented in such a way as to ensure the safety of the food. Food quality is also of great importance and so changes made may not compromise on quality and only ensure safe food. However, there appears to be relatively easy potential for the food industry to implement methods which would make their energy usage more sustainable. A large amount of the energy required is electric, which can simply be moved to renewable electrical sources. The major processing concern at the moment is the generation of steam from coal and gas boilers. As low-pressure and relatively low-temperature steam is needed, the boilers could be replaced with renewably powered heat pumps. For retorted batch processes, it may be possible to prioritise operation during the day when renewable energy is more readily available, though this will be difficult for continuous operation as is used for pasteurisation. A major concern for the canning industry, as it pertains to fruits and vegetables, is the harvest season. Factories may sit dormant for months in a year and then operate full time in other periods. This makes the operation of renewable energy challenging as there would be a need for large-scale investment for energy storage, which may only be useful for a few months. In the immediate future, it may be expected that the canning industry will stay on coal boilers (though these may start to be co-fed with biomass, or may even be moved over to biomass entirely) for the security they provide, with a slow move towards heat pumps and resistive heaters for products which are processed in a batch fashion, as well as for relatively expensive products in which the cost of renewable energy may not be too severe for the end consumers. The eventual solution to 24/7 operation during harvest season may need to come from renewable fuels, as it would likely be impractical to build storage facilities to only use part of the year (unless they may be implemented with grid storage in the non-harvest season times).

8.4 Pharmaceutical Drugs

Contribution by Julie-Ann Hoffman.

The first indication of the use of medicines dates as far back as the Stone Age. In a grave of a Neanderthal man in the Shanidar cave, on the northern border between Iraq and Iran, several medicinal herbs were placed alongside the body. Many millennia later, the medicinal use of plants was recorded in China, Greece, Rome, India and the Middle East. Specifically, in 2500 BCE, over 350 medicinal plants or herbs were well-documented by a Chinese herbalist, Shen Nung. Throughout the Middle Ages and well into the nineteenth century, herbal medicines further developed into prescription medication and the concept of pharmaceutical drugs as a principal part of primary health care was born [78]. In the late nineteenth century, this concept further evolved and produced the segment of the chemical sector known as the pharmaceutical industry. The 'golden age' and origin of the modern pharmaceutical industry took place post-world wars, around the 1940s, where new inventions and advances of medicines such as alkaloids, coal tar derivatives and sulphonamides were introduced into the industry [79].

Medication is typically administered as tablets, capsules, creams, ointments, aerosols, patches, oral liquids or injectables [80]. The different forms reflect the variability in the combination of the active pharmaceutical ingredient (API) with a delivery system. In all those systems, the API typically makes up only a small amount of the total administered medication. For instance, a tablet is made up of the API, filler (e.g. sugar or starch), a binder (e.g. corn syrup or starch) and possibly a lubricant (e.g. magnesium stearate or polyethylene glycol). The filler guarantees the correct concentration/dosage of the API, the binder bonds the filler and the API together and a lubricant helps to slow down the disintegration of the API during its final processing [80]. Creams and ointments consist of the API in a water-in-oil or oil-in-water emulsion and thus also contain substances to stabilise the emulsion (emulsifiers), waxes, emollients and lubricants. Medication taken orally in the liquid form contains the API dissolved or suspended in a suitable liquid (water, saline water, ethanol), to which stabilisers, thickening agents and even sweeteners and colourants may be added.

The pharmaceutical industry is nowadays an economic powerhouse, with one of the fastest-growing markets in the world; the CAGR of the APIs is ca. 5.8%, due to the high population growth rate and the increasing diagnoses of chronic diseases like diabetes and cancer [81]. This can be further seen in the European economy where over the last two decades, especially after COVID-19, the industry value has continuously grown in various sections of the industry, i.e. the market value of imports, exports and production (see Figure 8.10) [82].

In 2022, the global pharmaceutical market for prescription medication was approximately € 1 222 921 million at ex-factory prices. The largest market share of 52.3% belonging to the North American market (USA and Canada), ahead of the European market share of 22.4% and the Chinese market share of 8.1% [82] (see Figure 8.11).

8.4.1 Current Process

Many different types of pharmaceutical drugs are being produced worldwide with an annual production ranging from tonnes per annum for aspirin and ibuprofen

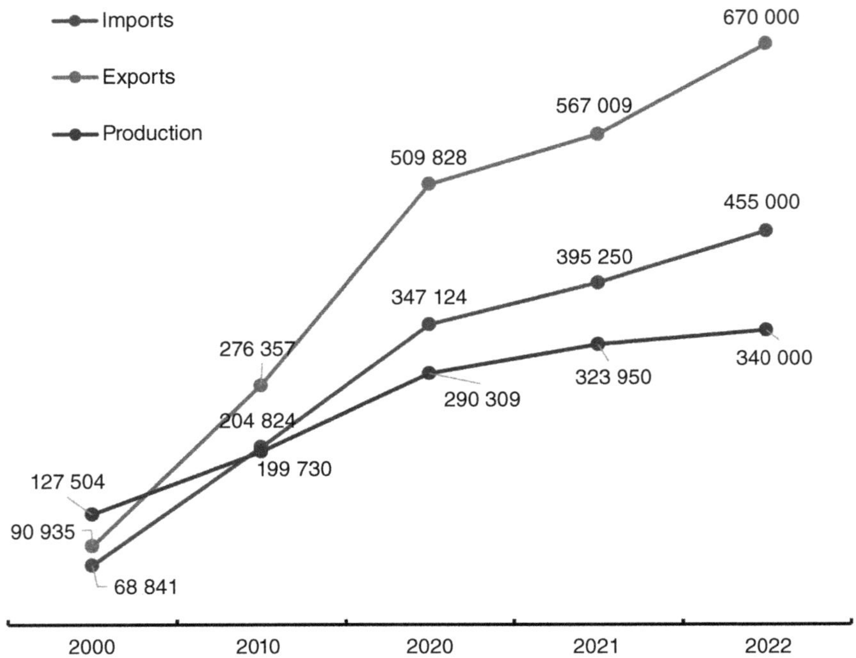

Figure 8.10 The pharmaceutical industry value (in EUR millions) in the European economy (adapted from The European Federation of Pharmaceutical Industries and Associations [82]; note: 2023 values are estimates, production value at factory price).

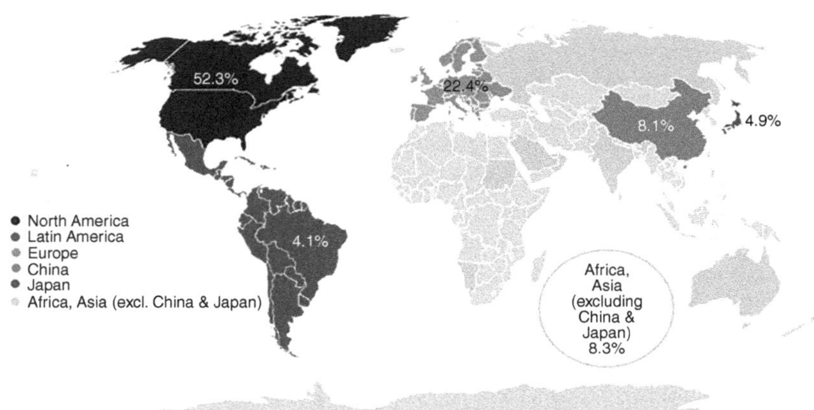

Figure 8.11 The global pharmaceutical sales market of prescription medication (2022) (adapted from The European Federation of Pharmaceutical Industries and Associations [82]).

to a few kilograms per annum for speciality drugs. Due to the diverse types of drugs, there are a variety of different processing routes. The production of drugs is generally divided into two main stages, viz. the primary stage consisting of manufacturing of the APIs, including synthesis, isolation and purification, and the secondary stage involves the drug formulation (or drug product manufacturing). These two stages may occur at different manufacturing sites, including locations in completely different countries. Hence, the production time can be extended by numerous weeks because of transportation between the two sites. Overall, the manufacturing production process can take up to 12–24 months, owing to the quality assurance measures that need to take place after each step [83, 84].

The production of APIs is via chemical synthesis, extraction, fermentation or a combination of these approaches [80], and takes place in batch processes (see Figure 8.12). The chemical synthesis produces drugs such as aspirin, ibuprofen, antihistamines, cardiovascular agents and central nervous system stimulants. On the other hand, pharmaceutical drugs such as insulin, some anti-cancer drugs and vaccines are extracted from naturally occurring substances. Fermentation is responsible for the production of various steroids and antibiotics. A combination of these processes is also used to produce antibiotics, antineoplastic agents, central nervous system depressants and vitamins [80].

The manufacturing of APIs using chemical synthesis involves several steps (see Figure 8.12). These include synthesis, separation, crystallisation, purification and drying. During the reaction step, raw materials, solvents and reagents partake in reactions such as alkylation, hydrogenation, oxidation or bromination. These reactions occur in reactor vessels [80]. The standard batch reactor or

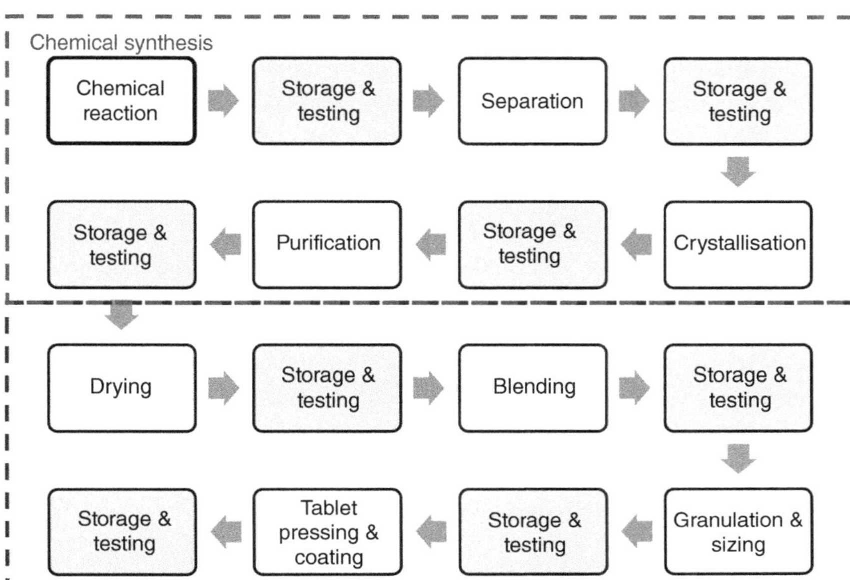

Figure 8.12 Block flow diagram representing the various steps in the production of pharmaceutical drugs as tablets via chemical synthesis highlighting the process quality control steps.

kettle-type reactor [80] has a capacity ranging from 500 to 10 000 l, and is made from glass-lined or stainless steel [85] with operating temperatures ranging from 20 to 120 °C [86]. Often the reactors are jacketed to allow coolant or steam to circulate to regulate the temperature within the reactor [85]. Both pressure and temperature are controlled and monitored within the reactor. Additionally, a condenser system may be necessary to control vent losses, as well as the addition of pollution control devices [80]. The separation step utilises a variety of different techniques, i.e. extraction, decanting, centrifuging, filtration and crystallisation. The most used technique is crystallisation, i.e. the formation of solid crystals from a supersaturated solution produced through cooling, solvent evaporation or chemical reactions. The crystals are then removed by filtration or centrifugation. The purification step generally occurs by recrystallisation, i.e. dissolving and crystallisation of the target compound. The last step is drying of the product or intermediates by evaporation of the residual solvent from the solid. There are various types of dryers that can be used such as tray dryers, rotary dryers, drum or tumble dryers or pressure filter dryers [80].

Extraction of the API is another method to isolate the pharmaceutical active compounds from natural materials. It is a costly process requiring suitable solvents and raw materials containing the active ingredient in sufficiently high concentration. Methods such as precipitation, recrystallisation and solvent extraction are commonly used to purify these natural active ingredients.

The production of penicillin proceeds via fermentation. Here, microorganisms are introduced into a liquid, which produce useful pharmaceutically active by-products [80]. Fermentation can be a slow process, with batch processing time ranging from hours to weeks, depending on the product.

Drug formulation is where the bulk pharmaceuticals are mixed with the delivery system ensuring that the pharmaceutical active ingredient is dosed in the correct concentration as well as other additives. Tablets are the most common form of medication, and the API is blended with a filler (e.g. sugar or starch), a binder (e.g. corn syrup or starch) and possibly a lubricant (e.g. magnesium stearate or polyethylene glycol). The purpose of a filler is to guarantee the API concentration, whereas the binder bonds them together and a lubricant helps to slow the disintegration of the API during its final processing [80]. Tablets are produced through compression, using methods such as wet or dry granulation. Wet granulation involves the API powder mixed with filler, wetted and blended with binder solution, mixed with the lubricants and pressed. Dry granulation is when coatings are applied to tablets that are sensitive to moisture or drying temperature. This process involves coating, drying and polishing in three separate rotary drums [80].

8.4.2 Energy Consumption in the Pharmaceutical Industry

The pharmaceutical industry is intertwined with its utilisation of fossil fuels. They are not only used as energy sources for electricity and process heat generation, but also as petroleum since it is a feedstock for many of its products [87, 88]. However, the pharmaceutical industry is also diverse, specifically

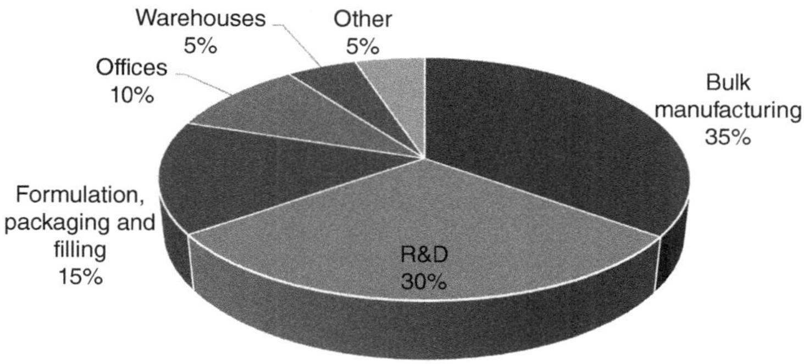

Figure 8.13 The energy use of the various facilities within a typical pharmaceutical plant (adapted from Galitsky et al. [80]).

when looking at the synthesis of the API. The average pharmaceutical plant (with a maximum productive surface area of about 15 000 m^2) consumes about 1926 kWh/m^2 per annum [89] where the heating, ventilation and air conditioning to maintain ambient air conditions are responsible for 65% of the energy consumption [80]. The need for climatisation of the process, laboratory and storage facility implies that energy input for heating, ventilation and air conditioning (HVAC) is required for 24/7.

The energy use is divided over several facilities within the pharmaceutical plant (see Figure 8.13) [90] with the bulk manufacturing section unit and the R&D (quality control section) each taking ca. 30–35% of the total energy usage. Formulation, packaging and filling typically use ca. 15% of the total energy usage.

8.4.2.1 Research and Development

The R&D process is comprised of four main stages: pre-clinical, clinical, review of new drug application and post-market surveillance [80]. The pre-clinical stages take up to six years; the substance is tested to determine if it is biologically safe and active, generally via animal testing. These results would be submitted to their governmental drug advisory boards, such as the U.S. Food and Drug Administration (FDA) and the European Medicines Agency (EMA), to receive approval for the start of clinical trials. The clinical trials are conducted on humans, who are identified by specific criteria. There are three phases of clinical trials, where each phase has a progressively larger pool of people tested. Phase 1 determines the safety of the drug, phase 2 determines the effectiveness of the drug and phase 3 is to confirm the results of phases 1 and 2 and further test for adverse reactions.

The next stage is the submission and waiting for approval for marketing of the new drug; this stage lasts for 1–2 years, depending on the regulatory advisory board. Lastly, the manufacturer evaluates and determines the optimum production for the drug formulation. Furthermore, the advisory board further monitors the ongoing safety of the drug once it's on the market [80].

Pharmaceutical raw materials can be divided into three main groups, which are APIs, intermediates (e.g. mostly solvents and reagents used in the manufacturing of APIs) and additives (e.g. binders, fillers and lubricants used in the formulation of the pharmaceutical drugs). These raw materials can include plant-derived materials, i.e. leaves, flowers, fruits, fungi and seeds as well as synthetic materials and biologics [91].

The pharmaceutical industry has long supply chains, where the final manufacturing plants are in the primary markets, e.g. the United States and Europe, whilst the raw materials are sourced from emerging economies, e.g. Brazil, China, India and Israel [91].

8.4.3 Life Cycle Analysis of the Pharmaceutical Industry

Life cycle assessments (LCAs) have been employed in assessing the production of pharmaceutical drugs throughout the supply chain [92]. LCAs of various APIs, such as ibuprofen [93], sildenafil citrate (API commonly used in Viagra™) [94], chlorhexidine gluconate [95], benzydamine hydrochloride [95], rufinamide [96], enrofloxacin [97], acetylsalicylic acid (API commonly known as aspirin) [98] and ciprofloxacin hydrochloride [99], indicated that greenhouse gas emissions, as well as freshwater toxicity are the main flags raised in the production of drugs. This could be driven by the numerous solvents used in drug manufacturing process and the high energy consumption for drying.

The production of the APIs can have a significant environmental impact on the carbon footprint of the produced drugs (see Table 8.3). The production of APIs can be associated with a very large carbon footprint (ranging from 1 to 600 kg $CO_{2,eq}$/kg API for simple APIs to various 1000s kg $CO_{2,eq}$/kg API for more complex products). The large variation in the carbon footprint to produce APIs can be attributed to the complexity of the synthesis route for the APIs and the carbon footprint can be correlated with the cost of the medication [105]. The high carbon footprint to produce APIs is associated with the low carbon efficiency of the overall process and the use of (large amounts of) solvents in the process. This can also be seen from frameworks and indicators such as Process Mass Intensity ([PMI]; total mass of process materials including water relative to the mass of product) and E-factor (mass of waste product relative to the mass of product) [93, 94] for the production of APIs. On an industrial average, the production of APIs contributes ca. 29% to the overall GHG emissions of medication with drug manufacturing (26%) and corporate emissions (35%), including R&D, sales and marketing and quality control being the other significant contributors [105]. These values will, of course, vary widely depending on the type of API present in the drug. Furthermore, in most cases, there is a lack of metrics such as ecotoxicity and end-of-life disposal, e.g. for the carbon footprint, for all manufacturing process steps and, e.g. APIs production for every pharmaceutical drug in all their dosage forms. These are needed to further evaluate the environmental impact of the manufacturing of pharmaceuticals. This limitation may be due to the substantial deficiency in data inventory of many pharmaceutical drugs, with respect to their environmental concerns [94].

Table 8.3 The related carbon footprint (in CO_2 equivalent) for specified active pharmaceutical ingredients (APIs).

APIs	Common uses	Synthesis route	Carbon footprint (kg $CO_{2,eq}$/kg dosage)	
Isostearic acids	Skin related conditions (cosmetic)	Extraction, followed by synthesis	1.9–3.8 (from soybean oil process) 1–1.5 (from tall oil process)	[92]
Paracetamol (acetaminophen)	Pain/fever relief	Chemical synthesis	38 (oral tablet) 151 (oral liquid) 310–628 (via i.v. dose including additional packaging and admin supplies)	[100]
Penicillin	Antibiotic	Fermentation	25.5–35.1[a)]	[101]
Tramadol (API commonly used in Tramacet®)	Pain relief	Chemical synthesis	450–500	[102]
Morphine	Severe pain relief	Extraction, followed by synthesis	60 000–18 000 (oral liquid) 80 000–155 000 (including mixing, filling and sterilisation of packaging)	[103]
Infliximab (a monoclonal antibody)	Anti-inflammatory	Fermentation	20 000–31 000	[104]

a) Estimated of 100 years of Global Warming Potential (GWP).

8.4.4 Making the Pharmaceutical Industry More Sustainable

The large environmental impact of the production of drugs originates from the inefficiencies in the synthesis of the APIs, with further contributions from so-called corporate emissions and drug manufacture [105]. Specifically, the use of large amounts of solvents in the numerous reaction/separation steps results in the large carbon footprint of these products and thus driving the need to reduce the environmental impact of the production of drugs.

8.4.4.1 Green Chemistry

The concept of green chemistry was introduced to emphasise the need for the creation of chemical products and processes that reduce the environmental impact of these processes, e.g. by minimising or avoiding the use and production of hazardous substances. There are 12 key contributing factors (see Figure 8.14) [107] to reduce negative environmental impacts (i.e. toxicity and waste generated from the synthesis of products); amongst others, the use of renewable feedstock, the use of safer chemicals and solvents, the reduction of derivatives and energy-efficient design. Although not all these principles are applicable to the pharmaceutical industry, an area that green chemistry would be valuable would be for the sourcing of alternative solvents used in the manufacturing of APIs, as currently solvents used in API manufacturing are responsible for 80% of the waste generated [108].

For instance, intermediates, APIs and excipients may be produced from biomass-derived residues known as lignocellulosic residues (LRs) [81], which originate from agricultural, forestry or wood waste products. Their chemical makeup consists largely of cellulose, lignin, tannin, saponin and hemicellulose [109], which can be depolymerised into polyols, acids, furans and phenolic compounds [81].

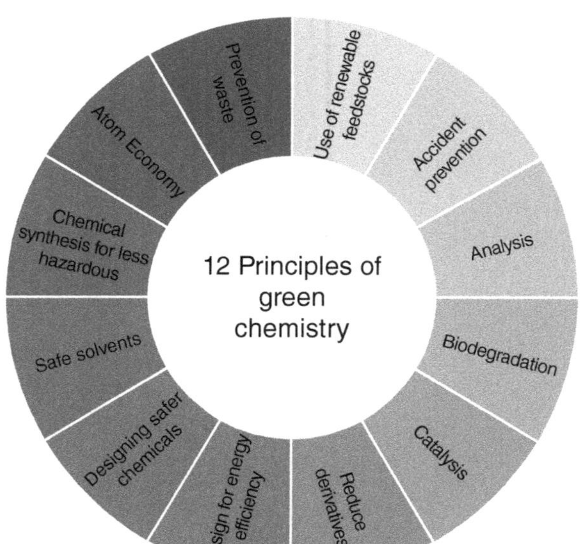

Figure 8.14 The principles of green chemistry (adapted from Abdussalam-Mohammeda et al. [106]).

For instance, polyols, such as sorbitol and xylitol, derived from cellulose and hemicellulose, are widely used as excipients or as APIs for the treatment of constipation. This is due to their sweet flavour, with the added advantage that they are not harmful to teeth [81]. Notably, the blockbuster API, paracetamol (acetaminophen) can be potentially derived from biomass, from the recently patented key biomass-derived intermediate, i.e. *p*-hydroxybenzamide [110] or via terpenes (derived from residues from the paper and pulp industries) [111]. Overall, LRs are one of the most abundant carbon sources on earth making it very appealing as a renewable feedstock, but good management of logistics must be undertaken to ensure that an efficient supply chain is utilised avoiding microbial degradation prior to its application. The major challenge that LRs must overcome is the stringent requirements of the purity and quality stipulated for the pharmaceutical industry. For the realisation of this sustainable route for APIs, further innovation and technologies will need to be in the forefront [81].

There are further other obstacles that hinder the implementation of green chemistry: the high cost, and the difficulties experienced when converting old, traditional synthesis methods to the 'green' product (not only due to lack of known alternatives to the raw materials [106]). Green chemistry falls short of its utilisation in the pharmaceutical industry due to the perceived uncertainties in introducing new or alternative synthesis methods in contrast to its established and approved methodology that has its guarantee of quality and safety [108]. Despite these shortcomings, the application of these green chemistry concepts, where possible, could be beneficial in reducing the environmental impact of the pharmaceutical industry [108].

8.4.4.2 Continuous Processing

Over the past two decades, the method of continuous manufacturing as an alternative to the traditional batch processes in the pharmaceutical industry has drawn attention from industry, academia and regulatory committees [83, 86]. Contrary to batch manufacturing, and indicative of its name, this process involves continuous feeding of the raw materials to the process and ejection of the products from the equipment (see Figure 8.15) [112]. This eliminates the numerous different time intervals for each of the batch processing step [84] as well as eradicating the expensive storage required to maintain the intermediates, thus concurrently reducing the pharmaceutical plant's carbon footprint [112]. Additionally, as this process is largely automated, the handling of chemical reagents and solvents is diminished,

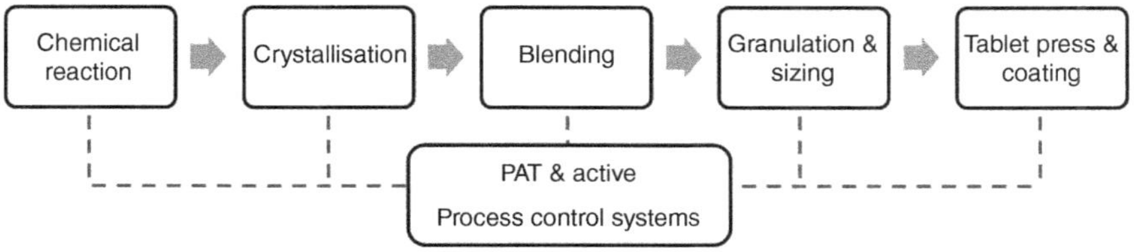

Figure 8.15 Representation of the continuous manufacturing process (adapted from Lee et al. [112]).

thus reducing hazard and safety issues [113]. Furthermore, capital costs can be decreased by 20–76% (depending on active ingredient loading, key intermediates expenditure and process chosen) and operating costs can be lowered to up to 40% [114].

The use of a regulatory framework known as Process Analytical Technologies (PATs) can ensure that quality assurance standards are kept in place. This real-time process monitoring and control system is integrated with advanced data processing [83]. Overall, this would significantly shorten the supply chain as it would lead to manufacturing of APIs and the drug formulation occurring at one site, as well as the use of PATs, which would reduce the length of the API production time from months to days [112].

Flow chemistry is a key concept in the development of continuous manufacturing in the pharmaceutical industry, although a relatively new development in the synthesis of APIs [115–118]. Relatively fast reactions can be carried out more readily and better controlled in a flow configuration than in the classical batch reactor set-up [116]. The main advantages of carrying out these multi-step reactions in a flow configuration are related to improved process safety [115, 116], improved control of reaction parameters [116], improved mixing [117] and reduced usage of solvents [118]. It should, however, be noted that challenges exist in the translation of the classical batch process into a flow process (e.g. reactions requiring concentrated, aqueous HCl-solutions may degas under the applied conditions [116], or with reactions, which may foul up the flow system [116, 117], as well as the strict process quality control required in this industry. The use of flow chemistry instead of a series of batch reactors would allow for heat integration in the overall process [116], thus reducing the energy usage even on the small scale of these processes [116]. Synthesis of API using flow chemistry may have an additional advantage, as the process scheduling may be more readily adapted to the availability of energy and thus to the intermittent nature of renewable energy.

8.4.4.3 Potential for Recycle

Many unused drugs are discarded, and they pollute the environment. The employment of a closed-loop supply chain in the pharmaceutical industry is a way to curb the economic burden of patients and promote environmental sustainability. This would be achieved by re-dispensing or recycling unused drugs [119]. Additionally, there are increasing concerns for the environmental impact of the packaging of pharmaceuticals, which are necessary for the safe dispensing of medications as well as ensuring the stability and effectiveness of the medications. The pharmaceutical industry is accountable for producing more than 300 million tonnes of waste annually, of which 50% of this is single-use plastics. Traditional alternative packaging materials, such as glass and metal, have their own environmental concerns; thus, the need for further research into biodegradable materials such as polymers, i.e. polylactic acid (PLA) and polyhydroxyalkanoates (PHA), is promising, even so the pharmaceutical regulatory landscape will also need to be navigated for its implementation [120].

8.4.4.4 Solar Thermal Systems

Solar thermal energy is a sustainable alternative for power generation, in particular for heat-related industrial processes [121]. The main benefits of this energy source are that it reduces reliance on fossil fuels and results in a decrease in greenhouse gas emissions [121]. A study conducted to demonstrate the viability of solar thermal systems for the Swiss pharmaceutical industry showed that the application of the system, where it meets between 28% and 38% energy requirements, results in natural gas savings that corresponds to 95–154 t $CO_{2,eq}$ per annum [122]. However, solar thermal energy is intermittent and is largely affected by the irregularity of weather patterns [123]; thus, it is necessary to integrate a solar thermal energy storage (STES) system to efficiently harness this renewable energy source [121].

8.4.5 Current Research Around Future Processes

Generally, the drug manufacturing industry is considered rigid in terms of production process; this is due to the need for strict quality assurance and good manufacturing practices that need to be adhered to strictly. However, in recent years, many countries have considered revaluating the long supply chains of the pharmaceutical industry and becoming more self-sufficient. This shift in perspective is particularly due to the impact of COVID-19, which led to sudden changes in regulations and disrupted supply chains that stretched over many countries' boundaries [83].

8.4.6 Outlook for the Manufacture of Pharmaceutical Drugs

Pharmaceutical drugs are a rapidly growing field and will be a necessity for humankind going forward. However, the environmental impact of manufacturing these products will need to be reduced to minimise their ultimate effect on the environment. A promising route is by altering the manufacturing processes, i.e. going from the classical batch processes to the continuous processing utilising flow chemistry, which would allow the reduction of solvents used in the process(es). The major obstacle here is a regulatory one, as the approval of drugs is linked to their processing route. Regulatory bodies have opened towards continuous processing over the last decade [124], but strict controls will have to be enforced regarding parameters such as physical attributes of the starting materials (variability in composition, hygroscopicity, specific area, viscosity, etc.) to ensure a consistent feed to the process, start-up and interruptions in the process, and physical drift of the process to ensure product quality and safety. The necessary loop to re-apply for regulatory access to the market may retard the introduction of these, from an environmental perspective, technologies to reduce the carbon footprint of the industry significantly, but they have to be followed to ensure that quality standards are maintained.

The energy consumption in the pharmaceutical industry is typically relatively modest, but maintaining a constant atmosphere in the manufacturing, storage and quality assurance laboratories would require that the energy is available at all times and intermittency in the power delivery should be balanced with dedicated power storage capacity.

8.5 Wastewater Treatment

Contribution by Mamello Maliehe and Petros Tefo Ndebele.

Water is one of the most critical materials needed for human survival, second to air [125]. In 2010, the United Nations General Assembly and the Human Rights Council recognised an individual's access to safe drinking water as a basic human right [126]. In 2015, the United Nations General Assembly released a comprehensive list of 17 sustainable development goals (SDGs) necessary for world transformation by 2030; amongst these was SDG 6: 'to ensure access to clean water and sanitation for all' [127]. This goal should be achieved by 2030 by increasing the global capacity of wastewater treatment and reuse whilst conserving ecosystems [128].

Although 71% of the Earth's surface is water [129], fresh water only accounts for less than 3% of the total volume of water in the world; the rest is seawater [128], and less than 1% of the globally available water is available for direct use [128]. Hence, clean water, although a necessity, is scarce. In 2017, approximately 660 million people did not have access to potable water and about 1.8 billion people drank contaminated water. As of 2023, about 1.5 billion people suffer from water scarcity and it is projected that in 2025 approximately 1.8 billion people are affected by water scarcity. Further, over 4 billion people experience a month of acute water scarcity each year [128]. It is estimated that in 2030, more than 160% of the total fresh water volume in the world will be needed to meet global fresh water requirements [130] (a more accurate statement would be that the annual fresh water consumption amounted in 2021 to ca. 9% of worldwide annual renewable internal fresh water flows [131]); the total global fresh water demand is estimated to rise from 4000 to 5500 billion m^3/yr. by 2050 [132], thus stressing the need of reuse/recycle of wastewater.

As of 2024, only Belarus, Andorra and Monaco have achieved SDG 6 with most countries either having challenges or major challenges remaining. Estimates indicate a global need of up to US$ 1.1 trillion to achieve this SDG, however, taxes, tariffs, etc. are generally unable to cover the costs required to provide clean drinking water, sanitation and hygiene. Other reasons range from weak governance in terms of policy adherence as well as policy challenges [133].

The use of water generates wastewater, which can be divided into four categories, namely domestic wastewater, normally discharged from residences, industrial wastewater, infiltration/inflow which is water that enters the sewer system indirectly, and stormwater which is runoff from rainfall [129].

Water reuse has therefore become a high-priority issue in the world with most continents affirming the reuse of potable water, having a preference for membrane-based treatment [128]. Water reuse has found not to have any negative health impacts [134]. The reuse of water from wastewater treatment plants (WWTPs) has come to light as a way of alleviating water scarcity. Singapore struggled to provide a sustainable water supply for decades, until 2003, when they opened their first plants to produce 'NEWater', recycled drinking water treated by advanced membrane separation methods. The water, added to reservoirs following treatment, passed World Health Organisation (WHO) drinking standards in addition to over 65 000 other scientific tests, and is clean enough to be used in the electronics industry and as drinking water. Since 1969, Namibia has drunk recycled water with 35% of the water for Windhoek, its capital city, coming from water reclamation plants [134].

Wastewater does contain a number of contaminants and reuse of potable water may expose humans to higher levels of pollutants. A study was conducted to analyse the risk level of drinking water from two potable reuse projects compared to a de facto scenario where 5% of the source water came from water discharged upstream after secondary treatment (chlorination/dichlorination; forming the base case/Scenario 1, [134]). In Scenario 2, the treated wastewater was filtered through surface solids into an aquifer before usage as potable water. In Scenario 3, treated wastewater was subjected to advanced water treatment techniques including microfiltration, reverse osmosis and advanced oxidation prior to injection of the treated water into an aquifer and used for drinking water. Figure 8.16 shows the reduction in the risks of exposure to four pathogens: norovirus, salmonella, cryptosporidium and adenovirus. Filtering wastewater and/or exposing it to advanced treatment processes reduces the risk of exposure to chemical or microbial contaminants when drinking reclaimed water.

It is important to remember that reuse water systems have a cost, which varies by location. Non-potable use may require less treatment than potable use, and thus the desired end use of the wastewater dictates the treatment steps and their cost. Treated wastewater has to be a high purity to be potable, this is a way of avoiding pollution expansion from wastewater, even from partially treated wastewater for its use in irrigation and industrial use [135].

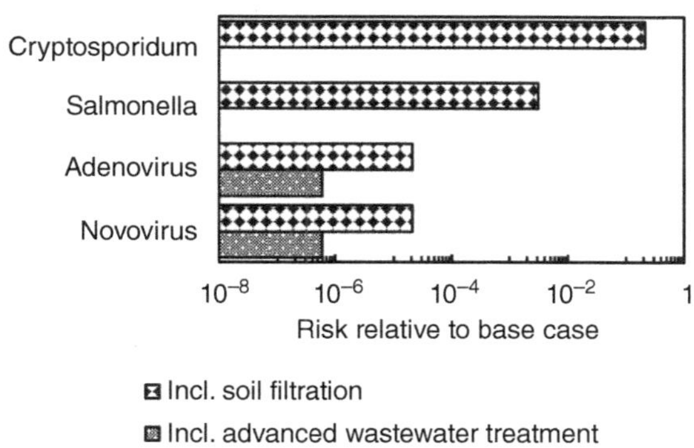

Figure 8.16 The risk of the selected pathogens relative to a base case, in which wastewater is 5% of the source water came from water discharged upstream after secondary treatment; Scenario 2: wastewater filtered in surface soil; Scenario 3: wastewater exposed to advanced treatments, such as microfiltration, reverse osmosis and oxidation (redrawn from Ghernaout [134]; the level of Cryptosporidum and Salmonella was below detection limit after advanced wastewater treatment).

8.5.1 Wastewater Treatment

The purpose of wastewater treatment can be to obtain a stream, which can be released into the environment, or to obtain potable water and is directly used for human consumption. Treatment of wastewater is achieved using physical, chemical and biological processes [136]. The purpose of wastewater treatment is to remove suspended solids in the water, reduce the oxygen demand and remove dissolved inorganic compounds and harmful bacteria and pathogens [129]. Various techniques are employed to treat the wastewater to a quality that ensures that it is safe for drinking and other activities or before it is discharged into water bodies such as rivers and lakes [137].

Wastewater treatment can be split into two main parts: (i) treatment of wastewater influent and (ii) treatment of sludge that is removed from the wastewater. In the treatment steps, non-dissolved solids are concentrated into sludge, which is removed in the primary and secondary stages, resulting in a 'clear' water effluent (see Figure 8.17). Dissolved pollutants are removed from the wastewater using biological or chemical processes, which transform the dissolved materials into solids/gases that can be separated from the water. In the treatment steps, materials that are dissolved in the wastewater are concentrated or transformed into a state that will be easy to separate physically from the water. The concentrated sludge from the primary and secondary treatment stages is separated from the wastewater using clarifiers. The sludge then undergoes further transformation (biological stabilisation) and concentration by thickening and dewatering [137].

Primary treatment is the first treatment stage that wastewater undergoes at the WWTP. Sewage will enter the WWTP through the inlet works. As it enters the WWTP, it either enters at the correct elevation where it then undergoes screening (the first treatment step) or may require lifting to the screening treatment step. The lifting can be achieved using an Archimedes screw (to handle the solid/liquid mixture) typically requiring 0.016–0.034 kWh/m^3 [140]. In this treatment stage, large objects, sand, grit and inorganic matter are removed from the wastewater by screening, comminution, grit removal, sedimentation and/or flotation [129]. The screens are used to remove debris and materials such as wood, rags, leaves, sticks, plastics and other undesired objects that could cause damage to pipes and pumps. The material screened from the wastewater is washed and disposed of [141, 142]. In the comminutor, material that was not screened is ground and shredded into debris that is later removed by sedimentation or flotation. Grit removal is facilitated by grit chambers which remove sand, coffee grounds, eggshells, various inorganic materials, etc. The material removed in the grit chambers is disposed of in landfills. Suspended solids (SSs) that pass through the screens and chambers are removed from the wastewater in the sedimentation tanks (primary clarifiers). The purpose of the primary sedimentation step is to remove some of the organic material present in the wastewater, with 60% of the suspended solids and 35% of the biochemical oxygen demand (BOD) possibly being removed at this stage [140]. In the sedimentation tanks, suspended materials settle to the bottom of the tank by gravity (the typical residence time of wastewater here is ca. two hours).

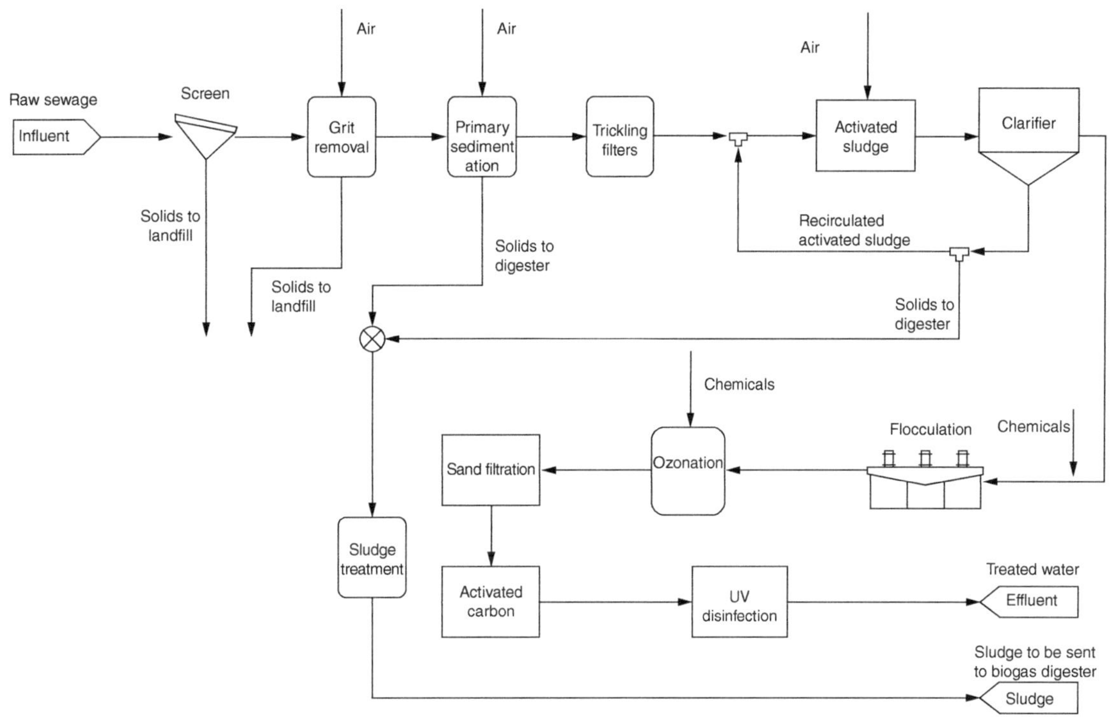

Figure 8.17 Flow diagram of a typical wastewater treatment process (redrawn from References [138, 139]).

The settled solids, also known as the primary sludge, are scraped from the bottom of the primary clarifiers using mechanical scrapers. The primary sludge then goes to the sludge processing steps [140, 141].

The energy requirement for the primary treatment step is low, with screening having an average energy requirement of 0.01 kWh/m^3, grit removal requiring 0.005–0.008 kWh/m^3 and the primary sedimentation step typically requiring ca. 0.01 kWh/m^3. [140].

In the secondary treatment stage, soluble organic material is removed using biological separation techniques. The purpose of the secondary treatment stage is to reduce the biochemical oxygen demand, remove suspended solids and remove nutrients such as nitrogen and phosphorus (typically present in wastewater from sewage and agricultural discharge).

In the biological treatment steps, biodegradable material dissolved in the wastewater is biodegraded by bacteria. The bacteria flocculate, allowing material that was dissolved in the wastewater to be extracted from the water by particulate and gravity settling. The biological reactions inducing the reduction in the concentration of dissolved organics are carried out, as introduction of oxygen promotes the growth of microorganisms [141].

The three biological treatment methods used include: trickling filter, the activated sludge process and aeration tanks. The trickling filter tanks are made up of a bed of stones. Settled sewage is sprayed onto the stone beds, and as this happens, the bacteria in the wastewater accumulate and multiply on the stones. An air draft supplies oxygen for the growth of the bacteria through the spaces between the stones. As the wastewater flows over the stones, microbes extract the dissolved organics out of the wastewater thus lowering the biochemical oxygen demand and purifying the wastewater. After the trickling filters, the wastewater passes through secondary clarifiers which remove the microbes that are removed by the trickling filter rocks [137].

The activated sludge treatment process comprises of an aeration tank and a secondary clarifier. Air is sparged through the aeration tank to provide oxygen, so that microorganisms can grow yielding the activated sludge. The mixture of settled solids and fresh sludge has a residence time of ca. six hours in the aeration tank which is adequate time for the dissolved organics to be absorbed by the microbes, thus further reducing the biochemical oxygen demand of the wastewater.

From the aeration tank, the mixture flows into the secondary clarifier. The activated sludge settles to the bottom by gravity and 'clear water' flows out as the secondary treatment effluent. The activated sludge is pumped out at the bottom of the tank with ca. 30% of the sludge being recirculated back to the aeration tank to mix with the primary effluent to ensure a healthy microbial population in the activated sludge treatment process. The remainder of the sludge (70%) is then treated prior to disposal or can be used as a feedstock for a biogas reactor to generate energy for the WWTP.

In the secondary treatment stage, the trickling filters require ca. 0.18 kWh/m^3. The activated sludge step has different energy requirements in the range of 0.27–0.41 kWh/m^3 depending on which technique is used. The sedimentation tanks have a low energy requirement of ca. 0.01 kWh/m^3 with the energy required for sewage recirculation between 0.03 and 0.23 kWh/m^3, depending on the recirculation rate [140].

The tertiary treatment stage involves the disinfection of secondary effluent using chlorine-based chemicals to remove any pathogens present in the water. Other treatment technologies employed at this stage include UV light, ozone and granular activated carbon (GAC). The disinfection step (chlorination) is associated with an energy consumption of ca. 0.02 kWh/m^3; the UV treatment step requires ca. 0.06 kWh/m^3, and ozonation ca. 0.08 kWh/m^3 [140]. Advanced treatment technologies can/need to be applied if necessary for the generation of potable water, such as reverse osmosis to remove any dissolved salts and other contaminants and advanced oxidation processes to break down complex organic molecules [140]. In this treatment stage, the sludge is also treated to be disposed of or used as an energy source in the biogas digester.

Sludge treatment technologies can be categorised into chemical (alkali, acid and oxidative), physical (mechanical, thermal, ultrasound, electrochemical) and biological (microbial and enzymes). The pre-treatment of sludge increases the yield of biogas from anaerobic digestion when utilising the sludge to generate power [143].

The energy consumption on WWTPs varies depending on the size, plant location, types of treatment processes employed, effluent quality standards, final usage (potable, agricultural or disposed into a water body) and the age of the plant. Table 8.4 shows the energy consumption of wastewater treatment plants utilising the activated sludge process of different sizes [140]. Approximately, three quarters of the energy consumption is related to aeration and sludge management. The consumption ratios show that energy consumption per volume of wastewater decreases as the size of the plant increases [144], and will reach a minimum of ca. 0.29 kWh/m^3 for very large plants.

Table 8.4 Energy consumption of different treatment steps for activated sludge process for wastewater treatment plants (WWTPs) treating different amounts of wastewater (adapted from GIZ & SAGEN [140]).

Process unit	**Energy requirement (kWh/day)**			
Plant size (10^3 m^3/day)	**2**	**10**	**25**	**100**
Pumping	114.1	507.7	1205.3	4336.4
Inlet works	26.9	48.6	89.8	334.2
Primary clarifiers	7.9	42.6	102.4	409.5
Aeration	281.1	1453.7	4705.6	17 908.2
Secondary clarifiers	7.9	42.6	102.4	409.5
Disinfection	0.5	2.7	17.8	70.0
Sludge management	637.2	1424.8	1194.7	4379.0
Lights & buildings	105.7	218.6	528.3	1585.0
Totals	**1181.3**	**3741.3**	**7946.3**	**29 431.8**
Consumption ratio (kWh/m^3)	0.59	0.37	0.32	0.29

Table 8.5 Energy demand of wastewater treatment plants (WWTPs) in different countries.

Country	Energy intensity (kWh/m^3)	
Australia	0.5–1.0	[144]
Brazil	0.57	[145]
China	0.31	[144]
Egypt	0.27–0.59	[146]
Germany	0.40–0.43	[144]
India	0.23	[147]
South Africa	0.08–0.41	[144]
South Korea	0.24	[144]
USA	0.52	[144]

The required energy consumption for the treatment of wastewater depends on the location and the country. Table 8.5 shows the energy consumption of wastewater treatment per cubic metre of wastewater treated. The United States shows an electricity consumption for WWTPs of 0.52 kWh/m^3 which accounts for 0.6% of their annual electricity consumption. Asian countries report a lower energy intensity than Western countries. It is important to note that the differences in energy consumption are due to different treatment levels employed [144].

8.5.2 Sewage Sludge

Sewage is a mixture of industrial and domestic wastes containing over 99% water [148]. Wastewater is usually treated physically, chemically or biologically to remove its contaminants. After treating sewage, a semi-solid or slurry waste by-product is produced; this is called sewage sludge. It is mainly formed during the primary treatment in the primary clarifiers, where precipitated solids form most of the primary sludge.

There are two main forms of sludge: sludge and secondary sludge. Secondary sludge is called activated sludge when formed in activated sludge processes. The sludge produced from biological processes is often referred to as wastewater biosolids [148]. Sewage sludge is rich in organic matter and nutrients such as nitrogen and phosphorus; Table 8.6 gives the typical composition of sewage sludge.

The organic content of dried sewage can range between 21% and 48%, giving a latent energy of 11–22 MJ/kg, a calorific value similar (if not better) than various other biomass materials [150]. Energy can be extracted from sludge via combustion or incineration, pyrolysis or anaerobic digestion (see Figure 8.18). The effectiveness of the energy recovery process is mainly influenced by heavy metal content, nitrogen and moisture. The most common pathways are anaerobic digestion and incineration.

Table 8.6 The composition of sewage sludge in terms of organic matter and nutrients (adapted from He et al. [149]).

Nutrient	Composition (%)
Proteins	≈40
Lipids	10–25
Carbohydrates	14
Lignin and ash	30–50

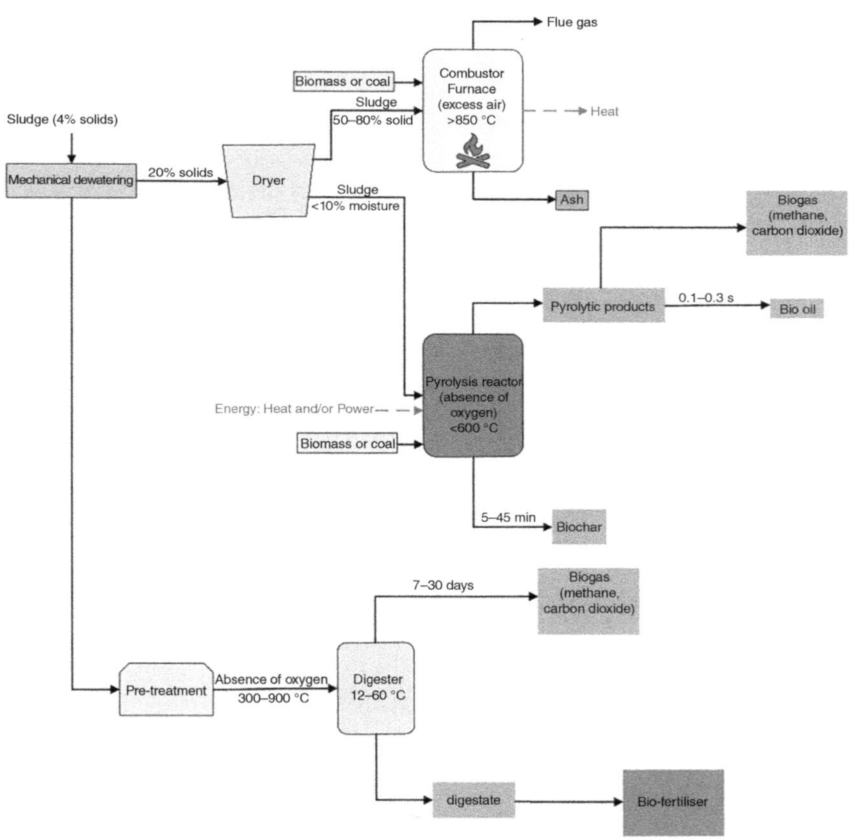

Figure 8.18 Sludge process routes (adapted from Oladejo et al. [150]).

Before processing the sludge, it is taken to a mechanical dewatering unit which increases the solid content from ca. 4% to ca. ca. 20%. A dryer can then take the solids content to 50–80% for combustion or <10% moisture for pyrolysis.

8.5.2.1 Combustion and Incineration

Combustion can be used for the generation of heat (conventional combustion) or for reducing the volume of waste materials (incineration). Incineration has been shown to reduce approximately 90% of sludge volume whilst terminating pathogens and is an effective strategy for clinical and municipal solid waste. The heat generated in the incineration may be used for power generation, but is generally used for burning harmful gases [150].

In conventional combustion, the latent energy is captured as heat through burning the dried sewage sludge. The heat can be converted to electrical energy through the use of combined heat and power (CHP) engines or used directly [150]. Conventional solid fuel combustion often involves drying after the dewatering step to reduce the moisture content of the sludge to less than 50 wt.%. The ash from incineration could be used in construction, whilst the ash produced through combustion can be used as fertiliser.

8.5.2.2 Pyrolysis

Pyrolysis is the thermal degradation or decomposition without using an oxidising agent in an inert environment. For sewage sludge, pyrolysis involves a thermal treatment at 350–600 °C in the absence oxygen. The resulting products are biogas, bio-oil and biochar, the relative amounts are controlled by the residence time and temperature in the reactor. For example, long residence time of the sludge/fuel in the reactor at low temperature and slow heating rates enhances char formation whilst low or short residence times at high temperature promotes bio-oil and biogas [150]. Biogas can be burned as fuel or reformed into syngas. Bio-oil can be used for heating or for generating electricity; it can also be refined to produce liquid fuels. Although bio-oil from sewage sludge has a heating value of about 33 MJ/kg, which is higher than that from biomass, it contains a high fraction of oxygen-containing compounds which limit its fuel quality and value. The yield of bio-oil is between 14% and 58% and is maximised at 450–550 °C, depending on the moisture content and the sewage sludge feed composition. Biochar can be used for direct combustion, as an adsorbent, and in agriculture.

Pyrolysis occurs in an inert environment requiring high temperature, and thus an external heat source is required. Heat can be produced by partially combusting the biogas or bio-oil produced to provide heat for the pyrolysis reaction in a self-sustaining manner.

8.5.2.3 Anaerobic Digestion

Anaerobic digestion of (activated) sludge produces biogas of a calorific value of 28–39 MJ/m^3 and composition 60–70% methane, 30–40% carbon dioxide, with traces of other gases. It is a slow process with a residence time of ca. 20–60 days. The obtained biogas can be scrubbed with water to bring the composition of methane up to 97% and the calorific value to 51 MJ/m^3. This biogas can be used to generate heat and electricity directly since it has a high methane content or scrubbed/upgraded to produce biomethane. The supernatant product has a high nutrient content, especially phosphorus and nitrogen, which can be recovered for use as fertiliser [150].

8.5.2.4 Wastewater and Sludge to Fertiliser and Irrigation

Reuse of stabilised sludge for agriculture is an attractive way of reusing sewage, though is subject to stringent sludge quality legislation. Usually, wastewater is the source of high-quality fertigation (fertiliser and irrigation), and reused effluents can provide water and nutrients to crops and urban green [151].

Wastewater needs to be treated before it is released into the environment to prevent the accumulation of its various pollutants in the receiving system or at least controlled so the maximum concentration of various elements and compounds in the environment does not exceed on the acceptable daily human intake based on the various ways that pollutants enter the food chain (see Table 8.7). It should be

Table 8.7 Maximum concentrations of toxic metals and organic compounds in soil, $C_{max,soil}$, based on acceptable daily human intake (from Chang et al. [152]).

Element	$C_{max,soil}$ (mg/kg)	Organic compound	$C_{max,soil}$ (mg/kg)
Antimony	36	Aldrin	0.48
Arsenic	8	Benzene	0.14
Barium	302	Chlordane	3
Beryllium	0.2	Chlorobenzene	211
Boron	1.7	Chloroform	0.47
Cadmium	4	2,4-D	0.25
Fluorine	635	DDT	1.54
Lead	84	Dichlorobenzene	15
Mercury	7	Dieldrin	0.17
Molybdenum	0.6	Dioxins	0.00012
Nickel	107	Heptachlor	0.18
Selenium	6	Hexachlorobenzene	1.40
Silver	3	Lindane	12
Thallium	0.3	Methoxychlor	4.27
Vanadium	47	PAH (as benzo [a]pyrene)	16
		PCB's	0.89
		Pentachlorophenol	14
		Phthalate	13 733
		Pyrene	41
		Styrene	0.68
		2,4,5 T	3.82
		Tetrachloroethane	1.25
		Tetrachloroethylene	0.54
		Toluene	12
		Toxaphene	0.0013
		Trichloroethane	0.68

noted that a similar consideration is valid for the dispersion of sludge obtained from wastewater into the environment. Wastewater is a valuable source of both water and nutrient content for crops, and the reuse of water for non-potable purposes such as irrigation is well-advanced (although requiring strict control and regulation).

Sewage sludge may be spread directly on to land. This minimises energy requirement, because fewer processing steps, such as heating, turbines and chemicals, are required. This is an attractive process route as it is inexpensive, and it eco-cycles nutrients so they are used in the soil, but sludge analysis has shown that 50% of samples have a very high concentration of at least one metal [153]. Applying the sludge directly onto farmland was found to be the most energy-efficient in terms of greenhouse gas emissions and energy, however, it may introduce metals such as cadmium to the soil [154]. This was when comparing mineral fertiliser, certified sewage sludge (struvite) precipitated from wastewater, and phosphorus recovered from the incineration of sludge. The consistent addition of cadmium and other heavy metals to the soil poses a high potential for long-term accumulation of toxic elements in the soil [155, 156]. China has reduced the direct application of sewage sludge to farmland applications from 60.9% in 2009 to 21.9% in 2017. This reduction may be due to the toxic characteristics of direct addition of sewage sludge to land, which adds heavy metals to the soil; this is adsorbed by plants and gets to humans as well [156].

It is recommended [156] that sewage sludge undergoes prior treatment before its use in agriculture, to reduce its potential risks with pathogens, heavy metals and other contaminants.

Sewage sludge is an organic waste containing high amounts of phosphorus and nitrogen, key for stimulating soil enzyme and microbial activity. However, it also contains heavy metals, pharmaceuticals, endocrine disruptors and hydrocarbons, which are dangerous for people and the environment [157]. The ash resulting from the combustion of sewage sludge is rich in phosphorus, which can be extracted from ash using thermochemical treatment, alkaline or acid leaching, precipitation with lime water or reaction with phosphoric acid. If leaching is used, the heavy metals can be removed by further using ion exchange or sulphide precipitation, The most commonly used technologies are the thermo-chemical treatment and wet extraction.

8.5.2.5 Energy Generation from Sludge

The WWTPs have the potential to be energy-neutral systems. Wastewater is considered to be an energy source [158]. Sludge from the wastewater can be used to produce biogas that can be used to generate heat and electricity. In countries, such as Austria, Germany and America, anaerobic digesters are coupled with CHP systems which produce electricity and heat for the wastewater treatment plant [158]. The efficiencies of biogas conversion in CHP systems are reported to be 25–45% [143] Studies have shown that a wastewater treatment plant could generate enough biogas to be entirely energetically self-sufficient [159]. In 2006, the Netherlands was producing 95 million Nm^3 of biogas from sludge digestion and using CHP to convert it into 143 MWh of electricity and heat [144].

Table 8.8 Mass balance and energy balance (adapted from Sarpong and Gude [160]) (assuming 1 g COD produces 13.9 kJ through formation of methane in AD).

	Inlet flow	AD inlet flow	CHP flow		
	COD (kg/day)	COD (kg/day)	COD (kg/day)	COD (g/m^3)	Energy (kWh/m^3)
PS	5000	2000	520	52	0.200
SS	3000	180	12.6	1.26	0.005
FOG	653	653	555	55.5	0.214
Total	8653	2833	1087.6	108.8	0.42

Wastewater treatment plants can be energy self-sufficient by harvesting the energy contained in the sludge and increasing the efficiency of energy utilisation. The challenges associated with running energy self-sufficient WWTPs are that there is a high upfront cost required to construct the energy systems. If the sludge does not undergo adequate anaerobic treatment, toxins and harmful gases may be released to neighbouring environments, leakage of the CH_4 and CO_2 may increase pollution of air and cause global warming [144].

Figure 8.19 shows an adapted conventional activated sludge wastewater treatment plant (CAS-WWTP) design that is energy self-sufficient [160]. The WWTP comprises of a design of a conventional plant that includes primary sedimentation (PS), activated sludge tank (AST), secondary sedimentation (SS), fats, oil and grease (FOG), anaerobic digester (AD) and a CHP [160]. The flow rate for the WWTP was scaled to a feed of 10 000 m^3/day (small/medium-sized plant) and the mass balance was conducted using a medium strength organic load of 500 ml/day, which gives a chemical oxygen demand (COD) of 5000 kg/day (Table 8.8).

Recoverable energy to electricity calculations:

$$\text{Sludge} = \text{PS} + \text{SS} = \frac{13.9\ \text{kJ}}{\text{g COD}} \cdot \frac{1\ \text{kWh}}{3600\ \text{kJ}} \cdot \frac{53.26\ \text{g}}{\text{m}^3} = 0.206\ \frac{\text{kWh}}{\text{m}^3}$$

$$\text{FOG} = \frac{13.9\ \text{kJ}}{\text{g COD}} \cdot \frac{1\ \text{kWh}}{3600\ \text{kJ}} \cdot \frac{55.5\ \text{g}}{\text{m}^3} = 0.214\ \frac{\text{kWh}}{\text{m}^3}$$

Total electricity produced = 0.42 kWh/m^3

A 10000 m^3/day WWTP requires 0.37 kWh of electrical power per m^3. From the energy calculations, this decreases to net 0.16 kWh/m^3 when digesting the sludge produced. With co-digestion of FOG, enough energy is produced to power the theoretical WWTP, with an excess of 0.05 kWh/m^3 [160].

The benefits of sludge-to-energy conversion in WWTPs include [161]

1) Production of energy from the sludge will render wastewater treatment plants less dependent on external energy sources (primarily in the form of electricity) and will thus be less affected by power outages.
2) The carbon footprint of WWTPs can be drastically reduced as biogas generated in the (anaerobic) decomposition of sludge generates methane, a much more powerful greenhouse gas than CO_2. Hence, converting the biogas into useful energy

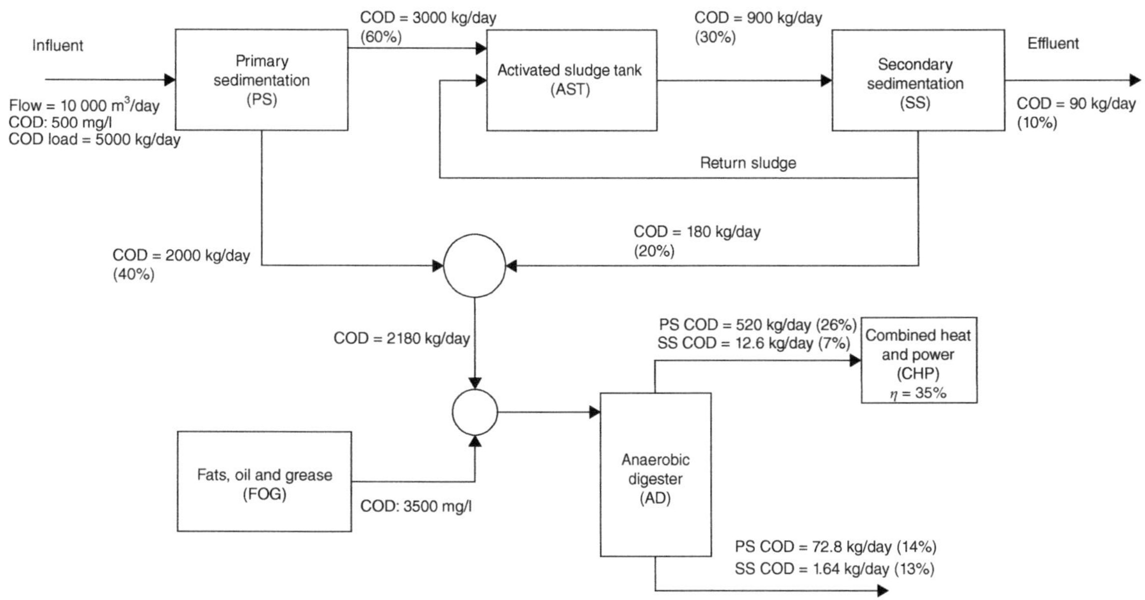

Figure 8.19 Energy-efficient CAS-WWTP block flow diagram (adapted from Sarpong and Gude [160]).

will reduce the equivalent CO_2 emissions, whilst recovering some of the energy in the form of electricity.

3) Sludge-to-energy allows for a reduction in the solid waste generation and thus reduces the landfill requirements.

8.5.3 Outlook

Water is essential to all life on Earth; fresh water is especially essential to all human processes. Freshwater, and moreover potable water, is not abundant in all areas, with notable water scarce regions in Africa and Asia, continents with large and rapidly growing populations and thus rapidly growing demand for clean water. As such, getting the most out of all water will be essential in the future. A way of achieving this will be the processing of wastewater into a re-useable form. Wastewater is available from multiple sources; these sources create a need for slightly different processes which can turn it into reusable products. However, the required quality of water for reuse also differs, with potable water having the most rigorous standards. It will become important to process wastewater for its reuse as water becomes increasingly scarce and therefore increasingly valuable.

The further treatment of sludge generated in wastewater treatment plants will become an essential part on these plants so that energy-autonomy on these plants may be obtained with minimal input from the grid required.

Whilst most industries are essential to modern society and everyday life, their absence would not necessarily collapse the global economy and potentially population. Water (and the food industry) differs from most industries here as it is absolutely essential. Treatment of wastewater will, therefore, become an essential part of a future society/economy (even one that is not renewable).

8.6 Outlook for Energy in the Consumer Goods Industry

The consumer goods industry typically does not have a high energy consumption per unit produced, and most of their greenhouse gas emissions may be imported through the feed. For instance, global greenhouse gas emissions from food production amounted to 13.4 Gton $CO_{2,eq}$ per annum in 2009–2011 [162] (ca. 28% of the anthropogenic CO_2-emissions in 2011). This can mainly be attributed to crop growth and harvesting and keeping livestock, which accounted for almost 82% of all emissions related to food production. Food processing, transport, packaging and retail came to 2.4 Gton $CO_{2,eq}$ per annum (18% of total greenhouse gas emissions), each contributing 25%, 33%, 26% and 16%, respectively. Hence, to improve the sustainability of the consumer goods industry, not only the energy usage in the actual processing should be considered, but also the environmental impact of the raw materials going into the final product (as well as the packaging used by the industry and transport of the product). For instance, the carbon footprint of the food industry will be greatly affected by using green ammonia instead of grey ammonia as a source for fertiliser (see Chapter 5).

Packaging is an essential aspect in some of the consumer goods industries, and greenhouse gas emissions associated with packaging may even dominate the carbon footprint of the product. For instance, one of the biggest problems facing the pharmaceutical industry is the necessity for single-use packaging as the products require sterilisation and isolation from moisture and oxygen, due to their sensitive nature. However, there are opportunities to reduce the carbon footprint of the used packaging by producing it from recycled/recyclable plastics.

Consumer goods may be controlled by stringent quality control, which must be met. This may be affected by environmental factors in the production environment as well as the product storage facility (and even the feed storage facility), such as temperature and humidity, which then have to be monitored and controlled accurately. Failure to maintain those conditions may result in degradation of the product, which then must be disposed of.

Electricity is one of the main energy sources used in this industry, and the transition towards the use of renewable energy will go with the transition of the overall electricity grid. The actual production run may even be scheduled around the availability of power, but maintaining the environmental conditions in the production and storage facility may need continuous power and would require a battery energy storage system (BESS).

The consumer goods industry typically utilises relatively low temperatures, which offer the opportunity to implement heat pumps to improve the utilisation of electric energy. Alternatively, heating of streams can easily be achieved using resistive heaters.

The relatively small scale and often batch processing nature of the consumer goods industries may make them prime candidates for implementation of renewable energy.

References

1 Shabir, I., Dash, K.K., Dar, A.H. et al. (2023). Carbon footprints evaluation for sustainable food processing system development: a comprehensive review. *Future Foods* 7: 100215. https://doi.org/10.1016/j.fufo.2023.100215.

2 Crippa, M., Solazzo, E., Guizzardi, D. et al. (2021). Food systems are responsible for a third of global anthropogenic GHG emissions. *Nature Food* 2 (3): 198–209. https://doi.org/10.1038/s43016-021-00225-9.

3 University of Michigan (2024). Carbon Footprint Factsheet. Center for Sustainable Systems [Online]. Available: https://css.umich.edu/publications/factsheets/sustainability-indicators/carbon-footprint-factsheet (accessed 8 May 2024).

4 Li, L., Awada, T., Shi, Y. et al. (2025). Global greenhouse gas emissions from agriculture: pathways to sustainable reductions. *Global Change Biology* 31 (1): e70015. https://doi.org/10.1111/gcb.70015.

5 Menegat, S., Ledo, A., and Tirado, R. (2022). Greenhouse gas emissions from global production and use of nitrogen synthetic fertilisers in agriculture. *Scientific Reports* 12 (1): 14490. https://doi.org/10.1038/s41598-022-18773-w.

6 O. US EPA (2024). Wastes [Online]. Available: https://www.epa.gov/report-environment/wastes (accessed 25 September 2024).

7 Sekerka, L.E. and Stimel, D. (2014). Embracing waste as a resource: insights from the informal sector. *Management Research Review* 37 (3): 241–260. https://doi.org/10.1108/MRR-11-2012-0254.

8 Wang, Y., Ying, H., Stefanovski, D. et al. (2025, https://doi.org/10.1038/s43016-025-01140-z). Food waste used as a resource can reduce climate and resource burdens in agrifood systems. *Nature Food* 6: 478–490.

9 Elroi, H., Zbigniew, G., Agnieszka, W.-C., and Piotr, S. (2023). Enhancing waste resource efficiency: circular economy for sustainability and energy conversion. *Frontiers in Environmental Science* 11: https://doi.org/10.3389/fenvs.2023.1303792.

10 Zavos, S. and Pyyhtinen, O. (2024). The limits of waste as a resource: a critique and a proposition towards a new scalar imagination for the circular economy model. *Cambridge Journal of Regions, Economy and Society* 17 (3): 683–697. https://doi.org/10.1093/cjres/rsae013.

11 Greer, R., von Wirth, T., and Loorbach, D. (2021). The Waste-Resource Paradox: practical dilemmas and societal implications in the transition to a circular economy. *Journal of Cleaner Production* 303: 126831. https://doi.org/10.1016/j.jclepro.2021.126831.

12 Greenflow (2024). What is hazardous waste and examples? Understanding dangerous waste materials [Online]. Available: https://www.greenflow.com/blog/what-is-hazardous-waste-and-examples-understanding-dangerous-waste-materials (accessed 25 September 2024).

13 Attrah, M., Elmanadely, A., Akter, D., and Rene, E.R. (2022). A review on medical waste management: treatment, recycling, and disposal options. *Environments* 9 (11): https://doi.org/10.3390/environments9110146.

14 Kvashnina, K., Claret, F., Clavier, N. et al. (2024). Long-term, sustainable solutions to radioactive waste management. *Scientific Reports* 14 (1): 5907. https://doi.org/10.1038/s41598-024-55911-y.

15 Clayton, R., Kirk, J., Banford, A., and Stamford, L. (2025, https://doi.org/10.1007/s10098-024-02998-6). A review of radioactive waste processing and disposal from a life cycle environmental perspective. *Clean Technologies and Environmental Policy* 27: 665–682.

16 von Sperling, M. and de Lemos Chernicharo, C.A. (2006). *Biological Wastewater Treatment in Warm Climate Regions*, vol. 1. London: IWA.

17 Cappelli, A., Oliva, N., and Cini, E. (2020). Stone milling versus roller milling: a systematic review of the effects on wheat flour quality, dough rheology, and bread characteristics. *Trends in Food Science & Technology* 97: 147–155. https://doi.org/10.1016/j.tifs.2020.01.008.

18 Cappelli, A., Bettaccini, L., and Cini, E. (2020). The kneading process: a systematic review of the effects on dough rheology and resulting bread characteristics, including improvement strategies. *Trends in Food Science & Technology* 104: 91–101. https://doi.org/10.1016/j.tifs.2020.08.008.

19 Rahman, M., Islam, R., Hasan, S. et al. (2022). A comprehensive review on bio-preservation of bread: an approach to adopt wholesome strategies. *Food* 11 (3): 319. https://doi.org/10.3390/foods11030319.

20 Myers, G. (2020). Bread factory bakes, ships out 10,000 loaves daily using old-world recipe. *Morristown Daily Record*, 2 February 2020 [Online]. Available: https://www.dailyrecord.com/story/news/2020/02/02/nj-bakery-makes-ships-out-10-000-bread-loaves-daily-using-old-world-recipe/4557385002/

21 Premier (2024). Bakeries [Online]. Available: https://www.premierfmcg.com/operations/bakeries (accessed 23 July 2024).

22 Hovis (2025). Our Business [Online]. Available: https://www.hovis.co.uk/our-business (accessed 17 September 2025).

23 Statista Market Insights (2024). Food - Worldwide [Online]. Available: https://www.statista.com/outlook/cmo/food/worldwide (accessed 23 July 2024).

24 The Business Research Company (2024). Grain Farming Global Market Report 2024 [Online]. Available: https://www.thebusinessresearchcompany.com/report/grain-farming-global-market-report (accessed 24 September 2024).

25 Vijayant Singh, K., Gupta, R., Aggarwal, A., and Agarwal, A. (2024). Evolving with steel: Future growth and opportunities, McKinsey and Company [Online]. Available at: https://www.mckinsey.com/in/our-insights/evolving-with-steel-future-growth-and-opportunities (accessed 9 September 2025).

26 Shahbandeh, M. (2024). Grain production worldwide 2023/24, by type, 6 February 2024 [Online]. Available: https://www.statista.com/statistics/263977/world-grain-production-by-type/ (accessed 24 September 2024).

27 Shahbandeh, M. (2024). Global wheat consumption 2023/24, by country, 22 May 2024 [Online]. Available: https://www.statista.com/statistics/1094065/total-global-wheat-consumption-by-country/ (accessed 24 September 2024).

28 Statista Market Insights (2024). Bread - Worldwide [Online]. Available: https://www.statista.com/outlook/cmo/food/bread-cereal-products/bread/worldwide#revenue (accessed 23 July 2024).

29 Singh, A. (2024). Bakery packaging market size, growth rate and industry analysis (2023–2032) [Online]. Available: https://www.towardspackaging.com/insights/bakery-packaging-market-sizing (accessed 22 October 2024).

30 Beech, G.A. and Crafts-Lighty, A.L. (1980). Energy use in flour production. *Journal of the Science of Food and Agriculture* 31 (8): 830–838. https://doi.org/10.1002/jsfa.2740310813.

31 Olaoye, O.S., Adefajo, A.A., and Ekundayo, S.O. (2014). Energy analysis of a wheat processing plant in Nigeria. *ARPN Journal of Engineering and Applied Sciences* 9 (9).

32 Odunfa, M. and Joel, O.L. (2022). Energy, exergy and cost analysis of flour production from wheat in Nigeria: case study of Nigeria Eagle Flour mills, Ibadan. *Global Journal of Engineering and Technology Advances* 13 (1): 067–097. https://doi.org/10.30574/gjeta.2022.13.1.0120.

33 Ahmed, Z., Nadulski, R., and Panasiewicz, M. (2017). The influence of wheat hardness on energy consumption during the wheat milling. *Dept. of Mach.*

Exploit. and Management of Prod. Processes, ULS in Lublin, Poland [Online]. Available: https://open.icm.edu.pl/handle/123456789/14846 (accessed 19 May 2025).

34 Federation of Bakers (2017). Factsheet No. 7 – How bread is made [Online]. Available: https://www.fob.uk.com/wp-content/uploads/2017/01/FS-7-How-bread-is-made.pdf (accessed 25 September 2024).

35 Maicas, S. (2020). The role of yeasts in fermentation processes. *Microorganisms* 8 (8): 1142. https://doi.org/10.3390/microorganisms8081142.

36 Paton, J.B. (2013). Energy utilisation in commercial bread baking. Ph.D. dissertation, School of Mechanical Engineering, University of Leeds [Online]. Available: https://core.ac.uk/download/pdf/18451907.pdf.

37 Lynn, A. (2002). Current practice in milling and baking. In: *Cereal Biotechnology* (ed. P. Morris and J. Bryce). Cambridge: Woodhead Publishing.

38 Condessa, B.M.B., da Silva, K.V., da Silva, J.F.M. et al. (2022). Performance of wild *Saccharomyces* and non-*Saccharomyces* yeasts as starter cultures in dough fermentation and bread making. *Int J of Food Sci Tech* 57 (5): 3046–3059. https://doi.org/10.1111/ijfs.15633.

39 Cappelli, A., Lupori, L., and Cini, E. (2021). Baking technology: a systematic review of machines and plants and their effect on final products, including improvement strategies. *Trends in Food Science & Technology* 115: 275–284. https://doi.org/10.1016/j.tifs.2021.06.048.

40 Wisconsin Oven Corporation (2023). Direct fired & indirect gas fired [Online]. Available: https://www.wisoven.com/support/technical-information/direct-fired-indirect-gas-fired (accessed 13 May 2023).

41 International Thermal Systems (2020). The 3 types of heating sources for industrial ovens [Online]. Available: https://www.internationalthermalsystems.com/wp-content/uploads/2020/05/heat-source-industrial-ovens.pdf (accessed 13 May 2023).

42 Lenexa Manufacturing (2022). Types of bread packaging [Online]. Available: https://lenexamc.com/2022/12/types-bread-packaging/ (accessed 22 October 2024).

43 Espinoza-Orias, N., Stichnothe, H., and Azapagic, A. (2011). The carbon footprint of bread. *International Journal of Life Cycle Assessment* 16 (4): 351–365. https://doi.org/10.1007/s11367-011-0271-0.

44 Department for Energy Security and Net Zero (2023). Greenhouse gas reporting: conversion factors 2023 [Online]. Available: https://www.gov.uk/government/publications/greenhouse-gas-reporting-conversion-factors-2023 (accessed 10 August 2024).

45 Zisopoulos, F.K., Moejes, S.N., Rossier-Miranda, F.J. et al. (2015). Exergetic comparison of food waste valorization in industrial bread production. *Energy* 82: 640–649. https://doi.org/10.1016/j.energy.2015.01.073.

46 Himmelblau, D.M. and Riggs, J.B. (2012). *Basic Principles and Calculations in Chemical Engineering*, Prentice Hall International Series in the Physical and Chemical Engineering Sciences, 8e. Upper Saddle River, NJ: Prentice Hall.

47 Istudor, A., Voicu, G., Muscalu, G., and Tudor, P. (2020). Evaluation of carbon dioxide released by bread dough during proving stage. *E3S Web Conf.* 180: 03012. https://doi.org/10.1051/e3sconf/202018003012.

48 Ashcroft, M. (2015). Improving the efficiency of bakery ovens: case study. *The Carbon Trust* [Online]. Available: https://unepccc.org/wp-content/uploads/sites/3/2016/03/cts402-improving-efficiency-of-bakery-ovens-0.pdf (accessed 10 October 2024).

49 Vandermeersch, T., Alvarenga, R.A.F., Ragaert, P., and Dewulf, J. (2014). Environmental sustainability assessment of food waste valorization options. *Resources, Conservation and Recycling* 87: 57–64. https://doi.org/10.1016/j.resconrec.2014.03.008.

50 Kumar, V., Brancoli, P., Narisetty, V. et al. (2023). Bread waste – a potential feedstock for sustainable circular biorefineries. *Bioresource Technology* 369: 128449. https://doi.org/10.1016/j.biortech.2022.128449.

51 Aniello, A., Poinsot, T., Selle, L., and Schuller, T. (2022). Hydrogen substitution of natural-gas in premixed burners and implications for blow-off and flashback limits. *International Journal of Hydrogen Energy* 47 (77): 33067–33081. https://doi.org/10.1016/j.ijhydene.2022.07.066.

52 Perry, R.H. and Green, D.W. (ed.) (1999). *Perry's Chemical Engineers' Handbook*, 7e, Perry's Chemical Engineers' Platinum Edition. New York, NY: McGraw-Hill.

53 Moseme Forsythe, H.D. and Madyira, D.M. (2019). Experimental performance assessment of a solar powered baking oven. *Procedia Manufacturing* 35: 535–540. https://doi.org/10.1016/j.promfg.2019.05.076.

54 Madsbøll, H. (2022). High temperature heat pump for tunnel oven. *Danish Technological Institute* [Online]. Available: https://energiforskning.dk/files/media/document/352-009%20Slutrapport.pdf (accessed 21 October 2024).

55 Le Page, M. (2017). A loaf of bread emits half a kilo of CO_2, mainly from fertiliser [Online]. Available: https://www.newscientist.com/article/2122857-a-loaf-of-bread-emits-half-a-kilo-of-co2-mainly-from-fertiliser/ (accessed 13 May 2023).

56 Smart, K. (2022). Review of recent progress in green ammonia synthesis: decarbonisation of fertiliser and fuels via green synthesis. *Johnson Matthey Technology Review* 66 (3): 230–244. https://doi.org/10.1595/205651322X16334238659301.

57 Tucker, G. and Featherstone, S. (2021). *Essentials of Thermal Processing*, 2e. Wiley.

58 Featherstone, S. (2015). *A Complete Course in Canning and Related Processes 1: Fundamental Information on Canning*, 3 vols, 14e. Woodhead Publishing.

59 Masanet, E. (2008). Energy efficiency improvement and cost saving opportunities for the Fruit and Vegetable Processing Industry. An ENERGY STAR guide for energy and plant managers, May 2008 [Online]. Available: https://escholarship.org/uc/item/8h25n5pr (accessed 11 August 2024).

60 World Health Organisation (2023). Botulism [Online]. Available: https://www.who.int/news-room/fact-sheets/detail/botulism#:~:text=The%20median%20lethal%20dose%20for,for%20lower%20levels%20of%20intoxication (accessed 29 April 2024).

61 Featherstone, S. (2015). *A Complete Course in Canning and Related Processes 3: Processing Procedures for Canned Food Products*, 3 vols, 14e. Woodhead Publishing.

62 BBC (2021). Kraft Heinz to invest in the UK to make tomato ketchup, 1 June 2021 [Online]. Available: https://www.bbc.com/news/business-57312162 (accessed 10 August 2024).

63 Wikimedia (1909). Cutaway diagram of the "solderless seam" used by H. J. Heinz's canning process in 1909 to attach the can ends to the cylindrical body. From the materials for the Alaska-Yukon-Pacific Exposition of 1909, held in Seattle [Online]. Available: https://commons.wikimedia.org/wiki/File:Can_solderless_seam_cutaway.jpg (accessed 8 August 2024).

64 Food and Agriculture Organization of the United Nations (1985). Planning and engineering data 2. Fish canning - Table of Contents, FAO Fisheries Circular No. 784 [Online]. Available: https://www.fao.org/4/R6918E/R6918E00.htm#Contents (accessed 19 May 2025).

65 Escaño, J.M., Bordons, C., Vilas, C. et al. (2009). Neurofuzzy model based predictive control for thermal batch processes. *Journal of Process Control* 19 (9): 1566–1575. https://doi.org/10.1016/j.jprocont.2009.07.016.

66 Maheswarappa, N.B., Mohan, K., and Jagadeesh, D.S. (2016). Meat products packaging. In: *Reference Module in Food Science*. Elsevier. https://doi.org/10.1016/B978-0-08-100596-5.03221-2.

67 Tavman, S., Kumcuoglu, S., and Gaukel, V. (2007). Apparent specific heat capacity of chilled and frozen meat products. *International Journal of Food Properties* 10 (1): 103–112. https://doi.org/10.1080/10942910600755151.

68 Barreiro, J.A., Perez, C.R., and Guariguata, C. (1984). Optimization of energy consumption during the heat processing of canned foods. *Journal of Food Engineering* 3 (1): 27–37. https://doi.org/10.1016/0260-8774(84)90005-0.

69 Classen, J., Siebenmorgen, T.J., Griffis, C.L., and Young, J.C. (1992). Technical Notes: processing energy requirements for several vegetables. *Transactions of the ASAE* 35 (3): 973–974. https://doi.org/10.13031/2013.28689.

70 IEA (2020). Motor-driven system electricity use as a share of electricity use by industry subsector – Charts – Data & Statistics [Online]. Available: https://www.iea.org/data-and-statistics/charts/motor-driven-system-electricity-use-as-a-share-of-electricity-use-by-industry-subsector (accessed 17 August 2024).

71 Worrell, E., Galitsky, C., and Martin, N. (2002). Energy efficiency opportunities in the brewery industry, LBNL--50979, 881595, June 2002, https://doi.org/10.2172/881595.

72 Chowdhury, J.I., Hu, Y., Haltas, I. et al. (2018). Reducing industrial energy demand in the UK: a review of energy efficiency technologies and energy saving potential in selected sectors. *Renewable and Sustainable Energy Reviews* 94: 1153–1178. https://doi.org/10.1016/j.rser.2018.06.040.

73 Teo, A. and Teo, S. (2016). Food canning waste in industrial processes. https://doi.org/10.13140/RG.2.1.2769.4487.

74 Dey, S. and Agrawal, D.M.K. (2016). Tinplate as a sustainable packaging material: recent innovation and developments to remain environment friendly and cost effective. *International Journal of Research in IT, Management and Engineering* 8: 9–22.

75 Ayres, R.U. (1997). Metals recycling: economic and environmental implications. *Resources, Conservation and Recycling* 21 (3): 145–173. https://doi.org/10.1016/S0921-3449(97)00033-5.

76 Hasan Bdeir, L. and Abood Mohammed, K. (2008). Recycling of aluminum beverage cans. *Engineering and Development* 12: 157–163.

77 Recyclops (2024). How long does it take to decompose? - Recyclops [Online]. Available: https://recyclops.com/how-long-does-it-take-to-decompose/ (accessed 18 October 2024).

78 Fowler, M.W. (2006). Plants, medicines and man. *Journal of the Science of Food and Agriculture* 86 (12): 1797–1804. https://doi.org/10.1002/jsfa.2598.

79 Malerba, F. and Orsenigo, L. (2015). The evolution of the pharmaceutical industry. *Business History* 57 (5): 664–687. https://doi.org/10.1080/00076791.2014.975119.

80 Galitsky, C., Galitsky, C., Chang, S. et al. (2008). Energy efficiency improvement and cost saving opportunities for the Pharmaceutical Industry. An ENERGY STAR guide for energy and plant managers, LBNL-57260-Revision, 929403, March 2008. https://doi.org/10.2172/929403.

81 Espro, C., Paone, E., Mauriello, F. et al. (2021). Sustainable production of pharmaceutical, nutraceutical and bioactive compounds from biomass and waste. *Chemical Society Reviews* 50 (20): 11191–11207. https://doi.org/10.1039/D1CS00524C.

82 The European Federation of Pharmaceutical Industries and Associations (2023). The Pharmaceutical Industry in Figures: Key Data: 2023, November 2023 [Online]. Available: https://efpia.eu/media/2rxdkn43/the-pharmaceutical-industry-in-figures-2024.pdf (accessed 16 July 2024).

83 Domokos, A., Nagy, B., Szilágyi, B. et al. (2021). Integrated continuous pharmaceutical technologies—a review. *Organic Process Research and Development* 25 (4): 721–739. https://doi.org/10.1021/acs.oprd.0c00504.

84 Plumb, K. (2005). Continuous processing in the pharmaceutical industry. *Chemical Engineering Research and Design* 83 (6): 730–738. https://doi.org/10.1205/cherd.04359.

85 Roy, J. (2011). Bulk drugs or active pharmaceutical ingredients. In: *An Introduction to Pharmaceutical Sciences: Production, Chemistry, Techniques and Technology*. O'Reilly Media, Inc.

86 Burcham, C.L., Florence, A.J., and Johnson, M.D. (2018). Continuous manufacturing in pharmaceutical process development and manufacturing. *Annual Review of Chemical and Biomolecular Engineering* 9 (1): 253–281. https://doi.org/10.1146/annurev-chembioeng-060817-084355.

87 Uhl, A., Schmidt, A., Jensch, C. et al. (2022). Development of concepts for a climate-neutral chemical–pharmaceutical industry in 2045. *Processes* 10 (7): 1289. https://doi.org/10.3390/pr10071289.

88 Wollensack, L., Budzinski, K., and Backmann, J. (2022). Defossilization of pharmaceutical manufacturing. *Current Opinion in Green and Sustainable Chemistry* 33: 100586. https://doi.org/10.1016/j.cogsc.2021.100586.

89 Bruni, G., Martini, C., Martini, F., and Salvio, M. (2023). On the energy performance and energy saving potential of the pharmaceutical industry: a study based on the Italian Energy Audits. *PRO* 11 (4): https://doi.org/10.3390/pr11041114.

90 Cushman-Roisin, B. and Cremonini, B.T. (2021). *Data, Statistics, and Useful Numbers for Environmental Sustainability: Bringing the Numbers to Life*. Elsevier.

91 SpendEdge (2025). The biggest problems with pharmaceutical raw materials market, and how you can fix it, SpendEdge Procurement Market Intelligence Solutions [Online]. Available: https://www.spendedge.com/blogs/problems-pharmaceutical-raw-materials/ (accessed 18 May 2025).

92 Riazi, B., Zhang, J., Yee, W. et al. (2019). Life cycle environmental and cost implications of isostearic acid production for pharmaceutical and personal care products. *ACS Sustainable Chemistry & Engineering* 7 (18): 15247–15258. https://doi.org/10.1021/acssuschemeng.9b02238.

93 Wang, D., Cheow, W.S., Amalina, N. et al. (2021). Selecting optimal pharmaceutical excipient formulation from life cycle assessment perspectives: a case study on ibuprofen tablet formulations. *Journal of Cleaner Production* 292: 126074. https://doi.org/10.1016/j.jclepro.2021.126074.

94 Cespi, D., Beach, E.S., Swarr, T.E. et al. (2015). Life cycle inventory improvement in the pharmaceutical sector: assessment of the sustainability combining PMI and LCA tools. *Green Chemistry* 17 (6): 3390–3400. https://doi.org/10.1039/C5GC00424A.

95 Güneş, Ç.E. and Şengül, H. (2022). Cradle-to-gate life cycle assessment of a liquid pharmaceutical product through analysis of chemical production pathways. *Clean Technologies and Environmental Policy* 24 (6): 1741–1755. https://doi.org/10.1007/s10098-022-02283-4.

96 Ott, D., Borukhova, S., and Hessel, V. (2016). Life cycle assessment of multi-step rufinamide synthesis – from isolated reactions in batch to continuous microreactor networks. *Green Chemistry* 18 (4): 1096–1116. https://doi.org/10.1039/C5GC01932J.

97 Kong, W., Lv, B., Yang, S. et al. (2021). Case study on environmental safety and sustainability of pharmaceutical production based on life cycle assessment of enrofloxacin. *Journal of Environmental Chemical Engineering* 9 (4): 105734. https://doi.org/10.1016/j.jece.2021.105734.

98 Amado Alviz, P.L. and Alvarez, A.J. (2017). Comparative life cycle assessment of the use of an ionic liquid ([Bmim]Br) versus a volatile organic solvent in the production of acetylsalicylic acid. *Journal of Cleaner Production* 168: 1614–1624. https://doi.org/10.1016/j.jclepro.2017.02.107.

99 Yang, K., Lv, B., Shen, H. et al. (2021). Life cycle assessment of pharmaceuticals: the ciprofloxacin hydrochloride case. *International Journal of Life Cycle Assessment* 26 (1): 64–75. https://doi.org/10.1007/s11367-020-01841-6.

100 Davies, J.F., McAlister, S., and Eckelman, M.J. (2024). Environmental and financial impacts of perioperative paracetamol use: a multicentre international life-cycle assessment. *British Journal of Anaesthesia* 133 (6): 1439–1448. https://doi.org/10.1016/j.bja.2023.11.053.

101 Harding, K.G., Dennis, J.S., and Harrison, S.T.L. (2018). Generic flowsheeting approach to generating first estimate material and energy balance data for Life Cycle Assessment (LCA) of Penicillin V production. *Sustainable Production and Consumption* 15: 89–95. https://doi.org/10.1016/j.spc.2018.05.004.

102 De Soete, W., Dewulf, J., Cappuyns, P. et al. (2013). Exergetic sustainability assessment of batch versus continuous wet granulation based pharmaceutical tablet manufacturing: a cohesive analysis at three different levels. *Green Chemistry* 15 (11): 3039. https://doi.org/10.1039/c3gc41185k.

103 McAlister, S., Ou, Y., Neff, E. et al. (2016). The Environmental footprint of morphine: a life cycle assessment from opium poppy farming to the packaged drug. *BMJ Open* 6 (10): e013302. https://doi.org/10.1136/bmjopen-2016-013302.

104 Renteria Gamiz, A.G., De Soete, W., Heirman, B. et al. (2019). Environmental sustainability assessment of the manufacturing process of a biological active pharmaceutical ingredient. *J of Chemical Tech & Biotech* 94 (6): 1937–1944. https://doi.org/10.1002/jctb.5975.

105 Piffoux, M., Le Tellier, A., Taillemite, Z. et al. (2024). Carbon footprint of oral medicines using hybrid life cycle assessment. *Journal of Cleaner Production* 475: 143576. https://doi.org/10.1016/j.jclepro.2024.143576.

106 Abdussalam-Mohammeda, W., Alia, A.Q., and Errayesb, A.O. (2020). Green chemistry: principles, applications, and disadvantages. *Chemistry Methodology* 4 (4): 408–423. https://doi.org/10.33945/SAMI/CHEMM.2020.4.4.

107 Singh, R.M., Pramanik, R., and Hazra, S. (2021). Role of green chemistry in pharmaceutical industry: a review. *JUSST* 23 (12): 291–299. https://doi.org/10.51201/JUSST/21/121018.

108 Tucker, J.L. (2006). Green Chemistry, a pharmaceutical perspective. *Organic Process Research and Development* 10 (2): 315–319. https://doi.org/10.1021/op050227k.

109 Adewuyi, A. (2022). Underutilized lignocellulosic waste as sources of feedstock for biofuel production in developing countries. *Frontiers in Energy Research* 10: 741570. https://doi.org/10.3389/fenrg.2022.741570.

110 Ralph, J., Karlen, S., and Mobley, J. (2019). Synthesis of paracetamol (acetaminophen) from biomass-derived p-hydroxybenzamide. US Patent 10286504B2.

111 Tsolakis, N. and Srai, J.S. (2018). Mapping supply dynamics in renewable feedstock enabled industries: a systems theory perspective on 'green' pharmaceuticals. *Operations Management Research* 11 (3–4): 83–104. https://doi.org/10.1007/s12063-018-0134-y.

112 Lee, S.L., O'Connor, T.F., Yang, X. et al. (2015). Modernizing pharmaceutical manufacturing: from batch to continuous production. *Journal of Pharmaceutical Innovation* 10 (3): 191–199. https://doi.org/10.1007/s12247-015-9215-8.

113 Mascia, S., Heider, P.L., Zhang, H. et al. (2013). End-to-end continuous manufacturing of pharmaceuticals: integrated synthesis, purification, and final dosage formation. *Angewandte Chemie, International Edition* 52 (47): 12359–12363. https://doi.org/10.1002/anie.201305429.

114 Schaber, S.D., Gerogiorgis, D.I., Ramachandran, R. et al. (2011). Economic analysis of integrated continuous and batch pharmaceutical manufacturing: a case study. *Industrial and Engineering Chemistry Research* 50 (17): 10083–10092. https://doi.org/10.1021/ie2006752.

115 Porta, R., Benaglia, M., and Puglisi, A. (2016). Flow chemistry: recent developments in the synthesis of pharmaceutical products. *Organic Process Research and Development* 20 (1): 2–25. https://doi.org/10.1021/acs.oprd.5b00325.

116 Burange, A.S., Osman, S.M., and Luque, R. (2022). Understanding flow chemistry for the production of active pharmaceutical ingredients. *iScience* 25 (3): 103892. https://doi.org/10.1016/j.isci.2022.103892.

117 Plutschack, M.B., Pieber, B., Gilmore, K., and Seeberger, P.H. (2017). The Hitchhiker's guide to flow chemistry. *Chemical Reviews* 117 (18): 11796–11893. https://doi.org/10.1021/acs.chemrev.7b00183.

118 Rabadán González, I., McLean, J.T., Ostrovitsa, N. et al. (2024). A thiol–ene mediated approach for peptide bioconjugation using 'green' solvents under continuous flow. *Organic & Biomolecular Chemistry* 22 (11): 2203–2210. https://doi.org/10.1039/D4OB00122B.

119 Suhandi, V. and Chen, P.-S. (2023). Closed-loop supply chain inventory model in the pharmaceutical industry toward a circular economy. *Journal of Cleaner Production* 383: 135474. https://doi.org/10.1016/j.jclepro.2022.135474.

120 Ashiwaju, B.I., Orikpete, O.F., Fawole, A.A. et al. (2023). A step toward sustainability: a review of biodegradable packaging in the pharmaceutical industry. *Matrix Science Pharma* 7 (3): 73–84. https://doi.org/10.4103/mtsp.mtsp_22_23.

121 Kumar, L., Hasanuzzaman, M., and Rahim, N.A. (2019). Global advancement of solar thermal energy technologies for industrial process heat and its future prospects: a review. *Energy Conversion and Management* 195: 885–908. https://doi.org/10.1016/j.enconman.2019.05.081.

122 Guillaume, M., Wagner, G., Jobard, X. et al. (2019). Solar thermal systems for the Swiss Pharmaceutical Industry Sector. *Proceedings of the ISES Solar World Congress 2019*, Santiago, Chile: International Solar Energy Society, pp. 1–10. https://doi.org/10.18086/swc.2019.12.06.

123 Engeland, K., Borga, M., Creutin, J.-D. et al. (2017). Space-time variability of climate variables and intermittent renewable electricity production – a review. *Renewable and Sustainable Energy Reviews* 79: 600–617. https://doi.org/10.1016/j.rser.2017.05.046.

124 U.S. Department of Health and Human Services Food and Drug Administration (2023). Q13 Continuous Manufacturing of Drug Substances and Drug Products, March 2023.

125 Kekes, T., Tzia, C., and Kolliopoulos, G. (2023). Drinking and natural mineral water: treatment and quality–safety assurance. *Water* 15 (13): 2325. https://doi.org/10.3390/w15132325.

126 United Nations (2023). Human rights to water and sanitation [Online]. Available: https://www.unwater.org/water-facts/human-rights-water-and-sanitation (accessed 10 October 2024).

127 United Nations (2024). Goal 6: Ensure access to water and sanitation for all, Sustainable Development Goals [Online]. Available: https://www.un.org/sustainabledevelopment/water-and-sanitation/ (accessed 8 October 2024).

128 Abou-Shady, A., Siddique, M.S., and Yu, W. (2023). A critical review of recent progress in global water reuse during 2019–2021 and perspectives to overcome future water crisis. *Environments* 10 (9): 159. https://doi.org/10.3390/environments10090159.

129 Sonune, A. and Ghate, R. (2004). Developments in wastewater treatment methods. *Desalination* 167: 55–63. https://doi.org/10.1016/j.desal.2004.06.113.

130 Ricart, S. and Rico, A.M. (2019). Assessing technical and social driving factors of water reuse in agriculture: a review on risks, regulation and the yuck factor. *Agricultural Water Management* 217: 426–439. https://doi.org/10.1016/j.agwat.2019.03.017.

131 Ritchie, H. and Roser, M. (2018). Water use and stress, OurWorldinData.org [Online Resource]. Retrieved from. https://ourworldindata.org/water-use-stress (accessed 10 September 2025).

132 Dalezios, N.R., Angelakis, A.N., and Eslamian, S.S. (2018). Water scarcity management: Part 1: methodological framework. *IJGENVI* 17 (1): 1. https://doi.org/10.1504/IJGENVI.2018.090629.

133 Evaristo, J., Jameel, Y., Tortajada, C. et al. (2023). Water woes: the institutional challenges in achieving SDG 6. *Sustain Earth Reviews* 6 (1): 13. https://doi.org/10.1186/s42055-023-00067-2.

134 Ghernaout, D. (2018). Increasing trends towards drinking water reclamation from treated wastewater. *WJAC* 3 (1): 1. https://doi.org/10.11648/j.wjac.20180301.11.

135 Ghernaout, D. and Elboughdiri, N. (2019). Upgrading wastewater treatment plant to obtain drinking water. *OALib* 06 (12): 1–14. https://doi.org/10.4236/oalib.1105959.

136 Crini, G. and Lichtfouse, E. (2019). Advantages and disadvantages of techniques used for wastewater treatment. *Environmental Chemistry Letters* 17 (1): 145–155. https://doi.org/10.1007/s10311-018-0785-9.

137 Watertechnologies (2023). Water Handbook - Wastewater Treatment | Veolia [Online]. Available: https://www.watertechnologies.com/handbook/chapter-37-wastewater-treatment (accessed 6 October 2024).

138 Ranieri, E., Leverenz, H., and Tchobanoglous, G. (2011). An examination of the factors involved in agricultural reuse: technologies, regulatory and social aspects. *JWARP* 03 (05): 300–310. https://doi.org/10.4236/jwarp.2011.35038.

139 Veolia (2023). Producing drinking water from wastewater, WINGOC - producing drinking water from wastewater Municipal solution [Online].

Available: https://www.veolia.co.za/producing-drinking-water-wastewater (accessed 14 October 2024).

140 GIZ & SAGEN (2020). A practical guideline for energy efficiency audits at wastewater treatment works. Deutsche Gesellschaft für Internationale Zusammenarbeit (GIZ) South African German Energy Programme (SAGEN), Hatfield, Pretoria, South Africa, May 2020 [Online]. Available: https://www.cityenergy.org.za/uploads/resource_490.pdf (accessed 10 October 2024).

141 Britannica (2024). Wastewater treatment - Sedimentation, Filtration, Clarification | Britannica [Online]. Available: https://www.britannica.com/technology/wastewater-treatment (accessed 14 October 2024).

142 SafeRack (2024). Steps of a wastewater treatment plant [Online]. Available: https://www.saferack.com/wastewater-treatment-plant/wastewater-treatment-process-illustrationv2/ (accessed 6 October 2024).

143 Gandiglio, M., Lanzini, A., Soto, A. et al. (2017). Enhancing the energy efficiency of wastewater treatment plants through co-digestion and fuel cell systems. *Frontiers in Environmental Science* 5: 70. https://doi.org/10.3389/fenvs.2017.00070.

144 Gu, Y., Li, Y., Li, X. et al. (2017). Energy self-sufficient wastewater treatment plants: feasibilities and challenges. *Energy Procedia* 105: 3741–3751. https://doi.org/10.1016/j.egypro.2017.03.868.

145 Vieira, A.S. and Ghisi, E. (2016). Water-energy nexus in low-income houses in Brazil: the influence of integrated on-site water and sewage management strategies on the energy consumption of water and sewerage services. *Journal of Cleaner Production* 133: 145–162. https://doi.org/10.1016/j.jclepro.2016.05.104.

146 Negi, R. and Chandel, M.K. (2021). Analysing water-energy-GHG nexus in a wastewater treatment plant of Mumbai Metropolitan Region, India. *Environmental Research* 196: 110931. https://doi.org/10.1016/j.envres.2021.110931.

147 Siddiqi, A. and Anadon, L.D. (2011). The water–energy nexus in Middle East and North Africa. *Energy Policy* 39 (8): 4529–4540. https://doi.org/10.1016/j.enpol.2011.04.023.

148 Demirbas, A., Edris, G., and Alalayah, W.M. (2017). Sludge production from municipal wastewater treatment in sewage treatment plant. *Energy Sources, Part A: Recovery, Utilization, and Environmental Effects* 39 (10): 999–1006. https://doi.org/10.1080/15567036.2017.1283551.

149 He, C., Chen, C.-L., Giannis, A. et al. (2014). Hydrothermal gasification of sewage sludge and model compounds for renewable hydrogen production: a review. *Renewable and Sustainable Energy Reviews* 39: 1127–1142. https://doi.org/10.1016/j.rser.2014.07.141.

150 Oladejo, J., Shi, K., Luo, X. et al. (2018). A review of sludge-to-energy recovery methods. *Energies* 12 (1): 60. https://doi.org/10.3390/en12010060.

151 Tzanakakis, V.A., Capodaglio, A.G., and Angelakis, A.N. (2023). Insights into global water reuse opportunities. *Sustainability* 15 (17): 13007. https://doi.org/10.3390/su151713007.

152 Chang, A., Pan, G., Page, A., and Asano, T. (2001). Developing human health-related chemical guidelines for reclaimed waster and sewage sludge applications in agriculture. *Water Science and Technology* 33 (10–11): 463–472.

153 Zsirai, I. (2011). Sewage sludge as renewable energy. *Journal of Residuals Science and Technology* 8 (4): 165–179.

154 Linderholm, K., Tillman, A.-M., and Mattsson, J.E. (2012). Life cycle assessment of phosphorus alternatives for Swedish agriculture. *Resources, Conservation and Recycling* 66: 27–39. https://doi.org/10.1016/j.resconrec.2012.04.006.

155 Hudcová, H., Vymazal, J., and Rozkošný, M. (2019). Present restrictions of sewage sludge application in agriculture within the European Union. *Soil and Water Research* 14 (2): 104–120. https://doi.org/10.17221/36/2018-SWR.

156 Giwa, A.S., Maurice, N.J., Luoyan, A. et al. (2023). Advances in sewage sludge application and treatment: process integration of plasma pyrolysis and anaerobic digestion with the resource recovery. *Heliyon* 9 (9): e19765. https://doi.org/10.1016/j.heliyon.2023.e19765.

157 Chew, K.W., Chia, S.R., Yen, H.-W. et al. (2019). Transformation of biomass waste into sustainable organic fertilizers. *Sustainability* 11 (8): 2266. https://doi.org/10.3390/su11082266.

158 Shen, Y., Linville, J.L., Urgun-Demirtas, M. et al. (2015). An overview of biogas production and utilization at full-scale wastewater treatment plants (WWTPs) in the United States: challenges and opportunities towards energy-neutral WWTPs. *Renewable and Sustainable Energy Reviews* 50: 346–362. https://doi.org/10.1016/j.rser.2015.04.129.

159 Burton, S., Cohen, B., Harrison, S. et al. Energy From Wastewater – A Feasibility Study Essence Report. WRC Report No. TT 399/09 July 2009, ISBN 978-1-77005-847-7. Available at: https://www.wrc.org.za/?mdocs-file=59280

160 Sarpong, G. and Gude, V.G. (2020). Near future energy self-sufficient wastewater treatment schemes. *International Journal of Environmental Research* 14 (4): 479–488. https://doi.org/10.1007/s41742-020-00262-5.

161 Smestad, G.P., Schlichting, M.L., Mohamed, T.A. et al. (ed.) (2021). *Distributed Renewable Energies For Off-Grid Communities: Empowering a Sustainable, Competitive, and Secure Twenty-First Century*, 2e, 427–428. Cambridge, MA: Brian Romer.

162 Poore, J. and Nemecek, T. (2018). Reducing food's environmental impacts through producers and consumers. *Science* 360 (6392): 987–992. https://doi.org/10.1126/science.aaq0216.

9

General Outlook

There is a need to limit greenhouse gas emissions (and specifically CO_2), so that the concentration of these gases in the atmosphere is reduced, which would thus limit the impact of climate change on society. However, the drive towards renewable energy goes further as society strives towards sustainable development, i.e. 'meeting the needs of the present without compromising the ability of future generations to meet their own needs' [1]. This implies that we should reduce usage of all limited resources, including energy resources, as well as minimise waste generation and the dispersion of waste. A consequence of this is to minimise the use of fossil fuels for energy generation, together with its waste disposal into the atmosphere.

The process industry is a major user of energy and thus, when using fossil fuel as the energy source, is a large emitter of greenhouse gases. Hence, societal pressure is on the process industry to reduce its emissions. This is valid not only for individual operations but also for industry sectors. The latter should consider that the industrial sector is expected to increase significantly when serving an increasing population striving for economic growth.

The process industry is diverse both in size and in extent of emissions per unit of product produced. Many different technologies are available or have been proposed for the various industries to minimise their emissions and to modernise the industry. Over the last few decades, energy savings in the process industry were achieved specifically through on-site heat integration, which at the same time has been a major driver to reduce operational cost. This came with the added benefit of reduced emissions and a reduced carbon footprint. Otherwise stated, the reduction in the emissions of greenhouse gases was 'financed' by the drive to reduce the cost for utilities and thus a reduction in the production costs. The drive towards renewable energy puts the process industry in a new position, where presently (in 2024) the cost associated with the reduction in emission of greenhouse gases by switching from fossil-fuel-based energy to renewable is not (yet) automatically offset by a decrease in the production cost or an increase in the selling price. However, appropriate costing of waste disposal should create drivers to minimise these emissions, as it will then reduce any penalties associated with emissions. Furthermore, it will allow the industry to continue to operate.

Renewable Energy in the Process Industry, First Edition.
Eric van Steen, Nicholas S. Featherstone, and Linda H. Callanan.

9.1 Drivers to Reduce Greenhouse Gas Emissions

The systematic reduction of emissions of greenhouse gases to minimise the impact on global warming is a large task, but humankind has been successful in minimising danger from emissions transcending national borders before, specifically by stopping the use of chlorofluorocarbons (CFCs) and hydrochlorofluorocarbons (HCFCs) as aerosol propellants to minimise the damage to the ozone layer (Montreal Protocol [2]) and replacing them with less harmful substances. The urge to reduce emissions of greenhouse gases does have similarities with the ban on CFCs and HCFCs in that it is a trans-border problem, but with a major difference. HCFCs as a propellant could readily be replaced by other, environmentally less damaging, substances having a similar functionality, such as light, saturated hydrocarbons.

Governments have, in principle, several mechanisms to enforce a change in behaviour, viz. carbon tax, carbon emission trading and incentives.

Taxation has its origin in the idea that producers will need to pay for the negative societal impact of their emissions [3]. The original idea of the Pigouvian taxation is to include the social marginal cost through taxation into the price to account for the negative externality imposed by the product, which would yield a new economic optimum for the quantity of greenhouse gases produced (see Figure 9.1). Here the polluter pays in principle for the marginal cost induced by the greenhouse gas emissions.

From a policy perspective, carbon tax is seen as a mechanism to influence companies to implement more efficient technologies to reduce their carbon footprint. Carbon taxation is a financial instrument incentivising companies to drive technologies aimed at mitigating greenhouse gas emissions by these companies, whilst at the same time by shifting consumer behaviour towards products with a reduced environmental impact [4]. However, it has been argued that this will only induce innovations in the large, productive industries, whereas smaller, more polluting industry will disappear [5]. Measures to enhance productivity targeting smaller companies may be needed to introduce more environmentally sound technologies in these companies.

Carbon tax may be compared to 'sin' taxes, such as those put on the sale of tobacco and alcoholic beverages to account for the negative externalities associated with those products. For instance, the consumption of tobacco and alcoholic beverages has been linked with increased hospitalisation costs incurred by the State. It may be argued that 'sin' tax is not only aimed at shifting consumer behaviour (by making these products financially less attractive), but that they are also transactional, i.e. society at large will take on the consequence, e.g. through hospital funding. In a similar manner, carbon tax may be considered transactional, where polluter pays for the waste emission with the understanding that the responsibility for carbon emissions is shifted from the producer to society at large. In this viewpoint, carbon tax should be ringfenced and used specifically for actions mitigating the effects of climate change, and, if possible, reducing the carbon emissions by society at large. However, equating carbon tax with the classical 'sin' tax is not quite correct due to the trans-border impact of greenhouse gas emissions, as they will not (necessarily) only

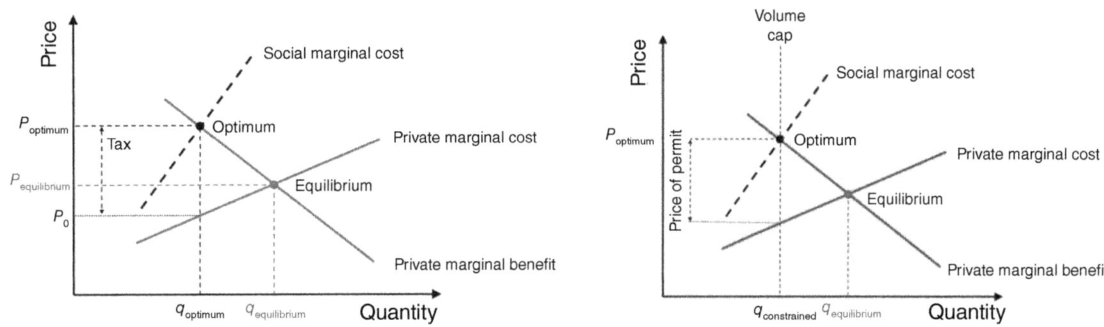

Figure 9.1 Paying for carbon emissions through Pigouvian taxation (a) and an emission trading system (b) (redrawn from Pan et al. [3]).

affect the country, where these gases are being emitted, nor just the country where the taxation is being levied. Hence, the social marginal cost is not paid, where the cost will be incurred. Hence, the direct transactional linkage in taxation is lost with national or even regional carbon taxation.

Only 5.65% of global emissions are subject to carbon tax [6]. Carbon tax is typically collected at the point of entry into a country or economic unit (either through import of fuel, or at the mine/well) [7]. The level at which carbon taxes should be set is intensely debated worldwide [7–9] often focusing on the macro-economic and direct social impact of carbon tax [10, 11]. They have often been introduced in a revenue-neutral way [7], thus ignoring the actual cost to mitigate the impact of global warming. The current taxation rate varies widely and is, in some countries, as high as US$(2024) 100/tonne $CO_{2,eq}$ (see Figure 9.2). Ideally, the level of carbon tax should be offset by the marginal social cost associated with the emission [12]. The level of carbon taxation levelled at the moment is still far off from the cost of carbon removal from the atmosphere, which can arguably be taken as a benchmark for the social marginal cost associated with the emissions. It has been estimated for direct air capture technology to cost at US$ 400–1350/tonne CO_2 for plants at a scale of

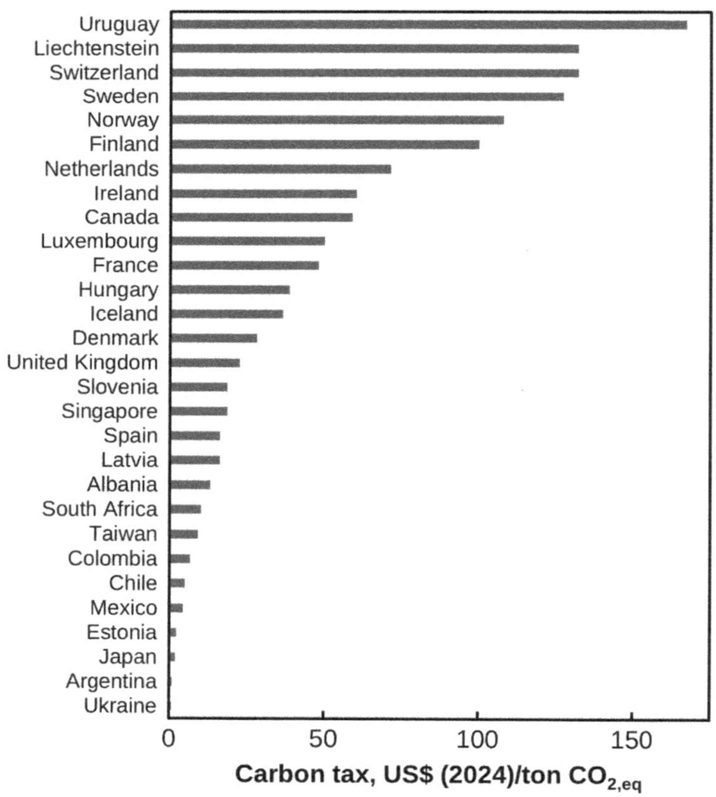

Figure 9.2 Carbon tax (in US$ [2024]) for carbon emissions (data from World Bank Group [6]).

Gigatonnes (with the associated high energy input) [13]. Hence, taxation aimed at reducing the concentration of greenhouse gases in the atmosphere should be seen as a driver for industry to reduce its emissions and thus reduce the cost of their product, rather than society taking over the responsibility for reducing the impact of global warming.

Emission trading systems (like carbon credits) are a different mechanism to give an economic incentive to reduce emissions. They are also known as cap-and-trade or a carbon market, with a finite limit (cap) set on the total greenhouse gas emissions that may be generated by an industry [14]. The emissions cap sets the maximum level of emissions allowed by a producer. This cap is set by the government, who is regarded as the custodian and ultimate owner for the environment (the rights of emissions can be given off by governments for free or auctioned off). This system may be criticised as aiming to maintain the status quo with the current players in an expanding market, thereby favouring established operators over newcomers in the market, and being influenced by lobbyists working for specific industries [15], although progressive reduction in the allowable emissions (cap) is a tool to attain sustainable development goals. Proponents of the emission trading system point out that this system allows for the required reduction in emissions by difficult-to-reduce industries to be achieved by trading/exchanging their required reduction (using certificates) with a reduction in an industry, whose emissions are more easily reduced. This also immediately links to a weakness in the system, as the industries covered must be diverse in their emission profile.

Worldwide emission trading systems cover 18.3% of the global emissions [6]. The pricing of the emission certificates can vary widely. The European Trading System (ETS) is possibly the best-known emission trading system covering CO_2 emissions for power and heat generation, aviation and maritime transport (to some extent) and the manufacture of cement/concrete, iron and steel, glass, ceramics, paper and pulp, acids and bulk organic chemicals in addition to covering N_2O emissions, e.g. from nitric acid production, and emissions of perfluorocarbons in the production of aluminium. It allows for a 'free' allocation (severely reduced in Phase III and Phase IV, see Figure 9.3) of the trading system in line with the targeted reduction in emissions, and emission allowances, which can be auctioned off. Not all emission allowances allocated to the auction volume will be auctioned off, to manage the supply of these allowances on the market and thus the price.

The price for the EU emission allowances was initially low (at ca. US$ 4–6/tonne CO_2) due to oversupply (see Figure 9.3). In 2018, the price started rising to above US$ 18/tonne CO_2 by constraining the supply of emission allowances due to placing some of them in the so-called Market Stability Reserve (MSR) instead of selling them [15]. Around 2021, a steep increase in the price of EU emission allowances was observed to more than US$ 80/tonne CO_2, because of the cold weather at the start of 2021 giving rise to an increased energy demand [17]. The higher gas price encouraged electricity suppliers to switch from gas back to coal, thus increasing the demand for the emission allowances, as well as the reduced (and shrinking) supply of allowances. In 2024/2025, the price for the EU emission allowances stabilised at ca. US$ 60/tonne CO_2.

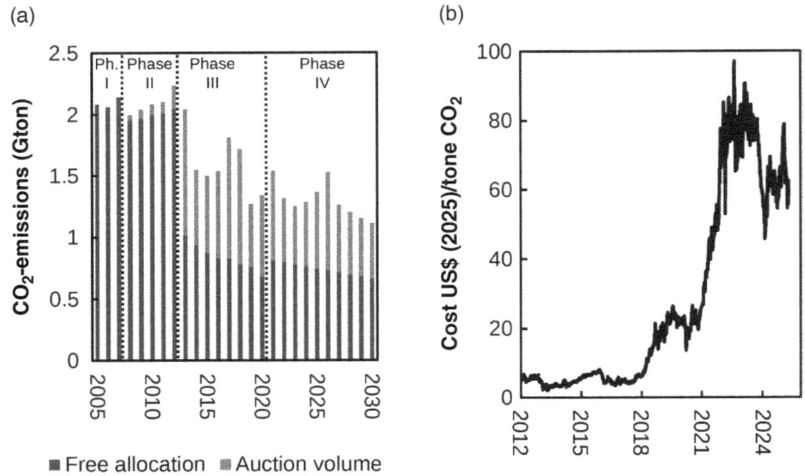

Figure 9.3 European Trading System: free allocation and auction volume (excl. aviation) for Phase I–IV (a; Sato et al. [15]) and price of EU emission allowances in US$ (2025)/tonne CO_2 (b; data from Instrat Foundation [16]).

A concern, when introducing carbon tax and/or carbon emission trading systems (carbon credits), is the so-called carbon leakage, i.e. the shifting of the production of products subject to carbon tax/carbon credits to so-called carbon tax havens, countries with minimal or no carbon taxation and not part of an emission trading scheme. This can be countered by a border tax adjustment, as the European Border Adjustment Mechanism (EBAM) [18] for imports on large amounts (currently larger than 50 tonne per annum) of products produced in processes with carbon emissions greater than the legislated limit, e.g. cement, iron and steel, aluminium, fertilisers and hydrogen, as well as the import of electricity.

Carbon taxation and carbon emission trading systems can be viewed as a method to price emissions and are thus essentially 'stick' methods. An alternative approach would be a 'carrot' method, where the introduction of novel, emission-reducing systems is rewarded with tax incentives (as e.g. in the Inflation Reduction Act introduced in the United States in 2022). There is a philosophical difference between these two methods, where carbon taxation and carbon emission trading systems recognise the cost associated with the reduction in carbon emissions, and a tax incentive scheme recognises that the initial investment needs to be supported, so the ultimate operational costs will be competitive in the market.

9.2 Accounting for Greenhouse Gas Emissions

The driving force for introducing renewable energy into the process industry is to reduce the emissions of greenhouse gases and specifically CO_2 into the atmosphere originating from its activities. Hence, it is important for the industry to account for its emissions and know its origins. This becomes even more important when products will have to be declared with their emissions.

Greenhouse gas emissions have been categorised as so-called Scope 1, Scope 2 and Scope 3 type of emissions. Scope 1 type emissions are emissions by the process itself/on the process site itself. Scope 2 type emissions are emissions associated with imported energy, in the form of electricity, heat, steam or cooling, used by the process. For instance, if a process uses electric power to drive the process, the direct greenhouse gas emissions associated with the generation of electricity would be classified as Scope 2 emissions.

Scope 3-type emissions are the so-called life cycle emissions [19] or non-energy indirect emissions, which are more difficult to characterise as they include emissions from a variety of different sources [20] both upstream and downstream of the actual process. The upstream emissions include emissions associated with the production of intermediates fed into the process. For example, soda ash is used as a fluxing agent in the glass industry (see Chapter 7.6), and the CO_2 emissions associated with the production of soda ash would count as Scope 3 emissions in the glass manufacture (whereas the CO_2 emissions caused by the decomposition of soda ash in glass manufacturing would count as Scope 1 emissions). Furthermore, emissions associated with waste generated by the process, the production of the assets required for the process, transport around the process (transport of raw materials, products, employees) and the processing, use and end-of-life treatment of sold products are all classified as Scope 3 emissions.

The classification of emissions into these groupings has been in use for some time claiming to give more insight in the origin of the emission by an organisation, as it conveniently divides emissions into emissions generated by the organisation, emissions associated with energy imported from other sources and lumping all other associated imported emissions into Scope 3. Furthermore, it can then be used as an indicator of the environmental impact of a product manufactured/delivered by a certain organisation. However, this grouping does not give an easy insight into how process modifications may affect the emissions.

A more suitable classification for the process industry in the view of the authors, more clearly outlining the impact of decisions on the choice of raw materials, type of process, source of energy used in the process, but also location of the process, on the emissions generated by the production process, would be based on the functionality of the operation. Scope A emissions can be assigned to emissions directly associated with the choice of process, and resulting from there the raw materials/intermediates required, the amount of energy required and the amount/type of waste generated. The choice of the energy source will also affect the overall Scope A emissions, as well as the possible generation of heat in the process. Scope B emissions are linked to the location of the process, i.e. close to the source of the feedstock, close to the market or even close to the required workforce. Scope C emissions originate from the fate of the product, and the overall emissions associated with a product are given by the sum of Scope A, Scope B and Scope C emissions. Scope D emissions are generated by having/leasing assets required for the production process. The classification can be refined as:

- Scope A emissions are linked to the chosen process and are under direct control by the production site
  - Scope A1 emissions are associated with the chemical conversion of materials in the process
  - Scope A2 emissions are associated with the production of the feed to the process and thus the choice of the feedstock
  - Scope A3 emissions are associated with the energy usage in the process, which is thus affected by the chosen energy source(s) for the process, and which can be offset by the sale of energy to alternative users
  - Scope A4 emissions are associated with waste generated by the process
- Scope B emissions are associated with location of the process with
  - Scope B1 emissions related to transport of feedstock and intermediates
  - Scope B2 emissions related to transport of products and waste
  - Scope B3 emissions related to transport of employees
- Scope C emissions are associated with the associated fate of the products in its life (and are thus harder to determine) with
  - Scope C1 emissions related to further processing of the product by third parties
  - Scope C2 emissions related to the use of the product and its ultimate disposal
- Scope D emissions are associated with the assets held by the organisation.

This type of classification will be more useful to process engineers. Figure 9.4 shows a linkage between this proposed classification scheme and the classically used Scope 1, Scope 2 and Scope 3 classification.

Calculating emissions for a given production process is straightforward, but considering the whole process value chain may result in double accounting.

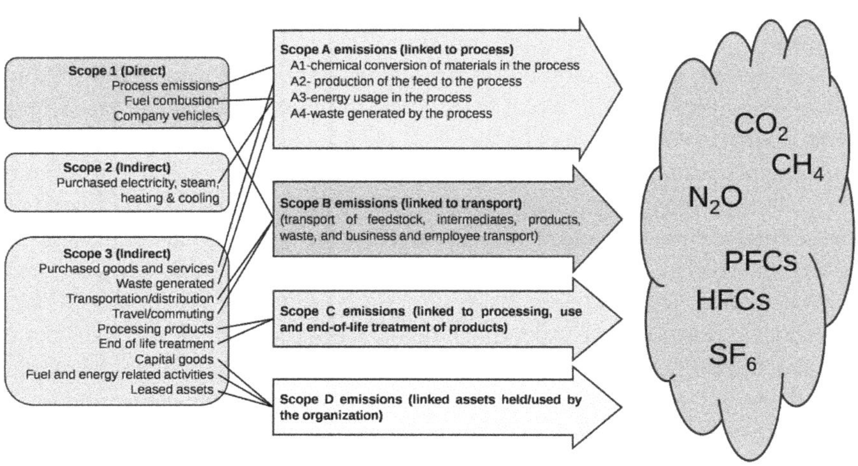

Figure 9.4 Correlating Scope 1, 2 and 3 emissions with Scope A, B, C and D emissions.

For instance, glass manufacturers use sodium carbonate as an essential ingredient. The emissions associated with sodium carbonate production can be classified as outlined in Figure 9.4, and include the emissions linked to the production of sodium carbonate (Scope A emissions) but also transport of sodium carbonate to its users (Scope B emissions), and emissions linked to processing of products (Scope C emissions). The transport of the feedstock (sodium carbonate) is also included in the emissions associated with the production of glass as Scope B emissions. Furthermore, the Scope A emissions associated with the production of glass are included as Scope C emissions for sodium carbonate.

These classifications are specific to a particular product associated with a specific process and cannot be taken as a measure of the overall emissions. These classification schemes may result in double accounting. This is specifically problematic, when dealing with reaction intermediates. For instance, emissions associated with the manufacture of an intermediate are taken over by the producer of the end-product, but transport of a reaction intermediate to the production site will need to be carefully considered as it contributes Scope B emissions to manufacturing of both the intermediate and the end-product. Hence, clear boundaries need to be drawn up when accounting for emissions so that double accounting is avoided to obtain an overview of the global greenhouse gas emissions, but more work will need to be done in this area.

9.3 Electrification

In 2024, ca. 42% of all primary energy was consumed to generate electricity and the availability of renewable energy, which is harnessed as electrical energy, will drive the further electrification of society at large. Electric power is not only used in households, but is also increasingly used for transport purposes using electric vehicles (EVs). Furthermore, data centres and AI-driven technologies will push the use of electrical energy up. This implies a large extension of the power grid to cater for all consumers.

Electric energy is also being used in the process industry (to run pumps, compressors, etc.), but the main energy usage in the process industry is for heating, which has been obtained more economically by combusting fossil fuels or from exothermic processes running on-site. The use of renewable energy for heating purposes would reduce the emission of greenhouse gases by the process industry significantly. The cost associated with the change in the heating method is determined by the capital cost of the equipment, but mainly by the operational cost, i.e. the cost of energy, and specifically the cost of electricity versus the cost of fossil fuels for heating purposes.

The globally averaged levelised cost of electricity (LCOE) for onshore wind power is ca. US$ 33/MWh (2023 values), US$ 75/MWh for offshore wind power, ca. US$ 44/MWh for PV-power and ca. US$ 117/MWh for concentrated solar power (CSP) (Table 9.1 shows the range of the LCOEs) [21]. This must be seen in the context of LCOE for power derived from fossil fuels (see Table 9.1). The straight comparison between the LCOE of renewable energy and the LCOE of fossil-fuel-derived power

Table 9.1 Levelised cost of electricity in US$/MWh (2023) from various sources (data from IRENA [21]).

Source	LCOE
Coal	78 – 293
Gas – OCGT	140 – 337
Gas – CCGT	48 – 168
Oil-fired	357
Wind power (onshore)	19 – 72
Wind power (offshore)	63 – 171
Solar power (PV)	31 – 110
Solar power (CSP)	107 – 139

generation is important for power generation as it indicates that power generated from renewable energy coming from new installations is cheaper than power generated from newly constructed fossil-fueldriven power stations.

The cost of electricity generated using renewable energy is expected to drop even more going forward. Figure 9.5 shows the installation cost of wind power (both onshore and offshore power [21, 22] and PV-installations [21, 23] over the last 25 years. Over the last 10 years, the price of both onshore and offshore wind power installations has come down by ca. 50%, and the price of PV-cells by more than 60%.

The predicted installation cost is typically depicted in the form of a learning curve. The double-logarithmic plot of the cost of wind power installations as a function of the cumulative installed capacity for onshore wind power shows a more rapid drop in the installation cost, when the cumulative installed capacity exceeded ca. 400 GW. It is estimated that the cost of onshore wind power installations will drop by ca. 40% if the installed capacity doubles and the installation costs can be extrapolated from the learning curve. Installation costs for offshore wind power installations are significantly higher but seem to follow a similar declining function. The installation costs for PV-cells are much lower than those for wind power (but will have a lower capacity factor) and its costs have been dropping consistently over a period of ca. 25 years, with a drop in the installation cost of ca. 40% upon doubling the installed capacity.

The LCOE for renewable energy is not only dependent on the installation costs, but also on the availability factor. Improving the operation of wind and solar farms may result in an improved availability factor, thus further reducing the LCOE for renewable energy.

The (anticipated) decrease in the cost of electric power from renewable energy could render heating using electric power economically more attractive than heating using fossil fuels. However, for this purpose, the cost of renewable electricity should not be compared to the cost of fossil electricity but rather to the lower heating value of fossil fuels. It can be estimated that using electricity to generate low temperature heat can be economically competitive with heating using fossil

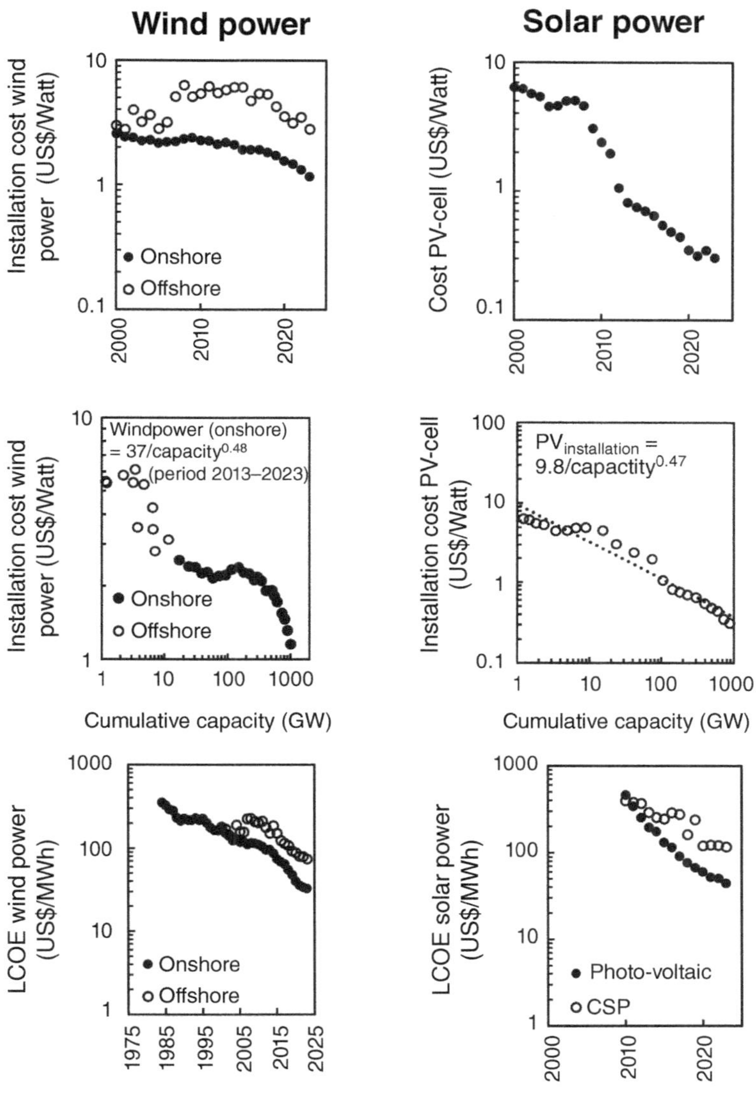

Figure 9.5 Installation cost (top), learning curve (middle) and levelised cost of electricity (bottom) in 2023 US$ for wind power (left) and solar power (right) (data from References [21, 22]).

fuels, with an average electricity price between US$ 30 and 70/MWh [24], which is in the range of the current LCOE for renewable electricity. This considers the use of heat pumps with a relatively high coefficient of performance (C.O.P.) of 3–6. This indicates that the cost of electric heating, to supply low temperature heat, already starts undercutting heating using fossil fuels.

Heating to higher temperatures (in the range of 100–1000 °C) using electricity would require a lower LCOE, as the C.O.P. becomes 1. Taking the price of natural gas at ca. US$ 2–4/MMBtu or US$ 6.8–13.6/MWh and a furnace efficiency of 80%, heating using electricity would become competitive at a cost of electricity of US$ 8.5–17/MWh (a value between US$ 15 and 25/MWh is quoted by [24]). This will become even more competitive when taking carbon taxes into account. For instance, a carbon tax of CO_2 of US$ 100/tonne of CO_2 would make electricity at a price of 28–37 US$/MWh well in the range of LCOE for energy generated from onshore wind power and in the range of the LCOE of PV-installations.

A similar comparison can be made to evaluate heating using electricity and heating using coal. Taking a coal at a price of US$ 100–150/tonne with a heating value of 28–30 GJ/tonne (corresponding US$ 13–18/MWh), and a furnace efficiency of 80%, heating using electricity would become competitive at a cost of electricity of US$ 16–23/MWh. Carbon taxation would again render electric heating more competitive (a price for CO_2 of US$ 100/ton of CO_2 would render electric heating more competitive at a cost of electricity of ca. US$ 56–79/MWh). With the expected decrease in the cost of renewable electricity, electric heating for higher temperatures will become more attractive than heating using fossil fuels (with carbon taxation aiding renewable electricity) provided heating elements can operate with the same level of confidence for prolonged times under industrial conditions.

The further electrification of the process industry does come with some caveats [25], i.e. intermittency in the power supply, sufficient capacity on the power lines to the industrial site (transmission), stability of power supply (able to adapt to rapid demand changes), availability of low-cost power, cost for switching heating technology and availability of alternative heating technologies. The further electrification of the process industry would require a significant extension of the current grid to transport the required amount of electric power. This is specifically the case in areas such as Southeast Asia, Africa and South and Middle America, where the energy usage at large is expected to increase strongly.

Intermittency of the power supply and stability of the grid are important aspects for the process industry, when considering further electrification of the process industry. Some industries will be able to schedule their operations with the availability of power and take advantage of low-cost power. This type of industry would thus rely on intimate interaction with the power producers and distributors on the (predicted) availability of power to schedule their operation. Intermittency of power supply could result in an increase in the production cost due to the underutilisation of the production units, which needs to be offset by a lower electricity cost, which may be minimised by having on-site power back-up.

Other industries will need a constant power supply, but even these industries will need to work out strategies on how to operate when power from the grid is not available (and may thus need larger energy back-up systems to ensure safe operation). For instance, furnaces in the glass industry will need to be maintained at temperatures higher than the melting temperature of glass to avoid solidification within the furnace and back-up power will be needed to maintain this temperature. Figure 9.6 shows the limited data on battery energy systems storage for utility-scale lithium battery systems over time. The costs for the battery cells have come down from over US$ 500/kWh in 2013 to US$ 108–120/kWh in the period 2019–2022

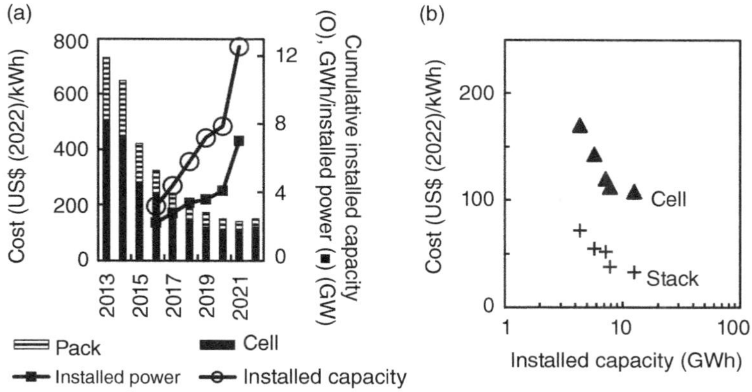

Figure 9.6 Installation cost for utility scale Li-battery storage and the global installed capacity of lithium-ion batteries (a), and learning curve for utility scale Li-battery energy storage systems (b) (data from References [26, 27]).

resulting in a total installation cost for utility battery energy storage systems of US$ 140–150/kWh [26]. This will bring the levelised cost of energy storage (LCOS) to ca. US$ 200/MWh [28, 29].

The total installed capacity of batteries has increased from 3.2 GWh in 2016 to over 12.5 GWh in 2021 [27]. The cost of the Li-battery cells appears to have stagnated, possibly caused by the rapid increase in demand. However, if the learning curve describing the cost of the Li-battery cells can be described using Wright's model, doubling the installed capacity could decrease the price of the cell to US$ 62/kWh and the price for the whole installation to US$ 80/kWh.

Extended electrification of the heating requirements in the process industry would require new learning in the industry. Over the last century, a wealth of knowledge has been developed on the operation of boilers and furnaces. The experience using electrical heating elements in industrial environments is still limited, especially with respect to the durability of these elements and how to deal with operational problems associated with their replacement.

9.4 Alternative Fuels

Hydrogen, ammonia, methanol and hydrocarbons are being discussed as alternative ways to store energy in the Power-to-X concept. These fuels could, in principle, also be used in the chemical industry. It should however be realised that these fuels will come at a premium as the transformation of electric power into these fuels will result in a loss of energy efficiency. Hence, the use of these fuels must come with a significant advantage in comparison to electric heating (possibly stabilised with some form of storage).

Another alternative is fuels from waste, and in particular biowaste and municipal waste, which, when put in landfill sites will decompose, releasing CO_2 and even more potent greenhouse gases such as methane. Humankind produces on average

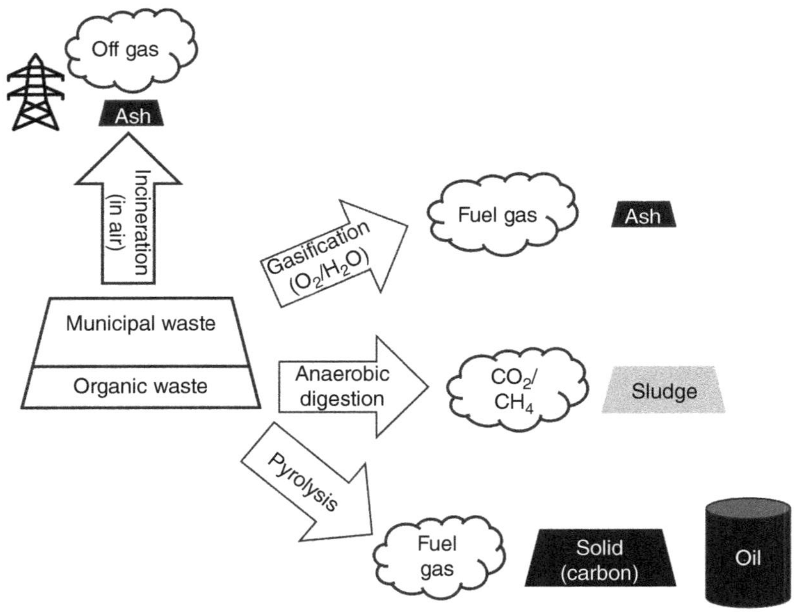

Figure 9.7 Possible ways to convert municipal waste (and organic waste) into fuel.

0.74 kg of waste per person per day (the actual value ranges from 0.1 to 4.5 kg per person per day with higher values in the more developed economies) [30], which could be used to extract energy/work. Dried municipal solid waste has a lower heating value (LHV) of 7–18 MJ/kg [31] strongly depending on its moisture content [32].

A variety of technologies are available to convert waste to energy (see Figure 9.7) [33, 34]. Incineration of waste can be used to generate electric power (potentially 2 MJ/tonne municipal waste [35]) by combusting municipal solid waste at temperatures in the range of 900–1100 °C for steam generation and use in a Rankine cycle, although its main objective is a reduction in the mass and volume of waste (with some possible recovery of metals from ash). Waste incineration reduces the greenhouse gas emissions associated with waste disposal, as disposing of it in a landfill site will generate landfill gas, a mixture of CO_2 and the potent greenhouse gas, methane, whilst incineration produces only CO_2. It should be recognised that the time span of this release changes, with immediate release in the incineration, whilst decomposition of waste on a landfill site to gas may take a long time.

Gasification of municipal solid waste with water and air (or oxygen) yields a synthesis gas, which can be cleaned and subsequently combusted to generate electric power with a reduced environmental impact in comparison to incineration (primarily due to higher temperatures that can be achieved and thus higher efficiency in the power generation step) [36]. Synthesis gas obtained by gasification of solid municipal waste with steam and supplying the heat externally is obtained at a rate of 0.7–1.5 Nm^3 (dry synthesis gas)/kg (dry waste) and contains (on a dry basis) hydrogen (27–53%), CO (9–16%), CO_2 (22–35%) and hydrocarbons (7.8–28.4%) with more hydrogen and CO being formed upon increasing temperature [37]. Supplying the required heat by gasification using a mixture of steam and air would decrease the concentration of hydrogen and CO due to the dilution with nitrogen and formation of CO_2. Clean synthesis gas is not only useful for its heating value but could also be used industrially as a reduction or synthesis gas.

Anaerobic digestion is an established technology to valorise pre-sorted biowaste [38] via the generation of biogas (typically 0.2–0.5 Nm^3/kg volatile or organic solids). Although, its widespread use may be limited due to the low methane productivity and thus the long hydraulic retention times, as well as the generation of compounds inhibiting the biological conversion [34, 39]. The process is mainly aimed at devolatilising waste and co-generating electric power through biogas combustion. The resulting sludge is then landfilled or distributed as fertiliser on land (see Section 8.5.2.4).

Heating up municipal waste in the absence of air will result in the formation of gases, a pyrolysis oil and a char. The amount of gas formed in this process is typically low and has thus become less popular as a method for waste treatment [34]. The amount of char formed, which can be maximised by controlling the pyrolysis temperature and heating rate, can reach up to ca. 40% of the carbon fed into the process [40]. However, from an industrial perspective, the char formed in this process may be an interesting alternative to coal used for reduction purposes. For instance, the waste produced in a metropolitan area of 10 million people would produce sufficient carbon to serve a steel plant producing almost 600 000 tonne of steel per annum if the population produces on average 1.5 kg of solid waste per person per day with a moisture content of 40% and a carbon content of 20% on a dry basis, which is pyrolysed and the solid yield is 35%. This shows that the amount of carbon produced in municipal waste could be used as a reducing agent, but the total amount of carbon used in the production of steel worldwide, is much larger than the amount of carbon recovered from municipal waste, and alternative reduction methods together with recycling will be necessary to fulfil the current per capita demand for steel. Furthermore, the effect of the minor components in the char produced from municipal waste on the reduction of process (and on the elimination of impurities in the reduction process) would need to be investigated. Also, the reduction process will then still emit CO_2, albeit CO_2 which would otherwise be emitted (at least in part) on the landfill site.

9.5 Carbon Capture and Storage

The chemical industry emits CO_2 due to its energy usage, but also because of its choice of reagents. Emissions associated with energy usage can be mitigated by switching to renewable energy sources. However, so-called process emissions, i.e. emissions associated with the choice of reactants and process, are more difficult to mitigate. In some cases, these emissions are inevitable, such as the use of thermodynamically stable carbonate minerals (e.g. calcite/limestone or dolomite), which decompose upon heating, releasing both the valuable elements and CO_2.

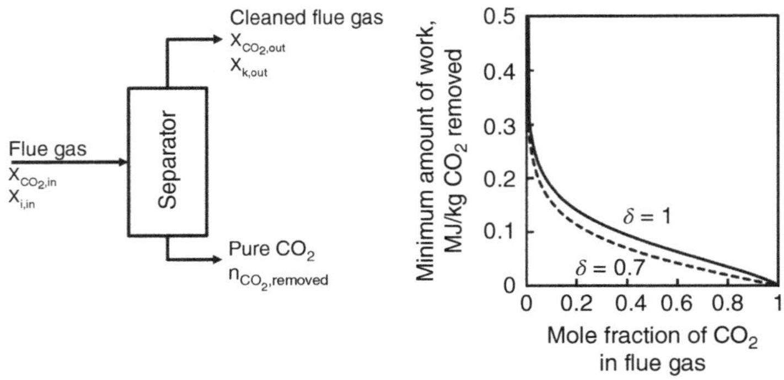

Figure 9.8 Isobaric and isothermal separation of CO_2 from flue gas treated as an ideal mixture and the minimum amount of work required for this separation at 25 °C.

Carbonates as a raw material often cannot be readily replaced (e.g. in cement, but also as a flux in steel and glass manufacture). In other cases, e.g. the replacement of carbon as a reducing agent yielding CO_2 may require a complete redesign of the process.

Carbon capture is a typical end-of-pipeline solution to remove CO_2 from flue gas before it is released into the atmosphere. It requires a process to remove CO_2 from the flue gas and a destination for the recovered CO_2.

The minimum amount of work and heat for the removal of CO_2 from flue gas is given by the change in the Gibbs free energy and entropy of mixing in the streams involved in the separation process (see Figure 9.8). Assuming that pure CO_2 is being extracted from an ideal mixture in an isothermal, isobaric separation process, the minimum amount of heat and work is given by:

$$\frac{\dot{W}_{\text{min}}}{RT \cdot \dot{n}_{\text{CO}_2,\text{removed}}} = \frac{-\dot{Q}_{\text{min}}}{RT \cdot \dot{n}_{\text{CO}_2,\text{removed}}}$$

$$= -\frac{1 - x_{\text{CO}_2}^{\text{in}}}{\delta \cdot x_{\text{CO}_2}^{\text{in}}} \cdot \ln\left(1 - \delta \cdot x_{\text{CO}_2}^{\text{in}}\right) + \frac{1}{\delta} \cdot \ln \frac{(1-\delta)}{1 - \delta \cdot x_{\text{CO}_2}^{\text{in}}}$$

$$- \ln \frac{(1-\delta) \cdot x_{\text{CO}_2}^{\text{in}}}{1 - \delta \cdot x_{\text{CO}_2}^{\text{in}}}$$

and thus the amount of work required per mole of CO_2 removed is only dependent on the mole fraction of CO_2 in the untreated flue gas ($x_{\text{CO}_2,\text{in}}$) and the fraction of CO_2 removed from the flue gas ($\delta = \frac{\dot{n}_{\text{CO}_2,\text{removed}}}{\dot{n}_{\text{CO}_2,\text{in}}}$). Figure 9.8 shows that the minimum amount of work required increases strongly upon decreasing the mole fraction of CO_2 in the flue gas; the minimum amount of work to recover CO_2 from a flue gas containing 15% CO_2 amounts to 0.159 MJ/kg (or 0.044 MWh/tonne) if all CO_2 is recovered, which decreases to 0.131 MJ/kg (or 0.036 MWh/tonne) if 70% of CO_2 is recovered.

A variety of technologies based on absorption, adsorption, chemical reaction (looping) and membrane separation have been developed to remove CO_2 from flue gas streams. In an absorption process, the flue gas is contacted with a liquid, which absorbs CO_2. Carbon dioxide is subsequently stripped from the liquid by raising the temperature. The liquid can be a non-selective liquid, e.g. chilled methanol (~ −60 °C to −50 °C), as in the Rectisol process, or a selective liquid, such as aqueous amine solutions. The non-selective Rectisol process is energetically quite efficient with an energy usage of ca. 1.9 MJ/kg (or 0.52 MWh/tonne CO_2 removed) for a flue gas containing 35% CO_2, but it does require a high initial capital investment [41]. The carbon capture using amine-containing solutions requires a higher energy input of ca. 3–5.5 MJ/kg (or 1–1.5 MWh/ton) [41, 42], due to the stronger bonding of CO_2 in the solution.

Carbon capture using adsorption involves passing CO_2-containing flue gas over a bed containing the solid adsorbent (zeolite, carbon-based materials or metal-organic frameworks (MOFs) [43]), which can be regenerated thermally and/or through depressurisation. The selectivity for the uptake of CO_2 on these adsorbents can be a limiting factor [44], thus requiring multiple units to obtain a relatively pure CO_2-stream [45, 46]. The energy requirements are relatively low for the processes with an energy requirement of 0.4–0.8 MJ/kg CO_2 or 0.1–0.2 MWh/tonne CO_2 recovered for a stream containing ca. 15% CO_2 using multiple separation units reaching an overall CO_2-recovery of more than 90% and a purity of the recovered CO_2 of greater than 95% [45, 46]. The use of more selective adsorbents in carbon capture, which adsorb CO_2 more strongly, is expected to be associated with a higher energy input.

Calcium looping is an alternative technology for carbon capture, where CO_2 reacts with slaked lime (hydrated lime or $Ca(OH)_2$) yielding $CaCO_3$ and water. CO_2 is then recovered by thermal decomposition of calcium carbonate at temperatures above 750 °C. The recovered lime is hydrated to recover the reactivity of the calcined lime [47]. This process utilises a cheap adsorbent, but the energy cost for removing CO_2 from a stream containing 12% CO_2 using calcium looping is still ca. 4.5 MJ/kg (or 1.25 MWh/tonne CO_2 captured; this value does not include additional energy cost required for the hydration and water removal step) [48].

In a membrane-based separation process, pressurised flue gas is passed over a membrane permeable for CO_2 and recovered as permeate. The selectivity for CO_2 can thus be problematic; to obtain a stream containing at least 95% CO_2, the permeate must be recompressed and passed over a second membrane [49]. The energy cost to capture 90% of CO_2 from a stream containing 14% CO_2 to obtain a stream containing at least 95% CO_2 was estimated to come to ca. 1.9 MJ/kg (0.5 MWh/tonne CO_2 captured).

The selective carbon capture processes typically require an energy input of ca. 1.9–5 MJ/kg CO_2. Putting these values in perspective, this energy is equivalent to 10.4–27.4% of the energy released by the heat of combustion of natural gas to generate this amount of CO_2 (or 21.2–55.9% of the energy released by combusting coal to generate the recovered amount of CO_2).

The energetic cost for the capture of CO_2 from flue gas typically decreases with increasing concentration of CO_2 in the flue gas. Carbon capture processes will thus become more readily feasible, if the concentration of CO_2 in the flue gas can be increased. In combustion processes, this may be achieved by combusting with oxygen or oxygen-enriched air. This will result in an increase in the combustion temperature, which could be reduced by co-feeding recycled flue gas/CO_2 to the furnace [50].

9.6 New Opportunities

The move to a hydrogen economy may offer opportunities for new ventures in the process industry involving the production of green hydrogen and its derivatives, such as ammonia, methanol and liquid hydrocarbons, in the power-to-X (PtX) value chain (see Chapter 5; Figure 9.9), as these are seen as attractive routes to store and transport energy over long distances and for prolonged periods of time (e.g. to counter the seasonal variation in energy supply). It should be realised that the conversion of energy into hydrogen (and its derivatives) is associated with a loss in energy, and thus this stored energy will be more expensive than the original renewable energy source. For instance, the conversion of electric power into hydrogen may proceed with an energetic efficiency of 61–69% (see Table 5.7) and the conversion of hydrogen back into energy using a fuel cell with an efficiency of close to 50% (see Section 5.2.5.2.) would thus imply that the so-called roundtrip energetic efficiency for the storage of energy in the form of hydrogen would be ca. 30–35% (this value will be lowered further due to efficiency loss during storage/transport). Tsiklios et al. [51] estimated a roundtrip efficiency when using hydrogen as the energy carrier of 36.9% (with power generation using a combined cycle gas turbine [CCGT]). Storing energy in the form of ammonia, methane or methanol would lower the roundtrip efficiency to 31.4%,

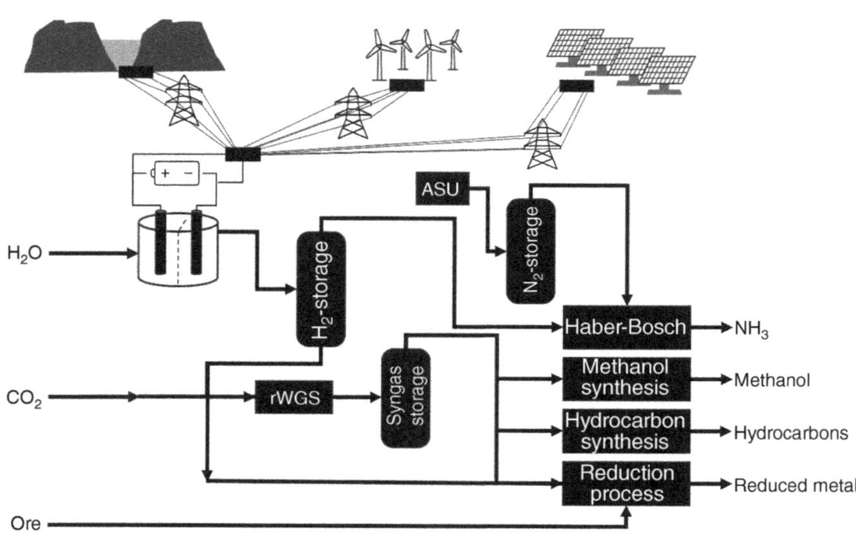

Figure 9.9 Conversion of renewable energy into the basic chemicals, hydrogen, ammonia, methanol or hydrocarbons, or reduced metals.

27.0%, 26.5%, respectively (with carbon originating from the capture of CO_2 from flue gas but not accounting for the energy required to capture CO_2). Hence, the cost of energy stored in this form will be at least ca. 3–4 times the cost of direct usage of renewable energy, if the production costs for these chemicals can be kept low.

The production of chemicals such as hydrogen, ammonia, methanol and liquid hydrocarbons starting with renewable energy will become a reality in the decade to come, specifically to green large-scale ammonia production, whereas the production of methanol and liquid hydrocarbons may be linked to inevitable CO_2-sources (e.g. the cement industry). However, the development of these processes will pose challenges in their design, process financing and operation of the ultimate plant.

The scale of these processes and specifically the amount of energy required for these processes imply that they will need to be intimately linked with a (large) renewable energy facility or facilities, since most or even all the energy generated will be consumed by the production site. The renewable energy facility may form a part of the production company, or it could be a separate company in a symbiotic relationship with the production process. Capturing renewable energy as a part of the production process will have the advantage of direct control of the energy supply, whereas supply of renewable energy from a closely aligned company may reduce the cost, as excess energy could be sold off to other consumers. It is not necessary to generate the renewable energy on the production site because it can be wheeled in over existing or dedicated transmission lines. Renewable energy could also be taken from a grid operator supplying renewable energy from a variety of different suppliers. The latter may result in a more stable supply of energy, albeit coming at a higher cost with pricing dependent on the time of the day as well as the season. The ultimate choice on how to structure the energy supply may depend on external factors, such as grid availability and grid stability.

The design of the plant will need to account for short-time fluctuations in the supply of energy and/or its cost, as well as the variability of these factors over long periods (typically a year assuming that the variability between various years over the lifespan of the plant can be neglected). The short-term variability may be negated by energy storage, as well as storage of the intermediate products, but the cost of energy storage and/or storage of reagents needs to be weighed against the cost of oversizing equipment and running the plant at variable load. The variability of the supply of energy as a feedstock to these processes puts a new dimension on the design of these processes, and the ultimate design of these plants will depend on the availability of renewable energy sources and the availability/feasibility of intermediate storage facilities.

The relative immaturity of these processes on an industrial scale (specifically the hydrogen generation, the reverse water-gas shift [rWGS] and even the green ammonia synthesis) renders this part of the process as a high-risk operation, thus pushing the discount rate up. On the other hand, the generation of renewable energy on a large scale, will be relatively low risk, as it may operate on its own by selling power to the grid. Financing these projects could be derisked by separating the project into different companies linked by various purchasing agreements, with separate companies responsible for energy generation, hydrogen generation and downstream processing. This type of solution may not be looked upon favourably by investors due to the possibility of profit-shifting. Hence, the overall process will need to be derisked with the techno-economic risk of the overall process being re-assessed by looking at its individual components, bearing in mind that the individual parts of the process may be able to continue to operate even if the overall project fails. Government grants in the form of subsidies may be an essential component in financing these projects in the transitionary period to overcome inequality in risk assessment (but also inequality in other areas such as carbon taxation).

The operation of these types of plants will have its challenges due to the necessary fluctuations in the energy supply (either based on the time-of-use cost or the actual availability of renewable energy). Chemical processes are not easy to turn on and off at short notice, not compared to charging a battery or pumping water at least, and thus chemical plants relying on renewable energy may need to incorporate large amounts of electrical storage, like battery storage or compressed hydrogen storage, to ensure safe operation. Hence, plants running on renewable energy will likely operate under a more variable load than plants currently operating in the chemical process industry. Consequently, secure communication between the energy generation section/companies and/or energy distribution companies and the process section of these plants will be critical to react to variations in the availability of the energy and feedstock. This should be enforced by daily forecasting models on the availability of energy and the feedstock required for the plant.

9.7 Concluding Remarks

The ultimate aim of the introduction of renewable energy into the process industry is to reduce anthropogenic greenhouse gas emissions. This must be seen in the context of an expanding industrial basis implying that even when the greenhouse gas emissions of the individual plants remain constant, the total greenhouse gas emissions from the process industry will increase.

Greenhouse gas emissions by the process industry are related to the so-called process emissions, i.e. emissions due to choice of reactants and process, and energy emissions, i.e. emissions due to the energy usage in the process industry. Process emissions are quite widespread in the metallurgical industry when using, e.g. carbothermal reductions, but also in food industry in fermentation processes. These may be avoided/reduced by changing the process and/or reactants, as proposed for the reduction of iron in the direct reduction of iron (DRI) process, or by implementing energy-intensive carbon capture processes. Often, both measures will be needed to decrease the concentration of CO_2 in the emissions.

Energy from fossil fuels is widely used in the process industry to provide heating (although several processes can be, at steady-state, net energy exporters), which can be replaced by renewable energy. Renewable energy can be supplied in the form of electric power, which can be used to power heat pumps or to provide heat through resistive heaters. The use of heat pumps to provide low-temperature heat (typically below 100–150 °C) should be competitive with heating by combusting fossil fuels directly when the cost of electricity from renewable energy matches the cost of electricity from fossil fuels, if the C.O.P. of the heat pump is at least 2–3. Resistive heaters can be used when requiring higher temperatures, although there are still questions on the durability of these heating elements, especially when exposed to specific atmospheres.

Introducing renewable energy for heating purposes in existing plants will be an ongoing process in the process industry for the next decade, when furnaces/boilers will need to be replaced. This process may be accelerated by incentives/carbon taxes on products, making the replacement of current equipment economically attractive prior to their replacement date. Depending on when this replacement occurs in the life cycle of the process plant, it may not necessarily lead to an overall reduction in greenhouse gas emissions over the plant lifetime.

The increased usage of electrical energy by the process industry should be matched by an extension of the capacity of the grid. The process industry will preferentially require a constant (base) load for its operation, and thus grid operators will face the challenge of supplying a much larger base load than can currently be supplied by variable supply sources of the day. Furthermore, the process industry will have to consider the consequence of time-of-use tariff. Specifically, the process industry will have to develop protocols on how to deal with requests to reduce energy usage (which may come as direct requests in situations with a high load on the grid). Furthermore, they will need to put processes in place ensuring safe operation/shutdown for unplanned power reduction and/or failure.

Industries relying on hydrogen may see a fundamental change in their mode of operation provided that the electricity cost comes down, as hydrogen is mainly obtained from natural gas and even coal nowadays. Hence, the drive to produce 'green' hydrogen will be assisted by carbon taxes/incentives. This will then have consequences for the industrial production of ammonia, one of the chemicals produced at the largest scale worldwide and could even affect industrial production of methanol and aviation fuel/diesel, when using captured CO_2 as the carbon source. The 'greening' of these products may occur simultaneously with the development of these product compounds as energy carriers.

Overall, the process industry stands at the eve of an interesting time with electrical energy having a much more dominant role. Abundance of cheap electricity from renewable sources will drive innovations in the industry replacing fossil fuels as the energy source of choice, as a necessary measure to control greenhouse gas emissions, thus confirming the statement '*The Stone Age came to an end, not because we had a **lack of stones,** and the oil age will come to an end not because we have a **lack of oil**.*' [52].

References

1 UN Secretary-General (1987). World Commission on Environment and Development. Report of the World Commission on Environment and Development: "Our common future", UN, New York.

2 United Nations Environment Programme (2003). Handbook for the International Treaties for the Protection of the Ozone Layer Sixth Edition: The Vienna Convention (1985); The Montreal Protocol (1987) [Online]. Available: https://wedocs.unep.org/20.500.11822/8045 (accessed 15 March 2025).
3 Pan, J., Cross, J., Zou, X., and Zhang, B. (2024). To tax or to trade? A global review of carbon emissions reduction strategies. *Energy Strategy Reviews* 55: 101508.
4 Manta, A., Doran, N., Bădîrcea, R. et al. (2023). Does the implementation of a Pigouvian tax be considered an effective approach to address climate change mitigation? *Economic Analysis and Policy* 80: 1719–1731.
5 Bertarelli, S. and Lodi, C. (2019). Heterogeneous firms, exports and Pigouvian pollution tax: does the abatement technology matter? *Journal of Cleaner Production* 228: 1099–1110.
6 World Bank Group (2025). State and trends of carbon pricing dashboard [Online]. Available: https://carbonpricingdashboard.worldbank.org (accessed 7 May 2025).
7 Parry, I. (2012). How to design a carbon tax. In: *Fiscal Policy to Mitigate Climate Change - A Guide for Policymakers*, 218. International Monetary Fund.
8 Marron, D. and Toder, E. (2014). Tax policy issues in designing a carbon tax. *American Economic Review* 104 (5): 563–568.
9 Gokhale, H. (2021). Japan's carbon tax policy: limitations and policy suggestions. *Current Research in Environmental Sustainability* 3: 100082.
10 Shi, Q., Ren, H., Cai, W., and Gao, J. (2019). How to set the proper level of carbon tax in the context of Chinese construction sector? A CGE analysis. *Journal of Cleaner Production* 240: 117955.
11 Wang, X., Li, J.F., and Zhang, Y.X. (2011). An analysis on the short-term sectoral competitiveness impact of carbon tax in China. *Energy Policy* 39: 4144–4152.
12 Gorbach, O., Kost, C., and Pickett, C. (2022). Review of internal carbon pricing and the development of a decision process for the identification of promising Internal Pricing Methods for an organisation. *Renewable and Sustainable Energy Reviews* 154: 111745.
13 Young, J., McQueen, N., Charalambous, C. et al. (2023). The cost of direct air capture and storage can be reduced via strategic deployment but is unlikely to fall below stated cost targets. *One Earth* 6 (7): 899–917.
14 Ahmad, M., Li, X., and Wu, Q. (2024). Carbon taxes and emission trading systems: which one is more effective in reducing carbon emissions? – A meta-analysis. *Journal of Cleaner Production* 476: 143761.
15 Sato, M., Rafaty, R., Calel, R., and Grubb, M. (2022). Allocation, allocation, allocation! The political economy of the development of the European Union Emissions Trading System. *WIREs Climate Change* 13 (5): e796.
16 Instrat Foundation (2025). CO_2 emission allowances price in EU ETS system [Online]. Available: https://energy.instrat.pl/en/prices/eu-ets (accessed 10 May 2025).
17 Ampudia, M., Bua, G., and Kapp, D.S.D. (2022). The role of speculation during the recent increase in EU emissions allowance prices, European Central

Bank – Eurosystem [Online]. Available: https://www.ecb.europa.eu/press/economic-bulletin/focus/2022/html/ecb.ebbox202203_06~ca1e9ea13e.en.html (accessed 10 May 2025].

18 Directorate-General for Taxation and Customs Union (EU) (2025). Carbon border adjustment mechanism, 28 March 2025 [Online]. Available: https://taxation-customs.ec.europa.eu/carbon-border-adjustment-mechanism_en (accessed 17 May 2025).

19 McKinsey & Company (2024). What are Scope 1, 2, and 3 emissions?, 17 September 2024 [Online]. Available: https://www.mckinsey.com/featured-insights/mckinsey-explainers/what-are-scope-1-2-and-3-emissions (accessed 17 May 2025).

20 Bourgon, T. (2024). Tracenable Academy, 2024 December 2024 [Online]. Available: https://academy.tracenable.com/the-complete-guide-to-scope-3-emissions-from-supply-chain-to-end-of-life-impact (accessed 12 March 2025).

21 IRENA (2024). Renewable Power Generation Costs in 2023, Knowledge Centre, Reports, 27 September 2024.

22 Our World in Data (2025). Installed wind energy capacity – Total wind capacity [Online]. Available: https://ourworldindata.org/grapher/cumulative-installed-wind-energy-capacity-gigawatts (accessed 15 March 2025).

23 Ritchie, H., Rosado, P., and Roser, M. (2024). Data Page: Solar photovoltaic module price [Online]. Available: https://ourworldindata.org/grapher/solar-pv-prices (accessed 15 March 2025).

24 Roelofson, O., Speelman, E., Witteveen, M., and Somers, K. (2020). Plugging in: what electrification can do for industry, McKinsey & Company, 28 May 2020 [Online]. Available: https://www.mckinsey.com/industries/electric-power-and-natural-gas/our-insights/plugging-in-what-electrification-can-do-for-industry (accessed 8 May 2025).

25 Wei, M., McMillan, C., and de la Rue du Can, S. (2019). Electrification of industry: potential, challenges and outlook. *Current Sustainable/Renewable Energy Reports* 6: 140–148.

26 BloombergNEF (2023). Top 10 Energy Storage Trends in 2023 [Online]. Available: https://about.bnef.com/blog/top-10-energy-storage-trends-in-2023 (accessed 16 May 2025).

27 Nguyen, T. and Tamrakar, U. (2024). DOE Global Energy Storage Database, National Technology & Engineering Sciences of Sandia, LLC (NTESS); US Department of Energy [Online]. Available: https://gesdb.sandia.gov/index.html (accessed 16 May 2025).

28 Schmidt, O., Melchior, S., Hawkes, A., and Staffell, I. (2019). Projecting the future levelized cost of electricity storage technologies. *Joule* 3: 81–100.

29 Steckel, T., Kendall, A., and Ambrose, H. (2021). Applying levelized cost of storage methodology to utility-scale second-life lithium-ion battery energy storage systems. *Applied Energy* 300: 117309.

30 Kaza, S., Yao, L.C., Bhada-Tata, P., and van Woerden, F. (2018). *What a Waste 2.0: A Global Snapshot of Solid Waste Management to 2050*. Urban Development.

31 Wang, B., Gupta, R., Bei, L. et al. (2023). A review on gasification of municipal solid waste (MSW): syngas production, tar formation, mineral transformation and industrial challenges. *International Journal of Hydrogen Energy* 48: 26676–26706.

32 Lin, X., Wang, F., Chi, Y. et al. (2015). A simple method for predicting the lower heating value of municipal solid waste in China based on wet physical composition. *Waste Management* 36: 24–32.

33 Beyene, H.D., Werkneh, A.A., and Ambaye, T. (2018). Current updates on waste to energy (WtE) technologies: a review. *Renewable Energy Focus* 24: 1–11.

34 Traven, L. (2023). Sustainable energy generation from municipal solid waste: a brief overview of existing technologies. *Case Studies in Chemical and Environmental Engineering* 8: 100491.

35 Alam, S., Rahman, K.S., Rokonuzzaman, M. et al. (2022). Selection of waste to energy technologies for municipal solid waste management – towards achieving sustainable development goals. *Sustainability* 14 (19): 11913.

36 Dong, J., Tang, Y., Nzihou, A. et al. (2018). Comparison of waste-to-energy technologies of gasification and incineration using life cycle assessment: case studies in Finland, France and China. *Journal of Cleaner Production* 203: 287–300.

37 He, M., Hu, Z., Xiao, B. et al. (1009). Hydrogen-rich gas from catalytic steam gasification of municipal solid waste (MSW): influence of catalyst and temperature on yield and product composition. *International Journal of Hydrogen Energy* 34: 95–103.

38 Saravanakumar, A., Sudha, M.R., Chen, W.-H. et al. (2023). Biomethane production as a green energy source from anaerobic digestion of municipal solid waste: a state-of-the-art review. *Biocatalysis and Agricultural Biotechnology* 53: 102866.

39 Nayeri, D., Mohammadi, P., Bashardoust, P., and Eshtiaghi, N. (2004). A comprehensive review on the recent development of anaerobic sludge digestions: performance, mechanism, operational factors, and future challenges. *Results in Engineering* 22: 102292.

40 Hwang, I.-H., Kobayashi, J., and Kawamoto, K. (2014). Characterization of products obtained from pyrolysis and steam gasification of wood waste, RDF, and RPF. *Waste Management* 34: 402–410.

41 Ma, Y., Liao, Y., Su, Y. et al. (2021). Comparative investigation of different CO_2 capture technologies for coal to ethylene glycol process. *PRO* 9 (2): 207.

42 Husebye, J., Brunsvold, A.L., Roussanaly, S., and Zhang, X. (2012). Techno economic evaluation of amine based CO_2 capture: impact of CO_2 concentration and steam supply. *Energy Procedia* 23: 381–390.

43 Das, A., Peu, S.D., Hossain, M.S. et al. (2023). Advancements in adsorption based carbon dioxide capture technologies – a comprehensive review. *Heliyon* 9 (12): e22341.

44 Boer, D.G., Langerak, J., and Pescarmona, P.P. (2023). Zeolites as selective adsorbents for CO_2 separation. *ACS Applied Energy Materials* 6 (5): 2634–2656.

45 Riboldia, L. and Bolland, O. (2017). Overview on Pressure Swing Adsorption (PSA) as CO_2 capture technology: state-of-the-art, limits and potentials. *Energy Procedia* 114: 2390–2400.

46 Liu, Z., Grande, C.A., Li, P. et al. (2011). Multi-bed Vacuum Pressure Swing Adsorption for carbon dioxide capture from flue gas. *Separation and Purification Technology* 81 (3): 307–317.

47 Rong, N., Wang, Q., Fang, M. et al. (2013). Steam hydration reactivation of CaO-based sorbent in cyclic carbonation/calcination for CO_2 capture. *Energy & Fuels* 27 (9): 5332–5340.

48 Mantripragada, H.C. and Rubin, E.S. (2014). Calcium looping cycle for CO_2 capture: performance, cost and feasibility analysis. *Energy Procedia* 63: 2199–2206.

49 Zhang, X., He, X., and Gundersen, T. (2013). Post-combustion carbon capture with a gas separation membrane: parametric study, capture cost, and exergy analysis. *Energy & Fuels* 27 (8): 4137–4149.

50 Chen, L., Yong, S., and Ghoniem, A. (2012). Oxy-fuel combustion of pulverized coal: characterization, fundamentals, stabilization and CFD modeling. *Progress in Energy and Combustion Science* 38 (2): 156–214.

51 Tsiklios, C., Schneider, S., Hermesmann, M., and Müller, T.E. (2023). Efficiency and optimal load capacity of E-Fuel-based energy storage systems. *Advances in Applied Energy* 10: 100140.

52 Quote Investigator (2018). The stone age did not end because the world ran out of stones, and the oil age will not end because we run out of oil, 1 January 2018 [Online]. Available: https://quoteinvestigator.com/2018/01/07/stone-age/#note-17685-1 (accessed 14 April 2021).

Index

Renewable Energy in the Process Industry, First Edition.
Eric van Steen, Nicholas S. Featherstone, and Linda H. Callanan.

f

h

i

j

k

l

m